Progress in Optical Science and Photonics

Volume 26

Indexed by Scopus

The purpose of the series Progress in Optical Science and Photonics is to provide a forum to disseminate the latest research findings in various areas of Optics and its applications. The intended audience are physicists, electrical and electronic engineers, applied mathematicians, biomedical engineers, and advanced graduate students.

Vijay Kumar · Irfan Ayoub · Vishal Sharma ·
Hendrik C. Swart
Editors

Optical Properties of Metal Oxide Nanostructures

Springer

Editors
Vijay Kumar
Department of Physics
National Institute of Technology Srinagar
Jammu and Kashmir, India

Irfan Ayoub
Department of Physics
National Institute of Technology Srinagar
Jammu and Kashmir, India

Vishal Sharma
Institute of Forensic Science
and Criminology
Panjab University
Chandigarh, India

Hendrik C. Swart
Department of Physics
University of the Free State
Bloemfontein, South Africa

ISSN 2363-5096 ISSN 2363-510X (electronic)
Progress in Optical Science and Photonics
ISBN 978-981-99-5639-5 ISBN 978-981-99-5640-1 (eBook)
https://doi.org/10.1007/978-981-99-5640-1

This Springer imprint is published by the registered company Springer Nature Singapore Pte Ltd.
The registered company address is: 152 Beach Road, #21-01/04 Gateway East, Singapore 189721, Singapore

Preface

We are delighted to present this edited volume on the interesting topic of the optical properties of Metal Oxide Nanostructures. This book brings together a collection of in-depth contributions from well-known experts in the field, providing a broad overview of the innovative advancements, research findings, and emerging trends in the study of metal oxide nanomaterials and their optical characteristics.

Nanotechnology has developed as a paradigm-shifting field during the last few decades, transforming a variety of scientific fields and creating exciting new opportunities for technological improvement. Metal oxides, particularly their nanostructures, have emerged as a vital class of materials with a rich spectrum of properties and a huge potential for device applications. Metal oxides show a sensitivity to stoichiometry that is unique among semiconductors, which allows the fine-tuning of these properties. In addition to bulk materials, nanostructured metal oxides with high surface-to-volume ratios are particularly interesting. An extensive variety of optical effects can be produced for various applications by simply manipulating the shape, size, and surface functionality. The possibility of controlling doping levels to tune the electronic and optical properties of these materials offers new avenues for potential applications.

The primary focus of this edited book is to explore many aspects related to the optical properties of metal oxides, including material design, structure-property relationships, different structural fabrications, and their applications. The fundamental science and mechanism behind optoelectronic phenomena are discussed to help researchers to understand the structure-property relationship in metal oxide nanostructures. Each chapter of the edited book comprises different metal oxide materials, such as TiO_2, ZnO, Fe_2O_3, SnO_2, and various mixed metal oxides. Each chapter explores the methods of synthesis and the structural and optical characterization techniques, particularly for the studied material under consideration, offering a detailed understanding of their optical behavior.

Studying optical phenomena in metal oxide nanostructures is interesting from scientific and technological viewpoints. The chapters of this book explore the principles of light-matter interaction and show the effects of size, shape, composition, and surface characteristics on the optical properties of metal oxide-based nanomaterials.

Many optical phenomena, such as scattering, reflection, transmission, absorption, luminescence, plasmonic effects, and luminescence, are discussed in this book. The unraveling of the complex links between the structural factors and the ensuing optical behavior lays the groundwork for the fabrication and modification of the properties of metal oxide nanostructures.

This book examines the different applications of metal oxide nanostructures made available by their outstanding optical properties. The chapters also cover the applications of metal oxides in solar cells, solar energy harvesting, energy storage devices, electrochemical sensors, electronics, memoirists, resistive switching, sensors, and detectors while emphasizing the methods employed to boost their efficiency and performance.

We express our gratitude to the distinguished authors, who contributed their wisdom and ideas to this book. Their invaluable contributions, in-depth study, and dedication to advancing the field of metal oxide nanostructures have impacted this volume. We hope that the many viewpoints and thorough coverage provided within these pages will motivate academics, scholars, and professionals to investigate and innovate metal oxide nanoparticles and their optical characteristics.

Finally, we want to express our sincere gratitude to the editorial and production staff, who have worked relentlessly to make this book a reality. The accuracy, coherence, and readability of the information have all been carefully preserved, making this book an invaluable tool for both academics and readers.

We cordially invite you to join us on this enthralling exploration of the optical properties of metal oxide nanostructures as we reveal the wonder and complexity that lie at the nexus of nanoscience and optics. May this book serve as a catalyst for additional research, creativity, and cooperation in this dynamic and quickly developing field.

Jammu and Kashmir, India — Vijay Kumar
Jammu and Kashmir, India — Irfan Ayoub
Chandigarh, India — Vishal Sharma
Bloemfontein, South Africa — Hendrik C. Swart

Contents

An Introduction to Metal Oxides

Suresh Chandra Baral, P. Maneesha, Srishti Sen, Sagnika Sen, and Somaditya Sen

Abstract Metal oxides (MOs) are important due to their tuneable electrical, magnetic and optical properties. These find a wide range of applications. MOs are stable and durable and are used widely in industries related to the environment, security, medicines, petrochemicals, and agricultural fields. The surface morphology and crystalline structure of MOs can be tailored by modifications in the synthesis routes and conditions. This opens a wide range of applicability. This chapter briefs about a structure correlated study of physical, chemical, and optical properties of several important MOs and their potential applications/scopes. There is a clear advantage of MOs over sophisticated electronic materials in terms of stability. The electronic/optoelectronic and magnetic properties bear a rich physics. Nanostructured and thin-film materials and their possible use in diodes, transistors, optoelectronic/electronics, magnetic storage media, microstrip antennas, catalysis, energy conversion, sensing properties, solar cells, batteries, and other energy devices, will be discussed. Different structures and their relevance to the functionalities will also be covered. This chapter provides an insight into contemporary research and developments, thus leading the reader to experience the structure correlated properties of the MOs.

Keywords Metal oxides · Opto electronics · Metal oxide synthesis · Perovskite oxides · Multifunctional oxides

---

Suresh Chandra Baral and P. Maneesha—Equal first authorship.
Srishti Sen and Sagnika Sen—Equal contribution.

S. C. Baral · P. Maneesha · S. Sen (✉)
Department of Physics, Indian Institute of Technology Indore, Indore 453552, India
e-mail: sens@iiti.ac.in

S. Sen
School of Bioscience, Engineering and Technology, VIT Bhopal University, Bhopal 466114, India

S. Sen
Rankers International School, Indore 452016, India

V. Kumar et al. (eds.), *Optical Properties of Metal Oxide Nanostructures*,
Progress in Optical Science and Photonics 26,
https://doi.org/10.1007/978-981-99-5640-1_1

1 Introduction

Reactive metals available on earth are mostly react with oxygen to form MOs. Compared to pure metals these MOs are having unique properties and high stability. The various exciting properties exhibited by these MOs can tune by various methods such as by doping with different metal ions, by varying the morphology, size, dimension etc. [1, 2]. Also, they form the basis for developing multiple theories of condensed matter physics. Again, metals can react with oxygen to produce Oxides: O^{2-}, peroxides: O_2^{2-}, superoxide: O_2^- out of which MOs thus typically contain an anion of oxygen in the oxidation state of −2. The properties exhibited by the MOs compounds are extensive. Various crystal structure, optical, magnetic and electrical properties with high stability, chemical and physical properties with different phase transitions are a few among them [3]. Due to these properties these oxides are used for different technological applications such as gas sensors, solar cell, battery, electronic components, transducers, optoelectronic devices, membrane reactors, catalytic, energy storage and bio applications [4–7]. If we talk about the types of metal oxides in general, the following two things can be considered.

I. Simple Metal Oxides:
A simple metal oxide is one carrying several oxygen atoms that the normal valency of its metal allows. Examples include MgO, CaO, CuO, ZnO, NiO, Al_2O_3, Fe_3O_4, BaO, etc.
II. Mixed metal oxides:
mixed metal oxides are produced when two or more simple metal oxides combine. For Example, Red lead (Pb_3O) is a mixed oxide of lead dioxide (PbO_2) and lead monoxide (PbO). Another example is ferro-ferric oxide (Fe_3O_4) which is a mixed oxide of two simple oxides—ferric oxide (Fe_2O_3) and ferrous oxide (FeO).

Minerals can naturally provide some metal oxides. However, these metal oxides are not readily available for application. Multicomponent dry or aqueous melts crystallizes to form naturally occurring MOs. The melts formed into rocks that underwent a weathering process for millions of years by the natural agencies involving prolonged interaction of many ionic constituents [8]. Compared to the above time frame of the formation of these natural MOs, laboratory synthesis of various metal oxides, such as TiO_2 [9, 10], ZnO [11], SnO_2 [12] and CuO [13], is a recent affair and has been extensively studied for practical use applications.

At room temperature metal oxides generally are brittle. But, as far as their electrical conductivity is concerned, they cover a wide range from insulators, semiconductors, and metallic conductors to superconductors [14]. Whereas the application of these oxides for electrical applications have not been exploited in comparison to silicon and copper. The primary reason is the problem to maintain the required level of oxygen content. The brittleness is another problem followed by crystalline phase transformations. Silicon has a typical carrier orbital of 30 Å. For doping concentrations above $4 \times 10^{18}/cm^3$ the orbitals overlap. However, in the case of transition metal oxides, typically an orbital is ~1 Å [15]. Only at higher doping concentrations ~$10^{22}/cm^3$ overlapping of orbitals occurs. Often, the crystalline structure of the host material is altered by such high doping levels, and modification of crystal structure happens in order to accommodate the dopants. Then concept of doping is no longer use in this situation and instead categorize oxides based on crystalline structure. The electrical properties are different in different crystal structure of the same MOs [16]. This phase change is a disadvantage for MOs application in comparison with Si. Whereas it can consider as an advantage to explore variety of properties of MOs over the various crystalline structures.

Metal-Insulator transition is an important electrical property exhibited by MOs [17]. In this type of transition an insulator turns to metal at a specific temperature or pressure. Even though this change occurs for a particular composition, the word transition is usually used. This field had undergone intensive research before high-temperature superconductivity related research area. Temperature dependent electrical conductivity at higher temperatures is another characteristic feature of MOs. Scattering by optical phonons plays an important role in this attribute. In TMOs, oxygen 2p- and metal d-orbitals forms the conduction band. Valence band constituted by the oxygen orbitals which are located lower in energy. The directionality of the p- and d-orbitals are correlated to the transport properties.

The localization of the electrons leads to the various magnetic and dielectric properties. Ferromagnetic or ferroelectric applications of MOs are frequent in technological applications. Whereas conductivity of these oxides can be detrimental for such applications [18]. Extensive studies are going on to suppress the conductivity of such materials. The hopping conduction studies are widely studied for such applications. The electrons and phonons in solids and their unavoidable interaction lays the roadmap of the modifications of the electronic states. Combined state of an electron and phonon is called a polaron. Very low temperature in the superconducting state these polarons becomes more stable two carrier state.

Synthesis of MOs can be done in different structures, morphologies and sizes (from bulk to nano). The properties of these different size, shape, form and morphology influence the properties of the materials modifying the crystal lattice parameters, bond lengths/angles, electronic and other properties. These modifications influence their applicability in components in electronics, optoelectronic devices [19], memory devices [20], energy storage devices, catalysts [21], solid state batteries [22], gas sensors [23] and bio applications [24].

When the lattice extends in a single unit beyond 100 nm it is customary to call such materials bulk. The properties of bulk are different from the sub-100 nm range called nanoparticles. Nanomaterials are defined as materials with one of the three external dimensions to be below 100 nm (nanoscale). Some of the materials have nanoscale internal structure or surface structure which are also categorized as nanomaterials (ISO 2015). There is a separate category of materials in the form of a nearly 2-D format called thin films which are actually a 3-D structure but with one of the dimensions much lesser than the other two. These are generally used in electronic, optical and mechanical devices and are generally considered to be a different category of materials. These three different forms of the MOs will be discussed in three different sections below.

Bulk MOs can be engineered by doping different materials or by treating at different atmospheres to obtain tailored electrical, mechanical, optical and magnetic, properties. These materials are very diverse compounds with various properties includes almost all fields of material science and physics. Based on the conductivity MOs can be superconductors, metallic, semiconductors, and insulators. Bonding characteristics also have variations from covalent to ionic. Bulk MOs has many applications in the fields of lasing materials, materials for nonlinear optics, sensors and catalysis to simple things like paint pigments. One good example of the use of a MO single crystal is in the Nd:YAG laser where Nd-doped $Y_3Al_5O_{12}$ (Yttrium Aluminium Garnet) is used to produce lasing action. BBO (β-barium borate), KDP (potassium dihydrogen phosphate), KTP (potassium titanyl phosphate), and lithium niobate are examples of materials exhibiting nonlinear optics [25]. There are more of these excellent examples where MO single crystals are used in physics. Bulk TiO_2 and ZnO are very well-known examples of sensors and catalytic materials along with being used in pigments. Bulk CuO has recently been used in catalysis using visible light.

MO thin films have a vast application in today's world. Thin films of MOs have modifiable properties depending on the dimensions and composition. In the nano range there is a huge scope of attaining modified characteristics and also of miniaturization of the device size thereby moving towards better electronics. Today, multilayer thin films of MOs are being used to fabricate heterojunction structures for modern electronic devices. These MO multilayer thin films are used to exhibit enhanced properties as compared to monolayer thin films. It is used for transistors, photovoltaic applications, gas sensors, and catalytic applications [26–28]. Oxides of ruthenium, iridium, manganese, cobalt, nickel, tin, iron, perovskites, ferrites etc. have been used for making MO thin film electrode in supercapacitors [29].

The fabrication techniques of these MO thin films differ from industry to industry, from research group to research group. Among these mostly used current techniques are sputtering, spray pyrolysis, sol-gel, thermal or electron-beam evaporation, pulsed laser deposition (PLD), rf/dc sputtering, hydrothermal, spin coating, drop coating, dip coating, screen printing, atomic layer deposition (ALD), and so on [29–34]. More details about the synthesis techniques will be discussed in the later sections below.

MO nanomaterials can be categorized as nanoparticles, nanofibers, nanotubes, nanocomposites and nanostructured materials. MO nanoparticles have unique physicochemical properties like large surface to-volume ratio, extremely small size and size-dependent optical properties. These enable them to be used in various products. TiO_2 and ZnO nanomaterials are commonly used in sun care products to provide protection for skin from UV rays and pleasant skin aesthetics [35]. Iron oxide nanoparticles is having superparamagnetic properties, which can be utilized for drug delivery, vivo cell tracking and magnetic separation of cells, magnetic resonance imaging [36, 37]. Similarly, aluminium oxide is having high hardness and mechanical strength which is being utilized in textiles industry, fuel cells, solar energy application, airbag propellants, paints, polymer coatings etc. [38, 39] Copper oxide nanoparticles (CuO NPs) are highly reactive and have antimicrobial activity which are used in various commercial applications such as, catalysts, gas sensors, antimicrobial agent, batteries and so on [40].

2 Different Classes of Metal Oxides

2.1 S Block Metal Oxides

The elements of the IA and IIA group of the periodic table are called s-block elements. IA [ns^1] group elements are called alkali metals and IIA [ns^2] group elements are called alkaline earth metals respectively. The different binary oxygen compounds formed by alkali metals and alkaline earth metals are (1) oxides, O^{2-} (2) peroxides, $O_2{}^{2-}$, with oxygen–oxygen covalent single bonds (3) superoxide, O_2^-, which also have oxygen–oxygen covalent bonds but with single negative charge [41]. Alkali metals forms all the three types oxides M_2O, peroxides, M_2O_2, and superoxide, MO_2^-. Whereas alkaline earth metals form only oxides, MO, and peroxides, MO_2. Both alkali metal oxides and alkaline earth metal oxides are ionic and react with water to form basic solutions of the metal hydroxide. Thus, these compounds are often called basic oxides.

$$M_2O + H_2O \rightarrow 2\,MOH\ (\text{where } M = \text{group 1 metal})$$

$$MO + H_2O \rightarrow M(OH)_2 (\text{where } M = \text{group 2 metal})$$

Group IA consists of lithium (Li), sodium (Na), potassium (K), rubidium (Rb), caesium (Cs) and francium (Fr) out of which only lithium (Li) produces oxide, sodium (Na) produce peroxides and rest of the elements produce superoxide. The alkali metal form oxide, M_2O by limiting the supply of oxygen. Peroxides M_2O_2, or superoxide, MO_2 is formed by alkali metals when there is excess oxygen. Sodium and Lithium reacts with oxygen to form sodium peroxide, Na_2O_2 and Lithium peroxide Li_2O_2 which are used in emergency breathing devices in spacecrafts and submarines. Other group members form (K, Rb, Cs, and Fr) the super oxide. Group IIA consists of Beryllium (Be), Magnesium (Mg), Calcium (Ca), Strontium (Sr), Barium (Ba), Radium (Ra). Except Ba and Ra, all other metals when burnt in oxygen form oxide of type MO. BeO is amphoteric in nature (i.e., can react chemically as a base and also as an acid) while other oxides are basic in nature (i.e., when it reacts with water gives bases).

$$2M + O_2 \rightarrow 2MO$$

2.2 D-Block Metal Oxides (D-Block MOs)

The d-block MOs include the Transition Metals (elements having a partially filled d or f subshell in any common oxidation state) Oxides (TMOs) and the 2B metal oxides (zinc, cadmium and mercury, which are usually included with the transition elements because of their similar properties). TMOs contain metals elements between titanium (Ti) and copper (Cu) in the 3d series, between Zirconium (Zr) and silver (Ag) in the 4d series, or between Hafnium (Hf) and Gold (Au) in the 5d series of the periodic table. Binary TMOs contain a transition metallic element and oxygen as in NiO, FeO, CuO and TiO_2 etc. In ternary and complex compounds additional metallic elements are also present. The additional metallic elements can be either transition metal, as in Fe_2CoO_4. But sometimes, the additional metallic elements can be pre-transition or post-transition groups as in $LaNiO_3$ and $PbTiO_3$. Most of the transition MOs are solid under normal conditions of temperature and pressure, except a few TMOs which are volatile (OsO_4). TMOs may be good metals (ReO_3), semiconductors (CuO), insulators (TiO_2) and superconductors ($YBa_2Cu_3O_7$). Many compounds show transitions from a metallic state to a non-metallic state as a function of temperature (VO_2), pressure (V_2O_3) or composition (Na_xWO_3). The electrical conductivity and related properties vary greatly for different TMOs [42]. TMOs possess many spectacular magnetic and electrical properties including high-temperature superconductivity and colossal magnetoresistance etc. This makes them particularly promising for nanoscience technological applications, where finite-size effects (quantum confinement), plays an important role compared to the bulk properties due to the reduction of the system dimensions [43]. Similarly, ultrathin MnO_2 nanosheets are capable of absorbing double-stranded DNA that can be used for gene delivery and TiO_2 have promising characteristics for efficient gene silencing [44]. TMOs can show multiple oxidation states, short diffusion path lengths and large surface area that make them extremely efficient to be used as cathode catalyst material in fuel cells. Some examples are molybdenum oxide (MoO_2), tungsten oxide (WO_3), manganese dioxide (MnO_2), vanadium oxide (V_2O_5), cobalt oxide (Co_3O_4), iron oxide (Fe_3O_4), nickel oxide (NiO), and zinc oxide(ZnO) [45]. TMO thin films of WO_3, MoO_3, V_2O_5, NiO, etc. exhibit a vibrant range of Electrochromic (EC) coloration and hence are extensively used in EC devices [46]. Furthermore, $LiMO_2$ (M = 3d transition metal) with a layered rock salt structure are widely used as insertion electrodes in several energy related devices e.g. rechargeable lithium-ion batteries along with $LiCoO_4$ [47], Na-Ion batteries [48] and supercapacitors [49]. TMOs possess important properties like visible light absorption potential, fast charge transfer capability, higher mechanical and thermal stability, non-toxicity, shorter energy band gap, etc. for which these are also used as a photocatalyst for removing organic dyes for water, photo electrochemical cells etc. [50, 51]. The fact that these are cheap materials enable their applicability for these purposes. The d-block metal oxides include zinc oxide (ZnO), cadmium oxide (CdO) and mercury oxide (HgO). Zinc oxide (ZnO) has been the most studied oxide in the last few decades. It has been used in diverse applications including a wide range of cosmetics and personal care products e.g., makeup, nail products, baby lotions, bath soaps and foot powders. ZnO is also used in skin protectants, such as diaper rash ointments and sunscreen products. ZnO has been widely explored for thousands of years. Hence, this material in particular is a mature engineering material of utmost importance [52]. Cadmium oxide (CdO) is an n-type semiconductor with a band gap of 2.5 eV. It has been widely used in cadmium plating baths, electrodes for storage batteries, cadmium salts, catalyst, ceramic glazes, phosphors, and nematocide. Major uses for cadmium oxide are as an ingredient for electroplating baths, optoelectronic devices, solar cells and in pigments [53]. Elemental mercury is provided by Mercury oxide (HgO) for various inorganic mercury salts and organic mercury compounds. This is also used as an electrode in mercury batteries and zinc-mercuric oxide electric cells by mixing it with graphite [54].

2.3 *Other Metal Oxides*

Except alkali metal oxides, alkaline earth metal oxides, transition metal oxides and rare earth metal oxides, other metals present in the periodic table can react with oxygen to form oxides. Most common examples known in science and technology are aluminium oxide (Al_2O_3), gallium oxide (Ga_2O_3), indium oxide (In_2O_3), tin oxide (SnO_2), thallium oxide (Tl_2O), lead oxide (PbO), bismuth oxide (Bi_2O_3), etc.

Al_2O_3 (Alumina) is an amphoteric oxide of aluminium. It is an electrical insulator but has a relatively high thermal conductivity (30 Wm^{-1} K^{-1}). Hence, it is used as a ceramic material. Alumina nanoparticles are stable over a wide range of temperatures. Alumina has many practical applications which includes antimicrobial properties to kill bacteria, drug delivery, biomolecular preservation and stabilisation, biosensing, cancer therapy and immunotherapy [55]. Last but not the least, the use of alumina templates for growing ordered arrangement of nanostructures is an extraordinary application [56].

Gallium oxides (Ga_2O_3) has been investigated from different perspectives. Its applicability is in microwave and optical maser studies. It is used as a phosphor and electroluminescent material. It is also used in chemical sensing and catalysis. Electronically, it is used as transparent conductive coatings, power and high voltage devices, Schottky diodes, field effect transistors, substrates for III-nitrides epitaxy, spintronic devices and many other applications [57].

Indium oxide (In_2O_3) is a transparent semiconducting oxide (TSO) that has been investigated for many decades. While trying to deposit indium thin films at high temperatures in air, it was observed that it oxidized to a transparent and conducting film of polycrystalline In_2O_3. Depending on the oxygen content of the film and the oxygen content of the ambience related to its conductivity. It was sensitive to vacuum and illumination. These properties enable it to use for gas sensing applications. The conductivity or Seebeck coefficient of this oxide is having significant dependence on the gas atmosphere. Since it is activated by illumination it is also known to be a prototype ozone sensor. The widespread application of In_2O_3 as transparent conducting oxide (TCO) is extensively studied. Indium-tin-oxide (ITO) is widely used as TCO in solar cell and active gas sensor material [58].

Tin oxide (SnO_2) finds its importance in applicability as a material for solid state gas sensing, magnetic data storage, magnetic resonance imaging, catalysis, energy-saving coatings, anti-static coatings, electrodes and anti-reflection coatings in solar cells, oxidation catalysis, transparent conductor, optoelectronic devices and resistors [59].

Thallium has two oxidation states +1 and +3 to form oxides. Thallium (III) oxide is a n-type semiconductor which is a highly insoluble thermally stable thallium source suitable for glass, optic and ceramic applications [60].

Lead has many oxide forms including PbO (α, β and amorphous), Pb_2O_3, Pb_3O_4, PbO_2 (α, β and amorphous). Although lead oxides are toxic to human health, it has many applications. It is used in batteries, gas sensors, pigments, ceramics, and glass industry. A recent notable usage of lead oxide has been reported in the field of nuclear radiation shielding (gamma rays shielding) using a composite with PVA, etc. [61].

Bismuth oxide, similar to lead oxides, has various polymorphs that included α-Bi_2O_3, β-Bi_2O_3, γ-Bi_2O_3, δ-Bi_2O_3, ε-Bi_2O_3, and ω-Bi_2O_3 based on synthesis temperature [62]. Amongst these, a stable polymorph is the monoclinic α-Bi_2O_3 phase synthesized at low temperature and cubic δ-Bi_2O_3 at high temperature. With increasing temperature tetragonal β-Bi_2O_3 nature decreases and monoclinic α-Bi_2O_3 dominates. Bismuth oxide has various applications for photovoltaic, photocatalyst, antibacterial and biomedical, gas sensor and solid oxide fuel cells, waste water treatment [63].

2.4 *Rare Earth Metal Oxides (REMOs)*

The rare earth metal oxides are lanthanum oxide (La_2O_3), yttrium oxide (Y_2O_3), cerium (CeO_2), gadolinium oxide (Gd_2O_3), neodymium oxide (Nd_2O_3), dysprosium oxide (Dy_2O_3), praseodymium oxide (Pr_6O_{11}), samarium oxide (Sm_2O_3), europium oxide (Eu_2O_3), terbium oxide (Tb_4O_7), holmium oxide (Ho_2O_3), erbium oxide (Er_2O_3), ytterbium oxide (Yb_2O_3), lutetium oxide (Lu_2O_3), thulium oxide (Tm_2O_3) and scandium oxide (Sc_2O_3).

Rare-earth metal oxides (REMOs) have been used as dopants for titania to enhance photocatalytic activity and photovoltaic performance of dye-sensitized solar cells (DSSCs) [64]. REMOs have many applications that include as a catalyst for oxidation of methane to form hydrocarbons, 1-butene isomerization, hydrogenation of 1,3-butadiene, acetone aldol addition and dehydration/dehydrogenation of alcohols and help to form methanol from CO_2 and H_2O. In the field of superconductors and ceramics, Y_2O_3, La_2O_3 and Er_2O_3 are important ingredients. In carbon arc electrodes the ceramic cores, in optical glass and calcium lights La_2O_3 is being widely use. RE metal oxides glasses are used for optical, electronic and medical purpose technological applications [65].

2.5 *Multifunctional Metal Oxides*

Multifunctional metal oxides are those who possess more than one simultaneous functional properties. They possess unique physical and chemical properties. Among all the multifunctional materials, double perovskites have attracted much scientific and technological interest. Ordered double perovskite structure is derived from single perovskite (ABO_3, where A is rare earth/alkaline earth Metal, B is a transition metal) where exactly half of the B-site cation is replaced by another B-cation and a rock salt ordering is achieved.

The properties of double perovskite depend upon the relative cation size and valency for which they possess the unique properties such as half-metallic ferromagnetism which are having high curie temperature T_C. These ferromagnetic half-metals exhibit large low field room temperature magnetoresistance (MR). La_2NiMnO_6 (LNMO) is an ordered double perovskite ferromagnetic semiconductor with a T_c = 280 K [66]. The electronic cloud distribution can be modified by the application of large magnetic field and hence a change in resistivity and dielectric properties of LNMO. Similarly, under high pressure the double perovskite Bi_2NiMnO_6 (BNMO) in the bulk format exhibits both ferroelectricity and ferromagnetism with transition temperature of 485 K and 140 K respectively [67]. Such materials with co-existing ferroic orders of magnetic and electric properties are referred to as "Multiferroic". These are very rare to find. B-site ordered Sr_2MgMoO_6, is also a mixed oxide-ion/electron conductor. It provides direct electrochemical oxidation in dry methane due to good sulfur-tolerance capacity. La-doped Sr_2MgMoO_6 exhibits better performance on natural gas [68]. Many of the double perovskites exhibits two or more functional properties with rich physics. This field of oxides opens up the doors of vast range of technological applications. Although the study of double perovskite is still limited, it has been reported that many of them display two or more functional properties with rich and exciting fundamental physics. This therefore opens up the doors for a wide range of potential applications.

A summary of different classes of metal oxides discussed above is tabulated with their examples and properties in Table1.

3 Different Synthesis Procedure of Metal Oxide

Almost 100 years back the investigation of various MO powders had been started for different applications. In order to prepare high quality MOs new methods and techniques had been reported by several research groups around the world. The main motivation for this intensive research in this area is due to the significant role of these oxides for many technological and industrial applications. Natural equivalents to these synthetic MOs are not really available. This is a prime reason for an explosion of research reports related to MO synthesis in the last few decades. Naturally occurring MOs are usually a compositional mixture of several components or structural phases

Table 1 Different classes of metal oxides with their properties

Metal oxides	S-block MOs		Alkali MOs	Li_2O, Li_2O_2	• Alkali metals +1 oxidation state form oxides, M_2O, peroxides, M_2O_2, and Super-oxides, MO_2^- (M represents a metal atom) • Na_2O_2 along with Li_2O_2 is used in emergency breathing devices in submarines and spacecrafts	[41, 69, 70]
				Na_2O_2		
				KO_2^-		
				RbO_2^-		
				CsO_2^-		
				FrO_2^-		
			Alkaline Earth MOs	BeO	• The alkaline earth metals (with a + 2 Oxidation state) form only oxides, MO, and peroxides, MO_2 • Except Ba and Ra, all other metals when burnt in oxygen form oxide of type MO • BeO is amphoteric in nature • All other oxides basic in nature	[72–74]
				MgO, MgO_2		
				CaO		
				SrO		
Metal oxides	D-block MOs	Transition MOs	Binary TMOs	TiO_2, TiO, Ti_2O_3 VO, VO_2, V_2O_3, V_2O_5 CrO, Cr_2O_3, CrO_2, CrO_3 MnO, Mn_3O_4 Mn_2O_3 MnO_2 MnO_3 Mn_2O_7 Etc	• Binary transition MOs contain a transition metallic element and oxygen • Depending upon various oxidation states, the oxides can be different	[43, 71, 75, 76]

(continued)

Table 1 (continued)

		Ternary TMOs	$MnFe_2O_4$ $CuFe_2O_4$ $CoFe_2O_4$ $NiFe_2O_4$ $TiFe_2O_4$ VFe_2O_4 Fe_2CoO_4 Etc.	• Ternary Transition MOs contain two transition metallic element and oxygen • Ternary Transition MOs has many practical applications that Energy storage applications and Electrochromic devices	[46–49]
		Complex MOs	$ZnNiCo_2O_4$ $ZnNiMnO_4$ $NiCo_2O_4$ $ZnCo_2O_4$ $LaNiO_3$ $PbTiO_3$	• The additional metallic elements can be a transition metal elements or pre-transition/ post-transition groups as in $LaNiO_3$ and $PbTiO_3$	[77–80]
	2B MOs		ZnO CdO HgO	• ZnO is a n-type semiconductor that has been widely used in cosmetics and personal care products • CdO is an n-type semiconductor that is used in electrodes for storage batteries, cadmium salts, catalyst, ceramic glazes etc • This red or yellow crystalline solid is also used as an electrode (mixed with graphite) in zinc-mercuric oxide electric cells and in mercury batteries	[52, 53, 81]

(continued)

Table 1 (continued)

	Other MOs	Al_2O_3 Ga_2O_3 In_2O_3 SnO_2 Tl_2O PbO Bi_2O_3	• Aluminium Oxide (Al_2O_3) is an amphoteric oxide of aluminium • Gallium oxide (Ga_2O_3) was investigated from the standpoints of microwave and optical maser studies, as a material for phosphors and electroluminescent devices • Indium oxide is widely used as transparent conducting oxide (TCO) indium-tin-oxide and to some extent as active gas-sensor material • Tin oxide (SnO_2) has applications in a solid-state gas sensor, magnetic data storage, magnetic resonance imaging, as catalysts, energy-saving coatings, anti-static coatings, as electrodes and anti-reflection coatings in solar cells • Thallium (III) oxide is a n-type semiconductor which is a highly insoluble thermally stable thallium source suitable for glass, optic and ceramic applications • Lead oxides have many applications as in batteries, gas sensors, pigments, ceramics, and glass industry. The notable application is that it is used in Nuclear Radiation Shielding (gamma rays shielding) • Bismuth oxide has various applications such as in preparation of nanostructures, photocatalyst, solid oxide fuel cell, gas sensor, catalyst for oxidation of hydrocarbons, catalytic performance for reduction, water purification, photovoltaic applications, biomedical, and antibacterial effect	[55–59, 63, 82, 83]

(continued)

Table 1 (continued)

	Rare earth MOs	La_2O_3 Y_2O_3 CeO_2 Gd_2O_3 Nd_2O_3 Dy_2O_3 Pr_6O_{11} Sm_2O_3 Eu_2O_3 Tb_4O_7 Ho_2O_3 Er_2O_3 Yb_2O_3 Lu_2O_3 Tm_2O_3 Sc_2O_3	• REMOs have been used as dopants for titania to enhance photocatalytic activity and photovoltaic performance of dye-sensitized solar cells (DSSCs) • REMOs have many applications that include as a catalyst for oxidation of methane to form hydrocarbons, 1-butene isomerization, hydrogenation of 1,3-butadiene, acetone aldol addition and dehydration/ dehydrogenation of alcohols and help to form methanol from CO_2 and H_2O • Y_2O_3, La_2O_3 and Er_2O_3 are important ingredients in the manufacture of superconductors and ceramics • La_2O_3 is also used in calcium lights, optical glass and in the formation of ceramic cores for carbon arc electrodes	[84–86]
	Multi-functional MOs	La_2NiMnO_6 (LNMO) Bi_2NiMnO_6 (BNMO) Sr_2MgMoO_6	• Multifunctional metal oxides are those who possess two or more intrinsic functional properties simultaneously • They possess the unique properties such as half-metallic ferromagnetism with relatively high curie temperature T_c • These ferromagnetic half-metals exhibit large low field room temperature magneto-resistance (MR)	[87–89]

with a combination of various properties. Thus, these naturally occurring MOs are unsuitable for industrial applications where a particular functionality with precise properties is desirable. The improvement in the single-phase nature of natural MO is highly expensive and the control of size distribution is also a barrier for their use in technological applications. Hence, technology demanded the need of synthetic MOs. Today, the synthetically processed materials vary from nano to bulk sized powders and single crystals or even thin films or composites with different morphology and varying physical/magnetic properties. The preparation techniques have improved from once highly difficult and costly affairs to effective, low-cost methods.

The MO synthesis methods can broadly classify into physical and chemical methods. The chemical reactions induced thermally or mechanically are called physical methods. Whereas the formation of new compounds due to the precipitation process or the decomposition process of the used precursors is called chemical methods. Other methods are biological which are fungi, bacteria and plant mediated synthesis. Top-down approach and bottom-up approach are involved in all the above methods for MO synthesis. The formation of desired structure by assembling small building blocks is the bottom-up approach. Examples for these approaches are coprecipitation, sol-gel, hydrothermal, solvothermal, microemulsion, and chemical vapor reactions. The removal of bulk matter in order to form a smaller desired structure by the usage of mechanical crushing, grinding, milling, pulverization, and mechanochemical reaction is the top-down approach. These two approaches are used in both physical and chemical methods for metal oxide synthesis. Different types of synthesis routes of metal oxide with examples and their advantage and disadvantage are tabulated in Table 2.

Table 2 Different synthesis methods of metal oxides their examples, advantages and disadvantages

Metal oxide synthesis methods

Methods	Physical methods	Chemical methods	Biological methods
Examples	Laser ablation, Ultra sonication, spray pyrolysis, vapour condensation, freeze drying	Sol gel, hydrothermal, coprecipitation	Bacteria, fungi, plant, algae, peptides
Advantages	• Simple • Low cost • Large scale production • Catalyst free	• Obtained nanoparticles, nanotubes • Various parameters can be modified to control the growth	• Non toxic • Inexpensive reagents • Eco friendly
Disadvantages	• Control of parameters is difficult • Energy use	• High cost of precursors • Chemical purification is needed • Use of surfactant	• Stability of the prepared MOs • Unclear reaction mechanism

3.1 Physical Methods

The larger solid particles reduce to smaller powders through the process of milling, grinding and crushing is known as comminution. Different types of milling processes like ball milling, vibratory milling, and attrition milling are used to produce metal oxide powders. These can be conducted in a dry or wet environment of reaction. For the preparation of different MOs high energy ball mill is the widely used milling process. By this method micron-sized to nanosized MO powders can be formed. In order to prepare nanosized powders extensive milling for a long time is required. The size and the size distribution, chemical composition, morphology of the prepared metal oxides samples will depend on the milling time. The variation of the grinding conditions can be greatly influenced by the grinding conditions.

In spray drying a slurry or emulsion is sprayed. Droplets containing multiple components of the anticipated product are sprayed on the clean surface of an appropriate substrate. The substrate is subject to a particular temperature and pressure where the sprayed material is converted to the desired metal oxide nano powders upon further calcination. The composition, crystallinity, size distribution, shape, etc., are controlled by the choice of different stages in the synthesis process, including the spray dryer system and the steps and amount of spray used.

Spray pyrolysis involves more processes compared to spray drying. This method involves decomposition of the precursor droplets through different reactions and formation of final MOs. The different steps involved in these processes are atomization of the liquid feedstock, evaporation followed by drying the sprayed droplets into solid particles. The dried precursor spheres undergo pyrolysis followed by sintering into MOs [90]. With the variation of the conditions of the pyrolysis reaction, the properties of the obtained metal oxides can be varied. Also, this method can be used for the preparation of metal oxides on a large scale.

Freeze drying is another process of metal oxide synthesis in which the desired cations salt solution is atomized using liquid nitrogen or liquid hexane into a liquid of low freezing point. From these atomized droplets the salt precipitates out during freezing. To preserve the solution, homogeneity freezing will be rapid. Porous salt particles in soft agglomerates can be produced in low vacuum or low-pressure conditions through sublimation of frozen solvent. By calcination metal oxide powders are obtained. The particle size will be uniformly distributed in this method. The sudden quenching of the precursor immobilizes the ions within the droplets and leads to the homogenous concentration.

Pulsed laser ablation (PLA) is an eco-friendly process. Very minimal use of chemicals is needed. The material in the target is removed impinging a pulsed laser beam. On the laser irradiated surface, the temperature increases rapidly during the operation. A plasma plume will be generated due to the excitation of electron state coupled with the light emission and formation of electrons and ions which result from the collision of the evaporated species. The metal oxides obtained in this way are highly pure without any impurities and do not need further purification. Low cost and easily controlled parameters make PLA a good method for metal oxide fabrication. Whereas the disadvantages are the agglomeration of the nano powders and low yield.

Vaporization-condensation method involves heating of the required metal target which initiates the vaporization followed by rapid condensation of the vapor of the metal on a substrate. The metal particles are either oxidised before being attached to the substrate or are oxidized on the substrate to produce MOs. This technique has been used for various thin film preparation, MOs nano powders and particles on substrate. It has been used for the preparation of various thin film and particles on substrates as well as metal oxide nano powders.

3.2 Chemical Methods

Chemical methods mainly involve hydrothermal methods and sol gel methods and Chemical Vapour Deposition (CVD). These are based on the precursor gel preparation of the desired component metals needed followed by the decomposition of the precursor gel towards crystallisation of MO structure at a particular convenient temperature.

Sol gel process also known as wet chemical or soft chemistry method. The main steps in sol gel are preparation of the colloidal solution known as sol, then the gelation of the sol followed by the exclusion of the solvent. Schematic of the different stages of sol gel synthesis routes is shown in the Fig. 1. The advantage of this method is the low cost, low temperature is needed to form the functional material. Also, a very small number of precursors are needed for the sol solution which gives maximum yield as well. However, there are certain disadvantages to this process. When it is dried there will be shrinkage of the product. Alkoxide precursors have different reactivity and hence during the sol formation there is a possibility of precipitation of specific oxides. The presence of hydroxyl group is difficult to avoid in this method. Further, the materials are agglomerated and hence residual porosity is high.

A widespread synthesis method for the fabrication of nanostructure MO with different morphology is called the Hydrothermal method. Thin film, single crystal, nanocrystals, and bulk powder can be made by using this method. The advantages of hydrothermal methods over other solution-based methods are the ability to control the size and morphology, flexibility, and energy saving. The disadvantage of these methods is the low yield of the product. The morphology of the prepared samples can be varied by changing the pH value of the solution, reactant concentrations, nature of the substrate, temperature of the deposition and time. As the morphology of the samples determine and monitor the functionality of the materials this method has emerged to be one of the most important techniques of modern-day research on nanomaterials.

Another example of a chemical synthesis method is the Chemical Vapour Deposition (CVD). Here in the first step the reaction chamber is filled by flowing gaseous raw materials. Through a complex reaction process solid material is deposited on the substrate. The product of this synthesise method is denoted as vapor ordered crystal

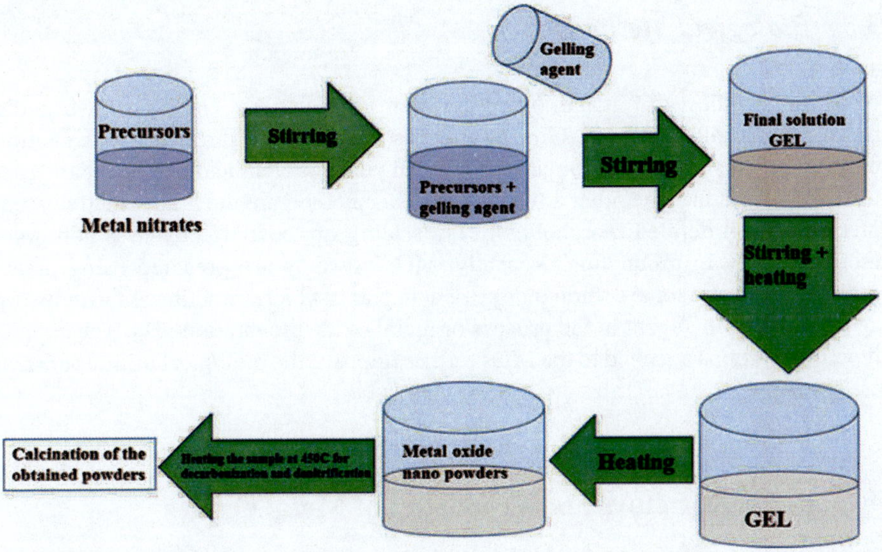

Fig. 1 Different stages of sol gel synthesis method

growth. Quality and purity will be very high for the samples prepared by these methods. For the deposition of thin films in the semiconductor industry mainly use this method of preparation. In the CVD process the first step is flowing the precursor into the reaction chamber which is in gas state. Chemical reactions will be taking place when precursor gas comes in contact with the heated substrate. The product of this chemical reaction forms a solid state and deposits to the substrate. Thermal decomposition or thermolysis involves high temperature chemical decomposition, such that the temperature of deposition will be the temperature of the decomposition. This is considered as a subclass of solution-phase synthesis.

Another important chemical method is sono-chemical method, which uses ultrasonic irradiation with high intensity. High purity crystalline nanomaterials are prepared by these methods. Molecules undergo many chemical reactions because of the powerful ultrasound radiation. Also, different steps such as generation, growth and collapse of bubbles in the liquid phase are influenced by these ultrasounds. Sono-chemical methods can be classified into homogeneous and heterogeneous. In homogenous sono-chemical methods metal oxides are synthesised through free radical formation. Whereas there are certain disadvantages as well such as need for the argon atmosphere for the proper functioning of the process, long processing time and heating during the reaction, random morphology of nanoparticles are the disadvantages of this method.

3.3 Biological Methods

Nontoxic synthesis of metal oxide powder with various environment-friendly processes can also be used. Green synthesis can be done through the mediation of microorganisms such as fungi, bacteria and yeast, protein and plants. Apart from the nontoxic nature, the other advantages of this method are the high selectivity and precision, well defined morphology, easy scaling up, cost effectiveness and good reproducibility in production. Recently, MOs powders are prepared using plants extracts. There are some compounds present in plants which are acting as the reducing agent and capping agent in the process of metal oxide preparation [91]. The capped functional groups attached to the MOs further improve the biological utilities of such materials.

4 Crystal Structure and Properties of Metal Oxides

Different properties and functionalities exhibited by the MOs can be correlated to the crystal structure of these materials. Diffraction techniques are used to look at the crystal structure of these oxide materials. Analysis of these metal oxide structures can explain the different exotic properties shown by them. The modification of crystal structure through different methods such as doping can greatly enhance the various properties of these materials.

Metal oxide crystal structure depends on the stoichiometry and coordination. In binary oxides the most common stoichiometries are MeO, MeO_2, Me_2O_3, and Me_3O_4. Ternary metal oxides appear as Me(I) Me(II)O_3, and Me(l)Me(Il)$_2O_4$. In all these cases the tetrahedral (surrounded by four oxygen) and octahedra coordination (surrounded by six oxygen) of metal ions are common.

When considering the crystal structure there can be simple binary monoxides (NiO, CoO) with NaCl structure or ternary oxides ($BaTiO_3$, $SrTiO_3$) with perovskite structure. They also can form complicated structures such as Co_3O_4 or Fe_3O_4 with the normal and/or inverse spinel structure. Examples of metal oxides for different crystal structures are tabulated in Table 3. The close-pack array of oxygen anions along with the metal cations at the lattice sites constitute metal oxide structures. These close-packed arrays consist of usually face-centred-cubic and hexagonal-close packed structures. The smaller metal cations are observed to be in an octahedral or tetrahedral crystal field with O-anions. A tetrahedral site with four neighbouring oxygen ions (ionic radius ~ 1.21 Å) can accommodate a spherical cation of radius 0.32 Å without any distortion. On the other hand, an octahedral site with six neighbouring oxygen ions can accommodate a larger ion of radius 0.58 Å. Hence, structures are influenced by the ionic size of the component metal ions. For each oxygen atom there are two tetrahedral and one octahedral interstice in the close packed lattice [92].

The crystal structure of metal oxides varies from cubic to triclinic. Most transition metal oxides have a cubic rock salt crystal structure, but ZnO and CoO are the only

Table 3 Different MO crystal structures and their examples

No	Crystal structure	Examples	References
1	Cubic	In_2O_3, CdO, NiO, CeO_2	[93–95]
2	Hexagonal	ZnO, CoO	[96]
3	Tetragonal	SnO_2	[97]
4	Orthorhombic	V_2O_5	[98]
5	Monoclinic	CuO, $Nb_{12}O_{29}$	[99]
6	Rhombohedral	Cr_2O_3	[100]
7	Spinel/inverse spinel	Fe_3O_4, Co_3O_4, $MgAl_2O_4$, $CuFe_2O_4$, $ZnCr_2O_4$	[101–105]
8	Rutile	TiO_2, NbO_2, RuO_2 and IrO_2	[10, 106–108]
9	Perovskite	$BaTiO_3$, $LaMnO_3$, $BiFeO_3$, $PrNiO_3$	[109–112]
10	Double Perovskite	La_2NiMnO_6, Sr_2FeCoO_6, $BiFeCrO_6$	[113–115]

stable transition metal oxides known to possess a hexagonal structure. ZnO has hexagonal wurtzite structure. The structure can be described as a formation of two interpenetrating hcp lattices. Both of these hcp lattices have the same axis with one of them displaced with respect to the other. In the hcp array of the lattice half of the tetrahedral sites are occupied by the metal cations. CuO is a famous metal oxide with monoclinic structure. The Cu atom in a CuO lattice is in an octahedral geometry with six O around it in a distorted octahedral form. However, the separation of two O atoms is too large to be considered as a bond. Hence, practically Cu is four-coordinated in CuO lattice. NiO has a cubic close packed structure or rock salt structure with Ni^{2+} and O^{2-} occupying the octahedral sites. The Ni^{2+} and O^{2-} ratio should be 1:1 ideally but it may not be always depending on the synthesis conditions. For 1:1 ratio of cation and anion NiO exhibits green colour when it deviates from the ideal behaviour it exhibits black colour.

A TiO_2 rutile structure is formed with Ti-cations in an octahedral field with six O-anions. This is a polymorph of TiO_2 which can also be present in anatase and brookite forms. All these polymorphs retain the octahedral field with differences in the octahedral arrangement in the lattice. In general, the rutile structure is a robust one which can be obtained due to a permanent phase transition from the other polymorphs on account of total structural reconstruction mediated by a thermodynamically assisted rotation of the octahedra in order to attain maximum density of the lattice. Other rutile phases can be observed for NbO_2, RuO_2 and IrO_2. In the MO_6 octahedra in rutile structure the adjacent octahedra are shared by the tetragonal c-axis. In NbO_2, RuO_2 and IrO_2, a d-orbital overlap is observed, which generates a metallic nature of these oxides.

Perovskite structures of metal oxides are investigated intensively for a wide variety of applications because of their tunability of compositions and structures. Perovskite family is having general formula ABO_3 where, lattice site A is for 12-fold coordinated cube-octahedral cages of the oxygen sublattice occupied by alkaline-earth or

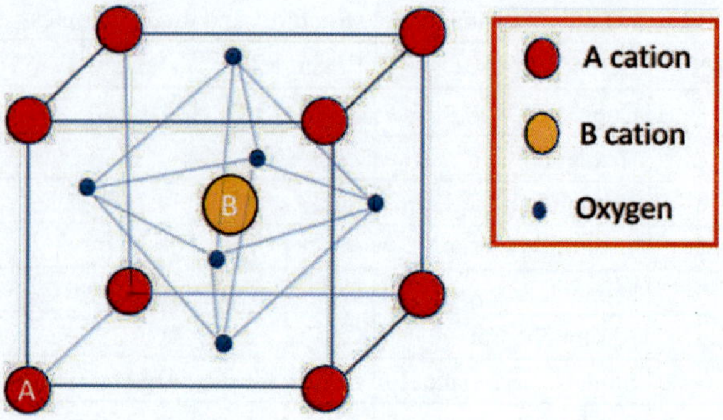

Fig. 2 Perovskite ABO_3 crystal structure

rare-earth metal cation. B lattice site surrounded by six oxygen atoms in an octahedral coordination occupied by a transition-metal cation. Figure 2 provides a general schematic of the ABO_3 perovskite structure. Where 12 coordinated A cations are represented by red balls and the octahedrally coordinated B cations are represented by yellow balls. Blue balls represented by the Oxygen anions. The difference in the ionic radii of different sites leads to the distortions in the perovskite structure which is expressed in terms of tolerance factor. Wide variety of A and B cations having different sites and oxidation states can occupy the lattice sites of ABO_3. This will helpful for tuning various properties of these compounds sites. The valence state of the constituent ions determines the oxygen stoichiometry often leading to a lattice with oxygen vacancies (deficient) or interstitials (excess).

Perovskite oxides may also be arranged as double perovskite or layered perovskite based on the composition or crystal structure. If the A site or B site is constituted with two different species of cations namely A', A'', B', B'' then one can achieve chemical compositions of the formulae $A_2B'B''O_6$, $A'A''BO_6$, $A'A''B'B''O_6$. These compounds retain the basic perovskite structure with a periodic arrangement of alternating cations and are called double perovskites. The physicochemical properties of perovskite oxides are based on their structure, composition and morphology [116]. Figure 3 shows the schematic of the periodic arrangement of the general double perovskite structure. Where violet and blue sites represent the ordered periodic arrangement of B' and B'' cation which gives the double perovskite structure. Green balls representing the A cation in the interstitial sites of theses ordered B', B'' cations. Oxygen anions are represented by red balls.

The chemical formula of spinel structure is AB_2O_4, where A is tetrahedrally coordinated and B ions are octahedrally coordinated cations with O-ions. Figure 4. shows the representation of the spinel AB_2O_4 structure with grey site represent tetrahedrally coordinated A cations and blue lattice sites represent the octahedrally coordinated B cations. The red sites are oxygen ions. The B-site is occupied by transition metal ions Fe^{3+}, Cr^{3+}, Mn^{3+} surrounded by a close-packed array of O-ions. The A-site is occupied by divalent cations such as Mn^{2+}, Fe^{2+}, Co^{2+}, Ni^{2+}, Cu^{2+}, Zn^{2+}, or Mg^{2+}, sometimes Li^+ or vacancies as well. An addition of a higher valent ion other than a trivalent cation at the B-site can lead to a vacancy at the A-site as a

Fig. 3 Double perovskite $A_2B'B''O_6$ crystal structure

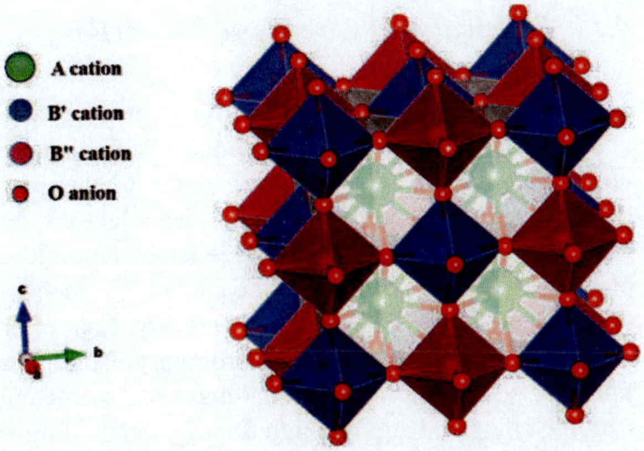

requirement of charge compensation. This can also be achieved by accommodating an O-interstitial. 1/8 of the tetrahedral sites are occupied by the A site cations and 1/2 of the octahedral sites are occupied by B site cations. The spinel lattice $A_8B_{16}O_{32}$ has 8 formula units which contains 32 oxygen anions. There are two sublattice in the spinel structure one with 8 tetrahedrally coordinated cations and other sublattice with 16 cations in octahedral sites. In an inverse spinel structure, the A-site ions and half of the B-site ions switch places but the same large unit cell as that of spinel. The chemical formula includes $B(AB)O_4$, in which the AB ions in the bracket occupy the octahedral sites and the B ions outside the bracket occupy the tetrahedral sites. The intermediate between the spinel and inverse spinel structure are called mixed spinels. The crystal field stabilization energy (CFSE) of ions in the tetrahedral and octahedral sites determine the choice of spinels and inverse spinels.

Fig. 4 AB_2O_4 spinel crystal structure

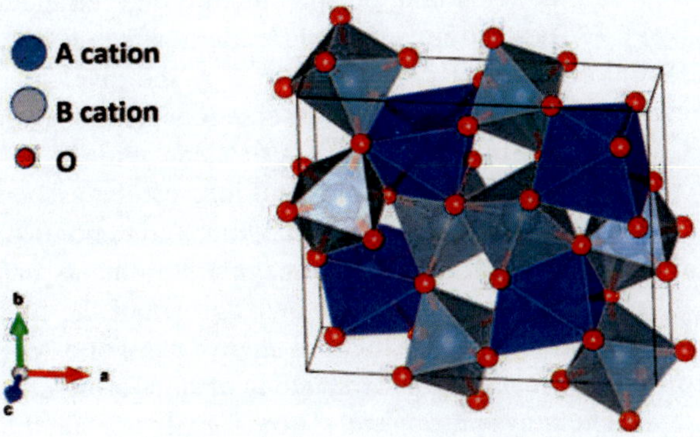

4.1 Physical and Chemical Properties

MOs show a rich variety of electronic properties, ranging from insulating to metallic and even superconducting behaviour. Oxide materials are of great scientific and commercial interest due to the various applications in high temperature superconductors, magneto resistive materials, piezoelectrics, ferroelectrics, multiferroic materials. Metal oxides and metal oxide-based heterojunction materials are interesting class of compounds because of the advantages such as optical transparency, mechanical stress tolerance, high carrier mobility, Cost effectiveness etc. MOs are having definite crystal structures. They are non-poisonous, and have greater stability, and interesting surface properties that make them efficient semiconductors. The modification of the bandgaps through doping, nano composites, formation of composites leads to the improvement of the physical and electronic properties of MOs. Also, this bandgap engineering will be beneficial for the enhanced light absorption in the visible region and widens the horizon of application of MOs. For example, MO heterojunctions on Si substrates have been studied over the years in order to fabricate modern electronic devices. These devices have efficient charge transport properties at the interface of different MOs.

4.2 Magnetic Properties

The magnetic properties exhibited by the MOs is due to the partially filled d shell metal ions such as Fe, Co, Mn, Ni, Cu etc. The first magnetic material to be discovered was lodestone, which is a well-known magnetic MO ferrite Fe_3O_4 which is having a cubic inverse spinel structure. Half of the octahedral sites are occupied by Fe^{2+} ions and the remaining octahedral sites and tetrahedral sites are evenly distributed by Fe^{3+}. These arrangement of Fe ions leads to ferrimagnetism in these compounds. Partial or full replacement of Fe^{2+} by the divalent ions such as Co, Mn, Zn, etc. can create various magnetic interactions. Cations such as Mn^{2+}, Co^{2+}, Ga^{3+}, In^{3+}, Cu^{+}, Mg^{2+}, Si^{4+}, Ge^{4+}, Zn^{2+}, Cd^{2+}, can replace Fe^{3+} ions in the tetrahedral site. The magnetism of free atoms and ions can be associated with angular momentum. For a free atom the spin and angular momentum of each electron obeys Hund's rule. Following this rule, the magnetic moments can be calculated easily without resorting to any particular knowledge about the individual atoms. Whereas when ions are placed in a crystal in an environment of oxygen neighbours the effect will be different. A complex understanding of spin–orbit interaction, exchange-integrals and crystal fielding splitting etc. is necessary to explain the magnetism of these materials. Depending on the exchange interaction the material will show Ferromagnetic (FM), Antiferromagnetic (AFM), Ferrimagnetic (FM)behaviour.

When considering the MO nanomaterials, the magnetic properties will be different from their bulk counterparts. Curie temperature, Neel temperature, coercive field etc. will vary from bulk to nano region [117]. In bulk MOs small regions of a particular spin-ordering can be observed called magnetic domains. Different energy phenomena like exchange energy, magneto crystalline anisotropy, and magnetostatic energy determine how the magnetic domains are formed and their orientation. In nano MOs the magnetic contribution is from two regions, surface spins component of the particle and core spin component of the particle. So the nanomagnetic interaction can be modelled via core- shell interaction. This interaction in AFM or FIM nanoparticle occurs at the interface between FM surface and the AFM (or FIM) core. This interaction can also called the exchange bias or exchange coupling [118].

4.3 *Electronic Properties*

The electronic structure of the metal in the MOs plays an important role in the properties exhibited by these materials. These MOs include wide band gap insulators, semiconductors, metals, and superconductors. There are various ways to change the electronic structure of MOs by doping, strain and defect engineering. Depending on the type of charge carriers metal oxides have p type and n type semiconductors. Doping metal oxides with acceptor or donor atoms leads to the formation of defect states in the electronic band structure. This may lead to the changes in the electronic properties exhibited by these oxides. The optical, magnetic properties exhibited by MOs are closely related to their electronic structure. Hence band structure modification with dopants leads to the modification of the other physical properties exhibited by them.

4.4 *Optical Properties*

Depending on the band gap of MOs these are semiconductors and larger band gaps are considered as insulators. Semiconducting metal oxides with a band gap below 3.5 eV, have an optical absorption in the range of near UV and visible electromagnetic spectrum. But the insulating MO has optical absorption only in the UV or higher energy. Depending on the metal the band gap of MO can vary. Also, with doping or impurity atoms in MO there is alteration of bandgap. This change in band gap with doping is due to the formation of defect states within the bandgap that leads to decrease of band gap or removal of inherent defect states in the sample that leads to increase of band gap. The colour exhibited by these MOs are depending on their band gap such as the Fe_2O_3 has rich colour due to its small bandgap while ZnO has no colour because of its relatively large bandgap.

4.5 *Catalytic Properties*

MO catalyst and catalytic reactions are essential in industrial processes, chemical synthesis, petrochemical processes and environmental issues. In the mid-1950s the importance of MOs became prominent when they were found to catalyse various acid base reactions. These are simple oxides like CuO, ZnO, Alumina, TiO_2, and other complex oxides, perovskite oxides etc. Research in the field of enhancing the selectivity of the catalyst with various preparation methods are going vigorously and many new strategies have resulted in the development of single sites with 100% selectivity of the metal oxide catalyst. Usefulness of a catalyst is mainly determined by the activity, selectivity, resistance to deactivation, and ability for regeneration. Catalytic activity includes oxidation (selective or total), acid and base catalysis, photocatalysis, depollution and biomass conversion. Apart from being a catalyst MOs are sometimes used as supports of active phases such as silica, alumina, silica-alumina, mesoporous oxides etc. The synergistic electron conductivity or/and thermal conductivity effects arises due to metal oxide support interaction influences the catalytic properties [119].

5 Applications

Due to their distinctive structure, catalytic capabilities, large surface area, and good mechanical stability, metal oxide nanoparticles play important roles in several diverse sectors. They are actively applied in gas sensing, energy storage, solar cells, catalysis, electronics, ceramic fabrication, etc. Some materials show excellent biocompatibility which poses a great role in the biomedical field. The diverse fields in which Mos are involved and examples for their utility is schematized in Fig. 5.

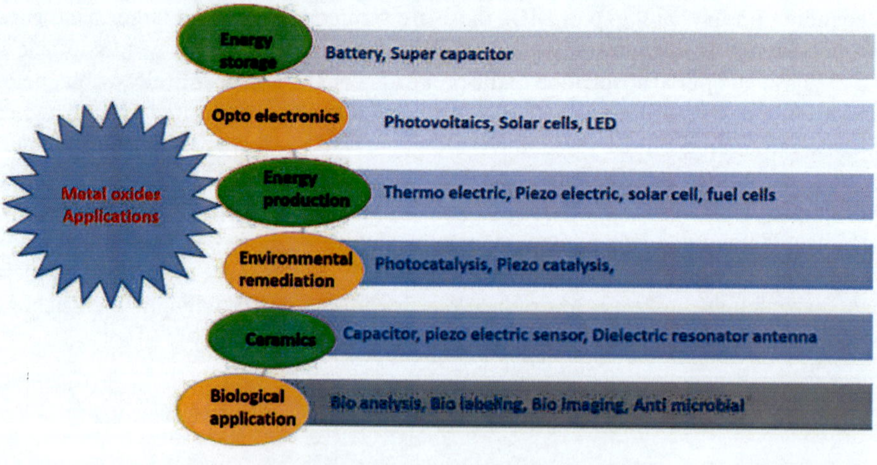

Fig. 5 Schematic of the various applications of MOs in different fields

There have been significant increases in the discharge of poisonous, explosive, and flammable gases in developed countries leading to increased environmental pollution, manufacturing wastes, and urbanization issues. Gas sensor applications play an important role in detecting the presence and quantity of such hazardous gases. Many metal oxides are ideal for using conductive measurements to detect combustible, reducing, or oxidizing gases. Various factors like chemical composition, surface modifications, microstructures, humidity, and temperature are important parameters for the sensitivity of the material. The electronic structures of the metal oxides play an important role in the selection of gas sensors. By using conductivity measurements, several metal oxides can detect flammable, reducing, or oxidizing gases. The conductivity of the following oxides shows a gas response: CuO, Ta_2O_5, Cr_2O_3, Nd_2O_3, SrO, Fe_2O_3, GeO_2, WO_3, CeO_3, Nb_2O_5, V_2O_3, NiO, Mn_2O_3, Co_3O_4, In_2O_3, MoO_3, and TiO_2.

Metal oxides are appealing cathode materials in electrochemical energy storage systems considering their relatively low molecular weight and negligible toxicity along with their good electrochemical and solid-state properties. Electricity supply, transmission, and distribution systems on the grid can benefit from energy storage. It removes the need for reliance on other options with a higher greenhouse gas footprint, which would otherwise be favoured with zero storage options.

The integrated circuits, diodes and thin-film transistors, field-effect transistors in the electronics are all made of metal oxide semiconductor materials. MO semiconductors are used because they are efficient in regulating the current flow in the device as they have highly desirable electron charge transport propertie. Nanoelectronics also see the usage of metal oxides rather than any other semiconductor, insulator or conductor because of the many electrical, chemical and physical properties that are ideal for use in electrical devices. Both n-type and p-type MOs semiconductors are observed. Amongst n-type MOs examples are tin oxide (SnO_2), indium oxide (In_2O_3), zinc oxide (ZnO), and titanium dioxide (TiO_2) while some prominent p-type MOs can be enlisted as nickel oxide (NiO) and copper oxide (CuO). These have been used in different applications. Zinc tin oxide (ZTO) [120], aluminium zinc oxide (AZO), magnesium zinc oxide (MgZO), gallium zinc oxide (GaZO), indium hafnium zinc oxide (InHfZO), and indium silicon zinc oxide (InSiZO) are important examples of functional complex MOs [121].

The use of metal oxides in Dielectric Resonator Antenna (DRA) applications has become important in the last decade. These antennas transmit and receive microwave frequencies and are generally made of ceramic materials. Indium tin oxide (ITO) [122] has been used in plasmonic nano antennas. The feature of MO here is that it acts as a TCO (Transparent conducting oxide). They have a large band gap dielectric. This enhances the usage of antennas for various purposes.

The property of MO to generate charge carriers when stimulated with suitable energy of light makes them important in technology and environment remediation. The charge transfer property followed by the light absorption makes these materials interesting for photocatalysis for environmental remediation. Considering economic and social development the MOs are efficient for organic dye removal and wastewater treatment processes.

Energy production and storage are the two main core areas of research to tie up the energy crisis that we are going to face in the near future. The application of MOs in supercapacitors and batteries is quite overwhelming. The property of variety of charge states in MOs allow them for the oxidation reduction charge transfer processes in supercapacitors. The blooming of nanotechnology and the research in the field of MO heterostructures paved the way for supercapacitors with good efficiency compared to the conventional ones. Device fabrication using MOs for thermoelectric and piezoelectric nanogenerators are becoming extremely important in recent times by making use of thermal and mechanical energy. The thin film preparation of these materials with different synthesis techniques are now used for the wearable nanogenerators (Table 4).

Table 4 Properties and applications of some important MOs

No	Important metal oxides	Properties	Applications	References
1	ZnO	High band gap, piezoelectric	Optical sensor, opto electronics, piezoelectric	[123]
2	CuO	Thermal stability, high reactivity, low band gap	Photocatalysis, antibacterial, photocells	[124]
3	NiO	Magnetic property, Capacitive property, electro active	Supercapacitor, sensor, photodetector	[125, 126]
4	TiO_2	Highly refractive, wide band gap	Pigment, photocatalysis	[127, 128]
5	Al_2O_3	Electrical insulator, high thermal conductivity	Gas purification, catalysis,	[129]
6	SnO_2	Magnetic property, wide band gap	Magnetic data storage, electrodes in solar cell, optoelectronic devices	[130, 131]
7	$BaTiO_3$	Dielectric, Piezoelectric	Capacitor, sensor	[132]
8	Co_3O_4	High specific capacitance, Magnetic property,	Supercapacitor, Bio medical application, Li-Ion batteries	[133–135]
9	Fe_3O_4	Magnetic property, antimicrobial property,	Biosensing, Antibacterial, Cancer theragnostic, catalysis	[136, 137]
10	La_2NiMnO_6	Ferromagnetism, spin phonon coupling	Spintronics, solar cells	[138]
11	$PbTiO_3$	Pyroelectric, transducer	Pyro detector, acoustic transducer	[139]
12	$LiNbO_3$	Piezoelectric	Pyro detector, surface acoustic wave guide	[140]

6 Conclusions

This chapter covers an overview of the MOs in terms of their structure, properties, synthesis methods and its wide applications. MOs have prime importance in the field of Physics, Chemistry, Biology and Material Science. Hence, understanding the science behind this important class of materials is of prime importance. Scientists explore the various properties of MOs for technological applications. Applications of MOs in energy storage, environmental remediation, biological field, sensors, optoelectronics are closely related to their structure correlated properties. Structural modification of these materials may unveil the unexplored properties of various MOs which may pave a novel pathway for different technological applications. Device fabrication based on MOs, MO thin film, MO nanomaterials, flexible MO films, MO heterojunctions are growing very fast for their practical applications and utility. Different research groups around the world are intensively exploring the exotic properties of these materials.

References

1. D. Smyth, The effects of dopants on the properties of metal oxides. Solid State Ionics **129**, 5–12 (2000). https://doi.org/10.1016/S0167-2738(99)00312-4
2. M. Sytnyk et al., Tuning the magnetic properties of metal oxide nanocrystal Heterostructures by cation exchange. Nano Lett. **13**(2), 586–593 (2013). https://doi.org/10.1021/nl304115r
3. H. van Gog, W.-F. Li, C. Fang, R. S. Koster, M. Dijkstra, M. van Huis, Thermal stability and electronic and magnetic properties of atomically thin 2D transition metal oxides. NPJ 2D Mater. Appl. **3**(1), Art. no. 1 (2019). https://doi.org/10.1038/s41699-019-0100-z
4. G. Kumar Dalapati et al., Tin oxide for optoelectronic, photovoltaic and energy storage devices: a review. J. Mater. Chem. A**9**(31), 16621–16684 (2021). https://doi.org/10.1039/D1TA01291F
5. Y. Yoon, P.L. Truong, D. Lee, S.H. Ko, Metal-oxide nanomaterials synthesis and applications in flexible and wearable sensors. ACS Nanosci. Au **2**(2), 64–92 (2022). https://doi.org/10.1021/acsnanoscienceau.1c00029
6. W. Smok, T. Tański, A short review on various engineering applications of electrospun one-dimensional metal oxides. Materials (Basel) **14**(18), 5139 (2021). https://doi.org/10.3390/ma14185139
7. N. Joshi, D. K. Pandey, B. G. Mistry, D. K. Singh, Metal oxide nanoparticles: synthesis, properties, characterization, and applications, in *Nanomaterials: advances and Applications*, eds. by D.K. Singh, S. Singh, P. Singh (Springer Nature, Singapore, 2023), pp. 103–144. https://doi.org/10.1007/978-981-19-7963-7_5
8. D. Ganguli, M. Chatterjee, *Ceramic Powder Preparation: a Handbook*, vol. 59 (Springer, 1997)
9. S. Jafari, B. Mahyad, H. Hashemzadeh, S. Janfaza, T. Gholikhani, L. Tayebi, Biomedical applications of TiO_2 nanostructures: recent advances. IJN **15**, 3447–3470 (2020). https://doi.org/10.2147/IJN.S249441
10. I. Ali, M. Suhail, Z.A. Alothman, A. Alwarthan, Recent advances in syntheses, properties and applications of TiO_2 nanostructures. RSC Adv. **8**(53), 30125–30147 (2018). https://doi.org/10.1039/C8RA06517A

11. S. Raha, M. Ahmaruzzaman, ZnO nanostructured materials and their potential applications: progress, challenges and perspectives. Nanoscale Adv. **4**(8), 1868–1925 (2022). https://doi.org/10.1039/D1NA00880C
12. Q. Zhao, L. Ma, Q. Zhang, C. Wang, X. Xu, SnO_2-based nanomaterials: synthesis and application in lithium-ion batteries and supercapacitors. J. Nanomater. **2015**, e850147 (2015). https://doi.org/10.1155/2015/850147
13. X. Wang, C. Hu, H. Liu, G. Du, X. He, Y. Xi, Synthesis of CuO nanostructures and their application for nonenzymatic glucose sensing. Sens. Actuator. B Chem. **144**(1), 220–225 (2010). https://doi.org/10.1016/j.snb.2009.09.067
14. N. Tsuda, K. Nasu, A. Yanase, K. Siratori, *Electronic Conduction in Oxides*, vol. 94 (Springer, Berlin, Heidelberg, 1991). https://doi.org/10.1007/978-3-662-02668-7
15. Springer Series in Solid-State Sciences. Springer. https://www.springer.com/series/682. Accessed 27 Jan. 2023
16. Structure–Property Relationship and Chemical Aspects of Oxide–Metal Hybrid Nanostructures | Chemical Reviews. https://pubs.acs.org/doi/full/https://doi.org/10.1021/cr300307n. Accessed 27 Jan. 2023
17. Z. Yang, C. Ko, S. Ramanathan, Oxide electronics utilizing ultrafast metal-insulator transitions. Annu. Rev. Mater. Res. **41**(1), 337–367 (2011). https://doi.org/10.1146/annurev-matsci-062910-100347
18. A.O. Chibirev, A.V. Leontyev, V.V. Kabanov, R.F. Mamin, Origin of negative photoconductivity at the interface of Ba0.8Sr0.2TiO3/LaMnO3/Ba0.8Sr0.2TiO3 Heterostructures. Nanomaterials **12**(21) (2022). https://doi.org/10.3390/nano12213774
19. S. David, A. Soosaimanickam, T. Sakthivel, B. Sambandam, S. Anandhi, Thin film metal oxides for displays and other optoelectronic applications, pp. 185–250 (2020). https://doi.org/10.1007/978-3-030-53065-5_6
20. D. Kumar, R. Aluguri, U. Chand, T. Tseng, Metal oxide resistive switching memory: materials, properties, and switching mechanisms. Ceram. Int. (2017)
21. J.C. Védrine, Metal oxides in heterogeneous oxidation catalysis: state of the art and challenges for a more sustainable world. Chemsuschem **12**(3), 577–588 (2019). https://doi.org/10.1002/cssc.201802248
22. S. Guo et al., Coordination-assisted precise construction of metal oxide nanofilms for high-performance solid-state batteries. J. Am. Chem. Soc. **144**(5), 2179–2188 (2022). https://doi.org/10.1021/jacs.1c10872
23. Metal Oxide Gas Sensors by Nanostructures | IntechOpen. https://www.intechopen.com/chapters/68941. Accessed 30 Jan. 2023
24. M.P. Nikolova. M.S. Chavali, Metal oxide nanoparticles as biomedical materials. Biomimetics **5**(2) (2020). https://doi.org/10.3390/biomimetics5020027
25. M. Roth, M. Tseitlin, N. Angert, Oxide crystals for electro-optic Q-switching of lasers. Glass Phys Chem **31**(1), 86–95 (2005). https://doi.org/10.1007/s10720-005-0028-6
26. G. Liu et al., Low-temperature, nontoxic water-induced metal-oxide thin films and their application in thin-film transistors. Adv. Func. Mater. **25**(17), 2564–2572 (2015). https://doi.org/10.1002/adfm.201500056
27. J. Bouclé, P. Ravirajan, J. Nelson, Hybrid polymer–metal oxide thin films for photovoltaic applications. J. Mater. Chem. **17**(30), 3141–3153 (2007). https://doi.org/10.1039/B706547G
28. H. Meixner, J. Gerblinger, U. Lampe, M. Fleischer, Thin-film gas sensors based on semiconducting metal oxides. Sens. Actuator. B Chem. **23**(2), 119–125 (1995). https://doi.org/10.1016/0925-4005(94)01266-K
29. C.D. Lokhande, D.P. Dubal, O.-S. Joo, Metal oxide thin film based supercapacitors. Curr. Appl. Phys. **11**(3), 255–270 (2011)
30. M.-G. Kim, M.G. Kanatzidis, A. Facchetti, T.J. Marks, Low-temperature fabrication of high-performance metal oxide thin-film electronics via combustion processing. Nat. Mater. **10**(5), 382–388 (2011)
31. A. Amri, Z.T. Jiang, T. Pryor, C.Y. Yin, Z. Xie, N. Mondinos, Optical and mechanical characterization of novel cobalt-based metal oxide thin films synthesized using sol-gel dip-coating method. Surface Coatings Technol. **207**, (2012)

32. H.Y. Sohn, A. Murali, Plasma synthesis of advanced metal oxide nanoparticles and their applications as transparent conducting oxide thin films. Molecules **26**(5), 1456 (2021)
33. M.G. Lee, J.S. Park, H.W. Jang, Solution-processed metal oxide thin film nanostructures for water splitting photoelectrodes: a review. J. Korean Ceram. Soc. **55**(3), 185–202 (2018)
34. W.-J. An, E. Thimsen, P. Biswas, Aerosol-chemical vapor deposition method for synthesis of nanostructured metal oxide thin films with controlled morphology. J. Phys. Chem. Lett. **1**(1), 249–253 (2010)
35. W.J. Stark, P.R. Stoessel, W. Wohlleben, A. Hafner, Industrial applications of nanoparticles. Chem. Soc. Rev. **44**(16), 5793–5805 (2015)
36. S. Naqvi et al., Concentration-dependent toxicity of iron oxide nanoparticles mediated by increased oxidative stress [Retraction]. Int. J. Nanomed. **17**, 1459–1460 (2022)
37. Q. Feng, Y. Liu, J. Huang, K. Chen, J. Huang, K. Xiao, Uptake, distribution, clearance, and toxicity of iron oxide nanoparticles with different sizes and coatings. Sci. Rep. **8**(1), 1–13 (2018)
38. I.-S. Kim, M. Baek, S.-J. Choi, Comparative cytotoxicity of Al2O3, CeO2, TiO2 and ZnO nanoparticles to human lung cells. J. Nanosci. Nanotechnol. **10**(5), 3453–3458 (2010)
39. Z. Poborilova, R. Opatrilova, P. Babula, Toxicity of aluminium oxide nanoparticles demonstrated using a BY-2 plant cell suspension culture model. Environ. Exp. Bot. **91**, 1–11 (2013)
40. G. Ren, D. Hu, E.W. Cheng, M.A. Vargas-Reus, P. Reip, R.P. Allaker, Characterisation of copper oxide nanoparticles for antimicrobial applications. Int. J. Antimicrob. Agents **33**(6), 587–590 (2009)
41. T.R. Krawietz, D.K. Murray, J.F. Haw, Alkali metal oxides, peroxides, and superoxides: a multinuclear MAS NMR study. J. Phys. Chem. A **102**(45), 8779–8785 (1998). https://doi.org/10.1021/jp9823190
42. P.A. Cox, *Transition Metal Oxides: an Introduction to their Electronic Structure and Properties*, vol. 27 (Oxford university press, 2010)
43. M. Finazzi, Magnetic properties of oxide ultrathin films on nonmagnetic substrates (2018)
44. M. Nurunnabi, J. McCarthy, *Biomedical Applications of Graphene and 2D Nanomaterials* (Elsevier, 2019)
45. Z. Zhang, J. Liu, J. Gu, L. Su, L. Cheng, An overview of metal oxide materials as electrocatalysts and supports for polymer electrolyte fuel cells. Energy Environ. Sci. **7**(8), 2535–2558 (2014). https://doi.org/10.1039/C3EE43886D
46. P. Ashrit, Electrochromic thin films and devices, in *Transition Metal Oxide Thin Film based Chromogenics and Devices* (Elsevier, 2017), pp. 73–151. https://doi.org/10.1016/B978-0-08-101747-0.00003-9
47. R. Kanno, Secondary batteries—Lithium rechargeable systems—Lithium-ION | positive electrode: lithium nickel oxide, in *Encyclopedia of Electrochemical Power Sources* (Elsevier), pp. 297–306. https://doi.org/10.1016/B978-044452745-5.00195-7
48. M. Mirzaeian, Q. Abbas, M.R. Hunt, A. Galeyeva, R. Raza, Na-ion batteries (Elsevier, 2021)
49. S.-J. Park, Y.-R. Son, Y.-J. Heo, Prospective synthesis approaches to emerging materials for supercapacitor, in *Emerging Materials for Energy Conversion and Storage* (Elsevier, 2018), pp. 185–208
50. S.R. Meher, Transition metal oxide-based materials for visible-light-photocatalysis, in *Nanostructured Materials for Visible Light Photocatalysis* (Elsevier, 2022), pp. 153–183
51. S. Sankaranarayanan, M. Hariram, S. Vivekanandhan, C. Ngamcharussrivichai, Biosynthesized transition metal oxide nanostructures for photocatalytic degradation of organic dyes, in *Green Functionalized Nanomaterials for Environmental Applications* (Elsevier, 2022), pp. 417–460
52. A. Moezzi, A.M. McDonagh, M.B. Cortie, Zinc oxide particles: synthesis, properties and applications. Chem. Eng. J. **185**, 1–22 (2012)
53. F. Yakuphanoglu, Synthesis and electro-optic properties of nanosized-boron doped cadmium oxide thin films for solar cell applications. Sol. Energy **85**(11), 2704–2709 (2011)

54. K. Kordesch, W. Taucher-Mautner, Zinc-Mercury, in *Encyclopedia of Electrochemical Power Sources* (Elsevier BV, 2009), pp. 62–67
55. P. Hassanpour et al., Biomedical applications of aluminium oxide nanoparticles. Micro Nano Lett. **13**(9), 1227–1231 (2018)
56. J. Sarkar, G.G. Khan, A. Basumallick, Nanowires: properties, applications and synthesis via porous anodic aluminium oxide template. Bull. Mater. Sci. **30**(3) (2007)
57. S. Stepanov, V. Nikolaev, V. Bougrov, A. Romanov, Gallium OXIDE: properties and applica 498 a review. Rev. Adv. Mater. Sci **44**, 63–86 (2016)
58. O. Bierwagen, Indium oxide—A transparent, wide-band gap semiconductor for (opto)electronic applications. Semicond. Sci. Technol. **30**, 024001 (2015). https://doi.org/10.1088/0268-1242/30/2/024001
59. M. Batzill, U. Diebold, The surface and materials science of tin oxide. Prog. Surf. Sci. **79**(2–4), 47–154 (2005). https://doi.org/10.1016/j.progsurf.2005.09.002
60. P. Sahoo, R.G. Shrestha, L.K. Shrestha, J.P. Hill, T. Takei, K. Ariga, Surface oxidized carbon nanotubes uniformly coated with nickel ferrite nanoparticles. J. Inorg. Organomet. Polym Mater. **26**, 1301–1308 (2016)
61. A. Hashim, K.H. Al-Attiyah, H. Obaid, Modern developments of polymer blend/oxide nanocomposites for biomedical applications as antibacterial and radiation shielding materials: a review. Res. J. Agric. Biol. Sci. **14**(1), 8–18 (2019)
62. G. Liu, S. Li, Y. Lu, J. Zhang, Z. Feng, C. Li, Controllable synthesis of α-Bi2O3 and γ-Bi2O3 with high photocatalytic activity by α-Bi2O3→γ-Bi2O3→α-Bi2O3 transformation in a facile precipitation method. J. Alloys Compd. **689** (2016). https://doi.org/10.1016/j.jallcom.2016.08.047
63. N. Motakef-Kazemi, M. Yaqoubi, Green synthesis and characterization of bismuth oxide nanoparticle using mentha pulegium extract. Iran. J. Pharm. Res.: IJPR **19**(2), 70 (2020)
64. P.L. Kharel, P.M. Cuillier, K. Fernando, F.P. Zamborini, B.W. Alphenaar, Effect of rare-earth metal oxide nanoparticles on the conductivity of nanocrystalline titanium dioxide: an electrical and electrochemical approach. J. Phys. Chem. C **122**(27), 15090–15096 (2018)
65. G.A. Hussein, Rare earth metal oxides: formation, characterization and catalytic activity thermoanalytical and applied pyrolysis review. J. Anal. Appl. Pyrol. **37**(2), 111–149 (1996)
66. X. Luo et al., The magnetic entropy change in the double perovskite La2NiMnO6 with strong spin–phonon coupling. Solid State Commun. **149**(19), 810–813 (2009). https://doi.org/10.1016/j.ssc.2009.02.020
67. Y. Shimakawa, M. Azuma, N. Ichikawa, Multiferroic compounds with double-perovskite structures. Materials (Basel) **4**(1), 153–168 (2011). https://doi.org/10.3390/ma4010153
68. Y. Ji, Y.-H. Huang, J.-R. Ying, J.B. Goodenough, Electrochemical performance of La-doped Sr2MgMoO6−δ in natural gas. Electrochem. Commun. **9**(8), 1881–1885 (2007). https://doi.org/10.1016/j.elecom.2007.04.006
69. K.P.C. Yao et al., Thermal stability of Li2O2 and Li2O for Li-Air batteries: in situ XRD and XPS studies. J. Electrochem. Soc.**160**(6), A824 (2013). https://doi.org/10.1149/2.069306jes.
70. N. Xiao, R.T. Rooney, A.A. Gewirth, Y. Wu, The long-term stability of KO2 in K-O2 batteries. Angew. Chem. Int. Ed. **57**(5), 1227–1231 (2018). https://doi.org/10.1002/anie.201710454
71. J.L. Hubisz, Transition metal oxides: an introduction to their electronic structure and properties. Phys. Teach. **52**(9), 574–574 (2014). https://doi.org/10.1119/1.4902220
72. Intrinsic Conductivity in Magnesium–Oxygen Battery Discharge Products: MgO and MgO2 | Chemistry of Materials. https://pubs.acs.org/doi/full/https://doi.org/10.1021/acs.chemmater.7b00217. Accessed 31 Jan. 2023
73. X. Liu, H. He, Y. Wang, S. Zhu, Transesterification of soybean oil to biodiesel using SrO as a solid base catalyst. Catal. Commun. **8**(7), 1107–1111 (2007). https://doi.org/10.1016/j.catcom.2006.10.026
74. B. Baumeier, P. Krüger, J. Pollmann, Structural, elastic, and electronic properties of SiC, BN, and BeO nanotubes. Phys. Rev. B **76**(8), 085407 (2007). https://doi.org/10.1103/PhysRevB.76.085407

75. M.N. Hasan, M. Nafiujjaman, Y.-K. Lee, Chapter 4—2D Nanomaterials for gene delivery, in *Biomedical Applications of Graphene and 2D Nanomaterials*, eds by M. Nurunnabi. J. R. McCarthy (Elsevier, 2019), pp. 87–104. https://doi.org/10.1016/B978-0-12-815889-0.00004-0
76. A. Šalak, M. Selecká, Thermodynamic conditions for the Mn–O system in sintering of manganese steels, in *Manganese in Powder Metallurgy Steels*, eds by A. Šalak, M. Selecká (International Science Publishing Ltd., Cambridge, 2012), pp. 5–21. https://doi.org/10.1007/978-1-907343-75-9_2
77. G. Blasse, Ferromagnetism and ferrimagnetism of oxygen spinels containing tetravalent manganese. J. Phys. Chem. Solids **27**(2), 383–389 (1966). https://doi.org/10.1016/0022-3697(66)90045-X
78. S. Zhang, F. Li, High performance ferroelectric relaxor-PbTiO3 single crystals: status and perspective. J. Appl. Phys. **111**(3), 031301 (2012). https://doi.org/10.1063/1.3679521
79. M. Golalikhani et al., Nature of the metal-insulator transition in few-unit-cell-thick LaNiO3 films. Nat. Commun. **9**(1) (2018). https://doi.org/10.1038/s41467-018-04546-5
80. S. Thoka et al., Spinel Zinc Cobalt Oxide (ZnCo2O4) porous nanorods as a cathode material for highly durable Li–CO2 batteries. ACS Appl. Mater. Interfaces **12**(15), 17353–17363 (2020). https://doi.org/10.1021/acsami.9b21347
81. K. Kordesch, W. Taucher-Mautner, Primary batteries—Aqueous systems | Zinc–Mercury, in *Encyclopedia of Electrochemical Power Sources* (Elsevier, 2009), pp. 62–67. https://doi.org/10.1016/B978-044452745-5.00103-9
82. M. Moeinian, K. Akhbari, Various methods for synthesis of bulk and nano Thallium(III) oxide. J. Inorgan. Organometall. Polym. Mater. **26** (2016). https://doi.org/10.1007/s10904-015-0289-z
83. K. Al-Attiyah, A. Hashim, S. Fadhil, Fabrication of novel (carboxy methyl cellulose-polyvinylpyrrolidone-polyvinyl alcohol)/lead oxide nanoparticles: structural and optical properties for gamma rays shielding applications, Int. J. Plast. Technol. (2019)
84. P.L. Kharel, P.M. Cuillier, K. Fernando, F.P. Zamborini, B.W. Alphenaar, Effect of rare-earth metal oxide nanoparticles on the conductivity of nanocrystalline titanium dioxide: an electrical and electrochemical approach. J. Phys. Chem. C **122**(27), 15090–15096 (2018). https://doi.org/10.1021/acs.jpcc.8b02971
85. G.A.M. Hussein, Rare earth metal oxides: formation, characterization and catalytic activity Thermoanalytical and applied pyrolysis review. J. Anal. Appl. Pyrol. **37**(2), 111–149 (1996). https://doi.org/10.1016/0165-2370(96)00941-2
86. G. Adachi, N. Imanaka, Z. C. Kang, *Binary Rare Earth Oxides* (Springer Science & Business Media, 2004)
87. J.J. Alvarado Flores, M.L. Ávalos Rodríguez, J.V. Alcaraz Vera, J.G. Rutiaga Quiñones, S.J. Guevara Martínez, R.A. Zarraga, Advances in the knowledge of the double perovskites derived from the conformation and substitution of the material Sr2MgMoO6-δ as anode with potential application in SOFC cell. Int. J. Hydr. Energy **46**(51), 26152–26162 (2021). https://doi.org/10.1016/j.ijhydene.2021.03.030
88. M. Azuma, K. Takata, T. Saito, S. Ishiwata, Y. Shimakawa, M. Takano, Designed ferromagnetic, ferroelectric Bi2NiMnO6. J. Am. Chem. Soc. **127**(24), 8889–8892 (2005). https://doi.org/10.1021/ja0512576
89. A. Hossain, A.K.M. Atique Ullah, P. Sarathi Guin, S. Roy, An overview of La2NiMnO6 double perovskites: synthesis, structure, properties, and applications. J. Sol-Gel Sci. Technol. **93**(3), 479–494 (2020). https://doi.org/10.1007/s10971-019-05054-8
90. M. Eslamian, N. Ashgriz, The effect of atomization method on the morphology of spray dried particles. ASME Int. Mech. Eng. Congr. Expos. **42223**, 969–974 (2005)
91. C.H. Voon, K.L. Foo, B.Y. Lim, S.C.B. Gopinath, Y. Al-Douri, Synthesis and preparation of metal oxide powders, in *Metal Oxide Powder Technologies* (Elsevier, 2020), pp. 31–65
92. K.M. Nam et al., New crystal structure: synthesis and characterization of hexagonal wurtzite MnO. J. Am. Chem. Soc. **134**(20), 8392–8395 (2012)

93. D. Liu et al., Large-scale synthesis of hexagonal corundum-type In2O3 by ball milling with enhanced lithium storage capabilities. J. Mater. Chem. A **1**(17), 5274 (2013). https://doi.org/10.1039/c3ta00182b
94. R.R. Mahdi, S.A. Makki, Synthesis and properties of cadmium oxide thin films prepared by simple chemical method. Energy Procedia **157**, 261–269 (2019). https://doi.org/10.1016/j.egypro.2018.11.189
95. A. Younis, D. Chu, S. Li, A. Younis, D. Chu, S. Li, Cerium oxide nanostructures and their applications. IntechOpen (2016). https://doi.org/10.5772/65937
96. T.V. Butkhuzi, T.G. Chelidze, A.N. Georgobiani, D.L. Jashiashvili, T.G. Khulordava, B.E. Tsekvava, Exciton photoluminescence of hexagonal ZnO. Phys. Rev. B **58**(16), 10692–10695 (1998). https://doi.org/10.1103/PhysRevB.58.10692
97. Q. Zhang, P. Liu, C. Miao, Z. Chen, C.M.L. Wu, C.-H. Shek, Formation of orthorhombic SnO2 originated from lattice distortion by Mn-doped tetragonal SnO2. RSC Adv. **5**(49), 39285–39290 (2015). https://doi.org/10.1039/C5RA04946F
98. A. Chakrabarti, K. Hermann, R. Druzinic, M. Witko, F. Wagner, M. Petersen, Geometric and electronic structure of vanadium pentoxide: a density functional bulk and surface study. Phys. Rev. B **59**(16), 10583–10590 (1999). https://doi.org/10.1103/PhysRevB.59.10583
99. H.-C. Lu, N. Katyal, G. Henkelman, D.J. Milliron, Controlling the Shape Anisotropy of monoclinic Nb12O29 nanocrystals enables tunable electrochromic spectral range. J. Am. Chem. Soc. **143**(38), 15745–15755 (2021). https://doi.org/10.1021/jacs.1c06901
100. S. Rajagopal, M. Bharaneswari, D. Nataraj, O.Y. Khyzhun, Y. Djaoued, Crystal structure and electronic properties of facile synthesized Cr_2 O_3 nanoparticles. Mater. Res. Express 3(9), 095019 (2016). https://doi.org/10.1088/2053-1591/3/9/095019
101. K.M.E. Miedzinska, B.R. Hollebone, J.G. Cook, An assigment of the optical absorption spectrum of mixed valence Co3O4 spinel films. J. Phys. Chem. Solids **48**(7), 649–656 (1987). https://doi.org/10.1016/0022-3697(87)90154-5
102. M. Anbarasu, M. Anandan, E. Chinnasamy, V. Gopinath, K. Balamurugan, Synthesis and characterization of polyethylene glycol (PEG) coated Fe3O4 nanoparticles by chemical co-precipitation method for biomedical applications. Spectrochim. Acta Part A Mol. Biomol. Spectrosc. **135**, 536–539 (2015). https://doi.org/10.1016/j.saa.2014.07.059
103. I. Ganesh, A review on magnesium aluminate (MgAl2O4) spinel: synthesis, processing and applications. Int. Mater. Rev. **58**(2), 63–112 (2013). https://doi.org/10.1179/1743280412Y.0000000001
104. K. Faungnawakij, N. Shimoda, T. Fukunaga, R. Kikuchi, K. Eguchi, Crystal structure and surface species of CuFe2O4 spinel catalysts in steam reforming of dimethyl ether. Appl. Catal. B **92**(3), 341–350 (2009). https://doi.org/10.1016/j.apcatb.2009.08.013
105. Z. Mousavi, F. Soofivand, M. Esmaeili-Zare, M. Salavati-Niasari, S. Bagheri, ZnCr2O4 nanoparticles: facile synthesis, characterization and photocatalytic properties. Sci Rep. **6**(1) (2016). https://doi.org/10.1038/srep20071
106. D. Mckeown, P. Hagans, L. Carette, A. Russell, K. Swider, D. Rolison, Structure of hydrous ruthenium oxides: implications for charge storage. J. Phys. Chem. B **103** (1999). https://doi.org/10.1021/jp990096n
107. N. Bestaoui, E. Prouzet, P. Deniard, R. Brec, Structural and analytical characterization of an iridium oxide thin layer. Thin Solid Films **235**(1), 35–42 (1993). https://doi.org/10.1016/0040-6090(93)90239-L
108. J. Zhang et al., Thermally induced crystallization in NbO2 thin films. Sci. Rep. **6**(1) (2016). https://doi.org/10.1038/srep34294
109. J. Silva, A. Reyes, H. Esparza, H. Camacho, L. Fuentes, BiFeO3: a review on synthesis, doping and crystal structure. Integr. Ferroelectr. **126**(1), 47–59 (2011). https://doi.org/10.1080/10584587.2011.574986
110. E.G. Rini, M. Gupta, R. Mittal, A. Mekki, M. Saeed, S. Sen, Pbnm to R3-c phase transformation in $(1-x)$LaFeO3.xLaMnO3 solid solution due to modifications in structure, octahedral tilt and valence states of Fe–Mn (2020)

111. R. Amin, N. Khatun, S. Sen, Optimization of Pb content in enhancing ferroelectricity and shifting the Tc of BaTiO3 to a higher temperature. J. Appl. Phys. **126**(17), 174105 (2019). https://doi.org/10.1063/1.5116651
112. K.S. Samantaray, R. Amin, M.P, I. Bhaumik, S. Sen, Effect of electrical poling on the structural, vibrational, and electrical properties of 0.94(Na0.5Bi0.5TiO3)-(0.06-x) CaTiO3-x (BaTiO3) lead-free ceramics. Ceram. Int. (2023). https://doi.org/10.1016/j.ceramint.2023.01.018
113. R. Pradheesh, H.S. Nair, V. Sankaranarayanan, K. Sethupathi, Exchange bias and memory effect in double perovskite Sr2FeCoO6. Appl. Phys. Lett. **101**(14), 142401 (2012). https://doi.org/10.1063/1.4756792
114. R. Nechache, F. Rosei, Recent progress in nanostructured multiferroic Bi2FeCrO6 thin films. J. Solid State Chem. **189**, 13–20 (2012). https://doi.org/10.1016/j.jssc.2012.01.022
115. M. Nasir et al., Role of antisite disorder, rare-earth size, and superexchange angle on band gap, curie temperature, and magnetization of R2NiMnO6 double perovskites. ACS Appl. Electron. Mater. **1**(1), 141–153 (2019). https://doi.org/10.1021/acsaelm.8b00062
116. X. Huang, G. Zhao, G. Wang, J.T. Irvine, Synthesis and applications of nanoporous perovskite metal oxides. Chem. Sci. **9**(15), 3623–3637 (2018)
117. A.P. Guimarães, A.P. Guimaraes, *Principles of Nanomagnetism*, vol. 7 (Springer, 2009)
118. J. Nogues et al., Exchange Bias in nanostructures. Phys. Rep.-Rev. Sect. Phys. Lett. **422**, 65–117 (2005). https://doi.org/10.1016/j.physrep.2005.08.004
119. S. Biswas, A. Pal, T. Pal, Supported metal and metal oxide particles with proximity effect for catalysis. RSC Adv. **10**(58), 35449–35472 (2020). https://doi.org/10.1039/D0RA06168A
120. J. Seo, H. Yoo, Zinc–Tin oxide film as an earth-abundant material and its versatile applications to electronic and energy materials. Membranes (Basel) **12**(5), 485 (2022). https://doi.org/10.3390/membranes12050485
121. M. Carofiglio, S. Barui, V. Cauda, M. Laurenti, Doped zinc oxide nanoparticles: synthesis, characterization and potential use in nanomedicine. Appl. Sci. **10**(15) (2020). https://doi.org/10.3390/app10155194
122. M. Abb, Y. Wang, N. Papasimakis, C.H. De Groot, O.L. Muskens, Surface-enhanced infrared spectroscopy using metal oxide plasmonic antenna arrays. Nano Lett. **14**(1), 346–352 (2014)
123. E.S. Nour, A. Echresh, X. Liu, E. Broitman, M. Willander, O. Nur, Piezoelectric and opto-electrical properties of silver-doped ZnO nanorods synthesized by low temperature aqueous chemical method. AIP Adv. **5**(7), 077163 (2015). https://doi.org/10.1063/1.4927510
124. P.K. Mishra, R. Dobhal, E.G. Rini, M. Kumar, S. Sen, Rapid organic dye degradation and wavelength dependent sensing study in Cu1–xFexO. Ceram. Int. **48**(5), 5995–6006 (2022). https://doi.org/10.1016/j.ceramint.2021.11.135
125. V. Manjunath et al., Perovskite-based facile NiO/CH3NH3PbI3 heterojunction self-powered broadband photodetector. ACS Appl. Electron. Mater. **3**(10), 4548–4557 (2021). https://doi.org/10.1021/acsaelm.1c00679
126. S.-I. Kim, J.-S. Lee, H.-J. Ahn, H.-K. Song, J.-H. Jang, Facile route to an efficient NiO supercapacitor with a three-dimensional nanonetwork morphology. ACS Appl. Mater. Interfaces **5**(5), 1596–1603 (2013). https://doi.org/10.1021/am3021894
127. Q. Guo, C. Zhou, Z. Ma, X. Yang, Fundamentals of TiO2 photocatalysis: concepts, mechanisms, and challenges. Adv. Mater. **31**(50), 1901997 (2019). https://doi.org/10.1002/adma.201901997
128. J. George, C.C. Gopalakrishnan, P.K. Manikuttan, K. Mukesh, S. Sreenish, Preparation of multi-purpose TiO2 pigment with improved properties for coating applications. Powder Technol. **377**, 269–273 (2021). https://doi.org/10.1016/j.powtec.2020.08.050
129. M.M. Sultonovich, I.J.R. Ogli, U.A. Abdurashidovich, A.T. Sirozhevich, Technology of modified sodium-aluminum catalysts for nitrogen gas purification systems. Am. J. Appl. Sci. **2**(09) (2020). https://doi.org/10.37547/tajas/Volume02Issue09-24
130. S. Pan, G. Li, Recent progress in p-type doping and optical properties of SnO2 nanostructures for optoelectronic device applications. Recent Pat. Nanotechnol. **5**(2), 138–161 (2011). https://doi.org/10.2174/187221011795909161

131. M. Sharma, S.N. Doila, S. Kumar, P.A. Alvi, Effect of Fe doping on magnetic behavior of SnO2 nanoparticles for spintronics applications. IOP Conf. Ser.: Mater. Sci. Eng.**594**(1), 012004 (2019). https://doi.org/10.1088/1757-899X/594/1/012004
132. W.-B. Li et al., BaTiO3-based multilayers with outstanding energy storage performance for high temperature capacitor applications. ACS Appl. Energy Mater. **2**(8), 5499–5506 (2019). https://doi.org/10.1021/acsaem.9b00664
133. S.K. Meher, G.R. Rao, Ultralayered Co3O4 for high-performance supercapacitor applications. J. Phys. Chem. C **115**(31), 15646–15654 (2011). https://doi.org/10.1021/jp201200e
134. Recent Advances in Co3O4 as anode materials for high-performance lithium-ion batteries. https://www.espublisher.com/journals/articledetails/307. Accessed 27 Jan. 2023
135. J. Xie et al., Co3O4-reduced graphene oxide nanocomposite as an effective peroxidase mimetic and its application in visual biosensing of glucose. Anal. Chim. Acta **796**, 92–100 (2013). https://doi.org/10.1016/j.aca.2013.08.008
136. M.I. Kim, J. Shim, T. Li, J. Lee, H.G. Park, Fabrication of nanoporous nanocomposites entrapping Fe3O4 magnetic nanoparticles and oxidases for colorimetric biosensing. Chem. A Euro. J. **17**(38), 10700–10707 (2011). https://doi.org/10.1002/chem.201101191
137. M.R. Ghazanfari, M. Kashefi, S.F. Shams, M.R. Jaafari, Perspective of Fe_3O_4 nanoparticles role in biomedical applications. Biochem. Res. Int. **2016**, e7840161 (2016). https://doi.org/10.1155/2016/7840161
138. A. Raj, A. Anshul, V. Tuli, P.K. Singh, R.C. Singh, M. Kumar, Effect of appropriate ETL on quantum efficiency of double perovskite La2NiMnO6 based solar cell device via SCAPS simulation. Mater. Today: Proc. **47**, 1656–1659 (2021). https://doi.org/10.1016/j.matpr.2021.04.581
139. G. Sághi-Szabó, R.E. Cohen, H. Krakauer, First-principles study of piezoelectricity in $PbTiO_3$. Phys. Rev. Lett. **80**(19), 4321–4324 (1998). https://doi.org/10.1103/PhysRevLett.80.4321
140. F. Chen et al., The electromechanical features of LiNbO3 crystal for potential high temperature piezoelectric applications. J. Materiom. **5**(1), 73–80 (2019). https://doi.org/10.1016/j.jmat.2018.10.001

Optical and Electrical Switching of Thermochromic Metal Oxide Nanostructures

Irfan Ayoub, Ab Mateen Tantray, Rishabh Sehgal, Vishal Sharma, Rakesh Sehgal, Hendrik C. Swart, and Vijay Kumar

Abstract Solid state thermochromic materials are a class of materials that, at some specific temperatures, undergo critical temperature changes from semiconductors to metals. These materials act as a class of smart materials that change colour because of a phase transition caused by the change in temperature. These materials are distinguished by the memory capabilities in response to the change in temperature. These materials have significant potential for a variety of industrial applications, such as smart windows, sensors, and anti-counterfeiting technologies. These materials also exhibit various decolorization mechanisms in a wide variety of forms. This chapter reviews the general concept of the chromogenic and their different forms, categorization of different thermochromic materials, decolorization mechanisms, and some prominent applications.

Keywords Optical · Electrical · Switching · Thermochromic · Metal oxides

I. Ayoub · A. M. Tantray · V. Kumar (✉)
Department of Physics, National Institute of Technology Srinagar, Srinagar, Jammu and Kashmir 190006, India
e-mail: vj.physics@gmail.com

R. Sehgal
Department of Electrical and Computer Engineering, University of Texas at Austin, Austin, TX 78751, USA

V. Sharma
Institute of Forensic Science & Criminology, Panjab University, Chandigarh 160014, India

R. Sehgal
Department of Mechanical Engineering, National Institute of Technology Srinagar, Srinagar, Jammu and Kashmir 190006, India

I. Ayoub · H. C. Swart · V. Kumar
Department of Physics, University of the Free State, P.O. Box 339, Bloemfontein ZA9300, South Africa

V. Kumar et al. (eds.), *Optical Properties of Metal Oxide Nanostructures*,
Progress in Optical Science and Photonics 26,
https://doi.org/10.1007/978-981-99-5640-1_2

1 Introduction

Over the last several decades, researchers have been increasingly interested in the investigation of metal oxides. Metal oxides belong to the class of ionic compounds, as they are composed of positive metallic and negative oxygen ions. The formation of the ionic bonds results from electrostatic interactions between the oppositely charged ions. Most of the metal oxides possess excellent thermal and chemical stability because of the presence of an entirely filled s-shell [1, 2]. However, partially filled d-shells provide them with a wide range of features and lend them the potential to be of considerable utility in different electronic devices. Besides electrochromic features, these one-of-a-kind qualities include broad bandgaps, high dielectric constants, reactive electronic transitions, and efficient electrical conductivity [3–17]. As a result of this, metal oxides are commonly regarded as among the most intriguing functional materials and are extensively used in a broad variety of technical applications. Continuous advancements in nanoscience and nanotechnology have inspired and encouraged the scientific community and industrialists to investigate new properties of both typical and novel materials at the nanoscale level. These materials are classified as having one of the three distinct morphologies or nanostructures: 0D, 1D, or 2D. The electrical properties of nanoparticles vary from those of their bulk counterparts because of the variation in their size and dimensionality [18, 19]. These disparities in the electronic structures result in a wide range of shifts in the physical and chemical characteristics of the substance [20, 21]. Nanomaterials with at least one dimension between 1 and 100 nm are classified as novel materials. 1D nanostructures offer the greatest potential for nanoelectronic and nano devices compared to other nanostructures (2D and 3D) because they all have a 1D morphology. The widths and thicknesses of the 1D nanostructures are restricted to the nanoscale range (1–100 nm), although the lengths vary from a few millimeters to several millimeters or even more. The millimeter length scale makes it possible for 1D nanostructures to be in touch with the macroscopic environment, which enables a wide range of physical measurements to be taken. As a result, the use of two terminals in the manufacture of 1D building blocks simplifies electrical measurement [22–27]. A significant amount of work has been done to date in the synthesis and characterization of various metal oxide nanostructures in various shapes, such as rods, wires, needles, belts, tubes, ribbons, coaxial cables, and many more [28–36]. As these nanostructures have potentially improved the different characteristic properties of materials, they are well-suited for a wide range of applications, such as sensor technology, smart windows, solar cells, supercapacitors, photodetectors, light-emitting diodes, transistors, etc. [37–44]. These metal oxide nanostructures have thus sparked a great deal of interest and enabled significant advances to be made in different technological fields for various functional applications. In the current technological domain, the area of functional materials acts as a focal point of research, and at least 85% of the current research is focused on them. Thermochromic compounds are a novel class of functional materials with distinctive color changes that have captured the interest of various scientists since the 1970s [45–47]. Thermochromic materials,

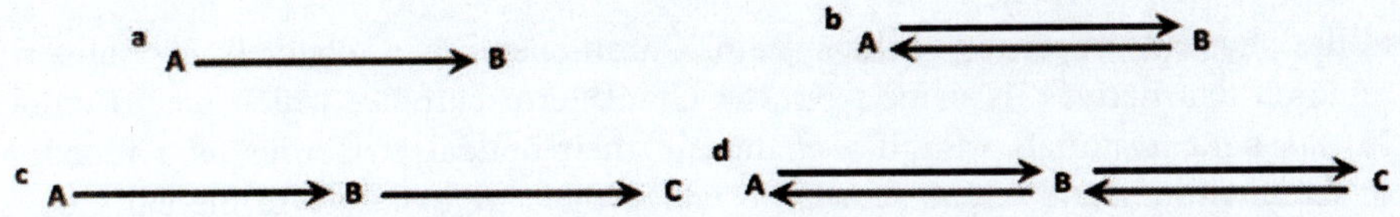

Fig. 1 Discoloration patterns for different thermochromic materials (A, B and C represents various colors)

commonly known as temperature-sensitive materials, comprise discoloring agents and other subsidiary components. The visible absorption spectra of discoloration substances or different mixtures may change on receiving heat or on cooling, hence resulting in the color change with the change in temperature up to a certain range [48, 49]. There are two types of thermochromic materials that have reversible and irreversible decolorization. There are many pathways of decolorization, as depicted in Fig. 1 [50]. In order to improve the usefulness and cognition of these materials, thermochromic discoloration often occurs with a photochromic reaction [51–53]. Many industries, including aviation, the military, intelligent windows, printing technology, textiles, and architectural coatings, amongst others, have investigated the potential uses of thermochromic materials [38, 54–57]. There have been several studies done on the use of thermochromic materials in anti-counterfeiting technologies and toys for children, but few of those studies have focused on the use of such materials in food packaging. Thermochromic materials are going to have tremendous potential in the realm of food packaging once people pay more attention to the safety of the food they eat.

This book chapter presents a thorough discussion of the phenomenon known as thermochromism, which comes into play when the optical characteristics of a substance undergo reversible transformation as a direct consequence of a change in temperature. Besides thermochromism, there are other types of chromic materials that are also discussed. In this chapter, emphasis is placed on the thermochromism shown by the metal oxide-based solid-state materials. Emphasis is also laid on discussing the basic concepts that explain the electronic structure and characteristics of solids, along with the chronological advancements in the theories underlying the thermochromic transitions that resulted in the semiconductor's discovery to metal transition. Last, besides having a brief discussion on the different synthesis methods, some applications exhibited by these materials are also discussed.

2 Chromogenics

A phase transition is a drastic change from one phase to another under the influence of some external stimulus, such as a thermal field, strain energy, surface energy, an external force, a magnetic field, etc. [58]. All these transitions are represented by a single word called "chromogenic". The changes are preceded by a considerable shift in the physical properties of the system. Chromogenic is a relatively new area of research that derives its name from the Greek term chromo, which means color. In this field, materials exhibit a change in their optical properties as a function of the environmental circumstances to which they are exposed. In the early days of photography, this term was used to describe light-sensitive materials. One of the prominent examples was photographic paper coated with an emulsion of silver halide. Light was the agent that brought about the changes in color of the paper. However, the color shift that occurred on the photographic paper was discovered to be permanent and irreversible. The term chromogenics is now used in a broader context. It refers to any changes that occur in any substance when subjected to some external stimuli, including light, an electric field, exposure to gas, and others, and can exhibit a change in the optical property that can be reversed [59]. These materials make up an important class of smart materials since they undergo reversible transformations in their physical characteristics in response to an external stimulus. The importance of chromogenic materials has increased because of their optically dynamic and interactive properties. This effective responsiveness to different external stimuli has led to the progressive development of different smart systems relying on the control of light. Depending on the type of external stimulus, there are different chromogenic materials such as electrochromic, thermochromic, piezochromic, etc. [60, 61]. A brief discussion on different types of chromogenic materials is given in the following sections.

2.1 *Electrochromic Materials*

In this class of materials, a modest electric field may elicit a bidirectional change in the optical characteristics. Owing to the simplicity of control and reversibility of optical characteristics with the help of an applied electric field, electrochromic coloration is the most explored and applicable type of chromogenic materials. In recent decades, liquid crystal displays have emerged as the most prevalent class of electrochromic devices [62]. The alignment of the liquid crystal molecules tends to cause a shift in the optical features from a transparent to an opaque state upon receiving an impulse from the applied electric field. This phenomenon has been used for a long time in a broad range of electronic display devices, such as watches and TVs. The use of liquid crystal relying on electrochromic materials has several drawbacks, the most significant of which is the extremely high-power consumption for keeping the device in the 'on' state along with the switching of the device between the 'on and off' states. Electrophoretic deposition also belongs to the class of electrochromic materials. In these systems, an electric field is applied in small amounts, and then metal particles suspended in the solution are deposited on a transparent electrode for the development of the film and thereby result in a change in optical properties [63]. Thus, for the operation of these devices, the presence of a liquid component is necessary. Till date, it has been observed that an enormous number of organic and inorganic materials possess electrochromic characteristics owing to

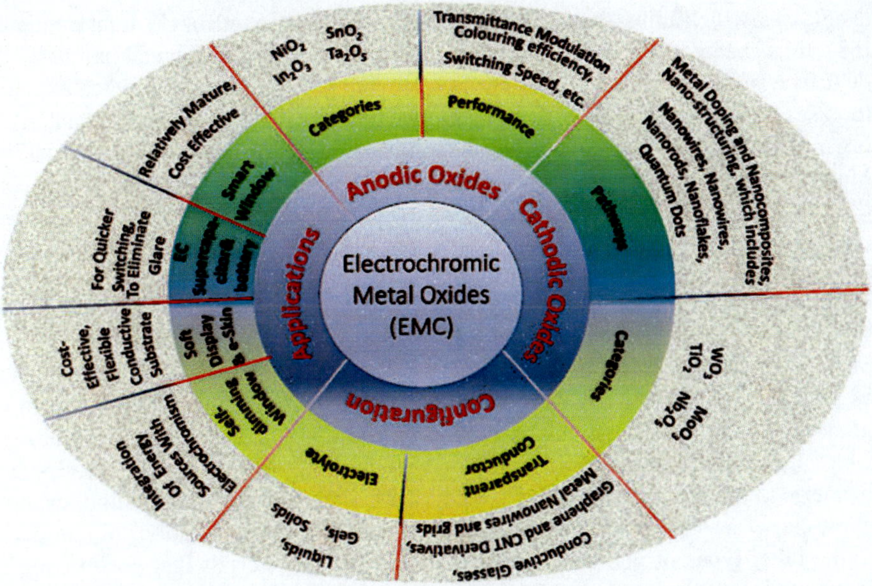

Fig. 2 Pictorial representation of different characteristic features and applications of electrochromic metal oxides

their different reversible chemical properties. Numerous metal oxides among the inorganic substances display an effective kind of electrochromic coloring [64–66].

Besides these, different conducting polymers and molecular dyes also act as effective electrochromic materials. The electrochromic colour exhibited by these different kinds of materials is caused by a variety of different phenomena, like redox reactions, variations in the oxidation state, the creation of polarons, and other similar processes [67]. An overview of the thermochromic materials is depicted in Fig. 2.

2.2 Photochromic Materials

When these materials are exposed to some kind of electromagnetic radiation, a reversible change in their optical characteristics takes place. The type of colouring produced and the metastable conditions of the final product are the parameters that determine the reversibility of the colour attained while going through the photochromic changes. Other types of activation sources, such as electric fields or heat, can be used to directly regulate the discoloration in addition to electromagnetic radiation. UV or visible light will always activate the photochromic colouring and result in changing its conductivity and other optical properties in the visible and infrared spectrums. These changes increased the domain of different applications for a wide range of real-world uses. A common example of this type of material is eyeglasses, which adjust their opacity depending on the strength of UV light irradiated [68]. It has been observed that on applying a small electric field, metal halide-based devices transform from a transparent to a reflecting mirror-like state when exposed to gases like hydrogen [69]. It has been revealed that a broad variety of materials, organic and inorganic, display the phenomenon of photochromic colouring [70]. Many organic compounds go through a process of change that involves the breakdown of their bonds when they are exposed to UV or blue light. In this case, it has been observed that when a single bond in a compound like spiropyrans, triarylmethane, or polymethine is broken, which causes colouring, the resulting photochromism remains very temporary since the system swiftly reverts back to its initial conditions. Under the different conditions of photo breakdown and coloration, the creation of pairs of radicals is brought about by hemolytic compounds. Compounds such as pyrroles, hydrazine, disulfides, etc. are able to exhibit this particular form of photochromism. Modifications brought about by this phenomenon are unstable as well as very unattractive from an application point of view. However, the cis-trans isomeric photochromism shown by compounds like urocanic acid, indigo, cyanine dyes, etc. was found to be more stable and interesting for applications as compared to the other types of photochromism [71]. Inorganic materials that possess significant bandgaps display metastable photochromic behavior. Observations have also revealed that the presence of defects like F centers, impurities, or dislocations in the materials is also responsible for the colour. Only additional activations, such as heat or an electric field, can reverse the coloration that is caused by photochromism. This type of photochromism effect has been exhibited by a few inorganic metal oxides and semi-conductors. The metastable colouring of these materials is what makes them more valuable from an application standpoint than other similar materials.

2.3 *Piezochromic Materials*

These are a class of materials that, on being subjected to pressure, undergo a reversible colour change. Due to their peculiar sensitivity to strain and variation of pressure, these materials are being extensively studied from a research point of view. This phenomenon is also sometimes referred to as mechanochromism. There are three different mechanisms through which the piezochromic materials can change their colour: (1) perturbation of the electronic energy levels in different ways depending on the pressure that is being applied to the piezochromic surfaces; (2) initiation of the first order phase transition as a consequence of changes occurring while applying the pressure; and (3) changes in the geometric configuration by applying the pressure to the surface of the piezochromic material, which in turn brings about a change in the colour of the material. Seeboth et al. [72] have created different piezochromic materials by combining various polymers among the derivatives of cholesteryl. These materials, when subjected to pressure, exhibit a colour shift that has the capability to move either from red to green or vice versa. For depicting the change in colour, different materials need to have activation pressures of different magnitudes, such as

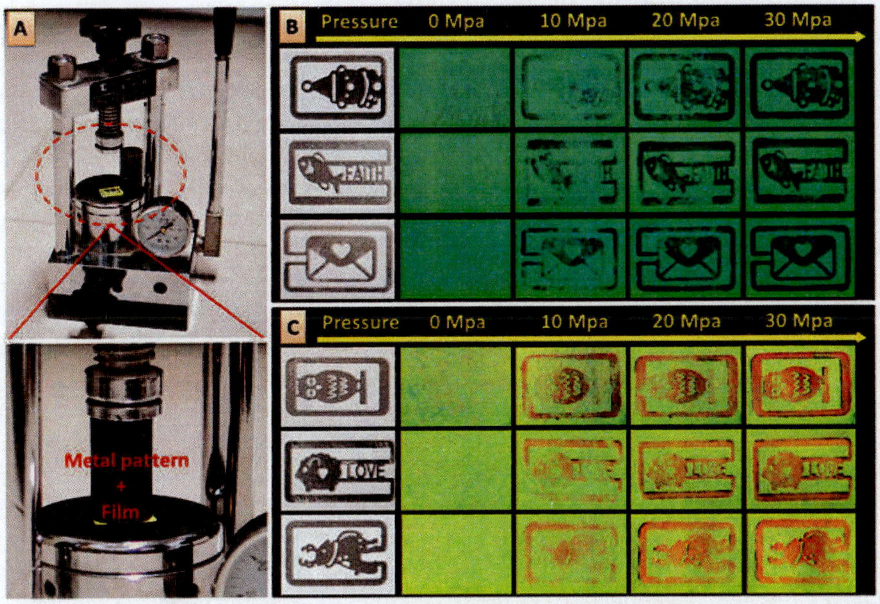

Fig. 3 **a** Depiction of the mechanical printing systems one with table press and other one with metal pattern moulds. **b**, **c** Emission of green and yellow light by the black paper due to variation in pressure along with the imprinting of cartoons along with visualization of quenching effect. Reused with permission from [74]

the dimethylglyoxime complex, which reveals the piezochromic response (change in colour) when subjected to a pressure of 63–150 Kbar. Similarly, the inorganic material samarium monosulfide changes from its semiconducting phase to a metal phase when subjected to a pressure of 6.5 bar [73]. These materials have an enormous number of applications, but the most prominent one is the detection of any variation in the surface pressure. Some of the others include offset printing inks, traditional flexo ink, screen printing ink, lamination, and piezochromic paints, which are being used on tennis balls, volleyballs, basketballs, bathrooms, and many more. An example of the piezochromic effect, that is, a change in colour as a result of a change in pressure, is depicted in Fig. 3, where the pressure has been varied from 0 to 30 MPa [74]. Observations have also revealed that the colour change in a few of the photochromic materials is irreversible at a certain pressure. These kinds of materials are often used in the detection of shock or in the monitoring of health-related factors.

2.4 *Magnetochromic Materials*

These are the materials that undergo an immediate and reversible change in colour when subjected to the influence of an external magnetic field. Magnetochromic is another name for the change in the optical constant of materials under the influence of a magnetic field. Because of the sensitivity of the material to an applied magnetic field from the outside, the spin-induced electric polarization can be regulated magnetically, and significant spin-charge interactions are evident. Materials which are considered to be efficient for the development of magnetochromic materials should possess high charge spin coupling with magnetic control. Spin-crossover complexes, colossal magnetoresistive oxides, and low-dimensional magnetics are some of the potential examples that possess the mentioned property and are being used for the development of magnetochromic materials [75]. Multiferroic compounds, which are distinguished by the simultaneous presence of ferromagnetism and ferroelectricity in the same phase, also reveal magnetochromic features.

An example of this kind of multiferroic magnetochromic material is $MnWO_4$, which exhibits spin-induced ferroelectricity in addition to substantial magnetoelectric reactions [76]. Another prominent example is the magnetochromic microspheres, which can be dispersed in a variety of media, including water, alcohol, polymers, etc. In comparison to other chromic materials, the colour change in these microspheres is caused by the action of the magnetic field on the orientation of the microspheres [77]. The shift in the colour attained due to the impact of the magnetic field does not affect the fundamental qualities or the basic structure of the materials. The rotation that occurs as a result of an applied external magnetic field is caused by the presence of iron oxide inside the microsphere. The ability to alter the colours can be achieved by rotating the microspheres in different directions. These materials possess an enormous number of applications, such as in display units, rewritable signs, sheets, and posters. Besides these applications they also act as effective materials for paints, printing inks, and anti-counterfeiting.

2.5 *Chemochromic Materials*

Materials are said to be chemochromic if they are capable of altering their colour when subjected to some chemical changes. Depending on the external agent used, different subclasses of chemochromic materials exist: (1) gasochromic, which refers to materials that typically exhibit reversible coloration when exposed to some adequate gas; (2) halochromic, which are materials that change their colour in response to pH variation; (3) ionochromic, which are materials that exhibit a colour change when placed in an ion-reaching medium; and (4) hydro or hygrochromic, which are materials that change either reversibly or permanently in response to exposure to water or liquid. These materials are also referred to as solvatochromic. Among the gasochromic materials, thin films made of WO_3 and V_2O_5 are two of the most well-known examples [78, 79]. When exposed to hydrogen gas, both of these materials change colour significantly. Observation has revealed that when WO_3 thin film is exposed to hydrogen, it changes its condition from being transparent to a state in which it is dark blue. It is being used in double-pane windows, where a thin layer of WO_3 is adhered to one of the inner sides of the window, and hydrogen gas is collected from the space between the two panes of glass. Besides this, the gasochromic devices relying on WO_3 are being used for the detection of hydrogen leakage, thereby protecting people from excessive exposure [78].

2.6 Thermochromic Materials

When subjected to high temperatures, these materials undergo reversible color shifts. Nearly every substance undergoes a reversible colour change when subjected to a particular temperature. However, thermochromism is used more explicitly for the materials that experience a sudden change in their optical characteristics when subjected to a temperature of a certain threshold value called the critical temperature (Tc). Thermochromism refers to the changes that occur in the optical characteristics of a substance as a function of temperature. This effect often occurs across a wide temperature range and manifests itself as a gradual colour shift. This type of thermochromism is referred to as continuous thermochromism. Furthermore, the structural shift that occurs at the transition temperature in this process is known as discontinuous thermochromism. The thermodynamics of the systems that govern whether the phase transition occurs are either first- or second order in nature, along with their reversible or irreversible nature [80]. Thermochromic materials possess the potential to be used in a variety of technological applications, such as in thermometers as fever indicators, gadgets, designing purposes, as temperature sensors for the purpose of safety, laser markings, and warning signals. In addition to inorganic oxides, many other diverse materials, including liquid crystals, conjugated oligomers, and leco dyes, are known to exhibit reversible colour change with temperature [81–83]. Apart from these thermochromic dyes, which rely on organic substances, they also undergo reversible colour change when subjected to the required amount of heat. The use of these materials has led to an enormous increase in the textile industry. Mao et al. [84] investigated the surreptitious capability of the W-doped vanadium oxide thin film coatings on cotton textiles employing semiconductor to metallic phase transition characteristics. They observed that the emissivity of coated fabrics changes abruptly with respect to its surroundings, indicating that W-doped vanadium oxide acts as an efficient and viable material for infrared stealth purposes. Because of this characteristic feature, these types of textiles are continuously used for security purposes in different uniforms, helmets, and shoes, and therefore act as military textiles. Many other researchers have also investigated these thermochromic materials for different purposes and revealed different properties; a few of the studies carried out are presented in Table 1.

Table 1 Different thermochromic materials and their application in different sectors

Chromic compounds	Textiles	Applications	References
Rose red TF-R1 and crystal violet lactone pigment	Polypropylene fibers	Temperature indicator	[85]
Crystal violet lactone	Poly(methyl methacrylate) microcapsules	Protective clothing	[86]
Cholesteric liquid crystals microcapsules	Polyvinylpyrrolidone fibers	General	[87]
Paste of blue and orange dyes	Cotton fabric	Military	[88]
Rear earth materials and thermochromic pigments	Polyacrylonitrile fiber	General	[89]
Light fiber containing $Sr_2ZnSi_2O_7$:Eu^{2+}, Dy^{3+} and thermochromic Y_2O_2S:Eu^{3+}, Mg^{2+}, Ti^{4+} pigment rose red TF-R1	Polyacrylonitrile fibers	Optical and thermal sensors	[90]
Thermochromic inks and graphene oxide Ultraviolet	Double-covered yarns	Multifunctional applications of elastic visual sensors and wearable visor	[91]
Ultraviolet curable ink	Polyester and cotton fabric thermochromic	Non-emissive displays, interior design, art and as indicators in interactive systems	[92]
Complex of doped transition metal with rare earths	Leather	Heat-resistant safety products	[93]
Nanofibers containing lauric acid	Polyester/cotton yarns	Temperature control	[94]

3 Thermochromic Transition Metal Oxides

With the discovery of the concept of phase transition, which takes place at a particular temperature known as critical temperature (T_c), many new things have evolved in science. The concept has become an ideal research point for studying the metallic and non-metallic states of a system due to the fact that at the phase transition temperature there occurs an abrupt change in the optical and electronic properties of the system. Since its establishment, transition metal oxide-based thermochromic materials have been investigated thoroughly. The main challenge in investigating these materials, according to Alder, was the complexity of developing pure and stoichiometric single crystals. He has observed that the electrical measurements performed were not associated with the intrinsic properties of the materials but rather were caused by the high concentration of lattice defects along with the different particulates present in the crystal. The investigations have led to advancements in crystal growth and instrumentation. Different advancements have led to the categorization of transition metal oxides into metals, semiconductors, and insulators that are capable of undergoing semiconductor to metal transition. As a consequence of this, metals are distinguished by a low resistivity of the order of 10^{-2} to 10^{-6} Ω cm at room temperature, while their resistivity rises linearly with the rise in temperature. Similarly, at the normal temperature, the resistivity of the semiconductors and insulators ranges from 10^3 to 10^{17} Ω cm, but this value falls off exponentially as the temperature increases [95]. Till date, the phenomenon of thermochromism has been studied in different transition metal oxides having different crystal structures, such as monoxides, dioxides, sesquioxides, spinels, and many more. The thermochromic effects have been evidently observed in different oxides of vanadium, tungsten, molybdenum, etc. These materials find use in a variety of technical applications, including electrical and optical switching devices, as well as other uses, some of which are covered in more detail later in this chapter. In the case of vanadium-based oxides, the primary focus has remained on studying their metal-insulator and phase transitions as a function of temperature. These thermochromic materials exhibit exceptional electrical, structural, and magnetic properties, but there is still considerable controversy about their theoretical implications, especially with regard to VO_2 and VO_3. Different researchers explored some of the enormous number of metal oxides available [96–100].

3.1 Vanadium Sesquioxide

Vanadium sesquioxide, also represented by the chemical formula V_2O_3, possesses the structure of corundum and holds a rhombohedral symmetry at its normal temperature, the crystal structure of which is depicted in Fig. 4 [101]. In this structure, the vanadium ions are organized into V–V pairs along the axis perpendicular to the honeycomb like structure. In the basal plane, there are three subsequent interaction spaces that remain occupied by V^{3+} ions. Due to the presence of one O^{2-} ion around each of the V^{3+} ions, the formation of deformed octahedrons around each cation occurs. Since the 1970s, experiments such as photoemission and x-ray absorption spectroscopy have been performed to reveal the electronic structure of V_2O_3. Besides this, there exist many computationally based studies such as dynamical mean field theory and different density functional methods, with the help of which different calculations related to the variation in the spectral characteristics of a material interaction in the field of strong electronic columbic potential are made. However, both the electronic structure and the contentious nature of the magnetic characteristics of V_2O_3 have continuously remained a subject of research discussion for the researchers [102]. Ever since the discovery of V_2O_3 in 1964 by Foex [103], it has been held in high esteem as the best illustration of a tightly correlated electron system that experiences a Mott-Hubbard transition. Recently, the notion of 3d-orbital hybridization and charge transfer systems are being considered more relevant, as the hybridization between 3d orbitals of vanadium and 2p orbitals of oxygen can vary during the phase transition, implying that the Hubbard model may not fit significantly. As a consequence of this, it is often hypothesized that the fundamental mechanism that will properly explain

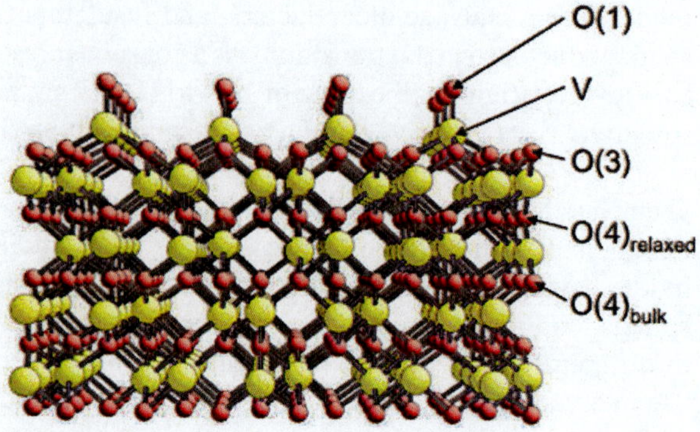

Fig. 4 Crystal structure of rhombohedral vanadium sesquioxide. Reused with permission from [101]

the transition from semiconductor to metal may reside somewhere in-between the Moot-Hubbard model and the charge transfer model [104, 105].

The initiation of a first-order phase change starts at a temperature of the order of T = 150 K. In this transition, the low-temperature antiferromagnetic monoclinic phase changes to its rhombohedral paramagnetic metallic phase [106]. Studies have revealed that the distortion of the monoclinic phase occurs because of the shifting of cation-ion pairs in the basal plane, which results in the effective rotation of the c-axis [107]. Observation has revealed that the distance between V–V bonds along the hexagonal c-axis is greater in the monoclinic insulating phase as compared to the rhombohedral metallic phase. The importance of this bond in the structure is still a matter of debate [108]. Despite this much explanation, there is widespread agreement that the single-band Hubbard model does not properly explain the transition from semiconductor to metal in V_2O_3 owing to the unique multi-orbital structure [109].

3.2 Vanadium Pentoxide

Vanadium pentoxide (V_2O_5) has been observed to display the thermochromic behaviour and, thereby, experience a phase shift from semiconductor to metal at a temperature of about 530 K [110]. Also because of the presence of intertwined optical and electrical properties, V_2O_5 exhibits the electrochromic properties, which cause its colour to shift from blue to green and then to yellow in a period of about two seconds [111]. These colour shifts are caused by the material's ability to change its electronic configuration. In response to the applied voltage, these materials possess the capability to alter their optical characteristics in a way that is both reversible and long-lasting. Thus, the phenomenon of electrochromism reasonably provides a straightforward approach for the generation of varying light transmission by applying different electrochromic processes. The structure of the V_2O_5 was figured out by Bachmann in 1961 [112]. The studies have revealed that it possesses an orthorhombic structure with a band gap energy of 2.24 eV below the transition temperature in the

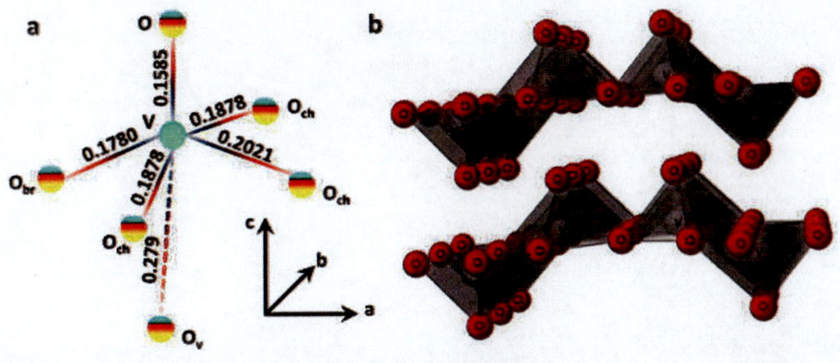

Fig. 5 **a** Coordination geometry of vanadium ion in V_2O_5 compound. In the b-direction there is a puckered chain due to the formation of V-O_{ch} bonds. These are connected to one another in the a-direction by O_{br} bridging bonds. Also, the interaction of the weak V-Ov vanadyl bonds in the c-direction results in the formation of layers. **b** Depiction of orthorhombic layered structure of VO_5. Reused with permission from [114]

semiconductor state. The orthorhombic structure is composed of a VO_6 octahedron, which shares its corners and edges with other electrochemical metal oxides. But the octahedron lacks the regular configuration because the distance between the V and O faces is not constant, as depicted in Fig. 5a [113].

Due to the significant distance between the V and O along the crystallographic c-axis, this orthorhombic structure can be characterized as a layer of square pyramids of VO_5 (Fig. 5b), where the vanadium atoms remain surrounded by five oxygen atoms. The layer structure possessed by V_2O_5 is well suited for the intercalation of cations, which is responsible for its prominent use in thermally activated thermochromic devices. The optical feature of the V_2O_5 at the low temperature state has been thoroughly investigated, but very little effort has been made in investigating the metallic state, which therefore needs more attention [115].

3.3 *Vanadium Dioxide*

The compound vanadium dioxide (VO_2) exists in a number of different phases with different space groups, that is VO_2 (M) [P21/c], VO_2 (R) [P 42/mnm], VO_2 (A) [P 42/nmc], VO_2 (B) [C 2/m], VO_2 (C) [I 4/mmm], VO_2 (D) [P 2/c] and VO_2 (P) [Pbnm], all of which are depicted in Fig. 6 [116–125]. These different phase structures can easily transform into one another under certain appropriate circumstances. The different possibilities and the meeting circumstances are depicted in Fig. 7 [120, 121, 126–128].

Amongst these different phases, VO_2 (M), VO_2 (R), VO_2 (A) and VO_2 (B) are found to have similar crystal structures with little variation. The main fundamental unit of these structures is the VO_6 octahedron, wherein the vanadium located at the centre is accompanied by the six surrounding oxygen atoms. These octahedral

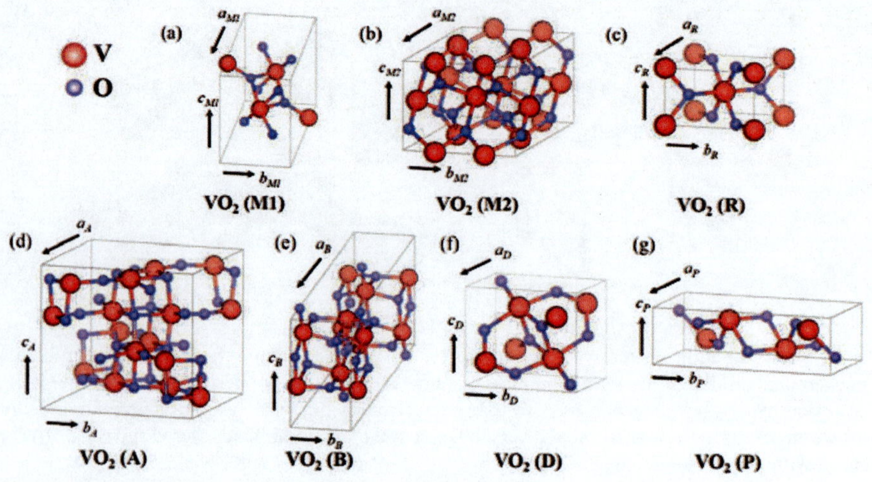

Fig. 6 Different polymorph crystalline structure of VO_2 [58]

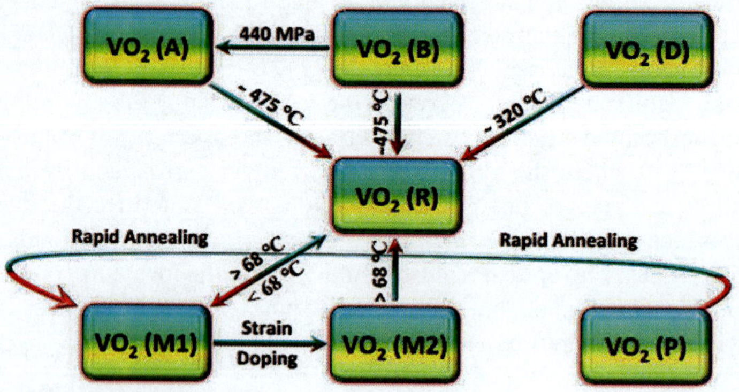

Fig. 7 Schematic representation of different possibilities and circumstance of phase transformation in VO_2

structures are formed by the sharing of edges and vertices with the neighbouring atoms, which results in the generation of a three-dimensional long-range ordered structure. The form of the VO_6 octahedron as well as their modes of connectivity with one another vary for VO_2 crystals holding distinct crystal structures, resulting in variance in the atomic coordinates as well as in their spatial symmetry groups. Phase VO_2 (M) is an insulating monoclinic in nature and contains two phases that are VO_2 (M1) and VO_2 (M2). In most cases of metal-insulator transitions (MIT) in VO_2, the transformation of VO_2 (M1) into VO_2 (R) takes place soon after the temperature exceeds the MIT temperature. Furthermore, the studies have revealed that the low temperature monoclinic phase is an insulating phase (P 21/c) called the M or M1 phase, while the high temperature rutile phase is metallic (P 42/mnm) in nature and designated as the R-phase [116, 117]. In the monoclinic phase, the vanadium atoms arrange themselves in a zigzag configuration along the c-axis. However, in the rutile phase, vanadium atoms arrange themselves along a line which is perpendicular to the c-axis and has a period of 12 C_M.

The octahedral structure composed of V–O bonds gets transformed from partial octahedral at low temperature to the standard octahedral structure at high temperature in the process of phase transition (insulating-metallic). Also, during this transformation, the bond angles of V–O get shifted from 90° to 78°–99°, as depicted in Fig. 6c. During the transformation, it has been observed that in comparison to the high temperature tetragonal phase, the lattice constants of the M-phase, that is, "a" and "b" are significantly raised by 0.6% and 0.4%, respectively, while the lattice constant "c" gets reduced by 1.0%. This variation in the lattice constants results in a change in the volume by 0.044. These structural changes eventually cause a change in the macroscopic properties such as electrical, optical, magnetic, mechanical etc. Both the phases of VO_2 (M), that is VO_2 (M1) and VO_2 (M2), have a monoclinic structure and are insulating. However, the vanadium atoms in the VO_2 (M1) phase arrange themselves into a zigzag chain of V-V pairs along the c-axis, while in the case of VO_2 (M2) the atoms arrange themselves into regular pairs. These regular pairs generally form two shapes that are composed of paired chains of vanadium atoms, but the atoms are not inclined and vice versa, as depicted in Fig. 6a, b. On exerting strain along 1–10 directions or doping with +3 valence ions, it is possible to convert the VO_2 (M1) phase to the VO_2 (M2) phase [126]. Besides this, the phase VO_2 (M2) is often considered an intermediary phase of the phase transition from VO_2 (M1) to VO_2 (R) [129]. In the c-plane of the VO_2 (A) phase, four sets of double-sided octahedrons couple with one another, resulting in the formation of a square of 22 units in size. The formation of a Z-shaped long chain of vanadium atoms is accomplished by staking blocks of 22 along the c-axis in the form of shared edges, as depicted in Fig. 6d. It has also been found that all the phases of VO_2 (A) have obvious MIT behaviour. VO_2 (A) exhibits a normal first-order phase transition with a critical temperature (T_c) of 435 K, which is significantly greater than the T_c of VO_2 (M). The phase transition of VO_2 (A) is accompanied by an abrupt shift in the resistance that ranges between 1 and 2 orders of magnitude [124]. In contrast to the monoclinic VO_2 (M) and VO (A), VO_2 (B) does not exhibit any apparent MIT characteristics and does not undergo any sudden changes in its characteristic properties. The crystal structure of VO_2 (B) is composed of two octahedrons which are connected to one another by the proper sharing of edges and vertices. Also, relative to the first layer, the second layer of the atoms is shifted by some fractional coordinate, that is (12, 12, 0) in the unit cell as depicted in Fig. 6e. Also, at an ambient temperature, VO_2 (B) possesses an adequate square sheet resistance of the order of 20–50 K along with an outstanding temperature coefficient resistance (TCR) of 7.0%/K [130]. As a result of this, VO_2 (B) is an excellent candidate for use in batteries as an electrode and thermally sensitive material. Also, VO_2 (B) can be potentially converted into VO_2 (A) when subjected to a pressure of about 440 MPa [127]. In addition, VO_2 (A) and VO_2 (B) may be converted into VO_2 (R) through an annealing process at a temperature of about 475 °C for at least an hour in the presence of argon gas [128]. VO_2 (C) has been first reported by Hagrman et al. [123] wherein they employed hydrothermal synthesis techniques for its fabrication. The structural configuration of VO_2 (C) is built up of VO_5 square pyramids, each of which is connected to the neighbouring four pyramids via the base edges. On the contrary, only a few studies have been conducted on it, and thus more attention is required to uncover its chemical and physical properties. Similarly, Liu et al. [120] developed a micro/nanostructure that is made up of nanosheets and possesses a novel phase of VO_2 (D). They have concluded from their studies that the formation energy of VO_2 (D) is very similar to that of rutile-type VO_2 (R). They have also proved that VO_2 (D) can be converted to VO_2 (R) by performing annealing at 320 °C for two hours in a gas medium. Following the mentioned procedure, VO_2 (M) can be acquired by performing the cooling of the sample after the annealing process. In addition to this, the Arrhenius-type behaviour has been revealed by the VO_2 (D), which possesses a bandgap of 0.33 eV. The magnetic characteristics of VO_2 (D) were determined by observing the variation of its magnetic susceptibility with temperature. Due to the presence of efficient magnetic properties in VO_2 (D), it acts as a potential candidate for being used in different electrical and magnetic devices. In addition, there is another new phase that is VO_2 (P), which has been produced by a straightforward chemical reaction pathway [121]. Finally, it has also been observed that by undergoing a quick annealing process, VO_2 (P) can be converted into VO_2 (M).

3.4 Tungsten Trioxide

The electrochromic nature of tungsten trioxide (WO_3) was discovered for the first time by Deb in 1969 [131]. Since then, it has become one of the most extensively studied transition metal oxides due to the fact that it possesses very effective electrochromic and photochromic capabilities. Studies have shown that the fundamental structures of ReO_3 (general structure formula) and perovskite have a great deal of bearing on the characteristics of WO_3 in both its natural form and its electrochemically switched state. In its natural state, WO_3 possesses a ReO_3-type structure that is composed of corner-sharing octahedral units of WO_6. In each of these octahedral units, each of the tungsten ions (W^{+6}) is surrounded by six oxygen ions (O^{2-}). The clear visualization can be obtained from Fig. 8a, which depicts the octahedral networking structure where each pair of the tungsten ions is separated by an oxygen ion. These octahedral units are considered the fundamental building block of WO_3, as the numerous configurations of them lead to the generation of its different forms.

The corner sharing that occurs at the oxygen site is what results in the generation of crystalline and stoichiometrically defined regular network structure. As depicted in Fig. 8a, the resultant network possesses an enormous vacant space that plays a crucial role in the electrochromic property of WO_3, as this vacant space is responsible for the functionalization through foreign ions. WO_3 possesses a monoclinic structure at ambient temperature. The variation in the temperature, bond lengths, and octahedral displacements results in the generation of different possible structures of WO_3, from

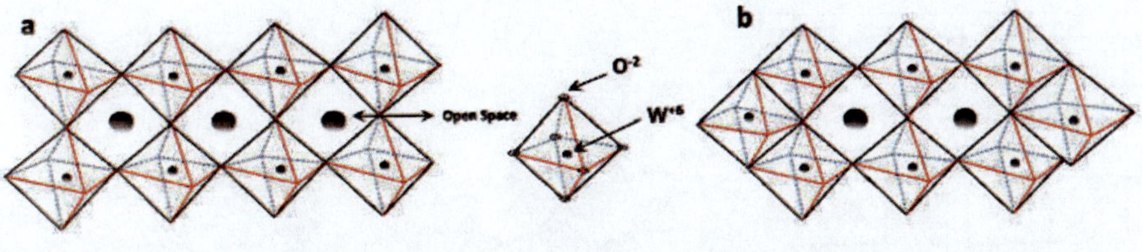

Fig. 8 Illustration of WO_3 structures. **a** Developed by sharing the octahedral units of WO_6 octahedral units. **b** Developed by sharing the edges of WO_6 octahedral units

orthorhombic to monoclinic to triclinic. There exists a sub-stoichiometric phase of WO_3, also known as the Magneli phase, which is depicted in Fig. 8b [132, 133]. The Magneli phase of transition metal oxides is generated from the original perovskite structures via sharing of the edges and by the structural deformation caused by the arranged oxygen deficits. In the perovskite structure of WO_3, the possibility of variation in the position of W as well as in the orientation of the WO_6 octahedron provides large space between the octahedral units of the network. The possibility of these variations results in the generation of eleven distinct structures of WO_3 [134]. Among these structures, besides the standard perovskite structure, two others, tetragonal and hexagonal, are depicted in Fig. 9. It has been observed that the free space depicted in Fig. 9a continuously increases when the perovskite evolves from the perovskite to the tetragonal to the pyrochlore and so on. Although stoichiometric WO_3 is optically transparent and electrically insulating, oxygen-deficient WO_3 may appear faintly blue to brown depending on the severity of the oxygen shortage. In the sub-stoichiometric configurations, the oxides exhibit the characteristics of both optical absorption and electrical conductivity. The fact that the electrical characteristics of the WO_3 structure arise from the orbitals of tungsten and oxygen can be clearly understood from the bandgap diagram depicted in Fig. 10 [135, 136]. As is evident from Fig. 10, there exist two separate bands: the valence band, which is totally filled with regard to the 2p bands of oxygen, and the vacant conduction band, composed of 5d orbitals of tungsten. The fermi levels (E_f) reside in between these two bands. The bandgap in the case of the WO_3 thin films has been estimated to reside at between 2.6 and 3 eV [137]. In contrast to the regular network, the amorphous phase of the polycrystalline WO3 is characterized by the absence of any long-range ordering of the octahedral network [138].

This is the case despite the fact that WO_6 octahedrons are distorted and there is a variation in bond lengths within the octahedrons. Thus, the structure of amorphous films is considered to be composed of a random network of microcrystals with WO_6 octahedrons as the basic building unit. The bandgap of the amorphous form of WO_3 was found to be greater than that of its crystalline form, with a value falling in the range of 3.2–3.5 eV. [139]. Besides the mentioned deficits, due to the amorphous nature of WO_3, it may also include additional defects such as interstitials, dangling bonds, vacancies, etc. [140]. These defects are often present in these structures and occur during the formation of WO_3 thin films by a variety of different processes. These defects play a significant role in enhancing the thermochromic behavior of

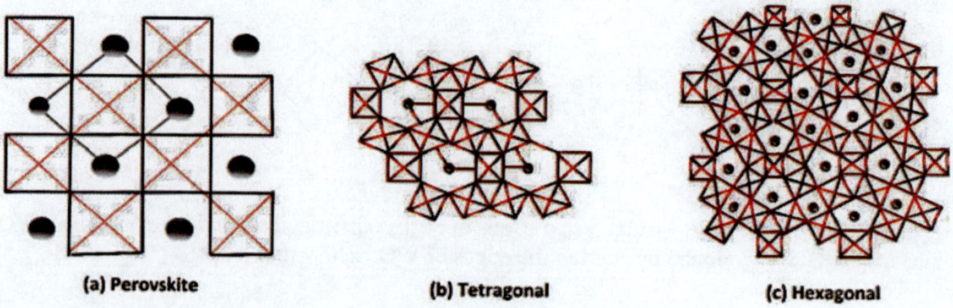

Fig. 9 Perovskite, tetragonal and hexagonal structures of WO_3 molecule

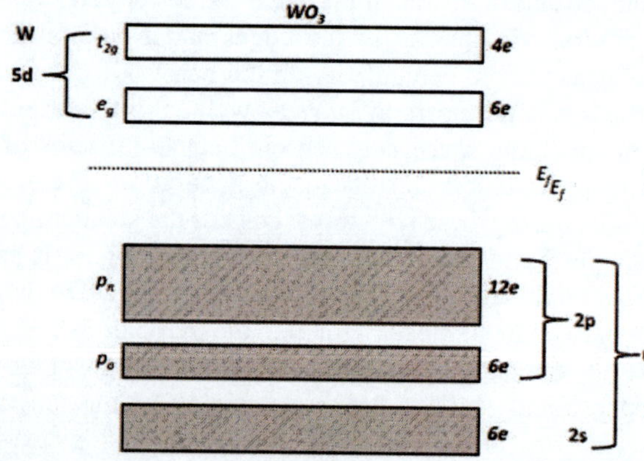

Fig. 10 Energy level band diagram of WO_3. Besides the fermi level (E_f), the unoccupied bands emerging from the tungsten 5d orbitals are depicted at the top and the occupied 2p orbitals arising from the oxygen are represented at the bottom

materials. It is anticipated that different deposition methods often employed for the creation of WO_3 thin films will produce a variety of species of W_xO_y form, where x and y vary over a large range [133].

It is envisioned that the electrochromic coloration of WO_3 will be brought about by the double insertion of either protons, electrons, or different alkali ions because of a reversible chemical reaction. This ca be symbolized through a straightforward transformation that is:

$$\underset{\text{(Transparent)}}{WO_3 + xM^+ + xe^-} \leftrightarrow \underset{\text{(Blue)}}{M_xWO_3}$$

In the above equation, M represents the particular kind of ion that was put in. As a result of the intercalation of a certain number of ions and electrons into the WO_3, the material's optical characteristics shift from the transparent blue to the invisible portion of the spectrum. This occurs concurrently with the reversible production of M_xWO_3, also known as tungsten bronze. Since WO_3 possesses an open structure similar to that of perovskites, different ions are easily able to intercalate inside the material. This causes vibrant colorization in the materials. This whole process shows its reversible nature until an optimum value of x is obtained. Continuing the insertion of ions beyond the optimum value of x, although technically feasible, results in the nullification of the reversible nature of the reaction. For instance, in the process of lithium-ion insertion into WO_3, the reversibility of the reaction is retained up to an x value of 0.7 [136]. It is not anticipated that the size of the inserted ion will have any effect on the value of x, which is the number of ions that can be intercalated into the tungsten bronze. In order to assess the impact of the inserted ions and electrons in defining the electrical, structural, and optical characteristics of the intercalated WO_3, a number of different models have been presented (M_xWO_3). These models cover a wide range of phenomena, from oxygen-vacancy-generated F centers to the intervalence charge transfer (IVCT) of inserted electrons to polaron absorption as well as oxygen extraction [135, 141–144]. However, as discussed, there exist many configurations of the octahedral units with different phases, which results in the generation of a large range of energy bandgaps and structures, some of which are even contrary to one another. Therefore, the search for an all-encompassing theoretical model capable of explaining the process responsible for the electrochromic coloration of WO_3 is still a challenge [71].

4 Decolorization Mechanisms of Thermochromic Materials

These materials are broadly classified into three different categories: organic, inorganic, and liquid crystals. Inorganic thermochromic materials primarily consist of metal iodide, double salt, transition metal oxides, metal alloys, etc., which possess different characteristic features such as being resistant to high temperatures, long-lasting, easy to fabricate, and not affected by light [50]. On the other hand, organic thermochromic compounds include most quantities and species such as spiropyrans, fluoranthene, triarylmethane, organic complexes, and many more. These materials possess different advantages, like optional and adjustable colours, being cost-effective, effective thermochromic behaviour at low temperatures, and high sensitivity for colour change. Because of this, the organic thermochromic materials offer substantial benefits over other ones, which are under consideration in the research community. The third category, that of liquid crystal, has been further classified into three distinct types, namely smectic, nematic, and cholesteric, on the basis of their molecular configurations. Although this class of thermochromic materials possesses good stability and efficient thermal stability, its applications are restricted due to its greater sensitivity to chemicals and high cost [50]. Thus, under these sections, the different decolorization mechanisms involved in reversible organic and inorganic materials are briefly discussed.

4.1 For Organic Materials

The decolorization mechanisms in reversible organic materials mainly occur via two different phenomena, which are intermolecular electron transfer or different structural changes occurring in the molecule. The first mechanism occurs in the thermochromic materials, which are composed of electron donors, electron acceptors, and some solvent compounds. In most cases, the oxidation-reduction potential of an electron donor is approximately equal to that of the electron acceptor. As a result, slight temperature variations cause electron transfer between them, changing the molecular structure of the electron donor. The direction of the reduction process is determined by the temperature, which in turn causes a change in the colour of the system [145]. The most common procedure involved in the transfer of electrons is depicted in Fig. 11a. The second mechanism occurs in such compounds, the molecular structures of which are sensitive to temperature changes. Due to the rise in temperature, these materials exhibit discoloration phenomena of their own. The decolorization occurs through different processes depending on the temperature, that is, (I) by intermolecular proton transfer (Example Fig. 11b), (II) through dimensional structural changes (Example Fig. 11c), (III) ring opening of molecules (Example Fig. 11d), and many more [10, 50, 146, 147].

Aside from the major factors mentioned above, there are a few other variables that contribute to the discoloration of organic compounds, such as electron spin state and pH value. Different researchers have successfully developed the spin-crosslinked

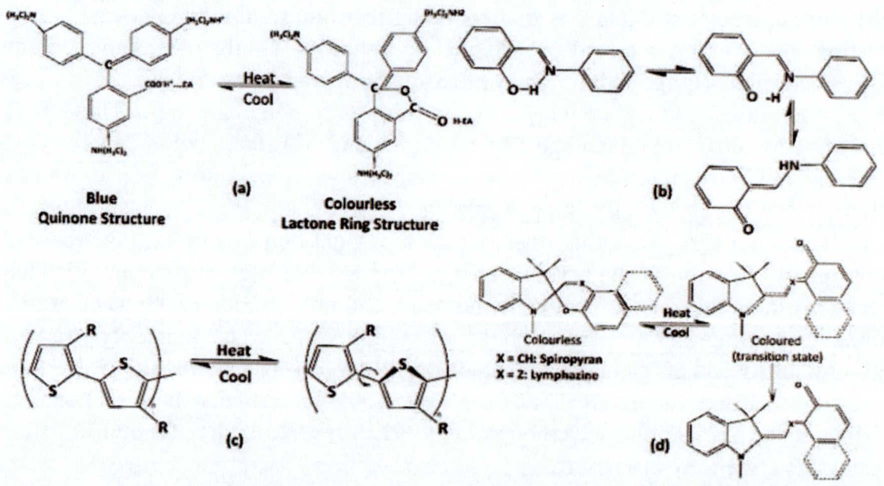

Fig. 11 Depiction of configurational changes in the molecule. **a** Transport of electrons in crystal violet alctone. **b** Intermolecular proton transfer in N-salicylideneaniline via keto-enol tautomerization. **c** Conformational variations of polythiophene. **d** Process of ring opening in spiropyrane

composite transparent films in a high spin state, which are colourless when maintained at 27 °C. It has been observed that as the temperature drops, the film attains a purple colour. In most discoloration systems, phenol red or phenolphthalein is used as an acid-base indicator. These systems also contain weak fatty acids, which assist in increasing the availability of protons. The carboxyl protons are triggered at a temperature where they can react with the affinity materials, thereby culminating in the gain and loss of protons. When the carboxyl cools, the proton will go back to how it was before, and the colour of the substance will return quickly [50].

4.2 *For Inorganic Materials*

Crystal lattice changes, ligand geometry, and crystal water are the primary factors that contribute to discoloration in inorganic thermochromic materials. The discoloration in the inorganic complexes occurs due to variation in the coordination structure or by changing the degree of hydration. On the other hand, in the case of inorganic oxides, the discoloration is due to changes in the crystal structure.

The crystal transition is responsible for the discolorations revealed by majority of the metal ion compounds. It has been observed that most of the discoloured crystalline materials on heating produce lattice displacement as a result of which the crystal structure gets changed. Soon after the crystal returns to the ambient temperature, the original crystal shape and colour are eventually recovered. For example, the compounds Cu_2HgI_4 and Ag_2HgI_4 possess a dual crystal structure that is positively tetrahedral at low temperatures and cubic at higher temperatures. The colour of these compounds fluctuates due to their dual crystal forms, which rely on temperature change [148]. Another factor involves the variation in the geometry of the ligands with the variation in the temperature. This type of thermochromism phenomenon has been observed in the Cr^{3+}, as the heating in it causes broadening in the ionic lattice, thus causing changes in its geometry. It has been discovered that the distance between the central atom and Cr^{3+} ions vary with temperature in compounds where an octahedral site is occupied by Cr^{3+} ions. Due to this variation, there occurs a change in the colour of the compound in accordance with the change in temperature [149]. Similarly, the modifications of the coordination number in the complex molecules are the fundamental cause of colour change. Thermochromism due to this factor has been mainly observed in crystallised inorganic salts. For example, iron phosphate dihydrate ($FePO_4$, $2H_2O$), which possesses either an orthogonal or monoclinic structure. In its pure form, it looks like white or yellowish white powder. However, the colour eventually becomes pure yellow due to loss in the crystallised water. Lastly, cholesteric liquid crystals are the most common form of reversible liquid crystal thermochromic materials. Liquid crystals made by cholesterol often have spiral shares. The variation in temperature results in shifting the pitch of the screw along with the variation in the wavelength of the reflected visible light, as a result of which the discoloration of the liquid crystal takes place [150]. The length of the screw pitch is proportional to the wavelength of the visible light, with both having the same magnitude. Because of this, some liquid crystal thermochromic materials have unique visual qualities in addition to a clear change in colour when the temperature changes.

5 Applications

As defined earlier that the Chromogenic substances are ones whose optical characteristics modify with response to external stimulation. Such substances are most frequently found to be photochromic, thermochromic, as well as electrochromic, in which the stimuli are, alternately, exposure to sunlight (photons), change in temperature, and then an applied electric field. As a result, when a thermochromic substance reaches a specific transition temperature, it changes colour [80]. Different transition-metal oxides including such as Ti_2O_3, V_2O_3, VO_2, as well as VO exhibit a transition together into metallic state at the Neel temperature and they are all semi-conducting under low temperatures. Morin conducted extensive thermo-conductive experiments on these oxides' electrical characteristics in 1959 [151]. The discovery of the phase transitions which take place at a critical temperature (T_c) in which an abrupt change throughout optical and electronic properties is observed, transition-metal oxides have attracted attention in research especially among solid-state thermochromic materials, making them excellent research subjects for examining both metallic and non-metallic states. They are currently being used for different applications in the field of optical and electrical switching devices as well as in memory devices because to their outstanding MIT (metal insulator transition) characteristics. In this section some of the major application that is smart windows, switches, field effect transistors (FET) etc., are briefly discussed.

5.1 Smart Windows

The adoption of the thermochromic materials for the mart windows has been a focused topic in recent years as a result of challenges of limited available energy and global warming. Smart windows assist the buildings in maintaining the high energy states by dynamically boosting and decreasing the heat gain cold and hot weather respectively [152]. By adjusting the incident solar heat flux, the adjustable absorbing layer on the glass surface possess the capability to alter the optical characteristics [153]. Among the different thermochromic materials, VO_2 are being considered as an effective material for the thermochromic smart windows that has proved to be energy efficient. This is due to VO_2's ability to modify the temperature to efficiently regulate thermal transfer and solar thermal transmission effectively [154–157]. Whenever MIT happens, VO_2 displays a significant shift in optical characteristics. Insulating VO_2 (M) has a high infrared light transmittance prior MIT. Because when temperature increases to the Tc, VO_2 (M) transforms into VO_2 (R), which leads to decreased infrared transmittance. It is really appealing when it comes to using thermochromic smart windows due to its distinctive property. At low temperatures, infrared rays may generally travel through the glass. The infrared beam is isolated and the inside temperature stays constant when the temperature exceeds a particular level. VO_2-based smart windows have been the subject of much research, although they have not yet been put to commercial use. The inadequate inherent characteristics of VO_2 smart windows severely restrict their practical uses. To create VO_2 thermochromic windows, researchers have been attempting to find solutions to these difficulties for the past few decades. Doping with high-valence tungsten elements is indeed a useful way to manage the phase transition temperature [158]. Since hydrogen is the smallest and lightest atomic element, it may easily change the spacing location of VO_2, therefore influences the behaviour of the MIT [159, 160]. Yoon et al. [160] revealed that even at low hydrogen concentrations, VO_2 stabilizes in the metal phase, whereas at high hydrogen concentrations, it stabilizes in the hydrogenated insulating phase yet again. Utilizing this unique capability, Chen et al. [161] created an electrochromic smart window system based on a VO_2 film under room-temperature solid-state electrolyte gate voltage control. The phase transition of VO_2 thin films might well be controlled by gate voltage at room temperature, and the gate-controlled VO_2 phase transition is a reversible process. During three separate VO_2 film stages, significant colour shifts could be seen. Although strongly hydrogen-doped VO_2 thin films were nearly transparent, pure VO_2 thin films and low hydrogen-doped VO_2 thin films (H_xVO_2) exhibit comparable yellow colours. At 650 nm, the original VO_2 film and metal H_xVO_2 have a transmittance of around 56%, while the insulating HVO_2 film seems to have a transmittance of up to 72%. According to the MIT procedure, the transmittance variation between a pure VO_2 film at 2000 nm is approximately 40.3%, as well as the gap between a metal H_xVO_2 film and an insulating HVO_2 film is 49.1% [161]. Outstanding performances of these gadgets surpass the potential of conventional VO_2 smart windows as well as shatter all prior records. The above implies that their technology holds promise for energy-saving use. The prototype for just an electrochromic smart window device based on VO_2 film for solid-state electrolyte gate voltage control is illustrated in Fig. 12.

5.2 *Electrical Switches*

One significant characteristic of VO_2 that considers this one a fantastic prospective substance for electrical switches is indeed the abrupt change in conductance throughout MIT [162, 163]. Typically, temperature or even the application of a static current as well as a static voltage might control the electronic switches relying on VO_2 [164, 165]. In terms of speed as well as broadband functionality, the electronic switches based on VO_2 perform well [166]. Whenever VO_2 is linked in series to the circuitry, the metal or insulating phase of the gas may be changed to regulate the circuit's connection and disconnection. Majority of the electronic switches require triggering by voltage-driven MIT of VO_2 to activate. A VO_2-based electrical switch

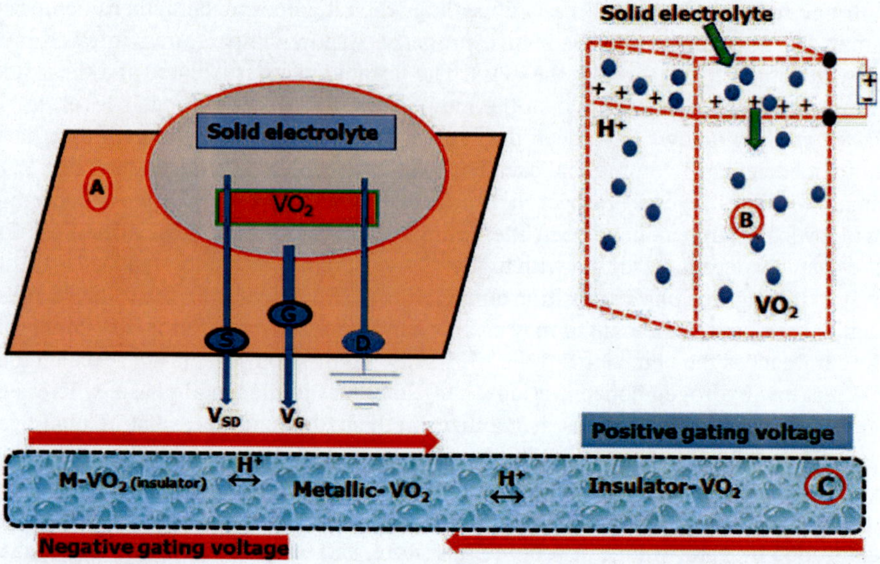

Fig. 12 **a** The source, drain, & gate electrodes on the VO_2 device's gating diagram. **b** Gating-controlled hydrogen ion migration. **c** The VO_2 hydrogenating content is tuned with either positive (blue) or negative (red) gating voltages, yielding into reversible insulator-metal-insulator tristate phase transformations

device was devised by Stefanovich and collaborators [164]. The I–V characteristics of VO_2 switching devices with various temperatures were also examined by Boriskov et al. [165] and they have observed that when the ambient temperature changes, different voltages are needed to trigger MIT. Crunteanu et al. [166] also investigated voltage mode and current-mode VO_2-based switches. They established that VO_2-based switches functioned for more than 260 million cycles without breakdown when propelled by current, which is a greater lifespan than that when powered by voltage (breakdown at about 16 million cycles). Especially devices with many reliable and reproducible switching cycles, the incorporation of VO_2 provides therefore a lot of promise. On the other hand, temperature is a simple way to achieve VO_2's resistance switching, that everyone is aware of. As a result, the accompanying resistance switching matches MIT behaviour.

5.3 *Memory Appliances*

Broadly speaking, in addition to temperature, an electric field is thought to be an additional efficient approach to influence MIT activity. The differing phase transition routes throughout the cooling and heating processes are typically blamed for hysteresis, which is typically present even during MIT. Strain, doping, and flaws in the

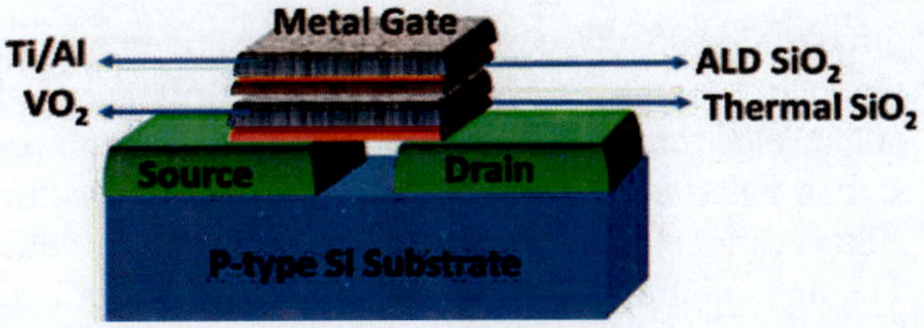

Fig. 13 Schematic representation of a ferroelectric memory device with a VO_2 layer operating as the gate insulator

lattice during the MIT process all impact this hysteresis. When grown on a substrate, VO_2 can get the mismatch strain from the substrate. Further, majority of the electrical equipment frequently accumulate VO_2 on the substrate, creating a significant external strain during MIT. Hysteresis causes the metallic phase of VO_2 to be held at a comparatively small temperature (or applied voltage). As a result, VO_2 can indeed be turned into an additional possible memory device application [167–169]. Another multistate memory with two-terminal current control was developed by Pellegrino et al. Investigators created multistate memory devices close to the T_c of 70 °C employing VO_2/TiO_2 free-standing thin film heterostructures as well as a Joule self-heating procedure. Optimizing the development and creation of metal clusters inside the micrometric zone by efficient Joule heating results in the memory effect [158, 170]. Also, Bae et al. [171] has reported a two-terminal memristor memory based on a single VO_2 NW at room temperature. Coy et al. [167] revealed the optoelectronic and so all multiple memory states in VO_2, which is noteworthy. Investigators examined the VO_2/SiO_2 specimen's resistance as well as near-infrared transmittance just at proper temperature inside the heating branches of the hysteresis loop. A variety of reading-writing modes may be produced by efficiently blending electrical and optical loads. Apparently numerous memory regions were discovered to be stable for at least a few hours, and in this case possibly permanently, if the proper temperatures were maintained. Furthermore, a well-developed chemical doping technique or strain effect may be employed in VO_2-based memory devices to modulate the hysteresis gap of VO_2 to produce a more stable memory effect [172]. Employing traditional CMOS processing technologies, ferroelectric memory devices with VO_2 films serving as insulators were manufactured. The architecture of the device is quite similar to a flash memory, as illustrated in Fig. 13.

6 Field-Effect Transistor (FET)

The gate voltage regulates the conductivity of field-effect transistors (FETs), which are conducting devices based on early triodes. FETs have many benefits over triodes, including a high input impedance, a low temperature effect, low noise, low power consumption, a high switching speed, as well as a simple manufacturing method. The metal-oxide-semiconductor FET (MOSFET) is perhaps the most frequently utilised semiconductor devices across all FETs [173]. While using drift as well as carrier diffusion, MOSFET develops current. Although in contemporary MOSFETs, which produce current by carrier diffusion and drift, silicon-based FETs have emerged as a crucial electronic component. However, in contemporary semiconductor technology, silicon-based FETs have turned into a crucial electrical component. Unfortunately, silicon cannot be used to create small FET devices due to its intrinsic size restriction. The maximum allowed sub-threshold swing rate is 60 mV/dec. If not, FET device integration, switching rate, and power consumption would all be negatively impacted. Therefore, the demand for silicon-alternative materials is critical. For nanoscale FETs, MIT materials are regarded as potential channel materials. The fluctuations in carrier concentration in MIT materials are greater than those in conventional silicon-based materials [173, 174]. Because of their exceptionally high MIT speeds as well as significant fluctuations in carrier concentration, VO_2, based FETs have received the most research attention among all FETs based on MIT substances. The concentration of free carriers produced by MIT in VO_2 were shown by Hormoz et al. to be many orders of magnitude higher than those in silicon. As a result, the VO_2-based FET can easily get beyond the classic MOSFETs carrier transit-time barrier [174].

7 Photodetector

In addition to smart windows, the distinctive optical characteristics of VO_2 discovered at MIT also indicate other intriguing properties for being used in photodetectors, tunable photonic crystals, and many other fields [160, 175, 176]. Wu et al. [176] constructed VO_2 nanodevices based on photoinduced phase transitions of VO_2 NWs under ultraviolet (UV) light at ambient temperature. They found that the synthesized gadgets had exceptional photo-responsiveness (R) and external quantum efficiency (EQE). The photodetector responded quite well to UV light. A heater based on Ag NW was used by Li et al. [177] to measure the infrared (IR) response of self-heating VO_2 NPs. IR photodetector technology is crucial for both civil and military applications. An IR photodetector constructed on a network of as-grown VO_2 (M1) NRs was also investigated by different researchers [184]. The device's I–V characteristics beneath various IR light intensities (980 nm) were evaluated. The gadget clearly showed a low-current condition in the dark and a high-current state when exposed to light. With rise and decaying periods of 1.6 and 1.0 s, respectively, and exceptional stability of much more than 150 cycles at room temperature, the gadget was employed with high photosensitivity. Their findings suggested that the networks of VO_2 (M1) NRs are the best method for using an IR photodetector [178].

8 Conclusion

In recent years, thermochromic materials have emerged as a popular research area; however, only a few studies have been published on practical applications. Its high transition temperature and high cost limit its application. In addition, other challenges that must be overcome with widespread use materials are finding ways to successfully lower the transition temperature and enhance the variation ranges of electrical and optical properties before and after transition. The trend in the growth of thermochromic materials reveals that the most important factor in improving the transformation properties is to optimize the synthesis parameters, and the doping of different elements has proven to be the most effective way to lower the temperature at which the phase transition occurs. The lowering of the phase transition temperature, which may be accomplished by doping with ions and annealing, is a significant step forward in broadening the range of applications of thermochromic materials. Using thermochromic materials also benefits from microencapsulation, which has become increasingly relevant in recent years. It is crucial to create innovative technologies to full use thermochromic materials since they play an important role in the growth of contemporary society. Functional materials play an important part in the progression of modern civilization. One proposal that may be considered is packaging, particularly for food packaging. People in today's contemporary society are much more concerned about the safety of the food they eat, and thermochromic materials will play a significant role in the testing of food quality and the enhancement of its shelf life in the future.

References

1. J. Lee, M. Christopher Orilall, S.C. Warren, M. Kamperman, F.J. Disalvo, U. Wiesner, Direct access to thermally stable and highly crystalline mesoporous transition-metal oxides with uniform pores. Nat. Mater. **7**, 222–228 (2008). https://doi.org/10.1038/nmat2111
2. H.J. Bolink, E. Coronado, J. Orozco, M. Sessolo, Efficient polymer light-emitting diode using air-stable metal oxides as electrodes. Adv. Mater. **21**, 79–82 (2009). https://doi.org/10.1002/adma.200802155
3. A.V. Emeline, G.V. Kataeva, A.V. Panasuk, V.K. Ryabchuk, N.V. Sheremetyeva, N. Serpone, Effect of surface photoreactions on the photocoloration of a wide band gap metal oxide: probing whether surface reactions are photocatalytic. J. Phys. Chem. B. **109**, 5175–5185 (2005). https://doi.org/10.1021/jp0452047
4. M. Kröger, S. Hamwi, J. Meyer, T. Riedl, W. Kowalsky, A. Kahn, P-type doping of organic wide band gap materials by transition metal oxides: a case-study on Molybdenum trioxide. Org. Electron. **10**, 932–938 (2009). https://doi.org/10.1016/j.orgel.2009.05.007
5. A. Dev, J.P. Richters, J. Sartor, H. Kalt, J. Gutowski, T. Voss, Enhancement of the near-band-edge photoluminescence of ZnO nanowires: important role of hydrogen incorporation versus plasmon resonances. Appl. Phys. Lett. **98**, 131111 (2011). https://doi.org/10.1063/1.3569951
6. D.R. Rosseinsky, R.J. Mortimer, Electrochromic systems and the prospects for devices. Adv. Mater. **13**, 783–793 (2001). https://doi.org/10.1002/1521-4095(200106)13:11%3c783::AID-ADMA783%3e3.0.CO;2-D

7. C.G. Granqvist, Oxide electrochromics: why, how, and whither. Sol. Energy Mater. Sol. Cells. **92**, 203–208 (2008). https://doi.org/10.1016/j.solmat.2006.10.027
8. K. Takada, H. Sakurai, E. Takayama-Muromachi, F. Izumi, R.A. Dilanian, T. Sasaki, Superconductivity in two-dimensional CoO2 layers. Nature **422**, 53–55 (2003). https://doi.org/10.1038/nature01450
9. Z.A. Ren, G.C. Che, X.L. Dong, J. Yang, W. Lu, W. Yi, X.L. Shen, Z.C. Li, L.L. Sun, F. Zhou, Z.X. Zhao, Superconductivity and phase diagram in iron-based arsenic-oxides ReFeAsO $1-\delta$ (Re = rare-earth metal) without fluorine doping. EPL Europhys. Lett. **83**, 17002 (2008). https://doi.org/10.1209/0295-5075/83/17002
10. K.S. Ahn, D.K. Lee, J.H. Kim, S.H. Kim, Domino-like thermal phase transition of 2,3-bis(phenylethenyl)-5,6-dicyanopyrazine crystal. Dye. Pigment. **89**, 93–95 (2011). https://doi.org/10.1016/j.dyepig.2010.08.016
11. G.M. Rignanese, Dielectric properties of crystalline and amorphous transition metal oxides and silicates as potential high-κ candidates: the contribution of density-functional theory. J. Phys. Condens. Matter. **17**, R357 (2005). https://doi.org/10.1088/0953-8984/17/7/R03
12. J. Robertson, High dielectric constant gate oxides for metal oxide Si transistors, reports. Prog. Phys. **69**, 327–396 (2006). https://doi.org/10.1088/0034-4885/69/2/R02
13. K. Chen, A.T. Bell, E. Iglesia, The relationship between the electronic and redox properties of dispersed metal oxides and their turnover rates in oxidative dehydrogenation reactions. J. Catal. **209**, 35–42 (2002). https://doi.org/10.1006/jcat.2002.3620
14. Sysoev, V.V., Button, B.K., Wepsiec, K., Dmitriev, S., Kolmakov, A. Toward the nanoscopic "electronic nose": hydrogen vs carbon monoxide discrimination with an array of individual metal oxide nano- and mesowire sensors (2006). https://doi.org/10.1021/NL060185T
15. G. Mavrou, S. Galata, P. Tsipas, A. Sotiropoulos, Y. Panayiotatos, A. Dimoulas, E.K. Evangelou, J.W. Seo, C. Dieker, Electrical properties of La2 O3 and $HfO_2La_2O_3$ gate dielectrics for germanium metal-oxide-semiconductor devices. J. Appl. Phys. **103**, 014506 (2008). https://doi.org/10.1063/1.2827499
16. M.J. Lee, S. Han, S.H. Jeon, B.H. Park, B.S. Kang, S.E. Ahn, K.H. Kim, C.B. Lee, C.J. Kim, I.K. Yoo, D.H. Seo, X.S. Li, J.B. Park, J.H. Lee, Y. Park, Electrical manipulation of nanofilaments in transition-metal oxides for resistance-based memory. Nano Lett. **9**, 1476–1481 (2009). https://doi.org/10.1021/nl803387q
17. X. Su, Z. Zhang, M. Zhu, Melting and optical properties of ZnO nanorods. Appl. Phys. Lett. **88**, 061913 (2006). https://doi.org/10.1063/1.2172716
18. A.P. Alivisatos, Semiconductor clusters, nanocrystals, and quantum dots. Science (80), 271, 933–937 (1996). https://doi.org/10.1126/science.271.5251.933
19. F.W. Wise, Lead salt quantum dots: the limit of strong quantum confinement. Acc. Chem. Res. **33**, 773–780 (2000). https://doi.org/10.1021/ar970220q
20. B. Wu, A. Heidelberg, J.J. Boland, Mechanical properties of ultrahigh-strength gold nanowires. Nat. Mater. **4**, 525–529 (2005). https://doi.org/10.1038/nmat1403
21. X. Duan, Y. Huang, Y. Cui, J. Wang, C.M. Lieber, Indium phosphide nanowires as building blocks for nanoscale electronic and optoelectronic devices. Nature **409**, 66–69 (2001). https://doi.org/10.1038/35051047
22. Y. Cui, C.M. Lieber, Functional nanoscale electronic devices assembled using silicon nanowire building blocks. Science (80), **291**, 851–853 (2001). https://doi.org/10.1126/science.291.5505.851
23. Y. Huang, X. Duan, Y. Cui, L.J. Lauhon, K.H. Kim, C.M. Lieber, Logic gates and computation from assembled nanowire building blocks. Science (80). **294**, 1313–1317 (2001). https://doi.org/10.1126/science.1066192
24. J. Andzane, N. Petkov, A.I. Livshits, J.J. Boland, J.D. Holmes, D. Erts, Two-terminal nanoelectromechanical devices based on germanium nanowires. Nano Lett. **9**, 1824–1829 (2009). https://doi.org/10.1021/nl8037807
25. C. Ke, H.D. Espinosa, In situ electron microscopy electromechanical characterization of a bistable NEMS device. Small **2**, 1484–1489 (2006). https://doi.org/10.1002/smll.200600271

26. X. Wang, C.J. Summers, Z.L. Wang, Large-scale hexagonal-patterned growth of aligned ZnO nanorods for nano-optoelectronics and nanosensor arrays. Nano Lett. **4**, 423–426 (2004). https://doi.org/10.1021/nl035102c
27. Y.R. Ma, C.M. Lin, C.L. Yeh, R.T. Huang, Synthesis and characterization of one-dimensional WO_2 nanorods. J. Vac. Sci. Technol. B Microelectron. Nanom. Struct. **23**, 2141 (2005). https://doi.org/10.1116/1.2050668
28. Y.R. Ma, C.C. Tsai, S.F. Lee, K.W. Cheng, Y. Liou, Y. Der Yao, Magnetic properties of large-area one-dimensional WO_2 and MoO_2 nanorods. J. Magn. Magn. Mater. **304**, e13–e15 (2006). https://doi.org/10.1016/j.jmmm.2006.02.025
29. R.S. Devan, W. Der Ho, J.H. Lin, S.Y. Wu, Y.R. Ma, P.C. Lee, Y. Liou, X-ray diffraction study of a large-scale and high-density array of one-dimensional crystalline tantalum pentoxide nanorods, in Cryst. Growth Des. Am. Chem. Soc. 4465–4468 (2008). https://doi.org/10.1021/cg800369s
30. L. Kumari, Y.R. Ma, C.C. Tsai, Y.W. Lin, S.Y. Wu, K.W. Cheng, Y. Liou, X-ray diffraction and Raman scattering studies on large-area array and nanobranched structure of 1D MoO2 nanorods. Nanotechnology **18**, 115717 (2007). https://doi.org/10.1088/0957-4484/18/11/115717
31. Y.W. Heo, D.P. Norton, L.C. Tien, Y. Kwon, B.S. Kang, F. Ren, S.J. Pearton, J.R. Laroche, ZnO nanowire growth and devices. Mater. Sci. Eng. R Rep. **47**, 1–47 (2004). https://doi.org/10.1016/j.mser.2004.09.001
32. J. Zhou, Y. Ding, S.Z. Deng, L. Gong, N.S. Xu, Z.L. Wang, Three-dimensional tungsten oxide nanowire networks. Adv. Mater. **17**, 2107–2110 (2005). https://doi.org/10.1002/adma.200500885
33. Z.L. Wang, J. Song, Piezoelectric nanogenerators based on zinc oxide nanowire arrays, Science (80-) **312**, 242–246 (2006). https://doi.org/10.1126/science.1124005
34. C.J. Park, D.K. Choi, J. Yoo, G.C. Yi, C.J. Lee, Enhanced field emission properties from well-aligned zinc oxide nanoneedles grown on the AuTin–Si substrate. Appl. Phys. Lett. **90**, 083107 (2007). https://doi.org/10.1063/1.2643979
35. P.M. Rao, X. Zheng, Rapid catalyst-free flame synthesis of dense, aligned α-Fe_2O_3 nanoflake and CuO nanoneedle arrays. Nano Lett. **9**, 3001–3006 (2009). https://doi.org/10.1021/nl901426t
36. S.M. Liu, L.M. Gan, L.H. Liu, W.D. Zhang, H.C. Zeng, Synthesis of single-crystalline TiO_2 nanotubes. Chem. Mater. **14**, 1391–1397 (2002). https://doi.org/10.1021/cm0115057
37. Direct electrochemistry and electrocatalysis of hemoglobin immobilized in TiO_2 nanotube films. Talanta **74**, 1414–1419 (2008). https://doi.org/10.1016/J.TALANTA.2007.09.017
38. Y. Wang, X. Jiang, Y. Xia, A. Solution-Phase, Precursor route to polycrystalline SnO_2 nanowires that can be used for gas sensing under ambient conditions. J. Am. Chem. Soc. **125**, 16176–16177 (2003). https://doi.org/10.1021/JA037743F/SUPPL_FILE/JA037743FSI20031117_042556.PDF
39. R. Rosario, D. Gust, A.A. Garcia, M. Hayes, J.L. Taraci, T. Clement, J.W. Dailey, S.T. Picraux, Lotus effect amplifies light-induced contact angle switching. J. Phys. Chem. B. **108**, 12640–12642 (2004). https://doi.org/10.1021/jp0473568
40. L.E. Greene, B.D. Yuhas, M. Law, D. Zitoun, P. Yang, Solution-grown zinc oxide nanowires. Inorg. Chem. **45**, 7535–7543 (2006). https://doi.org/10.1021/ic0601900
41. P.C. Chen, G. Shen, S. Sukcharoenchoke, C. Zhou, Flexible and transparent supercapacitor based on In_2O_3 nanowire/carbon nanotube heterogeneous films. Appl. Phys. Lett. **94**, 043113 (2009). https://doi.org/10.1063/1.3069277
42. J.D. Prades, R. Jimenez-Diaz, F. Hernandez-Ramirez, L. Fernandez-Romero, T. Andreu, A. Cirera, A. Romano-Rodriguez, A. Cornet, J.R. Morante, S. Barth, S. Mathur, Toward a systematic understanding of photodetectors based on individual metal oxide nanowires. J. Phys. Chem. C **112**, 14639–14644 (2008). https://doi.org/10.1021/jp804614q
43. S.Y. Li, P. Lin, C.Y. Lee, T.Y. Tseng, Field emission and photofluorescent characteristics of zinc oxide nanowires synthesized by a metal catalyzed vapor-liquid-solid process. J. Appl. Phys. **95**, 3711–3716 (2004). https://doi.org/10.1063/1.1655685

44. Y.W. Heo, L.C. Tien, Y. Kwon, D.P. Norton, S.J. Pearton, B.S. Kang, F. Ren, Depletion-mode ZnO nanowire field-effect transistor. Appl. Phys. Lett. **85**, 2274–2276 (2004). https://doi.org/10.1063/1.1794351
45. A.M. Donia, E.Z.M. Ebeid, Thermochromism in Ni^{2+} complexes with schiff base derivatives of 4-aminoantipyrine. Thermochim. Acta **131**, 1–6 (1988). https://doi.org/10.1016/0040-6031(88)80051-0
46. T. Inabe, N. Hoshino, T. Mitani, Y. Maruyama, Structure and optical properties of a Thermochromic Schiff base. low-temperature structural studies of the N, N′-Disalicylidene-p-phenylenediamine and N, N′-Disalicylidene-1,6-pyrenediamine crystals. Bull. Chem. Soc. Jpn. **62**, 2245–2251 (1989). https://doi.org/10.1246/bcsj.62.2245
47. M. Neebe, D. Rhinow, N. Schromczyk, N.A. Hampp, Thermochromism of bacteriorhodopsin and its pH dependence. J. Phys. Chem. B. **112**, 6946–6951 (2008). https://doi.org/10.1021/JP7111389
48. C.F. Zhu, A.B. Wu, Studies on the synthesis and thermochromic properties of crystal violet lactone and its reversible thermochromic complexes. Thermochim. Acta **425**, 7–12 (2005). https://doi.org/10.1016/j.tca.2003.08.001
49. C.G. Granqvist, S. Green, G.A. Niklasson, N.R. Mlyuka, S. von Kræmer, P. Georén, Advances in chromogenic materials and devices. Thin Solid Films **518**, 3046–3053 (2010). https://doi.org/10.1016/j.tsf.2009.08.058
50. Y. Cheng, X. Zhang, C. Fang, J. Chen, Z. Wang, Discoloration mechanism, structures and recent applications of thermochromic materials via different methods: a review. J. Mater. Sci. Technol. **34**, 2225–2234 (2018). https://doi.org/10.1016/j.jmst.2018.05.016
51. G. Gorgolis, D. Karamanis, Solar energy materials for glazing technologies. Sol. Energy Mater. Sol. Cells. **144**, 559–578 (2016). https://doi.org/10.1016/j.solmat.2015.09.040
52. D. Fan, Q. Li, Y. Xuan, Y. Xia, Thermal radiation from silicon microcavity coated with thermochromic film. Sol. Energy Mater. Sol. Cells. **144**, 331–338 (2016). https://doi.org/10.1016/j.solmat.2015.09.022
53. C.G. Granqvist, Electrochromics and thermochromics: towards a new paradigm for energy efficient buildings, in *Materials Today Proceedings* (Elsevier, 2016), pp. S2–S11. https://doi.org/10.1016/j.matpr.2016.01.002
54. W. Chen, J. Ketkaew, Z. Liu, R.M.O. Mota, K. O'Brien, C.S. Da Silva, J. Schroers, Does the fracture toughness of bulk metallic glasses scatter? Scr. Mater. **107**, 1–4 (2015). https://doi.org/10.1016/j.scriptamat.2015.05.003
55. G.Y. Sun, X. Cao, H. Zhou, S. Bao, P. Jin, A novel multifunctional thermochromic structure with skin comfort design for smart window application. Sol. Energy Mater. Sol. Cells. **159**, 553–559 (2017). https://doi.org/10.1016/j.solmat.2016.09.045
56. J. Zhang, J. Wang, C. Yang, H. Jia, X. Cui, S. Zhao, Y. Xu, Mesoporous SiO_2/VO_2 double-layer thermochromic coating with improved visible transmittance for smart window. Sol. Energy Mater. Sol. Cells **162**, 134–141 (2017). https://doi.org/10.1016/j.solmat.2016.12.048
57. S. Zheng, Y. Xu, Q. Shen, H. Yang, Preparation of thermochromic coatings and their energy saving analysis. Sol. Energy **112**, 263–271 (2015). https://doi.org/10.1016/j.solener.2014.09.049
58. Y. Zhang, W. Xiong, W. Chen, Y. Zheng, Recent progress on vanadium dioxide nanostructures and devices: fabrication, properties, applications and perspectives. Nanomaterials **11**, 1–48 (2021). https://doi.org/10.3390/nano11020338
59. C.M. Lampert, Chromogenic smart materials. Mater. Today **7**, 28–35 (2004). https://doi.org/10.1016/S1369-7021(04)00123-3
60. H.A. Sodano, D.J. Inman, G. Park, Comparison of piezoelectric energy harvesting devices for recharging batteries. J. Intell. Mater. Syst. Struct. **16**, 799–807 (2005). https://doi.org/10.1177/1045389X05056681
61. C.M.. Lampert, C.G. Granqvist, Large-area chromogenics: materials and devices for transmittance control. Soc. Photo Opt. 606 (1990)
62. T. Sluckin, D. Dunmur, H. Stegemeyer, Crystals that Flow (2004). https://doi.org/10.1201/9780203022658

63. J. Fransaer, P.M. Vereecken, G. Oskam, Semiconductors, metal oxides, and composites: metallization and electrodeposition of thin films and nanostructures (2015)
64. C.G. Granqvist, *Handbook of Inorganic Electrochromic Materials* (Elsevier, 1995). https://doi.org/10.1016/B978-0-444-89930-9.X5000-4
65. D.R. Rosseinsky, J.D. Slocombe, A.M. Soutar, P.M.S. Monk, A. Glidle, Simple diffuse-reflectance monitoring of emerging surface-attached species. Bipyridilium, Prussian-Blue and zinc-ferrocyanide voltammetry. J. Electroanal. Chem. **258**, 233–239 (1989). https://doi.org/10.1016/0022-0728(89)85176-9
66. B. Grant, N.J. Clecak, M. Oxsen, A. Jaffe, G.S. Keller, Study of the electrochromism of methoxyfluorene compounds. J. Org. Chem. **45**, 702–705 (1980). https://doi.org/10.1021/jo01292a030
67. P. Monk, R. Mortimer, D. Rosseinsky, *Electrochromism: fundamentals and Applications* (VCH, 2008)
68. R.C. Duncan, D.L. Staebler, Inorganic photochromic materials, 133–160 (1977). https://doi.org/10.1007/3-540-08293-x_45
69. W. Lohstroh, R.J. Westerwaal, B. Noheda, S. Enache, I.A.M.E. Giebels, B. Dam, R. Griessen, Self-organized layered hydrogenation in black Mg2NiH x switchable mirrors. Phys. Rev. Lett. **93**, 197404 (2004). https://doi.org/10.1103/PhysRevLett.93.197404
70. G.H. Brown, *Photochromism* (Wiley, 1971)
71. P. Ashrit, *Transition Metal Oxide Thin Film-Based Chromogenics and Devices* (2017)
72. A. Seeboth, D. Loetzsch, R. Ruhmann, Piezochromic polymer materials displaying pressure changes in bar-ranges. Am. J. Mater. Sci. **1**, 139–142 (2012). https://doi.org/10.5923/j.materials.20110102.23
73. P. Jin, S. Tanemura, Manufacturing methods of Samarium Sulfide thin films (2000)
74. P. Shi, D. Deng, C. He, L. Ji, Y. Duan, T. Han, B. Suo, W. Zou, Mechanochromic luminescent materials with aggregation-induced emission: mechanism study and application for pressure measuring and mechanical printing. Dye. Pigment. **173**, 107884 (2020). https://doi.org/10.1016/j.dyepig.2019.107884
75. N. Kida, S. Kumakura, S. Ishiwata, Y. Taguchi, Y. Tokura, Gigantic terahertz magnetochromism via electromagnons in the hexaferrite magnet $Ba_2Mg_2Fe_{12}O_{22}$. Phys. Rev. B Condens. Matter Mater. Phys. **83**, 064422 (2011). https://doi.org/10.1103/PhysRevB.83.064422
76. S. Toyoda, N. Abe, T. Arima, S. Kimura, Large magnetochromism in multiferroic $MnWO_4$. Phys. Rev. B Condens. Matter Mater. Phys. **91**, 054417 (2015). https://doi.org/10.1103/PhysRevB.91.054417
77. M. Bichurin, V. Petrov, V. Leontyev, A. Saplev, Two-range magnetoelectric sensor. AIP Adv. **7**, 015009 (2017). https://doi.org/10.1063/1.4973875
78. G. Korotcenkov, *Handbook of Gas Sensor Materials*, vol. 2, 181–184 (2014). https://doi.org/10.1007/978-1-4614-7165-3
79. P. Liu, S.H. Lee, C. Edwin Tracy, J.A. Turner, J.R. Pitts, S.K. Deb, Electrochromic and chemochromic performance of mesoporous thin-film vanadium oxide. Solid State Ionics **165**, 223–228 (2003). https://doi.org/10.1016/J.SSI.2003.08.044
80. N.F. Mott, L. Friedman, Metal-insulator transitions in VO_2, Ti_2O_3 and $Ti_{2-x}V_xO_3$. Philos. Mag. **30**, 389–402 (1974). https://doi.org/10.1080/14786439808206565
81. M. Seredyuk, A.B. Gaspar, V. Ksenofontov, S. Reiman, Y. Galyametdinov, W. Haase, E. Rentschler, P. Gütlich, Room temperature operational thermochromic liquid crystals. Chem. Mater. **18**, 2513–2519 (2006). https://doi.org/10.1021/cm052632w
82. H. Xie, S. O'Dwyer, J. Corish, D.A. Morton-Blake, The thermochromism of poly(3-alkylthiophene)s: the role of the side chains. Synth. Met. **122**, 287–296 (2001). https://doi.org/10.1016/S0379-6779(00)00305-2
83. A. Seeboth, A. Klukowska, R. Ruhmann, D. Lötzsch, Thermochromic polymer materials. Chin. J. Polym. Sci. (English Ed.) **25**, 123–135 (2007). https://doi.org/10.1142/S0256767907001923

84. Z. Mao, W. Wang, Y. Liu, L. Zhang, H. Xu, Y. Zhong, Infrared stealth property based on semiconductor (M)-to-metallic (R) phase transition characteristics of W-doped VO_2 thin films coated on cotton fabrics. Thin Solid Films **558**, 208–214 (2014). https://doi.org/10.1016/j.tsf.2014.02.055
85. Y. Zhang, Z. Hu, H. Xiang, G. Zhai, M. Zhu, Fabrication of visual textile temperature indicators based on reversible thermochromic fibers. Dye. Pigment. **162**, 705–711 (2019). https://doi.org/10.1016/j.dyepig.2018.11.007
86. M.S. Tözüm, S.A. Aksoy, C. Alkan, Microencapsulation of three-component thermochromic system for reversible color change and thermal energy storage. Fibers Polym. **19**, 660–669 (2018). https://doi.org/10.1007/s12221-018-7801-3
87. Y. Guan, L. Zhang, D. Wang, J.L. West, S. Fu, Preparation of thermochromic liquid crystal microcapsules for intelligent functional fiber. Mater. Des. **147**, 28–34 (2018). https://doi.org/10.1016/j.matdes.2018.03.030
88. K.R. Karpagam, K.S. Saranya, J. Gopinathan, A. Bhattacharyya, Development of smart clothing for military applications using thermochromic colorants. J. Text. Inst. **108**, 1122–1127 (2017). https://doi.org/10.1080/00405000.2016.1220818
89. Y. Jin, Y. Bai, Y. Zhu, X. Li, M. Ge, Thermosensitive luminous fiber based on reversible thermochromic crystal violet lactone pigment. Dye. Pigment. **146**, 567–575 (2017). https://doi.org/10.1016/j.dyepig.2017.07.062
90. Y. Jin, C. Shi, X. Li, Y. Wang, F. Wang, M. Ge, Preparation and luminescence studies of thermosensitive PAN luminous fiber based on the heat sensitive rose red TF-R1 thermochromic pigment. Dye. Pigment. **139**, 693–700 (2017). https://doi.org/10.1016/j.dyepig.2017.01.003
91. Q. Li, K. Li, H. Fan, C. Hou, Y. Li, Q. Zhang, H. Wang, Reduced graphene oxide functionalized stretchable and multicolor electrothermal chromatic fibers. J. Mater. Chem. C **5**, 11448–11453 (2017). https://doi.org/10.1039/c7tc02471a
92. Z. Ahmed, Y. Wei, R. Torah, J. Tudor, Actively actuated all dispenser printed thermochromic smart fabric device. Electron. Lett. **52**, 1601–1603 (2016). https://doi.org/10.1049/el.2016.1073
93. V. Tamilmani, D. Kanadasan, R. Muthazhagan, K.J. Sreeram, J.R. Rao, B.U. Nair, Thermochromism for smart Leathers. J. Am. Leather Chem. Assoc. **110**, 161–164 (2015). https://www.journals.uc.edu/index.php/JALCA/article/view/3615
94. X. Xia, Y. Zhang, Q.H. Li, C. Li, Preparation and characterization of thermochromic phase change nanofibers/woven composite material. Adv. Mater. Res. 427–431 (Trans Tech Publications Ltd., 2014). https://doi.org/10.4028/www.scientific.net/AMR.1048.427
95. D. Adler, Mechanisms for metal-nonmental transitions in transition-metal oxides and sulfides. Rev. Mod. Phys. **40**, 714–736 (1968). https://doi.org/10.1103/RevModPhys.40.714
96. A. Gupta, R. Aggarwal, P. Gupta, T. Dutta, R.J. Narayan, J. Narayan, Semiconductor to metal transition characteristics of VO_2 thin films grown epitaxially on Si (001). Appl. Phys. Lett. **95**, 111915 (2009). https://doi.org/10.1063/1.3232241
97. S. Surnev, M.G. Ramsey, F.P. Netzer, Vanadium oxide surface studies. Prog. Surf. Sci. **73**, 117–165 (2003). https://doi.org/10.1016/j.progsurf.2003.09.001
98. M. Imada, A. Fujimori, Y. Tokura, Metal-insulator transitions. Rev. Mod. Phys. **70**, 1039–1263 (1998). https://doi.org/10.1103/revmodphys.70.1039
99. K. Held, G. Keller, V. Eyert, D. Vollhardt, V.I. Anisimov, Mott-Hubbard metal-insulator transition in paramagnetic V_2O_3: an LDA+DMFT(QMC) study. Phys. Rev. Lett. **86**, 5345–5348 (2001). https://doi.org/10.1103/PhysRevLett.86.5345
100. C. Kolczewski, K. Hermann, S. Guimond, H. Kuhlenbeck, H.J. Freund, Identification of the vanadyl terminated V_2O_3 (0 0 0 1) surface by NEXAFS spectroscopy: a combined theoretical and experimental study. Surf. Sci. **601**, 5394–5402 (2007). https://doi.org/10.1016/j.susc.2007.09.033
101. M. Yethiraj, Pure and doped vanadium sesquioxide: a brief experimental review. J. Solid State Chem. **88**, 53–69 (1990). https://doi.org/10.1016/0022-4596(90)90205-C
102. M. Foëx, Dilatometric and electric studies of the anomaly of vanadium sesquioxide at low temperatures, C.R. Acad. Sci. **223**, 1126 (1946)

103. I. Pollini, A. Mosser, J.C. Parlebas, Electronic, spectroscopic and elastic properties of early transition metal compounds. Phys. Rep. **355**, 1–72 (2001). https://doi.org/10.1016/S0370-1573(01)00018-7
104. R. Zimmermann, R. Claessen, F. Reinert, P. Steiner, S. Hüfner, Strong hybridization in vanadium oxides: evidence from photoemission and absorption spectroscopy. J. Phys. Condens. Matter. **10**, 5697–5716 (1998). https://doi.org/10.1088/0953-8984/10/25/018
105. P.D. Dernier, M. Marezio, Crystal structure of the low-temperature antiferromagnetic phase of V_2O_3. Phys. Rev. B. **2**, 3771–3776 (1970). https://doi.org/10.1103/PhysRevB.2.3771
106. D. Adler, J. Feinleib, H. Brooks, W. Paul, Semiconductor-to-metal transitions in transition-metal compounds. Phys. Rev. **155**, 851–860 (1967). https://doi.org/10.1103/PhysRev.155.851
107. A. Tanaka, Electronic structure and phase transition in V_2O_3: importance of 3d spin-orbit interaction and lattice distortion. J. Phys. Soc. Japan. **71**, 1091–1107 (2002). https://doi.org/10.1143/JPSJ.71.1091
108. M.M. Qazilbash, A.A. Schafgans, K.S. Burch, S.J. Yun, B.G. Chae, B.J. Kim, H.T. Kim, D.N. Basov, Electrodynamics of the vanadium oxides VO_2 and V_2O_3, Phys. Rev. B Condens. Matter Mater. Phys. **77**, 115121 (2008). https://doi.org/10.1103/PhysRevB.77.115121
109. G.S. Nadkarni, V.S. Shirodkar, Experiment and theory for switching in Al/V_2O_5/Al devices. Thin Solid Films **105**, 115–129 (1983). https://doi.org/10.1016/0040-6090(83)90200-6
110. Y. Shimizu, K. Nagase, N. Miura, N. Yamazoe, Electrochromic properties of spin-coated V_2O_5 thin films. Solid State Ionics **53–56**, 490–495 (1992). https://doi.org/10.1016/0167-2738(92)90419-P
111. H.G. Bachmann, F.R. Ahmed, W.H. Barnes, The crystal structure of vanadium pentoxide. Zeitschrift Für Krist.—Cryst. Mater. **115**, 110–131 (1961). https://doi.org/10.1524/zkri.1961.115.16.110
112. A. Talledo, C.G. Granqvist, Electrochromic vanadium-pentoxide-based films: Structural, electrochemical, and optical properties. J. Appl. Phys. **77**, 4655–4666 (1995). https://doi.org/10.1063/1.359433
113. S. Beke, A review of the growth of V_2O_5 films from 1885 to 2010. Thin Solid Films. **519**, 1761–1771 (2011). https://doi.org/10.1016/j.tsf.2010.11.001
114. E.E. Chain, Optical properties of vanadium dioxide and vanadium pentoxide thin films. Appl. Opt. **30**, 2782 (1991). https://doi.org/10.1364/ao.30.002782
115. W. Li, S. Ji, Y. Li, A. Huang, H. Luo, P. Jin, Synthesis of VO_2 nanoparticles by a hydrothermal-assisted homogeneous precipitation approach for thermochromic applications. RSC Adv. **4**, 13026–13033 (2014). https://doi.org/10.1039/c3ra47666a
116. C. Cao, Y. Gao, H. Luo, Pure single-crystal rutile vanadium dioxide powders: synthesis, mechanism and phase-transformation property. J. Phys. Chem. C. **112**, 18810–18814 (2008). https://doi.org/10.1021/jp8073688
117. H. Yin, M. Luo, K. Yu, Y. Gao, R. Huang, Z. Zhang, M. Zeng, C. Cao, Z. Zhu, Fabrication and temperature-dependent field-emission properties of bundlelike VO_2 nanostructures. ACS Appl. Mater. Interfaces **3**, 2057–2062 (2011). https://doi.org/10.1021/am200291a
118. S.R. Popuri, M. Miclau, A. Artemenko, C. Labrugere, A. Villesuzanne, M. Pollet, Rapid hydrothermal synthesis of VO_2 (B) and its conversion to thermochromic VO_2 (M1). Inorg. Chem. **52**, 4780–4785 (2013). https://doi.org/10.1021/ic301201k
119. L. Liu, F. Cao, T. Yao, Y. Xu, M. Zhou, B. Qu, B. Pan, C. Wu, S. Wei, Y. Xie, New-phase VO_2 micro/nanostructures: investigation of phase transformation and magnetic property. New J. Chem. **36**, 619–625 (2012). https://doi.org/10.1039/c1nj20798a
120. C. Wu, Z. Hu, W. Wang, M. Zhang, J. Yang, Y. Xie, Synthetic paramontroseite VO_2 with good aqueous lithium-ion battery performance. Chem. Commun. 3891–3893 (2008). https://doi.org/10.1039/b806009f
121. Y. Oka, S. Sato, T. Yao, N. Yamamoto, Crystal structures and transition mechanism of VO_2(A). J. Solid State Chem. **141**, 594–598 (1998). https://doi.org/10.1006/jssc.1998.8025
122. D. Hagrman, J. Zubieta, C.J. Warren, L.M. Meyer, M.M.J. Treacy, R.C. Haushalter, A new polymorph of VO_2 prepared by soft chemical methods. J. Solid State Chem. **138**, 178–182 (1998). https://doi.org/10.1006/jssc.1997.7575

123. C. Wang, X. Liu, J. Shao, W. Xiong, W. Ma, Y. Zheng, Structural transition and temperature-driven conductivity switching of single crystalline VO_2(A) nanowires. RSC Adv. **4**, 64021–64026 (2014). https://doi.org/10.1039/c4ra12392a
124. J.S. Park, J.H. Jo, Y. Aniskevich, A. Bakavets, G. Ragoisha, E. Streltsov, J. Kim, S.T. Myung, Open-structured vanadium dioxide as an intercalation host for Zn Ions: investigation by first-principles calculation and experiments. Chem. Mater. **30**, 6777–6787 (2018). https://doi.org/10.1021/acs.chemmater.8b02679
125. J.P. Pouget, H. Launois, J.P. D'Haenens, P. Merenda, T.M. Rice, Electron localization induced by uniaxial stress in pure VO_2. Phys. Rev. Lett. **35**, 873–875 (1975). https://doi.org/10.1103/PhysRevLett.35.873
126. Y. Oka, T. Yao, N. Yamamoto, Structural phase transition of VO_2(B) to VO_2(A). J. Mater. Chem. **1**, 815–818 (1991). https://doi.org/10.1039/jm9910100815
127. G.A. Horrocks, S. Singh, M.F. Likely, G. Sambandamurthy, S. Banerjee, Scalable hydrothermal synthesis of free-standing VO_2 nanowires in the M1 phase. ACS Appl. Mater. Interfaces **6**, 15726–15732 (2014). https://doi.org/10.1021/am504372t
128. J.I. Sohn, H.J. Joo, D. Ahn, H.H. Lee, A.E. Porter, K. Kim, D.J. Kang, M.E. Wellandt, Surface-stress-induced Mott transition and nature of associated spatial phase transition in single crystalline VO_2 nanowires. Nano Lett. **9**, 3392–3397 (2009). https://doi.org/10.1021/nl900841k
129. S. Chen, J. Lai, J. Dai, H. Ma, H. Wang, X. Yi, Characterization of nanostructured VO_2 thin films grown by magnetron controlled sputtering deposition and post annealing method. Opt. Express. **17**, 24153 (2009). https://doi.org/10.1364/oe.17.024153
130. S.K. Deb, A novel electrophotographic system. Appl. Opt. **8**, 192 (1969). https://doi.org/10.1364/ao.8.s1.000192
131. A. Magnéli, Structures of the ReO 3-type with recurrent dislocations of atoms: 'homologous series' of molybdenum and tungsten oxides. Acta Crystallogr. **6**, 495–500 (1953). https://doi.org/10.1107/s0365110x53001381
132. K. Bange, Colouration of tungsten oxide films: a model for optically active coatings. Sol. Energy Mater. Sol. Cells. **58**, 1–131 (1999). https://doi.org/10.1016/S0927-0248(98)00196-2
133. H.A. Wriedt, The O-W (oxygen-tungsten) system. Bull. Alloy Phase Diagr. **10**, 368–384 (1989). https://doi.org/10.1007/BF02877593
134. S.K. Deb, Optical and photoelectric properties and colour centres in thin films of tungsten oxide. Philos. Mag. **27**, 801–822 (1973). https://doi.org/10.1080/14786437308227562
135. L. Berggren, Optical absorption and electrical conductivity in lithium intercalated amorphous tungsten oxide films (Acta Universitatis Upsaliensis 2004)
136. J.M. Berak, M.J. Sienko, Effect of oxygen-deficiency on electrical transport properties of tungsten trioxide crystals. J. Solid State Chem. **2**, 109–133 (1970). https://doi.org/10.1016/0022-4596(70)90040-X
137. A. Nakamura, S. Yamada, Fundamental absorption edge of evaporated amorphous WO_3 films. Appl. Phys. **24**, 55–59 (1981). https://doi.org/10.1007/BF00900398
138. G.A. Niklasson, C.G. Granqvist, Electrochromics for smart windows: thin films of tungsten oxide and nickel oxide, and devices based on these. J. Mater. Chem. **17**, 127–156 (2007). https://doi.org/10.1039/b612174h
139. B.W. Faughnan, R.S. Crandall, P.M. Heyman, Electrochromism in WO_3 Amorphous Films. R.C.A. Rev. **36**, 177–197 (1975)
140. I.G. Austin, N.F. Mott, Polarons in crystalline and non-crystalline materials. Adv. Phys. **18**, 41–102 (1969). https://doi.org/10.1080/00018736900101267
141. I.F. Chang, B.L. Gilbert, T.I. Sun, Electrochemichromic systems for display applications. J. Electrochem. Soc. **122**, 955–962 (1975). https://doi.org/10.1149/1.2134377
142. C. Sunseri, F. Di Quarto, A. Di Paola, Kinetics of coloration of anodic electrochromic films of $WO_3{\cdot}H_2O$. J. Appl. Electrochem. **10**, 669–675 (1980). https://doi.org/10.1007/BF00615491
143. P. Hao, H. Li, T. Yu, Y. Fu, Pyrazinium iodoargentate with versatile photo- and thermo-chromism. Dye. Pigment. **136**, 825–829 (2017). https://doi.org/10.1016/j.dyepig.2016.09.045

144. V. Lukeš, M. Breza, Thermochromism of bithiophenes and internal aromatic chain rotation. J. Mol. Struct. Theochem. **820**, 35–39 (2007). https://doi.org/10.1016/j.theochem.2007.06.003
145. D. Chalasani, J.K. Potvin, B.L. Lucht, W.B. Euler, Two-step thermochromism in poly(3-docosoxy-4-methylthiophene): mechanistic similarity to poly(3-docosylthiophene). J. Polym. Sci. Part A Polym. Chem. **48**, 4370–4373 (2010). https://doi.org/10.1002/pola.24236
146. H.R.C. Jaw, M.A. Mooney, T. Novinson, W.C. Kaska, J.I. Zink, Optical properties of the thermochromic compounds Ag_2HgI_4 and Cu_2HgI_4. Inorg. Chem. **26**, 1387–1391 (1987). https://doi.org/10.1021/ic00256a011
147. J. Heiras, E. Pichardo, A. Mahmood, T. López, R. Pérez-Salas, J.M. Siqueiros, M. Castellanos, Thermochromism in (Ba, Sr)–Mn oxides. J. Phys. Chem. Solids. **63**, 591–595 (2002). https://doi.org/10.1016/S0022-3697(01)00198-6
148. W. Li, C. Zhu, W. Wang, J. Wu, Reversible thermochromism materials. Gongneng Cailiao/J. Funct. Mater. **28**, 337–341 (1997)
149. F.J. Morin, Oxides which show a metal-to-insulator transition at the Neel temperature. Phys. Rev. Lett. **3**, 34 (1959). https://doi.org/10.1103/PhysRevLett.3.34
150. F. Asdrubali, G. Baldinelli, Theoretical modelling and experimental evaluation of the optical properties of glazing systems with selective films. Build. Simul. **2**, 75–84 (2009). https://doi.org/10.1007/S12273-009-9112-5
151. J.M. Dussault, L. Gosselin, T. Galstian, Integration of smart windows into building design for reduction of yearly overall energy consumption and peak loads. Sol. Energy. **86**, 3405–3416 (2012). https://doi.org/10.1016/j.solener.2012.07.016
152. M.E.A. Warwick, R. Binions, Chemical vapour deposition of thermochromic vanadium dioxide thin films for energy efficient glazing. J. Solid State Chem. **214**, 53–66 (2014). https://doi.org/10.1016/j.jssc.2013.10.040
153. M.J. Miller, J. Wang, Influence of Na diffusion on thermochromism of vanadium oxide films and suppression through mixed-alkali effect. Mater. Sci. Eng. B Solid-State Mater. Adv. Technol. **200**, 50–58 (2015). https://doi.org/10.1016/j.mseb.2015.06.006.
154. J.P. Fortier, B. Baloukas, O. Zabeida, J.E. Klemberg-Sapieha, L. Martinu, Thermochromic VO2 thin films deposited by HiPIMS. Sol. Energy Mater. Sol. Cells **125**, 291–296 (2014). https://doi.org/10.1016/j.solmat.2014.03.007
155. A. Aijaz, Y.X. Ji, J. Montero, G.A. Niklasson, C.G. Granqvist, T. Kubart, Low-temperature synthesis of thermochromic vanadium dioxide thin films by reactive high power impulse magnetron sputtering. Sol. Energy Mater. Sol. Cells **149**, 137–144 (2016). https://doi.org/10.1016/j.solmat.2016.01.009
156. L. Hu, H. Tao, G. Chen, R. Pan, M. Wan, D. Xiong, X. Zhao, Porous W-doped VO_2 films with simultaneously enhanced visible transparency and thermochromic properties. J. Sol-Gel Sci. Technol. **77**, 85–93 (2016). https://doi.org/10.1007/s10971-015-3832-z
157. M. Jo, H.J. Lee, C. Oh, H. Yoon, J.Y. Jo, J. Son, Gate-induced massive and reversible phase transition of VO_2 channels using solid-state proton electrolytes. Adv. Funct. Mater. **28**, 1802003 (2018). https://doi.org/10.1002/adfm.201802003
158. H. Yoon, M. Choi, T.W. Lim, H. Kwon, K. Ihm, J.K. Kim, S.Y. Choi, J. Son, Reversible phase modulation and hydrogen storage in multivalent VO_2 epitaxial thin films. Nat. Mater. **15**, 1113–1119 (2016). https://doi.org/10.1038/nmat4692
159. S. Chen, Z. Wang, H. Ren, Y. Chen, W. Yan, C. Wang, B. Li, J. Jiang, C. Zou, Gate-controlled VO_2 phase transition for high-performance smart windows. Sci. Adv. **5** (2019). https://doi.org/10.1126/sciadv.aav6815
160. D.A. Mazurenko, R. Kerst, J.I. Dijkhuis, A.V. Akimov, V.G. Golubev, A.A. Kaplyanskii, D.A. Kurdyukov, A.B. Pevtsov, Subpicosecond shifting of the photonic band gap in a three-dimensional photonic crystal. Appl. Phys. Lett. **86**, 041114 (2005). https://doi.org/10.1063/1.1856687
161. B.G. Chae, H.T. Kim, D.H. Youn, K.Y. Kang, Abrupt metal-insulator transition observed in VO_2 thin films induced by a switching voltage pulse. Phys. B Condens. Matter. **369**, 76–80 (2005). https://doi.org/10.1016/j.physb.2005.07.032

162. G. Stefanovich, A. Pergament, D. Stefanovich, Electrical switching and Mott transition in VO_2. J. Phys. Condens. Matter. **12**, 8837–8845 (2000). https://doi.org/10.1088/0953-8984/12/41/310
163. P.P. Boriskov, A.A. Velichko, A.L. Pergament, G.B. Stefanovich, D.G. Stefanovich, The effect of electric field on metal-insulator phase transition in vanadium dioxide. Tech. Phys. Lett. **28**, 406–408 (2002). https://doi.org/10.1134/1.1482750
164. A. Crunteanu, J. Givernaud, J. Leroy, D. Mardivirin, C. Champeaux, J.C. Orlianges, A. Catherinot, P. Blondy, Voltage- and current-activated metal-insulator transition in VO_2-based electrical switches: a lifetime operation analysis. Sci. Technol. Adv. Mater. **11**, 6 (2010). https://doi.org/10.1088/1468-6996/11/6/065002
165. J. Cao, E. Ertekin, V. Srinivasan, W. Fan, S. Huang, H. Zheng, J.W.L. Yim, D.R. Khanal, D.F. Ogletree, J.C. Grossman, J. Wu, Strain engineering and one-dimensional organization of metal-insulator domains in single-crystal vanadium dioxide beams. Nat. Nanotechnol. **4**, 732–737 (2009). https://doi.org/10.1038/nnano.2009.266
166. T. Driscoll, H.T. Kim, B.G. Chae, M. Di Ventra, D.N. Basov, Phase-transition driven memristive system. Appl. Phys. Lett. **95**, 043503 (2009). https://doi.org/10.1063/1.3187531
167. H. Coy, R. Cabrera, N. Sepúlveda, F.E. Fernández, Optoelectronic and all-optical multiple memory states in vanadium dioxide. J. Appl. Phys. **108**, 113115 (2010). https://doi.org/10.1063/1.3518508
168. L. Pellegrino, N. Manca, T. Kanki, H. Tanaka, M. Biasotti, E. Bellingeri, A.S. Siri, D. Marré, Multistate memory devices based on free-standing VO_2/TiO_2 microstructures driven by Joule self-heating. Adv. Mater. **24**, 2929–2934 (2012). https://doi.org/10.1002/adma.201104669
169. S.H. Bae, S. Lee, H. Koo, L. Lin, B.H. Jo, C. Park, Z.L. Wang, The memristive properties of a single VO_2 nanowire with switching controlled by self-heating. Adv. Mater. **25**, 5098–5103 (2013). https://doi.org/10.1002/adma.201302511
170. S.H. Lee, M.K. Kim, J.W. Lee, Z. Yang, S. Ramanathan, S. Tiwari, Vanadium dioxide (VO_2) is also a ferroelectric: properties from memory structures, in *Proceedings of the IEEE Conference on Nanotechnology*, pp. 735–739 (2011). https://doi.org/10.1109/NANO.2011.6144602
171. Y.C. Li, H.M. Zhang, Y.M. Zhang, H.Y. Hu, B. Wang, Y. Le Lou, C.Y. Zhou, Two-dimensional threshold voltage model of a nanoscale silicon-on-insulator tunneling field-effect transistor. Chinese Phys. B. **22**, 038501 (2013). https://doi.org/10.1088/1674-1056/22/3/038501
172. D.M. Newns, J.A. Misewich, C.C. Tsuei, A. Gupta, B.A. Scott, A. Schrott, Mott transition field effect transistor. Appl. Phys. Lett. **73**, 780–782 (1998). https://doi.org/10.1063/1.121999
173. C. Zhou, D.M. Newns, J.A. Misewich, P.C. Pattnaik, A field effect transistor based on the Mott transition in a molecular layer. Appl. Phys. Lett. **70**, 598 (1998). https://doi.org/10.1063/1.118285
174. S. Hormoz, S. Ramanathan, Limits on vanadium oxide Mott metal-insulator transition field-effect transistors. Solid. State. Electron. **54**, 654–659 (2010). https://doi.org/10.1016/j.sse.2010.01.006
175. P. Markov, R.E. Marvel, H.J. Conley, K.J. Miller, R. F. Haglund Jr, S.M. Weiss, Optically monitored electrical switching in VO_2. Acs Photon. **2**, 1175–1182 (2015)
176. J.M. Wu, L.B. Liou, Room temperature photo-induced phase transitions of VO2 nanodevices. J. Mater. Chem. **21**, 5499–5504 (2011)
177. M. Li, S. Ji, J. Pan, et al., Infrared response of self-heating VO_2 nanoparticles film based on Ag nanowires heater. J. Mater. Chem. A, **2**, 20470–20473 (2014)
178. A. Joushaghani, J. Jeong, S. Paradis, D. Alain, J.S. Aitchison, J.K. Poon, Wavelength-size hybrid Si-VO_2 waveguide electroabsorption optical switches and photodetectors. Opt. Express **23**, 3657–3668 (2015)

Optical Properties of Metal Oxide-Based Perovskite Structures

Neeraj Rathee and Nirat Ray

Abstract Perovskite materials with the composition ABO_3, where A and B are cations and O is the oxygen anion, have been studied extensively for their intriguing physical properties as well as the potential for practical applications. Perovskites have been at the forefront of revolutionary discoveries ranging from ceramic high temperature superconductors to high-efficiency photovoltaics. Many different types of lattice distortions can occur owing to the flexibility of bond angles within the ideal perovskite structure. A broad range of novel functional materials and device concepts have been envisaged through fundamental understanding of the relationships between the structural and chemical compatibility, thermal stability, solid solubility, and lattice strain. In this chapter, we review some of the fundamental optical properties of metal oxide perovskites. We highlight the role of defects, and shape anisotropy to show the variation of optical properties from nanocrystals to nanowires and nanosheets. We discuss the advancements in photovoltaics and other optoelectronic devices, and conclude with a comparison of optical properties of metal oxide perovskites with halide perovskites.

Keywords Optical properties · Band gap · Photovoltaics · Photocatalysis

1 Introduction

Metal oxide perovskites, denoted by the chemical formula ABO_3, can be understood as a simple cubic Bravais lattice superimposed with a 5-atom motif. The A cation (typically an alkaline earth metal or rare earth metal) occupies the sites at the corners of the cube. The B cation (typically a transition metals) occupies the body centre and is co-ordinated by 6 oxygen atoms (O) in an octahedral geometry [1]. Figure 1 shows a schematic of the ideal perovskite structure. Given the large number of permutations

N. Rathee · N. Ray (✉)
Department of Materials Science and Engineering, Indian Institute of Technology Delhi, Hauz Khas, New Delhi 110016, India
e-mail: nirat@iitd.ac.in

V. Kumar et al. (eds.), *Optical Properties of Metal Oxide Nanostructures*,
Progress in Optical Science and Photonics 26,
https://doi.org/10.1007/978-981-99-5640-1_3

Fig. 1 An ideal unit cubic perovskite crystal structure of metal oxide perovskites

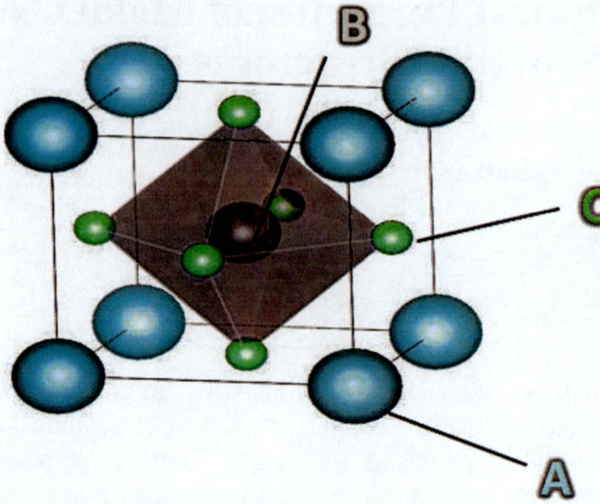

that can be created with different A and B atoms, an extraordinarily wide variety of nanostructures and phases with entirely distinct properties can be produced. Possible variations also include double perovskites or stacked perovskites also known as layered perovskites.

Perovskite oxides can be synthesized using a variety of routes, but the most commonly used methods include soft and hard template assisted synthesis, colloidal route, electrospinning technique, and hydrothermal method. The heat released during synthesis processes is quite advantageous for synthesis of phase-pure perovskites [2]. Defect formation does not include the partial or complete substitution of anion or cation but can also be considered if both processes are occurring simultaneously. If the A^{3+} ion is replaced by A^{2+} ion, then this substitution leads to a distorted lattice structure, and the creation of oxygen vacancies [3]. On the other hand, if allowed, the B component may also undergo partial oxidation [4, 5]. The defect-rich structures are particularly relevant for application of perovskites in heterogeneous catalysis [6, 7].

Perovskites are also widely used in photocatalysis in addition to their growing popularity as catalyst for oxygen reduction and oxygen evolution reactions [8, 9]. They are effective catalysts for low-temperature oxidations [10, 11], particularly for the oxidative dehydrogenation of propane [12]. Certain variables [13], like the chemical composition, the number of defects present, or the fractional reduction of metal ions, can have an effect on the photocatalysis characteristics [12, 14]. Several experimental and theoretical have shown how the catalytic performance of perovskite oxides can be dynamically modified through the inductive change of oxygen stoichiometry to produce oxygen-vacant structures for a variety of applications. One method that is commonly used to enhance the electrocatalytic activity of a complex oxide is the modification in surface anionic defects, specifically creating oxygen vacancies [15]. Oxygen deficiencies can significantly improve catalytic efficiency by modifying the electric structures and charge transport, with the change in electronic structure the absorption and desorption behaviour changes The band structure alignment of the absorber close to the charge transport layers could be affected by the defects generated in metal oxide perovskites [16]. Therefore, it is essential to eliminate or passivate these defects in order to achieve high efficiency.

Metal oxide perovskites are also known to exhibit luminescence, and while a variety of advanced theories have validated various mechanisms, new research on the luminescence mechanism is constantly providing different insights [17]. Photoluminescence (PL) spectroscopy is among the most significant and effective techniques for investigating the optoelectronic characteristics of broad band gap materials. Recently, in $SrTiO_3$, a typical perovskite material, unique PL bands have been found [18]. After doping of $SrTiO_3$ perovskites the PL spectra lies in the blue region wavelength provides a chance to deeply understand the charge carrier dynamics and electron characteristics [19, 20].

In this chapter, we first begin by describing the various optical transitions in these materials. We highlight the role of defects, quantum confinement, excitons in shaping the optical properties. We also discuss PL reports in metal oxide perovskites. We then focus on specific metal oxide perovskites, and discuss photovoltaic and optoelectronic device applications. We conclude our discussion by exploring the newly emerging metal halide perovskites (ABX_3).

2 Optical Transitions in Metal Oxide Perovskites

The wavelength of light absorbed by a semiconductor material is controlled by its energy band gap. Table 1 lists optical bandgaps of few metal oxide perovskites, along with the observed morphology and possible synthesis routes. We note that typical values of bandgap range from 2 to 4 eV, primarily due to the large difference in electronegativity between the oxygen and metal atoms. Systematic studies have investigated bandgap variations with change in B cations. An important example is the study of optical band gap in RMO_3 (R = La, Y) where M is the series of transition metals Sc, Ti, V, Cr, Mn, Fe, Co, Ni and Cu. The authors reported an almost linear decrease in the optical bandgap from Sc to Cu. Interestingly, the authors also reported a crossover between Mott insulator, where the charge gap is determined by the Hubbard splitting (U) of the d band and a charge transfer (CT) insulator, where the charge gap is typically an energy difference between filled p bands of ligand anions and unoccupied upper Hubbard 3d band. This crossover was reported at M = Cr [21]. The optical bandgap decreases from 3.4 to 3.2 eV as we move from $CaTiO_3$ to $SrTiO_3$ (going down the group). Similarly, the bandgap of $MSnO_3$ [22] perovskite stannate were found to be 4.44, 4.06, 3.10 eV for $CaSnO_3$, $SrSnO_3$, $BaSnO_3$, confirming the trend of decrease in the bandgap down a group in the periodic table. However, an increase in the bandgap from 2 to 2.4 eV was reported for $LaFeO_3$ to $BiFeO_3$ (same period).

Yousef et al. [34] studied the structural and optical properties of double perovskites, A_2YVO_6 (where A = Mg, Sr with different ionic radii). The band gap for Mg_2YVO_6 was 2.9 eV while for Sr_2YVO_6 it was around 2.4 eV. This trend in

Table 1 Bandgap of some perovskite materials

Material	Band gap (ev)	Morphology	Synthesis route	References
$CaTiO_3$	3.41	Nano cuboids	Hydrothermal route	[23]
$SrTiO_3$	3.20	–	Solid state reaction	[24]
$MSnO_3$ (M = Ca, Sr and Ba) $CaSnO_3$, $SrSnO_3$, and $BaSnO_3$	4.44, 4.06, and 3.10	–	Solid state reaction	[22]
$NaTaO_3$	4.0	Microspheres	Two-step synthesis approach-molten salt in combination with hydrothermal processing	[25]
$CaTiO_3$	3.62	Nanostructures	Wet chemical synthesis	[26]
$BiFeO_3$	2.40	Nanoparticles	Sol-gel technique	[27]
$LaFeO_3$	2.00	Nanoparticles	Sol-gel technique	[28]
$NaNbO_3$	3.48	Nanowires	Hydrothermal	[29]
Bi_2WO_6	2.70	Nanostructures	Ultrasonic solvothermal treatment	[30]
$In_2S_3/NaTaO_3$	2.15–2.52	Composite nanostructures	Hydrothermal	[31]
$NaTaO_3$/reduced graphene oxide	3.87	Composite nanostructures	Hydrothermal	[32]
Bi_2MoO_6	2.81–2.85	Nanosheets/ microrods	Hydrothermal	[33]

the band gap was attributed to a reduction in the transition distance between valence electrons with increasing ionic radius.

2.1 *Role of Defects*

Defects in metal oxide perovskites are inevitable, regardless of their material processing conditions, and play a critical role in the functional behaviour of the material. The most common defect in these materials are oxygen vacancies. However, grain boundaries, dislocations etc. are also found. The nature, concentration, and arrangement of these defects strongly influences the material behaviour and can result into significant enhancement/suppression of their electronic, optoelectronic, and magnetic device properties. A consequence of existence of point defects on the band structure of the material is the formation of additional energy levels known as defect states near conduction and valence band and in the forbidden energy gap. These extra energy levels may act as traps (deep or shallow) for the charge carriers (electrons and holes) hindering the process of optical and electronic transitions.

Oxygen vacancies may be created even during the synthesis of the material, and therefore depend on the synthesis route to some extent. Oxygen defects may be generated post-synthesis by treating with Ar and H_2 plasma. The incorporation of oxygen vacancies can be used as a tool to enhance adsorption of molecules on the oxygen deficient surface [35]. Yunmin et al. [36] have developed oxygen defects in $PrBa_{0.5}$ $Sr_{0.5}Co_{1.5}Fe_{0.5}O_{5+\delta}$ (PBSCF), in a controlled manner to improve Oxygen Evolution Reaction (OER) in electrochemical water splitting processes.

Substitutional defects, where another atom replaces either the A or B cation have also been explored. As an example, the introduction of noble metals (Pd, Pt and Rh) in $LaFeO_3$ show potential for catalysis of n-propane [37]. While pure $LaFeO_3$ sample exhibited a propane conversion of 18 with 42% olefin selectivity, after doping, the samples showed an improved 85% propane conversion and 79% olefin selectivity. Another interesting implication of substitutional defects is revealed by first-principles investigations. The study of Cu doping at Sr site in $SrTiO_3$ affects the electronic band structure significantly by introducing new states at Gamma point. The consequence is that the bandgap of the material transitions from indirect to direct, making it an appealing candidate for optoelectronic devices. In transition metal oxide perovskites defects and impurities can modify the electronic and physical level both by disrupting the crystal lattice normal periodicity and, more crucially, by introducing extra energy levels to the band gap [38]. Metal oxide perovskites will become more conducting in nature when more electrons or holes are introduced into the conduction band or valance band [39].

The role of oxygen vacancies of $LaCrO_3$ and $LaMnO_3$ on their electronic and optical properties have already discussed by Park and his co-workers [40]. By introducing oxygen single vacancies, the band gap reduces in $LaCrO_3$ while in case of $LaMnO_3$ the band gap completely vanishes. These band gap assessments were used for optimizing the role of oxygen vacancies in perovskites oxides with reference to optoelectronic parameters that can have a potential in optical sensors. The generation of oxygen vacancies have improved the catalytic properties in energy conversion processes [41].

2.2 *Shape Anisotropy and Quantum Confinement Effects*

Quantum confinement, or the restriction of motion of charge carriers in 1, 2 or 3-dimensions is known to have a strong impact on the band gap and consequently optical properties. The simplest model to understand these effects is that of a particle-in-a-box. Quantum theory tells us that the energy levels of a particle confined to move in a box are quantized, and quantization effects can be observed clearly as the size of the box becomes comparable to the de Broglie wavelength of the particle. Experimentally, confinement can be achieved in multiple ways. One route harnesses the shape anisotropy effects, and uses different nanoscale shapes such as the creation of nanorods, spherical nanocrystals or dots to achieve confinement. The other approach may use single crystal growth to create heterostructures and the band alignment can be used to create quantum wells, which can be further confined with electric fields.

One interesting recent development with 2D oxide perovskites is the achievement of freestanding $SrTiO_3$ and $BiFeO_3$ ultrathin crystalline films down to a thickness of 1–4 unit cells [42].

With careful calibration and understanding of the properties, these could be now considered to be part of the wider family of 2D materials, and the reduction in dimensionality could be used as a knob to modulate properties for optoelectronic applications. A key fact to consider here would be the dangling bonds in oxides, which could impact the properties. 1D oxide perovskites, such as $BaMnO_3$ nanorods, $La_{0.5}Sr_{0.5}MnO_3$ (LSMO) nanowires, $La_{0.8}Sr_{0.2}MnO_3$ nanofibers, and $La_{0.7}Ca_{0.3}MnO_3$ and $La_{0.325}Pr_{0.300}Ca_{0.375}MnO_3$ nanotubes, have been synthesized and investigated for applications in transistors and spintronics [43].

The impact of quantum confinement in atomic layered perovskites in various heterosystems have already been explored [44]. Miri and co-workers [45] used quantum wells created by growing $SrTiO_3$ on $LaAlO_3$ substrate. The thickness of the quantum wells was studied by varying the growth thickness and the resulting band gap variation could easily be measured. Other studies include quantum confinement behaviour of metal oxide $SrVO_3$ where the perovskites structures were embedded in an insulating layer [46]. In another study [47] formation of 2D electron-liquid states was carried out by growing $SrVO_3$ thin films on Nb: $SrVO_3$ substrate. The quantum well states showed a orbital dependent quantization derived out of anisotropic orbitals of 3d electronic states.

A lot of research has recently been devoted to nano porous perovskite metal oxides because of their distinctive appearance and characteristics in addition to their better productivity. When nano porous design techniques are employed on perovskite metal oxides, these techniques have proved to be better for characteristics and high performance in a variety of applications. Grosso et al. [48] created nano islands and mesoporous perovskite oxide layer of $SrTiO_3$ using a quasi-organic template. Brezesinki and co-workers [49, 50] have developed nano porous layered films by using evaporation-induced self-assembly method. This method includes the fabrication of $NaTaO_3$, $BiFeO_3$ by using co-polymers for high temperature stability along with 3-D honeycomb structures for their application in phototcatalysis.

Recently [51], novel core-shell nanostructures of $PbTiO_3$ were fabricated and utilized in reduction in presence of light from CO_2 to CH_4 by introducing Basalt fibre. With the introduction of sensitizer, a narrow band gap of 1.92 eV was generated for the perovskite, and photoreduction was improved. While $CaTiO_3$ have already been reported as a perovskite with wide band gap ranging from 3.0 to 3.6 eV. This material usually referred as photocatalytic material where the catalysis depends upon the ratio of Ca and Ti ions. Additionally $NaTaO_3$ nanostructures [52] reported to have a broad energy band gap in a range from 4.0 to 4.2 eV and accepted as an appropriate material for photocatalysis activity and splitting of water.

2.3 Excitons

The generation of excitons is one of the most typical defects in insulating or semiconducting structures. An exciton is basically a bound electron-hole pair and is generated whenever photon with energy higher than the energy band gap are absorbed by a material. Theoretical studies on metal oxide perovskites have estimated the exciton binding energies for perovskite oxides. The exciton binding energies for $SrTiO_3$, $SrZrO_3$, $SrHfO_3$ and $KTaO_3$ range from 0.2 to 0.3 eV. The exciton binding energies for La series perovskites fall within the 120–190 meV range and are overall smaller than those of their cubic nonmagnetic counterparts [43].

Excitons in the ferroelectric perovskite $BaTiO_3$ are of special interest. In the ferroelectric tetragonal phase, it was found that the hole and electron parts were well localized on oxygen and titanium atoms, respectively, with an interparticle distance of 7.0 Å. The exciton-induced dipole moment, however, was opposite to the existing ferroelectric polarization, and can lead to the loss of bulk ferroelectricity in the material [53]. Deexcitation of the exciton can gives rise to luminescence, and the luminescence energy estimated was ~0.94 eV.

2.4 Photoluminescence (PL) and Quenching

To investigate the optoelectronic features of broad band gap materials, luminescence spectroscopy is one of the most important tools. Extensive research has been done to understand the characteristics of luminescence in a variety of perovskites. Many broad-gap perovskite TMOs do not exhibit efficient photoluminescence (PL) at room temperature, however, a broad PL band can be seen at low temperatures. Recently, Yamada et al. [17] worked on $SrTiO_3$ which exhibited a novel emission spectrum in pure form under continuous wave excitation while the doped $SrTiO_3$ crystals showed a broad band. Band edge emission was observed at low temperature for the doped and undoped crystals. The broad band gap for other TMO perovskites $KTaO_3$, $BaTiO_3$, $LiNbO_3$, $SrTiO_3$ and $LiTaO_3$ showed weak or no luminescent bands at room temperature while if the temperature was lowered a broad PL band was obtained. After illuminating with Ar ions at room temperature, the bands were enhanced.

Another study was carried on $CaZrO_3{:}Eu^{3+}$ for its application in White LED's [54]. The doping concentration was varied and at 5 mol% maximum emission intensity was recorded. Zhang [55] and co-workers synthesised Bi^{3+} and Eu^{3+} doped Ca_2LaTaO_6 double perovskites and discussed their energy transfer mechanism. From the luminescence studies it was found that Ca_2LaTaO_6 can act as a good host for phosphors. The luminescence studies showed a broad band spectrum corresponding to charge transfer from Eu^{3+} to O^{2-} ions. The prominent peaks in the UV region demonstrates that the doped nanophosphors have their excitation wavelength in the UV and blue region which is suitable for applications in LED devices.

Another strategy has been to employ co-doping; Priti et al. [56] co-doped Eu^{3+} ions along with Li^+, Na^+ and K^+ ions in $CaTiO_3$. The doped perovskites emit at 613 nm which lies in the red region upon excitation at 398 and 466 nm of wavelength. 12 Eu^{3+}/10 Na^+ co-doped $CaTiO_3$ perovskites showed the maximum emission intensity, and showed red emission in warm light area. Ho^{3+} and Yb^{3+} ions have been employed as co-dopants in $CaZrO_3$, where Yb^{3+} ions transfer their excitation energy progressively and cooperatively across various energy levels of Ho^{3+} ions, significantly enhancing the emission intensity (Fig. 2a). The presence of Mg^{2+} ions further multiplies the emission intensity of the phosphor (Fig. 2b). Consequently, the Ho^{3+}/Yb^{3+}/Mg^{2+} co-doped $CaZrO_3$ phosphor holds potential as a candidate for generating intense monochromatic green light [57]. Rare earth ions are often used for doping/co-doping as they possess a unique ability to transform low-energy near-infrared (NIR) photons into higher-energy visible photons, known as up conversion (UC). This non-linear process involves either the sequential absorption of multiple low-energy photons through intermediate meta-stable levels or the cooperative absorption of two photons, resulting in the emission of a photon with higher energy.

Yan et al. [58] have studied the synthesis of double perovskites for its application in white LED's. The Sr_2InSbO_6 perovskites were activated by Europium ions, the emission spectrum of the perovskites exhibited an orange emission in the region 594 nm wavelength. The nature of the perovskites was converted to water insolvent with a coating of Oleic Acid. The luminescent studies showed that Eu activated double perovskites have a great possibility for its application in white LED's and latent fingerprints. Sanni and co-workers [59] have synthesised novel double perovskites for its application in LEDs. Mn doped Ba_2CaWO_6 nanostructures were fabricated by solid reaction method. The emission spectra for the doped and undoped structures were observed and it was found that for doped nanostructures under UV excitation

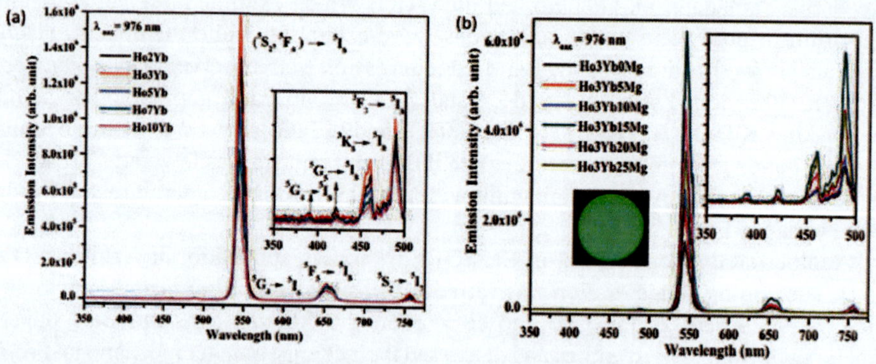

Fig. 2 **a** Up conversion emission spectra for Ho^{3+}/Yb^{3+} co doped $CaZrO_3$ metal oxide perovskites with different concentration of Yb^{3+} ions at an excitation of 976 nm wavelength. Reproduced with permission from Ref. [57] © 2016 Royal Society of Chemistry. **b** Up conversion emission spectra for Ho^{3+}/Yb^{3+} co doped $CaZrO_3$ metal oxide perovskites with different concentration of Mg^{2+} ions at an excitation of 976 nm wavelength. Reproduced with permission from ref [57] © 2016 Royal Society of Chemistry

the emission was noted at 695 nm wavelength which lies in the deep red region. Temperature dependent photoluminescence spectra have shown a potential for its application in plant cultivation red region LED's.

3 Materials and Devices

Historically, perovskite research has focused on oxide compounds, such as $CaTiO_3$ (discovered by Rose in 1839 and named after Lev Perovskiy) [60] $BaTiO_3$, $PbTiO_3$, $SrTiO_3$, $BiFeO_3$, and others, and the targeted applications have been ferroelectric, piezoelectric and pyroelectric. They have also found applications in photovoltaics (PVs), but the wide band gap of typical perovskite oxides has historically limited efficiencies. Metal oxide perovskites' potential catalytic activity has been displayed in a variety of ways. The abundance of oxygen vacancies or deficiencies on the surface has allowed scientists to utilize them in the reaction-based sensory readings [9, 61, 62].

Numerous mixed-metal oxides in the dense perovskite phase can transport significant amount of oxygen at high temperatures. At higher temperatures, the permeability of oxygen can exceed than that of porous membranes. Perovskite metal oxides that have been effectively used for oxygen transport includes $LiNbO_3$, $SrCeO_3$, and $SrTiO_3$ [63, 64]. These properties of more oxygen flux transport at higher temperatures have potential application in in several high-temperature membrane reactor. Due to the remarkable hole conductivity ($\sigma = 100$ S/cm), perovskite oxides such as $LaCoO_3$, $LaFeO_3$, and $LaMnO_3$ were just used as cathode materials in solid oxide fuel cells (SOFCs) [41]. Table 2 lists some of the applications of perovskite oxides along with the materials studied for those applications.

Table 2 Perovskite materials with their applications

Materials	Applications	References
$LaFeO_3$	Photocatalytic	[65]
$LaVO_3$	Solar cells	[66]
$BaSnO_3$	Solar cells	[67]
Pt/LaCoMnO/Nb:$SrTiO_3$	Magnetic memory devices	[68]
$PbZrTiO_3$	Sensors and actuators	[69]
Ho^{3+}/Yb^{3+}/Mg^{2+} doped $CaZrO_3$	Phosphor materials in photoluminescence	[57]
$SrMgTiO_3$	Oxygen gas sensor	[70]
$GdCaCoMnO_3$	Fuel cells	[71]
$CsPb_2Br_5$/$BaTiO_3$	Humidity sensor	[72]
$LaSrMnO_3$ and $LaBaMnO_3$	Magnetic field sensors	[73]
CuO/$BaTiO_3$	Carbon di-oxide sensor	[74]

3.1 Optical and Excitonic Properties of Transition Metal Oxide Perovskites

Transition metal oxides (TMO) are those where the B cation contains 4d or 5d electrons. The optical band gap can be calculated by a numerous technique experimentally including absorption spectroscopy, inverse photoemission while theoretical measurements of band gap can be done by using DFT. The band gap can further be modulated by doping with transition metals. We consider a few examples here. Doping with Cr, Ti, V, Mn, Fe has been used to tailor the band gap of $KNbO_3$ [75]. The band gap of $KNbO_3$ can be reduced from 3.3 to 1.1 eV when co-doped with Ba and Ni ions and the doped structures also showed multiferroic characteristics [76]. Hiroshi et al. [77] studied the room temperature luminescent spectra for $BaSnO_3$ perovskites. The sample was excited at 380 nm and the emission was recorded in near infrared region at wavelength of 905 nm. The broad band with tail at 760 nm suggests the presence of defects in the energy states. At lower temperature up to 77 k the luminescent spectra showed a sharp and broad band. When Sr ions were added as impurities the band gap and luminescence spectra were shifted.

3.2 Perovskite Based Photovoltaics

The interest in development of green technologies, and harnessing renewable energy sources such as solar and wind energy, has greatly expanded during the past few years. Due to their wide availability and high stability, metal oxide perovskites have been considered as candidates for replacing conventional amorphous Si based solar cells.

The first effort to fabricate perovskite based solar cell was held in 2013 with an working efficiency of 2.6% [78]. A rapid enhancement in the efficiency was observed in the perovskite solar cells due to their amazing charge transport properties. When stable metal oxides were mixed together with inverted device structure, the efficiency was recorded to be 17.7% while the flexible perovskite solar cell with a triple-cation arrangement resulted an efficiency of 18.6% [79]. Liu et al. has reported an efficiency of 17.3% for perovskites solar cells with metal oxide SnO_2 nanoparticles as the electron transport layers [80]. Figure 3 represents the power conversion efficiency over the time for metal oxide and halide perovskites and their comparison with the conventional Silicon solar cells.

With the development in the methods of fabrications of perovskites including interface engineering, creating oxygen defects, charge transport compound optimisation the efficiency of metal oxide perovskites has started from 3.4% reached up to 25.7% in just few years [81, 82]. Seong and co-workers have studied La doped $BaSnO_3$ metal oxide perovskite for their application in solar cells [83]. The doped nanoparticles were synthesised by a colloidal route at not so high temperature of 300 °C. The modified perovskites showed an efficiency of 21.2%. Perovskites $BaSnO_3$ were

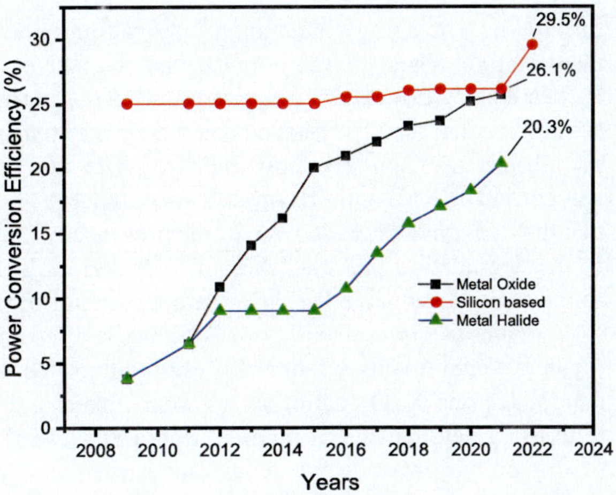

Fig. 3 Evolution of the power conversion efficiency of solar cells with time. Red markers denote efficiency for silicon based solar cells, black markers are for metal oxide and green for metal halide

treated with $TiCl_4$ to study the changes in their physical, chemical and charge transport properties. After treating with $TiCl_4$ a very thin layer of TiO_2 got deposited on the perovskites where the thickness of the layer deposited depends on the run time for deposition. The photo current density was increased and a maximum was achieved after 3 min of deposition with a power conversion efficiency of 5.5%.

3.3 *Other Opto-Electronic Device Applications*

Besides the broad categories discussed above, there are many other applications for perovskite oxides. Electro-optic modulators based on the perovskite $KNbO_3$ rely on the polarization optic coefficients [84]. Studies on different phases of $AgTaO_3$ [85] revealed that $AgTaO_3$ is a potential material for practical application in thermoelectric devices with low electrical conductivity and Seebeck coefficient of 2955 μV/K and power factor of 1.17×10^{12} W/K^{2}ms.

A key application of MOPs is in photocatalysis. Photocatalysis is viewed as a promising remedy for a clean and sustainable future. Significant work has gone into generating highly effective photo catalytic materials for several uses, including the reduction of Carbon dioxide, water pollutant degradation and the elimination of nitrogen from the environment. The photo generated electron–hole pair's short recombination time and the photocatalyst material's specific absorption wavelength are the significant barriers in photocatalysis. The majority of compounds are UV triggered, yet the UV band makes up just 5% of the spectrum. As a result, for increased photocatalysis, the compound energy band gap should be sensitive to visible and near-infrared radiations as well as the long electron–hole pair recombination time. In Perovskites compounds there are three different active sites (A, B and O) available for doping. The energy band gap modification induced by doping at these sites can increase light absorption in both the visible and Near infrared bands.

Yeqiu et al. [86] have synthesised Ag–$SrTiO_3$ perovskite nanoparticles by electrospinning technique. They have heated the nanofibers up to 700 °C and observed that the fabricated nanofibers show the perovskite structure. The photocatalytic activity was carried on RhB and MB organic dyes under visible light irradiation. It was found that the dyes have degraded due to the presence of plasmonic effect generated in Ag. In the degradation process loads of hydroxyl radicals were generated which were found to be responsible for the decomposition of organic dye molecules.

Venkatesan and co-workers [87] have studied the photocatalysis for different organic contaminants by utilizing the composites of two nanomaterials. $BiFeO_3$ nanoplates and V_2O_5 nanoflowers were mixed and annealed in different concentrations to degrade the pollutants. Zahra et al. [88] have studied the degradation of organic compounds by using rare earth metal oxide perovskites CuO(wt%)/$SmFeO_3$ composites with metal oxide. The nanocomposites were synthesised in different concentrations by simple physical mixing followed by calcination. The absorbance spectra were recorded with different concentration of organic pollutant and it was found that there is a decrease in intensity of absorbance with increase in organic pollutant. Also, the role of varying pH was measured for the nanocomposite material. The maximum catalytic efficiency of 65% was attained under the influence of mercury lamp irradiation for 10% CuO composite with $SmFeO_3$ for 8 mg L^{-1} of RhB dye. An enhancement in the photocatalytic efficiency up to 85% was observed when 0.05 ml H_2O_2 was added to the nanocomposite material. The presence of H_2O_2 in the composite played a great role for photo catalytic activity along with the presence of Fenton-like oxidation. Thus, the hybrid nanocomposite proved a potential material for water treatment processes.

Arabi and co-workers presented the degradation of MB under visible light by utilizing doped perovskite nanowires [89]. $LaMnO_3$ was doped with Calcium and its morphological and optical changes were recorded. XRD results showed that only single-phase nanowires were present. The photocatalytic efficiency came to be 73% under 360 min of illumination under visible light. The results were stable after six repetitions. Parvaneh et al. [90] synthesised novel material by doping Mn and N in rare-earth $LaCoO_3$ perovskites. Upon doping, the band gap of the doped materials decreased due to the formation of impurity bands. Malachite green dye was degraded under visible light illumination. It was found that the degradation of dye was increased two-fold due to the presence of Mn dopant. For N doped perovskite the catalytic activity was increased due to the presence of more oxygen vacancies. Similar study was also carried out by Jayapandi et al. [91] where MB was degraded by $LaCoO_3$ activated by Ag ions. Due to the presence of Ag ions the electron transport kinetics was found to be increased and in just 10 min the photocatalytic efficiency reached to 99% which was 75% for bare perovskites. Table 3 summarizes the photocatalytic activity, linking the materials and their morphology with the targeted pollutant.

Table 3 Photocatalytic activity of some perovskite materials

Material	Morphology	Pollutant	Synthesis route	Light source	References
$La_{1-x}Ca_xMnO_3$ [89]	Nanowires	Methylene blue	Hydrothermal	Visible	[89]
$CuO/SmFeO_3$ [88]	Nanocomposites	RhB phenol	Sol gel	Mercury lamp	[88]
$BaZrO_3$ [92]	–	Photoelectrochemical water splitting	Pechini-type process		[92]
$BiFeO_3/BiVO_4$ [93]	Nanocomposite	Tetracycline	Ultrasonic/hydrothermal route	Visible	[93]
$CaTiO_3$ [23]	Nano cuboids	Rhodamine B	Hydrothermal route	Ultraviolet	[23]
S–$SrTiO_3$ [94]	−18 eV	2-Propanol	Physical mixing	Xenon	[94]
$NaTaO_3$ and graphene oxide (GO) [32]	Nanostructures-hexahedral structure	Methylene blue	Hydrothermal method	Ultraviolet	[32]
$LaCoO_3$ [90]	Nanospheres	Malachite green	Sol-gel method	Visible	[90]
$Bi_2O_4/NaBiO_3$ [95]	Composite nanostructures	Methyl orange (Mo) and phenol	Hydrothermal method	–	[95]
$BiBaMFeO_3$(M = Na and K) [96]	Nanoparticles	Methylene blue	Sol-gel method	Visible	[96]
$Au/LaFeO_3/Cu_2O$ [97]	Composites	Rhodamine	Sol gel	Xenon	[97]

4 The Rise of Halide Perovskites

While oxide perovskites show high stability, they are limited by typical high temperature synthesis processes required to fabricate these materials Halide perovskites, with the flexibility of low temperature synthesis routes, have seen tremendous interest over the past few years. In general, the crystal structure for metal halide perovskites is represented as—ABX_3 (where X = Cl^-, Br^-, I^-). Based on the conventional crystallisation methods, various mixed-metal and mixed-halide perovskites [98, 99] have already been designed utilising components with tunable bandgap, resulting in various application [100]. Besides metal halide perovskites, organic–inorganic hybrid perovskites have also been explored and allow for much larger range in terms of materials.

Metal Halide Perovskites were primarily utilized as sensitising substances in dye-sensitized photovoltaic cells; however, it was quickly discovered that, in addition to increasing the absorbance, these materials have outstanding charge transfer characteristics. Metal halide perovskites due to their tunable size properties have been found to be the promising materials for their application in LEDs, Lasers, Opto-electronic materials. They have unusually high photoluminescence quantum yield. Unfortunately bulk metal halides showed very low quantum yields [101]. Imran and co-workers have studied the effect of benzoyl halides used as an alternate precursor for the preparation of different in-organic and hybrid metal halides. These prepared hybrid structures possess purity in the crystal phase, tunable size [102]. Nowadays researchers are more focused on the metal halide nanocrystals for better photoluminescence quantum yield and tunable size variation. The very first metal halide nanostructures were fabricated by Schmidt [103] in year 2014. Since then, the study of these metal halide compounds has practically exploded. Shi and co-workers [104] have studied the optical properties of $MAPbBr_3$ and $MAPbI_3$ and studied the energy band gap of the materials. Both compounds display a significantly narrow bandgap in single-crystalline form than those in equivalent films. A narrow bandgap can increase photon absorption and results in improvement of photocurrent generation. Both the materials show a narrow photoluminescence spectrum and a small hump at 790 nm is also visible in $MAPbI_3$ which matches with the literature results [105] of a thin film for $MAPbI_3$ material. The tunning of luminescence in the entire visible region can be achieved by varying the sizes of these nanocrystals and their initial ration of precursors. The mixed metal halides can be synthesised by just simply varying the concentration of the halides salts [106]. The mixed-perovskite nanocrystals also exhibit remarkable both optical and electrical characteristics with increased stability, including a very low trapping density and significant mobilities [107, 108]. Nevertheless, the impacts of defects in perovskite materials are inevitable at the interface and grain boundaries, even though the single-crystal structure has smaller grain boundaries and can thus go away with the ion migration channel. Therefore, it is important to research how imperfections affect perovskite single crystals in order to stop the electrical loss due to charge carrier recombination.

Peng and co-workers [109] have studied the solar cell characteristics of $MAPbBr_3$ using ITO substrate. The PCE of the fabricated solar cells was recorded to be 6.53%. Ni et al. [110] have studied the PCE for trapped states in polycrystalline perovskites. The results for the thin film of $MAPbI_3$ fabricated by DLCP method was found to be 20.8%. Yang and co-workers [111] have developed single crystal thin film $MAPbBr_3$ photodetector where they have utilized ITO as a substrate and Au as an electrode. When negative voltage was applied a Schottky barrier was formed at the barrier and the device was open where the value of current increases. The increase in current was proportional to the incident light. Thus, the resulting photoconductive gain value comes to be 5.0 × 107, it came out to be the higher value when compared with the literature pervious results. Chen et. al. [112] have developed $MAPbBr_3$ based Light Emitting Diode deposited on ITO substrate with an Au electrode. When the voltage was applied there was an increase in emission intensity but at 4.0 V a decrease was recorded in the emission intensity. The decrease was due to the reduction in light collecting efficiency as the photons were emitted from the side surface. The electroluminescence spectrum was studied and two different peaks at 542–544 and 548–551 nm wavelength was recorded. With the increase in bias voltage the intensity for the second peak lying at the higher wavelength was increased due to the more injection current density and a red shift in the wavelength was also noted.

Li and co-workers [113] have designed $FAPbBr_3$ micro disks for laser application. The emission spectrum was recorded at 542 nm wavelength at a lower pumping rate. When the pumping rate was increased there is a red shift in the emission peak at 553 nm. The red shift was due to the reabsorption held in one photon pumping. The intensity of the emission spectra also depends on the pumping rate. Anna and co-workers have studied the quantum confinement behaviour of metal halide perovskite materials for their application in optoelectronic devices. They have used water as an additive in the chemical reaction with the precursor material methylammonium lead bromide nanoparticles to study the mobility of species and their variation in lattice parameters. This method has regulated the size of fabricated metal halide quantum dots in the range of 4–7 nm. Due to quantum confinement effect with the addition of water the band gap decreases while the quantum yield was increased [114]. In the past ten years, a significant amount of research has been conducted on the photovoltaic properties of metal halide perovskites, and its efficiency for the application solar cells have been reported up to 25.8% [115].

5 Conclusions

In conclusion, perovskites are a highly versatile family, exhibiting a wide range of properties with the potential to target a far wider range of applications. Most single oxide perovskites have a wide band gap (~2–4 eV), and are therefore poor solar absorbers, but may be more suitable for water splitting. Single perovskites have been widely investigated for photocatalysis (PC), where electrocatalysts are deposited on a solar absorber material and immersed in an electrolyte. Again, the main concern arises as most single perovskites are only capable of absorbing in the UV part of the solar spectrum. Perovskites, such as $SrTiO_3$, $NaTaO_3$, $AgNbO_3$, $KNbO_3$, $NaNbO_3$, and $BiFeO_3$ have been studied for PC applications, with dopant strategies being used to reduce bandgaps. The more recently trending halide perovskites offer advantages but also disadvantages compared to the oxide counterparts. The biggest advantage stems from ease of processing for halide perovskites, which can be done at relatively low temperatures. The biggest disadvantage is the limited thermal and environmental stability of halide perovskites.

Overall, the perovskite field offers rich opportunities for development of materials for the future. However, the optical properties are not independent of the electrical properties and attention needs to be paid towards the electronic band structure and its modulation with doping and/or defects. While the introduction of defects or dopants may help tailor the bandgap, increasing relevance for solar absorber or water splitting applications, they may also influence trap states and care then needs to be paid to non-radiative processes. With advances in nanofabrication, and our understanding of phenomenon at those length scales, dimensionality may also be used as another knob to tailor the electronic and optical properties. This is definitely an avenue where more work continues to be done, and insights gained may guide future efforts in these materials.

References

1. M.A. Peña, J.L.G. Fierro, Chemical structures and performance of perovskite oxides. Chem. Rev. **101**, 1981–2017 (2001)
2. X. Huang, G. Zhao, G. Wang, J.T.S. Irvine, Synthesis and applications of nanoporous perovskite metal oxides. Chem. Sci. **9**, 3623–3637 (2018)
3. A.I. Kurbakov, Electronic, structural and magnetic phase diagram of Sm1–xSrxMnO3 manganites. J. Magn. Magn. Mater. **322**, 967–972 (2010). https://doi.org/10.1016/j.jmmm.2009.11.034
4. P. Ciambelli, S. Cimino, S. De Rossi et al., AMnO3 (A = LA, Nd, Sm) and Sm(1–x)Sr(x)MnO3 perovskites as combustion catalysts: structural, redox and catalytic properties. Appl. Catal. B Environ. **24**, 243–253 (2000). https://doi.org/10.1016/S0926-3373(99)00110-1
5. F.S. Toniolo, M. Schmal, Improvement of catalytic performance of perovskites by partial substitution of cations and supporting on high surface area materials, in *Perovskite Materials—Synthesis, Characterisation, Properties, and Applications* (2016)
6. S. Royer, D. Duprez, F. Can et al., Perovskites as substitutes of noble metals for heterogeneous catalysis: dream or reality. Chem. Rev. **114**, 10292–10368 (2014)
7. H. Najjar, H. Batis, Development of Mn-based perovskite materials: chemical structure and applications. Catal. Rev. Sci. Eng. **58**, 371–438 (2016). https://doi.org/10.1080/01614940.2016.1198203
8. S. Tasleem, M. Tahir, Recent progress in structural development and band engineering of perovskites materials for photocatalytic solar hydrogen production: a review. Int. J. Hydrog. Energy **45**, 19078–19111 (2020)
9. E. Grabowska, Selected perovskite oxides: characterization, preparation and photocatalytic properties-A review. Appl. Catal. B Environ. **186**, 97–126 (2016)
10. F. Polo-Garzon, Z. Wu, Acid-base catalysis over perovskites: a review. J. Mater. Chem. A **6**, 2877–2894 (2018)

11. K. Kamata, Perovskite oxide catalysts for liquid-phase organic reactions. Bull. Chem. Soc. Jpn. **92**, 133–151 (2019). https://doi.org/10.1246/bcsj.20180260
12. G. Koch, M. Hävecker, D. Teschner et al., Surface conditions that constrain alkane oxidation on perovskites. ACS Catal. **10**, 7007–7020 (2020). https://doi.org/10.1021/acscatal.0c01289
13. J.L.G. Fierro, L.G. Tejuca, Non-stoichiometric surface behaviour of LaMO3 oxides as evidenced by XPS. Appl. Surf. Sci. **27**, 453–457 (1987). https://doi.org/10.1016/0169-4332(87)90154-1
14. R. Ignatans, G. Mallia, E.A. Ahmad et al., The effect of surface reconstruction on the oxygen reduction reaction properties of LaMnO3. J. Phys. Chem. C **123**, 11621–11627 (2019). https://doi.org/10.1021/acs.jpcc.9b00458
15. A. Badreldin, A.E. Abusrafa, A. Abdel-Wahab, Oxygen-deficient perovskites for oxygen evolution reaction in alkaline media: a review. Emergent Mater. **3**, 567–590 (2020)
16. G. Koch, M. Hävecker, P. Kube et al., The influence of the chemical potential on defects and function of perovskites in catalysis. Front. Chem. **9**(2021). https://doi.org/10.3389/fchem.2021.746229
17. Y. Yamada, Y. Kanemitsu, Photoluminescence spectra of perovskite oxide semiconductors. J. Lumin. 30–34 (2013)
18. H. Yasuda, Y. Yamada, T. Tayagaki, Y. Kanemitsu, Spatial distribution of carriers in SrTiO3 revealed by photoluminescence dynamics measurements. Phys. Rev. B—Condens. Matter. Mater. Phys. **78**(2008). https://doi.org/10.1103/PhysRevB.78.233202
19. H. Yasuda, Y. Kanemitsu, Dynamics of nonlinear blue photoluminescence and Auger recombination in $SrTiO_3$. Phys. Rev. B **77**, 193202 (2008)
20. Y. Yamada, H. Yasuda, T. Tayagaki, Y. Kanemitsu, Temperature dependence of photoluminescence spectra of nondoped and electron-doped SrTiO3: crossover from auger recombination to single-carrier trapping. Phys. Rev. Lett. **102**(2009). https://doi.org/10.1103/PhysRevLett.102.247401
21. T. Arima, Y. Tokura, J.B. Torrance, Variation of optical gaps in perovskite-type 3d transition-metal oxides. Phys. Rev. B **48**, 17006–17009 (1993). https://doi.org/10.1103/PhysRevB.48.17006
22. W. Zhang, J. Tang, J. Ye, Structural, photocatalytic, and photophysical properties of perovskite MSnO3 (M = Ca, Sr, and Ba) photocatalysts. J. Mater. Res. **22**, 1859–1871 (2007). https://doi.org/10.1557/jmr.2007.0259
23. Y. Yan, H. Yang, X. Zhao et al., Enhanced photocatalytic activity of surface disorder-engineered CaTiO3. Mater. Res. Bull. **105**, 286–290 (2018). https://doi.org/10.1016/j.materresbull.2018.05.008
24. S. Suzuki, A. Iwase, A. Kudo, Long wavelength visible light-responsive SrTiO3photocatalysts doped with valence-controlled Ru for sacrificial H2and O2evolution. Catal. Sci. Technol. **10**, 4912–4916 (2020). https://doi.org/10.1039/d0cy00600a
25. Y. Li, H. Gou, J. Lu, C. Wang, A two-step synthesis of NaTaO3 microspheres for photocatalytic water splitting. Int. J. Hydrog. Energy. 13481–13485 (2014)
26. L. Ernawati, R.A. Wahyuono, H. Widiyandari et al., Experimental data of CaTiO3 photocatalyst for degradation of organic pollutants (Brilliant green dye)—Green synthesis, characterization and kinetic study. Data Br. **32**(2020). https://doi.org/10.1016/j.dib.2020.106099
27. S. Sharma, M. Kumar, Band gap tuning and optical properties of BiFeO3 nanoparticles. Mater. Today Proc. **28**, 168–171 (2020). https://doi.org/10.1016/j.matpr.2020.01.496
28. M. Ismael, M. Wark, Perovskite-type LaFeO3: photoelectrochemical properties and photocatalytic degradation of organic pollutants under visible light irradiation. Catalysts **9**(2019). https://doi.org/10.3390/catal9040342
29. Q. Liu, Y. Chai, L. Zhang et al., Highly efficient Pt/NaNbO3 nanowire photocatalyst: its morphology effect and application in water purification and H2 production. Appl. Catal. B Environ. **205**, 505–513 (2017). https://doi.org/10.1016/j.apcatb.2016.12.065
30. C. Huang, L. Chen, H. Li et al., Synthesis and application of Bi2WO6 for the photocatalytic degradation of two typical fluoroquinolones under visible light irradiation. RSC Adv. **9**, 27768–27779 (2019). https://doi.org/10.1039/c9ra04445k

31. J. Xu, B. Luo, W. Gu et al., Fabrication of In2S3/NaTaO3 composites for enhancing the photocatalytic activity toward the degradation of tetracycline. New J. Chem. **42**, 5052–5058 (2018). https://doi.org/10.1039/c7nj05123a
32. Z. Fu, S. Zhang, Z. Fu, Hydrothermal preparation of NaTaO3/rGO composite photocatalyst to enhance UV photocatalytic activity. Results Phys. **15**(2019). https://doi.org/10.1016/j.rinp.2019.102669
33. L. Zhang, T. Xu, X. Zhao, Y. Zhu, Controllable synthesis of Bi2MoO6 and effect of morphology and variation in local structure on photocatalytic activities. Appl. Catal. B Environ. **98**, 138–146 (2010). https://doi.org/10.1016/j.apcatb.2010.05.022
34. Y.A. Alsabah, A.T. Elden, M.S. AlSalhi et al., Structural and optical properties of A2YVO6 (A = Mg, Sr) double perovskite oxides. Results Phys. **15**(2019). https://doi.org/10.1016/j.rinp.2019.102589
35. T. Ling, D.Y. Yan, Y. Jiao et al., Engineering surface atomic structure of single-crystal cobalt (II) oxide nanorods for superior electrocatalysis. Nat. Commun. **7**(2016). https://doi.org/10.1038/ncomms12876
36. Y. Zhu, X. Zhong, S. Jin et al., Oxygen defect engineering in double perovskite oxides for effective water oxidation. J. Mater. Chem. A **8**, 10957–10965 (2020). https://doi.org/10.1039/d0ta04362a
37. S.N. Basahel, A.H.A. Medkhali, M. Mokhtar, K. Narasimharao, Noble metal (Pd, Pt and Rh) incorporated LaFeO3 perovskite oxides for catalytic oxidative cracking of n-propane. Catal. Today **397–399**, 81–93 (2022). https://doi.org/10.1016/j.cattod.2021.11.032
38. W. Li, J. Shi, K.H.L. Zhang, J.L. Macmanus-Driscoll, Defects in complex oxide thin films for electronics and energy applications: challenges and opportunities. Mater. Horizons **7**, 2832–2859 (2020)
39. A. Herklotz, D. Lee, E.J. Guo, et al., Strain coupling of oxygen non-stoichiometry in perovskite thin films. J. Phys. Condens. Matter **29** (2017)
40. J. Park, Y.N. Wu, W.A. Saidi et al., First-principles exploration of oxygen vacancy impact on electronic and optical properties of ABO3: δ(A = La, Sr; B = Cr, Mn) perovskites. Phys. Chem. Chem. Phys. **22**, 27163–27172 (2020). https://doi.org/10.1039/d0cp05445c
41. Q. Ji, L. Bi, J. Zhang et al., The role of oxygen vacancies of ABO3perovskite oxides in the oxygen reduction reaction. Energy Environ. Sci. **13**, 1408–1428 (2020)
42. D. Ji, S. Cai, T.R. Paudel et al., Freestanding crystalline oxide perovskites down to the monolayer limit. Nature **570**, 87–90 (2019). https://doi.org/10.1038/s41586-019-1255-7
43. L. Li, L. Liang, H. Wu, X. Zhu, One-dimensional perovskite manganite oxide nanostructures: recent developments in synthesis, characterization, transport properties, and applications. Nanoscale Res. Lett. **11** (2016)
44. S. Stemmer, A.J. Millis, Quantum confinement in oxide quantum wells. MRS Bull. **38**, 1032–1039 (2013). https://doi.org/10.1557/mrs.2013.265
45. M. Choi, C. Lin, M. Butcher et al., Quantum confinement in transition metal oxide quantum wells. Appl. Phys. Lett. **106**(2015). https://doi.org/10.1063/1.4921013
46. K. Yoshimatsu, T. Okabe, H. Kumigashira et al., dimensional-crossover-driven metal-insulator transition in $SrVO_3$ ultrathin films. Phys. Rev. Lett. **104**, 147601 (2010)
47. K. Yoshimatsu, K. Horiba, H. Kumigashira, et al., Metallic quantum well states in artificial structures of strongly correlated oxide. Science (80) **333**, 319–322 (2011). https://doi.org/10.1126/science.1205771
48. D. Grosso, C. Boissiere, B. Smarsly et al., Periodically ordered nanoscale islands and mesoporous films composed of nanocrystalline multimetallic oxides. Nat. Mater. **3**, 787–792 (2004). https://doi.org/10.1038/nmat1206
49. C. Reitz, K. Brezesinski, J. Haetge et al., Nanocrystalline NaTaO3 thin film materials with ordered 3D mesoporous and nanopillar-like structures through PIB-b-PEO polymer templating: towards high-performance UV-light photocatalysts. RSC Adv **2**, 5130–5133 (2012). https://doi.org/10.1039/c2ra20203d

50. C. Reitz, C. Suchomski, C. Weidmann, T. Brezesinski, Block copolymer-templated BiFeO3 nanoarchitectures composed of phase-pure crystallites intermingled with a continuous mesoporosity: effective visible-light photocatalysts? Nano Res. **4**, 414–424 (2011). https://doi.org/10.1007/s12274-011-0096-y
51. J.Y. Do, Y. Im, B.S. Kwak et al., Preparation of basalt fiber@perovskite PbTiO3 core-shell composites and their effects on CH4 production from CO2 photoreduction. Ceram. Int. **42**, 5942–5951 (2016). https://doi.org/10.1016/j.ceramint.2015.12.142
52. Y.L. Liu, C.L. Yang, M.S. Wang et al., Te-doped perovskite NaTaO3 as a promising photocatalytic material for hydrogen production from water splitting driven by visible light. Mater. Res. Bull. **107**, 125–131 (2018). https://doi.org/10.1016/j.materresbull.2018.06.040
53. L.M. Prócel, F. Tipán, A. Stashans, Mott-Wannier excitons in the tetragonal BaTiO3 lattice. Int. J. Quantum Chem. **91**, 586–590 (2003). https://doi.org/10.1002/qua.10471
54. X. Li, L. Guan, J.Y. An et al., Synthesis of red phosphor CaZrO3:Eu3+ for white light-emitting diodes. Chinese Phys. Lett. **28**(2011). https://doi.org/10.1088/0256-307X/28/2/027805
55. L. Zhang, X. Sun, Enhanced Eu3+ emission in Ca2LaTaO6:Eu3+ phosphors by codoping Bi3+. J. Mater. Sci. Mater. Electron. **31**, 9160–9166 (2020). https://doi.org/10.1007/s10854-020-03445-0
56. P. Singh, R.S. Yadav, S.B. Rai, Enhanced photoluminescence in a Eu3+ doped CaTiO3 perovskite phosphor via incorporation of alkali ions for white LEDs. J. Phys. Chem. Solids **151**(2021). https://doi.org/10.1016/j.jpcs.2020.109916
57. A. Maurya, R.S. Yadav, R.V. Yadav et al., Enhanced green upconversion photoluminescence from Ho3+/Yb3+ co-doped CaZrO3 phosphor via Mg2+ doping. RSC Adv. **6**, 113469–113477 (2016). https://doi.org/10.1039/c6ra23835a
58. Y. Xie, X. Geng, J. Guo et al., Luminescence of a novel double-perovskite Sr2InSbO6:Eu3+ orange-red-emitting phosphor for white LEDs and visualization of latent fingerprints. Mater. Res. Bull. **146**(2022). https://doi.org/10.1016/j.materresbull.2021.111574
59. S. Zhao, J. Xiang, M.H. Fang et al., A novel high thermal stability Ba2CaWO6: Mn4+ far-red emitting phosphor with a double-perovskite structure for plant growth LEDs. Opt. Mater. (Amst) **124**(2022). https://doi.org/10.1016/j.optmat.2022.112052
60. S.C. Watthage, Z. Song, A.B. Phillips, M.J. Heben, Evolution of perovskite solar cells, in *Perovskite Photovoltaics: basic to Advanced Concepts and Implementation* (2018), pp 43–88
61. X.Z. Jin, Y.L. Shao, Y. Zheng, T. Zhang, Progress in modification of strontium Titanate photocatalyst. J. Mol. Catal. **34**, 559–568 (2020)
62. E. García-López, G. Marcì, B. Megna et al., SrTiO3-based perovskites: Preparation, characterization and photocatalytic activity in gas-solid regime under simulated solar irradiation. J Catal **321**, 13–22 (2015). https://doi.org/10.1016/j.jcat.2014.10.014
63. R. Ran, X. Wu, D. Weng, J. Fan, Oxygen storage capacity and structural properties of Ni-doped LaMnO 3 perovskites. J. Alloys Compd. **577**, 288–294 (2013). https://doi.org/10.1016/j.jallcom.2013.05.041
64. M.L. Crespillo, J.T. Graham, F. Agulló-López et al., Real-Time identification of oxygen vacancy centers in LiNbO3 and SrTiO3 during irradiation with high energy particles. Crystals **11**(2021). https://doi.org/10.3390/cryst11030315
65. N. Afifah, R. Saleh, Enhancement of photocatalytic activities of perovskite LaFeO3 composite by incorporating nanographene platelets. IOP Conf. Ser.: Mater. Sci. Eng.
66. L. Wang, Y. Li, A. Bera et al., Device performance of the Mott insulator LaVO3 as a photovoltaic material. Phys. Rev. Appl. **3**(2015). https://doi.org/10.1103/PhysRevApplied.3.064015
67. J. Dong, J. Wu, J. Jia et al., Efficient perovskite solar cells employing a simply-processed CdS electron transport layer. J. Mater. Chem. C **5**, 10023–10028 (2017). https://doi.org/10.1039/c7tc03343e
68. L. López-Mir, C. Frontera, H. Aramberri et al., Anisotropic sensor and memory device with a ferromagnetic tunnel barrier as the only magnetic element. Sci. Rep. **8**(2018). https://doi.org/10.1038/s41598-017-19129-5
69. K. Uchino, Glory of piezoelectric perovskites. Sci. Technol. Adv. Mater. **16** (2015)

70. C. Yu, Y. Shimizu, H. Arai, Investigation on a lean-burn oxygen sensor using perovskite-type oxides. Chem. Lett. **15**, 563–566 (1986). https://doi.org/10.1246/cl.1986.563
71. S.J. Skinner, Recent advances in perovskite-type materials for solid oxide fuel cell cathodes. Int. J. Inorg. Mater. **3**, 113–121 (2001). https://doi.org/10.1016/S1466-6049(01)00004-6
72. M.Y. Cho, S. Kim, I.S. Kim et al., Perovskite-induced ultrasensitive and highly stable humidity sensor systems prepared by aerosol deposition at room temperature. Adv. Funct. Mater. **30**(2020). https://doi.org/10.1002/adfm.201907449
73. Y. Xu, U. Memmert, U. Hartmann, Magnetic field sensors from polycrystalline manganites. Sens. Actuat. A Phys. **91**, 26–29 (2001). https://doi.org/10.1016/S0924-4247(01)00493-9
74. J. Herrán, G.G. Mandayo, E. Castaño, Solid state gas sensor for fast carbon dioxide detection. Sens. Actuat. B Chem. **129**, 705–709 (2008). https://doi.org/10.1016/j.snb.2007.09.028
75. Y. Liang, G. Shao, First principles study for band engineering of KNbO3 with 3d transition metal substitution. RSC Adv. **9**, 7551–7559 (2019). https://doi.org/10.1039/c9ra00289h
76. J. Tang, Y. Wang, L. Zhang et al., Multiferroic properties of Ba/Ni co-doped KNbO3with narrow band-gap. J. Alloys Compd. **703**, 67–72 (2017). https://doi.org/10.1016/j.jallcom.2017.01.180
77. H. Mizoguchi, P.M. Woodward, C.H. Park, D.A. Keszler, Strong near-infrared luminescence in BaSnO3. J. Am. Chem. Soc. **126**, 9796–9800 (2004). https://doi.org/10.1021/ja048866i
78. P. Docampo, J.M. Ball, M. Darwich et al., Efficient organometal trihalide perovskite planar-heterojunction solar cells on flexible polymer substrates. Nat. Commun. **4**(2013). https://doi.org/10.1038/ncomms3761
79. M. Najafi, F. Di Giacomo, D. Zhang et al., Highly efficient and stable flexible perovskite solar cells with metal oxides nanoparticle charge extraction layers. Small **14**(2018). https://doi.org/10.1002/smll.201702775
80. C. Liu, L. Zhang, X. Zhou, et al., Hydrothermally treated SnO2 as the electron transport layer in high-efficiency flexible perovskite solar cells with a certificated efficiency of 17.3%. Adv. Funct. Mater. **29** (2019). https://doi.org/10.1002/adfm.201807604
81. K. Valadi, S. Gharibi, R. Taheri-Ledari et al., Metal oxide electron transport materials for perovskite solar cells: a review. Environ. Chem. Lett. **19**, 2185–2207 (2021)
82. J.S. Shaikh, N.S. Shaikh, S.S. Mali et al., Nanoarchitectures in dye-sensitized solar cells: metal oxides, oxide perovskites and carbon-based materials. Nanoscale **10**, 4987–5034 (2018)
83. S.S. Shin, E.J. Yeom, W.S. Yang et al., Colloidally prepared La-doped BaSnO3 electrodes for efficient, photostable perovskite solar cells. Science (80) **356**, 167–171 (2017). https://doi.org/10.1126/science.aam6620
84. P. Günter, Electro-optical properties of KNbO3. Opt. Commun. **11**, 285–290 (1974). https://doi.org/10.1016/0030-4018(74)90183-7
85. S.A. Khan, H.U. Khan, S. Mehmood, Z. Ali, Structural, electronic, optical and thermoelectric properties in the phases of AgTaO3. Mater. Sci. Semicond. Process. **122**(2021). https://doi.org/10.1016/j.mssp.2020.105467
86. Y. Wu, T. He, Ag loading induced visible light photocatalytic activity for pervoskite SrTiO3 nanofibers. Spectrochim. Acta—Part A Mol. Biomol. Spectrosc. **199**, 283–289 (2018). https://doi.org/10.1016/j.saa.2018.03.078
87. V. Jayaraman, D. Sarkar, R. Rajendran et al., Synergistic effect of band edge potentials on BiFeO3/V2O5 composite: enhanced photo catalytic activity. J. Environ. Manag. **247**, 104–114 (2019). https://doi.org/10.1016/j.jenvman.2019.06.041
88. Z. Behzadifard, Z. Shariatinia, M. Jourshabani, Novel visible light driven CuO/SmFeO3 nanocomposite photocatalysts with enhanced photocatalytic activities for degradation of organic pollutants. J. Mol. Liq. **262**, 533–548 (2018). https://doi.org/10.1016/j.molliq.2018.04.126
89. A. Arabi, M. Fazli, M.H. Ehsani, Synthesis and characterization of calcium-doped lanthanum manganite nanowires as a photocatalyst for degradation of methylene blue solution under visible light irradiation. Bull. Mater. Sci. **41**(2018). https://doi.org/10.1007/s12034-018-1590-6

90. P. Nakhostin Panahi, M.H. Rasoulifard, S. Babaei, Photocatalytic activity of cation (Mn) and anion (N) substitution in LaCoO3 nanoperovskite under visible light. Rare Met. **39**, 139–146 (2020). https://doi.org/10.1007/s12598-019-01329-9
91. S. Jayapandi, D. Lakshmi, S. Premkumar et al., Augmented photocatalytic and electrochemical activities of Ag tailored LaCoO3 perovskite semiconductor. Mater. Lett. **218**, 205–208 (2018). https://doi.org/10.1016/j.matlet.2018.02.015
92. Y. Yuan, X. Zhang, L. Liu et al., Synthesis and photocatalytic characterization of a new photocatalyst BaZrO3. Int. J. Hydrog. Energy **33**, 5941–5946 (2008). https://doi.org/10.1016/j.ijhydene.2008.07.052
93. T. Soltani, A. Tayyebi, B.K. Lee, BiFeO3/BiVO4 p−n heterojunction for efficient and stable photocatalytic and photoelectrochemical water splitting under visible-light irradiation. Catal. Today **340**, 188–196 (2020). https://doi.org/10.1016/j.cattod.2018.09.030
94. T. Ohno, T. Tsubota, Y. Nakamura, K. Sayama, Preparation of S, C cation-codoped SrTiO3 and its photocatalytic activity under visible light. Appl. Catal. A Gen. **288**, 74–79 (2005). https://doi.org/10.1016/j.apcata.2005.04.035
95. Y. Zhang, W.X. Wang, Y.S. Guo et al., Hydrothermal synthesis of Bi2O4/NaBiO3 heterostructures with enhanced visible light photocatalytic properties. J. Phys. Chem. Solids **149**(2021). https://doi.org/10.1016/j.jpcs.2020.109766
96. A. Haruna, I. Abdulkadir, S.O. Idris, Synthesis, characterization and photocatalytic properties of Bi0.85−XMXBa0.15FeO3 (M = Na and K, X = 0, 0.1) perovskite-like nanoparticles using the sol-gel method. J. King Saud. Univ. Sci. **32**, 896–903 (2020). https://doi.org/10.1016/j.jksus.2019.05.005
97. S. Guan, R. Li, X. Sun et al., Construction of novel ternary Au/LaFeO3/Cu2O composite photocatalysts for RhB degradation via photo-Fenton catalysis. Mater. Technol. **36**, 603–615 (2021). https://doi.org/10.1080/10667857.2020.1782062
98. C.H. Ng, T.S. Ripolles, K. Hamada et al., Tunable open circuit voltage by engineering inorganic Cesium lead bromide/iodide perovskite solar cells. Sci. Rep. **8**(2018). https://doi.org/10.1038/s41598-018-20228-0
99. Y. Liu, Y. Zhang, X. Zhu et al., Triple-cation and mixed-halide perovskite single crystal for high-performance X-ray imaging. Adv. Mater. **33**(2021). https://doi.org/10.1002/adma.202006010
100. K. Poorkazem, T.L. Kelly, Improving the stability and decreasing the trap state density of mixed-cation perovskite solar cells through compositional engineering. Sustain. Energy Fuels **2**, 1332–1341 (2018). https://doi.org/10.1039/c8se00127h
101. W.J. Yin, T. Shi, Y. Yan, Unusual defect physics in CH3NH3PbI3 perovskite solar cell absorber. Appl. Phys. Lett. **104**(2014). https://doi.org/10.1063/1.4864778
102. M. Imran, V. Caligiuri, M. Wang et al., Benzoyl halides as alternative precursors for the colloidal synthesis of lead-based halide perovskite nanocrystals. J. Am. Chem. Soc. **140**, 2656–2664 (2018). https://doi.org/10.1021/jacs.7b13477
103. L.C. Schmidt, A. Pertegás, S. González-Carrero et al., Nontemplate synthesis of CH3NH3PbBr 3 perovskite nanoparticles. J. Am. Chem. Soc. **136**, 850–853 (2014). https://doi.org/10.1021/ja4109209
104. D. Shi, V. Adinolfi, R. Comin et al., Low trap-state density and long carrier diffusion in organolead trihalide perovskite single crystals. Science (80) **347**, 519–522 (2015). https://doi.org/10.1126/science.aaa2725
105. S.D. Stranks, G.E. Eperon, G. Grancini et al., Electron-hole diffusion lengths exceeding 1 micrometer in an organometal trihalide perovskite absorber. Science (80) **342**, 341–344 (2013). https://doi.org/10.1126/science.1243982
106. J. Shamsi, A.S. Urban, M. Imran et al., Metal halide perovskite nanocrystals: synthesis, post-synthesis modifications, and their optical properties. Chem. Rev. **119**, 3296–3348 (2019). https://doi.org/10.1021/acs.chemrev.8b00644
107. Z. Fan, J. Liu, W. Zuo et al., Mixed-cation MAxCs1−xPbBr 3 Perovskite single crystals with composition management for high-sensitivity X-ray detection. Phys. Status Solidi—Rapid Res. Lett. **14**(2020). https://doi.org/10.1002/pssr.202000226

108. L. Ma, Z. Yan, X. Zhou et al., A polymer controlled nucleation route towards the generalized growth of organic-inorganic perovskite single crystals. Nat. Commun. **12**(2021). https://doi.org/10.1038/s41467-021-22193-1
109. W. Peng, L. Wang, B. Murali et al., Solution-grown monocrystalline hybrid perovskite films for hole-transporter-free solar cells. Adv. Mater. **28**, 3383–3390 (2016). https://doi.org/10.1002/adma.201506292
110. Z. Ni, C. Bao, Y. Liu et al., Resolving spatial and energetic distributions of trap states in metal halide perovskite solar cells. Science (80) **367**, 1352–1358 (2020). https://doi.org/10.1126/science.aba0893
111. Z. Yang, Y. Deng, X. Zhang et al., High-performance single-crystalline perovskite thin-film photodetector. Adv. Mater. **30**(2018). https://doi.org/10.1002/adma.201704333
112. M. Chen, X. Shan, T. Geske et al., Manipulating ion migration for highly stable light-emitting diodes with single-crystalline organometal halide perovskite microplatelets. ACS Nano **11**, 6312–6318 (2017). https://doi.org/10.1021/acsnano.7b02629
113. X. Li, K. Wang, M. Chen et al., Stable whispering gallery mode lasing from solution-processed formamidinium lead bromide perovskite microdisks. Adv. Opt. Mater. **8**(2020). https://doi.org/10.1002/adom.202000030
114. A. Jancik Prochazkova, Y. Salinas, C. Yumusak et al., Controlling quantum confinement in luminescent perovskite nanoparticles for optoelectronic devices by the addition of water. ACS Appl. Nano Mater. **3**, 1242–1249 (2020). https://doi.org/10.1021/acsanm.9b01857
115. N.K. Tailor, S. Satapathi, The impact of Cs3Bi2Cl9 single crystal growth modality on its symmetry and morphology. J. Mater. Res. Technol. **9**, 7149–7157 (2020). https://doi.org/10.1016/j.jmrt.2020.04.097

Optical Behavior of Metal-Oxide-Based Nanofluids

A. S. Abdelrazik

Abstract How to make the best use of the unique properties of nanomaterials, in different fluidic applications, has gained a lot of attention in the last few decades. Dispersing the nanomaterials in a conventional fluid can provide a wide range of alternative thermal, rheological, and optical properties for the new nano-enhanced fluids (nanofluids), depending on different affecting parameters. Different metal, metal oxide, and carbon-based nanomaterials were used for this purpose. This chapter provides a step forward in understanding the optical behavior of metal-oxide-based nanofluids and how metal oxides play a major role in controlling the optical response of several nanofluids. This flexible controlled optical behavior has met great interest to increase the absorption of solar energy from one side and work as a spectral-selective filter for solar radiation from another side. This chapter, in different sections, deals with a number of topics in this regard starting from the techniques used for the synthesis of nanofluids, passing by the parameters influencing the optical behavior of nanofluids, the analytical evaluation of optical properties, the optical behavior of metal-oxide-based nanofluids, and ending by their applications in different solar-harvesting-related systems.

1 Introduction

The thermal behavior of nanofluids gained high attention from scholars and researchers during the last 3 decades, due to the unique thermal properties of the induced nanomaterials, but what about the optical behavior? The optical behavior of nanofluids did not take that much attention at that time, since the main scope of using these new emerging kinds of fluids was only to enhance the thermal properties and consequently the heat transfer capabilities of the conventional existing fluids.

A. S. Abdelrazik (✉)
Interdisciplinary Research Center for Renewable Energy and Power Systems (IRC-REPS), King Fahd University of Petroleum & Minerals, Dhahran 31261, Saudi Arabia
e-mail: ahmeds@kfupm.edu.sa

V. Kumar et al. (eds.), *Optical Properties of Metal Oxide Nanostructures*,
Progress in Optical Science and Photonics 26,
https://doi.org/10.1007/978-981-99-5640-1_4

Choi and Eastman [1], in the annual meeting of the American Society of Mechanical Engineers (ASME), were the first to propose the nanofluid term to the world. Nanofluid is a solution of conventional fluid and nanomaterials, that be synthesized either using one-step or two-step methods. Since then, strong research has been conducted on this new emerging fluid to present its high potential with future projections. "Mechanisms of heat flow in suspensions of nano-sized particles (nanofluids)" by Keblinski et al. [2] in 2001, "Nanofluids for thermal transport" by Keblinski et al. [3] in 2005, "Nanofluids: From Vision to Reality Through Research" by Choi in 2009, "Applications of Nanofluids: Current and Future" by Wong et al. in 2010, "Nanofluids: A New Field of Scientific Research and Innovative Applications" by Choi [4] in 2011, and many other studies that have been proposed on nanofluids in the last 25 years.

The talk about the optical, rather than the thermal, behavior of the nanofluids have started in 2006 through a study proposed by Kim et al. [5]. The researchers have examined that the nanofluids can provide unique optical properties, besides their unique thermal properties, that can be useful for solar harvesting applications. It was useful for solar absorption and solar filtration.

Introducing nanomaterials with unique optical properties spectrums was able to change the absorption and transmittance spectrum of any conventional base fluid and come up with new fluids that have different optical behavior compared to the base fluid. The nanomaterial type [6–8], size [8–13], shape [8–10, 13, 14], concentration [8, 9, 15–25], basefluid [8, 23–30] and surfactant types [8, 20, 31], and in-suspension stability are parameters that were found to highly affect the ending nanofluid optical behavior.

Different nanofluid mixtures existed in the literature as a result of dispersing various nanomaterials in different basefluids. However, the nanofluids can be classified as per their basefluid or nanomaterial constituents. Carbon, metal, and metal-oxide-based nanomaterials are the common nanomaterials classifications, while water, glycols, refrigerants, and oils are the main categories of basefluids.

While the oxide-based nanomaterials are one of several nanomaterials that have been introduced for the formation of nanofluids, they have shown a relatively better behavior in terms of in-suspension stability compared to the other nanomaterials [19, 32]. On the other side, in terms of better thermal properties, carbon and metal-based nanomaterials were showing significant improvements in the thermal properties of the formed nanofluids [33–37]. For solar absorption, carbon-based nanofluids were the best to satisfy this need because of their color and optical behavior nature [38–43].

Metal-oxide-based nanofluids have gained a lot of attention in the last few years due to the raised concept of spectrally-selective optical filtration. This concept depends on mainly manipulating the optical behavior of a basefluid using nanomaterials to achieve a completely different optical spectrum that allows the transmission of a specific solar-radiation-spectrum range. This transmitted radiation should reach the photovoltaic (PV) cells and be in line with the electrically-efficient spectral range, depending on the PV cell type. Metal-oxide-based nanofluids with their high stability and optical properties were in the spot of light to reach the optimized spectrally-selective optical-filtration window.

In this chapter, we are discussing the optical behavior of metal-oxide-based nanofluids. We will start in the next Sect. 2 with more explanation of the metal-oxide-based nanofluids and the techniques of preparation and characterization of nanofluids, followed by the parameters that significantly affect the optical behavior of nanofluids (Sect. 3). After that Sect. 4 discusses the analytical ways for the evaluation of the optical behavior of nanofluids as a two-phase fluid. The optical behavior of the metal-oxide-based nanofluids under the influence of the different affecting parameters explained in Sect. 3, is presented in Sect. 5. Section 6 demonstrates the different applications of the optical behavior of metal-oxide-based nanofluids in solar absorption and filtration. Finally, the conclusions of the chapter and future prospectives are summarized in Sects. 7 and 8.

2 Metal-Oxide-Based Nanofluids (Preparation and Characterization)

A nanofluid is simply a mixture of a basefluid and a nanomaterial that has been synthesized using specific techniques to fulfill certain purposes. These purposes were mainly serving toward improving the thermophysical and optical properties of the fluid. Several nanomaterials are used for this purpose; such as carbon, metal, and metal-oxide-based nanomaterials. Additionally, different basefluids are used to prepare these nanofluids. Water, refrigerants, oils, and glycols are the common fluids used as basefluids for these nanofluids.

Carbon-based nanofluids were the magic of nanofluids, which researchers have conducted several studies on due to their high thermal and absorption capabilities. However, the stability of these nanofluids remained a big challenge that needs to be sorted out before commercialization of these nanofluids. On the other side, oxide-based nanofluids have got more attention compared to plain nanomaterial-based nanofluids due to their comparatively higher stability which make them more feasible and applicable.

Metal-oxide-based nanofluids were the highest nanofluids of interest in the past few years due to their high spectral-selective-transmittance capabilities. Introducing metal-oxide-based nanomaterials with unique optical properties spectrums was able to change the absorption and transmittance spectrum of the basefluids to reach new fluids that have different optical behavior compared to the base fluid. Different metal-oxide-based nanofluids exist in the literature, however, most of them were not yet examined. Examples of metal-oxide-based nanomaterials are Aluminum oxide or Alumina (Al_2O_3), Titanium dioxide or Titania (TiO_2), Copper oxide (CuO), Iron oxide (Fe_3O_4), Tungsten oxide (WO_3), and Cerium oxide (CeO_2).

2.1 Preparation

There are two approaches for synthesizing nanofluids: one-step and two-step.

2.1.1 One-Step Method

Is described as preparing nanoparticles by chemical or physical techniques and dispersing them in a specified nanofluid in a single process. When compared to the two-step procedure, the one-step method provides greater dispersion of nanoparticles in based fluids [44]. This process gives the produced nanoparticles a good desperation and a very stable suspension in the fluid. The one-step procedure produced a molten salt nanofluid with improved physical and chemical characteristics over the two-step method [45].

2.1.2 Two-Step Method

The two-step process is the most commonly utilized method in nanofluid production. Two processes are carried out sequentially. The first step is to prepare dry nanomaterial powders using a physical or chemical process, and the second step is to disperse the nanoparticle powder in the based fluid. This approach had poor control over nanoparticles, which hampered stability. The use of surfactants is critical in obtaining more stable nanofluids and the stirring speed and sonication and stirring times, during preparation, are found as dependent on the surfactant type [46]. The two-step approach was discovered to be an effective cost-cutting solution in industrial applications [47].

2.2 Characterization

Characterization of a nanofluid simply refers to conducting several measurements to end up with a full description of the synthesized nanofluid. Description of nanofluid includes its homogeneity, thermal and dynamic stability, particle size distribution, and nanoparticles distribution, in addition to its thermal, rheological, and optical properties. In the following few subsections, different characterization parameters and techniques are discussed.

2.2.1 Nanoparticles' Morphology and Distribution

Both Scanning Electron Microscopy (SEM) and Transmission Electron Microscopy (TEM) are used to examine the morphology and distribution of the nanomaterials, in wet (dispersed) and dry conditions. This helps in having a better understanding of the effects of these parameters on the different resulting properties and nanofluid stability. Different morphologies can be achieved under different preparation conditions (sonication and stirring time and power, temperature [48], surfactant existence …etc.). Discovering higher sizes of nanomaterials dispersed in the nanofluid after preparation reflects a high degree of agglomeration [49] and low stability for this nanomaterial when dispersed in that basefluid in the applied synthesis conditions.

Several other pieces of equipment besides the SEM and the TEM are used for the assessment of the quality of the nanomaterial and nanofluid preparation and stability. Field Emission Scanning Electron Microscopy (FE-SEM), High-resolution Transmission Electron Microscopy (HR-TEM), Fourier Transform Infrared spectroscopy (FTIR), Energy-Dispersive X-ray spectroscopy (EDX), X-ray diffractometer (XRD), X-ray Photoelectron Spectroscopy (XPS), and Raman spectroscopy are some of the equipment used for this purpose as per [50–53].

2.2.2 Nanoparticles' Size Distribution and Stability

The nanoparticles' size distribution is usually measured using the Dynamic Light Scattering (DLS) method. The DLS method evaluates the particle size distribution by correlating it to the Brownian motion of the particles [54]. The distribution of the nanomaterial sizes while dispersed in the nanofluid usually reflects the homogeneity of the nanomaterial distribution in the nanofluid and any agglomeration tendency that exists.

Several parameters could be used to evaluate the stability of nanofluids. pH value and zeta potential are used to evaluate the physical stability of nanofluids, while time-dependent visual inspection and properties degradation are used for the assessment of their chemical stability [55]. The zeta potential represents the potential difference between the dispersion medium and a stationary layer attached to a nanoparticle. Electrically stabilized nanofluids, with less tendency of agglomeration, are usually found with high absolute zeta potential values [46]. Suspensions with a zeta potential of >60 mV, 30 mV–60 mV, <20 mV, and <5 mV reflect excellent, physical, limited, and low stability conditions, respectively [56]. Homogeneous nanoparticles' distribution, high zeta potentials, and lower pH values, with negligible degradation in the visual inspection and nanofluid properties, altogether is a strong sign of a stable nanofluid suspension.

2.2.3 Thermal Properties

Thermal conductivity and specific heat are the main thermal properties of concern in the characterization of nanofluids because of their importance as a sign of the improved thermal behavior of nanofluids compared to basefluids. Different types of equipment were found as capable of measuring thermal conductivity, however, Differential Scanning Calorimetry (DSC) is the equipment that is normally used for specific heat measurements. On the other side, the Thermogravimetric analysis (TGA) is used to evaluate the thermal stability, decomposition, and purity of the nanomaterial before dispersing it into the nanofluids [57, 58]. Very high temperatures are applied to the nanomaterials and the weight degradation with temperature, due to evaporation, is monitored.

2.2.4 Optical Properties

As our main concern in this chapter is to discuss the optical behavior of metal-oxide-based nanofluids, we are going to dig more into the optical properties measurements.

The nanofluid optical properties are measured using Ultraviolet-Visible (UV-Vis) Spectroscopy equipment. UV-Vis Spectroscopy (or Spectrophotometry) is a quantitative method for measuring the amount of light absorbed by a chemical compound. This is accomplished by comparing the intensity of light passing through a sample to the intensity of light passing through a standard sample or blank. As a function of wavelength, it quantifies the absorption/transmission or reflection of a substance. Despite being referred to as UV-Vis, the commonly employed wavelength range extends from 190 to 1100 nm in the near-infrared (NIR) [59].

In the following couple of sections, we are going to put more focus on the optical behavior of the metal-oxide-based nanofluids in terms of their absorptance and transmittance capabilities and applications in different solar systems. Recent research achievements are going to be discussed in a short, interesting, and informative way.

3 Parameters Affecting Optical Behavior of Metal-Oxide-Based Nanofluids

Nanofluids interact with the incident radiations by either absorption, transmittance, or reflection. Therefore, several nanomaterial-based parameters affect the optical behavior of the nanofluids. From these parameters: the base fluid and nanomaterial types, nanomaterial size, shape, concentration, and stability in the nanofluid solution. All the aforementioned parameters formulate the shape of the nanomaterials content in the nanofluid solution, which in turn affects its optical behavior.

3.1 Nanomaterial Type

Dispersing different nanomaterial **types**, separately, in the same base fluid results in nanofluid samples that are different in optical and thermophysical behavior [6–8]. These differentiations come as a result of different interactions between the nanomaterials and base fluids on the molecular level, which affects their dispersibility in the fluid. In addition, each single nanomaterial type has its optical imprint, which is integrated with the optical imprint of the base fluid to come up with a nanofluid with a new optical behavior imprint.

3.2 Nanomaterial Size

The nanomaterial **size** means diameter, length, and/or surface area, which all formulate the volume of the nanoparticle that finally affects the optical behavior of the synthesis nanofluid [8–13]. The size if large or small affects the direct absorption of the incident radiation and the homogeneity of optical behavior across the nanofluid. Assume that 2 mg of a metal-oxide-based nanomaterial has been dispersed in two cups of water. One time, it was dispersed as one large nanoparticle weighing 2 mg, while for the other cup, it was powdered into thousands of smaller nanoparticles. Logically, the distribution in the 2nd case will be better and the optical impact of the nanomaterial will be equally distributed among the fluid, which will not happen in the 1st case. Figure 1a, b illustrate how the nanomaterial type and size affect the optical behavior of different nanofluids.

3.3 Nanomaterial Shape

The **shape** plays another important role that influences the optical behavior of the nanofluid [8–10, 13, 14]. The nanomaterial might be circular, tubular, or sheet shape. Similar to the size, the nanomaterial shape affects the homogeneity of the optical behavior as well as, sometimes, the stability of the nanofluid. Specific shapes for some nanomaterials have been proven as of high probability to result in better suspension for the nanomaterials in the fluid that affects the current and long-run optical behavior of the system.

3.4 Nanomaterial Concentration

Without a doubt, the nanomaterials' **concentration** is one of the key parameters that affect the optical behavior of the constituting nanofluid as demonstrated in different studies [8, 9, 15–25]. Higher concentration means more opaque solid particles that are suspended in the nanofluid, which in consequence decreases the transmissivity of the nanofluid and increases its absorptivity. The rate of change of the optical behavior with concentration is not fixed but is dependent mainly on the nanomaterial type and solution stability. Figure 1c, d illustrate how the nanomaterial concentration affects the transmittance of nanofluids.

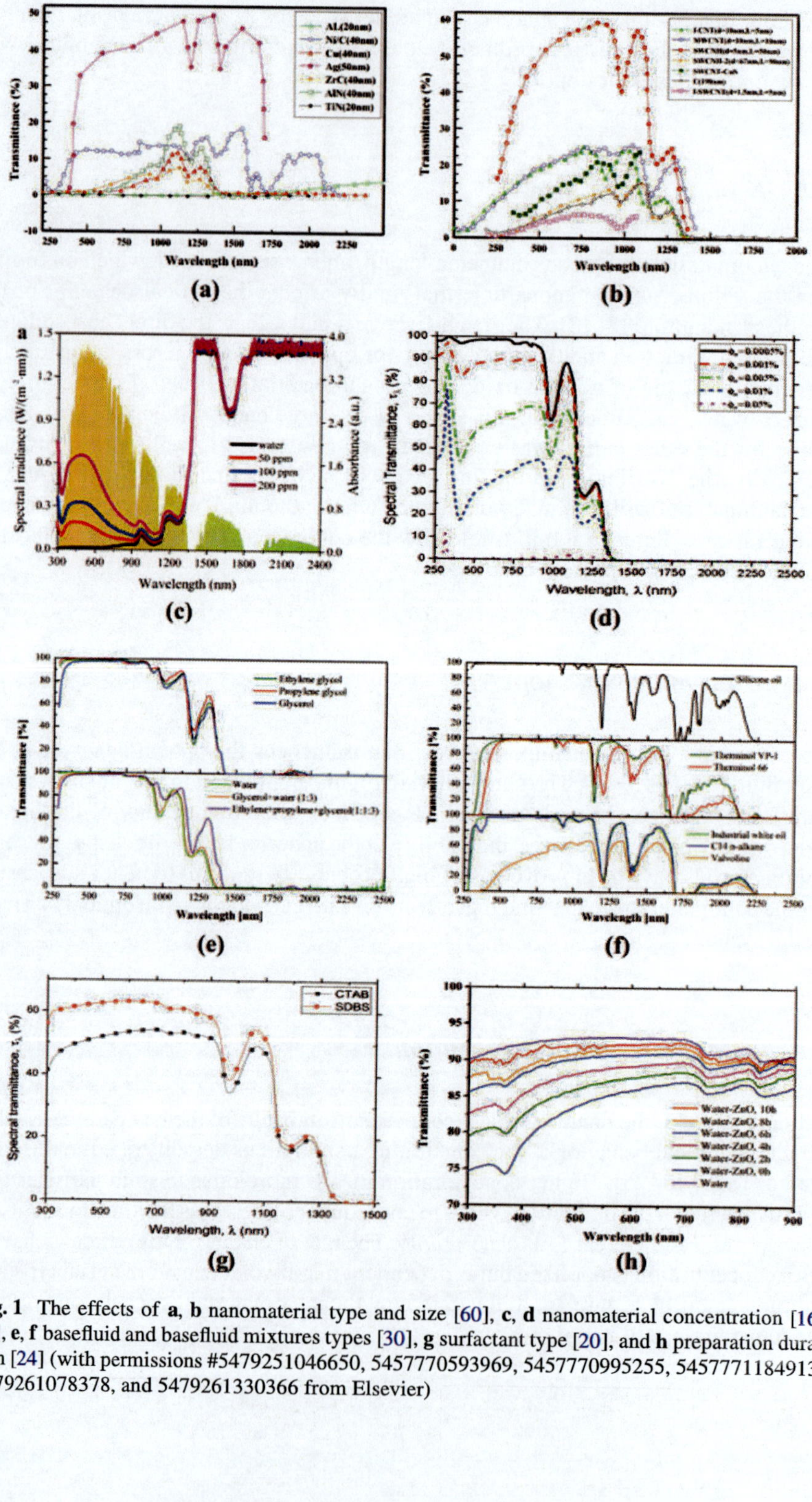

Fig. 1 The effects of **a**, **b** nanomaterial type and size [60], **c**, **d** nanomaterial concentration [16, 22], **e**, **f** basefluid and basefluid mixtures types [30], **g** surfactant type [20], and **h** preparation duration [24] (with permissions #5479251046650, 5457770593969, 5457770995255, 5457771184913, 5479261078378, and 5479261330366 from Elsevier)

3.5 Basefluid Type

Additionally, the **basefluid (or basefluid mixture) type** is one of the major parameters that affect the optical behavior of nanofluids [8, 23–30]. Each basefluid has its spectral optical properties that in case of integration with a specific nanomaterial type mush affects the resulting nanofluid optical behavior. The spectral transmittances of several basefluids and basefluid mixtures are depicted in Fig. 1e, f.

3.6 Surfactant Type

Surfactants are usually added to the nanofluid mixtures, prepared using the two-step method, for the sake of increasing the suspension stability. As a constituent, in this case, its optical properties that vary based on the surfactant type contributes also to the final nanofluid optical shape [8, 20, 31]. The influence of different surfactants were examined in the literature, such as the sodium dodecylbenzenesulfonate (SDBS), cetyltrimethylammonium bromide (CTAB), polyvinylpyrrolidone (PVP), and gum arabic (GA) [20, 31]. Figure 1g illustrates how the SDBS and CTAB surfactants affect the optical behavior of nanofluids.

3.7 Nanofluid Stability

Last and not least, the nanofluid **stability**. It is considered as of playing a magic role in the optical behavior of nanofluids. Nanofluids with similar constituents (base fluid, nanomaterial type, size, shape, and concentration) may have different optical behaviors if their stability and the suspension of the nanomaterials in their solutions are different. Hence, sustaining stability became of high importance to achieve the expected optical behavior and unify it among similar nanofluid samples. Several parameters were found as affecting the stability of nanofluids, such as the synthesis method, pH value, ultrasonication and stirring periods [24], existence of surfactants [20, 31], and ionic length. The impact of the preparation period on the nanofluid transmittance is demonstrated in Fig. 1h, while Fig. 2 presents the apparent nanofluid stability using different surfactants and for several nanoparticles' concentrations, immediately after preparation and after 7 days from preparation.

The measurement of nanofluid properties is a straightforward way to evaluate its optical behavior. However, as nanofluids are mixtures of liquids and nanomaterials, which can be considered as simply a two-phase fluid, different correlations were proposed to evaluate its optical properties as explained in the next section.

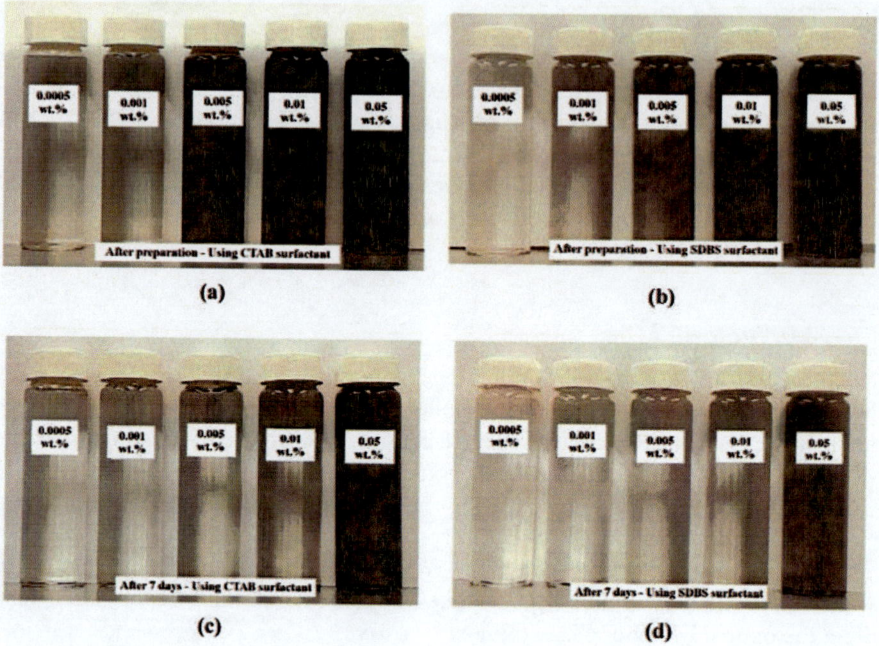

Fig. 2 Effect of **a**, **c** CTAB and **b**, **d** SDBS surfactants on the apparent stability of nanofluid at different concentrations and examined, **a**, **b** immediately after preparation and **c**, **d** after 7 days [20] (with permission 5480651210006 from Elsevier)

4 Evaluation of the Optical Properties of Nanofluid

This section provides the mathematical equations, based on the Rayleigh method, that are implemented to analytically evaluate the transmittance and absorptance of nanofluids.

The dimensionless equation below defines the nanoparticle size parameter:

$$\alpha = \frac{\pi d_{np}}{\lambda} \tag{1}$$

where, d_{np} is the diameter of the nanoparticle in nm, and λ is the wavelength in nm.

The incoming radiation intensity is calculated using the following equation:

$$\frac{\partial I_\lambda}{\partial y} = -(K_{of,\lambda} + \sigma_{of,\lambda})I_\lambda = -\beta_{of,\lambda} I_\lambda \tag{2}$$

where, I_λ is the spectral solar irradiance in W/m^2 nm, $K_{\text{of},\lambda}$, $\sigma_{\text{of},\lambda}$ and $\beta_{\text{of},\lambda}$ are the nanofluids' linear absorption, scattering, and extinction coefficients, respectively.

Due to the fact that they are independent, the extinction coefficient of the nanofluid is the total of the absorption coefficient of the base fluid and the extinction coefficient of the nanoparticles (Eq. 3) [61].

$$\beta_{of,\lambda} = \beta_{bf,\lambda} + \beta_{np,\lambda} \tag{3}$$

where, $\beta_{bf,\lambda}$ and $\beta_{np,\lambda}$ are the extinction coefficients of the base fluid and nanofluid, respectively, evaluated using the equations from Eqs. 4 to 11.

For the extinction coefficient of the base fluid:

$$\beta_{bf,\lambda} = K_{bf,\lambda} + \sigma_{bf,\lambda} \tag{4}$$

where, $K_{bf,\lambda}$ is the linear absorption of base fluid estimated as follows [62]:

$$K_{bf,\lambda} = \frac{4\pi k_{bf,\lambda}}{\lambda} \tag{5}$$

where, $k_{bf,\lambda}$ is the imaginary part of the refractive index of water.

Equation 6 [63] defines the extinction coefficient of nanoparticles as the sum of the linear absorption and scattering coefficients of the nanoparticles.

$$\beta_{np,\lambda} = \frac{3}{2}\frac{\varnothing Q_{ext,\lambda}}{d_{np}} \tag{6}$$

where, $Q_{ext,\lambda}$ is the extinction effectiveness of the nanoparticles as derived by Eq. 7 [64] using the Mie theory.

$$Q_{ext,\lambda} = Q_{abs,\lambda} + Q_{scat,\lambda} \tag{7}$$

where, $Q_{abs,\lambda}$ and $Q_{scat,\lambda}$ are the absorption and scattering efficiency of the nanoparticles, which can be determined using Rayleigh's approximation equations as stated in a number of classic texts by Eqs. 8 and 9.

$$Q_{abs,\lambda} = 4\alpha * Im\left\{\frac{m^2-1}{m^2+2}\left[1+\frac{\alpha^2}{15}\left(\frac{m^2-1}{m^2+2}\right)\times\frac{m^4+27m^2+38}{2m^2+3}\right]\right\} \tag{8}$$

where *Im* {} refers to the imaginary part for the number between brackets.

$$Q_{scat,\lambda} = \frac{8}{3}\alpha^4\left|\frac{m^2-1}{m^2+2}\right|^2 \tag{9}$$

In this equation, m represents the relative complex refractive index of the nanofluid, which may be determined by applying the following connection to the formula:

$$m = \frac{m_{np}}{n_{bf}} \tag{10}$$

where, n_{bf} is the real component of the refractive index of the base fluid and m_{np} is the complex refractive index of the nanoparticles calculated by Eq. 11.

$$m_{np} = n_{np} + ik_{np} \quad (11)$$

where, n_{np} and k_{np} are the real and imaginary components of the refractive index of the nanoparticles.

Following this step, the extinction coefficient of the nanofluid is computed, and the spectral transmittance of the nanofluid is obtained by applying the Beer-Lambert–Bouguer rule in the manner described in [65]:

$$\tau_{nf,\lambda} = \frac{I_\lambda}{I_{0,\lambda}} = e^{-h_{nf}\beta_{nf,\lambda}} \quad (12)$$

where I_λ is the transmitted irradiation, $I_{0,\lambda}$ is the incident irradiation (AM1.5 Global [ASTM G-173]), and h_{nf} is the nanofluid thickness, penetration length, or optical path length of the nanofluid.

The total transmittance of the nanofluid-based is then calculated using Eq. 13.

$$\tau_{nf} = \frac{I}{I_0} = \frac{\int_{0.28\mu m}^{2.5\mu m} \tau_{of,\lambda} I_{0,\lambda} d\lambda}{\int_{0.28\mu m}^{2.5\mu m} I_{0,\lambda} d\lambda} \quad (13)$$

As a result and neglecting the reflectance of the nanofluid, the total absorbance of the nanofluid is then calculated using Eq. 14.

$$\alpha_{nf} = 1 - \tau_{nf} \quad (14)$$

5 The Optical Behavior of Metal-Oxide-Based Nanofluids

Different properties evaluate the optical behavior of nanofluids and how it is influenced by the aforementioned parameters. The transmittance, absorptance, and extinction coefficient are considered the most commonly used parameters for this purpose, however, other several parameters are also used, such as refractive index, scattering coefficient, and illuminance.

The first application that attracted the researchers towards interest in studying the optical behavior of nanofluids was using them in the direct absorption of solar radiation. Therefore, the main focus was on studying the absorptance of nanofluids and how to increase it as a function of the different nanomaterial-base fluid individual properties and constituents. This in consequence led to more studies on the high-absorption-based carbon-based nanofluids. Few studies were conducted to evaluate the absorptance of non-carbon-based nanofluids. Recently, and with the release of a new direction toward using nanofluids in spectral-selective filtration of solar radiation, different studies were proposed to evaluate the optical behavior of metal and metal-oxide-based nanofluids.

For the metal-oxide-based nanofluids, the optical properties of Al_2O_3 and TiO_2-based nanofluids have been investigated by Leong et al. [31]. The authors used UV-Vis spectroscopy to measure both the transmittance and absorptance of nanofluids under variant nanomaterial and surfactant types, besides the pH of the prepared solutions. The measurements declared a high dependency for the transmittance of nanofluids on the examined parameters. Higher absorbtance was reported for the TiO_2 compared to the Al_2O_3-based nanofluids, but with lower nanofluid stability. Higher surfactant influence and nanofluid absorptance were observed at higher nanomaterial sizes. Yaseen et al. [66] reported a degradation in the refractive index and transmittance of cerium oxide (CeO_2)-based nanofluid after 7 days from preparation. Higher absorption was also reported at higher nanomaterial sizes.

Karami et al. [67] observed higher absorption with the dispersion of CuO into a DW-EG (70:30) basefluid mixture compared to the parent case without nanomaterial. Lower transmittance and higher extinction coefficients were achieved at higher CuO concentrations. In addition, thicker layers resulted in more energy absorption. Applying the theoretical approaches of Rayleigh, Maxwell-Garnett, and Lambert-Beer, besides the experimental measurements, Said et al. [6] evaluated the optical behavior of Al_2O_3/DW and TiO_2/DW nanofluids. Higher extinction coefficients and refractive indexes were generally recorded for the titania nanofluid compared to the alumina nanofluid. It was concluded that in terms of solar absorption, titania nanofluid will be the best choice.

Milanese et al. [68] conducted an intensive study on the optical behavior of six different metal-oxide-based nanofluids including Al_2O_3, CuO, TiO, ZnO, CeO_2, and Fe_2O_3 nanomaterials at different concentrations. Different optical behaviors have been observed depending on the nanomaterial type as presented in Fig. 3a, b). The highest and lowest solar absorption percentages were achieved using the Fe_2O_3 and Al_2O_3-based nanofluids and the difference was more distinguished at lower concentrations. On the other side, the spectral transmittances of the six nanofluid groups illustrated in Fig. 3c–h declared that the concentration negatively impacted the nanofluid transparency.

Crisostomo et al. [10] examined the core-shell Ag–SiO_2/water nanofluid at different concentrations of 0.006–0.025 wt.%. Increasing the concentration resulted in more energy absorption. The absorption was more pronounced in the VIS range (Fig. 4a). Negligible changes were observed above 950 nm. Li et al. [22] evaluated the optical behavior of Ag–TiO_2/water nanofluid. The measurements uncovered a high absorption coefficient in the visible range, with high absorptance at higher nanomaterial concentrations (Fig. 4b). More energy from the incident radiation could be absorbed using high-concentration nanofluid samples (Fig. 4c).

The optical behavior of magnetic electrolytes, a recently developed nanofluid, was assessed by Jin and Jing [12]. Fe_3O_4 was dispersed in a water-EG mixture (1:1). Two nanofluids (NF1 and NF2) were produced by modifying the aforementioned mixture with either methylene blue (MB) or copper sulfate (CS), which caused the difference

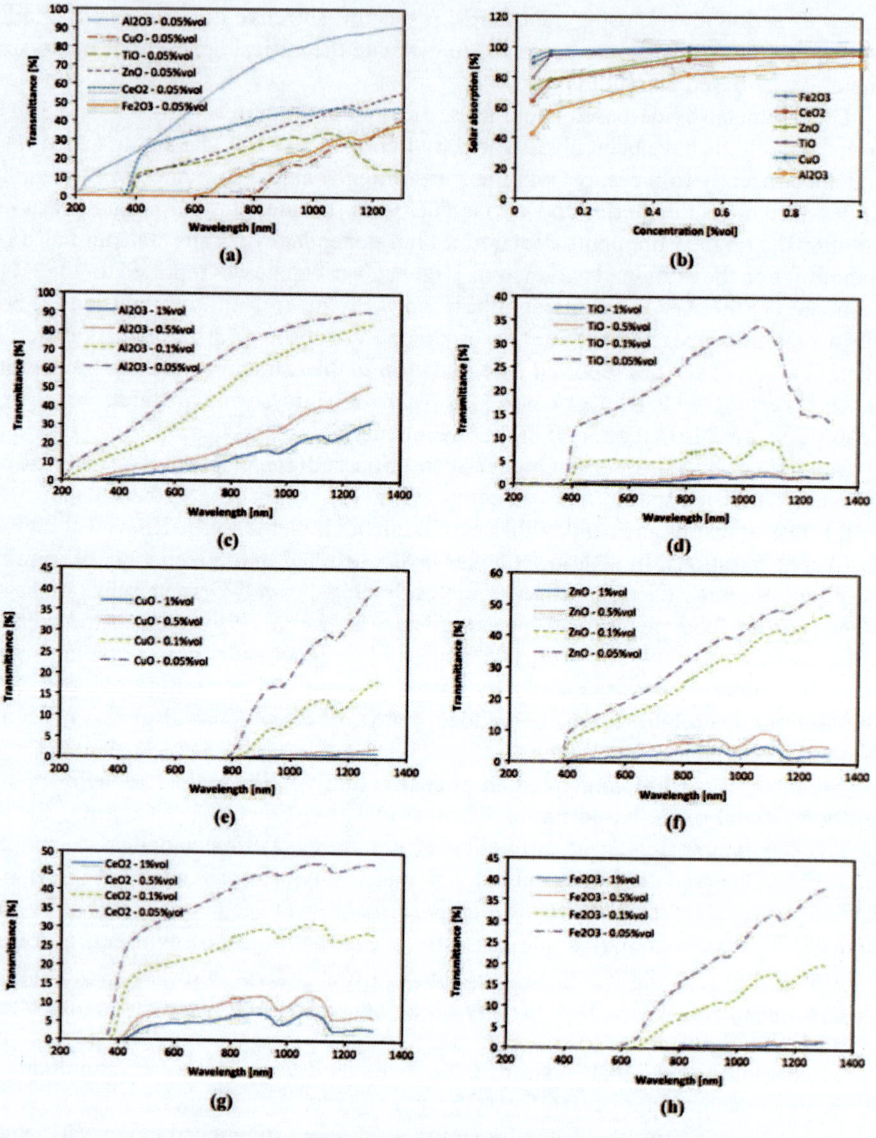

Fig. 3 **a** Spectral transmittance at 0.05 vol.% and **b** Solar absorption at different concentrations, of six metal-oxide-based nanofluids [68], and Spectral transmittance at different concentrations of **c** Al_2O_3, **d** TiO, **e** CuO, **f** ZnO, **g** CeO_2, and **h** Fe_2O_3-based nanofluids [68] (with permission #5457780471082 from Elsevier)

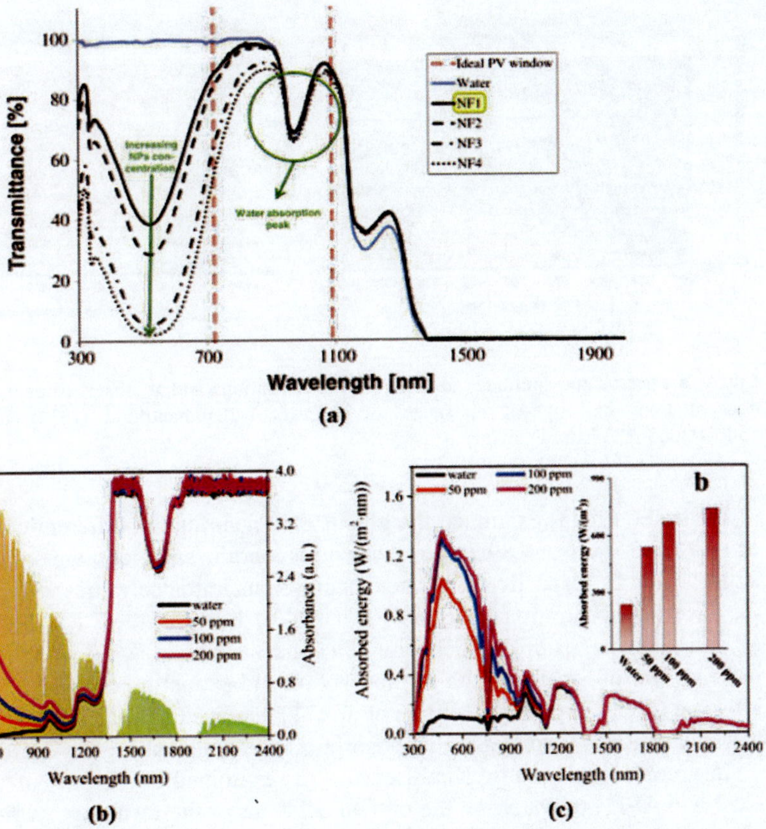

Fig. 4 **a** The optical characteristics of the Ag–Si core-shell nanofluids [10], **b** Absorbance, and **c** the absorbed energy of beam splitters [22] (with permissions # 5458120446083 & 5458130745477 from Elsevier)

in their absorptance schemes. Few negative changes in absorptance were reported as well with time. The absorptances of the synthesized nanofluids were found as negatively affected by the increase in their temperature and the corresponding heating periods.

Research on the stability and optical characteristics of water/SiO_2 nanofluid at extremely high temperatures has been conducted by Adam et al. [69]. Increasing the sonication temperature was negatively affecting the nanofluid absorptance. The nanoparticles had a strong propensity to agglomerate at higher exposure temperatures. On the other side, a little reduction in the spectral absorptance, with time, could be noticed in the VIS range after the preparation. It was revealed that only at very short wavelengths there were discernible differences in transmittance across the spectrum as a result of concentration increase. Additionally, between 25 and 90 °C of exposure, the nanofluid's absorption coefficient decreased.

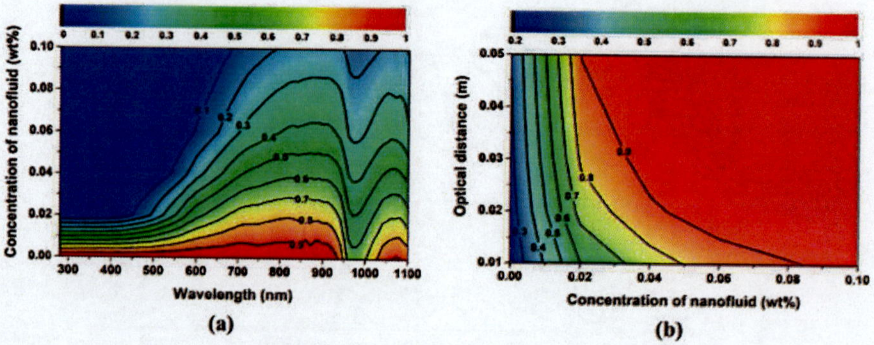

Fig. 5 **a** Spectral transmittance at different concentrations and **b** Absorbed energy at different concentrations and optical distances, of a Fe_3O_4/water nanofluid [73] (with permission #5460030056133 from Elsevier)

Huaxu et al. [23] examined the glycol/ZnO nanofluid at different concentrations of 11.2, 22.3, 44.6, and 89.2 ppm. The corresponding samples were named solutions A, B, C, and D, respectively. Higher spectral transmittance values were reported for the lowest-concentration solutions. Xu et al. [70] dispersed the CuO into the synthetic oil and evaluated its optical behavior at a concentration of 0.055 wt.%. A significant increase was observed in both the absorption and extinction coefficients of the fluid after the CuO dispersion. Menbari et al. [71] achieved similar results for the CuO/water nanofluid showing a higher absorption coefficient at higher CuO loadings.

In a comparable study, Khalil et al. [72] examined the impact of adding both CuO and Al_2O_3 to water, on the optical behavior of the resulting hybrid nanofluid. The measurements revealed a significant difference between the spectral absorptance distribution of the hybrid nanofluid, in comparison to the individual nanofluids. Ham et al. [73] examined the optical behavior of Fe_3O_4/water nanofluid at different concentrations and optical distances. The contours presented in Fig. 5 reveal lower spectral transmittance values and higher energy absorption at higher concentrations and optical distances.

6 Applications of the Optical Behavior of Metal-Oxide-Based Nanofluids

Two main applications are targeted by manipulating the optical behavior of nanofluids. The first is related to maximizing the absorption of incident solar radiation in direct absorption solar collectors (DASC). The second is related to allowing for a defined spectrum range from the incident solar radiation to reach the PV panel while absorbing the remaining spectrum inside the nanofluid itself and this is applied in PV-based hybrid solar systems.

In the following sections, the applications that benefit from the unique optical behavior of nano-enhanced fluids are discussed.

6.1 Direct Absorption Solar Collector (Absorptance Behavior)

The direct absorption solar collector (DASC) is a solar harvesting system that aims at converting solar energy into thermal energy through absorption of the incident solar radiation. Solar energy is usually absorbed in the operating fluid of the DASC and is used for several heat supplies. Increasing the absorption capability of the DASC is of utmost importance to increase the overall effectiveness of the system and this is the targeted impact of using nanofluids in this application.

Figure 6 illustrates the difference between the conventional absorption solar collector that was mainly depending on absorbing the incident radiation in the collector tube itself. High-absorption tubing materials with/without coating were the main parameters evaluating the functionality and solar-harvesting capabilities of the absorption solar collector (Fig. 6a). At that time, the nanofluids were of interest, but only with their thermal properties to increase the heat collection from the surface of the collector tubes. Later and with increasing interest in the optical behavior of nanofluids and that they can replace the high-absorption tubes and absorb the incident radiation directly, the conventional tubes have been replaced by glass tubes without coatings (Fig. 6b) and we started to have the DASC.

6.1.1 Experimental Efforts

Several experimental setups were established to examine the optical capabilities of nanofluids to suffice for the high absorption needs in direct absorption solar collectors (DASC). The experimental setups aimed at maximizing the solar energy absorption through the DASC by examining the deed capability of lab-synthesized nanofluids,

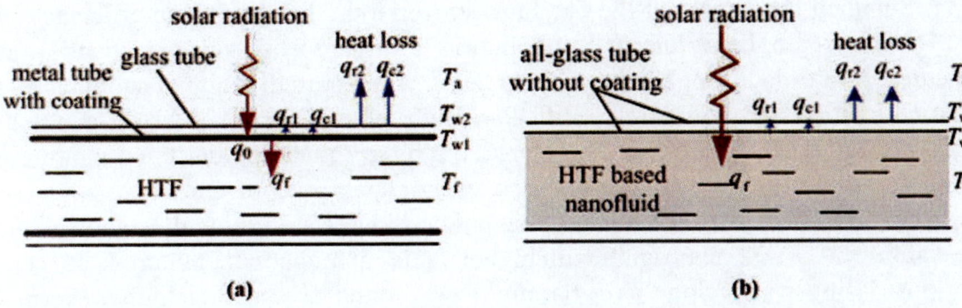

Fig. 6 **a** Old indirect absorption solar collector and **b** Direct absorption solar collector [70] (no permission is needed)

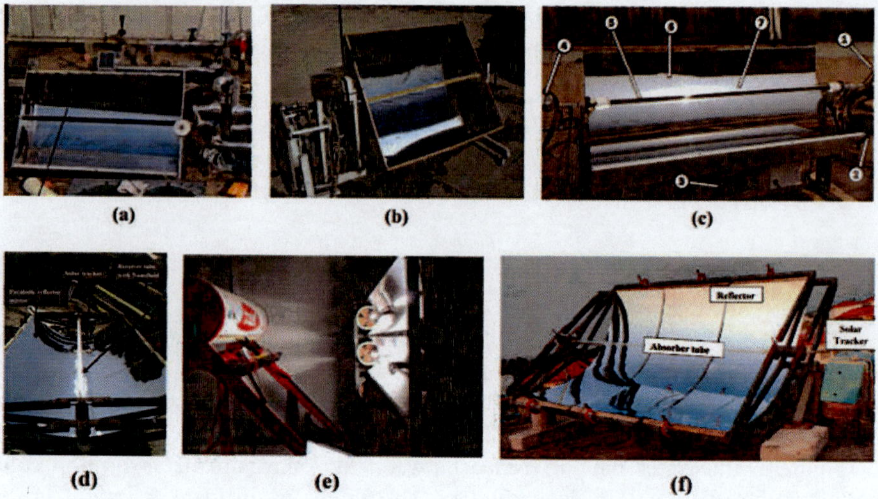

Fig. 7 Samples of the DASC experimental setups: **a** [70], **b** [71], **c** [74], **d** [75], **e** [76], and **f** [77] (with permission #5457780828691, 5457781041593, 5480671389889, 5457790097404, and 5457790235343 from Elsevier)

with unique absorption properties, to absorb the maximum spectrum of the incident solar radiation. Different nanofluids and system designs with and without concentrators were examined. Samples of these experimental setups as schematics or real systems are illustrated in Fig. 7.

Considering the high potential of nanomaterial addition in maximizing the optical absorption of the incident radiation on the DASCs, several experiments over years were conducted using various kinds of nanofluids. For sure, metal-oxide-based nanofluids have contributed extensively to these studies.

Xu et al. [70] have utilized the CuO/oil nanofluid as a novel nanofluid for solar absorption in a DASC with a parabolic trough concentrator. Better overall temperature distribution along the system's nanofluid was reported with higher absorption efficiency in comparison to the conventional distribution using the conventional indirect absorption solar collector. Mixing it with water instead of oil, Menbari et al. [71] examined the impact of the CuO nanomaterial on the absorption efficiency of the DASC. Again, better thermal distribution of the absorbed solar irradiance was detected. Additionally, higher thermal efficiency was reported; at higher nanomaterial volume fraction and flow rate specifically. With the addition of γ-Al_2O_3 nanomaterials to CuO, Menbari et al. [78] prepared a binary metal-oxide-based nanofluid to combine the high absorption and scattering capabilities of CuO and γ-Al_2O_3, respectively. Their finding revealed a significant increase in the collected sensible heat along the DASC pipe, again with higher values at higher nanomaterials concentrations. Using water alone as a basefluid was found as showing higher thermal efficiency compared to the case of a water-EG mixture. Similar results were reported for the DAPTC [71].

In a comparable study, Khalil et al. [72] examined the impact of adding both CuO and Al_2O_3 to water, to form a hybrid nanofluid, on the performance of a direct absorption parabolic trough collector (DAPTC). The CuO was selected for its high absorbtivity, while the Al_2O_3 was chosen for its high scattering capabilities. Using the developed hybrid nanofluid, the thermal efficiency increased to 31% compared to the basefluid condition, with a 19% difference between the DAPTC and the conventional one. Additionally, increasing the concentration of the nanomaterial content resulted in higher efficiency values. Gupta et al. [79] employed the Al_2O_3/water alone as the absorption nanofluid. In all the studied nanomaterial loadings, the results revealed efficiency enhancements, however, it was not correlated with the loading percentage. It peaked at certain loading and decreased for the other ones.

Kasaeian et al. [80] in their study, have examined the influence of several nanofluid combinations on the outlet fluid temperature and the DASC efficiency. Higher fluid temperatures were recorded at higher silica (SiO_2) concentrations when dispersed in Ethylene Glycol (EG). The temperature has increased by 2.2, 5.3, and 7.7 K at 0.1, 0.2, and 0.3% of SiO_2, respectively. The reported thermal efficiencies were ~19, 22.5, and 24.5% respectively, compared to 16.5% for the pure EG.

Evaluating the potential of metal-oxide-based hybrid nanofluid, Joseph et al. [77], prepared a SiO_2/Ag–CuO plasmonic nanofluid and studied how its optical properties affected the functionality of the DASC. A maximum enhancement in the thermal efficiency of 48.19%, compared to water fluid. Additionally, the exergy efficiency has increased by 9.4%, in comparison to water. Karami et al. [81] changed the basefluid and studied the influence of the CuO loadings on the optical and overall functionality of the DASC. A water-EG mixture (70:30) was used for this purpose. The existence of nanoparticles boosted the efficiency by 9–17% depending on the volume fraction and flow rate. Higher values were achieved at higher volume fractions.

Using different nanomaterials and basefluids, Subramaniyan and Rajangam [82] dispersed a TiO_2 and Carbon-doped TiO_2 (C–TiO_2) in water and EG, respectively. Carbon doping was performed to enhance the absorptivity of the nanomaterial. Higher solar absorption was achieved using the C–TiO_2/EG nanofluid, but with lower suspension stability compared to the TiO_2/water nanofluid. C–TiO_2 was recommended for utilization but at low volume fractions, to assure good stability and absorption altogether.

6.1.2 Numerical and Theoretical Efforts

Conducting experimental testing of nanofluids usually faces several challenges, starting from the availability of the materials, the preparation of enough samples, having these samples in high stability conditions, and ending with the difficulty of examining different structures of nanofluids, in terms of the nanomaterial concentrations, sizes, and shapes.

Therefore, using the numerical techniques was of high importance to have a better understanding of the potential of these new coming nanofluids, especially since they provide testing varieties, with some approximations. The main and commonly used approximation is examining the nanofluid' impact on different systems in the shape of a single-phase fluid, but with equivalent thermal, rheological, and optical properties.

Very few studies were conducted to examine the impact of the metal-oxide-based nanofluids, used for optical filtration, on the performance of DASCs. Kasaeian et al. [83] developed a mathematical model to evaluate the impact of SiO_2/EG nanofluid on the functionality of the DASC. They have reported the optimum nanomaterial concentration for high efficiency and less agglomeration of 0.4% with a thermal efficiency of 70.9%. Saray and Heyhat [84] proposed a numerical model for the DAPTC operated with either a CuO/water nanofluid or a CuO-MWCNT/water hybrid nanofluid. In comparison to several nanofluid solutions, the hybrid nanofluid at high CuO concentration provided the highest energy and exergy efficiencies (Fig. 8). The results indicated that using CuO/water nanofluid 30.8 GJ of embodied energy could be saved, while CuO-MWCNT/water hybrid nanofluid could save 40.44 GJ. This was attributed to the high thermal and absorption properties of the MWCNT.

The impacts of the size and concentration of TiO_2 on the extinction coefficient of a water-based nanofluid were examined by Said et al. [85] using the Rayleigh method. The authors aimed at assessing the factors affecting the increase in solar absorption. At TiO_2 sizes of less than 20 nm, a negligible impact on the optical properties was noticed, while with higher concentrations, more solar radiation could be absorbed.

Table 1 presents a summary of the experimental and numerical studies conducted on the DASCs operated by metal-oxide-based nanofluids, highlighting the examination conditions and the resulting impacts on temperature and efficiency, as provided by Sainz-Mañas [86].

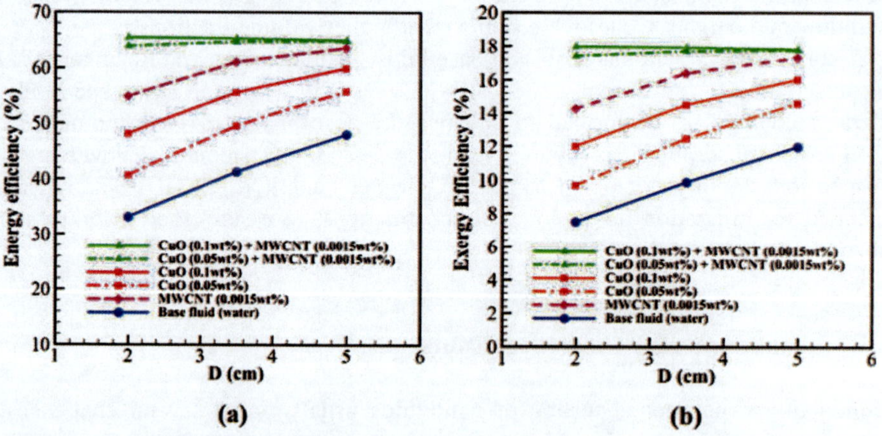

Fig. 8 Comparison between the impact of several nanofluids on the **a** energy and **b** exergy efficiencies of a DAPTC at different piping diameters [84] (with permission #5460030383680 from Elsevier)

Table 1 Summary of studies on DASCs with concentrators and operated by metal-oxide-based nanofluids [86]

Approach	Geometry	Irradiation	Depth	Nanofluid				Results		References
				Nanomaterial	Basefluid	Size	Concentration	Temperature	Efficiency (%)	
NUM	PTC—Tube receiver	5152	45 mm		WD SO			$\Delta T = 125$ °C	η_{col} = 42–52.5	[70]
				CuO	WD SO	D = 200 nm	0.055 wt.%	$\Delta T = 105$ °C	η_{col} = 40–62	
NUM	PTC—Tube receiver	4560	45 mm		WD-350 SO			T_{in} = 40–110 °C	η_{col} = 64	[87]
				CuO	WD-350 SO	D = 200 nm	0.05 wt.%	T_{in} = 40–110 °C	η_{col} = 63–70	
							0.06 wt.%	T_{in} = 40–110 °C	η_{col} = 66–72	
							0.075 wt.%	T_{in} = 40–110 °C	η_{col} = 65–72	
							0.1 wt.%	T_{in} = 40–110 °C	η_{col} = 64–71	
NUM/EXP	PTC—Tube receiver	~44600	20 mm	CuO	Water	D = 100 nm	0.002 vol.%	T_{out} = 28–36 °C	η_{col} = 28–33	[71]
							0.004 vol.%	T_{out} = 30–39 °C	η_{th} = 36–40	
							0.008 vol.%	T_{out} = 31–42 °C	η_{th} = 41–49	
NUM	PTC—Tube receiver	6531	26 mm		EG			T_{out} = 44–48 °C	η_{th} = 52.2–59.5	[83]
				SiO_2	EG	D = 6-15 nm	0.2–0.4 vol.%	T_{out} = 44–55 °C	η_{th} = 55.9–70.9	

(continued)

Table 1 (continued)

Approach	Geometry	Irradiation	Depth	Nanofluid				Results		References
				Nanomaterial	Basefluid	Size	Concentration	Temperature	Efficiency (%)	
EXP	PTC—Tube receiver	7464–9330	24 mm		EG			$T_{out} = 48.8$ °C	$\eta_{th} = 55.8$	[80]
				SiO_2	EG	D = 15 nm L = 10 μm	0.1–0.2–0.3 vol.%	$T_{out} = 38–59$ °C	$\eta_{th} =$ 58–60–64	
EXP	PTC—Tube receiver	~42500	20 mm	CuO + Al_2O_3	Water	CuO: D < 100 nm Al_2O_3: D < 40 nm	0.008 vol.% CuO + 0.2 vol.% Al_2O_3	$T_{out} = 38–59$ °C	$\eta_{th} =$ 36–48	[78]
					Water + EG			$T_{out} = 24–45$ °C	$\eta_{th} =$ 35–45	
					Water		0.004 vol.% CuO + 0.1 vol.% Al_2O_3	$T_{out} = 30–46$ °C	$\eta_{th} =$ 31–39	
					Water + EG			$T_{out} = 31–50$ °C	$\eta_{th} =$ 29–37	
					Water		0.002 vol.% CuO + 0.05 vol.% Al_2O_3	$T_{out} = 29–49$ °C	$\eta_{th} =$ 23–31	
					Water + EG			$T_{out} = 28–42$ °C	$\eta_{th} =$ 22–30	
EXP	CPC—Tube receiver	N/A	N/A	CuO	WD TO	L = 0.5–2 μm	0.06 mg/ml	N/A	$\eta_{col} = 68$	[76]
EXP	PTC—Tube receiver	25500–28500	22 mm	CuO	Water	D = 30 nm	0.01–0.1 vol.%	N/A	$\eta_{th} =$ 53.5–64.4	[74]

(continued)

Table 1 (continued)

Approach	Geometry	Irradiation	Depth	Nanofluid				Results		References
				Nanomaterial	Basefluid	Size	Concentration	Temperature	Efficiency (%)	
EXP	PTC—Tube receiver	~63000	15 mm	SiO_2/Ag–CuO	Water	SiO_2/Ag: D = 300 nm CuO: D < 50 nm	SiO_2/Ag: 0.21 g/l CuO: 0.86 g/l	ΔT = 10.8–8.4 °C	η_{th} = 55–64.1	[77]
EXP	PTC—Tube receiver	15000–19000	15 nm	Al_2O_3-CuO	Water	Al_2O_3: D < 100 nm CuO: D < 100 nm	0.11–0.55 vol.%	ΔT = 4–6 °C	η_{th} = 30.6–58.4	[72]
EXP	CPC—Tube receiver	1015	18 mm		Water			N/A	η_{th} = 42	[73]
				Fe_3O_4	Water	N/A	0.001–0.1 wt.%	N/A	η_{th} = 36–38	
NUM	PTC—Tube receiver	~28,700	20–50 mm		Water			N/A	η_{th} = 30–34	[84]
				CuO	Water	N/A	0.1–0.05 wt.%	N/A	η_{th} = 39–48	

6.2 Spectral-Selective Hybrid PV/Thermal (Transmittance Behavior)

Spectral-selective optical Filtration, often known as SSOF, is a novel method that was created to mitigate the negative effects of exposing photovoltaic (PV) cells directly to the full spectrum of solar radiation that is incident onto them. The optical filtering method is predicated on the capability of the filter, which is placed in front of the photovoltaic panel, to transmit a specific spectral range of the incident light while absorbing the remainder spectrum. This range corresponds to the range in which the cells are most capable of producing electricity. Spectrally selective optical filtration is used to absorb the solar radiation spectrum ranges that are not efficiently utilized by the PV cells, while the remaining spectrum is sent to the PV cells for the production of energy [13]. This allows the PV cells to more effectively utilize the solar radiation spectrum ranges that are not efficiently utilized by them. The optical properties of the filter are chosen so that it will transmit the solar radiation in the spectral range that will be utilized by the PV cells (visible and near-infrared), while the remaining solar radiation will be absorbed by the thermal receiver. This will result in the fluid being able to collect a greater quantity of heat. Several distinct methods of filtering, including those based on solids as well as liquids, have been suggested for the achievement of this goal.

Figure 9a, b illustrate a schematic illustration of an OF channel in conjunction with a typical PV system. On the other hand, it displays the expected interaction function that the OF channel should have in terms of attenuating the inefficient solar radiation spectrum. In Fig. 9c, we have a representation of the expected interaction participation of the OF channel in the poor absorption of solar energy. A wide variety of liquids, ranging from base fluids with traditional optical properties to nano-enhanced fluids (also known as nanofluids) with advanced optical properties, have been used for filtration purposes. In particular, nanofluids have a greater optical tenability [88–91] and more desirable thermal characteristics [92, 93] than pure fluids. In recent years, there has been a significant uptick in the number of technological endeavors that make use of nanomaterials. The use of nanomaterials to transmit and absorb light based on a selection criterion designed for a certain wavelength spectrum [10] is one of the most exciting new applications that has recently been developed. When nanoparticles are dispersed in a variety of base fluids, their optical properties can be more easily customized to get the required spectrum response of PV cells [10, 13]. This can be accomplished by tailoring the sizes and shapes of the nanomaterials.

It is possible to think of nanofluids as a low-cost alternative to spectrum-splitting mirrors due to the fact that they can work as both a cooler and an optical filter at the same time [25]. Again, nanofluids are considered to be powerful spectral beam splitters (optical filters) that have the capability of enhancing the functionality of hybrid PVT systems [21]. Nanofluid-based optical filters have been employed to provide efficient operation for standard PVT systems without causing any side effects [22]. This is possible due to the remarkably tuned optical properties that these filters possess. According to An et al. [94], the majority of the nanofluids that have been

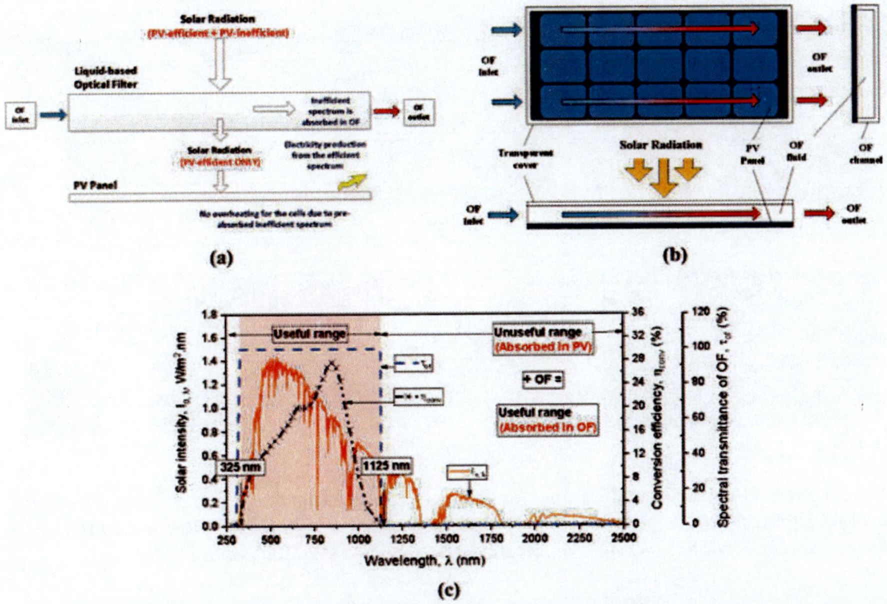

Fig. 9 Schematic diagrams for the liquid-based optical filtration (OF) arrangement attached to the conventional PV system: **a** Explanation, **b** Integration [8], and **c** The anticipated interaction effect of the OF channel in absorbing the inefficient solar radiation that is received by cSi PV cells [8] (with permission 5457790499320 from Elsevier)

produced for use in solar collectors have been utilized at moderate temperatures. However, it is possible to design additional nanofluids that can be employed at high temperatures. Finding the appropriate nanofluid that can transmit the spectrum that is theoretically expected while also matching the requirements of the PV cell continues to be a difficult task [12].

6.2.1 Experimental Efforts

In order to investigate the actual behavior of nanofluids for the purpose of spectral-splitting filtration, several different experimental settings were developed. The primary objective of the experimental designs was to enhance the performance of hybrid photovoltaic (PV) solar systems. This objective was to be accomplished by investigating the ability of lab-synthesized nanofluids, which have distinctive optical properties, to fully transmit the beneficial spectral range while absorbing the remaining portion of the light. Various designs for the channels, including single and double channels, with and without concentrators, an air gap or being directly coupled to the PV panel, and either still or moving filtration fluid. Figure 10 provides illustrations of several of these experimental settings, either in the form of schematics or actual systems.

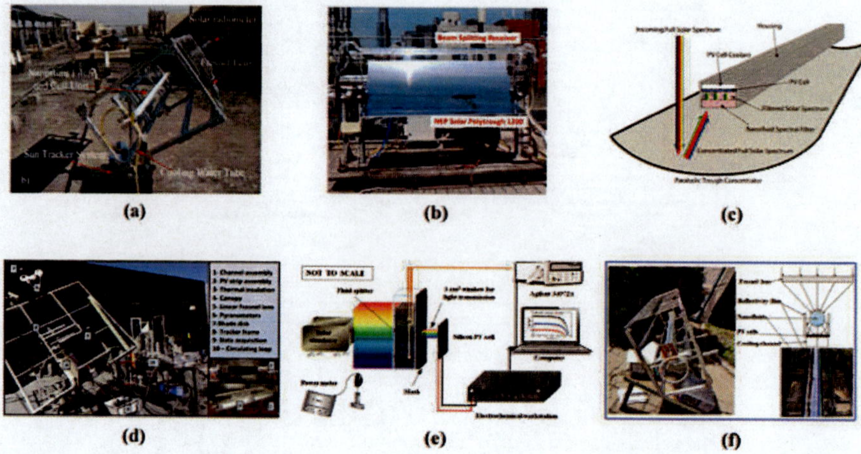

Fig. 10 Sort of experimental setups for the sake of investigating optical filtration: **a** [95], **b** [96], **c** [97], **d** [10], **e** [22], and **f** [23] (with permissions #5458110979702, 5458110614994, 5458111225464, 5458111405112, 5458120042511, and 5458120209542 from Elsevier)

Due to the high potential of nanomaterial addition in optimizing the optical behavior of the basefluids to fit the spectrally-selective needs of the optical filtration channel attached to the PV-based hybrid systems, a few experiments were carried out utilizing a variety of nanofluids. These experiments were successful in meeting the spectrally-selective requirements of the optical filtration channel. There is little doubt that nanofluids based on metal oxides have made a significant contribution to these investigations.

Experimental research was carried out by Hjerrild et al. [98] to investigate the effectiveness of a combined photovoltaic and thermal system that was also equipped with an optical filter. Both Ag–SiO_2/water and carbon nanotube/water were used to create core-shell nanofluids for the experiment. The rate of heating rose when carbon nanotubes were coupled with water nanofluid; nevertheless, this resulted in a decrease in the amount of electrical output. When a conventional water filter was substituted by an Ag–SiO_2/water nanofluid, there was an increase in performance across the board. The authors demonstrated that the increased costs associated with the nanoparticles were negligible (1 USD per liter).

Crisostomo et al. [10], who also employed core-shell Ag–SiO_2 dispersed in water, studied the effect that optical filtering had on the overall performance of a photovoltaic/thermal prototype system. Their findings showed that optical filtering had a negative influence. The tool was put through its paces in the natural environment it was designed for. The authors developed a hybrid configuration, which consisted of a thermal receiver channel and a PV module that were physically separated. The impacts of different nanoparticle loadings as well as different intensities of solar radiation were investigated. In comparison to the PV system that was operating on its own, this solution resulted in a 12% increase in the overall weighted energy output. The created nanofluid samples transmitted light with a spectrum that was almost exactly the same as what was required by the solar panel, with the exception of the infrared and short wavelength regions. It follows that utilizing the nanofluid samples in the settings that were tested resulted in a reduction in the output of electrical power while simultaneously increasing the output of thermal power and the overall efficiency of the system.

Jin and Jing [12] investigated the possibility of using the magnetic electrolyte, a nanofluid that was only recently produced, as an optical filter in a hybrid PVT system. The magnetic electrolyte nanofluid was created by dispersing Fe_3O_4 in a mixture of water and EG, in which EG contributed a share of 50%. The previously described mixture was modified with either methylene blue (MB) or copper sulfate, which resulted in the production of two different magnetic electrolyte nanofluids (CS). The ability of the created nanofluids to satisfy the spectrum requirements of cSi and GaAs solar cell materials was evaluated using these nanofluids. The spectral requirements of the ideal filter were more closely satisfied by the recently discovered nanofluids. It was discovered that an increase in the temperature of the newly optimized nanofluid filters, as well as the associated heating times, had a detrimental impact on the absorptance of the filters.

Experiments with nanofluid optical filtering were carried out by Hjerrild et al. [25] on three different types of PV cell materials: c-Si, GaAs, and Ge solar cells. The optical parameters of the fluid filter need to be tuned such that they are compatible with the spectrum responses of GaAs, Ge, and Si PV cells. The nanofluid was produced through the combination of nanoparticles. The mixture included gold and gold-copper nanorods that had silica coatings on them, as well as silver-silica (Ag–SiO_2) nanocores that were encased in silica shells. Because it can take in both visible and infrared rays, we decided to go with it. In order to get the desired spectrum for each cell, the researchers used a variety of loading percentages when dispersing nanoparticles in either water or Multi-Therm PG-1 oil. The most inefficient cells were those that used a Multi-Therm PG-1-based fluid with Ge PV. The water/Ag–SiO_2 nanofluid and the c-Si PV cells had the highest degree of productivity. Because of its relatively low price and high spectrum-splitting efficiency, this hybrid PV-nanofluid system produced some encouraging results (optical filtration).

For the purpose of optical filtration, Huaxu et al. [23] have introduced a nanofluid made of glycol and ZnO that is relatively affordable. The authors have generated glycol/ZnO nanofluids with varied amounts of nanoparticles in each fluid. The decrease in transmittance that occurred at higher concentrations brought in a reduction in the amount of irradiance and electrical output that was transferred (Fig. 11a). When the loading concentration of ZnO nanoparticles was increased from 11.2 to 89.2 ppm, there was a 47% increase in the photothermal efficiency (Fig. 11b). In general, the system was very effective at larger concentrations, despite the fact that it required a less amount of electrical contribution (Fig. 11c). When compared to the cost of gold, silver, and polypyrrole nanoparticles, ZnO was found to be an extremely cost-effective alternative.

According to Li et al.'s research [22], the presence of Ag–TiO_2 in water functions as an optical filter when PVT systems are also present. Because of its high selective absorptivity, Ag–TiO_2 was chosen as the material of choice. An investigation into

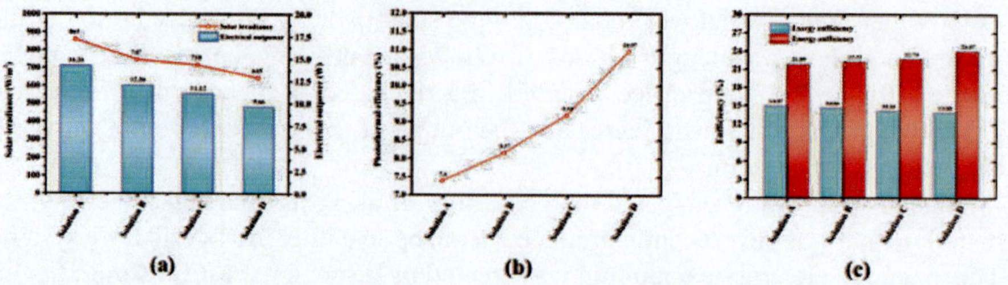

Fig. 11 **a** Solar irradiance and electrical outpower, **b** Photothermal efficiency, and **c** Efficiency of the system, under different glycol/ZnO solutions [23] (with permission #5460030729816 from Elsevier)

the optical performance of the nanofluid revealed a high transmittance coefficient in the region of the spectral spectrum where the Si PV cells were most electrically efficient. Figure 12a shows that increasing the nanoparticle loading led to improved photothermal conversion efficiencies but decreased the PV conversion efficiencies (0–200 ppm). It can be seen in Fig. 12b that the loading concentration of 100 ppm resulted in the best possible merit function being obtained. Because an increase in the nanoparticle loading led to an increase in the thermal energy efficiency but a decrease in the electrical efficiency, and vice versa, the loading can be controlled depending on how the hybrid system is being utilized.

Hashemian et al. [15] published their findings on a new nanofluid for optical filtering in 2022. The nanofluid was created by dispersing Ag–Cr_2O_3 nanoparticles in glycerol. The researchers believe that their recently developed nanofluid is unique in that it possesses the optimal transmittance spectrum, high nanofluid stability, and a low cost of manufacture all at the same time. In order to achieve the best possible spectrum response from the Si PV cells, it was necessary to tune the nanofluid for transmission between 650 and 1075 nm. The authors measured the overall exergy efficiency, merit function, and conversion efficiency of the nanofluid while testing its effectiveness as a filter by adjusting the loading on it from 10 to 80 ppm and changing the range over which it was evaluated. As can be observed in Fig. 12c, the temperatures of the PV devices got lower as the number of nanoparticles that were introduced got higher. The conversion efficiency reached its highest point with the 80 ppm loaded glycerol/Ag–Cr_2O_3 nanofluid and reached 31.55%, which is an increase from 13.25% without filtration. Figure 12d demonstrates that the nanofluid filter loaded with 40 ppm attained the highest merit function of 1.558 and the highest overall exergy efficiency of 41%.

6.2.2 Numerical and Theoretical Efforts

Conducting experimental testing of nanofluids usually faces several challenges, starting from the availability of the materials, the preparation of enough samples, having these samples in high stability conditions, and ending with the difficulty

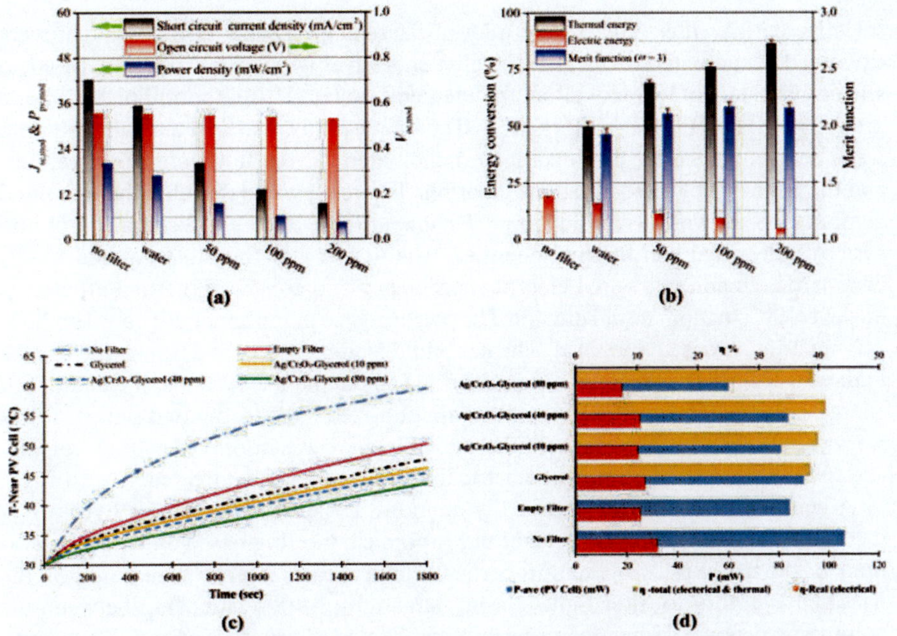

Fig. 12 **a** PV conversion results under different nanoparticle concentrations and **b** the power conversion efficiency and merit function without and with nanofluid splitters [22], and Comparison between the **c** temperature near the PV cells and **d** generated power and system efficiency, for different examined systems with and without optical filtration [15] (with permissions #5460030582280 & 5460030936113 from Elsevier)

of examining different structures of nanofluids, in terms of the nanomaterial concentrations, sizes, and shapes.

Therefore, using the numerical techniques was of high importance to have a better understanding of the potential of these new coming nanofluids, especially since they provide testing varieties, with some approximations. The main and commonly used approximation is examining the nanofluid' impact on different systems in the shape of a single-phase fluid, but with equivalent thermal, rheological, and optical properties.

Crisostomo et al. [10] evaluated the impact of core-shell Ag–SiO_2, distributed in water, optical filtration on the overall performance of a photovoltaic/thermal system. The high potential of this material was demonstrated in a number of different experimental studies, and the authors used a numerical technique that has been demonstrated to be effective in experimental settings. When the loading of nanoparticles was increased, the thermal performance of the material as a whole was improved. With higher loading, the samples might be able to absorb sunlight in the near visual area of the spectrum (300–700 nm). However, because of the excessive loads, the system's electrical output has been significantly reduced.

A numerical CFD analysis has been provided by Abd El-Samie et al. [17] in order to evaluate a hybrid PVT system that makes use of a nanofluid-based spectrally selective optical filter. There were a lot of nanomaterials and base fluids that were examined. Nanomaterials include the elements silver (Ag), iron oxide (Fe_3O_4), and silicon dioxide (SiO_2), as well as the elements water (H_2O), therminol VP-1, and propylene glycol (PG). The exergetic and energetic performance of the hybrid system were both impacted by the loading of nanomaterials, the flow rate of nanofluids, and the intensity of incident solar radiation. The results declared that the nanofluid combination as well as the loading of nanoparticles had an effect on the hybrid system's electrical and thermal energies. The figures illustrate that Ag and Fe_3O_4 nanoparticle loading lowered electrical efficiency and enhanced thermal efficiency, however, SiO_2 nanomaterial did not. The performance of the nanofluid based on SiO_2 was superior across all loadings. The nanofluid made of water and iron oxide has the highest thermal efficiency. In comparison to standalone PV systems, the amount of energy produced by hybrid PV systems that utilized nanofluid-based optical filters was between 79 and 140% higher. Thermal efficiency was improved by flow velocity and irradiation from the sun, but electrical efficiency was not affected.

Adam et al. [69] have theoretically analyzed the effectiveness of a hybrid PV/thermal system employing the nanofluid for optical filtering based on their observations of the nanofluid's optical properties. These measurements were based on the nanofluid's ability to filter light. An investigation of the relationship between the volume percentage of nanoparticles and performance was carried out. The examination of the optical filtering that was attached to the hybrid CPVT found that it promoted a greater overall performance at higher loadings while simultaneously reducing the amount of electrical sharing. The findings of the scientists indicate that the filter's absorbance can be increased by the addition of nanoparticles, and as a consequence, the filter's total efficiency can be improved in comparison to the filter designed for pure water.

Table 2 provides a summary of several studies that were conducted to evaluate the optical behavior impact of metal-oxide-based nanofluids on the performance of spectrally-selective optical filtration hybrid PV/thermal sstems.

7 Conclusions

Understanding the optical behavior of metal-oxide-based nanofluids and the crucial role of metal oxides in influencing the optical response of numerous nanofluids was advanced by this chapter. To maximize solar energy absorption from one side and act as a spectral-selective filter for solar radiation from the other side, this adaptable controlled optical behavior has garnered a lot of attention. This chapter covered a wide range of relevant topics, from the methods used to create nanofluids to the factors that affect their optical properties to how metal oxides' optical properties are affected, and finally to how these metal-oxide-based nanofluids are put to use in various solar-harvesting systems. Several salient conclusions can be extracted as follows, while more conclusions can be extracted from the chapter discussions:

Table 2 Summary of the experimental and numerical research articles published on liquid-based optical filtration

Approach	Filter fluid	Main findings	References
EXP	Magnetic electrolyte (Fe_3O_4/ 50%water + 50%EG) nanofluid	• Compared to water/core-shell nanofluids, the spectrum matches the ideal filter • Stable, thermally conducting nanofluids	[12]
NUM	Nanofluid	• Higher nanoparticle loadings increased thermal output • Nanofluid temperatures increased cut-on wavelengths • 1100 nm cutoff frequency • Nanofluid temperatures and sunlight concentrations increased efficiency	[94]
THE	Magnetic electrolyte (Fe_3O_4/ 50%water + 50%EG) nanofluid	• Higher merit functions than water/ core-shell nanofluids • Hybrid PV/T with OF system outperforms standalone PV system	[12]
NUM	Different mixtures (Water, Therminol VP-1, and Glycol) + (Ag, Fe_3O_4, and SiO_2)	• Nanofluid type affected performance • Ag and Fe_3O_4 loading lowered electrical efficiency but boosted thermal efficiency • SiO_2 loading had little effect • SiO_2 was the most electrically efficient nanomaterial • Water/Fe_3O_4 produced the most heat • Hybrid systems collected 79–140% more energy than standalone PV systems	[17]
EXP	Water/Ag–TiO_2	• High VR absorption • High electrically-efficient spectrum transmittance • Higher photothermal conversion efficiency, poorer electrical efficiency • High-loadings' thermal efficiency	[22]
EXP	$CoSO_4$/Ag	• Higher UV and V absorption than water-based nanofluids • Lower electrical output than water/ Ag nanofluid • Higher merit function than water/ Ag–SiO_2/Ag nanofluids	[21]

(continued)

Table 2 (continued)

Approach	Filter fluid	Main findings	References
EXP	Water/ZnO and Glycol/ZnO nanofluids	• Glycol/ZnO was more stable and homogeneous than water/ZnO • At wavelengths below 800 nm, nanoparticle diameter affected optical characteristics • Lowering nanoparticle diameter and loading increased Glycol/ZnO UV-Vis transmittance • Lower path lengths increased Glycol/ZnO transmittance • The optical path length worked better around 850–1400 nm	[24]
EXP	$(CoSO_4 + PG)/Ag$	• High absorption outside 325–670 nm • Highly efficient 615–970 nm transmittance • 500% more efficient than a PV system alone • 9% more efficient than $(CoSO_4 +$ water)/Ag	[97]
NUM	Ideal filter, water, Therminol 66, Valvoline, and Nanofluid	• Si and CdTe PV cells might absorb 27 and 39% of unnecessary incident radiation with a perfect filter • CdTe cells increased electrical efficiency	[98]

- Using the two-step method in nanofluid preparation resulted in a need to use surfactant for stability assuring. This impacted the optical behavior of nanofluids as an affecting constituent.
- Several parameters, such as the basefluid, nanomaterial, and surfactant types, nanomaterial size, shape, and concentration, besides nanofluid stability, play important roles in formulating the final optical shape of nanofluids.
- The optical behavior of nanofluids is an integration of the optical specifications of the constituting basefluid, nanomaterial, and surfactant (if any).
- Various basefluid and nanomaterial combinations showed varying degrees of filtration performance. They were, however, unified in their reduced transmittance and thus electrical output as compared to pure basefluid.
- Conclusions of optical behavior
- It has been demonstrated that lower nanomaterial loadings, smaller diameters, and shorter optical lengths create a more efficient electrical output (for SSOF). Nevertheless, the contrary was more successful from a thermal standpoint (for DASC).
- Nanofluids were initially proposed to improve solar absorption in different direction absorption solar systems and therefore, more focus was directed towards the carbon-based nanofluids, due to their high absorption capabilities. Few studies were conducted on metal-oxide-based nanofluids.

- The launch of the new application of nanofluid as a spectrally-selective optical filter provided a high potential solution to the unnecessary absorption of the out-of-PV-spectral-response solar radiation. The metal-oxide-based nanofluids were a challenging applicant for this target that has gained more research attention to evaluate the optical behavior of several PV-based solar systems with integrated spectrally-selective optical filtration channels placed on the front side of the PV panel.
- Among the DIW, IPA, IBA, DMSO, and GLY basefluids, DIW performed best for heat extraction and transmittance, while IPA/DIW and DMSO/DIW mixes performed best for SSOF.
- For SSOF, among the water-based nanofluids containing Ag, SiO_2, and Fe_3O_4, the Fe_3O_4 nanomaterial produced the maximum thermal output, whereas SiO_2 produced the highest electrical efficiency with negligible dependence on loading %.
- For SSOF, when the water/CNT and water/Ag–SiO_2 nanofluids were compared, the first produced a faster heating rate, while the second demonstrated higher electrical and overall efficiency. In general, the water/core-shell Ag–SiO_2 transmittance was quite close to the efficient spectral range of the cells.
- For SSOF, lower nanomaterial loadings, smaller diameters, and shorter optical lengths have all been shown to provide more efficient electrical output. However, from a thermal standpoint, the reverse proved more successful.

The SSOF fluid's side-cooling capabilities can compensate for the poor transmittance and, as a result, the low radiation received by the panel via extra-cooling of the PV cells.

8 Future Perspectives

However the optical behavior of nanofluids has been studied numerously, the optical behavior of metal-oxide-based nanofluids still has a lot to provide to the research and application community. Many research ideas and floors are still open in this regard. A few of these future research perspective ideas can be listed as follows:

- Doping the metal-oxide-based nanomaterials with carbons provides a high-potential solution for the lack of absorption in some spectral ranges of the metal-oxide-based nanofluids.
- The parameters impacting the optical behavior of metal-oxide-based nanofluids have not been optimized until now. Providing the optimized ranges of parameters' values will help in better controlling the optical behavior for various applications.
- More understanding of the optical behavior of metal-oxide-based nanofluids can be achieved using molecular dynamics simulations, which did not break into this topic to date.

- The application of spectrally-selective optical filtration using nanofluids, with PV-based systems is still a very hot topic. Balancing between reducing the PV temperature increase, increasing the heat absorbed in the filtration nanofluid, and assuring no obstruction of incident solar radiation, to maintain the electrical efficiency of the system, has not been achieved.

References

1. S.U.S. Choi, J.A. Eastman, Enhancing thermal conductivity of fluids with nanoparticles. ASME Int. Mech. Eng. Congr. Expo. **66**, 99–105 (1995). https://doi.org/10.1115/1.1532008
2. P. Keblinski, S.R. Phillpot, S.U.S. Choi, J.A. Eastman, Mechanisms of heat flow in suspensions of nano-sized particles (nanofluids). Int. J. Heat Mass Transf. **45**, 855–863 (2001). https://doi.org/10.1016/S0017-9310(01)00175-2
3. P. Keblinski, J.A. Eastman, D.G. Cahill, Nanofluids for thermal transport. Mater. Today **8**, 36–44 (2005). https://doi.org/10.1016/S1369-7021(05)70936-6
4. Choi SUS. Nanofluids: a new field of scientific research and innovative applications **29**, 429–431 (2011). https://doi.org/10.1080/01457630701850778
5. J.-K. Kim, J.Y. Jung, Y.T. Kang, The effect of nano-particles on the bubble absorption performance in a binary nanofluid. Int. J. Refrig. **29**, 22–29 (2006). https://doi.org/10.1016/j.ijrefrig.2005.08.006
6. Z. Said, R. Saidur, N.A. Rahim, Optical properties of metal oxides based nano fluids. Int. Commun. Heat Mass Transf. **59**, 46–54 (2014). https://doi.org/10.1016/j.icheatmasstransfer.2014.10.010
7. M.A. Alim, Z. Abdin, R. Saidur, A. Hepbasli, M.A. Khairul, N.A. Rahim, Analyses of entropy generation and pressure drop for a conventional flat plate solar collector using different types of metal oxide nanofluids. Energy Build. **66**, 289–296 (2013). https://doi.org/10.1016/j.enbuild.2013.07.027
8. A.S. Abdelrazik, The potential of liquid-based spectrally-selective optical filtration and its use in hybrid photovoltaic/thermal solar systems. Sol. Energy **249**, 569–605 (2023). https://doi.org/10.1016/J.SOLENER.2022.12.010
9. S.K.S.K. Verma, A.K.A.K. Tiwari, M. Tripathi, An evaluative observation on impact of optical properties of nanofluids in performance of photo-thermal concentrating systems. Sol. Energy **176**, 709–724 (2018). https://doi.org/10.1016/j.solener.2018.10.084
10. F. Crisostomo, N. Hjerrild, S. Mesgari, Q. Li, R.A. Taylor, A hybrid PV/T collector using spectrally selective absorbing nanofluids. Appl. Energy **193**, 1–14 (2017). https://doi.org/10.1016/j.apenergy.2017.02.028
11. D. Jing, Y. Hu, M. Liu, J. Wei, L. Guo, Preparation of highly dispersed nanofluid and CFD study of its utilization in a concentrating PV/T system. Sol. Energy **112**, 30–40 (2015). https://doi.org/10.1016/j.solener.2014.11.008
12. J. Jin, D. Jing, A novel liquid optical filter based on magnetic electrolyte nanofluids for hybrid photovoltaic/thermal solar collector application. Sol. Energy **155**, 51–61 (2017). https://doi.org/10.1016/j.solener.2017.06.030
13. N. Brekke, J. Dale, D. Dejarnette, P. Hari, M. Orosz, K. Roberts et al., Detailed performance model of a hybrid photovoltaic/thermal system utilizing selective spectral nano fluid absorption. Renew. Energy **123**, 683–693 (2018). https://doi.org/10.1016/j.renene.2018.01.025
14. M.U. Sajid, Y. Bicer, Nanofluids as solar spectrum splitters: a critical review. Sol. Energy **207**, 974–1001 (2020). https://doi.org/10.1016/j.solener.2020.07.009
15. M. Hashemian, S. Jafarmadar, S. Khalilarya, M. Faraji, Energy harvesting feasibility from photovoltaic/thermal (PV/T) hybrid system with Ag/Cr2O3-glycerol nanofluid optical filter. Renew. Energy **198**, 426–439 (2022). https://doi.org/10.1016/j.renene.2022.07.153

16. A.S.S. Abdelrazik, F.A.A. Al-Sulaiman, R. Saidur, Optical behavior of a water/silver nanofluid and their influence on the performance of a photovoltaic-thermal collector. Sol. Energy Mater. Sol. Cells **201**, 110054 (2019). https://doi.org/10.1016/j.solmat.2019.110054
17. M.M. Abd El-Samie, X. Ju, C. Xu, X. Du, Q. Zhu, Numerical study of a photovoltaic/thermal hybrid system with nanofluid based spectral beam filters. Energy Convers. Manag. **174**, 686–704 (2018). https://doi.org/10.1016/j.enconman.2018.08.083
18. W. An, J. Zhang, T. Zhu, N. Gao, Investigation on a spectral splitting photovoltaic/thermal hybrid system based on polypyrrole nanofluid: preliminary test. Renew. Energy **86**, 633–642 (2016). https://doi.org/10.1016/j.renene.2015.08.080
19. A.S.S. Abdelrazik, K.H.H. Tan, N. Aslfattahi, R. Saidur, F.A. Al-Sulaiman, F.A. Al-Sulaiman, Optical properties and stability of water-based nanofluids mixed with reduced graphene oxide decorated with silver and energy performance investigation in hybrid photovoltaic/thermal solar systems. Int. J. Energy Res. er.5770 (2020). https://doi.org/10.1002/er.5770
20. A.S. Abdelrazik, K.H. Tan, N. Aslfattahi, A. Arifutzzaman, R. Saidur, F.A. Al-sulaiman, Optical, stability and energy performance of water-based MXene nanofluids in hybrid PV/thermal solar systems. Sol. Energy **204**, 32–47 (2020). https://doi.org/10.1016/j.solener.2020.04.063
21. X. Han, X. Chen, Q. Wang, S.M. Alelyani, J. Qu, Investigation of CoSO4-based Ag nanofluids as spectral beam splitters for hybrid PV/T applications. Sol. Energy **177**, 387–394 (2019). https://doi.org/10.1016/j.solener.2018.11.037
22. H. Li, Y. He, C. Wang, X. Wang, Y. Hu, Tunable thermal and electricity generation enabled by spectrally selective absorption nanoparticles for photovoltaic/thermal applications. Appl. Energy **236**, 117–126 (2019). https://doi.org/10.1016/j.apenergy.2018.11.085
23. L. Huaxu, W. Fuqiang, Z. Dong, C. Ziming, Z. Chuanxin, L. Bo et al., Experimental investigation of cost-effective ZnO nanofluid based spectral splitting CPV/T system. Energy **194**, 116913 (2020). https://doi.org/10.1016/j.energy.2020.116913
24. L. Huaxu, W. Fuqiang, L. Dong, Z. Jie, T. Jianyu, Optical properties and transmittances of ZnO-containing nanofluids in spectral splitting photovoltaic/thermal systems. Int. J. Heat Mass Transf. **128**, 668–678 (2019). https://doi.org/10.1016/j.ijheatmasstransfer.2018.09.032
25. N.E. Hjerrild, F. Crisostomo, R.L. Chin, J.A. Scott, R. Amal, R.A. Taylor, Experimental results for tailored spectrum splitting metallic nanofluids for c-Si, GaAs, and Ge solar cells. IEEE J. Photovolt. **9**, 385–390 (2019). https://doi.org/10.1109/JPHOTOV.2018.2883626
26. X.-Y. Han, D.-S. Xue, Y.-J. Guo, J. Qu, Optical properties analysis of liquid based selective beam-splitting filter for concentrating PV/T beam splitting applications. K Cheng Je Wu Li Hsueh Pao/J. Eng. Thermophys. **38**, 2313–2319 (2017)
27. D. Chemisana, E.F. Fernandez, A. Riverola, A. Moreno, Fluid-based spectrally selective fi lters for direct immersed PVT solar systems in building applications. Renew. Energy **123**, 263–272 (2018). https://doi.org/10.1016/j.renene.2018.02.018
28. X. Han, D. Xue, X. Chen, Spectral characterisation and evaluation of liquid based beam splitters for CPV/T solar systems. Int. Heat Transf. Conf. (2018). https://doi.org/10.1615/ihtc16.nee.022475
29. X. Han, D. Xue, J. Zheng, S.M. Alelyani, X. Chen, Spectral characterization of spectrally selective liquid absorption fi lters and exploring their effects on concentrator solar cells. Renew. Energy **131**, 938–945 (2019). https://doi.org/10.1016/j.renene.2018.07.125
30. K.Z.K. Ahmad, H. Chyuan, Investigation on stability and optical properties of titanium dioxide and aluminum oxide water-based nanofluids. Int. J. Thermophys. (2017). https://doi.org/10.1007/s10765-017-2218-6
31. L. Chen, C. Xu, J. Liu, X. Fang, Z. Zhang, Optical absorption property and photo-thermal conversion performance of graphene oxide/water nanofluids with excellent dispersion stability. Sol. Energy **148**, 17–24 (2017). https://doi.org/10.1016/j.solener.2017.03.073
32. Ç. Demirkır, H. Ertürk, Rheological and thermal characterization of graphene-water nanofluids: hysteresis phenomenon. Int. J. Heat Mass Transf. **149** (2020). https://doi.org/10.1016/j.ijheatmasstransfer.2019.119113

33. W. Yu, H. Xie, X. Wang, X. Wang, Significant thermal conductivity enhancement for nanofluids containing graphene nanosheets. Phys. Lett. A **375**, 1323–1328 (2011). https://doi.org/10.1016/j.physleta.2011.01.040
34. M.Z. Saghir, A. Ahadi, A. Mohamad, S. Srinivasan, Water aluminum oxide nanofluid benchmark model. Int. J. Therm. Sci. **109**, 148–158 (2016). https://doi.org/10.1016/j.ijthermalsci.2016.06.002
35. R. Taherialekouhi, S. Rasouli, A. Khosravi, An experimental study on stability and thermal conductivity of water-graphene oxide/aluminum oxide nanoparticles as a cooling hybrid nanofluid. Int. J. Heat Mass Transf. **145** (2019). https://doi.org/10.1016/j.ijheatmasstransfer.2019.118751
36. R. Essajai, A. Mzerd, N. Hassanain, M. Qjani, Thermal conductivity enhancement of nano fluids composed of rod-shaped gold nanoparticles: insights from molecular dynamics. J. Mol. Liq. **293**, 111494 (2019). https://doi.org/10.1016/j.molliq.2019.111494
37. T.B. Gorji, A.A. Ranjbar, S.N. Mirzababaei, Optical properties of carboxyl functionalized carbon nanotube aqueous nanofluids as direct solar thermal energy absorbers. Sol. Energy **119**, 332–342 (2015). https://doi.org/10.1016/j.solener.2015.07.012
38. M. Simonetti, F. Restagno, E. Sani, M. Noussan, Numerical investigation of direct absorption solar collectors (DASC), based on carbon-nanohorn nano fluids, for low temperature applications. Sol. Energy **195**, 166–175 (2020). https://doi.org/10.1016/j.solener.2019.11.044
39. R.C. Shende, S. Ramaprabhu, Thermo-optical properties of partially unzipped multiwalled carbon nanotubes dispersed nano fl uids for direct absorption solar thermal energy systems. Sol. Energy Mater. Sol. Cells **157**, 117–125 (2016). https://doi.org/10.1016/j.solmat.2016.05.037
40. S. Delfani, M. Karami, M.A. Akhavan-Behabadi, Performance characteristics of a residential-type direct absorption solar collector using MWCNT nanofluid. Renew. Energy **87**, 754–764 (2016). https://doi.org/10.1016/j.renene.2015.11.004
41. E. Sani, L. Mercatelli, S. Barison, C. Pagura, F. Agresti, L. Colla et al., Potential of carbon nanohorn-based suspensions for solar thermal collectors. Sol. Energy Mater. Sol. Cells **95**, 2994–3000 (2011). https://doi.org/10.1016/j.solmat.2011.06.011
42. R.M. Cheriyan, I. Anila, M.J. Mathew, Characterization and magnetic phase resolution of CoFe2O4 nanocubes and nanospheres. AIP Conf. Proc. **2162**, 79–87 (2019). https://doi.org/10.1063/1.5130308
43. B. Ma, D. Shin, D. Banerjee, One-step synthesis of molten salt nanofluid for thermal energy storage application—A comprehensive analysis on thermophysical property, corrosion behavior, and economic benefit. J. Energy Storage **35**, 102278 (2021). https://doi.org/10.1016/j.est.2021.102278
44. A. Abadeh, M. Passandideh-Fard, M.J. Maghrebi, M. Mohammadi, Stability and magnetization of Fe3O4/water nanofluid preparation characteristics using Taguchi method. J. Therm. Anal. Calorim. **135**, 1323–1334 (2019). https://doi.org/10.1007/s10973-018-7662-4
45. A.K. Tiwari, N.S. Pandya, Z. Said, H.F. Öztop, N. Abu-Hamdeh, 4S consideration (synthesis, sonication, surfactant, stability) for the thermal conductivity of CeO2 with MWCNT and water based hybrid nanofluid: an experimental assessment. Colloids Surf. A Physicochem. Eng. Asp. **610** (2021). https://doi.org/10.1016/j.colsurfa.2020.125918
46. J.H. Gu, L. Long, Z. Lu, Z.Y. Zhong, Optical, electrical and structural properties of aluminum-doped nano-zinc oxide thin films deposited by magnetron sputtering. J. Mater. Sci. Mater. Electron. **26**, 734–741 (2015). https://doi.org/10.1007/s10854-014-2457-2
47. S. Ambika, S. Gopinath, K. Saravanan, K. Sivakumar, C. Ragupathi, T.A. Sukantha, Structural, morphological and optical properties and solar cell applications of thioglycolic routed nano cobalt oxide material. Energy Rep. **5**, 305–309 (2019). https://doi.org/10.1016/j.egyr.2019.02.005
48. N. Gupta, A. Kumar, H. Dhasmana, A. Kumar, A. Verma, P. Shukla et al., Effect of shape and size of carbon materials on the thermophysical properties of magnesium nitrate hexahydrate for solar thermal energy storage applications. J. Energy Storage **41**, 102899 (2021). https://doi.org/10.1016/j.est.2021.102899

49. U.S. Shenoy, A.N. Shetty, Simple glucose reduction route for one-step synthesis of copper nanofluids. Appl. Nanosci. **4**, 47–54 (2014). https://doi.org/10.1007/s13204-012-0169-6
50. S. Chen, X. Zeng, Q. Yuan, Effect of carbon nanotube concentration on cooling behaviors of oil-based nanofluids during the immersion quenching. J. Shanghai Jiaotong Univ. **22**, 395–401 (2017). https://doi.org/10.1007/s12204-017-1852-2
51. R. Kathiravan, R. Kumar, A. Gupta, R. Chandra, Preparation and pool boiling characteristics of copper nanofluids over a flat plate heater. Int. J. Heat Mass Transf. **53**, 1673–1681 (2010). https://doi.org/10.1016/j.ijheatmasstransfer.2010.01.022
52. S. Ponmani, J. Karen, M. William, R. Samuel, R. Nagarajan, J.S. Sangwai, Colloids and surfaces a: physicochemical and engineering aspects formation and characterization of thermal and electrical properties of CuO and ZnO nanofluids in xanthan gum. Colloids Surf. A Physicochem. Eng. Asp. **443**, 37–43 (2014). https://doi.org/10.1016/j.colsurfa.2013.10.048
53. R. Gómez-Villarejo, E.I. Martín, A. Sánchez-Coronilla, T. Aguilar, M. Teruel, R. Alcántara et al., Experimental characterization and theoretical modelling of Ag and Au-nanofluids: a comparative study of their thermal properties. J. Nanofluids **7**, 1059–1068 (2018). https://doi.org/10.1166/jon.2018.1544
54. R.H. Muller, *Zetapotential und Partikelladung in der Laborpraxis* (Wissenschaftliche Verlagsgesellschaft, Stuttgart, n.d.)
55. R.C. Shende, S. Ramaprabhu, Thermo-optical properties of partially unzipped multiwalled carbon nanotubes dispersed nanofluids for direct absorption solar thermal energy systems. Sol. Energy Mater. Sol. Cells **157**, 117–125 (2016). https://doi.org/10.1016/j.solmat.2016.05.037
56. Z. Qin, L. Tan, Z. Liu, S. Chen, J. Qin, J. Tang et al., One-pot synthesis of ultrafine TiO2 nanoparticles with enhanced thermal conductivity for nanofluid applications. Adv. Powder Technol. **27**, 299–304 (2016). https://doi.org/10.1016/j.apt.2015.12.018
57. https://www.edinst.com/techniques/uv-vis-spectroscopy/#:~:text=UV%2DVis%20Spectroscopy%20(or%20Spectrophotometry,a%20reference%20sample%20or%20blank (n.d.). Accessed 29th Dec. 2022
58. S.H.A. Ahmad, R. Saidur, I.M. Mahbubul, F.A. Al-Sulaiman, Optical properties of various nanofluids used in solar collector: a review. Renew. Sustain. Energy Rev. **73**, 1014–1030 (2017). https://doi.org/10.1016/j.rser.2017.01.173
59. R.A. Taylor, T. Otanicar, G. Rosengarten, Nanofluid-based optical filter optimization for PV/T systems 1–7 (2012). https://doi.org/10.1038/lsa.2012.34
60. H. Tyagi, P. Phelan, R. Prasher, Predicted efficiency of a low-temperature nanofluid-based direct absorption solar collector. J. Sol. Energy Eng. **131**, 041004 (2009). https://doi.org/10.1115/1.3197562
61. R.A. Taylor, T.P. Otanicar, Y. Herukerrupu, F. Bremond, G. Rosengarten, E.R. Hawkes et al., Feasibility of nanofluid-based optical filters. Appl. Opt. **52**, 1413–1422 (2013). https://doi.org/10.1364/AO.52.001413
62. C.F. Bohren, D.R. Huffman, *Absorption and Scattering of Light by Small Particles* (1998). https://doi.org/10.1002/9783527618156
63. R. Siegel, J.R. Howell, Thermal radiation heat transfer. Library (Lond) **1**, 134 (2002). https://doi.org/10.1002/apj.5500020220
64. S.A. Yaseen, A.S. Alameen, F.A. Saif, S.B. Undre, P.B. Undre, Study of volumetric and optical properties of cerium oxide nano-fluid. J. Phys. Conf. Ser. **1644** (2020). https://doi.org/10.1088/1742-6596/1644/1/012030
65. M. Karami, M.A. Akhavan-Behabadi, M. Raisee Dehkordi, S. Delfani, Thermo-optical properties of copper oxide nanofluids for direct absorption of solar radiation. Sol. Energy Mater. Sol. Cells **144**, 136–142 (2016). https://doi.org/10.1016/j.solmat.2015.08.018
66. M. Milanese, G. Colangelo, A. Creti, M. Lomascolo, F. Iacobazzi, A. De Risi, Optical absorption measurements of oxide nanoparticles for application as nanofluid in direct absorption solar power systems—Part I: water-based nanofluids behavior. Sol. Energy Mater. Sol. Cells **147**, 315–320 (2016). https://doi.org/10.1016/j.solmat.2015.12.027
67. S.A. Adam, X. Ju, Z. Zhang, J. Lin, M.M.M.M.M. Abd El-Samie, C. Xu, Effect of temperature on the stability and optical properties of SiO2-water nanofluids for hybrid photovoltaic/thermal

applications. Appl. Therm. Eng. **175**, 1–11 (2020). https://doi.org/10.1016/j.applthermaleng.2020.115394
68. G. Xu, W. Chen, S. Deng, X. Zhang, S. Zhao, Performance evaluation of a nanofluid-based direct absorption solar collector with parabolic trough concentrator. Nanomater **5**, 2131–2147 (2015). https://doi.org/10.3390/NANO5042131
69. A. Menbari, A.A. Alemrajabi, A. Rezaei, Heat transfer analysis and the effect of CuO/Water nanofluid on direct absorption concentrating solar collector. Appl. Therm. Eng. **104**, 176–183 (2016). https://doi.org/10.1016/J.APPLTHERMALENG.2016.05.064
70. A. Khalil, M. Amjad, F. Noor, A. Hussain, S. Nawaz, E.P.B. Filho et al., Performance analysis of direct absorption-based parabolic trough solar collector using hybrid nanofluids. J. Brazil. Soc. Mech. Sci. Eng. **42**, 1–10 (2020). https://doi.org/10.1007/S40430-020-02654-2/FIGURES/8
71. J. Ham, Y. Shin, H. Cho, Comparison of thermal performance between a surface and a volumetric absorption solar collector using water and Fe3O4 nanofluid. Energy **239**, 122282 (2022). https://doi.org/10.1016/J.ENERGY.2021.122282
72. M.M. Heyhat, M. Valizade, S. Abdolahzade, M. Maerefat, Thermal efficiency enhancement of direct absorption parabolic trough solar collector (DAPTSC) by using nanofluid and metal foam. Energy **192**, 116662 (2020). https://doi.org/10.1016/J.ENERGY.2019.116662
73. S. Sreekumar, A. Joseph, C.S. Sujith Kumar, S. Thomas, Investigation on influence of antimony tin oxide/silver nanofluid on direct absorption parabolic solar collector. J. Clean. Prod. **249**, 119378 (2020). https://doi.org/10.1016/j.jclepro.2019.119378
74. N. Wang, G. Xu, S. Li, X. Zhang, Thermal properties and solar collection characteristics of oil-based nanofluids with low graphene concentration. Energy Procedia **105**, 194–199 (2017). https://doi.org/10.1016/J.EGYPRO.2017.03.301
75. A. Joseph, S. Sreekumar, S. Thomas, Energy and exergy analysis of SiO2/Ag–CuO plasmonic nanofluid on direct absorption parabolic solar collector. Renew. Energy **162**, 1655–1664 (2020). https://doi.org/10.1016/J.RENENE.2020.09.139
76. A. Menbari, A.A. Alemrajabi, A. Rezaei, Experimental investigation of thermal performance for direct absorption solar parabolic trough collector (DASPTC) based on binary nanofluids. Exp. Therm. Fluid Sci. **80**, 218–227 (2017). https://doi.org/10.1016/J.EXPTHERMFLUSCI.2016.08.023
77. H.K. Gupta, G.D. Agrawal, J. Mathur, An experimental investigation of a low temperature Al2O3–H2O nanofluid based direct absorption solar collector. Sol. Energy **118**, 390–396 (2015). https://doi.org/10.1016/j.solener.2015.04.041
78. A. Kasaeian, R. Daneshazarian, R. Rezaei, F. Pourfayaz, G. Kasaeian, Experimental investigation on the thermal behavior of nanofluid direct absorption in a trough collector. J. Clean. Prod. **158**, 276–284 (2017). https://doi.org/10.1016/J.JCLEPRO.2017.04.131
79. M. Karami, M.A. Akhavan-Bahabadi, S. Delfani, M. Raisee, Experimental investigation of CuO nanofluid-based direct absorption solar collector for residential applications. Renew. Sustain. Energy Rev. **52**, 793–801 (2015). https://doi.org/10.1016/j.rser.2015.07.131
80. A.L. Subramaniyan, P.S. Lakshmi, R. Ilangovan, Energy harvesting through optical properties of TiO2 and C–TiO2 nanofluid for direct absorption solar collectors. Int. J. Renew. Energy Res. **5**, 542–547(2015). https://doi.org/10.20508/ijrer.v5i2.2177.g6609
81. A. Kasaeian, R. Daneshazarian, F. Pourfayaz, Comparative study of different nanofluids applied in a trough collector with glass-glass absorber tube. J. Mol. Liq. **234**, 315–323 (2017). https://doi.org/10.1016/J.MOLLIQ.2017.03.096
82. J. Ahbabi Saray, M.M. Heyhat, Modeling of a direct absorption parabolic trough collector based on using nanofluid: 4E assessment and water-energy nexus analysis. Energy **244**, 123170 (2022). https://doi.org/10.1016/J.ENERGY.2022.123170
83. Z. Said, M.H. Sajid, R. Saidur, G.A. Mahdiraji, N.A. Rahim, Evaluating the optical properties of TiO2 nanofluid for a direct absorption solar collector. Numer. Heat Transf. Part A Appl. **67**, 1010–1027 (2015). https://doi.org/10.1080/10407782.2014.955344
84. M. Sainz-Mañas, F. Bataille, C. Caliot, A. Vossier, G. Flamant, Direct absorption nanofluid-based solar collectors for low and medium temperatures. A review. Energy **260** (2022). https://doi.org/10.1016/j.energy.2022.124916

85. W. Chen, G. Xu, S. Zhao, X. Zhang, Numerical simulation on the performance of nanofluid-based direct absorption solar collector with parabolic trough concentrator, in *ASME 2016 5th International Conference on Micro/Nanoscale Heat and Mass Transfer MNHMT 2016*, vol. 1 (2016). https://doi.org/10.1115/MNHMT2016-6647
86. T.P. Otanicar, P.E. Phelan, R.S. Prasher, G. Rosengarten, R.A. Taylor, Nanofluid-based direct absorption solar collector. J. Renew. Sustain. Energy **2**(3), 033102 (n.d.)
87. Q. He, S. Wang, S. Zeng, Z. Zheng, Experimental investigation on photothermal properties of nanofluids for direct absorption solar thermal energy systems. Energy Convers. Manag. **73**, 150–157 (n.d.)
88. R. Taylor, S. Coulombe, T. Otanicar, P. Phelan, A. Gunawan, W. Lv, G. Rosengarten, R. Prasher, H. Tyagi, Small particles, big impacts: a review of the diverse applications of nanofluids. J. Appl. Phys. **113**(1), 1 (n.d.)
89. M.S. Hossain, R. Saidur, M.F. Sabri, Z. Said, S. Hassani, Spotlight on available optical properties and models of nanofluids: a review. Renew. Sustain. Energy Rev. **43**, 750–762 (n.d.)
90. H. Younes, G. Christensen, D. Li, H. Hong, A.A. Ghaferi, Thermal conductivity of nanofluids. J. Nanofluids **4**(2), 107–132 (n.d.)
91. R. Saidur, K.Y. Leong, H.A. Mohammed, A review on applications and challenges of nanofluids. Renew. Sustain. Energy Rev. **15**(3), 1646–1668 (n.d.)
92. W. An, J. Li, J. Ni, R.A. Taylor, T. Zhu, Analysis of a temperature dependent optical window for nanofluid-based spectral splitting in PV/T power generation applications. Energy Convers. Manag. **151**, 23–31 (2017). https://doi.org/10.1016/j.enconman.2017.08.080
93. W. An, J. Zhang, T. Zhu, N. Gao, Investigation on a spectral splitting photovoltaic/thermal hybrid system based on polypyrrole nano fluid: preliminary test. Renew. Energy **86**, 633–642 (2016). https://doi.org/10.1016/j.renene.2015.08.080
94. C. Stanley, A. Mojiri, M. Rahat, A. Blakers, G. Rosengarten, Performance testing of a spectral beam splitting hybrid PVT solar receiver for linear concentrators. Appl. Energy **168**, 303–313 (2016). https://doi.org/10.1016/j.apenergy.2016.01.112
95. T. Otanicar, J. Dale, M. Orosz, N. Brekke, D. DeJarnette, E. Tunkara et al., Experimental evaluation of a prototype hybrid CPV/T system utilizing a nanoparticle fluid absorber at elevated temperatures. Appl. Energy **228**, 1531–1539 (2018). https://doi.org/10.1016/j.apenergy.2018.07.055
96. N.E. Hjerrild, S. Mesgari, F. Crisostomo, J.A. Scott, R. Amal, R.A. Taylor, Hybrid PV/T enhancement using selectively absorbing Ag–SiO2/carbon nanofluids. Sol. Energy Mater. Sol. Cells **147**, 281–287 (2016). https://doi.org/10.1016/j.solmat.2015.12.010
97. X. Han, X. Chen, Y. Sun, J. Qu, Performance improvement of a PV/T system utilizing Ag/CoSO4-propylene glycol nanofluid optical filter. Energy **192**, 116611 (2020). https://doi.org/10.1016/j.energy.2019.116611
98. G. Huang, K. Wang, S. Riera et al., Performance analysis of fluid-based spectral- splitting hybrid photovoltaic-thermal solar collectors, in *Proceedings of the International Conference on Efficiency, Cost, Optimization, Simulation and Environmental Impact of Energy Systems 2020*

Nonlinear Optical Properties of Metal Oxide Nanostructures

Vineet Gunwant, Preeti Gahtori, and Ravindra Pandey

Abstract Metal oxide nanostructures (MONS) have attracted substantial research interest due to their unique possibility of engineering their electronic and optical properties through material design. MONS, such as $BaTiO_3$, $LiNbO_3$, ZnO, CuO, $BiFeO_3$, etc., have shown promising applications in optoelectronics, photonics, and biomedical devices. This chapter provides insight into specific nonlinear optical properties, such as Second Harmonic Generation, Z-scan, Hyper Rayleigh scattering, two-photon absorption, optical limiting, etc. In addition, we have summarized recent results of several nonlinear optical properties and potential applications of MONS in cell labeling, fluorescence excitation, sensing, and optoelectronics.

Keywords Metal oxide nanostructures (MONS) · Non-linear optics (NLO) · Second harmonic generation (SHG) · Cell labeling · Optoelectronics

1 Introduction

Metal-oxide nanostructures (MONS) have great potential for several applications in energy storage [1], biosensing [2], optoelectronics [3], and light emission due to their ability to interact with light [4]. The resonant wavelength at which MONS interact with light depends upon metals [5], their size, shape [6], the distance between neighboring objects, the surrounding dielectric medium, etc. [7]. These nanostructures can act as antennae to harvest light surrounding them by enhancing the intensity of the coupled electric field [5, 8, 9].

V. Gunwant · P. Gahtori · R. Pandey (✉)
Department of Chemistry, Indian Institute of Technology Roorkee, Roorkee, Uttarakhand 247667, India
e-mail: rpandey@cy.iitr.ac.in

V. Gunwant
e-mail: v_gunwant@cy.iitr.ac.in

P. Gahtori
e-mail: pgahtori@cy.iitr.ac.in

V. Kumar et al. (eds.), *Optical Properties of Metal Oxide Nanostructures*,
Progress in Optical Science and Photonics 26,
https://doi.org/10.1007/978-981-99-5640-1_5

In a linear optical regime, the optical properties of MONS depend linearly on the incident electric field. However, this linear variation of their properties is broken when a strong electric field is applied, which results in nonlinearity. Various optical devices have continuously shaped the advancement in nonlinear optics (NLO), ranging from optical fibers to photonic crystals [10–12]. The unique property of MONS is to localize electromagnetic fields at nanoscale volumes which helps to control the properties of light at dimensions much smaller than their wavelength. Moreover, the localized electric field at the metal surface enhances the weak Raman process, which is used in surface-enhanced Raman scattering [7].

Nonlinear optical effects arise when the motion of electrons under a strong electromagnetic field cannot be considered harmonic. As a result, the anharmonicity in the field strength mixes the incident fields. It produces new fields that oscillate either with the same frequency (optical rectification) or at the sums or differences of the incident frequencies [13, 14]. Moreover, the nonlinear phenomena respond on the time scale of picosecond and femtosecond, allowing ultrafast processing of optical signals [7], which enables nonlinear phenomena in MONS to be applicable for biosensing and monitoring chemical growth [6]. Apart from these applications, MONS are also used as fluorescent probes for tracking biological processes to study cellular morphology, functions, cell targeting, etc. [9] which are advantageous with respect to organic fluorescent probes that are highly susceptible to photobleaching [15]. MONS such as potassium niobate ($KNbO_3$), lithium niobate ($LiNbO_3$), Barium titanate ($BaTiO_3$), potassium titanyl phosphate ($KTiOPO_4$), and zinc oxide (ZnO) have higher cell imaging than organic chromophores [5]. Also, to achieve the desired optical and nonlinear optical properties, it is crucial to have reasonable control over the synthesis of MONS of precisely controlled size and shapes [16]. This book chapter covers the following sections, from explaining the nature of the interaction of light with MONS to their nonlinear properties and applications. Section 2 describes the interaction of light with MONS. Section 3 discusses various nonlinear optical techniques used to study MONS, such as second harmonic generation (SHG), Z-scan, Hyper Rayleigh scattering, etc. Section 4 presents several properties of MONS such as optical limiting effect, imaging, and photostability, wavelength tunability, nonlinear wave mixing, etc. Section 5 presents various applications of MONS e.g., cell labeling, fluorescence excitation, sensing, optoelectronics, etc. In Sect. 6, the effect of conjugating metal to MONS is explained. Section 7 explains challenges and gives future perspectives for MONS.

2 Interaction of Metal Oxide Nanostructures (MONS) with Light

A ray of light reaching at an interface between two materials is reflected and transmitted. The properties of reflection and transmittance are governed by Snell's law and Fresnel coefficients. These laws are valid for the case when the size of the material is greater than the wavelength of light [1]. For those materials which are having subwavelength size or structures smaller a size than the wavelength of the light, these laws are not valid. In these cases, the light is diffracted, absorbed and scattered by subwavelength structures. These nanostructures have optical properties that are direct consequence of their energy band structures [2]. Metals have free electrons due to partially filled conduction band, which can transport energy and electric current efficiently. For such structures, the electrons oscillate due to the external excitation, from the entire volume (Fig. 1). All free electrons in the nanostructures are displaced along the polarization of the incident electric field, that forms a dipole which oscillates with the incident frequency [4, 7]. With the help of these oscillations, nanostructures can interact with light efficiently [4].

The incident field couples with the dipole at a frequency which depends upon the nanostructure type, size, shape, surrounding dielectric medium etc. By varying these properties, MONS of potential applications can be made. Some MONS show size dependent tunability of their optical properties [5], the representative examples are given in Fig. 2. In addition, the nanostructures concentrate the light energy which intensifies the electric field between them. The property of these nanostructures to absorb, scatter visible light, act as an electron sink enables them to be used for photovoltaics, biosensors and photocatalysts applications [1, 6].

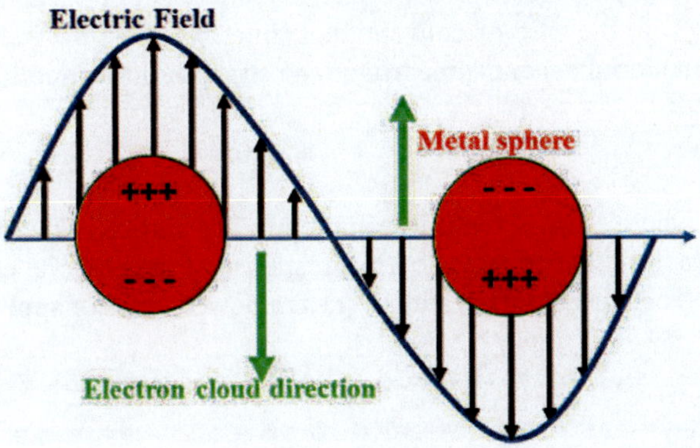

Fig. 1 Illustration of oscillation of metal electron cloud when electric field is applied. All the electrons are displaced along the polarization of incident electric field

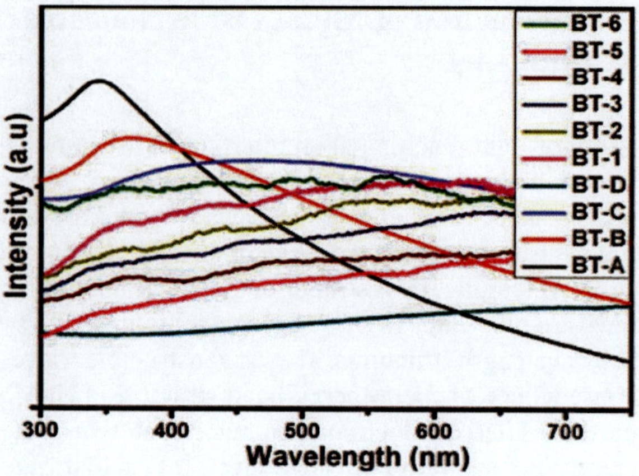

Fig. 2 Absorptin spectra of $BaTiO_3$ (BT) nanoparticles of different size. BT 1–6 were prepared by varying citric acid percentage at same temperature. BT, A–D were prepared at different temperatures. The broadening of absorption peaks are due to increase in band gap energy with decrease in grain size of the nanoparticles. Reprinted with permission from Ref. [5]. Copyright 2015 Elsevier

3 Nonlinear Optics (NLO)

NLO is a subfield of optics, that occur as a consequence of the modification of the optical properties of a material by light. They are nonlinear in the sense that they occur when the response of the material to an applied optical field varies in a nonlinear manner, that depends on the strength of the optical field. In most of the cases, the laser light of very high intensity can modify the optical properties of the material by mixing incident pulses or exciting higher order frequencies from the material. The study of NLO properties of materials started in the late 19th century, with the discovery of the Pockels or electro-optic effects. This effect refers to the change in refractive index by an external electric field applied to the material. As light propagates through such crystals, their phase and polarization can be tuned [7].

In the case of conventional (linear) optics, the induced polarization $P(t)$ of the material depends linearly on the strength of the applied electric field $E(t)$,

$$P(t) = \epsilon_0 \chi^{(1)} E(t)$$

where, $\chi^{(1)}$ is the linear susceptibility and ϵ_0 is the permittivity of free space. For nonlinear optics, the optical response of the material can be described by expressing the induced polarization $P(t)$ as a power series of applied electric field strength $E(t)$.

$$P(t) = \epsilon_0[\chi^{(1)} E(t) + \chi^{(2)} E^2(t) + \chi^{(3)} E^3(t) + \ldots]$$

$$P(t) = P^{(1)}(t) + P^{(2)}(t) + P^{(3)}(t) + \ldots$$

$\chi^{(2)}$ and $\chi^{(3)}$ denote second and third order nonlinear susceptibilities, respectively. $P^{(2)}$ and $P^{(3)}$ are second and third order induced polarization of the material [8]. The second order nonlinear interactions occur only in the crystals that are noncentrosymmetric in nature. On the other hand, third order nonlinear interactions can occur in both centrosymmetric as well as noncentrosymmetric crystals.

The NLO processes like multiphoton absorption, where several photons are absorbed can be used for laser cutting and ablation. Nonlinear phenomena are useful as they allow rapid change of the material properties by using only light [9]. Nonlinear techniques such as second harmonic generation, Z-scan, hyper-Rayleigh scattering etc. have become useful to characterize MONS [20].

3.1 Second Harmonic Generation

On atomic scale, atoms develop a dipole moment which oscillates with the same frequency as the applied electric field and emit secondary radiation. If magnitude of applied electric field is large enough, like in case of laser, the emitted radiation can contain higher frequencies compared to the incident one. For example, in the case of second harmonic generation (SHG), two photons with the same energy are annihilated at the same time and one photon with the double energy is emitted from the atom (Fig. 3). The excitation of the atom leads it to the virtual energy state from where, it instantaneously relaxes back to the ground state. The excitation of atoms in the virtual state leads to the three useful characteristics of the SHG, (I) The SHG is not a resonant process, it can occur at any wavelength. (II) The emission of photons in the SHG process occurs instantaneously compared to the resonant processes such as two photon absorption, which has finite lifetime in the excited state. (III) The bandwidth of the emitted SHG signal is narrower than the bandwidth of the excitation radiation, whereas in the case of the resonant process, the bandwidth is broader due to non-radiative losses [8, 10].

SHG being a second order nonlinear process, the induced polarization is given as:

$$P^{(2)}(2\omega) = \epsilon_0 \chi^{(2)}(\omega) E^2(\omega)$$

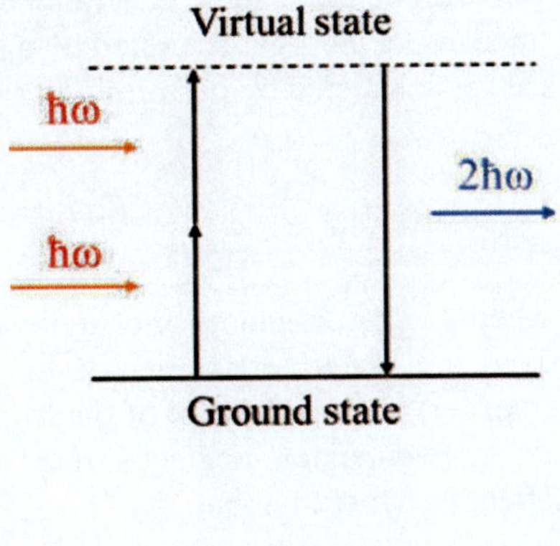

Fig. 3 Energy level diagram of SHG process

where, $P^{(2)}(2\omega)$ is second order induced polarization, $\chi^{(2)}(\omega)$ and $E^2(\omega)$ are second order susceptibility and applied electric field at frequency ω, respectively.

Application of various MONS uses the advantage of SHG process such as nanostructures with high refractive index e.g. silicon, gallium, arsenide etc. are good nonlinear emitters but these materials have significant losses in the visible and ultraviolet region [1]. Therefore, new materials are required for applications in the near ultraviolet-visible range. Ferroelectric metal oxide lithium niobate ($LiNbO_3$) nanocubes show applications in nonlinear photonics in the near ultraviolet-visible range [11]. These $LiNbO_3$ nanocubes are efficient SHG emitters as a result of the strong electric field confinement at the SHG wavelength. Hence, they emit second harmonic signal in the near ultraviolet range and therefore useful for near ultraviolet photonic applications. Some other materials e.g. $BiFeO_3$ [12], $KTiOPO_4$ [13] etc. have been utilized in cell imaging and photo stability.

3.2 *Hyper-Rayleigh Scattering*

In nonlinear optics, the frequency, phase of an incident electromagnetic field is modified through its nonlinear interaction with matter. The interaction leads to the induced dipole moment which depends on the power series of the applied electric field (E).

$$\mu = \alpha.E + \beta.E^2 + \ldots$$

where, α is linear polarizability and β is the first hyper polarizability. The macroscopic nonlinear response of materials is proportional to the molecular hyper polarizabilities, which is directly related to the performance of MONS [14]. In 1991, the Hyper-Rayleigh Scattering (HRS) technique was developed for the measurement of molecular hyperpolarizabilities. It can measure β, directly of all molecules, irrespective of symmetry or charge. For dipolar molecules, the hyperpolarizability can be calculated without the knowledge of the dipole moment. The HRS can be measured by measuring the intensity of the incoherently scattered light, at the second-harmonic wavelength, that can be generated by an intense laser beam in a nanostructure. The intensity of the light scattered by a single molecule at the harmonic wavelength 2ω can be calculated by performing orientational average over β [15].

$$I_s(2\omega) = \frac{32\pi^2}{c\varepsilon_0^3\lambda^4 r^2}\langle\beta_{HRS}^2\rangle I(\omega)^2$$

where, $I_s(2\omega)$ is intensity of light scattered by MONS at harmonic frequency, and $I(\omega)$ is intensity of incident light. The effective hyperpolarizability for a MONS ($\langle\beta_{HRS}\rangle$) is characteristic of the MONS SH generation efficiency. If nanostructures are large enough to neglect surface contribution of the scattered SH signal, then the effective hyperpolarizability $\langle\beta_{HRS}\rangle$ has the same properties as $\chi^{(2)}$ and is related to the averaged SH coefficient <d> by mean particle volume V.

$$\langle \beta_{HRS} \rangle = \langle d \rangle V$$

<d> is an intensive physical property of each material. Consistently, large size nanostructures (ZnO 90–200 nm) have a higher hyperpolarizability than small nanostructures (ZnO 20 nm). Application of the HRS technique on nanostructures allows one to achieve a reliable assessment of the SH efficiencies of noncentrosymmetric nanomaterials for applications in bio-imaging as nonlinear optical markers [15, 16].

3.3 Z-Scan

Materials having third-order optical nonlinearities have been explored for their application to high-speed optical switching devices. To assess a material for this application, nonlinear refraction index (NLR) and two-photon absorption coefficient (TPA) should be known. These parameters can be determined by Z-scan technique, in which a sample is scanned near the focal region of a laser beam [17]. As the sample is moved along the propagation direction of the laser beam, it experiences modulation in phase and intensity, that can be measured by its transmittance as a function of the sample position. If all the transmitted light is measured, then only TPA affects the Z-scan, which is known as open -aperture Z-scan. On the other hand, if some of the transmitted light is detected due to the presence of an aperture in front of the detector, both NLR and TPA contributes to the Z-scan, which is referred as closed aperture Z-scan [18].

In the experiment, a continuous variation of the input fluence can be achieved by moving the sample through the focal region of a focused laser beam. As the position of sample is taken as z with $z = 0$ for the focal point of incident laser beam, hence named as Z-scan.

Z-scan curves of graphene and graphene-CuO nanocomposites are shown in Fig. 4. The NLO properties of graphene-CuO nanocomposites are larger compared with graphene, due to doping of CuO, which increases the time of electron transfer and the photon transition. Also, it suppresses the recombination between electrons and holes that ultimately provides longer time for free carriers to work. This leads to enhancement in the NLO properties of the nanocomposites [19]. These nanocomposites have potential applications in various nonlinear optical devices such as optical switches [20], optical sensors [6] and so forth. These applications of MONS are discussed in later sections.

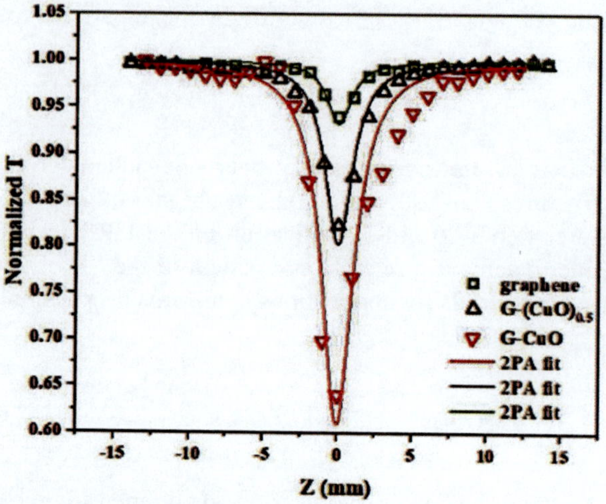

Fig. 4 Normalized Z-scan curves of multi-photon absorption of graphene and graphene-CuO nanocomposites. Reprinted with permission from Ref. [19]. Copyright 2019 Elsevier

3.4 Multiphoton Excitation Processes

Multiphoton absorption is simultaneous absorption of several photons. Multiphoton based processes are applicable in many scientific areas such as confocal scanning microscopy, 3D data storage etc. It provides tunable excitation wavelength in the near infrared (NIR) range in which absorption probability depends nonlinearly on the excitation intensity [21]. The confocal scanning microscopy based on multi-photon absorption is considered as promising luminescent technique for imaging living systems which allows subcellular resolution without photodamage [1].

The efficiency of multiphoton absorption of a compound is determined by its absorption cross section. The three-photon absorption cross section is defined by:

$$dN = \sigma^{(3)} p(t, z)^3 c dz$$

Here, dN denotes the number of absorption events along the interval dz, c is the concentration and $p(t, z)$ is the time and z-dependent photon flux. Compared to the one photon absorption, there are sufficiently small number of photons in three photon absorption process during propagation of the laser beam through the sample. Hence, the z-dependence can be neglected for three photon absorption. The photon flux can be calculated by considering only temporal dependence:

$$p(t) = \frac{E/h\nu}{2\pi w_{x,y}} \frac{1}{w_t\sqrt{2\pi}} e^{-\frac{t^2}{2w_t^2}}$$

The photon flux is given for laser pulse with energy E, frequency ν, beam width of circular gaussian beam $w_{x,y}$ and temporal width w_t [21].

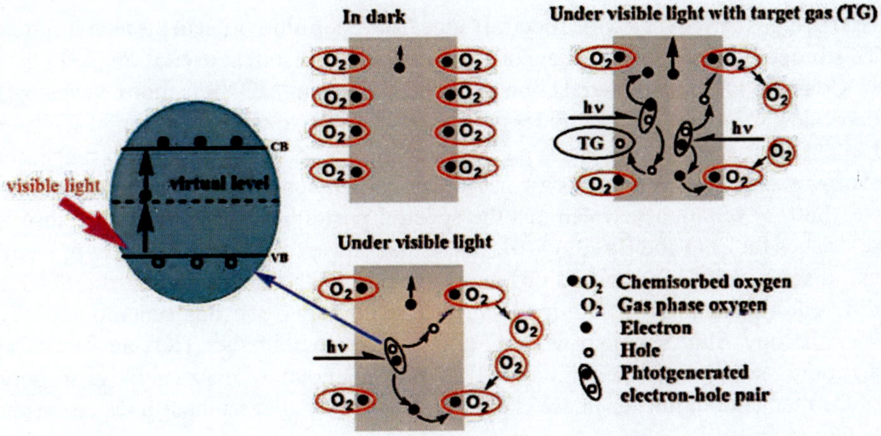

Fig. 5 Visible light-activated gas sensing mechanism of ZnO sensor, induced by multiphoton excitation process. Reprinted with permission from Ref. [22]. Copyright 2013 Elsevier

The photo response in ZnO and TiO_2 MONS can be used in gas sensing applications which can be induced by two-photon or multiphoton excitation processes. The photo responded property of ZnO and its sensing response were observed under visible light irradiation (two photon absorption), while in absence of visible light it shows no sensing response (Fig. 5). It is suggested that native defects of ZnO are responsible for photo-responded behavior under visible light irradiation [22].

4 Nonlinear Optical Properties

4.1 Size and Excitation Wavelength Tunability

MONS have strong dependency on their size for nonlinear absorption. For $BaTiO_3$ nanoparticles the size ranging from 16 to 26 nm have nonlinear absorption coefficient which ranges from 11×10^{-11} to 6.2×10^{-11} m/W. The nonlinear absorption coefficient is larger for particles having smaller size and vice-versa. For a spherical nanostructure surface area to volume ratio is given by 3/r, where r is the radius of spherical nanostructure. For the smaller size MONS the surface to volume ratio is high, which implies that there is maximum distribution of particles on the surface [5]. Hence, the smaller size MONS are having large quantum confinement which can enhance the nonlinear absorption. By optimizing nonlinearity and size dependency, the nonlinear absorption in MONS can be tuned. The MONS are smaller than the coherence length of the material, hence no destructive interference can be possible between the emitted signals from different locations within the nanostructures. Also, laser sources can cover a wide spectral range from deep ultraviolet to the near infrared. This property becomes advantageous with respect to resonant excitation [23].

Compared to bulk materials, the SHG intensity from MONS is more wavelength dependent. Also, the MONS are less sensitive towards excitation wavelengths, except for ZnO and $BaTiO_3$. For these particles, in the 760–970 nm excitation range, the shorter excitation wavelengths are having higher efficiency than larger ones because for shorter excitation wavelengths the spectral position of the bandgap is close to resonance for ZnO and $BaTiO_3$ [10]. These material specific characteristics of signal can discriminate between two different types of MONS by recording the SHG at different wavelengths. In addition to this, using the sum frequency generation (SFG) spectroscopy, Shultz and co-workers [24] showed that smaller TiO_2 nanoparticles are more reactive than the larger on. The relative amount of dissociative adsorption of the methanol in the gas phase is larger in case of smaller nanoparticles compared to larger one.

Apart from MONS, nonlinear optical techniques can be applied to study the size dependency of metal nanostructures. Chandra et al. [25] used Hyper-Rayleigh scattering (HRS) to show the size to wavelength ratio (r/λ) affects the SH response from Cu nanoparticles. It was shown that hyperpolarizability values of the Cu nanoparticles vary with particle size. It was concluded that SH response from nanoparticles were electric dipolar in origin if r/λ was 1/20 or less. For larger sized nanoparticles The SH response was disturbed by quadrupolar contribution as a result of retardation effect.

4.2 Optical Limiting Effect

Optical limiting is a nonlinear effect in which transmittance of a sample decreases on applying high intensity light source. Optical limiting property of MONS can be utilized to protect the delicate optical instruments and human eye from intense light sources. They strongly reduce the intense laser pulse thereby transmitting low ambient light. Optical limiting in MONS such as Fe_2O_3 [26], Fe_3O_4 [27], spinel ferrites [28], graphene oxide-Fe_3O_4 [29], $BaTiO_3$ [5] have been reported. The large nonlinear optical response of these materials makes them as potential materials for optical limiting and optical switching applications [30]. Cd-Fe co-doped CuO MONS absorb more photon in the visible region and shows lower onset optical limit threshold, therefore higher nonlinear coefficients. These properties of doped CuO MONS can be used as a potential candidate for the development of laser safety devices [31]. Spinel ferrites are versatile MONS, having applications in magnetic and biological fields. Panit et al. [28] investigated the optical limiting properties of $ZnFe_2O_4$ and its Ni/Cu mixed forms and compared the results with C_{60}, which was extensively studied for its optical limiting effect [33, 37]. The ferrite MONS show a slight increase in transmittance in low fluence regions, while efficient limiting at higher fluences. The depth of the valley in the z-scan curve directly indicates the optical limiting efficiency of a material.

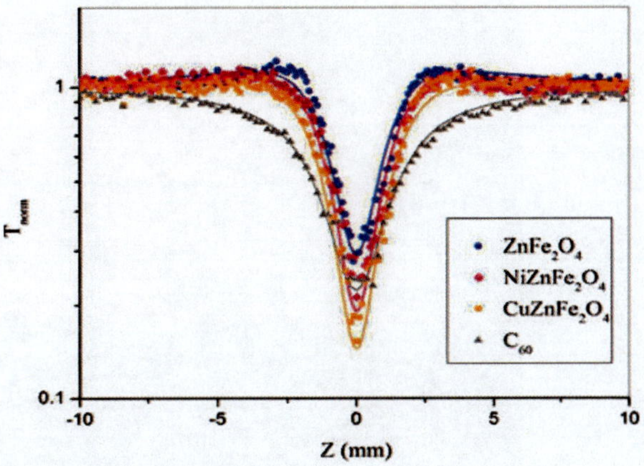

Fig. 6 Z-scans of different MONS, compared with C_{60}. The depth in the valley is indicative of optical limiting by each MONS. Reprinted with permission from Ref. [28]. Copyright 2012 American Institute of Physics

Figure 6 shows that C_{60} exhibits optical limiting throughout the range of the measurement while the MONS show optical limiting for a specific fluence region. However, for application purposes these MONS have unique advantage compared to C_{60} as they can increase the sensitivity of a detector by switching suddenly to optical limiting region when input fluence exceeds the threshold limit of the detector [32]. The increase in optical limiting in the case of Ni/Cu mixed ferrite MONS can be attributed to self-trapping of charge transfer states as there is increase in excited state absorption due to charge transfer between 2p orbital of oxygen and 3d orbital of metal. Further, the trapping of charge transfer transition is dependent on lattice strain. In the case of Ni and Cu zinc ferrite, introduction of these new ions create distortion and produces strain in the nanostructure which further increases the excited state absorption thereby enhancing optical limiting of Ni and Cu mixed zinc ferrite MONS [28]. In addition, the optical limiting increases with increase in the grain size of MONS and has dependence on their structural phase e.g. when MONS undergo transition from cubic to tetragonal phase, an increase in the nonlinear optical absorption has been observed [5, 17].

4.3 *MONS as Imaging Agents*

Several medical diagnoses rely on imaging by MONS. MONS based imaging agents with non-invasive optical techniques are essential to obtain high contrast images for applications in drug delivery [33], biomedical therapies [34] etc. Label free optical techniques such as high resolution optical coherence tomography [35], harmonic generation microscopy, two photon fluorescence [36] have been used for 3-dimensional imaging with enough depth penetration in biological species [37]. Moger et al. [36] showed that contrast from TiO_2, CeO_2 and ZnO can be used to image low particle concentration within biological species.

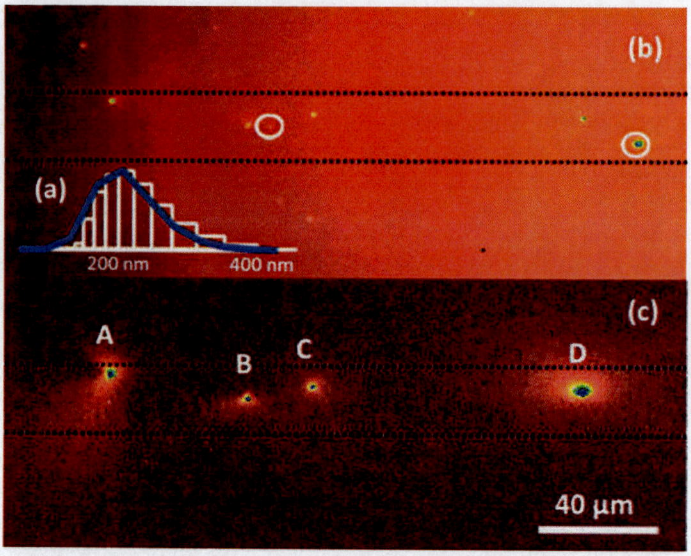

Fig. 7 **a** Size distribution of KTP MONS centered at 185 nm. **b** White light image of the KTP MONS on the waveguide. The dashed lines indicate the extension of the waveguide mode. **c** SH image of the same sample region. Reprinted with permission from Ref. [13]. Copyright 2010 Optical Society

Ronja et al. [13] used Potassium-Titanyl-Phosphate ($KTiOPO_4$, KTP) MONS, spread on a waveguide, to show their imaging property, as shown in Fig. 7. When these MONS are excited with laser radiation, they show enhanced contrast. The region (b) represents the nanoparticles under white light illumination. The differences in intensity originate from the size dispersion of MONS in the sample. In region (c), they are excited with intense laser source. Their positions (A-D) spatially correlate with those of MONS observed under white light illumination. It provides the possibility to employ MONS as molecular probes in experiments based on evanescent excitation.

4.4 *Enhancement of Nonlinear Efficiency of Nanostructures*

Several research efforts have been focused on increasing the conversion efficiency of incident light into the nonlinear signal. Pure metallic nanostructures emit strong nonlinear signals despite having symmetric structure, the reason being shape asymmetry [38], surface contributions [7] etc. However, the nonlinear signal emitted by the pure metallic nanostructures suffers from high losses in the visible range and small volumes which requires complex fabrication methods. On the other hand, hybrid nanostructures such as $BaTiO_3$ nanoparticles [39], $KNbO_3$ nanowires [40], $LiNbO_3$ disks surrounded by gold nanorings [41] enhance SHG signal.

Gold nanorings filled with lithium niobate ($LiNbO_3$) generates SH signal which comes out to be larger than unpatterned $LiNbO_3$ MONS. The larger refractive index of the substrate (MONS) than the surrounding medium provides enhanced field in close vicinity to the particle-substrate interface [42]. Hence, the nonlinear field generated at the interface region is stronger than the same substrate without resonant nanostructure [41]. These nanorings offer a self-organized core-shell model and work as nano-optical device to enhance SHG inside their core. Enhancing nonlinear optical response can contribute to fabricating efficient nanoscopic light sources.

Another approach to enhance the SHG response of nanostructures is to take advantage of intrinsic Mie scattering resonances. Timpu et al. [43] utilized mie scattering resonances of the nanostructures to enhance the SHG signal from individual $BaTiO_3$ nanostructures. They have used $BaTiO_3$ nanoparticles, which are transparent in visible and infrared range. In addition, $BaTiO_3$ has a noncentrosymmetric crystal structure and due to its large refractive index, $BaTiO_3$ nanostructure has several resonances in their linear scattering cross section, which helps a strong SHG response even at nanoscale. Due to the large refractive index and noncentrosymmetric structure, $BaTiO_3$ nanostructures show strong bulk SHG signal.

4.5 Nonlinear Wave-Mixing by MONS

Wave-mixing by MONS could lead to develop applications such as nanoscale light sources, optical switching [20, 44], and pulse characterization [45]. MONS having varied chemical compositions, size can readily generate multiple signals such as various harmonic generation, sum or difference frequency combinations. Liu et al. [46] used GaAs based dielectric metasurface to report simultaneous emission of harmonics and wave-mixing signals. Campargue et al. [47] reported multiple signals from $KNbO_3$ nanoparticles, upon two-color excitation at 1045 nm and 1300 nm that are resolved spectrally from deep UV (260 nm) to near infrared (874 nm). Excitation at 1045 nm produces two harmonics (523 and 348 nm) and four (650, 433, 325 and 260 nm) at excitation of 1300 nm (Fig. 8). Apart from this, sum frequency involving three wave mixing (3WM, 586 nm) and four wave mixing (373, 401 nm) associated with $\chi^{(2)}$ and $\chi^{(3)}$, respectively [47]. This response of $KNbO_3$ nanoparticles was shown to be conserved, when they are embedded in biological medium, which can be useful for bioimaging and photo interaction applications.

It is also interesting to study other nonlinear effects such as sum frequency generation (SFG). $LiNbO_3$ nanowires have been reported to generate and guide second-harmonic (SH) signals. As, nanowires provide superior surface properties and stronger optical confinement due to extended length of light matter interactions, they can be used for understanding optical switching.

Sergeyev et al. [48] used $LiNbO_3$ nanowires to generate sum frequency signal by overlapping the laser beams. To confirm that the generated signal was from SFG, they took spectral measurements of the signal that depend linearly on the power and wavelength of the incident laser source. Based on their results, they concluded that

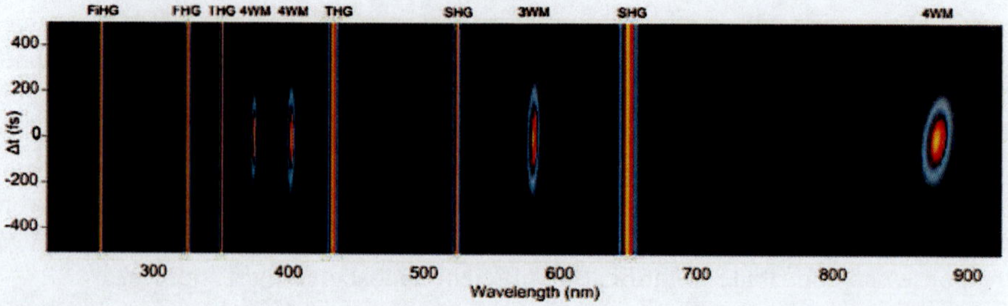

Fig. 8 Spectrograms of $KNbO_3$ nanoparticles under 1300 and 1045 nm excitation. Harmonic signals appear as continuous vertical strips contrary to superposition (mixing) of the two pulses. Reprinted with permission from Ref. [47]. Copyright 2020 American Chemical Society

phase matching of the guided SFG signal is possible in the nanowires that can be used to enhance the SFG signal. The signal enhancement due to phase matching condition was also shown in the case of SHG in $LiNbO_3$ nanowires [49].

4.6 MONS Show Larger Non-linear Response as Compared to Non-oxidized Nanostructures

MONS having good chemical stability and mechanical strength have been used for investigation of their non-linear optical properties [50]. MONS such as CuO, V_2O_5, Co_3O_4, Cr_2O_3, Mn_3O_4 have shown higher nonlinear susceptibility than their non-oxidized counterparts [51, 52]. The optical properties of nanostructures can be perturbed by changing their surface area to volume ratio. It means a decrease in the radius of sphere increases the ratio (the quantum confinement effect). In other words, the concentration of carriers increases with decrease in the size of nanostructures. Further, the higher concentration of carriers can have direct implications on the non-linear response of the nanostructure. Boltaev et al. [53] showed that red shift of the band gap of oxidized nanostructure (CuO) can increase its non-linear efficiency compared to non-oxidized one (Cu nanoparticles). In the case of Cu nanoparticles, the SPR was observed at 620 nm while a broad SPR was observed at 300 nm, which is characteristic of nanoellipsoids. The shift in the SPR band is a direct consequence of quantum confinement effect, which arises due to the presence of smaller nanoparticles and nanoellipsoids. Having the idea of quantum confinement, the non-linear response of both Cu nanoparticles and CuO nanoellipsoids was measured by z-scan technique. In the case of CuO nanoellipsoids, positive non-linear refractive properties were observed at 800 nm probe pulse. In addition, open aperture z-scan was used to measure the optical limiting properties of both materials at 800 nm, 60 fs probe pulses. The presence of optical limiting effect in CuO nanoellipsoids was attributed to the two-photon absorption process. However, no such optical limiting was observed in the case of Cu nanoparticles. The observed nonlinear response in CuO nanoellipsoids was attributed to be associated with decrease in the band gap of CuO nanoellipsoids. Apart from it, CuO nanoellipsoids are having 3D shape, which makes the nanostructures to have tunable properties as compared to 2D shaped Cu nanoparticles, which have inherent properties.

5 Applications of Metal-Oxide Nanostructures

5.1 MONS for Cell-Labeling

As discussed in the previous section, MONS have high photostability, nonlinear efficiency, biocompatibility, which makes them excellent for cell labeling applications. For example, MONS, which are capable of showing non-linearity, have many advantages over conventional fluorescent materials. Such MONS maximize imaging depth and maintain high specificity. Bismuth ferrite ($BiFeO_3$) MONS can monitor pulmonary macrophages in mice. Macrophages are immune cells which evolve as a result of lung diseases such as asthma. $BiFeO_3$ stably label exogenous or endogenous macrophages and produce excellent SHG signal after their in vivo application to mice [54]. The externally applied $BiFeO_3$ loaded macrophages can be tracked in the lung, allowing precise localization of the cells and monitoring their distribution over the time. In addition to this, $BiFeO_3$ MONS have shown long term photostability in vivo, because of their nonlinear interaction with the excitation field. The excitation wavelength flexibility in MONS widens the scope of molecular probes that can be used for fluorescence staining (Fig. 9). It is likely that longer wavelength excitation can enable visualization of deeper tissue segments, which can be helpful for the exploration of third harmonic generation (THG) and fourth harmonic generation (FHG) produced by tissue structures thereby minimizing the need of tissue labeling [54, 55].

Emission in ZnO nanoparticles lies in visible range which makes them very useful for cell labeling applications. The successful attachment of ZnO nanoparticles to the cell can be ascribed to the green color emission from the cells (Fig. 10). As the particle size increases, ZnO nanoparticles undergo a red shift due to quantum confinement effect. The visible emission in ZnO is a result of transitions that involve trapped states. These trapped levels are formed by the oxygen vacancies present on the surface of nanoparticles [9]. Also, by varying pH conditions, they can show different color emissions with good intensity. Zeta potential study shows that at pH 12, the surface charge is dominated by O^{2-} i.e., surface is having less vacancy of oxygen, while at pH 6, it is dominated by Zn^{2+} i.e., surface is having more vacancy of oxygen. Hence, due to presence and absence of oxygen vacancies, color emission from ZnO varies in the visible range. In the case of ZnO nanoparticles, with the increase in surface oxygen vacancies, the wavelength in the visible range increases from blue to orange color [56]. Although, we have discussed two MONS $BiFeO_3$, and ZnO, there are several other MONS which have been used for cell labeling applications such as $BaTiO_3$ [57], iron oxides [58], $KNbO_3$ [59] etc.

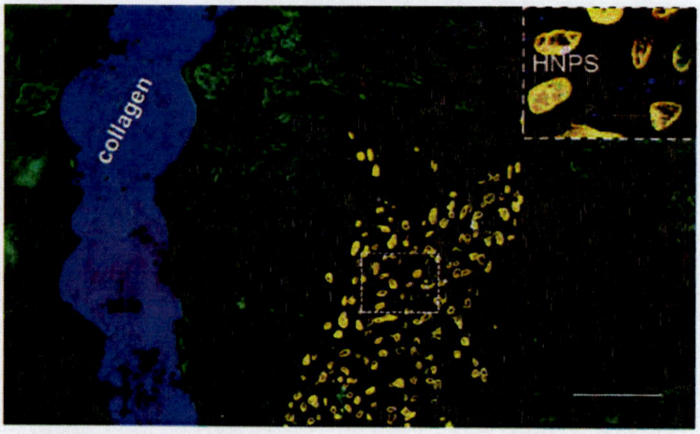

Fig. 9 Illustration of MONS-based labeling of stem cells which can be used to track cells or tissues. SHG (in blue color) from $BiFeO_3$ (inset) and collagen. Reprinted with permission from Ref. [55]. Copyright 2017 American Chemical Society

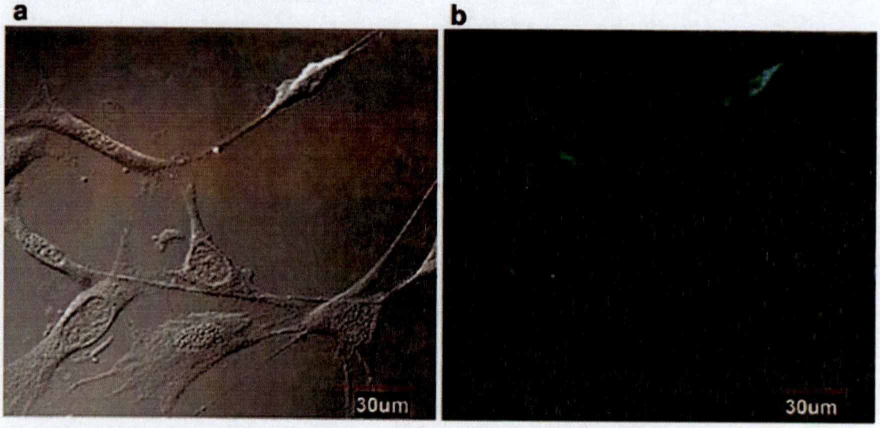

Fig. 10 **a** Differential interference contrast photograph and **b** fluorescent image of NIH/3T3 cells labeled with ZnO nanoparticles with green color emission. Reprinted with permission from Ref. [56]. Copyright 2010 American Chemical Society

5.2 *Applications in Monitoring Chemical Growth of Formation of MONS*

Nonlinear optics can be efficiently used to monitor the formation and growth of the nanostructures that are having significant hyperpolarizability. Hyper-Rayleigh scattering (HRS) is a nonlinear optical tool to monitor MONS formation in liquid [15]. Segets et al. [60] observed HRS from ZnO nanoparticles. The hyperpolarizabilities, β_{ZnO}, of different sized particles was shown to increase with particle size.

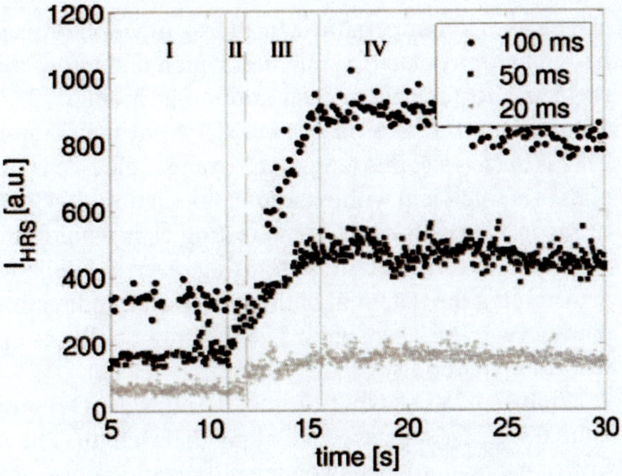

Fig. 11 Hyper rayleigh scattering (HRS) signals monitored for three different time resolutions. Four different regions are (I) only one reactant; (II) nucleation; (III) growth; (IV) ripening. Reprinted with permission from Ref. [60]. Copyright 2009 American Chemical Society

A long-term time resolved observation of ZnO nanoparticles allowed the different steps of the particle formation process and to monitor nucleation, growth, and ripening rates from the nonlinear optical signals. Figure 11 shows the evolution of the HRS signal as a function of time for three different time resolutions of 20, 50 and 100 ms. Region I represent, the observed signals for solvent and zinc acetate. Region II, III, and IV represent the nucleation, growth and ripening, respectively [60]. These time dependent studies clearly show the capability of the MONS based nonlinear optical technique to study and monitor growth of a chemical reaction. Sahyun et al. [61] observed significant hyperpolarizability from TiO_2 nanoparticles. The generated second harmonic could be used to monitor the formation and growth of these particles. Further, HRS can also be used to quantify the origin of NLO properties in such particles [62].

5.3 *MONS in Sensing Applications*

Optical sensors have been developed for a number of applications ranging from industry, defense to health monitoring and medicine. The popularity of these sensors can be attributed to their small size, light weight and immunity to electromagnetic interference, temperature dependence, higher sensitivity [10]. $LiNbO_3$ is an anisotropic material which shows spontaneous polarization dependent on temperature. As $LiNbO_3$ is heated, there is movement of lithium and niobium ions, and the crystal expands those results in change in the polarization. As the polarization changes, there is change in refractive index of the material. Hence, temperature variation causes refractive index to change therefore $LiNbO_3$ can be used as temperature sensor [63].

Distinct crystalline structures may lead to various optical properties. Sodium niobate ($NaNbO_3$) is another MONS having seven structural phase-transitions which can be used as temperature sensor as a function of its phase transition. Pure $NaNbO_3$ has an antiferroelectric structure, which undergoes phase transition from an antiferroelectric to ferroelectric on cooling at around 73 K and the reverse transition at 245 K, here the two phases coexist from temperature ranging from 12 to 280 K. This is because at this temperature range, the free energy difference between the two phases is small and within thermal fluctuations [64]. The small free energy difference between the two phases is interesting as it would allow to switch from the antiferroelectric to ferroelectric state using electric fields, which in turn will be helpful in determining the potential of this material in applications such as high-density optical data storage [65], sensing [6], enhancing nonlinear optical properties and hologram recording materials [66].

Niobate NWs ($KNbO_3$, $LiNbO_3$ and $NaNbO_3$) propagate and maintain nonlinear optical signals, such as SHG, along their length. The maximum SH signal generated from a nanowire depends on the polarization of incident laser light, crystal structure orientation and design of NW. SH signal propagates in periodic manner inside the NW as the coherence length of the SH process is smaller than the NW length. Also, the cross section design of NW influences the conversion efficiency between the fundamental harmonic and second harmonic signals that propagate inside the NW. Sergeyev et al. [67] showed that at the $LiNbO_3$ NW output, the average SH signal power was 347 ± 40 pW for the incident power of laser beam (85 ± 3 mW). It was stated that the propagation process inside the NW may be a combined form of wave guiding and scattering along the length of the NW.

The tunability of size and shape of MONS makes them a good candidate for sensing and catalytic applications [68]. Lee et al. [69] used an electrode made up of metal oxide to show electrochemical behavior for glucose biosensor. As the biosensing behavior often relies on the free active site and surface area of a nanostructure, tuning the size of MONS can help in improving the biosensing ability. Sun et al. [70] showed that cupric oxide (CuO) nanourchins having size smaller than 100 nm are more effective than the bigger ones for glucose biosensing applications. The CuO samples were employed as glucose-sensing anode materials. The glucose biosensing activity was measured by linear-sweep voltammograms to examine the glucose oxidation process at the CuO anode. As the concentration of glucose was increased, there was step-wise increment in the current, flowing through the cell. Here, the CuO anode acts as a sensor for glucose such a way that a small amount (~100 μM) of glucose increases the current in the cell. Also, the sensing process was increased due to the smaller size of CuO nanourchins. The smaller sized nanourchins facilitate larger surface to volume ratio which increases the quantum confinement, ultimately increasing the surface area so that more number of molecules can be adsorbed. MONS can also be used to quantify organic dyes, which can be used in biomedical applications. Rao et al. [63] used $LiNbO_3$ NWs for localized delivery of light. At the NW output, fluorescence signal detected as 17.6 ± 1.9 fW for 1 μg/ml concentration of dye. This dye concentration range is the same as is used for cell staining which indicates that using the NW (of appropriate crystal structure and orientation) for fluorescence excitation requires low dye concentration and can be used for cell studies.

5.4 MONS in Optoelectronics

The term optoelectronics refers to a branch of electronics in which electronic devices are used for sensing, emitting and transmitting light. One-dimensional (1D) MONS such as ZnO nanowires [71], TiO_2 nanorods [72], Ga_2O_3, Nb_2O_5 nanobelts [73, 74] with varying morphology have been used in nanoscale optoelectronics. 1D MONS have large surface to volume ratio, which results in increasing the lifetime of photocarriers thereby increasing the photoconductivity of MONS. Jin et al. [71] fabricated a flexible photoconductor based on ZnO nanostructures. When these were combined with 8-hydroxy-quinoline, the device displayed high responsivity at moderate electric field. Li et al. [73] used individual Ga_2O_3 nanobelts to monitor ozone layer. These nanobelts have deep ultraviolet (DUV) sensitivity (200–290 nm) which makes them suitable for monitoring ozone layer, since ozone layer absorbs radiation with wavelengths shorter than 290 nm.

Fang et al. [74] showed the photoconductivity in Nb_2O_5 nanobelts. The I-V curves for the device were taken at different wavelengths (320, 400, 500, 600 nm) and in the dark. Under dark condition, the current was very low (15.5 pA at 1.0 V). Also, illumination under 500 and 600 nm light, there was very less change in photoresponsivity, while large photocurrent was monitored for 320 nm light. Moreover, there was very small variation in the device performance which exhibits reproducibility of the device. These findings suggest that Nb_2O_5 nanobelts have excellent performance with respect to sensitivity, spectral selectivity and stability [75].

5.5 MONS as Ultraviolet and Near Ultraviolet Nanoprobes

Certain MONS emit nonlinear signal in Ultraviolet (UV) or Near Ultraviolet (NUV) range such as $LiNbO_3$ nanocubes [11], ZnO nanowires [76], SnO_2 nanowires [77], Ga_2O_3 nanobelts [73] etc. UV photodetectors are applicable to be utilized in high temperature conditions such as for fire detection purposes. The conventional semiconductor (Si) based UV photodetectors are prone to degradation when exposed to high temperature conditions [75]. Hence, it is required to develop such detectors which can resist the temperature variation. For this purpose, Hu et al. [78] synthesized MONS based on Ga_2O_3 nanobelts and studied UV photodetection performance at high temperature (433 K) with respect to sensitivity, stability and response speed.

Timpu et al. [11] showed that $LiNbO_3$ nanocubes emit sufficient SHG signal in the NUV range which can be used in NUV photonic applications. The lower refractive index of these nanocubes makes them good candidate for confining electromagnetic field inside them compared to semiconductor materials such as silicon, germanium, gallium arsenide etc., which have higher refractive index. Moreover, $LiNbO_3$ nanocubes with small size (below 250 nm) show efficient SHG emission in NUV range than the visible range due to increased confinement of electromagnetic field inside the nanocubes which enables the electromagnetic field to couple strongly

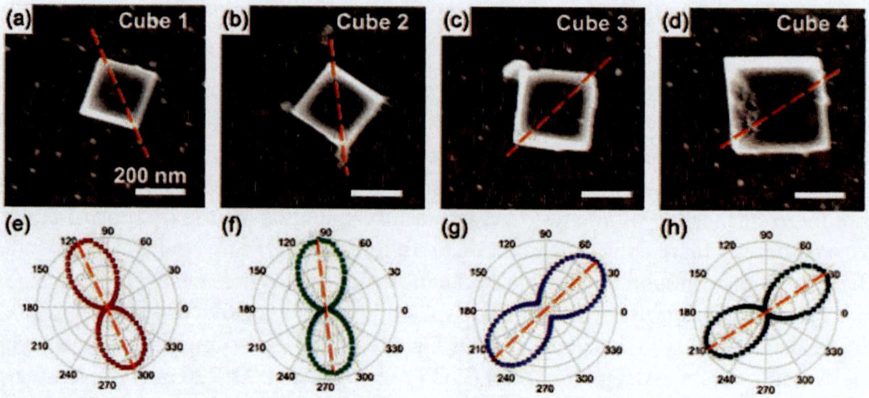

Fig. 12 $LiNbO_3$ nanocubes. **a–d** SEM images of nanocubes and **e–h** corresponding SHG polarization dependence at 400 nm. Reprinted with permission from Ref. [11]. Copyright 2019 American Chemical Society

with the nanocubes. Further, the dependency of SHG signal on the polarization of incident light was shown to predict the crystal orientation which is shown in Fig. 12. Hence, orientational details of the nanocube can be predicted, which is an important requirement for photonic device fabrication [11, 47]. These nanocubes serve as multifunctional materials for various NUV range applications such as sum-frequency generation, single photon sources etc. [75].

6 Effect of Conjugating Metal to Metal-Oxide Nanostructures

Nonlinear optical phenomena require a large distance of interaction between light and nonlinear material, devices which accompany this process tend to be bulky. For many applications such as gas sensing [79], environmental analysis, diagnosis [75] etc., the nonlinear light sources are required to be confined in three dimensional volumes at subwavelength scale [80]. One of the excellent choices is to place metallic nanostructures on top of the nonlinear substance. Choi et al. [81] synthesized a CuO/Cu_2O nanowire sensor for sensitivity towards acetone analytes. These nanowires were incorporated with silver nanoparticles which resulted in greater sensitivity towards the analyte than CuO/Cu_2O nanowires alone. Apart from it, Pt incorporated PdO nanowires exhibited enhanced gas sensing responses than PdO sensor [82]. Pt is known to have better performance for hydrogen sensing applications at room temperature. Pt nanowires conjugated to In_2O_3 nanoparticles significantly improved hydrogen sensing response [79].

High loading ability and surface properties of MONS makes them an excellent candidate for probe applications. Metos et al. [54] synthesized a nanoconjugate based on $LiNbO_3$ core chelated by gadolinium ions (Gd^{3+}) to showcase the ability of hybrid nanoprobes for near infrared (NIR) fluorescence, MRI, tumor imaging. MONS shown to be efficient for NIR probes having various advantages such as excitation wavelength flexibility, non-saturation, absence of bleaching/blinking, and localization of emitted second and third harmonic signals to identify malignant cells such as tumors. Dubreil et al. [55] showed cell labeling and tracking with the help of bismuth ferrite ($BiFeO_3$) MONS.

Timpu et al. [83] showed that the nanodimer, consisting of barium titanate ($BaTiO_3$) conjugated to gold nanoparticles ($BaTiO_3$–Au) leads to enhancement in second harmonic generation efficiency. The spectrum of the nanodimer shows pronounced resonance at 580 nm which is red shifted compared to the gold nanoparticles plasmonic resonance (Fig. 13a). The red shift is due to the $BaTiO_3$ nanoparticles attached to gold nanoparticles. The maximum SHG signal was obtained at 580 nm (Fig. 13b), which is close to the measured linear resonance for the nanodimer. The enhancement in SHG efficiency is due to the localized surface plasmon resonance of gold nanoparticles which couples with the resonance from $BaTiO_3$ nanoparticles. It was shown up to 15-fold enhancement compared to a single $BaTiO_3$ nanoparticles. Here, gold nanoparticles act as nanoantenna at the second harmonic wavelength [6].

The composite materials are useful for optical limiters and high-performance optical switchers. The applicability of these composites relies on their non-linear efficiency. It is established that metals in conjugation with MONS tend to increase the non-linear efficiency which, in turn, can increase the performance of MONS. Among the most suitable metals for composite formation are gold, silver, copper, cobalt, tin etc. which induce high values of nonlinear susceptibilities in MONS [52].

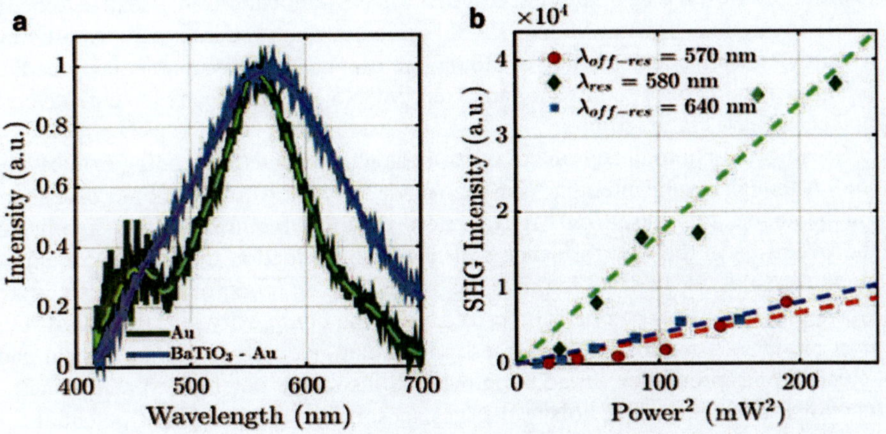

Fig. 13 **a** Linear spectra of gold nanoparticles (green line) and gold-$BaTiO_3$ nanodimer (blue line). Blue line is red shifted to 580 nm. **b** SHG intensity at 580 nm (resonance wavelength, green line) and 570, 640 nm (off resonance wavelengths). SHG signal of the nanodimer at linear resonance wavelength (580 nm) is enhanced compared to the off-resonance wavelengths. Reprinted with permission from ref [83]. Copyright 2017 American Chemical Society

Ryasnyansky et al. [84] studied the non-linear optical properties of composite made from Cu nanoparticles fabricated in indium titanium oxide (ITO) matrix. The non-linear optical phenomena and parameters were measured by z-scan technique. It was observed that the absorption of 600 nm light increased with the increment in Cu nanoparticles concentration, which implies that the nanoparticles are fabricated on the ITO matrix. The enhanced absorption is due to the surface plasmon resonance of Cu nanoparticles. Using non-linear optical technique i.e., close aperture z-scan, they have demonstrated that for unimplanted ITO, there was no transmittance in the z-scan profile. However, in the presence of Cu nanoparticles the transmittance increases with particle concentration. The enhancement of nonlinearity in presence of Cu nanoparticles was attributed to the interband transitions and surface plasmon resonance field enhancement in Cu nanoparticles.

7 Challenges and Future Outlook

In past years, several advances have been made to design MONS of different sizes, compositions, and symmetry to control nonlinear optical properties for various applications. Several studies have shown that nonlinear optical properties from MONS are governed by the surface dipole excitations, magnetic dipole interactions, electric quadrupoles, etc. One of the potential applications of MONS is to use them as cell labeling agents for therapeutic and diagnostic purposes. These applications may be controlled by the design of MONS (e.g., coating materials, size, shape, and surface charges). However, there are some challenges in the applications of MONS in therapeutic applications. One of the concerns that MONS applications may lead to toxicity in a living system. Hence, MONS should be removable systematically from living systems. Also, an understanding of biological response and environmental remediation is necessary. Once such issues are fixed, MONS can be extended to cure diverse medical diseases in the future.

Sensing and photo detection are other areas where MONS are being extensively used. Although several attempts have been made to develop different kinds of sensors and photodetectors based on MONS, there is still a requirement for significant improvements in their performance and low-cost production to meet future application demands. Besides, to improve the optoelectronic properties of MONS, more hybrid nanostructures can be synthesized since the complexity in these nanostructures produces a number of interfaces, which can give rise to unique optical and optoelectronic properties. In addition, MONS also have potential to act as multifunctional devices such as solar cells, gas sensors etc. to have better potential for practical applications and enhanced economic viability.

8 Conclusion

In this chapter, we have provided a comprehensive overview of the significant NLO properties inherent in MONS. These properties have been instrumental in enabling diverse applications across various fields such as cell imaging, sensing, and optoelectronics. The exceptional high optical contrast, photostability, and wavelength tunable properties of MONS makes them a potenial substitute for organic dyes based fluorescent probes. Expanding beyond these applications, the distinctive optical properties of MONS has led to the creation of innovative hybrid materials, such as metal conjugated to MONS and multifunctional nanodevices, aimed at enhancing the performance of optoelectronic devices. Furthermore, the remarkable electrochemical and physical properties exhibited by MONS render them suitable candidates for electrode materials in the development of next-generation supercapacitors.

References

1. S.T. Kochuveedu, Y.H. Jang, D.H. Kim, A study on the mechanism for the interaction of light with noble metal-metal oxide semiconductor nanostructures for various photophysical applications. Chem. Soc. Rev. **42**, 8467–8493 (2013)
2. I. Concina, Z.H. Ibupoto, A. Vomiero, Semiconducting metal oxide nanostructures for water splitting and photovoltaics. Adv. Energy Mater. **7**, 1–29 (2017)
3. A. Davoyan, N. Engheta, Plasmonic nanostructures. Nat. Commun. **5**, 1–5 (2014)
4. K. Wang, J. Zhou, L. Yuan et al., Anisotropic third-order optical nonlinearity of a single ZnO micro/nanowire. Nano Lett. **12**, 833–838 (2012)
5. T. Woldu, B. Raneesh, P. Sreekanth et al., Size dependent nonlinear optical absorption in BaTiO3 nanoparticles. Chem. Phys. Lett. **625**, 58–63 (2015)
6. W.T. Koo, H.J. Cho, D.H. Kim et al., Chemiresistive hydrogen sensors: fundamentals, recent advances, and challenges. ACS Nano **14**, 14284–14322 (2020)
7. F. Claudia, No ETH linear and nonlinear optics with metal oxides: from single nanoparticles to metasurfaces
8. Y.X. Zhang, Y.H. Wang, Nonlinear optical properties of metal nanoparticles: a review. RSC Adv. **7**, 45129–45144 (2017)
9. D. Staedler, T. Magouroux, R. Hadji et al., Harmonic nanocrystals for biolabeling: a survey of optical properties and biocompatibility. ACS Nano **6**, 2542–2549 (2012)
10. P.C. Ray, Size and shape dependent second order nonlinear optical properties of nanomaterials and their application in biological and chemical sensing. Chem. Rev. **110**, 5332–5365 (2010)
11. F. Timpu, J. Sendra, C. Renaut et al., Lithium niobate nanocubes as linear and nonlinear ultraviolet Mie resonators. ACS Photonics **6**, 545–552 (2019)
12. J. Riporto, M. Urbain, Y. Mugnier et al., Second harmonic spectroscopy of ZnO, BiFeO3 and LiNbO3 nanocrystals. Opt. Mater. Express **9**, 1955 (2019)
13. R. Bäumner, L. Bonacina, J. Enderlein et al., Evanescent-field-induced second harmonic generation by noncentrosymmetric nanoparticles. Opt. Express **18**, 23218 (2010)
14. E.V. Rodriguez, C.B.D. Araújo, A.M. Brito-Silva et al., Hyper-Rayleigh scattering from BaTiO3 and PbTiO3 nanocrystals. Chem. Phys. Lett. **467**, 335–338 (2009)
15. E. Hendrickx, K. Clays, A. Persoons, Hyper-rayleigh scattering in isotropic solution. Acc. Chem. Res. **31**, 675–683 (1998)
16. R. Le Dantec, Y. Mugnier, G. Djanta et al., Ensemble and individual characterization of the nonlinear optical properties of ZnO and BaTiO3 nanocrystals. J. Phys. Chem. C **115**, 15140–15146 (2011)

17. R. Podila, B. Anand, J.P. West et al., Evidence for surface states in pristine and Co-doped ZnO nanostructures: magnetization and nonlinear optical studies. Nanotechnology **22** (2011)
18. M. Yin, H.P. Li, S.H. Tang, W. Ji, Determination of nonlinear absorption and refraction by single Z-scan method. Appl. Phys. B Lasers Opt. **70**, 587–591 (2000)
19. L.F. Wu, Y.H. Wang, P.L. Li et al., Enhanced nonlinear optical behavior of graphene-CuO nanocomposites investigated by Z-scan technique. J. Alloys Compd. **777**, 759–766 (2019)
20. B. Piccione, C.H. Cho, L.K. Van Vugt, R. Agarwal, All-optical active switching in individual semiconductor nanowires. Nat. Nanotechnol. **7**, 640–645 (2012)
21. S.V. Eliseeva, G. Auböck, F. Van Mourik et al., Multiphoton-excited luminescent lanthanide bioprobes: two- and three-photon cross sections of dipicolinate derivatives and binuclear helicates. J. Phys. Chem. B **114**, 2932–2937 (2010)
22. Q. Geng, Z. He, X. Chen et al., Gas sensing property of ZnO under visible light irradiation at room temperature. Sens. Actuators B Chem. **188**, 293–297 (2013)
23. F. Particles, Optical absorption in sol-gel-derived crystalline barium titanium **12**, 3010–3012 (1998)
24. O. Crespo-Biel, B. Dordi, D.N. Reinhoudt, J. Huskens, Supramolecular layer-by-layer assembly: alternating adsorptions of guest- and host-functionalized molecules and particles using multivalent supramolecular interactions. J. Am. Chem. Soc. **127**, 7594–7600 (2005)
25. M. Chandra, S.S. Indi, P.K. Das, Depolarized hyper-Rayleigh scattering from copper nanoparticles. J. Phys. Chem. C **111**, 10652–10656 (2007)
26. C.P. Singh, K.S. Bindra, G.M. Bhalerao, S.M. Oak, Investigation of optical limiting in iron oxide nanoparticles. Opt. Express **16**, 8440 (2008)
27. S.S. Nair, J. Thomas, C.S. Suchand Sandeep et al., An optical limiter based on ferrofluids. Appl. Phys. Lett. **92** (2008)
28. P. Chantharasupawong, R. Philip, T. Endo, J. Thomas, Enhanced optical limiting in nanosized mixed zinc ferrites. Appl. Phys. Lett. **100** (2012)
29. X.L. Zhang, X. Zhao, Z.B. Liu et al., Nonlinear optical and optical limiting properties of graphene oxide-Fe3O4 hybrid material. J. Opt. **13** (2011)
30. L.A. Gómez, C.B. de Araújo, L.M. Rossi et al., Third-order nonlinearity of nickel oxide nanoparticles in toluene. Opt. Lett. **32**, 1435 (2007)
31. A. Pramothkumar, N. Senthilkumar, R.M. Jenila et al., A study on the electrical, magnetic and optical limiting behaviour of Pure and Cd-Fe co-doped CuO NPs. J. Alloys Compd. **878**, 160332 (2021)
32. D.M. Guldi, M. Prato, Excited-state properties of C60 fullerene derivatives. Acc. Chem. Res. **33**, 695–703 (2000)
33. A.K. Gupta, M. Gupta, Synthesis and surface engineering of iron oxide nanoparticles for biomedical applications. Biomaterials **26**, 3995–4021 (2005)
34. M. Mahmoudi, S. Sant, B. Wang et al., Superparamagnetic iron oxide nanoparticles (SPIONs): development, surface modification and applications in chemotherapy. Adv. Drug. Deliv. Rev. **63**, 24–46 (2011)
35. R.A. Leitgeb, W. Drexler, A. Unterhuber et al., Ultrahigh resolution Fourier domain optical coherence tomography. Opt. Express **12**, 2156 (2004)
36. J. Moger, B.D. Johnston, C.R. Tyler, Imaging metal oxide nanoparticles in biological structures with CARS microscopy. Opt. Express **16**, 3408 (2008)
37. F. Stracke, B. Weiss, C.M. Lehr et al., Multiphoton microscopy for the investigation of dermal penetration of nanoparticle-borne drugs. J. Invest. Dermatol. **126**, 2224–2233 (2006)
38. B.K. Canfield, H. Husu, M. Kauranen et al., Second-harmonic generation driven by local field asymmetry in noncentrosymmetric gold nano-Ts. Opt. InfoBase Conf. Pap. (2007)
39. Y. Pu, R. Grange, C.L. Hsieh, D. Psaltis, Nonlinear optical properties of core-shell nanocavities for enhanced second-harmonic generation. Phys. Rev. Lett. **104**, 1–4 (2010)
40. J. Richter, A. Steinbrück, M. Zilk et al., Core-shell potassium niobate nanowires for enhanced nonlinear optical effects. Nanoscale **6**, 5200–5207 (2014)
41. D. Lehr, J. Reinhold, I. Thiele et al., Enhancing second harmonic generation in gold nanoring resonators filled with lithium niobate. Nano Lett. **15**, 1025–1030 (2015)

42. L. Arizmendi, Photonic applications of lithium niobate crystals. Phys. Status Solidi Appl. Res. **201**, 253–283 (2004)
43. F. Timpu, A. Sergeyev, N.R. Hendricks, R. Grange, Second-harmonic enhancement with Mie resonances in perovskite nanoparticles. ACS Photonics **4**, 76–84 (2017)
44. C.J. Barrelet, H.S. Ee, S.H. Kwon, H.G. Park, Nonlinear mixing in nanowire subwavelength waveguides. Nano Lett. **11**, 3022–3025 (2011)
45. X. Zhang, H. He, J. Fan et al., Sum frequency generation in pure zinc-blende GaAs nanowires. Opt. Express **21**, 28432 (2013)
46. S. Liu, P.P. Vabishchevich, A. Vaskin et al., An all-dielectric metasurface as a broadband optical frequency mixer. Nat. Commun. **9**, 1–6 (2018)
47. G. Campargue, L. La Volpe, G. Giardina et al., Multiorder nonlinear mixing in metal oxide nanoparticles. Nano Lett. **20**, 8725–8732 (2020)
48. A. Sergeyev, M. Reig Escalé, R. Grange, Generation and tunable enhancement of a sum-frequency signal in lithium niobate nanowires. J. Phys. D Appl. Phys. **50**, 1–8 (2017)
49. A. Sergeyev, R. Geiss, A.S. Solntsev et al., Enhancing guided second-harmonic light in lithium niobate nanowires. ACS Photonics **2**, 687–691 (2015)
50. Q. Zhang, K. Zhang, D. Xu et al., CuO nanostructures: Synthesis, characterization, growth mechanisms, fundamental properties, and applications. Prog. Mater. Sci. **60**, 208–337 (2014)
51. M. Ando, K. Kadono, M. Haruta et al., Large third-order optical nonlinearities in transition-metal oxides. Nature **374**, 625–627 (1995)
52. S. Debrus, J. Lafait, M. May, Z-scan determination of the third-order optical nonlinearity of gold: silica nanocomposites **4469** (2008)
53. G.S. Boltaev, R.A. Ganeev, P.S. Krishnendu et al., Nonlinear optical characterization of copper oxide nanoellipsoids. Sci. Rep. **9**, 1–8 (2019)
54. F. Ramos-Gomes, W. Möbius, L. Bonacina et al., Bismuth ferrite second harmonic nanoparticles for pulmonary macrophage tracking. Small **15**, 1–10 (2019)
55. L. Dubreil, I. Leroux, M. Ledevin et al., Multi-harmonic imaging in the second near-infrared window of nanoparticle-labeled stem cells as a monitoring tool in tissue depth. ACS Nano **11**, 6672–6681 (2017)
56. X. Tang, E.S.G. Choo, L. Li et al., Synthesis of zno nanoparticles with tunable emission colors and their cell labeling applications. Chem. Mater. **22**, 3383–3388 (2010)
57. P. Pantazis, J. Maloney, D. Wu, S.E. Fraser, Second harmonic generating (SHG) nanoprobes for in vivo imaging. Proc. Natl. Acad. Sci. U S A **107**, 14535–14540 (2010)
58. S. Pillarisetti, S. Uthaman, K.M. Huh et al., Multimodal composite iron oxide nanoparticles for biomedical applications. Tissue Eng. Regen. Med. **16**, 451–465 (2019)
59. A. Rogov, Y. Mugnier, L. Bonacina, Harmonic nanoparticles: Noncentrosymmetric metal oxides for nonlinear optics. J. Opt. (United Kingdom) **17** (2015)
60. D. Segets, L.M. Tomalino, J. Gradl, W. Peukert, Real-time monitoring of the nucleation and growth of Zno nanoparticles using an optical hyper-rayleigh scattering method. J. Phys. Chem. C **113**, 11995–12001 (2009)
61. M.R.V. Sahyun, Hyper-Rayleigh scattering (HRS) spectroscopy applied to nanoparticulate TiO2. Spectrochim. Acta—Part A Mol. Biomol. Spectrosc. **58**, 3149–3157 (2002)
62. P.K. Das, Chemical applications of hyper-rayleigh scattering in solution. J. Phys. Chem. B **110**, 7621–7630 (2006)
63. C.N. Rao, A. Bharadwaj, S. Datar, S.N. Kale, Lithium niobate nanoparticulate clad on the core of single mode optical fiber for temperature and magnetic field sensing. Appl. Phys. Lett. **101** (2012)
64. S.K. Mishra, N. Choudhury, S.L. Chaplot et al., Competing antiferroelectric and ferroelectric interactions in NaNbO3: neutron diffraction and theoretical studies. Phys. Rev. B—Condens. Matter. Mater. Phys. **76**, 1–8 (2007)
65. E. Valdez, C.B. De Araújo, A.A. Lipovskii, Second harmonic scattered light from a transparent glass-ceramic containing sodium niobate nanocrystals. Appl. Phys. Lett. **89**, 1–4 (2006)
66. E.L. Falcão-Filho, C.A.C. Bosco, G.S. Maciel et al., Third-order optical nonlinearity of a transparent glass ceramic containing sodium niobate nanocrystals. Phys. Rev. B—Condens. Matter. Mater. Phys. **69**, 1–8 (2004)

67. A. Sergeyev, R. Geiss, A.S. Solntsev et al., Second-harmonic generation in lithium niobate nanowires for local fluorescence excitation. Opt. Express **21**, 19012 (2013)
68. Y. Jun, J. Choi, J. Cheon, Shape control of semiconductor and metal oxide nanocrystals through nonhydrolytic colloidal routes. Angew. Chem. Int. Ed. Engl. **45**, 3414–3439 (2006)
69. M.M. Rahman, A.J.S. Ahammad, J.-H. Jin et al., A comprehensive review of glucose biosensors based on nanostructured metal-oxides. Sensors (Basel) **10**, 4855–4886 (2010)
70. S. Sun, X. Zhang, Y. Sun et al., A facile strategy for the synthesis of hierarchical CuO nanourchins and their application as non-enzymatic glucose sensors. RSC Adv. **3**, 13712–13719 (2013)
71. Z. Jin, J. Wang, Flexible high-performance ultraviolet photoconductor with zinc oxide nanorods and 8-hydroxyquinoline. J. Mater. Chem. C **2**, 1966–1970 (2014)
72. R.S. Chen, C.A. Chen, H.Y. Tsai et al., Photoconduction properties in single-crystalline titanium dioxide nanorods with ultrahigh normalized gain. J. Phys. Chem. C **116**, 4267–4272 (2012)
73. L. Li, E. Auer, M. Liao et al., Deep-ultraviolet solar-blind photoconductivity of individual gallium oxide nanobelts. Nanoscale **3**, 1120–1126 (2011)
74. X. Fang, L. Hu, K. Huo et al., New ultraviolet photodetector based on individual Nb2O5 nanobelts. Adv. Funct. Mater. **21**, 3907–3915 (2011)
75. W. Tian, H. Lu, L. Li, Nanoscale ultraviolet photodetectors based on onedimensional metal oxide nanostructures. Nano Res. **8**, 382–405 (2015)
76. K. Liu, M. Sakurai, M. Liao, M. Aono, Giant improvement of the performance of ZnO nanowire photodetectors by Au nanoparticles. J. Phys. Chem. C **114**, 19835–19839 (2010)
77. L. Hu, J. Yan, M. Liao et al., Ultrahigh external quantum efficiency from thin SnO2 nanowire ultraviolet photodetectors. Small **7**, 1012–1017 (2011)
78. R. Zou, Z. Zhang, Q. Liu et al., High detectivity solar-blind high-temperature deep-ultraviolet photodetector based on multi-layered (l00) facet-oriented β-Ga 2O3 nanobelts. Small **10**, 1848–1856 (2014)
79. Y. Fang, Y. Liu, L. Li et al., Simultaneously-engineered composition and spatial position of metal/metal-oxide nanowires for hydrogen sensing applications. ACS Appl. Nano. Mater. **5**, 3667–3675 (2022)
80. F. Dutto, M. Heiss, A. Lovera et al., Enhancement of second harmonic signal in nanofabricated cones. Nano Lett. **13**, 6048–6054 (2013)
81. Y.M. Choi, S.Y. Cho, D. Jang et al., Ultrasensitive detection of VOCs using a high-resolution CuO/Cu2O/Ag nanopattern sensor. Adv. Funct. Mater. **29**, 1–9 (2019)
82. H.J. Cho, V.T. Chen, S. Qiao et al., Pt-functionalized PdO nanowires for room temperature hydrogen gas sensors. ACS Sens. **3**, 2152–2158 (2018)
83. F. Timpu, N.R. Hendricks, M. Petrov et al., Enhanced second-harmonic generation from sequential capillarity-assisted particle assembly of hybrid nanodimers. Nano Lett. **17**, 5381–5388 (2017)
84. A.I. Ryasnyansky, B. Palpant, S. Debrus et al., Nonlinear optical properties of copper nanoparticles synthesized in indium tin oxide matrix by ion implantation. J. Opt. Soc. Am. B **23**, 1348–1353 (2006)

Functionality of Metal Oxide-Based Core-Shell Nanoparticles

Umer Mushtaq, Vishal Sharma, Hendrik C. Swart, and Vijay Kumar

Abstract In recognition of their distinctive properties and diverse spectrum of applications in sectors such as catalytic processes, photocatalysis, biomedicine, biology, delivery of drugs, and sensing devices, core-shell nanostructures are currently attracting increased attention. The chemical composition and morphology of the core along with the shell of such nanomaterials can be transformed, leading to a wide range of tunable characteristics that may be essential in the development of catalyst and photocatalytic, and provide potential better and more efficient platforms for biomedical research. Due to their adjustable nanoscale functionality in both the core and shell parts, metallic materials such as Au, Ag, and Zn, along with metal oxide semiconductors such as ZnO, TiO_2, SiO_2, and SnO_2, have generated a lot of research attention in areas such as biomedicine, catalysis, and photocatalysis. In this chapter, the main focus was on some major synthesis techniques for the fabrication of core-shell nanostructures and various techniques employed for functionalizing the core, shell, or both in order to achieve improved and enhanced performance in different applications.

Keywords Functionalization · Metal oxides · Core-shell · Nanoparticles

U. Mushtaq · V. Kumar (✉)
Department of Physics, National Institute of Technology Srinagar, Srinagar, Jammu and Kashmir 190006, India
e-mail: vj.physics@gmail.com

V. Sharma
Institute of Forensic Science & Criminology, Panjab University, Chandigarh 160014, India

U. Mushtaq · H. C. Swart · V. Kumar
Department of Physics, University of the Free State, P.O. Box 339, Bloemfontein ZA9300, South Africa

V. Kumar et al. (eds.), *Optical Properties of Metal Oxide Nanostructures*,
Progress in Optical Science and Photonics 26,
https://doi.org/10.1007/978-981-99-5640-1_6

1 Introduction

Nanotechnology can be understood as a branch of modern science that deals with materials having sizes in the range of 1–100 nm [1]. This technology has brought about advancement in many aspects of our day-to-day lives, such as biomedical, pharmacological, agricultural, environmental, chemical science, physics, electronics, and information technology. The size of this kind of range has revolutionized most materials and technologies and thus has huge significance in a number of biomedical domains, including bioimaging, drug carrier systems, diagnostics, and many other domains that assist in achieving advancements in many fields and at the same time save lives. Because of the discovery of numerous novel characteristics such as flexibility of functionalization, and coupling of proteins, nanomaterials have proved to be valuable in almost all fields of modern technology, and their reduced dimension along with increased surface-to-volume ratio are the major characteristics that contribute to this. During the early 90 s, very few research articles were documented, which increased exponentially after that [2]. Our comprehension of intricate biological mechanisms that are directly accountable for illness diagnosis and therapy has been greatly expanded by the use of such novel nanostructured materials. While a number of techniques are still in the early stages of development, several are already being used in routine clinical settings. For many applications, including cellular healing, drug distribution, therapeutic uses, and diagnostic tools, the effective surface functionalization of newly produced nanoparticles is also crucial [3, 4]. Core-shell nanostructures can be understood as a family of particles that are made of two or more distinct layers of materials with distinct properties. Amongst them, the central part is known as the core, while the rest of the layers constitute the outer part of this core/shell nanostructure [5, 6]. Such an architecture offers the chance to synthesize a material that demonstrates traits and behaviors not possible with the core or shell components alone. Based on the synthesis parameters as well as the specific requirement, the central part can be in a solid, liquid, or gaseous state, whereas the exterior shell is often in solid form and synthesized using either organic or inorganic components [6]. Core-shell nanostructures could be divided into four classes on the basis of components used to synthesize core and shell material; these four groups can be identified as a combination of materials with core and shell made from inorganic/organic, organic/inorganic, organic/organic, and inorganic/inorganic nanoparticle components. Several biological, chemical, magnetic, and optical characteristics of core-shell nanoparticles are affected by changing the constituents of the core and shell. Core-shell nanomaterials are used in a variety of industries, including beverage and beauty products, biomedicine, healthcare, and material science, due to their distinctive qualities. Core/shell materials have been used as bioreactors, sustained release technologies, packaging systems to shield active payloads from deterioration and chemical change, as well as storage devices for heat energy [7, 8]. Additionally, medicines may be placed in several layers in a core/shell nanostructure to allow for their successive delivery inside the body [9, 10]. These types of structures can contain materials with hydrophobic properties as well as substances with hydrophilic characteristics [11]. Cellular encapsulation during the process of transplantation of cells to a wounded region for restoring damaged tissue is among the main applications of core/shell-structured materials. Because of the presence of shell material, these cells are shielded from the surrounding environment, allowing them to have a prolonged life and hence eventually adapt to the host tissue [12].

While using nanomaterials in the biomedical field, morphological properties must be taken into account when talking about surface functionalization. While taking into account the magnetic properties of a nanostructure, its size and the surface proportion of atoms within are crucial factors. As a result, the nanoparticle itself as well as the oxide of a nanomaterial generally possesses ferromagnetic properties. Research has shown that some nanomaterials, such as niobium nitride, possess ferromagnetic properties, while non-magnetic nanomaterials such as cerium oxide and aluminum oxide exhibit magnetic hysteresis at standard temperatures. This information tends to help further understand the attributes of ferromagnetism. Due to their small size, nanostructured substances have enhanced ferromagnetic characteristics [13]. A nanomaterial's total magnetization is comprised of two consequences: one that takes place on the exterior and the other that takes place within the inner core of the core/shell nanostructure. This study indicates that the presence of surface defects tends to encourage a magnetic perturbation that persists inside the nearest stratum. Both the surface effect as well as the anisotropic magnetic property of a nanoparticle are the most important properties to comprehend; as a result, comprehension of these properties is essential in the creation of nanomaterials having use in healthcare, such as magnetic resonance imaging and magnetic hyperthermia [13, 14]. Investigators have enhanced the attributes of nanomaterials, particularly stability, in a novel and important way through the process of surface modification or functionalization [15, 16]. Due to its superior stability against deterioration in comparison to biological-based shells, the silica covering can be considered one of the optimal solutions for surface modification. According to the study, functionalized silica has better characteristics than unfunctionalized silica, and to enhance the stability of the nanomaterial, biological substances such as enzymes and medications have been immobilized onto the porous morphology of silica [17]. Regarding various biological and ecological uses, silica comprises units of silanol on its exterior, and their existence enhances their ability for modification, biocompatibility, and hydrophilic-hydrophobic proportion [18]. Throughout the tests, Hui et al. [19] coated silica on nanomaterials using the Stöber technique, and Roca et al. [20] covered silica over magnetite nanostructures using the sol-gel procedure. The research material contains numerous publications involving microemulsion and reversed microemulsion that outlines both silica and iron oxide coating techniques. Such fabrication techniques have important ramifications in the biomedical field and provide key benefits for generating preferred nanostructures. To create a novel substance for biological uses, scientists have to successfully integrate the architecture of cubic magnetic nanomaterials into the framework of magnetite and then functionalize it with silica. Via this investigation, it emerged how easily and creatively the tunable silica thickness could be developed, as well as how the stability and biocompatibility of Fe_3O_4/SiO_2 core/shell nanocubes could be improved. They are interesting candidates for a variety of biosensing applications because they exhibit outstanding magnetic properties and sensitivity for attaching molecules to the synthesized core/shell structure [21]. Numerous studies demonstrate how to employ magnetism in various specialized applications, such as the elimination of metal objects such as safety pins [22, 23], gunshots, grenade shrapnel, dental prostheses, and catheters moving within the body and brain [14, 24].

2 Synthesis of Metal Oxide Core-shell Nanoparticles

During the past few years, metal oxides (MO) have proved to be an important group of chemicals that have been extensively used to generate core as well as shell materials in the case of core/shell nanomaterials because of their peculiar properties. Based on the manner of transition, the fabrication of MO can be divided into two groups: the first being the transition to the solid phase from the gas phase, and the second being the transition from the liquid phase to the solid phase [25]. While performing the transition from gas to solid phase, mostly two techniques are employed: chemical vapor deposition (CVD) and pulse laser deposition (PLD). Generally, compared to the gas-to-solid phase transition, liquid-to-solid phase transformation is the more prevalent mode for the synthesis of core or shell material in the process of generation of metal oxide-based core-shell nanostructures. The most prominent techniques in the case of liquid-to-solid phase types of synthesis are either sol-gel types of synthesis or co-precipitation methods. Among the two synthesis methods mentioned, the co-precipitation method has been more prevalent for the fabrication of metal oxide core-shell nanoparticles, while the sol-gel process is most frequently used to create silicon and titanium oxides.

2.1 General Overview

The sol-gel method of synthesis (Fig. 1) can be described as a wet chemical procedure that is frequently employed in material fabrication techniques that can be used for a variety of applications in several associated fields of science and technology such as biomedicine, sensors, water purification, and much more. This synthesis method has been extensively used for the fabrication of metal oxide nanostructures. The progressive creation of the solid-phase network is achieved through the integration of two stages, hydrolysis, and polycondensation, of specific molecular constituents. Sol-gel synthesis has attracted attention since the mid-1800s, after which this technique was later modified by Werner Stober in 1968 [26]. Studies in this field have greatly expanded since the 1990s, which resulted in more than 35 k research articles up to the year 2000 wherein different nanostructures were synthesized via this method [27, 28]. Improved sol-gel techniques were employed by multiple research teams to synthesize various types of metal oxides [29–32]. For the synthesis of MOs, the fabrication method generally employed is hydrolysis and polycondensation of

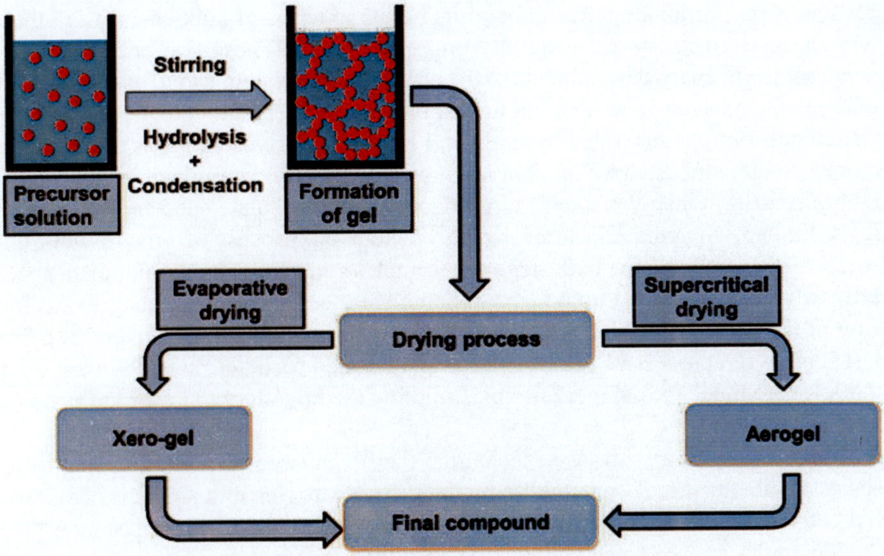

Fig. 1 Schematic diagram representing the sol-gel synthesis technique

metal alkoxide precursors, followed by aging and drying with the help of ambient conditions in the surrounding atmosphere.

Another synthesis technique employed to synthesize metal oxide core-shell nanoparticles is the co-precipitation method. Due to its adaptability and affordability, the precipitation reaction technique is one of the most popular ways to make core-shell nanoparticles. In this reaction, at least one water-insoluble salt forms a precipitate out of the medium after being mixed with two or more water-soluble salts. The most crucial factor that needs to be taken into account is the solubility product of the precipitated compound, which is essential in deciding whether the reaction is successful. The reaction product initially forms in the liquid phase; however, as soon as its content exceeds the compound's solubility product value in the reaction medium, particles begin to form through nucleation and growth. Nucleation, growth, and agglomeration are three separate steps in the procedure of nanoparticle fabrication by precipitation. Reactions such as acid-base processes, which take place in the liquid phase but produce solid-state particulates, can start precipitation reactions, including redox precipitation. A number of energy factors have an impact on this technique, which is also referred to as reactive crystallization. The activation energy of the reaction in the liquid phase is the first stage in the procedure, and lower activation energy levels are associated with faster reaction rates. After the reaction is finished, a few embryos in the bulk liquid phase are produced that contain atoms. These embryos change from a liquid to a solid state when they reach a critical radius or nucleus. New solid–liquid interfaces with particular surface energies inversely correlated to their radii are produced as a result of the generation of solid particles in the bulk phase. The energy increase that results from the formation of a solid particle is proportionate to the cube of its radius in terms of bulk energy, and these two energy terms have diametrically opposed natures. These two energy components are the primary determinants of the change in the system's total energy, though other elements such as particle motion in the medium, crystal defects, and electrostatic contributions may also have a small impact [33]. Nucleation is typically the quicker stage, and the development time is subsequently determined by the overall rate of reaction when the solubility product of the particulate substance is exceedingly limited. Nanomaterials develop throughout the process of growth while the reaction diffuses from the bulk stage to the nuclei's surface. The development speed primarily depends on the time taken to complete the reaction, and the rate of diffusion of molecules from the bulk phase to the nuclear interface is delayed [34]. The last stage is the process of agglomerating synthesized particles, in which very small particles are made into bigger ones by shrinking the boundary to lower the system's total energy level.

Spray-drying is an advanced technique that is capable of scaling up for large-scale manufacturing of core-shell-structured nanoparticles in a single process. But typically, the consequent product fragments aggregate into a solid dispersion [35]. Such particles are produced by pumping a solution of liquid or a suspension through a spray-drying device, in which it is divided up into tiny drops because of heated nitrogen gas [36]. After being separated in a cylindrical device, drops are dried and then gathered in a collector. Particulate matter possessing a particular shape and size could be generated by means of capillary forces, and to some extent, agglomeration is caused by the evaporation-driven contraction of the enclosed droplets. The composition of the precursor solution, along with the drying procedure, has a big effect on the surface morphological characteristics [37]. Large-scale core-shell nanoparticles that are sensitive to heat can be produced via this approach in an accurate and simple manner [38]. For example, Li et al. [39] created silicon/carbon hybrid microspheres to be utilized as anodes in lithium-ion batteries using a spray drying technique and surface coating process. These microspheres have a hierarchy-type core-shell structure. Spray drying has also been employed by Bruinsma et al. [37] to produce hollow, spherical particulates.

A popular approach for producing polymer-type core-shell nanoparticles with dimensions starting from tens of nanometers is the electrohydrodynamic atomization process, additionally referred to as electro-spraying. In order to create a jet and droplets, this technique involves feeding a high voltage to a conductive polymer-based sludge-type mixture at the top of a needle [40]. Surface tension is surmounted by the force of electrostatics, which also produces monodisperse drops at a regulated pace [41]. While electrospinning and electrospraying operate on essentially the same principles, they vary in terms of the degree of concentration of the chemical solution employed and the chain attachment density of the polymeric slurry [41, 42]. While electrospinning uses an increased concentration of solution to create fibers, electrospraying uses a lesser concentration to generate nano- and micro-sized particles [43]. In a single process, electrospraying can produce droplets without using a lot of solvents or templates, and by adjusting the needle diameter, it is simple to change the dimensions of the droplets from nanoscale to microscale [44]. The dimensions and form of the droplets generated through this technique can be regulated by changing the variables that are being experimented with, and they typically exhibit a narrow size distribution. The fabrication parameters for electro-spraying equipment are more straightforward and affordable than those for traditional mechanical spraying setups [45]. As an example, using helical electro-spraying, Hwang et al. [46] created core-shell capsule-type structures with uniformly sized polymethylmethacrylate (PMMA) core and PCL shell and polystyrene (PS) core and PCL shell. Owing to the autonomous flow of solutions from an inner or exterior capillary nozzle, the combination of polymeric species in the core and shell can be tailored. Micro-sized capsules have been created by introducing different polymeric solutions in a one-step procedure. In accordance with some research, corona discharge can be reduced and, at the same time, the size of particles can also be controlled by combining single-nozzle electro-spraying with spray drying [47, 48]. The coaxial electro-spraying and spray-drying processes, on the other hand, produce particulates that possess a higher degree of uniformity in size and structure. Additionally, the agglomeration of particles can be avoided through the use of electrical potential [47, 49]. In order to create core-shell nanoparticles for protein drugs, Ho et al. combined spray drying and electrohydrodynamic coaxial jetting, and as a consequence, the electrohydrodynamic force reduced agglomeration and the size of the particles. The particle and release analysis of a model protein (lysozyme) demonstrated that the use of an electric field also increased its encapsulating effectiveness [47]. To create polymer or ceramic-type hollow spheres, other studies combined the sol-gel synthesis technique with hetero-phase polymerization [50].

Many different types of compounds and elements used as a core in core-shell nanoparticles, such as cholesteric fluid crystalline structures [51], dyes [52], oils [53], and peptides, along with protein molecules [54], have been successfully encapsulated using the process of interfacial polymerization or polycondensation. The moderate encapsulating properties of this technique enable the protein's biological properties to be preserved [53]. This kind of polymerization takes place once monomer compounds are polymerized at an emulsion surface, causing the shell to expand quickly. At first, an aqueous solution is used to emulsify a phase that is organic and made up of reactive monomers that are solvable in oil and core components, resulting in the formation of droplets. Isocyanate or acidic chloride are the typical reacting monomers utilized in the aqueous phase. When a highly reactive monomeric compound that degrades in water is added, the reaction among the two different monomers at the boundary of the core particles distributed in the liquid phase causes the polymer to expand [54]. Interfacial polymerization has been used by Shenoy et al. [55] to produce core-shell nanoparticles possessing liquid centers. The interfacial structure begins via an enzymatic interaction between glucose and glucose oxidase, which results in the formation of hydrogen peroxide that, when present with iron (Fe^{2+}), generates hydroxyl radicals. The methacrylate monomer strands are polymerized as a result of the hydroxyl radicals. In order to prevent the initiation reaction from spreading, one or more of the starting materials are kept in a hydrogel center while the remaining components and monomers are kept in a shell solution.

In-situ polymerization has recently attracted more attention as a technique for altering the exterior of inorganic particulates such as silica and copper sulfide and nanolayers such as carbon nanotubes. This can be attributed to the strong interfacial contact, broad coverage, and durability it provides [56, 57]. The explosive crystalline material and polymeric bonding agent have weakened van der Waals interactions, which means that this method can't fully fix imperfections in the surface covering [56]. In-situ polymerization frequently results in imperfect polymerization and cannot be applied to all polymers [58]. The primary distinction between in-situ polymerization and interfacial polymerization is that the former does not involve reactive monomers in the core structure. A number of research studies have been done during the past few years on the development of core-shell particles employing in-situ polymerization [48]. Lin et al. [59] developed core-shell 1,3,5-triamino-2,4,6-trinitrobenzene/polydopamine (PDA) particles by employing this technique. In both instances, the teams altered the coating duration to modify the quantity of PDA covering on the exterior layer. Nabavi et al. [60] also documented synthesis of Fe_3O_4@polyaniline core-shell microspheres with clearly established blackberry-such as morphological characteristics, and Xuan et al. [61] published the fabrication of microcapsules with an n-hexadecane core and melamine formaldehyde as the outer covering.

Suspension polymerization is a form of chemical reaction employed to fabricate core-shell nanoparticles via encapsulation methods. Although the monomers and initiator elements are both insoluble in the solution, this technique includes starting polymerization in monomers using a dissolvable initiator. The monomeric stage, which comprises monomers, a bursting substance, and an initiator, is suspended in the solution alongside a stabilizing substance, employing a stirring device in the initial phase to create microdroplets. Inside the drops, the subsequent step starts the procedure of polymerization, which eventually results in the formation of identical-sized polymeric microbeads. Monomer, polymer, and water are the three distinct stages that make up the final granules [62, 63]. In order to expand using heat-core-shell frameworks with hollow interiors that contained an inert hydrocarbon of about 10–20 m in dimension, Jonsson et al. [63] utilized the suspension polymerization technique. Particulate surface morphology is greatly influenced by the monomer input makeup and polymerization temperature. Identical to this, Ting et al. [64] produced core-shell structures made of poly(vinyl neo-decanoate) interior cross-linked using poly (ethylene glycol di-methacrylate) and a poly (ethylene glycol methacrylate) exterior using suspension polymerization and a traditional radical-free polymerization method. To generate polymeric bead-such as core-shell structures, Besteti et al. [65] and Lenzi et al. [66] utilized a mix of suspension and emulsion polymerization techniques. Both of them started by using the common suspension polymerization procedure to produce a non-porous polymeric core material. Consequently, throughout the suspension phase, they incorporated all the elements required for conventional emulsion polymerization into the reaction solution. The emulsified polymeric particles gathered on the outside of the suspended polymeric granules due to the elevated viscosity and adhesion of the suspended polymer particulates, which resulted in the development of a core-shell bead.

Precipitation polymerization is an effective method for creating microscopic core-shell particles. In contrast to the dispersion polymerization technique, this technique starts out with a solution that is uniform in consistency; however, it's starting particulates do not expand in the solution, and hence the procedure known as "precipitation polymerization" is a consequence [62]. Precipitation polymerization, in contrast with alternative techniques that employ inhibitors or emulsifiers, yields polymeric microscopic particles with spotless interfaces and functional groups [67]. Li et al. [67] have effectively produced core and shell polymeric particles lacking stabilizers using the two-step precipitation polymerization technique. Fabricated particles are comprised of an inner core of divinylbenzenes with a diameter of between two and eight m and a shell material consisting of functional monomers such as mono-vinyl, chloromethyl-styrene, or divinyl methacrylic. The reaction substrate and the chosen monomer determine whether the shell is porous or non-porous. Acetonitrile along with toluene/acetonitrile combinations have been used as the reaction mediums, and monodisperse copolymer particles were generated by adding functional co-monomers such as maleic anhydride, chloromethyl-styrene, and methacrylates. Toluene, for example, is an appropriate co-solvent that produces permeable spheres with sizes ranging from nanometers to micrometers. Precipitation polymerization has additionally been effectively utilized in other investigations to create core- or shell-structured materials [68]. Through precipitation polymerizing styrene and divinylbenzene in the midst of titanium dioxide, Werts et al. [69] were able to cover titanium dioxide using cross-linked polystyrene shells, resulting in inorganic and organic core and shell particulates. They observed that, as a pretreatment procedure, 3-(trimethoxy-silyl) propyl methacrylate enhanced the titanium dioxide particulate's ability to be encapsulated. Precipitation polymerization was used in a two-step procedure by Barahona et al. [70] to produce microscopic molecularly imprinted polymeric (MIP) particles that possess a core-shell design. They created impermeable polymeric cores initially by precipitation polymerizing divinylbenzene-80 (DVB-80) in acetone, which they subsequently utilized as seeds for the useful molecularly imprinted polymeric shell materials. The shells were developed by copolymerizing DVB-80, methacrylic acid (MAA), and thiabendazole (TBZ) with each other in a mixture of the solvents toluene and acetonitrile to create a permeable shell.

The fabrication of core-shell particles containing a polymer-based core and shell utilizing a variety of polymerization techniques has been described in multiple scientific articles. Rudin et al. [71] used a series of seeded growth emulsified polymerizations to generate mono-disperse 3 m diameter rubber core-shell microparticles. Laus et al. [72] created monodisperse polystyrene microparticles with sizes varying from 3 to 10 m by stabilizing their outer layer by utilizing poly-carboxylic acid or poly-epichlorohydrin steric. Polystyrene seeds have been subjected to a two-step polymerization procedure by Okubo et al. [73]. In the initial phase, they used dispersion polymerization within an ethanol–water mixture having 2,20-azobisisobutyronitrile as the initiator and poly (acrylic acid) as the stabilizing agent in accordance with different circumstances to create monodisperse polystyrene particles possessing micrometer dimensions. Consequently, in the presence of 1.9-Mm monodisperse polystyrene particulates, seeded copolymerization of styrene and chloromethyl styrene was observed to take place.

Typically, agitation by mechanical means has been used to produce intense shear in two distinct emulsification stages in an attempt to produce several low-order emulsifications [60, 74]. Initially, two solutions that do not get mixed well are thoroughly combined to create an initial single emulsion. A double emulsion is generated by carefully combining the initial emulsion with a sizable amount of a solution that doesn't mix with the other two [75, 76]. In comparison with alternative techniques, emulsification is a relatively straightforward technique for creating numerous emulsions; however, bulk combining has several disadvantages, such as poor batch-to-batch reproducibility, increased consumption of materials and energy, uncertain regulation of processing variables, and challenges in managing the amount and size of encapsulated core droplets [60]. The generation of core-shell particles that serve as prototypes for core-shell particulates using the microfluidic approach is a substitute for conventional techniques. The drawbacks of techniques involving the mixing of large amounts could be solved by the use of microfluidics. Microfluid-based systems are compact and have an affordable production cost [77]. The ability to regulate small amounts of fluid in microchannels with extreme precision with microfluidic technology makes it possible to precisely alter the dimension of the core material along with the thickness of the shell [76]. This technique is capable of producing particles with a small distribution of size periodically. The ability to create microemulsions using two, three, or multiple cores represents one of the most noteworthy benefits of microfluidics [78]. Microemulsion-based cell packaging has advanced significantly in recent years owing to advancements in microfluidic technology [78]. Due to the fact that the complicated procedures involved, including coating, modifying the surface, chemical responses, centrifuging, cleansing, and dispersing, may damage the synthesized particles, traditional emulsifying techniques for encasing live cells have proven to be problematic. Microfluidics, on the other hand, can preserve consistency and cell viability [79].

2.2 Other Factors Affecting Core-shell Nanoparticle Synthesis

Complete and total control over synthesis variables, viz., nucleation and growth of semiconducting shell nanomaterials enclosing the metal core-based nanostructure, is necessary for the development of metal core-metal oxide shell nanostructured materials. Multiple experimental variables, including heat provided during synthesis, duration of synthesis, solution concentration, mismatches in the lattice characteristics, availability of surfactants, and interfacial energy, are used to modulate such pathways. A large amount of interfacial energy is typically generated between the metal oxide and metal as a result of this discrepancy in structural characteristics and the lack of chemical interaction among metal and metal oxide surfaces. However, it would be extremely difficult to produce the core-shell design if the two-component materials had a structural mismatch. To change interfacial energy among these two-component materials, however, an appropriate ligand or surfactant could be useful. Although the SiO_2 shell and noble metal-based core nanomaterial combination are very well known, this type of method hasn't been frequently used for other oxide shells, primarily due to the agglomeration caused within the solution [80]. Additionally, the use of metallic precursor salt is required for the synthesis of many metal oxides, with the exception of SiO_2, which might cause the seeds of nanomaterials to aggregate. However, if using metallic salt as a starting material is mandatory, incorporating the right surfactant, and ligand, could be a more advantageous substitute for developing core-shell structured materials with nanosized morphology. Because of this, even if their corresponding pure oxides are frequently accessible, there is still relatively little research material addressing the fabrication of core-shell nanomaterials using metal oxides. The option of a suitable ligand or surfactant remains intriguing. This is probably the most probable reason why core-shell nanostructures made of metal and metal oxides are unique in this circumstance. Both TiO_2 and SnO_2 have cube-type crystalline structures similar to noble metals such as Ag, Au, and Pt, and have their crystallite properties for core-shell building well matched, enabling the synthesis of TiO_2 and SnO_2 core-shell nanostructures to be significantly simpler. Moreover, because ZnO possesses a hexagonal crystal structure and gets generally developed unevenly along the c-axis of a crystal structure, it gets exceedingly challenging to build a ZnO-based shell around a noble metal to generate a core-shell type design. Due to this result, the development of a core-shell structured material by the use of metal and ZnO has proved to be difficult due to their mismatched lattices, and unfortunately, a surfactant or a ligand could not assist in solving the problem of encapsulation throughout the fabrication procedure [81]. In order to address this issue, Zhao et al. [82] successfully developed Ag–ZnO core-shell nanostructures using the laser ablation method. They suggested that the manufactured nanomaterials could possess exceptional potential for use in the creation of micro- and nano-sized optoelectronic devices. These researchers successfully created a core-shell nanostructure with a silver core and zinc oxide shell for the first time by laser ablation with the help of a 248 nm KrF excimer pulsed laser within a liquid solution. The thickness of the ZnO-based shell could be adjusted to vary the Ag–ZnO core-shell nanomaterial's strong surface plasma resonance absorbing property. The metal-based core with ZnO shell architecture and a unique structural design was effectively synthesized by Sun et al. [83], thus allowing a thorough understanding of the growth mechanism. Regarding encapsulating ZnO over different types of cores such as metals, oxides, polymeric nanostructures, graphene oxide, and carbon nanotubes, these scientists developed a universal method in which they also demonstrated that such a method is effective and can be expanded to use a wide range of different semiconducting materials as the shell, including Fe_3O_4, MnO_2, MnO_4, MnO, Co_2O_3, TiO_2, Eu_2O_3, Gd_2O_3, ZnS, Tb_2O_3, CdS, and β-Ni $(OH)_2$. They discovered that 4-mercaptobenzoic acid and polyvinylpyrrolidone (PVP) was crucial for the production of Au–ZnO core-shell nanostructures. Researchers have also demonstrated that the interfacial energy between Au and ZnO could be significantly reduced by the use of 4-mercaptobenzoic acid. The amphiphilic property of PVP is essential for encapsulating ZnO over a metallic core containing hydrophobic ligands, as indicated by the fact that ZnO doesn't interact effectively with the hydrophobic binding sites present on the exterior of the gold core in the absence of the PVP polymeric substance. Xu et al. [84] employed a hydrothermal technique to fabricate an innovative core-shell type of structure using Ag metal along with WO_3 metal-oxide nanoparticles for efficient localized surface plasmon transmission. Efficient preparation of Ag nanoparticles having a diameter between 25 and 60 nm is primarily dependent on morphology and dimensions. Ag nanostructures were first dissolved in a Na_2WO_4 liquid, and then nitric acid was added. This solution was observed to precipitate Ag_x–H_2WO_4. This precipitation was then calcinated at high temperatures, through which the excess water was removed, thus allowing the generation of an Ag_x–WO_3 core-shell nanostructure with a shell width of ~60 nm and a total particle size of ~200 nm. Figure 1 [85] shows the same core/shell nanoparticle synthesis by two different techniques, viz., the facile synthesis technique and the laser ablation method, and clearly highlights some of the very minute variations present in their structure. Figure 2a displays a TEM image of the zinc oxide outer shell covering the silver nanomaterial-based core, where it can be very clearly seen that these Ag ZnO nanostructures appear to be largely spherical with a diameter of ~90 $\pm$ 10 nm on the TEM image, along with having a characteristic core-shell type structure. The size of the Ag core was found to be in the range of 10–15 nm, while the width of the ZnO shell was calculated to be ~45 $\pm$ 5 nm. The formation of this core-shell structure was further supported by the STEM image in Fig. 2b. The inter-planar distance in the case of the silver core was revealed with the help of HR-TEM imaging, as shown in Fig. 2c, to be 0.235 nm, which was found to be consistent with the (111) plane of the FCC crystal structure. The wurtzite hexagonal structures (002) and (101) planes match the zinc oxide shell's interplanar spacing's of 0.271 and 0.256 nm, respectively.

Ag/ZnO core/shell nanostructures have also been synthesized via the laser ablation technique, which displayed a slight variation in the structural properties of such nanomaterials as observed from Fig. 2d–f [82]. Figure 2a displays the TEM image of nanostructures with a silver core and zinc oxide outer shell, which showed the diameter of such nanomaterials to be ~50–60 nm. High-resolution TEM images (Fig. 2e) of these nanostructures clearly show the shell thickness of ~5 nm covering a silver core of diameter ~15 nm, and the structural characteristics of this core-shell nanostructure have been shown in Fig. 2f. In order to create hollow Au–Cu_2O core-shell structured nanomaterials, Lu et al. [86] developed a reliable and room-temperature wet-chemical approach in which, by adjusting the nanostructured dielectric copper oxide shells, they are able to regulate and enhance the plasmonic characteristics of the core, composed of hollow Au nanomaterials. It became clear that the strong affinity of copper oxide for the Au surface, along with their corresponding crystallographic symmetry, promoted the epitaxial development of Cu_2O on the Au nanomaterial surface. Very little is known about the fabrication of core-shell nanostructures utilizing Co_3O_4 as the shell component, but by using the solvothermal fabrication technique Hu et al. [87] tried to develop Au–Co_3O_4 core-shell nanostructures with

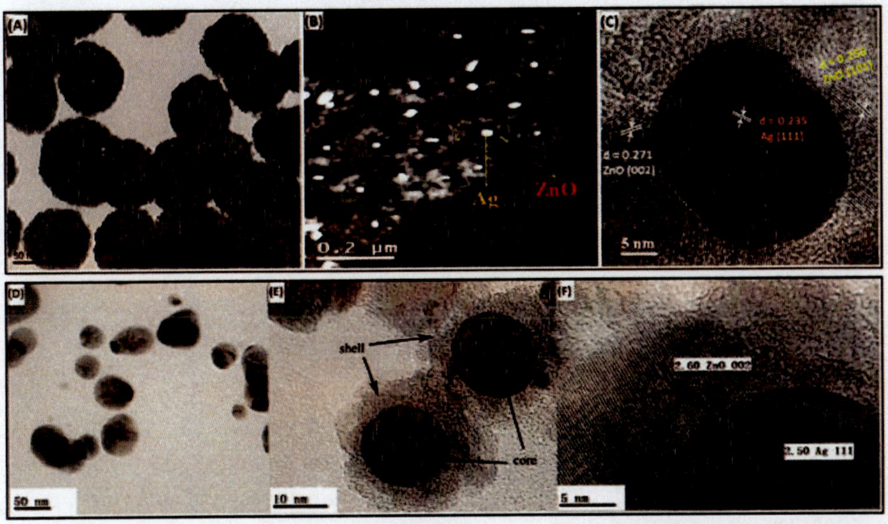

Fig. 2 **a** TEM image of Ag/ZnO core/shell nanostructures synthesized using facile synthesis technique, **b** high-angle annular dark field scanning TEM image, **c** corresponding high-resolution TEM of Ag–ZnO core-shell nanomaterials. Reproduced with permission from Ref. [85], Copyright 2018, Elsevier. **d** TEM image of Ag/ZnO core/shell nanostructures synthesized using laser ablation technique. **e**, **f** High-resolution TEM images of such nanostructures confirming their structural characteristics Reproduced with permission from Ref. [82]. Copyright 2015, APL materials

a cube-such as design; however, the fabricated nanostructures were not observed to possess a well-defined core-shell type structure.

3 Surface Functionalization of Metal Oxide Core/Shell Nanoparticles

Functional nanoparticles are becoming recognized as a distinct type of nanosystem because of their exceptional benefits, such as their large surface area and ease of surface modification and functionalization. These benefits make it possible for core-shell nanoparticles to combine multiple functional properties in a single nanosystem, resulting in the desired physicochemical properties needed for efficient targeted drug delivery [88]. A graphic representation for functionalizing core-shell nanostructure is given in Fig. 3 [89].

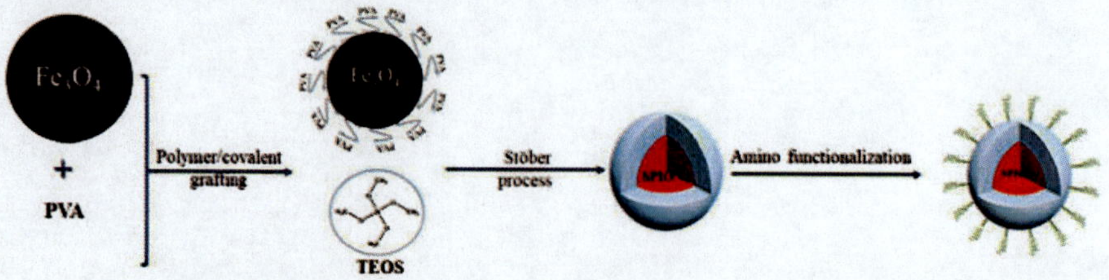

Fig. 3 General schematic representation for functionalization of core/shell nanomaterial [89]

3.1 Functionalization Via Covalent Bond

Noble metals covered with silica-based shells, along with a variety of silane compounds, can be used for coupling chemical reactions and are commonly linked to the idea of covalent binding. Spots for encapsulating biological molecules, medications, and coordination compounds can be generated through the condensation of silane linkers onto the outermost layer of a silica shell via alkoxysilane groups [90, 91]. Traditional conjugation procedures were originally used for covalent bond attachments because biological molecules and organic ligands have functional groups such as amines, carboxylic acids, and thiols (Fig. 4) [92].

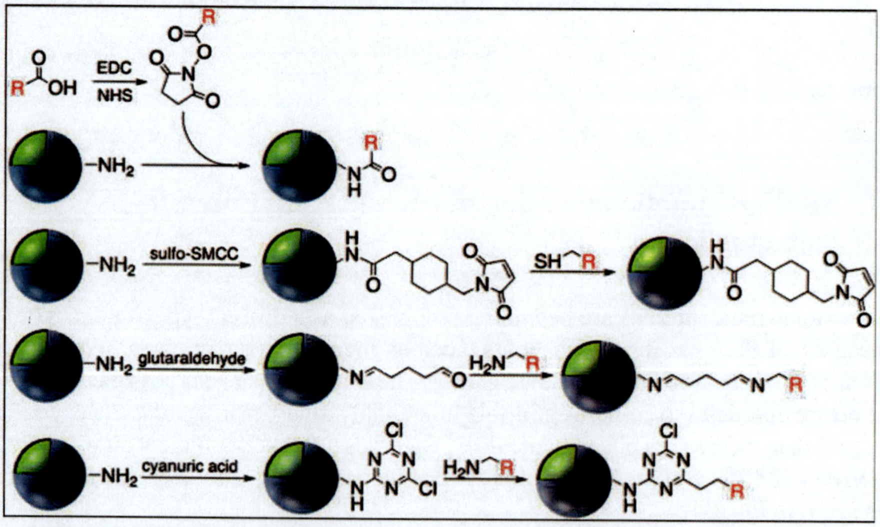

Fig. 4 Instances of modifying the surface of core-shell M@SiO_2 (where M represents Au or Ag) using organosilane agents [92]

3.1.1 Formation of Amide Bond

Carbodiimide crosslinking chemistry, utilizing a water-soluble compound called 1-ethyl-3-(3-dimethylaminopropyl) carbodiimide (EDC) to generate bonds of amide, is the most efficient and practical way of attaching chemical modifying agents to the exterior of noble metal nanostructures covered with silica (Fig. 2). The carboxylate compound (regardless of being on the exterior of the silica shell or a component of the chemical modifying agent) is initially coupled with EDC before undergoing a reaction with N-hydroxysuccinimide (NHS) to create a stable intermediary compound, that proceeds to react with an appropriate amine [93, 94]. Li and colleagues [95] designed Au@SiO_2 nanostructures featuring different core morphologies and coupled them to H_2N–DNA via EDC/NHS. The outcome of the experiment served as a probing device for surface-enhanced Raman scattering (SERS), which was employed to keep track of DNA hybridization. Similar techniques were used for attaching DNA with one strand to the exterior of Ag@SiO_2@SiO_2 core-shell nanomaterials, in which the energy-donating ligand complex RuBpy was incorporated together with the external silica coating. The DNA was used as a fluorescent detector to measure PSA levels after being hybridized with a prostate-specific antigen (PSA) aptamer [96]. By using the plasmon-enhanced fluorescent property of the silver core, Brouard, and coworkers [97, 98] also used this method to create Ag@SiO_2–DNA, enabling the identification of the human SRY gene. Prior to bioconjugation, eosin dyes were integrated using a covalent bond into the exterior silica layer. They also attached DNA and a cationic polymer-based transducer onto Ag@SiO_2 nanostructure exteriors to build a highly accurate metal-enhanced fluorescent (MEF) detector for DNA identification [97]. A sandwich-such as architecture was created in another investigation using a DNA-functionalized gold nano-star, a Raman reporter, and SiO_2. The resulting structure was employed for the SERS detection of small levels of heavy metallic contaminants in human saliva, especially silver and mercury released by dental fillings [99]. Finally, using a modified Stober process, Guo and associates developed Ag@SiO_2 nanostructures with varying shell thicknesses, to which they later attached rhodamine B isothiocyanate via covalent bonds [100]. A number of investigations have employed the EDC/NHS binding approach to link various compounds to core-shell nanostructures. For instance, in a particular investigation, hemoglobin was bound to gold nanomaterials encased in an amino group-containing titanium dioxide casing [101]. Following that, a biological sensor was created using this framework to identify the presence of hydrogen peroxide. In an alternative investigation, gold core-shell nanostructures were combined with rhodamine ligands to produce a luminous substrate for microbial identification [102]. The molecules were attached to the nanostructures using EDC/NHS, and the resultant substrate was subsequently applied to E. coli to produce sharp pictures utilizing laser fluorescence microscopy. For quicker evaluation, an identical procedure was executed with a microfluidic device. A separate study functionalized silver-titanium oxide core-shell nanomaterials with vancomycin for antibacterial action [101]. Dopamine was applied to the surface of the nanomaterial in order to produce primary amine groups, and the unbound amine groups on the nanostructures were used to interact with the antibiotic vancomycin. It was demonstrated that the resultant nanostructures with antibiotic functionalities, specifically photo-kill harmful microorganisms.

3.1.2 Imine (Schiff Base) Formation

An organic amine and an appropriate aldehyde can easily be combined to generate Schiff bases with the standardized formula R′-N]CR$_2$. Over the exterior of M@SiO$_2$, multiple molecules possessing main amine groups have been linked together through Schiff base formation, for which aldehyde-modified core-shell nanostructures have been employed. The conjugates created by attaching NH$_2$–DNA to M@SiO$_2$ nanostructures functionalized using aldehydes are currently being utilized for medical uses [103–105]. Anti-h-IgG was immobilized on the exterior of amine-functionalized Ag@SiO$_2$ and Au@SiO$_2$ to create a range of SERS biosensors. Before adding the amine-functionalized antibodies, glutaraldehyde (GA) was applied to the outside of the nano-composites [106–108]. For instance, Gong and colleagues utilized an identical method to attach anti-h-IgG to the outer layer of Ag@SiO$_2$ using a covalent bond, while Wang et al. [108] utilized imine production to produce a colorimetric biological sensor for the identification of h-IgG antigens. As a dual-mode biosensor, an anti-h-IgG antibody was incorporated into fluorescent dye-doped silica-coated gold nanomaterial aggregates. The improved nanocomposite with strong surface-plasma bands and SERS activity, combined with a coupled antibody, was utilized in multiplexed bio-detection. Biotinylated antibodies have been grafted onto the exterior of Ag@SiO$_2$ employing the identical method, and improved nanocomposite structures, that had been earlier doped by the lanthanide metal europium in the shape of a chelate complex (BHHCT-Eu-DPBT) have been proven to be appropriate for fluorescent biological imaging of cells [107, 109, 110]. Ag@SiO$_2$ was modified by Yin and colleagues [111] via a method including the application of amine-reactive 4-formylphenylboronic acid. A stable C=N imine bond was created by reacting the amino groups on the external layer of Ag@SiO$_2$ with the CHO functional group. Using this link and the boronic acid –B(OH)$_2$ functionality group's selectivity for sugar sialic acid as a substrate, an efficient ion-imprinted substrate for SERS monitoring of cancerous cells and tissue was designed. Bovine serum albumin (BSA), a protein, was cross-linked to the outside of Au@SiO$_2$@NH$_2$ nanostructures by Narayan and colleagues [112] by employing a glutaraldehyde cross-linker. They made use of the amine and aldehyde groups' reactivity as well as the imine linkage stability, coupled BSA via EDC/NHS couplings, and evaluated the altered Au@SiO$_2$ nanostructures using electrostatic absorbance relying on their physicochemical composition. In a different work, glutaraldehyde conjugation was used to covalently bond the metalloproteinase myoglobin (Mb) to amine-Ag@SiO$_2$ SERS tags. In order to identify proteins in biological research, functionalized AuNP with 2-(iminodiacetic acid) ethanethiol (IDA) was added to a colloidal solution of Mb-conjugated Ag@SiO$_2$ [113]. Ag@SiO$_2$ nanostructures functionalized with anti-HER$_2$ were produced by Xia and associates [114] utilizing a Schiff base formation procedure. They used this technique to covalently couple H$_2$N-antibody to oxidize dextran 500 altered Ag@SiO_2. The resultant compound was used as a SERS tag for detecting tumor cells and contained 4-MBA. To sum up, the formation of imine (Schiff base) has been identified as an efficient method for linking molecules to the exterior of core-shell nanostructures with a metal-based core and shell made from a metal oxide-based material. However, the primary challenge that remains in synthesizing these structures is the insufficient availability of appropriate amine and aldehyde precursors.

3.1.3 NHS Ester-Maleimide Conjugation

Proteins and oligonucleotides, which contain thiol and amine groups, are capable of being joined to the outside of silica-coated noble metal core-shell nanostructures using maleimide compounds as linking agents. For the purpose of identifying heavy metals and protein molecules, scientists have developed a maleimide-activated Ag@SiO_2 substrate using sulfosuccinimidyl-4-(N-maleimidomethyl)-cyclohexane-1-carboxylate (sulfo-SMCC) [115, 116]. Comparable to this, SH-ssDNA was applied to Au@SiO_2 nanostructures, which were then employed as SERS detectors for identifying DNA [117, 118]. The Au-nanocage@SiO_2 functionalized using amine groups was additionally customized with peptide tags possessing thiol endings utilizing sulfo-SMCC cross-linkers, and employed as a flexible SERS nanoprobe for imaging, drug administration, and photothermal treatment [118]. A comparable strategy was used to effectively cross-link DNA to the outside of silver nanoprisms covered with silica, utilizing the cross-linker succinimidyl 4-(p-malaminophenyl) butyrate (SMPB) [119]. Utilizing monoclonal antibodies that identify antigens, Kustner and associates [120] changed the exterior of silica-coated gold and silver nanostructures. A maleimide dual-functional cross-linker combined with a polyethylene glycol spacer was utilized to bind amine-functionalized nanomaterials with a core-shell configuration. The functionalized core-shell nanostructure was next joined to an antibody that had undergone thiol group modification. Additionally, they used a maleimide coupling to adhere a paramagnetic contrast substance known as 1,4,7,10-tetraazacyclododecane-1,4,7,10-tetraacetic acid (DOTA)-Gd^{3+} to the exterior of Au@SiO_2. The magnetic resonance imaging (MRI) procedure utilized customized core-shell nanomaterials with Raman receptors [121]. Maleimide (MAL)-discretePEG3-biotin and NHS-d-PEG3TM-biotin were the two reagents used to bind biotin to amine and thiol functionalized-Ag@SiO_2. The changed tags showed improved fluorescent characteristics, suggesting a possible application in biological imaging [122]. A 51 kDa protein called CD4 is present on the outside of T-cells. Antibodies against CD4 can be used to research how cells communicate. Sulfo-SMCC was employed by Dong et al. [123] for attaching CD4 antibodies to the surface of Ag@SiO_2 nanostructures, which were subsequently utilized for labeling cells. Maleimide-PEG-NHS compounds of varying molecular weights were used to generate reactive Au-nanorod@SiO_2 substrates, and they were further attached with peptide moieties due to thiols such as tLyP-1 peptide and SH-RGD [124, 125]. Targeted imaging, photothermal treatment (PPT), and drug delivery were all possible with these tailored nanomaterials. Jokerst et al. [126] covered $Au@SiO_2$ with 1,11-bismaleimido-triethyleneglycol (BM(PEG)) and succinimidyl-ester (SM(PEG)), proving that a number of linkers can be used for coupling. The adaptability of the approach was demonstrated by the subsequent conjugation of its altered shell to numerous biological molecules with sulfhydryl and amine moieties. The amine-end-encapsulated antibodies were attached to $Au@SiO_2$ nanostructures using the maleimide-thiol linking technique, and the resulting particles were subsequently used as SERS sensors for identifying biological indicators. These detectors were made by modifying monoclonal antibodies (mAbs) and fluorescent compounds with various excitation wavelengths on gold nanomaterials using a Raman reporter and a shell of silica incorporating reactive thiols. The aforementioned platforms have been employed to image molecules on the exteriors of new tissue [110, 127]. In conclusion, the maleimide-thiol linking method offers a flexible and successful method for covalently coupling molecules to the outside of core-shell nanostructures. By including a PDI dye containing an alkoxysilane molecule that interacted with silanol molecules on the silicon oxide shell, Ribeiro and colleagues [128] altered $Au@SiO_2$ nanostructures, and then their impact on dye discharge was investigated. Utilizing the fluorescein monomer (FITC) doped with silica in a single-particle study, Yan et al. [129] conducted research on the effect of $Ag@SiO_2$ nanostructures on fluorescent compounds. Furthermore, they grafted europium compounds onto $Au@SiO_2$ nanomaterials for identifying singlet oxygen, and they attached samarium and dysprosium benzoate compounds to the exterior of $Ag@SiO_2$ nanomaterials to explore their luminescent characteristics [130]. Furthermore, phenylazathiacrown monomers containing a silane group had been added to $Au@SiO_2$ nanostructures for usage in surface-enhanced Raman spectroscopy (SERS) for identifying mercury ions [130].

3.1.4 Miscellaneous Methods

Apart from the techniques explained in the previous sections, there are other ways of modifying the surface of core-shell nanostructures. For instance, a reactive compound containing an epoxy group called (3-glycidol-propyl) triethoxysilane has been utilized to attach thiol, amine, or hydroxy compounds to the outside of a SiO2 shell. In accordance with Krishnan et al. [131], an epoxy-amino process was implemented for attaching anti-E. coli McAb to $Ag@SiO_2$ core-shell nanoparticle surfaces, and this bioconjugate was then used for precisely identifying E. coli bacterial cells. The outer layer of the nanomaterials was functionalized through the use of (3-glycidoxypropyl) triethoxysilane before the antibody was linked to it. Epoxide chemical processes were used by Zhang and his associates [132] to investigate the bonding of a europium chelate to the outside of $Ag@SiO_2$ core-shell nanomaterials. They used the phenanthroline derivative epoxyphen to modify the exterior of amine-functionalized nanostructures. subsequently using Eu(ttfa)3, they introduced europium (III) ions to epoxyphen on the external layer of $Ag@SiO_2$. The created core-shell nanostructures with luminescence were employed in optical multiplexing devices. The rhodamine B iso-thiocyanate (RITC) cyanate compound responded to amine-functionalized core-shell nanostructures in a simple method to produce fluorescent-labeled Ag@SiO_2 nanomaterials. The resultant framework was used to look into the ways the silver core's ability to glow is affected by the degree of thickness of the covering. The hydroxyl compounds of Ag@SiO_2 were salinized with N-ethylenediamine (TMPED) and subsequently coupled with RITC to produce terminal primary amines on the outermost layer. Finally, tert-butyldimethylsilyl chloride (TBDMS) was used to preserve the free phenolic OH in order to create a metal-enhanced fluorescent (MEF) biological detector [133, 134]. Furthermore, monomers were successfully joined to the −OH group using Ag@TiO_2 nanomaterials that have been SH-functionalized. Ag@SiO_2 nanostructures coated with quantum dots (QDs) and containing the Raman reporter p-amino thiophenol (PATP) have been developed and employed for concurrent SERS and SEF immunoassays [135]. Identical techniques were employed to create CD47 antibody-SERS nanomaterials for the detection of breast cancer [136]. The Ag@SiO_2 interface has been attached to several vitamins, such as nicotinic acid, pantothenic acid, and biotin, using N-ethoxycarbonyl-2-ethoxy-1,2-dihydroquinoline (EEDQ) as a binding agent [137]. Europium complexes (III) were affixed onto the outer layer of Au@SiO_2 through the reaction of siloxane compounds from N-(4-benzoicacidyl) and N-(propyltriethoxysilyl) urea (RABI), followed by interaction with europium [138]. Luminescent nanostructures with a dual shell and silver shell identified as Ag@SiO_2@SiO_2-RuBpy were developed by Xu et al. [139] for the fluorescent-based detection of prostate-specific antigen (PSA), a tumor marker for identifying prostate tumors. The greatest enhancement in phosphorescence occurred when the silica shell doped with RuBpy was positioned around 10 nm away from the surface of the silver core. This was achieved by encapsulating RuBpy in the external silica layer of the nanomaterials. The study demonstrated the ability of Au@SiO_2@pMMA nanostructures to form a thick film on the interface of water and air; however, the resulting layer lacked the plasmonic characteristics of free nanomaterials in a solution. Metal oxide core-shell nanomaterials have been used for a variety of biomedical applications, as represented in the schematic representation given in Fig. 5 [89].

3.2 *Functionalization Via Electrostatic Interaction*

Whenever nanomaterials possessing a positive or negative charge come into contact with comparable cations or anions, electrostatic adsorption occurs. Surface interaction occurs more quickly with electrostatic modification of the surface than with covalent alterations to the interface, which is its main benefit. The primary drawback of this approach is that the molecule's link is rather weak, making it simple for different molecules and ions to take their positions. Recent research contains an extensive number of examples of electrostatic modification of surfaces. For

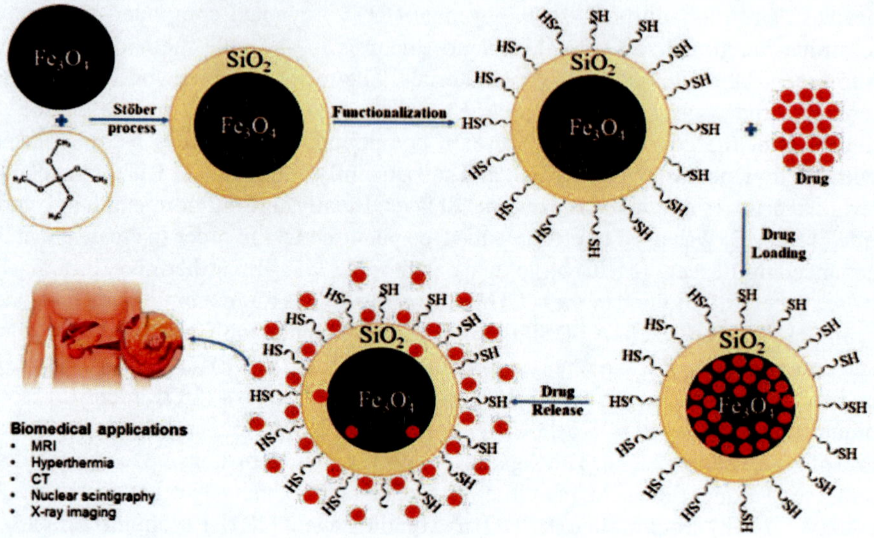

Fig. 5 Graphical representation of biomedical applications [89]

instance, a new kind of biological sensor has been created by combining positively charged conjugated polymers (CPs) such as bis(6-(bromohexyl) fluorene-2,7-ylenevinylene-co-alt-1,4-phenylene) (PFV) and poly(1H-imidazolium,1-methyl-3-[2-[(4-methyl-3-thienyl)oxy]ethyl]-chloride) (CCP) onto the surface of negatively charged Ag@SiO_2 [140, 141]. Yang et al. [142] electrostatically plated the exterior layer of fluorophore-doped mesoporous silica-covered AgNP by using poly (allylamine hydrochloride, or PAH). In another example, negatively charged Ag@SiO_2 became absorbed by polydimethyldiallylammonium chloride (PDDA). IgG-FITC was subsequently adsorbed onto Ag@SiO_2@PDDA, where it was transformed and used as a biomarker [143]. In a comparable manner, an identical method was utilized for immobilizing the negative-charged Au@SiO_2 interface with the positively charged cancer-fighting drug doxorubicin hydrochloride (DOX) [144, 145]. Because DOX's electrostatic adsorption is permanent, it is suited for delivering drugs that need simple therapeutic release. Owing to DOX being absorbed onto the hollow/rattle mesoporous Au@SiO_2 surface, it exhibited an elevated DOX loading potential. According to Sudeep et al. [144], adding the anionic tricarbocyanine dye IR-125 to Ag@TiO_2 core-shell nanostructures improved their photoresponse, enabling the development of dye-sensitized solar cells. The coupling of IR-125 to Au-nanostar@SiO_2 significantly increased the intensity of the fluorescence [147]. Ag@SiO_2 has been immobilized with 1-hydroxypyrene-3,6,8-trisulfonic acid (HPTS) by Bai et al. [146] by means of electrostatic absorption, creating a pH-sensitive probing device exhibiting beneficial uses as a pH sensor in biological structures. The efficient antibacterial fluoroquinolone levofloxacin hydrochloride (LEVO), which is effective against gram-negative and gram-positive microbes, was electrostatically attached to the exterior of a mesoporous silica-coated silver nanoparticle interface due to its zwitterionic molecular composition [147]. The primary goal of the work was to determine how the electric fields on the surfaces of silver nanoparticles can improve pyrene's Raman outputs. The silica shell's width was adjusted by the scientists in order to accomplish this. Furthermore, they altered the exterior of Ag@SiO_2 using Py-A, pyrene methylammonium. This alteration utilized the electrostatic bond that exists between the cationic ammonium group of the pyrene derivatives and the negatively charged surface of the core-shell nanomaterial [148].

4 Conclusion

Metal oxide core-shell nanostructures belong to a novel class of metal-semiconductor hybrid materials that have special optical electrical properties that render these materials as highly valuable in a wide range of operations, such as photocatalysis, and biomedical applications. The development of effective photocatalytic reactors, medical devices, systems for drug delivery, and biological engineering have been significantly influenced by core-shell nanostructures, along with their added functionalities. They may additionally be employed in environmental rehabilitation and bacterial decontamination. Owing to their distinctive physicochemical and electrical characteristics, metal nanoparticles, including Ag, and Au, have recently attracted a lot of curiosity as core elements. Metallic oxides have proven to be an excellent choice as a semiconductor shell material because of their availability and simplicity of manufacturing. Core-shell nanostructures can be used to create highly effective transistors, sensors, drug delivery systems, bioimaging systems, and photovoltaic devices; because of their superior electrical, magnetic, optoelectronic, photocatalytic, and photon-absorbing properties. Due to their ability to carry out effective solar light-driven photocatalysis, such nanoparticles exhibit significant promise for restoring the environment via the photocatalytic treatment of wastewater. Core-shell nanostructures with a semiconducting shell and metal core are a good contender for microbial decontamination along with decontamination of air since metal-core nanoparticles such as Ag and Au also have antibacterial capabilities, although metal-oxide photocatalysts are recognized for their ability to eliminate microorganisms through the use of UV light. Core-shell nano-catalysts are simple to recycle and reuse because of their distinctive physicochemical characteristics, such as their nanoscale shape, porosity, and magnetic properties, thus making them of particular importance to researchers.

References

1. M. Nasrollahzadeh, S.M. Sajadi, M. Sajjadi, Z. Issaabadi, An introduction to nanotechnology, interface. Sci. Technol. **28**, 1–27 (2019). https://doi.org/10.1016/B978-0-12-813586-0.00001-8
2. K. Chatterjee, S. Sarkar, K. Jagajjanani Rao, S. Paria, Core/shell nanoparticles in biomedical applications. Adv. Colloid Interface Sci. **209**, 8–39 (2014). https://doi.org/10.1016/j.cis.2013.12.008.
3. H.L. Liu, P. Hou, W.X. Zhang, J.H. Wu, Synthesis of monosized core–shell Fe3O4/Au multifunctional nanoparticles by PVP-assisted nanoemulsion process. Colloids Surf. A Physicochem. Eng. Asp. **356**, 21–27 (2010). https://doi.org/10.1016/J.COLSURFA.2009.12.023
4. G. Liu, M. Swierczewska, S. Lee, X. Chen, Functional nanoparticles for molecular imaging guided gene delivery. Nano Today **5**, 524–539 (2010). https://doi.org/10.1016/J.NANTOD.2010.10.005
5. R. Hayes, A. Ahmed, T. Edge, H. Zhang, Core–shell particles: preparation, fundamentals and applications in high performance liquid chromatography. J. Chromatogr. A. **1357**, 36–52 (2014). https://doi.org/10.1016/J.CHROMA.2014.05.010
6. R.A. Ramli, W.A. Laftah, S. Hashim, Core–shell polymers: a review. RSC Adv. **3**, 15543–15565 (2013). https://doi.org/10.1039/C3RA41296B
7. N. Sun, Z. Xiao, Synthesis and performances of phase change materials microcapsules with a polymer/BN/TiO2 hybrid shell for thermal energy storage. Energy Fuels **31**, 10186–10195 (2017). https://doi.org/10.1021/ACS.ENERGYFUELS.7B01271
8. K. Akamatsu, T. Yamaguchi, Novel preparation method for obtaining pH-responsive core−Shell microcapsule reactors. Ind. Eng. Chem. Res. **46**, 124–130 (2006). https://doi.org/10.1021/IE060857Q
9. L.J. De Cock, S. De Koker, B.G. De Geest, J. Grooten, C. Vervaet, J.P. Remon, G.B. Sukhorukov, M.N. Antipina, Polymeric multilayer capsules in drug delivery. Angew. Chemie Int. Ed. **49**, 6954–6973 (2010). https://doi.org/10.1002/ANIE.200906266
10. X.L. Yang, X.J. Ju, X.T. Mu, W. Wang, R. Xie, Z. Liu, L.Y. Chu, Core–Shell chitosan microcapsules for programmed sequential drug release. ACS Appl. Mater. Interfaces **8**, 10524–10534 (2016). https://doi.org/10.1021/ACSAMI.6B01277/SUPPL_FILE/AM6B01277_SI_003.AVI
11. W. Wang, T. Luo, X.J. Ju, R. Xie, L. Liu, L.Y. Chu, Microfluidic preparation of multicompartment microcapsules for isolated co-encapsulation and controlled release of diverse components. Int. J. Nonlinear Sci. Numer. Simul. **13**, 325–332 (2012). https://doi.org/10.1515/IJNSNS-2012-0402/MACHINEREADABLECITATION/RIS
12. D.K. Nguyen, Y.M. Son, N.E. Lee, Hydrogel encapsulation of cells in core-shell microcapsules for cell delivery. Adv. Healthc. Mater. **4**, 1537–1544 (2015). https://doi.org/10.1002/ADHM.201500133
13. M. Rajagopalan, K.A. Darling, C. Kale, S.A. Turnage, R.K. Koju, B.C. Hornbuckle, Y. Mishin, K.N. Solanki, Nanotechnology enabled design of a structural material with extreme strength as well as thermal and electrical properties. Mater. Today **31**, 10–20 (2019). https://doi.org/10.1016/J.MATTOD.2019.09.024
14. S.F. Medeiros, A.M. Santos, H. Fessi, A. Elaissari, Stimuli-responsive magnetic particles for biomedical applications. Int. J. Pharm. **403**, 139–161 (2011). https://doi.org/10.1016/J.IJPHARM.2010.10.011
15. X. Chu et al., Surface modification of magnetic nanoparticles in biomedicine*. Chin. Phys. B. **24**, 014704 (2015). https://doi.org/10.1088/1674-1056/24/1/014704
16. E.D. Smolensky, H.Y.E. Park, T.S. Berquó, V.C. Pierre, Surface functionalization of magnetic iron oxide nanoparticles for MRI applications—Effect of anchoring group and ligand exchange protocol. Contrast Media Mol. Imaging **6**, 189–199 (2011). https://doi.org/10.1002/CMMI.417

17. B. Sun, G. Zhou, H. Zhang, Synthesis, functionalization, and applications of morphology-controllable silica-based nanostructures: a review. Prog. Solid State Chem. **44**, 1–19 (2016). https://doi.org/10.1016/J.PROGSOLIDSTCHEM.2016.01.001
18. S. Dagher, A. Soliman, A. Ziout, N. Tit, A. Hilal-Alnaqbi, S. Khashan, F. Alnaimat, J.A. Qudeiri, Photocatalytic removal of methylene blue using titania- and silica-coated magnetic nanoparticles. Mater. Res. Express. **5**, 065518 (2018). https://doi.org/10.1088/2053-1591/aacad4
19. C. Hui, C. Shen, J. Tian, L. Bao, H. Ding, C. Li, Y. Tian, X. Shi, H.J. Gao, Core-shell Fe3O4@SiO2 nanoparticles synthesized with well-dispersed hydrophilic Fe3O4 seeds. Nanoscale **3**, 701–705 (2011). https://doi.org/10.1039/C0NR00497A
20. A.G. Roca, D. Carmona, N. Miguel-Sancho, O. Bomatí-Miguel, F. Balas, C. Piquer, J. Santamaría, Surface functionalization for tailoring the aggregation and magnetic behaviour of silica-coated iron oxide nanostructures. Nanotechnology **23**, 155603 (2012). https://doi.org/10.1088/0957-4484/23/15/155603
21. M. Abbas, Fe3O4/SiO2 core/shell nanocubes: novel coating approach with tunable silica thickness and enhancement in stability and biocompatibility. J. Nanomed. Nanotechnol. **05** (2014). https://doi.org/10.4172/2157-7439.1000244
22. M. Equen, G. Roach, R. Brown, T. Bennett, Magnetic removal of foreign bodies from the esophagus, stomach, and duodenum, AMA. Arch. Otolaryngol. **66**, 698–706 (1957). https://doi.org/10.1001/archotol.1957.03830300078007
23. S.A. Douglas, S. Mirza, F.W. Stafford, Magnetic removal of a nasal foreign body. Int. J. Pediatr. Otorhinolaryngol. **62**, 165–167 (2002). https://doi.org/10.1016/S0165-5876(01)00609-7
24. G.T. Gillies, R.C. Ritter, W.C. Broaddus, M.S. Grady, M.A. Howard, R.G. McNeil, Magnetic manipulation instrumentation for medical physics research. Rev. Sci. Instrum. **65**, 533 (1998). https://doi.org/10.1063/1.1145242
25. J.A. Rodríguez, M. Fernández Garcia, *Synthesis, Properties, and Applications of Oxide Nanomaterials*, vol. 717 (2007)
26. W. Stöber, A. Fink, E. Bohn, Controlled growth of monodisperse silica spheres in the micron size range. J. Colloid Interface Sci. **26**, 62–69 (1968). https://doi.org/10.1016/0021-9797(68)90272-5
27. J. Zha, H. Roggendorf, Sol-gel science, the physics and chemistry of sol-gel processing. Eds. by C.J. Brinker, G.W. Scherer, Adv. Mater. **3**, 522–522 (1991). https://doi.org/10.1002/adma.19910031025
28. L. Klein, *Sol-gel Optics: processing and Applications* (2013)
29. H. Sertchook, D. Avnir, Submicron silica/polystyrene composite particles prepared by a one-step sol-gel process. Chem. Mater. **15**, 1690–1694 (2003). https://doi.org/10.1021/cm020980h
30. A. Maliakal, H. Katz, P.M. Cotts, S. Subramoney, P. Mirau, Inorganic oxide core, polymer shell nanocomposite as a high K gate dielectric for flexible electronics applications. J. Am. Chem. Soc. **127**, 14655–14662 (2005). https://doi.org/10.1021/ja052035a
31. D. Wang, C. Song, Y. Lin, Z. Hu, Preparation and characterization of TiO2 hollow spheres. Mater. Lett. **60**, 77–80 (2006). https://doi.org/10.1016/j.matlet.2005.07.076
32. M. Ocaña, W.P. Hsu, E. Matijevic, Preparation and properties of uniform-coated colloidal particles. 6. Titania on Zinc Oxide. Langmuir **7**, 2911–2916 (1991). https://doi.org/10.1021/la00060a008
33. R. Smoluchowski, A review of: "The science of crystallization, microscopic interfacial phenomena." Radiat. Eff. Defects Solids. **118**, 191–191 (1991). https://doi.org/10.1080/10420159108220635
34. C.N.R. Rao, A. Müller, A.K. Cheetham, *Nanomaterials Chemistry: Recent Developments and New Directions* (2007). https://doi.org/10.1002/9783527611362
35. S. Wang, M. Zhang, D. Wang, W. Zhang, S. Liu, Synthesis of hollow mesoporous silica microspheres through surface sol-gel process on polystyrene-co-poly(4-vinylpyridine) core-shell microspheres. Microporous Mesoporous Mater. **139**, 1–7 (2011). https://doi.org/10.1016/j.micromeso.2010.10.002

36. A. Imhof, Preparation and characterization of titania-coated polystyrene spheres and hollow titania shells. Langmuir **17**, 3579–3585 (2002). https://doi.org/10.1021/la001604j
37. P.J. Bruinsma, A.Y. Kim, J. Liu, S. Baskaran, Mesoporous silica synthesized by solvent evaporation: spun fibers and spray-dried hollow spheres. Chem. Mater. **9**, 2507–2512 (1997). https://doi.org/10.1021/cm970282a
38. Z. Ma, B. Gao, P. Wu, J. Shi, Z. Qiao, Z. Yang, G. Yang, B. Huang, F. Nie, Facile, continuous and large-scale production of core-shell HMX@TATB composites with superior mechanical properties by a spray-drying process. RSC Adv. **5**, 21042–21049 (2015). https://doi.org/10.1039/c4ra16527f
39. S. Li, X. Qin, H. Zhang, J. Wu, Y.B. He, B. Li, F. Kang, Silicon/carbon composite microspheres with hierarchical core-shell structure as anode for lithium ion batteries. Electrochem. Commun. **49**, 98–102 (2014). https://doi.org/10.1016/j.elecom.2014.10.013
40. L. Zhang, S. Hao, B. Liu, H.C. Shum, J. Li, H. Chen, Fabrication of ceramic microspheres by diffusion-induced sol-gel reaction in double emulsions. ACS Appl. Mater. Interfaces **5**, 11489–11493 (2013). https://doi.org/10.1021/am402028w
41. W. Pu, D. Fu, H. Xia, Z. Wang, Preparation of hollow polyurethane microspheres with tunable surface structures via electrospraying technology. RSC Adv. **7**, 49828–49837 (2017). https://doi.org/10.1039/c7ra09831f
42. S.L. Shenoy, W.D. Bates, H.L. Frisch, G.E. Wnek, Role of chain entanglements on fiber formation during electrospinning of polymer solutions: good solvent, non-specific polymer-polymer interaction limit. Polymer (Guildf) **46**, 3372–3384 (2005). https://doi.org/10.1016/j.polymer.2005.03.011
43. P. Fattahi, A. Borhan, M.R. Abidian, Microencapsulation of chemotherapeutics into monodisperse and tunable biodegradable polymers via electrified liquid jets: Control of size, shape, and drug release. Adv. Mater. **25**, 4555–4560 (2013). https://doi.org/10.1002/adma.201301033
44. M. Cloupeau, B. Prunet-Foch, Electrohydrodynamic spraying functioning modes: a critical review. J. Aerosol. Sci. **25**, 1021–1036 (1994). https://doi.org/10.1016/0021-8502(94)90199-6
45. X.X. Wang, X.J. Ju, S.X. Sun, R. Xie, W. Wang, Z. Liu, L.Y. Chu, Monodisperse erythrocyte-sized and acid-soluble chitosan microspheres prepared via electrospraying. RSC Adv. **5**, 34243–34250 (2015). https://doi.org/10.1039/c5ra04726a
46. K.H. Yoon, J. Unyong, C.C. Eun, Production of uniform-sized polymer core-shell microcapsules by coaxial electrospraying. Langmuir **24**, 2446–2451 (2008). https://doi.org/10.1021/la703546f
47. H. Ho, J. Lee, PEG/PLA core/shell particles from coaxial electrohydrodynamic spray drying. Macromol. Res. **19**, 815–821 (2011). https://doi.org/10.1007/s13233-011-0814-z
48. M. Cloupeau, B. Prunet-Foch, Electrostatic spraying of liquids in cone-jet mode. J. Electrostat. **22**, 135–159 (1989). https://doi.org/10.1016/0304-3886(89)90081-8
49. S. Park, S. Hwang, J. Lee, PH-responsive hydrogels from moldable composite microparticles prepared by coaxial electro-spray drying. Chem. Eng. J. **169**, 348–357 (2011). https://doi.org/10.1016/j.cej.2011.02.063
50. F. He, M.J. Zhang, W. Wang, Q.W. Cai, Y.Y. Su, Z. Liu, Y. Faraj, X.J. Ju, R. Xie, L.Y. Chu, Designable polymeric microparticles from droplet microfluidics for controlled drug release. Adv. Mater. Technol. **4**, 1800687 (2019). https://doi.org/10.1002/ADMT.201800687
51. S.S. Lee, S.H. Kim, Controlled encapsulation of cholesteric liquid crystals using emulsion templates. Macromol. Res. **26**, 1054–1065 (2018). https://doi.org/10.1007/s13233-018-6148-3
52. H.S. Tan, T.H. Ng, H.K. Mahabadi, Interfacial polymerization encapsulation of a viscous pigment mix: emulsification conditions and particle size distribution. J. Microencapsul. **8**, 525–536 (1991). https://doi.org/10.3109/02652049109021876
53. W.-H. Li, H.D.H. Stöver, Monodisperse cross-linked core−shell polymer microspheres by precipitation polymerization. Macromolecules **33**, 4354–4360 (2000). https://doi.org/10.1021/ma9920691

54. R. Al-Shannaq, M.M. Farid, Microencapsulation of phase change materials for thermal energy storage systems, in *Advanced Thermal Energy Storage Systems Methods Application*, pp. 269–329 (2020). https://doi.org/10.1016/B978-0-12-819885-8.00010-3
55. R. Shenoy, M.W. Tibbitt, K.S. Anseth, C.N. Bowman, Formation of core-shell particles by interfacial radical polymerization initiated by a glucose oxidase-mediated redox system. Chem. Mater. **25**, 761–767 (2013). https://doi.org/10.1021/cm303913f
56. C. Lin, F. Gong, Z. Yang, L. Pan, S. Liu, J. Li, S. Guo, Bio-inspired fabrication of core@shell structured TATB/polydopamine microparticles via in situ polymerization with tunable mechanical properties. Polym. Test. **68**, 126–134 (2018). https://doi.org/10.1016/j.polymertesting.2018.03.046
57. M.J. Almasi, T. FanaeiSheikholeslami, M.R. Naghdi, Band gap study of polyaniline and polyaniline/MWNT nanocomposites with in situ polymerization method. Compos. Part B Eng. **96**, 63–68 (2016). https://doi.org/10.1016/j.compositesb.2016.04.032
58. M. Bin Ahmad, Y. Gharayebi, M.S. Salit, M.Z. Hussein, K. Shameli, Comparison of in situ polymerization and solution-dispersion techniques in the preparation of polyimide/montmorillonite (MMT) nanocomposites. Int. J. Mol. Sci. **12**, 6040–6050 (2011). https://doi.org/10.3390/ijms12096040
59. C. Lin, F. Gong, Z. Yang, X. Zhao, Y. Li, C. Zeng, J. Li, S. Guo, Core-shell structured HMX@Polydopamine energetic microspheres: Synergistically enhanced mechanical, thermal, and safety performances. Polymers (Basel) **11**, 568 (2019). https://doi.org/10.3390/polym11030568
60. S.A. Nabavi, G.T. Vladisavljević, V. Manović, Mechanisms and control of single-step microfluidic generation of multi-core double emulsion droplets. Chem. Eng. J. **322**, 140–148 (2017). https://doi.org/10.1016/j.cej.2017.04.008
61. S. Xuan, Y.X.J. Wang, K.C.F. Leung, K. Shu, Synthesis of Fe3O4@polyaniline core/shell microspheres with well-defined blackberry-such as morphology. J. Phys. Chem. C **112**, 18804–18809 (2008). https://doi.org/10.1021/JP807124Z/ASSET/IMAGES/LARGE/JP-2008-07124Z_0009.JPEG
62. R. Arshady, Suspension, emulsion, and dispersion polymerization: a methodological survey. Colloid Polym. Sci. **270**, 717–732 (1992). https://link.springer.com/article/https://doi.org/10.1007/BF00776142. Accessed 6 Apr. 2023
63. M. Jonsson, O. Nordin, E. Malmström, C. Hammer, Suspension polymerization of thermally expandable core/shell particles. Polymer (Guildf) **47**, 3315–3324 (2006). https://doi.org/10.1016/j.polymer.2006.03.013
64. S.R.S. Ting, T.L.U. Nguyen, M.H. Stenzel, One pot synthesis of surface PEGylated core-shell microparticles by suspension polymerization with surface enrichment of biotin/avidin conjugation. Macromol. Biosci. **9**, 211–220 (2009). https://doi.org/10.1002/MABI.200800161
65. M.D. Besteti, D.M.G. Freire, J.C. Pinto, Production of core–shell particles by combined semibatch emulsion/suspension polymerizations. Macromol. React. Eng. **5**, 518–532 (2011). https://doi.org/10.1002/MREN.201100012
66. M.K. Lenzi, F.M. Silva, E.L. Lima, J.C. Pinto, Semibatch styrene suspension polymerization processes. J. Appl. Polym. Sci. **89**, 3021–3038 (2003). https://doi.org/10.1002/APP.12443
67. G.L. Li, H. Möhwald, D.G. Shchukin, Precipitation polymerization for fabrication of complex core–shell hybrid particles and hollow structures. Chem. Soc. Rev. **42**, 3628 (2013). https://doi.org/10.1039/c3cs35517a
68. M.P.L. Werts, M. Badila, C. Brochon, A. Hébraud, G. Hadziioannou, Titanium dioxide—Polymer core-shell particles dispersions as electronic inks for electrophoretic displays. Chem. Mater. **20**, 1292–1298 (2008). https://pubs.acs.org/doi/full/https://doi.org/10.1021/cm071197y. Accessed 6 Apr. 2023
69. F. Barahona, E. Turiel, P.A.G.G. Cormack, A. Martín-Esteban, Chromatographic performance of molecularly imprinted polymers: core-shell microspheres by precipitation polymerization and grafted MIP films via iniferter-modified silica beads. J. Polym. Sci. Part A Polym. Chem. **48**, 1058–1066 (2010). https://doi.org/10.1002/pola.23860

70. K.J. O'Callaghan, A.J. Paine, A. Rudin, Emulsion polymerization of supermicron, monodisperse acrylic copolymer particles with core-shell structures. J. Polym. Sci. Part A Polym. Chem. **33**, 1849–1857 (1995)
71. M. Laus, C. Dinnella, G. Lanzarini, A. Casagrande, Core-shell functional microspheres by dispersion polymerization: 2. Synthesis and characterization. Polymer (Guildf) **37**, 343–347 (1996). https://doi.org/10.1016/0032-3861(96)81108-6
72. M. Okubo, Y. Iwasaki, Y. Yamamoto, Preparation of micron-size monodisperse polymer microspheres having cationic groups. Colloid Polym. Sci. **270**, 733–737 (1992). https://doi.org/10.1007/BF00776143
73. Z. Che, T.N. Wong, N.T. Nguyen, A simple method for the formation of water-in-oil-in-water (W/O/W) double emulsions. Microfluid. Nanofluidics**21**, 1–9 (2017). https://link.springer.com/article/https://doi.org/10.1007/s10404-016-1845-8. Accessed 6 Apr. 2023
74. L. Zhang, J. Huang, T. Si, R.X. Xu, Coaxial electrospray of microparticles and nanoparticles for biomedical applications. Expert Rev. Med. Device. **9**, 595–612 (2012). https://doi.org/10.1586/erd.12.58
75. Y. Yamagishi, T. Sugeno, T. Ishige, H. Takeuchi, A.T. Pyatenko, An evaluation of microencapsulated PCM for use in cold energy transportation medium, in IECEC 96. Proceedings of the 31st Intersociety Energy Conversion Engineering Conference (IEEE, 1996), pp. 2077–2083. https://doi.org/10.1109/IECEC.1996.553442
76. A.J.T. Teo, M. Yan, J. Dong, H.D. Xi, Y. Fu, S.H. Tan, N.T. Nguyen, Controllable droplet generation at a microfluidic T-junction using AC electric field. Microfluid. Nanofluidics**24**, 1–9 (2020). https://link.springer.com/article/https://doi.org/10.1007/s10404-020-2327-6. Accessed 6 Apr. 2023
77. K. Moshksayan, N. Kashaninejad, M.E. Warkiani, J.G. Lock, H. Moghadas, B. Firoozabadi, M.S. Saidi, N.-T.T. Nguyen, Spheroids-on-a-chip: recent advances and design considerations in microfluidic platforms for spheroid formation and culture. Sens. Actuators B Chem. **263**, 151–176 (2018). https://doi.org/10.1016/j.snb.2018.01.223
78. C. Kim, S. Chung, Y.E. Kim, K.S. Lee, S.H. Lee, K.W. Oh, J.Y. Kang, Generation of core-shell microcapsules with three-dimensional focusing device for efficient formation of cell spheroid. Lab Chip **11**, 246–252 (2011). https://doi.org/10.1039/c0lc00036a
79. A. Guerrero-Martínez, J. Pérez-Juste, L.M. Liz-Marzán, Recent progress on silica coating of nanoparticles and related nanomaterials. Adv. Mater. **22**, 1182–1195 (2010). https://doi.org/10.1002/adma.200901263
80. Z.L. Wang, Piezopotential gated nanowire devices: Piezotronics and piezo-phototronics. Nano Today **5**, 540–552 (2010). https://doi.org/10.1016/j.nantod.2010.10.008
81. Y. Zhao, S. Li, Y. Zeng, Y. Jiang, Synthesis and properties of Ag/ZnO core/shell nanostructures prepared by excimer laser ablation in liquid. APL Mater. **3**, 086103 (2015). https://doi.org/10.1063/1.4928287
82. H. Sun, J. He, J. Wang, S.Y. Zhang, C. Liu, T. Sritharan, S. Mhaisalkar, M.Y. Han, D. Wang, H. Chen, Investigating the multiple roles of polyvinylpyrrolidone for a general methodology of oxide encapsulation. J. Am. Chem. Soc. **135**, 9099–9110 (2013). https://doi.org/10.1021/JA4035335/SUPPL_FILE/JA4035335_SI_001.PDF
83. L. Xu, M.L. Yin, S. Liu, Superior sensor performance from Ag@WO3 core-shell nanostructure. J. Alloys Compd. **623**, 127–131 (2015). https://doi.org/10.1016/j.jallcom.2014.10.103
84. A.N. Kadam, D.P. Bhopate, V.V. Kondalkar, S.M. Majhi, C.D. Bathula, A.V. Tran, S.W. Lee, Facile synthesis of Ag–ZnO core–shell nanostructures with enhanced photocatalytic activity. J. Ind. Eng. Chem. **61**, 78–86 (2018). https://doi.org/10.1016/j.jiec.2017.12.003
85. B. Lu, A. Liu, H. Wu, Q. Shen, T. Zhao, J. Wang, Hollow Au–Cu2O core-shell nanoparticles with geometry-dependent optical properties as efficient plasmonic photocatalysts under visible light. Langmuir **32**, 3085–3094 (2016). https://doi.org/10.1021/ACS.LANGMUIR.6B00331
86. J. Hu, Z. Wen, Q. Wang, X. Yao, Q. Zhang, J. Zhou, J. Li, Controllable synthesis and enhanced electrochemical properties of multifunctional AucoreCo3O4shell nanocubes. J. Phys. Chem. B **110**, 24305–24310 (2006). https://doi.org/10.1021/JP063216H

87. R. Kumar, K. Mondal, P.K. Panda, A. Kaushik, R. Abolhassani, R. Ahuja, H.G. Rubahn, Y.K. Mishra, Core-shell nanostructures: perspectives towards drug delivery applications. J. Mater. Chem. B **8**, 8992–9027 (2020). https://doi.org/10.1039/d0tb01559h
88. A. Spoială, C.-I. Ilie, L.N. Crăciun, D. Ficai, A. Ficai, E. Andronescu, Magnetite-silica core/shell nanostructures: from surface functionalization towards biomedical applications—A review. Appl. Sci. **11**, 11075 (2021). https://doi.org/10.3390/app112211075
89. G. Steinberg, K. Stromsborg, L. Thomas, D. Barker, C. Zhao, Strategies for covalent attachment of DNA to beads. Biopolymers **73**, 597–605 (2004). https://doi.org/10.1002/bip.20006
90. R.A. Sperling, W.J. Parak, Surface modification, functionalization and bioconjugation of colloidal Inorganic nanoparticles. Philos. Trans. R. Soc. A Math. Phys. Eng. Sci. **368**, 1333–1383 (2010). https://doi.org/10.1098/rsta.2009.0273
91. S. Talebzadeh, C. Queffélec, D.A. Knight, Surface modification of plasmonic noble metal–metal oxide core–shell nanoparticles. Nanosc. Adv. **1**, 4578–4591 (2019). https://doi.org/10.1039/C9NA00581A
92. C.A.G.N. Montalbetti, V. Falque, Amide bond formation and peptide coupling. Tetrahedron **61**, 10827–10852 (2005). https://doi.org/10.1016/j.tet.2005.08.031
93. G.F. Luo, W.H. Chen, Q. Lei, W.X. Qiu, Y.X. Liu, Y.J. Cheng, X.Z. Zhang, A triple-collaborative strategy for high-performance tumor therapy by multifunctional mesoporous silica-coated gold nanorods. Adv. Funct. Mater. **26**, 4339–4350 (2016). https://doi.org/10.1002/adfm.201505175
94. M. Li, S.K. Cushing, J. Zhang, J. Lankford, Z.P. Aguilar, D. Ma, N. Wu, Shape-dependent surface-enhanced Raman scattering in gold-Raman-probe-silica sandwiched nanoparticles for biocompatible applications. Nanotechnology **23**, 115501 (2012). https://doi.org/10.1088/0957-4484/23/11/115501
95. Y.L. Deng, D.D. Xu, D.W. Pang, H.W. Tang, Target-triggered signal turn-on detection of prostate specific antigen based on metal-enhanced fluorescence of Ag@SiO2@SiO2-RuBpy composite nanoparticles. Nanotechnology **28**, 065501 (2017). https://doi.org/10.1088/1361-6528/28/6/065501
96. D. Brouard, O. Ratelle, A.G. Bracamonte, M. St-Louis, D. Boudreau, Direct molecular detection of SRY gene from unamplified genomic DNA by metal-enhanced fluorescence and FRET. Anal. Methods **5**, 6896–6899 (2013). https://doi.org/10.1039/c3ay41428k
97. D. Brouard, M.L. Viger, A.G. Bracamonte, D. Boudreau, Label-free biosensing based on multilayer fluorescent nanocomposites and a cationic polymeric transducer. ACS Nano **5**, 1888–1896 (2011). https://doi.org/10.1021/nn102776m
98. P. Zheng, M. Li, R. Jurevic, S.K. Cushing, Y. Liu, N. Wu, A gold nanohole array based surface-enhanced Raman scattering biosensor for detection of silver(i) and mercury(ii) in human saliva. Nanoscale **7**, 11005–11012 (2015). https://doi.org/10.1039/c5nr02142a
99. L. Guo, A. Guan, X. Lin, C. Zhang, G. Chen, Preparation of a new core-shell Ag@SiO2 nanocomposite and its application for fluorescence enhancement. Talanta **82**, 1696–1700 (2010). https://doi.org/10.1016/j.talanta.2010.07.051
100. N.H.A. Halim, Y.H. Lee, R.S.P.M. Marugan, U. Hashim, Mediatorless impedance studies with titanium dioxide conjugated gold nanoparticles for hydrogen peroxide detection. Biosensors **7** (2017). https://doi.org/10.3390/bios7030038
101. D. Gontero, A.V. Veglia, D. Boudreau, A.G. Bracamonte, Ultraluminescent gold core–shell nanoparticles applied to individual bacterial detection based on metal-enhanced fluorescence nanoimaging. J. Nanophotonics. **12**, 012505 (2017). https://doi.org/10.1117/1.jnp.12.012505
102. I.E. Sendroiu, M.E. Warner, R.M. Corn, Fabrication of silica-coated gold nanorods functionalized with DNA for enhanced surface plasmon resonance imaging biosensing applications. Langmuir **25**, 11282–11284 (2009). https://doi.org/10.1021/la902675s
103. S. Liu, Z. Zhang, M. Han, Gram-scale synthesis and biofunctionalization of silica-coated silver nanoparticles for fast colorimetric DNA detection. Anal. Chem. **77**, 2595–2600 (2005). https://doi.org/10.1021/ac0482864

104. P.P. Hu, L.L. Zheng, L. Zhan, J.Y. Li, S.J. Zhen, H. Liu, L.F. Luo, G.F. Xiao, C.Z. Huang, Metal-enhanced fluorescence of nano-core-shell structure used for sensitive detection of prion protein with a dual-aptamer strategy. Anal. Chim. Acta **787**, 239–245 (2013). https://doi.org/10.1016/j.aca.2013.05.061
105. C. Wang, Z. Ma, T. Wang, Z. Su, Synthesis, assembly, and biofunctionalization of silica-coated gold nanorods for colorimetric biosensing. Adv. Funct. Mater. **16**, 1673–1678 (2006). https://doi.org/10.1002/adfm.200500898
106. C. Wei, M.M. Xu, C.W. Fang, Q. Jin, Y.X. Yuan, J.L. Yao, Improving the sensitivity of immunoassay based on MBA-embedded Au@SiO2 nanoparticles and surface enhanced Raman spectroscopy. Spectrochim. Acta—Part A Mol. Biomol. Spectrosc. **175**, 262–268 (2017). https://doi.org/10.1016/j.saa.2016.12.036
107. Z. Wang, S. Zong, H. Chen, H. Wu, Y. Cui, Silica coated gold nanoaggregates prepared by reverse microemulsion method: dual mode probes for multiplex immunoassay using SERS and fluorescence. Talanta **86**, 170–177 (2011). https://doi.org/10.1016/j.talanta.2011.08.054
108. W. Deng, D. Jin, K. Drozdowicz-Tomsia, J. Yuan, J. Wu, E.M. Goldys, Ultrabright Eu-doped plasmonic Ag@SiO 2 nanostructures: time-gated bioprobes with single particle sensitivity and negligible background. Adv. Mater. **23**, 4649–4654 (2011). https://doi.org/10.1002/adma.201102027
109. D. Yin, S. Wang, Y. He, J. Liu, M. Zhou, J. Ouyang, B. Liu, H.Y. Chen, Z. Liu, Surface-enhanced Raman scattering imaging of cancer cells and tissues via sialic acid-imprinted nanotags. Chem. Commun. **51**, 17696–17699 (2015). https://doi.org/10.1039/c5cc05174f
110. S. Narayan, A. Rajagopalan, J.S. Reddy, A. Chadha, BSA binding to silica capped gold nanostructures: effect of surface cap and conjugation design on nanostructure-BSA interface. RSC Adv. **4**, 1412–1420 (2014). https://doi.org/10.1039/c3ra45887c
111. X. Kong, Q. Yu, X. Zhang, X. Du, H. Gong, H. Jiang, Synthesis and application of surface enhanced Raman scattering (SERS) tags of Ag@SiO 2 core/shell nanoparticles in protein detection. J. Mater. Chem. **22**, 7767–7774 (2012). https://doi.org/10.1039/c2jm16397g
112. X. Xia, W. Li, Y. Zhang, Y. Xia, Silica-coated dimers of silver nanospheres as surface-enhanced Raman scattering tags for imaging cancer cells. Interface Focus **3** (2013). https://doi.org/10.1098/rsfs.2012.0092
113. N. Sui, L. Wang, T. Yan, F. Liu, J. Sui, Y. Jiang, J. Wan, M. Liu, W.W. Yu, Selective and sensitive biosensors based on metal-enhanced fluorescence. Sens. Actuators B Chem. **202**, 1148–1153 (2014). https://doi.org/10.1016/j.snb.2014.05.122
114. N. Sui, L. Wang, F. Xie, F. Liu, H. Xiao, M. Liu, W.W. Yu, Ultrasensitive aptamer-based thrombin assay based on metal enhanced fluorescence resonance energy transfer. Microchim. Acta **183**, 1563–1570 (2016). https://doi.org/10.1007/s00604-016-1774-y
115. M.G. Cerruti, M. Sauthier, D. Leonard, D. Liu, G. Duscher, D.L. Feldheim, S. Franzen, Gold and silica-coated gold nanoparticles as thermographic labels for DNA detection. Anal. Chem. **78**, 3282–3288 (2006). https://doi.org/10.1021/ac0600555
116. F. Hu, Y. Zhang, G. Chen, C. Li, Q. Wang, Double-walled Au nanocage/SiO2 nanorattles: Integrating SERS imaging, drug delivery and photothermal therapy. Small **11**, 985–993 (2015). https://doi.org/10.1002/smll.201401360
117. C. Xue, X. Chen, S.J. Hurst, C.A. Mirkin, Self-assembled monolayer mediated silica coating of silver triangular nanoprisms. Adv. Mater. **19**, 4071–4074 (2007). https://doi.org/10.1002/adma.200701506
118. B. Küstner, M. Gellner, M. Schütz, F. Schöppler, A. Marx, P. Ströbel, P. Adam, C. Schmuck, S. Schlücker, SERS labels for red laser excitation: silica-encapsulated SAMs on tunable gold/silver nanoshells. Angew. Chem. Int. Ed. **48**, 1950–1953 (2009). https://doi.org/10.1002/anie.200804518
119. M.F. Kircher, A. De La Zerda, J.V. Jokerst, C.L. Zavaleta, P.J. Kempen, E. Mittra, K. Pitter, R. Huang, C. Campos, F. Habte, R. Sinclair, C.W. Brennan, I.K. Mellinghoff, E.C. Holland, S.S. Gambhir, A brain tumor molecular imaging strategy using a new triple-modality MRI-photoacoustic-Raman nanoparticle. Nat. Med. **18**, 829–834 (2012). https://doi.org/10.1038/nm.2721

120. S.M. Kang, W.J. Kim, I.S. Choi, Biomimetic approach to the formation of magnetic nanoparticle/silica core/shell structures. J. Nanosci. Nanotechnol. 5347–5350 (2008). https://doi.org/10.1166/jnn.2008.1286
121. M. Dong, Y. Tian, D. Pappas, Synthesis of a red fluorescent dye-conjugated Ag@SiO2 nanocomposite for cell immunofluorescence. Appl. Spectrosc. **69**, 215–221 (2015). https://doi.org/10.1366/14-07615
122. Y. Liu, M. Xu, Q. Chen, G. Guan, W. Hu, X. Zhao, M. Qiao, H. Hu, Y. Liang, H. Zhu, D. Chen, Gold nanorods/mesoporous silica-based nanocomposite as theranostic agents for targeting near-infrared imaging and photothermal therapy induced with laser. Int. J. Nanomed. **10**, 4747–4761 (2015). https://doi.org/10.2147/IJN.S82940
123. S. Shen, H. Tang, X. Zhang, J. Ren, Z. Pang, D. Wang, H. Gao, Y. Qian, X. Jiang, W. Yang, Targeting mesoporous silica-encapsulated gold nanorods for chemo-photothermal therapy with near-infrared radiation. Biomaterials **34**, 3150–3158 (2013). https://doi.org/10.1016/j.biomaterials.2013.01.051
124. J.V. Jokerst, Z. Miao, C. Zavaleta, Z. Cheng, S.S. Gambhir, Affibody-functionalized gold-silica nanoparticles for raman molecular imaging of the epidermal growth factor receptor. Small **7**, 625–633 (2011). https://doi.org/10.1002/smll.201002291
125. S. Kang, Y. "Winston" Wang, X. Xu, E. Navarro, K.M. Tichauer, J.T.C. Liu, Microscopic investigation of topically applied nanoparticles for molecular imaging of fresh tissue surfaces. J. Biophotonics **11** (2018). https://doi.org/10.1002/jbio.201700246
126. T. Ribeiro, C. Baleizão, J.P.S. Farinha, Artefact-free evaluation of metal enhanced fluorescence in silica coated gold nanoparticles. Sci. Rep. **7** (2017). https://doi.org/10.1038/s41598-017-02678-0
127. Y. Yan, L. Meng, W. Zhang, Y. Zheng, S. Wang, B. Ren, Z. Yang, X. Yan, High-throughput single-particle analysis of metal-enhanced fluorescence in free solution using Ag@SiO2 core-shell nanoparticles. ACS Sens. **2**, 1369–1376 (2017). https://doi.org/10.1021/acssensors.7b00522
128. D. Hu, Y. Wang, J. Liu, Y. Mao, X. Chang, Y. Zhu, Light-driven sequential shape transformation of block copolymer particles through three-dimensional confined self-assembly. Nanoscale (2022). https://doi.org/10.1039/d2nr01172g
129. S. Krishnan, T. Chinnasamy, S. Veerappan, K. Senthilkumar, D. Kannaiyan, Dual labeled Ag@SiO2 core-shell nanoparticle based optical immunosensor for sensitive detection of E. coli. Mater. Sci. Eng. C **45**, 337–342 (2014). https://doi.org/10.1016/j.msec.2014.09.028
130. J. Zhang, F. Song, Z. He, Y. Liu, Z. Chen, S. Lin, L. Huang, W. Huang, Wide-range tunable fluorescence lifetime and ultrabright luminescence of Eu-grafted plasmonic core-shell nanoparticles for multiplexing. Small **12**, 397–404 (2016). https://doi.org/10.1002/smll.201502107
131. R. Appiah-Ntiamoah, A. Chinnappan, † Ab, H. Kim, Erratum: a facile method to effectively combine plasmon enhanced fluorescence (PEF) and fluoride-Lewis acid based reactions to detect low concentrations of fluoride in solution. RSC Adv. **6**(82), 78918–78929 (2016). https://doi.org/10.1039/c6ra90080a
132. S. Li, T. Zhang, Z. Zhu, N. Gao, Q.H. Xu, Lighting up the gold nanoparticles quenched fluorescence by silver nanoparticles: a separation distance study. RSC Adv. **6**, 58566–58572 (2016). https://doi.org/10.1039/c6ra11265j
133. X. Zhang, X. Kong, Z. Lv, S. Zhou, X. Du, Bifunctional quantum dot-decorated Ag@SiO2 nanostructures for simultaneous immunoassays of surface-enhanced Raman scattering (SERS) and surface-enhanced fluorescence (SEF). J. Mater. Chem. B **1**, 2198–2204 (2013). https://doi.org/10.1039/c3tb20069h
134. R.M. Davis, J.L. Campbell, S. Burkitt, Z. Qiu, S. Kang, M. Mehraein, D. Miyasato, H. Salinas, J.T.C. Liu, C. Zavaleta, A raman imaging approach using CD47 antibody-labeled SERS nanoparticles for identifying breast cancer and its potential to guide surgical resection. Nanomaterials **8**, 953 (2018). https://doi.org/10.3390/nano8110953
135. M. Tudose, D.C. Culita, P. Ionita, M.C. Chifiriuc, Silver nanoparticles embedded into silica functionalized with vitamins as biological active materials. Ceram. Int. **41**, 4460–4467 (2015). https://doi.org/10.1016/j.ceramint.2014.11.138

136. H. Li, J. Kang, J. Yang, B. Wu, Distance dependence of fluorescence enhancement in Au Nanoparticle@Mesoporous Silica@Europium complex. J. Phys. Chem. C **120**, 16907–16912 (2016). https://doi.org/10.1021/acs.jpcc.6b01312
137. D.D. Xu, Y.L. Deng, C.Y. Li, Y. Lin, H.W. Tang, Metal-enhanced fluorescent dye-doped silica nanoparticles and magnetic separation: a sensitive platform for one-step fluorescence detection of prostate specific antigen. Biosens. Bioelectron. **87**, 881–887 (2017). https://doi.org/10.1016/j.bios.2016.09.034
138. F. Tang, F. He, H. Cheng, L. Li, Self-assembly of conjugated polymer-Ag@SiO 2 hybrid fluorescent nanoparticles for application to cellular imaging. Langmuir **26**, 11774–11778 (2010). https://doi.org/10.1021/la101714q
139. M.L. Viger, D. Brouard, D. Boudreau, Plasmon-enhanced resonance energy transfer from a conjugated polymer to fluorescent multilayer core-shell nanoparticles: a photophysical study. J. Phys. Chem. C **115**, 2974–2981 (2011). https://doi.org/10.1021/jp109993a
140. J. Yang, F. Zhang, Y. Chen, S. Qian, P. Hu, W. Li, Y. Deng, Y. Fang, L. Han, M. Luqman, D. Zhao, Core-shell Ag@SiO2@mSiO2 mesoporous nanocarriers for metal-enhanced fluorescence. Chem. Commun. **47**, 11618 (2011). https://doi.org/10.1039/c1cc15304h
141. Z.Y. Hu, J.H. Tang, Z.L. Ge, T.M. Song, T. Qiu, H.P. Xie, L.P. Guo, Fluorescence enhancement of IgG-FITC based on surface plasmon resonance of Ag@SiO2 nanoparticles as application of biomarker. Anal. Lett. **46**, 844–855 (2013). https://doi.org/10.1080/00032719.2012.738342
142. Z. Zhang, L. Wang, J. Wang, X. Jiang, X. Li, Z. Hu, Y. Ji, X. Wu, C. Chen, Mesoporous silica-coated gold nanorods as a light-mediated multifunctional theranostic platform for cancer treatment. Adv. Mater. **24**, 1418–1423 (2012). https://doi.org/10.1002/adma.201104714
143. A.S. Monem, N. Elbialy, N. Mohamed, Mesoporous silica coated gold nanorods loaded doxorubicin for combined chemo-photothermal therapy. Int. J. Pharm. **470**, 1–7 (2014). https://doi.org/10.1016/j.ijpharm.2014.04.067
144. P.K. Sudeep, K. Takechi, P.V. Kamat, Harvesting photons in the infrared. Electron injection from excited tricarbocyanine dye (IR-125) into TiO2 and Ag@TiO2 core-shell nanoparticles. J. Phys. Chem. C **111**, 488–494 (2007). https://doi.org/10.1021/jp0665022
145. R. Bardhan, N.K. Grady, N.J. Halas, Nanoscale control of near-infrared fluorescence enhancement using au nanoshells. Small **4**, 1716–1722 (2008). https://doi.org/10.1002/smll.200800405
146. Z. Bai, R. Chen, P. Si, Y. Huang, H. Sun, D.H. Kim, Fluorescent pH sensor based on Ag@SiO2 core-shell nanoparticle. ACS Appl. Mater. Interfaces **5**, 5856–5860 (2013). https://doi.org/10.1021/am401528w
147. Y. Wang, X. Ding, Y. Chen, M. Guo, Y. Zhang, X. Guo, H. Gu, Antibiotic-loaded, silver core-embedded mesoporous silica nanovehicles as a synergistic antibacterial agent for the treatment of drug-resistant infections. Biomaterials **101**, 207–216 (2016). https://doi.org/10.1016/j.biomaterials.2016.06.004
148. M. Shanthil, R. Thomas, R.S. Swathi, T.K. George, Ag@SiO 2 core-shell nanostructures: distance-dependent plasmon coupling and SERS investigation. J. Phys. Chem. Lett. **3**, 1459–1464 (2012). https://doi.org/10.1021/jz3004014

Role of Defects in Metal Oxide Nanostructures

Sayak Datta, P. Maneesha, Prashant Kumar Mishra, and Somaditya Sen

Abstract Metal-oxide nanostructures (MOs) are ideal systems for exploring a large number of novel phenomena. The intrinsic defects and the induced defects in MOs can have a major role in the various properties exhibited by them. The modification of defects in a MO lattice can be done through different ways such as by varying synthesis conditions, by varying dopant, with changes in the morphology, and by various thermal treatments. This article provides a comprehensive review on basic principle and the state of art research activities related to role of defects in controlling the various parameters suitable for application of the MOs. The construction and integration of the metal oxide-based devices such as optoelectronic devices, photodetectors, sensors, memory devices and for energy storage devices must need knowledge about the role of defects in MOs. This chapter is dealing with the general concept of defects in solids, its formation and related changes in the properties exhibited by them. It further deals about the different types of defects and correlated properties of some of the well-known MOs such as ZnO, CuO/Cu_2O, NiO and perovskite oxides. The defects modified properties and related applications are also given in detail. This book provides readers about the different pathways of modification of defects in MOs, and ways of identifying the defects in MOs and their characterization. This chapter will be beneficial for the readers to understand the rich physics of defects in MOs.

Keywords Metal oxides · Nanostructures · Defects · Oxides applications · Perovskites

---

S. Datta · P. Maneesha · P. K. Mishra · S. Sen (✉)
Department of Physics, Indian Institute of Technology Indore, Indore 453552, India
e-mail: sens@iiti.ac.in

V. Kumar et al. (eds.), *Optical Properties of Metal Oxide Nanostructures*,
Progress in Optical Science and Photonics 26,
https://doi.org/10.1007/978-981-99-5640-1_7

1 Simple Oxides and Their Functionalities

Nanoparticle research is now a hot topic in science due to its various possible uses in several sectors like material science, agriculture, biomedical engineering, IT, optics, electronics, energy, catalysis, and sensing [1–4]. The electronic band structure of nanoparticles can be modified by changing the crystallite size or by doping or by the application of external forces like a magnetic field, pressure, temperature etc. The nature of the materials can vary from being a semiconductor, a metal, or an insulator. In simple oxides nanoparticles of materials like ZnO, CuO, NiO, TiO_2, MgO, Al_2O_3, SnO_2, AgO, CeO_2, ZrO_2, etc., can exhibit size-related structural abnormalities [5–8]. The difference in size means differences in the lattice strain and hence is correlated to the structure, i.e., the bond lengths and bond angles involved in the lattice. Such changes can be a resultant of any minor changes in the purity of the materials or be caused by the modifications in the synthesis conditions or be a consequence of the conditions at which the material is being subjected to. These changes in the atomic level of interaction of the electronic clouds of the ions composing the lattice can modify the band structure of the materials. Such modifications in the band structure can affect the electronic, optoelectronic, magnetic, and other properties of the materials. Hence, such modifications lead to applicability of such materials in optoelectronics, optical and magnetic sensors. Further these modifications can lead to usage in wastewater treatment, and antibacterial characteristics, which have recently received greater attention.

Morphology of these metal oxide nanostructures have received an emphatic emerging interest and is due to the fact that some properties are all highly sensitive to sample shape, i.e., the terminating surfaces of the crystallites. This allows the formation of new and enhanced characteristics that can be significantly altered by finite size effects in systems with shorter length scales. The high surface-to-volume ratio of nanomaterials, which is most pronounced in nanoparticles, can also have a significant impact on the optoelectronic properties. In such materials due to their finite size the possibilities of surface defects are high and hence, the defect concentration in MO nanostructures is often higher than that in bulk systems. This hence correlates their significantly larger surface to volume ratio where the enhanced surface states are more prone to defects [9–11]. Apart from the surface defects, the termination of the lattice at a shorter distance affects the regularity of the lattice and hence the bond lengths. Such changes in the bond lengths often result in absence or excess of constituent ions known as vacancies and interstitials [12]. Occasionally there are anti-site defects due to one type of cation or anion being misplaced at the site of another type anion or cation. All such defects often create distinct energy states due to the different formation energies of such defects. These are intrinsic defects due to modifications of the constituent ions and hence are called latent defects. Apart from these latent defects there can be other defects due to doping which generally are extrinsic in nature [13]. When describing metal oxide nanostructures, it is necessary to take these defects into account which significantly impacts the physical characteristics of nanomaterials.

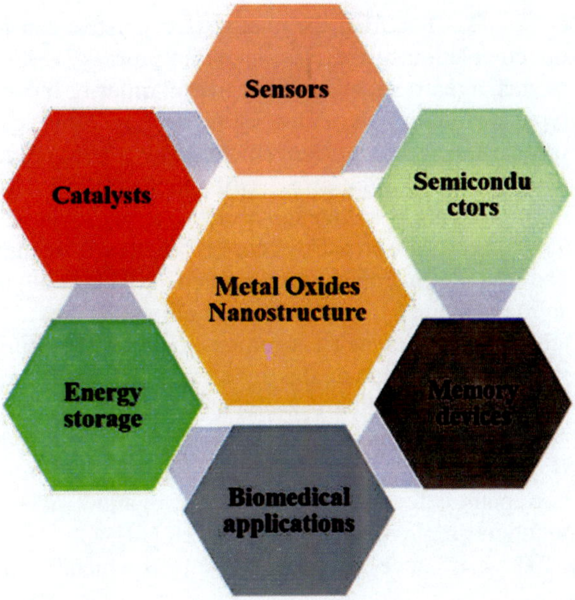

Fig. 1 Flow chart representations indicate multi-functionalities of nanomaterials

In this chapter we will briefly go through the different types of defects such as stoichiometric, non-stoichiometric and other noticeable defects which have a major impact on metal oxide properties and related applications. However, our focus will be defect states found in different metal oxide nanostructures. These will be discussed in detail separately for a few important simple oxides, like ZnO, CuO, NiO and some complex perovskite oxides as well. These examples may help us better understand how the defect states affect different oxides in a unique way and in turn contribute to a wide range of applications in several domains of life (Fig. 1).

2 Defect States in Solids

A solid is created from a variety of tiny crystals. However, defects found in solids arise due to fast and moderate rates of crystallization process. Larger crystalline solids are created by the accumulation or joining of several tiny crystals together thereby initially forming polycrystalline samples. Defects arise at the interface of these crystallites often leading to planar defects or much more complicated forms of defects. These defects are an arbitrary arrangement of those ions at the interface of neighbouring crystallites. The larger crystallites grow with time due to a continuous evolution of the entire assembly of the crystallites when the smaller particles subside and are absorbed into the larger ones by the process of diffusion at the interface of the two crystallites from the smaller to the larger crystallite. In the process, the defect states at the interface undergoes modifications with time as the diffusion continues. Therefore, the synthesized crystals can have a variety of defects after the completion of the crystallization process [13]. In such a process the imprints of such a rearrangement of the irregularities or inconsistencies can be of numerous types including point defects, line defects, planar defects etc. When there are any inconsistencies in the arrangement of any atom or point in an ideal crystal, it is referred to as a deviation point defect. Different forms of point defects can be observed. When the process of crystallization occurs at a very rapid rate, point defects are considered. These defects are primarily caused by variations in the arrangement of the constituent particles. A point defect occurs when the ideal configuration of solids in a crystalline solid is distorted around an atom or point. Like this, a line defect occurs when there is any divergence in the arrangement of the crystal's entire row of lattice points. Below is a short discussion on the same as per the literature [14].

Point defects, line defects, surface defects, and volume defects are the four categories into which defects or imperfections in crystalline solids can be subdivided. Historically at the very beginning of defect study, ionic crystals were first known to have point defects. With further investigation and research, such defects were also found in simple metal crystals as well [14].

There are three types of point defects which are mainly observed such as:

- **Stoichiometric defect**
- **Frenkel defect**
- **Schottky defect**.

Besides these, we also have impurity defects, non-stoichiometric defects, metal deficiency defects, etc.

2.1 *Stoichiometric Defects*

The defect in which the compound's stoichiometry, as represented by their chemical formula, remains unchanged is known as stoichiometric defect. Keeping the solid's electrical neutrality in mind, this type of point defect maintains the stoichiometric ratio of positive and negative ions. It is well known as intrinsic or thermodynamic defects. These defects basically fall into two categories:

- ***Vacancy defect***: When an atom is absent from one of its lattice sites, leaving the site unoccupied, then a vacancy defect is created. This is associated with another vacancy of an ion of opposite polarity. As a result, the density of the substance is expected to reduce. In a different situation such an ion may be displaced to an intermolecular space of the crystal and thereby a charge neutrality is maintained without the loss of any other ion. In such cases the density may or may not be affected [15].
- ***Interstitial defect***: When an atom or molecule occupies the intermolecular spaces of crystals, then interstitial defects are created. Such defects are created due to the presence of some other ion with a larger opposite polarity than the regular

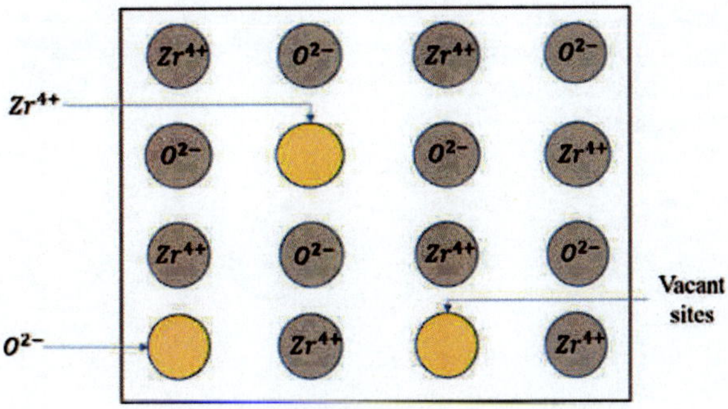

Fig. 2 Schottky Defect [Here one cation (Zr^{4+}) and two anions (O^{2-}) are missing from the lattice]

type, thereby requiring more quantity of such an ion. The density of the substance increases due to this defect. However, if this interstitial ion is due to a misplacement of the ion the density remains unaltered [15].

2.2 *Schottky Defects*

This kind of defect arises when an equal number of cations and anions get missing from the lattice site. The material having a high coordination number, shows this type of defect. In ionic compounds, the Schottky defect typically appears when the radius ratio of the cation and anion [r+/r−] is not too far below unity. It is a type of vacancy defect in which the crystal retains its neutrality despite having an equal number of cations and anions missing from the lattice. A representation of the Schottky defect in ZrO_2 crystal structure is shown in the Fig. 2. The density of the crystal decreases as a result of this defect. Examples include CeO_2, cubic ZrO_2, UO_2, ThO_2 and PuO_2 [15].

2.3 *Frenkel Defects*

Such a defect arises when some cations are absent from their original lattice site and instead occupy the interstitial sites of the lattice. Generally, substances having low coordination number show this type of defect. The dielectric constant of the crystal increases when similar charges come close to each other.

Frenkel defect is a type of interstitial defect or dislocation defect in which the smaller ions, mostly cations, dislocate from their original position and occupy the interstitial site of the crystal lattice. Since there is no ion movement outside the crystal, the density of such crystals remains constant. The ionic compounds in which the radius ratio[r+/r−] is low exhibit such kinds of defects. Example: Lu_2O_3, MgO, etc. [15]. A schematic of the Frenkel defect in Lu_2O_3 crystal structure is shown in the Fig. 3.

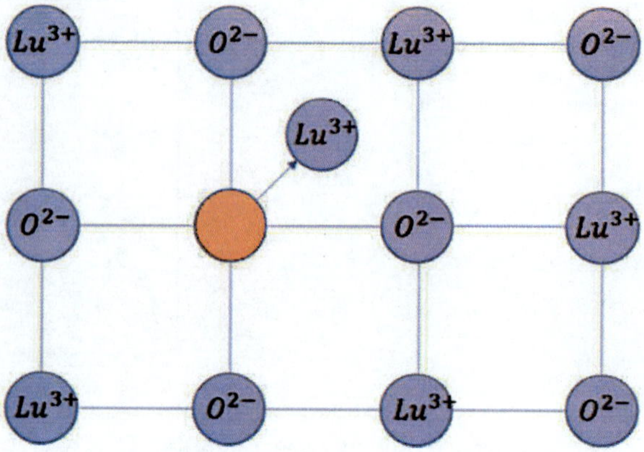

Fig. 3 Frenkel defect [One cation (Lu^{3+}) missing from the lattice occupying the interstitial site]

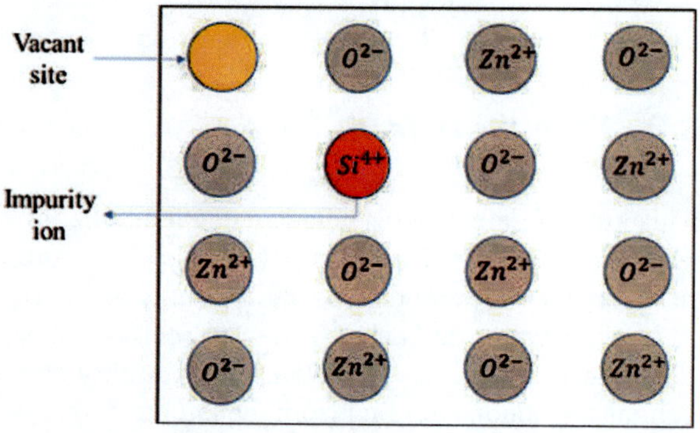

Fig. 4 Impurity defect (crystal of ZnO containing impurity of Si^{4+})

2.4 *Impurity Defects*

Sometimes, simple compounds can contain small amounts of impurities of some other compounds in them. Such defects are known as impurity defects. For example, as in Fig. 4, ZnO can contain a small amount of Si^{4+} as impurity in the crystal. Some of the Zn^{2+} ion positions can be replaced by one Si^{4+} ion. Sometimes, two Zn^{2+} ions can get replaced by one Si^{4+} ion to maintain charge neutrality of the crystal leaving one site of Zn^{2+} as vacant. In another condition the extra charge of Si^{4+} can be compensated by a O^{2-} interstitial [11].

2.5 *Non-stoichiometric Defects*

Compounds of d-block elements (i.e., transition metals) of the periodic table generally demonstrate this type of defect. These defects are described as metal excess defects which can be mainly of two types i.e., (i) metal excess defect due to anionic

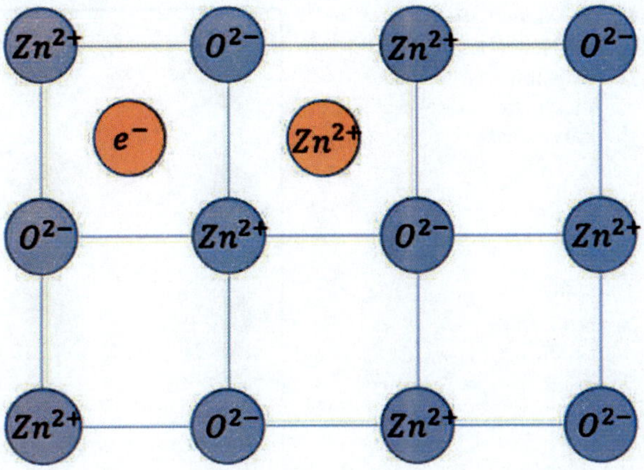

Fig. 5 Metal excess defect owing to presence of interstitial cation

vacancy and (ii) metal excess defect owing to presence of interstitial cation. Metal oxides generally show the latter type of defect which is briefly discussed below.

2.5.1 Metal Excess Defect

When a compound contains an excess of cations, it releases these extra cations upon heating it. These cations then occupy the interstitial sites in crystals while the neighbouring interstitial sites are also occupied by an equal number of electrons. For example, ZnO exhibits this type of defect. ZnO loses oxygen reversibly on heating. For maintaining electrical neutrality, extra Zn^{2+} goes into the interstitial sites, while neighbouring interstitial sites are occupied by electrons as shown in Figure 5. ZnO changes from white to yellow colour upon heating due to removal of oxygen [11].

$$ZnO \xrightarrow{\text{Heating}} Zn^{2+} + \frac{1}{2}O_2 + 2e^-$$

2.5.2 Metal Deficiency Defect

This defect is caused when some cations are missing from the lattice site and another cation still present increases its valency to keep the lattice electrically neutral. Consider the FeO mineral, which has a composition of $Fe_{0.95}O$. It might range from $Fe_{0.93}O$ to $Fe_{0.96}O$. Some Fe^{2+} cations are absent in the FeO crystal but the presence of the necessary amount of Fe^{3+} ions make up for the loss of positive charge. The metal deficiency defect in FeO is represented in the Fig. 6. The removal of one Fe^{2+} from the FeO crystal structure leads to change of oxidation states of two Fe^{2+} ions (from 2+ to 3+) to form Fe^{3+} ions in neighbouring sites to maintain the electrical neutrality [12].

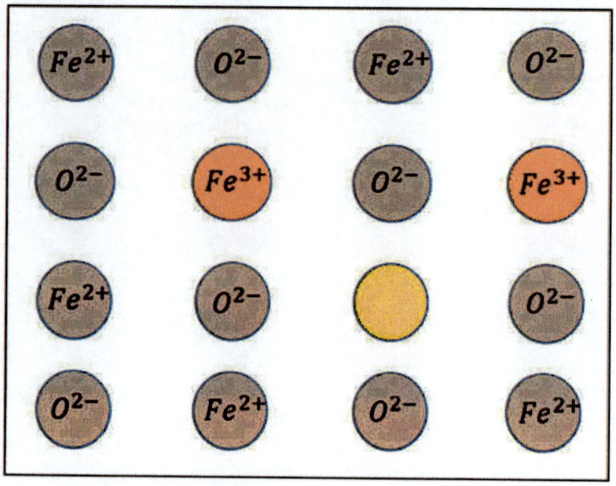

Fig. 6 Metal deficiency defect [Here one cation (Fe^{2+}) is missing but two Fe^{3+} ion compensates for this cation loss]

3 Defects in Some Important Metal Oxides

By moderating the defects within the semiconductor, the properties of the MOs can be varied. The roles of defects differ from application to application. In microelectronics, intrinsic defects such as interstitials, intrinsic, or vacancies mediate the diffusion of dopant atoms. As the feature size of the device becomes smaller the importance of the defects will become more pronounced. For example, in nanometer scale structures, the effect of a single point defect cannot be ignored since the size of the structure is small enough. Hence, in the device development the ability to control the defects is as much important for the proper functioning. Here we have discussed a few important MOs and their defects and correlated properties [9, 16–19].

4 Zinc Oxide (ZnO)

ZnO has three possible crystalline structures: Rock salt, Zinc blende, and Wurtzite. Amongst the three, the highest stability of the hexagonal ZnO wurtzite phase makes it a technologically important material [6]. The wurtzite structure of ZnO in a ball and stick representation (Fig. 7), can be illustrated as below.

Wurtzite ZnO is a tetrahedrally oriented II–VI semiconductor with a direct band gap of ~3.37 eV and an exciton binding energy of ~60 meV [6]. Each anion is surrounded by four cations in a tetrahedral geometry. This is a result of the sp^3 covalent bonding. A strong ionicity of ZnO from the covalent nature of the bonding is observed which shapes the bandgap of the material. Hence, small changes in the bond length and bond angles in these materials due to whatever reason, be it synthesis parameters or doping or any other physical changes in its applicability, can modify the covalent bonding nature and hence the electronic localization of the lattice and hence

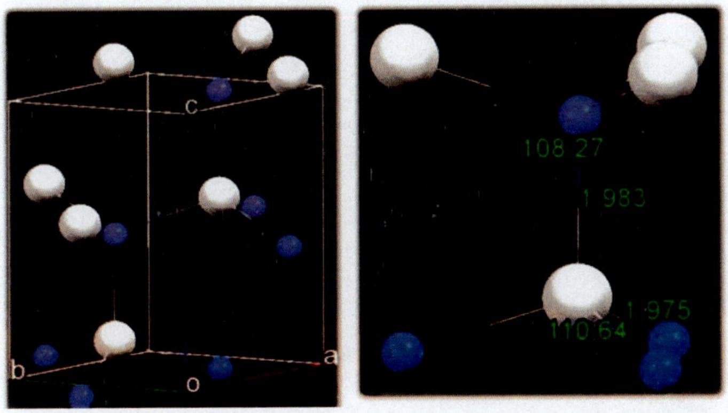

Fig. 7 Wurtzite form of ZnO and its tetrahedral Zn and O atoms arrangement in detail

modify the band structure and transport properties of the materials [6]. Thermally stable wurtzite ZnO nanostructures have excellent optoelectronic and piezoelectric properties and are biocompatible. ZnO is environmentally friendly in nature. This makes them the most potential candidate for optoelectronic devices such as solar cells, ultra-violet laser diodes, solar cells, light-emitting diodes (LEDs), etc. Other devices like field-emission devices, actuators, sensors, spintronics and piezoelectric devices are also important. ZnO is also used as a potential photocatalyst [9].

4.1 *Defects in ZnO*

The structure of ZnO is known to have inherent point flaws. Oxygen vacancies (V_O), oxygen interstitial (O_i), zinc vacancies (V_{Zn}), zinc interstitial (Zn_i), oxygen antisite (O_{Zn}), and zinc antisite (Zn_O) are the inherent point defects of ZnO. As illustrated in the image, these defects have formation energies corresponding to defect levels inside the bandgap of ZnO. The transition from conduction bands to these levels causes green, blue, violet, yellow, and orange-red emission (Fig. 8). The following can be formulated for different emissions [17, 20–22].

- Violet emission → 3 eV (413 nm) ~ 2.75 eV (450 nm) → Zn_i
- Blue emission → 2.75 eV (450 nm) ~ 2.50 eV (496 nm) → V_{Zn}
- Yellow and orange-red emission → 1.65 eV (750 nm) ~ 2.17 eV (571 nm) → O_i
- Green emission → 2.17 eV (570 nm) ~ 2.50 eV (496 nm) → V_O.

Besides these intrinsic or native defects, ZnO is also known to observe some extended defects. These defects can be found almost throughout the crystal but they remain largely confined along at least one direction and can have a profound effect on the properties of the crystal. Similarly, a high density of grain boundaries can increase the material's hardness and hence can also modify the conductivity of the ZnO crystal

- Violet emission → 3 eV (413 nm) ~ 2.75 eV (450nm) → Zn_i
- Blue emission → 2.75 eV (450 nm) ~2.50 eV (496 nm) → V_{Zn}
- Yellow and orange-red emission → 1.65 eV (750 nm) ~ 2.17 eV (571 nm) → O_i
- Green emission → 2.17 eV (570 nm) ~ 2.50 eV (496 nm) → V_O.

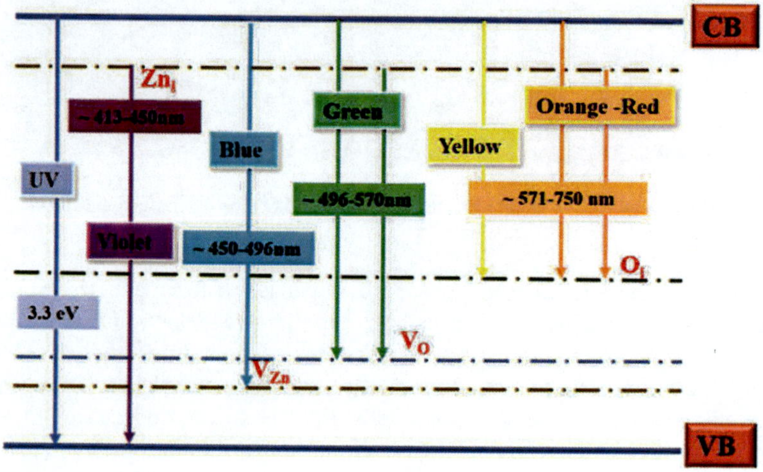

Fig. 8 Energy level diagram illustrating major defect levels in ZnO

which may result in an increase in the rate of diffusion of the charge carriers [23]. Sometimes, these defects can be in the form of surface conduction where only the material surface exhibits conductivity and very less conductivity is observed in the bulk material [24]. Also, these kinds of defects can arise from dislocations in the crystal lattice as well as from the synthesis route used for the preparation of these materials that leads to stacking faults in the lattice [25].

4.1.1 Oxygen Vacancies (V_O)

Oxygen vacancies, V_O, are loss of O^{2-} ion from the lattice. The removed O^{2-} had electrons bonded to four neighbouring Zn^{2+} ions. For an uncharged defect, the electrons will persist even after the removal of the O^{2-} ion, resulting in an abundance of electrons. Electron donor states are chemical defects that possess extra electrons and therefore can donate the excess electrons to an absorption process. By donating electrons, it is itself oxidized but reduces other entities involved in the process. Therefore, the V_O defects function as donors. The defect state energy is mostly reported below the CB boundary at 2.2–2.5 eV, i.e., a transition is possible from these defects to the VB emitting a wavelength of 496–563 nm, which corresponds to green luminescence

[17, 20, 22, 26]. Exceptions to this trend are also found in literature. Janotti et al. [27] found states belonging to V_O which are ~1.0 eV lower than the CB minimum.

4.1.2 Oxygen Interstitial (O_i)

In the ZnO lattice, two probable configurations for oxygen interstitial (O_i) have been proposed. The first is a tetrahedral arrangement, while the other is octahedral. The tetrahedral O_i was found to be unstable, while the octahedral was stable. In a semi-insulating and p-type ZnO, these O_i can be electrically inactive while in n-type materials these can remain as deep acceptors in an octahedral configuration. In fact, the O_i species may be held responsible for defect recombination on the O-sub lattice for n-type ZnO materials. The O_i levels are placed above the VB maxima. A literature reports the O_i defect states at 1.93 eV higher than the VB while these were observed at 1.96 eV above the VB. Others report these states to be 1.70–2.10 eV above the VB. When electrons transit from the CB or Zn_i–O_i, a yellow and orange-red color emission is obtained. An O_i to VB transition results in an orange emission. Hence, either way, the O_i defects are responsible for either yellow or orange or orange-red emissions [21, 28–31].

4.1.3 Zinc Vacancies (V_{Zn})

The removal of a Zn^{2+} from the lattice results in four dangling oxygen bonds, which are known as zinc vacancies (V_{Zn}). The four oxygen bonds combine to form a doubly occupied symmetric state in the valence band, whereas, they form three almost degenerate states close to the VB maxima in the bandgap. These three almost degenerate states contain four electrons. However, they can only take two electrons. Electron acceptor states are chemical defects that can accept electrons transferred to it from other sources. Therefore, these states can accept electrons from electron donors or VB and result in absorption. By accepting electrons, it is itself reduced but oxidizes other entities involved in the process. Hence, V_{Zn} shows an acceptor behaviour. Above the VB, two charge states are possible for V_{Zn}. These states have a ~0.8 eV energy difference. V_{Zn} states were found at 2.51–2.60 eV below CB maxima. These were also found at a lower range of 2.60–2.77 eV below CB maxima. These were even found as low as ~2.84 eV below CB maxima. Hence, a range of ~2.5 eV (460 nm) ~2.84 eV (387 nm) can be safely allotted to a CB to V_{Zn} transition, thereby providing a blue emission [27, 32–34].

4.1.4 Zinc Interstitial (Zn_i)

There are two probable places for zinc interstitial in ZnO: one as a tetrahedral site and the other as an octahedral site. According to reports, the octahedral Zn_i site is the most stable, whereas the tetrahedral Zn_i site has larger energy and is more unstable.

The nearest neighbours at the tetrahedral site are one Zn and one O atom, whereas the octahedral site includes three Zn and three O atoms. In ZnO lattice, the Zn-ion is in a predominant Zn^{2+} state. To maintain charge neutrality, an introduced extra Zn^{2+} ion will need two electrons to leave a Zn_i site uncharged. This introduces a two-electron state addition. These states are generally above the conduction band minimum (Fig. 8). With the supply of a very small amount of energy, these two electrons can be transferred to the CB. Hence, Zn_i is a shallow donor. This allows $Zn_i \rightarrow$ VB transitions of energy range, 2.75 eV (387 nm)–3 eV (460 nm). Zn_i can also be a defect state located deep inside the bandgap. It can be located at ~0.5 eV above the VB (deep donor). Zn_i is also discussed just 0.22 eV below CB (i.e., shallow donor; this is the most accepted position). The most accepted value of Zn_i is a shallow donor providing a luminescence that is violet in color [17, 22, 27, 34].

However, some literature also reports these states in the CB, ~1.2 eV above CB minima. This model ensures the n-type nature of the materials.

4.1.5 Zinc Antisites (Zn_O)

When a Zn^{2+} ion is replaced at the site of O^{2-} ion, a zinc antisite, Zn_O, defect is created. The formation energy of this type of defect is high and hence at equilibrium condition or proximity to the steady state, the contribution from these states is unlikely. Even if this state is a double donor state the Zn_O antisite in an n-type ZnO cannot take part in the n-type conductivity [27]. Oba et al. [34], on the other hand, discovered that Zn_O are equivalent to V_O defects in a Zn-rich lattice. The distinct charges of these states might result in the alteration of the energy levels. Zn_O has high formation energy. Equilibrium conditions with Fermi energy in the bandgap are difficult. Therefore, like Zn_i the Zn_O defects can exist in the CB, ~1.2 eV above CB minima. This enables an n-type nature [34, 35].

4.1.6 Oxygen Antisite (O_{Zn})

The oxygen antisite (O_{Zn}) defect occurs when an O^{2-} is substituted at the Zn^{2+} site. Despite being an acceptor-type defect, the O_{Zn} has an extremely high energy demand for production and is electrically inactive even under favorable O-rich circumstances. As a result, the concentration of O_{Zn} defects in ZnO at equilibrium is quite low. O_{Zn}, on the other hand, is a point defect of the acceptor type. It requires very high formation energy and hence these states are difficult to form even in the excess of O_2 in the synthesis process. As a result, antisite faults of this sort play a little role in deep level emission [27, 34, 36]. The various kind of defects found in ZnO can be listed as (Table 1).

Kung et al. [37] reported that ZnO and ZnO: *Y* nanorods have been successfully prepared at low temperature using the hydrothermal method. A detailed study was done on the relationship between the *Y*-doped induced defects and their structural,

Table 1 Tabulation of intrinsic defects or native defects in ZnO for different colors and centers

Deep level transitions	Wavelength range (nm)	Energy (eV)	Emission color	References
$Zn_i \rightarrow$ VB	~385–435	~3.22–2.85	Violet	[34]
$Zn_i \rightarrow V_{Zn}$ or CB $\rightarrow V_{Zn}$	~435–495	~2.85–2.50 ~0.21 eV near to CB	Blue	[34]
CB $\rightarrow V_O$ or V_{Zn} or CB $\rightarrow V_O$ and V_{Zn}	~495–570 532	~2.50–2.18 ~2.32	Green	[34]
CB $\rightarrow O_i$	~570–590	~ 2.18–2.10	Yellow	[34]
CB $\rightarrow O_i$ or $Zn_i \rightarrow O_i$	~590–620	~ 2.10–2.00	Orange/Red	[34]

optical, and magnetic properties in ZnO: *Y* nanorods. It was discovered that *Y* doping causes an increase in defects: V_O^+, V_O^{++}, and O_i.

According to Srivastava et al. [38], the additional charge of Si^{4+} in the ZnO lattice lowers the amount of V_O. Owing to the fact that Si^{4+} is smaller in size than Zn^{2+}, defects like V_{Zn} and O_i reduce to give space for Si^{4+}. Si doping causes nominal distortions that increase Zn interstitials. As a result, orange-red luminescence now reduces the dominance of green luminescence. Hence, in order to enable a more regular lattice, Si^{4+} substitution enhances the bandgap in modified ZnO and correlates to a decrease in oxygen-related defects.

Bajpai et al. [39] performed Si^{4+} substitution in Fe^{3+}-doped ZnO to reduce oxygen vacancies. This makes the lattice more crystalline. However, with the gradual substitution of Fe^{3+} in Si^{4+}-substituted ZnO, O_i and Zn_i as well as V_O's increase. This results in increased lattice strain. The presence of Si^{4+} in Fe^{3+}-doped ZnO lattice seems to facilitate oxygen interstitials.

Mishra et al. [20] showed a decrease in the oxygen-related defects for Fe/Ga co-doped samples. Such a reduction of V_O and O_i can result in an increase in V_{Zn}, which emits blue light. Mishra et al. [40] further reported that for Ga/Si co-doped samples, the PL spectra show that the NBE and DLE peak intensities are lower in the doped samples compared to pure ZnO. Hence, surface defects decrease with substitution of Ga/Si in ZnO. The orange–red shift in the CIE plot is consistent with a decrease in V_O and a rise in O_i in the lattice.

According to Ayaz et al. [21], samples with increased Li^+ content showed enhanced green emission, which suggests that there are more oxygen vacancies in the ZnO lattice.

Zong et al. [41] reported that Co^{2+} doping can significantly enhance the oxygen defects as well as improve the magnetic behaviour of ZnO nanoparticles while having minimal impact on its structural and morphological properties. Air annealing experiment serves as an additional confirmation to the significant role of oxygen defects on the magnetic properties. From the DFT (Density Functional Theory) studies, it has been calculated that Co doped ZnO nanoparticles have long range magnetic ordering due to the localized spin moments of 3d electrons of Co^{2+} ions. According to theoretical and experimental findings, in Co-doped ZnO system presence of oxygen defect-induced bound magnetic polarons in which coupling $Co^{2+} - O_V^{+} - Co^{2+}$ leads to the ferromagnetic behaviour [41].

Okada et al. reports that concentrations of defects such as (Zn_i, V_O and V_O^{+}) increase with increase of Al doping which provide more electrons to the conduction band of ZnO crystal.

Apart from this, these formed defect levels resist the direct transfer of electrons from conduction band to the valence band. This enhances the effective number of holes and electrons. Hence, the reduction of electron hole recombination leads to the increase in the photocatalytic activity [42].

The electronic charge distribution in the vicinity of Zn in the ZnO lattice is modified by the change of the local charges brought in by a foreign dopant's influence on the O-neighbourhood due to the differences in size and valence state of the dopant. Hence, the electronic band structure gets modified, thereby modifying the amount and nature of the latent crystal defects. The ionic size difference between the dopant ion and the host ion can modify the strain in the lattice and thereby modify the crystallite size, amount and nature of the defects and also affect the morphology, surface charge, etc. The role of V_O, as an effective electron trap, increases the efficiency of separation of the electron/hole pairs generated through photons [17, 26–28, 32].

Pure ZnO is known to be a dominant n-type semiconductor. In search of a p-type ZnO, a majority of research has been focused on the production of p-type ZnO thin films and the impact of acceptor–donor co-doping on physical properties. V_O is a common hindrance in achieving a solution-processed p-type ZnO thin film. Hence excess oxygen supply was required for the fabrication of such thin films. Such techniques have helped remove V_O or introduction of O_i as acceptor level in ZnO and enable the formation of a possible p-type ZnO.

ZnO latent defects can be a source of n-type nature:

The microscopic equilibrium processes of intrinsic doping asymmetry occurring in ZnO were investigated by Zhang et al. ZnO was assumed to be in equilibrium with a hypothetical reservoir containing Zn and O atoms. With this assumption, the enthalpy of formation of a defect α of charge q and defect transition energy was computed using Eqs. (1) and (3) respectively [43].

$$\Delta H^{(q,\alpha)} = \Delta E^{(q,\alpha)} + n_\alpha \mu_\alpha + q E_F \tag{1}$$

where

$$\Delta E^{(q,\alpha)} = \Delta E^{(q,\alpha)}(\text{defect} + \text{host}) - E(\text{host only}) + n_\alpha \mu_\alpha(\text{solid}) + q E_v \tag{2}$$

Here, $\Delta E^{(q,\alpha)}$(defect + host) represents the cell's total energy containing both the host and defect α. The total energy of the cell is represented by E (host only). E_F denotes Fermi energy, E_v denotes the host crystal's valence band maximum, n_α is the number of atoms that are removed from the host crystal to the atomic reservoir during the defect formation stage.

$$\varepsilon(q/q') = [\Delta E(q,\alpha) - \Delta E(q',\alpha)]/(q' - q) \quad (3)$$

where q and $q\prime$ are the two different states of defect α. The defect is in state q i.e., higher charge state when $E_F < E(q, q\prime)$ as opposed $q\prime$ i.e., lower charge state when $E_F > E(q, q\prime)$. From the above calculation in Eqs. (1) and (3), it can be observed that acceptor-like defects such as V_{Zn} and O_i in ZnO have quite high formation energies which clearly indicates a low equilibrium concentration [44].

Creation of shallow donor states like Zn_i and antisite Zn_O close to the VB maxima or CB minima can act as donors since they readily produce electrons. Increase in donor concentration by possible reduction of the formation enthalpy ΔH is possible by synthesis in an O-deficit environment resulting in the creation of V_O and Zn_i. Even if E_F is high in the gap, donors must have low formation enthalpy ΔH, for them to become abundant. It is applicable for V_O and Zn_i in Zn-rich conditions whereas electron-killer centers (O_i, V_{Zn}) must have a high formation enthalpy to form. This is favorable for the Zn-rich conditions and unfavorable for O-rich circumstances. Therefore, Zn_i can be the root cause of an intrinsic n-type ZnO because electron-killer centers (O_i, V_{Zn}) should possess high formation enthalpy despite E_F being high in the gap, so that they do not form. It is thus approximately valid for Zn-rich but not O-rich conditions [45].

During the p-type doping scenario, acceptors (O_i, V_{Zn}) need to have shallow levels in respect to V_B maxima to conveniently produce holes. O_i, on the other hand, is not a shallow acceptor. For acceptors to become widespread, they should have a low formation enthalpy ΔH, even when the gap's E_F is low. It only applies to circumstances that are O-rich; it does not apply to situations that are Zn-rich. As a result, the inherent n-type in ZnO in a Zn-rich environment is due to Zn_i, which accords well with experiment results. It is therefore concluded that due to lower formation enthalpies of hole-killer defects, ZnO could not be a p-type semiconductor in thermal equilibrium and Zn rich circumstances. Acceptor levels like O_i, V_{Zn} conveniently produce holes being close to the VB maxima (shallow levels). V_{Zn} is a shallow acceptor, but not O_i. Hence, they should have a low formation enthalpy ΔH, even when the gap's E_F is low, for them to become common. It only operates in O-rich settings, not in Zn-rich ones. To avoid forming, hole-killer centers (V_O, Zn_i, and ZnO) must have a high formation enthalpy, even if the gap's E_F is low [45].

Pure ZnO has higher resistivity than doped and co-doped ZnO due to the lower carrier density. The electronic and band structures of ZnO can be modified in order to increase the conductivity of ZnO.

The following three types of electronic transitions are possible to exist in ZnO nanostructures:

(i) Electron excitation from the ZnO host's VB to CB.
(ii) Dopant-induced d–d internal transitions.
(iii) Charge transfer transitions between the host CB and the dopant d-state, as well as between the host VB and the dopant d-state.

The type (ii) and (iii) transitions can appear in the visible range. This means that introducing foreign elements changes the energy states, which has a significant influence on aspects like optoelectronics, photo-sensing, and photocatalysis. Based on the change in functionalities of ZnO, the dopants were chosen and employed to examine the functional views of the parameters [46].

4.2 Applications of ZnO

The intrinsic defects developed in ZnO due to the various fabrication methods leads to the enhancement of properties that can be used for different applications. Eisenberger et al. [47] developed a new synthesis procedure that controls certain defects in pure ZnO nanoparticles. The created defects lead to the decrease in the bandgap and contribute to the n type conductivity. The control of the defects which are responsible for the green emissions leads to the strong emission in the violet to blue region. This can be applicable to bioimaging using fluorescent microscopy techniques. Most of the ZnO nanoparticles requires UV sources for excitation, this work allows 405 nm laser source for live-cell imaging experiments using a confocal microscope. The high sensitivity of ZnO can be utilized for application as sensors. The O_v on the surface which are responsible for the changes in the electronics property are involved for sensing action. The adsorption of gases like NO_2 on the vacancy site through oxidation leads to the withdrawal of electrons and depleted from the conduction band and hence reduction in the conductivity. Whereas gases like H_2 can react with surface adsorbed oxygen and undergo reduction process and leads to the increase in the conductivity. These can be utilized for the sensing of different types of gases. The major challenge is the selective sensing of various gases [48].

The modulation of intrinsic defects in ZnO can be utilized for the Resistive switching memory device application. The concentration of defects and the microstructural properties can modulate the area of switching region. In order to get promising memory effect, high resistivity ZnO film is required. Less leakage current and high resistivity can be obtained through the suppression of the native defects. Thin switching layer and lower current operation can be achieved with high resistive ZnO. One of the promising approach for scalable memory devices with low power is ZnO nanoisland-based switching memory device [49]. Room temperature ferromagnetism in pure ZnO is another field of research interest. Many researchers have proved that this magnetism is due to the V_O present in ZnO [50, 51]. The investigations have revealed that the magnetic property of the ZnO is strongly correlated with the concentration of Vo at the surface [52]. This observed ferromagnetism can be utilized for the magnetic memory devices applications.

For the photocatalytic waste water treatment application, the defects present in ZnO play a major role. The modification of the surface defects will help for the adsorption of dye molecule. The intrinsic defects will modify the band gap and act as charge trap centres lead to decrease of electron hole recombination rate. These defects are helpful for enhancing the photodegradation of the dye molecule present in the waste Water. Heavily doped n-type ZnO is currently a particularly desirable transparent conductive oxide material due to its high conductivity, and it can be utilized for solar cells and light emitters. There is a global scarcity of the industry standard ITO. Hence, industry has started searching for replacements in a possible substitute: Ga and Al-doped ZnO [53].

5 Cupric Oxide (CuO) and Cuprous Oxide (Cu_2O)

CuO having a narrow indirect bandgap (~1.4 eV), belongs to the transition metal oxide group and shows p-type behaviour. It possesses a monoclinic structure and has many interesting characteristics including high stability, super thermal conductivity, photovoltaic properties, and antimicrobial activity. It is widely used in many technological applications as solar cell devices, gas sensors, magnetic memory devices, active catalysts [54–61]. CuO is a C_62h symmetric monoclinic structure (Fig. 9). Cupric oxide which has four formula units, has a coordination number of 4 i.e., having square planar configuration (110) plane. 1.95 Å and 1.96 Å are the Cu–O bond lengths in this plane. The lattice parameters of cupric oxide are $a = 4.6837$ Å, $b = 3.4226$ Å, $c = 5.1288$ Å, $\beta = 99.54^{\circ}$, and $\alpha = \gamma = 90^{\circ}$.

The space group of Cu_2O is Pn3m (Fig. 10). All oxygen atoms have a tetrahedral coordination with Cu atoms and each Cu atom is linearly coordinated to oxygen atoms. The lattice parameters of the Cu_2O unit cell are (a = b = c = 4.252 Å). The calculated direct band gap of Cu_2O is 0.46 eV. The point defects dominant in Cu_2O in various oxygen conditions, obtained from experimental and theoretical results are both neutral and single negatively charged V_{Cu}. The growth conditions and sub-stoichiometry will influence the electronic and magnetic properties of Cu_2O.

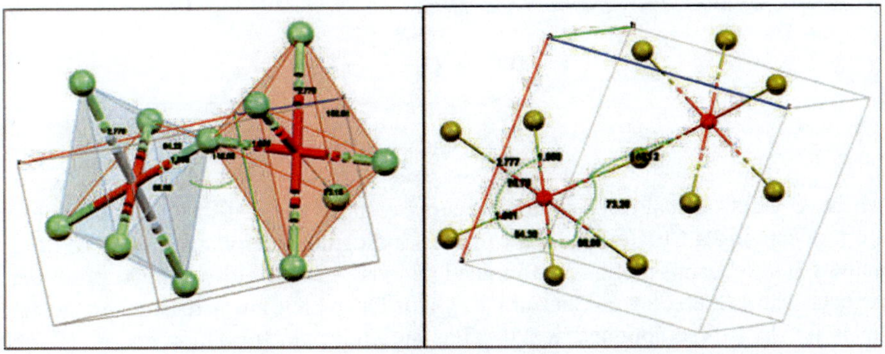

Fig. 9 CuO unit cell (Red ball = "Cu", Yellow ball = "O")

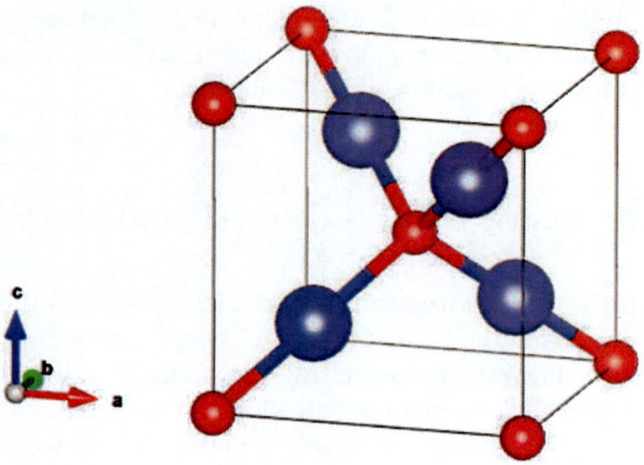

Fig. 10 Crystal structure of Cu_2O (Red Ball = "O", Blue Ball = "Cu")

5.1 *Defects in CuO and Cu₂O*

In intrinsic CuO semiconductors, the different types of defects are oxygen vacancy (V_O), copper vacancy (V_{Cu}), copper interstitials (Cu_i) and oxygen interstitial (O_i), Oxygen antisites (O_{Cu}) [62–64]. They are generated through the following defect reactions:

$$O_o \rightarrow V_o^{x+} + xe^- + \left(\frac{1}{2}\right)O_2$$

$$\left(\frac{1}{2}\right)O_2 \rightarrow V_{Cu}^{x-} + xh^+ + O_o$$

$$\left(\frac{1}{2}\right)O_2 \rightarrow O_i^{x-} + xh^+$$

$$Cu_i^{x+} \rightarrow Cu_{Cu} + xh^+$$

where, O_O is a normal oxygen atom occupying an O-site, Cu_{Cu} is a normal copper atom occupying a Cu-site, and x (= 0, 1, 2) indicating the ionization degree of the defects (i.e., neutral, mono-/dual-ionized defect). Accordingly, the two prominent defects (cation defects) in CuO are V_{Cu} and O_i, which are a reason for the free holes for the p-type conductivity. V_{Cu} has the minimum formation energy (O-rich condition) which indicates its p-type character. V_{Cu} prefers to capture an electron. Therefore, copper vacancy acts as an acceptor in CuO. V_O has the minimum formation energy (O-poor condition). V_O is a kind of anion defect which produces electrons and hence annihilates the holes, thereby being termed as a hole-killer. The optoelectronic property of CuO is determined by the most probable reaction above.

CuO has a complex structure consisting of a chain of tilted CuO_6 octahedra connected through an O-atom (Fig. 9). Compared to Cu_2O (octahedral symmetry with space group Pn3m), CuO crystallizes in a monoclinic crystal structure having less symmetry (space group C_2/c). The typical unit cell of CuO consists of eight atoms where four O atoms surround a Cu atom thereby forming a square planar configuration, whereas each O atom is situated at the center of four Cu atoms in the shape of a distorted tetrahedron. The lattice parameters of a CuO unit cell are a = 4.6837 Å, b = 3.4226 Å, c = 5.1288 Å. There are a substantial number of publications on theoretical study related to ground state properties, whereas the intrinsic defects have never been explored much. Nolan and Elliott [65] have investigated the defects in the simple antiferromagnetic configuration and conventional monoclinic cell. However, study on the experimentally observed ground state has not been done.

Aleksandar Zivkovic et al. [63] studied the dependence of growth condition on the defect's formation and the nature of (p type or n type) CuO semiconductor. CuO can either show p-type (O-rich/Cu-poor conditions) nature or n-type (Cu-rich/O-poor conditions) nature by varying the growth conditions. To analyze the intrinsic defects in CuO, they have considered a $2 \times 3 \times 2$ supercell (antiferromagnetic in nature) which ensures minimal interaction between introduced defects with an assumption that these defects are situated at least 10 Å apart in each crystallographic direction for calculation. They have obtained the result that the formation energy of defects shows similar trends between HSE and DFT + U despite the fact that it depends on the growth environment. The most favorable neutral defects are Cu_i and V_O, under Cu-rich growth environments, whereas p-type defects such as V_{Cu}, O_{Cu}, and O_i are favored under O-rich conditions over all n-type defects. Undoped CuO with high performance and conductivity can be created with different formation energies through distinct growth conditions [63].

The defect states formed within the bandgap of CuO are far from the CB and VB edges. While considering the device fabrication these defects are unfavorable for the proper working of devices which rely on electronic excitation through photon absorption. Because these states will act as centers for recombination rather than helping in the increment of carrier concentration. V_O and Cu_i with formation energies around 1 eV are the major intrinsic donors in the Cu rich environment. However, V_O functions as a deep donor, having a transition level at about 0.69 eV above the valence band maximum (VBM). The low carrier mobility (~0.48–1.73) $cm^2\ s^{-1}\ V^{-1}$ observed in n-type CuO is due to the recombination centers formed due to V_O [19]. On the contrary, Cu_i, even though required less energy than V_O, leads to neutral defects having ionization level situated in the VBM, therefore does not effectively compensate for the charge in CuO. V_{Cu} is the majority defect in a Cu-rich environment. Its energy level is too high to make up for the most apparent donor impurities. Acceptor defects outweigh the intrinsic impurities under O-rich conditions. O_i, along with the O_{Cu} and V_{Cu}, is the lowest energy acceptor and is observed to form up to 1 eV. These are not offset by any other defects throughout the full range of electronic band gap, making the material exclusively p-type. When used as an acceptor, V_{Cu} exhibits a relatively shallow transition at 0.17 eV above the VBM and a deeper transition at 0.28 eV lower than the conduction band minimum (CBM). O_{Cu} exhibits two deep acceptor levels, one at 0.49 eV and another at 1.07 eV above the VBM. As a result, the O_i's transition level is observed to be at 1.14 eV above the VBM, which restricts the otherwise strong conductivity which could be predicted, given the extremely low energy required in the defect formation.

Apart from the above-mentioned theoretical studies there are numerous experimental studies on the various defects related properties of CuO. Doping in CuO with various elements can create different types of defects in CuO lattice which can lead to modification of the properties exhibited by CuO. In case of La doped CuO, photoluminescence spectra clearly indicated increased peak intensities associated with defects as a result of doping and annealing without any change in peak position. According to positron studies, it is mostly caused by mismatch in the lattice and difference in oxidation state between Cu^{2+} and La^{3+} ions in correlation with positron studies. More quenching sites for non-radiative transition were created by the defects in La doped CuO [66]. By increasing the Fe doping content, Fe-doped CuO nanotubes (NT) showed a considerable blue shift in the band gap as compared to the un-doped CuO NTs. This could happen because of the Burstein-Moss Shift. The crystallinity of the CuO NTs can be increased by increasing the Fe content which in turn helps to remove the deep trap levels as proved by the steady-state photoluminescence studies [67]. In case of Li doped CuO, Zgair et al. [68] suggest that un-doped CuO showed the strongest intensity whereas with increase in Li doping concentration, it decreases. The near band gap emission (NBE) peak in the PL spectra corresponds to red emission as it is observed in the visible region at around 690 nm (1.797 eV). This can be explained by defects like copper interstitials (Cu_i) in the samples. Also, in another case the interchange of Li^+ with Cu^{2+} ions within CuO transforms the synthesized material from flakes-like morphology to flower-like morphology as reported by Siddiqui et al. [69]. Such structural modification causes changes in the optical properties due to annihilation of the parent phase. As the Li content increases, the band gap is observed to decrease and hence the green-yellow emission in PL spectra was much improved when Li was substituted in place of Cu, which is favorable for optoelectronic applications. Manoharan et al. [70] reports that green emission is associated with the PL peak at 546 nm and can be caused due to defects like copper interstitials. Comparing the pure and Mn/Nd co-doped CuO, it is observed that no new emissions are created because of doping but the intensities of the prominent peaks show a considerable change. Also, a small red shift is observed in the PL peak towards higher wavelengths with the co-doping of Mn and Nd. This red shift is indicative of changes of the defect states in terms of electronic hybridisation due to the presence of Mn and Nd. Hence, this is a nice example of modifications of the electronic clouds in the presence of impurities, even for the imperfections of the lattice. PL bands for the powders are not due to the band gap emission, but can be attributed to various structural defects, such as oxygen vacancy, which generally acts as deep defect donors in semiconductor oxide, and would induce the formation of new energy levels in the band gap present in the prepared pure and Co-doped CuO nanostructures. The green emission bands are reported at 489 nm (2.54 eV), 525 nm (2.37 eV) and 585–625 nm (1.99 eV) for deep level defects of CuO [71].

There is intensive study on the defect's formation in another oxide of Cu, cuprous oxide Cu_2O which is a competent phase during the formation of CuO. Wright et al. [72] studied the energetics of V_{Cu} in Cu_2O using density-functional theory (DFT)

Table 2 Cu defects formation energies values under Cu-rich conditions

Reported work	V_{Cu} (eV)	$V_{\text{split Cu}}$ (eV)
Wright et al. [72]	0.92	1.66
Raebiger et al. [73]	0.30	0.31
David et al. [74]	1.15	1.14
Persson et al. [75]	0.7	1.0
Nolan et al. [65]	0.41	1.24

Table 3 Cu defects formation energies values under O-rich conditions

Reported work	$O_{i(\text{octa})}$ (eV)	$O_{i(\text{tetra})}$ (eV)
Wright et al. [72]	2.07	1.97
Soon et al. [76]	1.36	1.65
David et al. [74]	1.94	1.87
Raebiger et al. [73]	1.3	1.8

and the local density approximation (LDA). A "simple" copper vacancy (V_{Cu}) is a type of vacancy in which one Cu is removed leaving tri-coordinated oxygens. There is another type of defect related to cations, where a "split" vacancy geometry, (V^{split}Cu), can be defined as a situation where the remaining Cu moves toward the vacancy site, to achieve tetrahedral coordination. The "split" V^{split}Cu, changes the six-coordinated Cu states to an anomalous four coordinated Cu state, by moving the remaining Cu ion toward the Cu-vacancy site. This transforms the octahedral coordination of Cu to a tetrahedral coordination. There are several other theoretical reports on this aspect. DFT-generalized gradient approximation (GGA) and DFT-GGA + U techniques were used by Nolan and Elliott, who reported that V_{Cu} is more favorable than V^{split}Cu [54, 63, 64]. The O_i can occupy octahedral interstitials ($O_{i(\text{oct})}$) or tetrahedral interstitials ($O_{i(\text{tetra})}$) and their corresponding formation energies will also vary. The formation energies of the V_{Cu} and V^{split}Cu defects and interstitial defects $O_{i(\text{oct})}$ and $O_{i(\text{tetra})}$ defects in Cu_2O reported in literature are tabulated in Tables 2, and 3.

5.2 *Properties and Applications of CuO*

Depending on the various properties such as size, surface characteristics, optical and magnetic properties, CuO nanoparticles may be utilized in a variety of applications. The synthesis technique plays a key role for regulating all these and consequently their biological properties. Doping materials in semiconductors are used in several of these applications, such as: catalysts, chemical sensors, photo detectors, antimicrobial agents, and suitable material for semiconductor resistive gas sensing applications due to its surface conductivity. CuO has been widely used in photovoltaic applications. CuO thin films are potential materials for solar dye-sensitized cells and light-emitting diodes. In the search for a clean and economical energy source, CuO is receiving growing interest from the scientific community. CuO can be used as an absorber in solar cells for direct solar energy transformation into electricity [55, 59, 61, 77–80].

6 Nickel Oxide (NiO)

NiO is naturally found in its mineralogical form known as bunsenite which is very rare. Nickel oxide nanoparticles appear in green powder form. It is a wide bandgap semiconductor (3.5–4 eV) with p type behaviour [81]. The high durability and the good chemical stability of NiO leads to the application in solar cells and UV photo-detectors. Also, the unique optical and magnetic properties find applications as electrochromic devices, anti-ferromagnetic layers, and chemical sensors [82].

NiO nano powders are black in color and have a cubic structure (Fig. 11) with Fm $\overline{3}$ m symmetry. The structure of NiO is like that of NaCl or rock-salt structure with Ni^{2+} and O^{2-} occupying the octahedral sites of the crystal lattice. Nickel atoms are located at the corners and at the face centers of the cube, while oxygen atoms are located at the center of the cube and at the edge centers, following the typical fcc-packed structure. The structural properties reveal that NiO nanoparticles have single phase face centered cubic structure with negligible strain and without any detectable impurity phases. Defects present in the crystal structure leads to non-stoichiometric nickel oxide which shows p-type semiconductor behaviour. Hence, NiO like many other binary metal oxides, is frequently non-stoichiometric, which means that the Ni:O ratio is not always 1:1 which can be due to the presence of excess oxygen atoms and vacancies in the nickel structure. This non-stoichiometry in NiO is accompanied by a color change. The stoichiometrically correct NiO is green while the non-stoichiometric NiO is black. It is a unique 3d transition-metal monoxide, which crystallizes in the rock salt structure with much higher symmetry in contrast to CuO or CoO. This reveals that the electronic structure NiO is fundamentally different from CuO or CoO.

It has four formula units per unit cell. Both Ni^{2+} and O^{2-} ions are octahedrally coordinated. Ni^{2+} ions are six-fold coordinated with six O^{2-} ions situated at the corners of an octahedral shape. The Ni^{2+} and O^{2-} ions occupy 4b (0, 1/2, 0) and 4e (0, y, 1/4) Wykoff's sites, respectively. The unit cell parameters are $a = b = c = 4.178$ Å, $\alpha = \beta = \gamma = 90°$ with volume = 72.92 Å. The Ni–O, bond distance is 2.0842 Å [35, 83].

6.1 *Defects in NiO*

NiO has both direct and indirect bandgap energies of values ~3.5 eV which is suggested by Density Functional Theory (DFT). The band gap is understood in

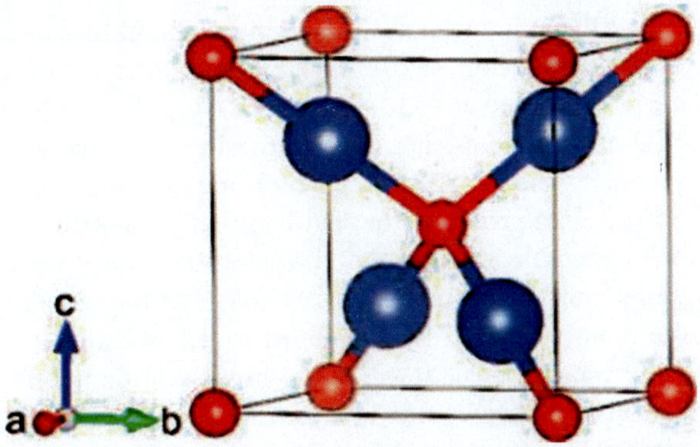

Fig. 11 NiO cubic structure arrangement (Red Ball = "O", Blue Ball = "Ni")

terms of the difference between the energy states of maxima of valence band and minima of conduction band. The experimental values of band gap are in the range of 3.5–4.0 eV. It hints at theoretical and experimental agreement in some cases. There are several reports that native defects in semiconductors play an important role in modification of electronic and optical properties. Literature suggests comparative studies of these native defects such as: Ni-rich NiO with O vacancies (NiO:V_O), and oxygen-rich NiO with Ni vacancies (NiO:V_{Ni}). The theoretical calculations result that defect-induced gap states in NiO are present in form: NiO:V_O are majority Ni-*3d* states, whereas both Ni-*3d* and O-*2p* states present in NiO:V_{Ni}. In comparison with NiO and NiO:V_O samples, the absorption spectra of NiO:V_{Ni} shows the significant defects below 3.0 eV. The enhancement of sub-gap absorptions in NiO:V_{Ni} is due to the gap states in the electronic density of states. The optical constants of NiO:V_{Ni} shows significant deviation from NiO:V_O at similar vacancy concentration while correlating native vacancy defects and optoelectronic properties. There are several reports for the defects formation energies calculated theoretically using GGA + U calculations [16, 35, 84, 85]. The experimental as well as theoretical studies supported Fermi energies up to 4 eV and band gap energies of NiO for both O-rich and Ni-rich limits. The Ni vacancies dominate in both conditions (O-rich and Ni-rich), which is consistent with the observed p-type behaviour and Ni deficiency in NiO.

These results are in agreement with the previous results using GGA + U calculations which demonstrate unstable interstitial defects. However, for the majority of E_F, Ni vacancies are the lowest energy defects. On the other hand, V_O dominates at low E_F (Ni-rich (O-poor)) limits. This denotes the defect formation energies are significantly influenced by the band's widening from GGA + U calculations. The defects such as V_O and Ni interstitials reduced the formation energies at the VB maxima while the defects such as O_i and Ni vacancies increased the formation energies, and vice versa. From the GGA + U calculations, the vacancies of Ni and O at the NiO (001) surface have a large increase in the energy for a Ni vacancy (1.9 eV) and a small reduction in energy for an V_O (0.07 eV) were found. It suggests that across a wider range of E_F in the Ni-rich limit, V_O's have the lowest energy defect than Ni vacancies. The authors claim that in this scenario, a DLE associated to green emission ~520 nm (2.38 eV) along with a weak near band edge emission in the UV range at ~333–357 nm (3.47–3.72 eV) dominates the luminescence of the NiO nanoparticles. The green emission is related to structural defects, such as O/Ni vacancies and interstitials, whereas the UV emission is caused by exciton recombination in close proximity to near band-edge transitions in NiO. The possible transition of trapped electrons at the Ni interstitial to the valence band is responsible for the violet emission to occur at ~2.97 eV. The radiative recombination of electrons from the doubly ionized Ni vacancies to the holes in the valence band causes the blue luminescence peak at ~2.70 eV. In view of this, different authors have reported that various synthesis routes, post treatments, and different types of dopants can be used to modify the structural defects in NiO that may lead to photoluminescence. Although Ni vacancies are responsible for the near-IR emission, they are also involved in the transitions of 3d energy levels (2–2.5 eV). The PL spectrum involves high energy emissions which are mainly due to the near band edge emissions and *d–d* charge transfer excitons in NiO. As a result, NiO's inherent low intensity can therefore be overcome to functionalize it for the luminescent devices, and the luminescence from NiO can be controlled by modifying it using doping elements. NiO microstructures have a broad luminescence spectrum that includes emissions in the near-IR (1.46 eV), the visible (2–2.5 eV) and the UV region (3.5 and 4.5 eV) [16, 85–87].

Gandhi et al. [88] reported that decreasing the particle size leads to decrease in the UV emission band. This leads to the ionic Ni–O and covalent Ni–Ni bond weakening through electron density calculations. This is another way in a nanostructure system to control the UV emission band by modifying the strength of Ni–O and Ni–Ni bonds.

The photogenerated holes from the deep level V_O^{2+} states and electrons in shallow levels at 2.37 eV just below the conducting band results in a green band emission spectrum. This seemingly anomalous phenomenon in the visible region shows weak violet, blue, orange, and strong green band emissions as a result of defects. These can be attributed to nickel interstitials, nickel vacancies, oxygen interstitials, and oxygen vacancies, respectively, in the deep level emission [88]. Another study by Ramesh et al. reports NiO nanocrystals having N and/or Fe doping can result in structural changes such as the incorporation of cationic impurities and surface vacancies as additional defects. Instead of the presence of defect states such as oxygen vacancies, N and/or Fe-doped nanocrystals show a decrease in the PL strength mainly due to increase in cation vacancies and surface related defects, which serves as an alternative non-radiative pathway to PL emission [38, 39]. However, a few cation vacancies in the N doped nanocrystals results in a partial drop in the luminescence intensity [89]. The recombination rate of photoinduced electron–hole pairs will increase with the increase of PL intensity of NiO nanoparticles. The holes localized at deep traps have low oxidizing potential and easily react with physically adsorbed substances. The stoichiometry of NiO may change with the heat treatment V_{Ni}, O_i, in NiO lattice leads to two wide emission peaks at 420 nm (2.95 eV) and 440 nm (2.82 eV) and a strong—broad peak at 460 nm (2.70 eV) in violet emission band. These peaks give the confirmation of such defects in NiO lattice. The charge transfer between Ni^{+2}

and Ni^{+3} can lead to V_{Ni}. The PL emission band at 460 nm results from the surface oxygen vacancies of NiO nanoparticles [90]. It has also been reported that Mg doped NiO exhibit bands in the visible (at ~410, 438, 457 and 490 nm) and UV (at ~362, 376, 385 and 390 nm) regions. The distinct feature of near band edge emission is the UV region band. Out of the several UV bands, the band located at 362 nm is associated with the excitonic recombination and its improvement in intensity as doping concentration of Mg increases. This might imply that *Mg* is substituted successfully in NiO. The reduction in defect density along with subsequent quality improvement of the nanostructure can be because of the increased intensity of UV emission in NiO with Mg doping. The presence of trapped electrons at Ni interstitial to the valence band was suggested by the violet emission bands in the samples at 410 and 438 nm. The recombination of electrons from Ni^{2+} vacancy to the holes in the valence band may have been the cause of the blue emission band with low intensity at ~457 and ~490 nm [91].

Bhatt et al. [92] reports that the Li^{+} concentration in NiO in Li doped NiO caused the formation of numerous shoulder peaks at wavelengths of around 402 and 422 nm (violet emission), 452 nm (blue emission) and 508 nm (green emission). The transition of trapped electrons at Ni interstitial to the valence band is the most probable cause of the violet emission peaks and subsequently, blue emission is due to the recombination of electrons from the Ni^{2+} vacancy and the holes. Another important factor in modifying the electronic structure of parent oxide materials is the size of the dopant. Numerous factors can modify the defect states of NiO, which has a significant impact on the luminescence and consequently optoelectronic properties. The dominant point defects in NiO are observed to be Nickel vacancies. From a thermodynamic approach. As a result, the radiative emissions promoted for the presence of Ni vacancies should dominate the optical properties of NiO. However, the presence of O_V and O_i defects are promoted through controlled experimental conditions [61, 80, 93–96].

6.2 Properties and Applications

NiO is an anti-ferromagnetic material with a cubic structure and a band gap of ~3.5–4.0 eV which can be adjusted by reducing the size/charge compensation of dopants by doping. Hence NiO finds applications as anti-ferromagnetic materials, electrochromic display materials and chemical sensors, p-type transparent conducting films, high-T_C superconductors, and giant magneto resistance compounds. NiO nanoparticles are widely employed in Li-ion batteries, supercapacitors, field emissions, photocatalysis, gas/biosensors, optoelectronics, and photodetectors [83, 84, 93, 94, 97–100]. These properties show the uniqueness of NiO among all transition metal oxides and have attracted a great deal of scientific attention. However, there are numerous competing theories regarding the origin of ferromagnetism even at room temperature. The magnetic and other properties of NiO can be tailored by adjusting the synthesis route, annealing temperature and processing conditions. Therefore,

Table 4 Summary of T_r and T_f calculated from the response cycles for self-powered NiO/MAPI heterojunction

Probing wavelength (nm)	T_r (ms)	T_f (ms)
290 ± 5 (UV)	452 ± 10	363 ± 10
450 ± 5 (Blue)	345 ± 10	290 ± 10
540 ± 5 (Green)	332 ± 10	319 ± 10
640 ± 5 (Red)	369 ± 10	320 ± 10

variation in synthesis methods could be responsible for conflicting reports on the magnetic properties of NiO. Also, NiO layered with perovskite material can show broadband photo sensing applications by Sen et al. [101] has been published. The response and recovery time in ms of self-powered photo-sensor is tabulated showing the sensing performance of NiO based device [101] (Table 4).

7 Perovskites Metal Oxides (ABO_3)

Perovskite are compounds having the structure ABO_3 where A stands for alkaline-earth or rare-earth metal cation which occupies the 12-fold coordinated cube-octahedral cages of the oxygen sublattice, and B signifies a transition-metal cation surrounded by six oxygen atoms in an octahedral coordination. A schematic of the perovskite ABO_3 crystal structure is shown in Fig. 12. The perovskite structure shows distortions as a result of difference in the ionic radii of different sites which is expressed in terms of the tolerance factor. The fascinating features, such as ferroelectricity, magnetism, superconductivity, ionic conductivity, etc., are governed by the corresponding local coordination environments and partially occupied d-orbitals. These are extremely sensitive to the defects in the lattice structure, compositions, surface, interface, and grain boundaries of the perovskites.

The defects mentioned in the simple oxides like anion defects O_V, O_i, and cation defects A_v, A_i, B_v, B_i plays a major role in the properties exhibited by these oxides. These defects modify the regular periodicity of the crystal lattice, and more importantly add extra energy levels into the band gap, changing the physical properties and electronic structure of perovskite MOs. Due to the strong correlation between crystal structure and electronic properties novel functionalities can be explored by tuning the defects in the parent lattice. The impurity cations added to the perovskite will have a site preference according to their charge and ionic radii, large cations replace larger ions and vice versa. Though the concentration of defects in the perovskite is usually very low, the effect of defects on the properties will be notable. These defects can be point defects, charged defects, defect dipoles, and ordered defect dipoles. The defect dipole creation in the perovskite lattice affects the oxidation states of neighbouring cations or leads to the formation of ionic vacancies. These changes will bring ferroelectricity, magnetism, and conductivity in nonmagnetic/

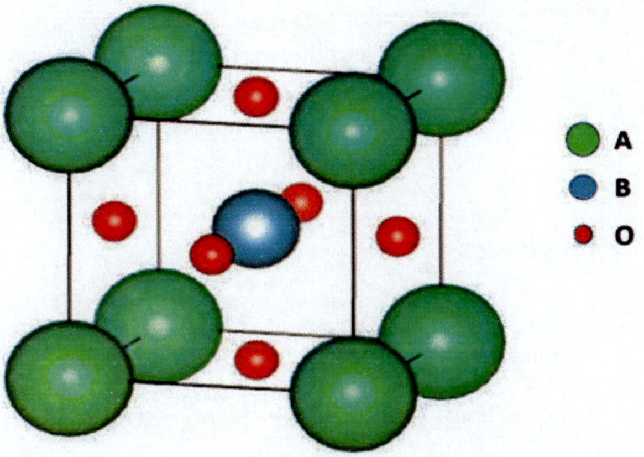

Fig. 12 ABO_3 perovskite structure

non ferroelectric/insulator materials. Reduction of the mobility of oxygen vacancies, enhancement of the breakdown field strength with the moderation of leakage current will occur with the formation of defect dipoles between dopants and oxygen vacancies. Coupling between defects, coupling of defects and other parameters like strain, surface, interfaces, external fields will also lead to the modification of the properties. For instance, while interfaces can offer a favorable site for defect formation, strain can modify the formation energies of defect dipoles and alter their spatial orientations. These will lead to the existence of new phenomena which are not present in the uncoupled systems of defects and strain/surface, etc. [102]. Regarding magnetic properties, it has been reported that V_O, BO_6 octahedral distortions, misfit dislocations and grain boundaries are efficient ways to modify the magnetism in these perovskite materials. Because of the unique interaction between spin, charge, orbital, and lattice degrees of freedom in perovskite manganites, $Ln_{1-x}A_xMnO_3$ (Ln = La, Pr, Bi; A = Ba, Sr, Ca) there has been a great deal of attention. Defects can also result in charge compensation, act as scattering sites to influence the transport properties like conductivity and mobility, and even cause lattice distortions and induce strain.

8 Conclusion

For application of metal oxides in the various technological fields, in-depth knowledge about the defects is very important for exact implementation of the device. In the case of MO photo (electro) catalysts, that needs high concentrations of charged point defects that are electronically active. In photovoltaic materials, defects are considered as detrimental apart from some selected dopants and should be eliminated to minimize the charge recombination. However, defects can have an active role in the stabilization of charge separation. It also mediates rate limiting catalytic steps. The strong correlation of structure and electronic properties of these materials can be used to make novel functionalities by changing the defects in them. Tuning the defect by doping impurities with different oxidation states and ionic radii, different synthesis techniques and thermal treatment will cause the modifications of the properties exhibited by these MOs. We hope this chapter gave an overview of the defects in MOs in a detailed way to the reader.

References

1. M.S. Chavali, M.P. Nikolova, Metal oxide nanoparticles and their applications in nanotechnology. SN Appl. Sci. **1**(6), 607 (2019). https://doi.org/10.1007/s42452-019-0592-3
2. A.B. Djurišić, X. Chen, Y.H. Leung, A.M.C. Ng, ZnO nanostructures: growth, properties and applications. J. Mater. Chem. **22**(14), 6526–6535 (2012). https://doi.org/10.1039/C2JM15548F
3. M.R. Bindhu, M. Umadevi, M. Kavin Micheal, M.V. Arasu, N. Abdullah Al-Dhabi, Structural, morphological and optical properties of MgO nanoparticles for antibacterial applications. Mater. Lett. **166**, 9–22 (2016). https://doi.org/10.1016/j.matlet.2015.12.020.
4. Z. Yu et al., Noninvasively modifying band structures of wide-bandgap metal oxides to boost photocatalytic activity. Adv. Mater. Deerfield Beach Fla **30**(14), e1706259 (2018). https://doi.org/10.1002/adma.201706259
5. P. Bindu, S. Thomas, Estimation of lattice strain in ZnO nanoparticles: X-ray peak profile analysis. J. Theor. Appl. Phys. **8**(4), 123–134 (2014). https://doi.org/10.1007/s40094-014-0141-9
6. R. Bhardwaj, A. Bharti, J.P. Singh, K.H. Chae, N. Goyal, S. Gautam, Structural and electronic investigation of ZnO nanostructures synthesized under different environments. Heliyon **4**(4), e00594 (2018). https://doi.org/10.1016/j.heliyon.2018.e00594
7. M. Bushell et al., Characterization of commercial metal oxide nanomaterials: crystalline phase, particle size and specific surface area. Nanomaterials **10**(9) (2020). https://doi.org/10.3390/nano10091812
8. S. Sahoo, G. Sharma, R. Katiyar, Raman spectroscopy to probe residual stress in ZnO nanowire. J. Raman Spectrosc. **43**, 72–75 (2012). https://doi.org/10.1002/jrs.3004
9. J. Al-Sabahi, T. Bora, M. Al-Abri, J. Dutta, Controlled defects of zinc oxide nanorods for efficient visible light photocatalytic degradation of phenol. Materials **9**(4) (2016). https://doi.org/10.3390/ma9040238
10. J.A. Dawson, Y. Guo, J. Robertson, Energetics of intrinsic defects in NiO and the consequences for its resistive random access memory performance (2015). 10/253626
11. P. Pratim Das et al., Redistribution of native defects and photoconductivity in ZnO under pressure. RSC Adv. **9**(8), 4303–4313 (2019). https://doi.org/10.1039/C8RA10219H
12. P.K. Mishra, R. Dobhal, E.G. Rini, M. Kumar, S. Sen, Rapid organic dye degradation and wavelength dependent sensing study in Cu1–xFexO. Ceram. Int. **48**(5), 5995–6006 (2022). https://doi.org/10.1016/j.ceramint.2021.11.135
13. S. Ayaz, R. Amin, K. Samantray, A. Dasgupta, S. Sen, Tunable ultraviolet sensing performance of Al-modified ZnO nanoparticles. J. Alloys Compd. **884**, 161113 (2021). https://doi.org/10.1016/j.jallcom.2021.161113
14. S. Jiao, X. Fu, L. Zhang, Y.-J. Zeng, H. Huang, Point-defect-optimized electron distribution for enhanced electrocatalysis: towards the perfection of the imperfections. Nano Today **31**, 100833 (2020). https://doi.org/10.1016/j.nantod.2019.100833
15. E. Pastor, M. Sachs, S. Selim, J.R. Durrant, A.A. Bakulin, A. Walsh, Electronic defects in metal oxide photocatalysts. Nat. Rev. Mater. **7**(7), 503–521 (2022). https://doi.org/10.1038/s41578-022-00433-0
16. H.J. Van Daal, A.J. Bosman, Influence of native defects on transport properties of Li-doped NiO. Phys. Lett. **23**(9), 525–526 (1966). https://doi.org/10.1016/0031-9163(66)90389-1

17. A. Bera, D. Basak, Role of defects in the anomalous photoconductivity in ZnO nanowires. Appl. Phys. Lett. **94**, 163119 (2009). https://doi.org/10.1063/1.3123167
18. H. Chen et al., A copper-based sorbent with oxygen-vacancy defect from mechanochemical reduction for carbon disulfide absorption. J. Mater. Chem. A **4** (2016). https://doi.org/10.1039/C6TA07134A
19. F. Yang, Y. Choi, P. Liu, D. Stacchiola, J. Hrbek, J.A. Rodriguez, Identification of 5–7 defects in a copper oxide surface. J. Am. Chem. Soc. **133**(30), 11474–11477 (2011). https://doi.org/10.1021/ja204652v
20. T. Srivastava, S. Kumar, P. Shirage, S. Sen, Reduction of O^{2-} related defect states related to increased bandgap in Si^{4+} substituted ZnO. Scr. Mater. **124**, 11–14 (2016). https://doi.org/10.1016/j.scriptamat.2016.06.038
21. P.K. Mishra, S. Ayaz, B. Kissinquinker, S. Sen, Defects assisted visible light sensing in Zn1 − x(GaSi)x/2O. J. Appl. Phys. **127**(15), 154501 (2020). https://doi.org/10.1063/1.5139896
22. P.K. Mishra et al., Role of Ga-substitution in ZnO on defect states, carrier density, mobility and UV sensing. J. Mater. Sci. Mater. Electron. **30**(20), 18686–18695 (2019). https://doi.org/10.1007/s10854-019-02221-z
23. W.B. Fowler, R. Phillips, A.E. Carlsson, Point and extended defects in crystals. Digital encyclopedia of applied physics (Wiley-VCH Verlag GmbH & Co. KgaA, Weinheim, Germany, 2003). https://doi.org/10.1002/3527600434.eap344
24. D.C. Look, H.L. Mosbacker, Y.M. Strzhemechny, L.J. Brillson, Effects of surface conduction on hall-effect measurements in ZnO. Superlattices Microstruct. **38**, 406–412 (2005)
25. P. Venńgùs, J.M. Chauveau, M. Korytov, C. Deparis, J. Zuniga-Perez, C. Morhain, Interfacial structure and defect analysis of nonpolar ZnO films grown on R-plane sapphire by molecular beam epitaxy. J Appl Phys. **103**, 083525 (2008). https://doi.org/10.1063/1.2905220
26. Oxygen vacancy induced band gap narrowing of ZnO nanostructures by an electrochemically active biofilm. Nanoscale (RSC Publishing). https://pubs.rsc.org/en/content/articlelanding/2013/NR/c3nr02678g
27. A. Janotti, C.G. Van de Walle, Native point defects in ZnO. Phys. Rev. B **76**(16), 165202 (2007). https://doi.org/10.1103/PhysRevB.76.165202
28. M. Saha, S. Ghosh, V.D. Ashok, S.K. De, Carrier concentration dependent optical and electrical properties of Ga doped ZnO hexagonal nanocrystals. Phys. Chem. Chem. Phys. **17**(24), 16067–16079 (2015). https://doi.org/10.1039/C4CP05480F
29. S.S. Gaikwad et al., Oxygen induced strained ZnO nanoparticles: an investigation of Raman scattering and visible photoluminescence. J. Mater. Chem. C **2**(35), 7264–7274 (2014). https://doi.org/10.1039/C4TC00566J
30. S.H. Skjærvø et al., Interstitial oxygen as a source of p-type conductivity in hexagonal manganites. Nat. Commun. **7**(1) (2016). https://doi.org/10.1038/ncomms13745
31. T. Srivastava et al., Opto-electronic properties of Zn(1 – x)VxO: Green emission enhancement due to V4+ state. J. Appl. Phys. **122**(2), 025106 (2017). https://doi.org/10.1063/1.4992087
32. Zinc Vacancy–Hydrogen Complexes as Major Defects in ZnO Nanowires Grown by Chemical Bath Deposition. J. Phys. Chem. C. https://pubs.acs.org/doi/10.1021/acs.jpcc.0c04264
33. F. Kayaci, S. Vempati, I. Donmez, N. Biyikli, T. Uyar, Role of zinc interstitials and oxygen vacancies of ZnO in photocatalysis: a bottom-up approach to control defect density. Nanoscale **6**(17), 10224–10234 (2014). https://doi.org/10.1039/C4NR01887G
34. F. Oba, M. Choi, A. Togo, I. Tanaka, Point defects in ZnO: an approach from first principles. Sci. Technol. Adv. Mater. **12**(3), 034302 (2011). https://doi.org/10.1088/1468-6996/12/3/034302
35. Point defects in sputtered NiO films. Appl. Phys. Lett. **94**(6). https://aip.scitation.org/doi.org/10.1063/1.3081025
36. J. Smalc-Koziorowska, Defects in crystals. Crystals **10**(10) (2020). https://doi.org/10.3390/cryst10100915
37. C.-Y. Kung et al., Influence of Y-doped induced defects on the optical and magnetic properties of ZnO nanorod arrays prepared by low-temperature hydrothermal process. Nanosc. Res. Lett. **7**(1), 372 (2012). https://doi.org/10.1186/1556-276X-7-372

38. Improved ultraviolet sensing, photo-stabilized visible transmission, and electrical conductance in Zn1 − xGax/2Fex/2O | SpringerLink. https://link.springer.com/article/10.1557/jmr.2020.82
39. G. Bajpai, T. Srivastava, F. Husian, S. Kumar, S. Biring, S. Sen, Enhanced red emission from Fe/Si co-doped ZnO nano-particles. Scr. Mater. **144**, 27–30 (2018). https://doi.org/10.1016/j.scriptamat.2017.09.041
40. Structure correlated optoelectronic and electrochemical properties of Al/Li modified ZnO. J. Appl. Phys. **126**(2). https://aip.scitation.org/doi/10.1063/1.5099894
41. Doping effect and oxygen defects boost room temperature ferromagnetism of Co-doped ZnO nanoparticles: experimental and theoretical studies—RSC Advances (RSC Publishing). https://pubs.rsc.org/en/content/articlelanding/2019/RA/C9RA03620B
42. F. Yu et al., Effects of Al doping on defect behaviours of ZnO thin film as a photocatalyst. Mater. Sci.-Pol. **37**(3), 437–445 (2019). https://doi.org/10.2478/msp-2019-0050
43. J.C. Fan, K.M. Sreekanth, Z. Xie, S.L. Chang, K.V. Rao, p-Type ZnO materials: theory, growth, properties and devices. Prog. Mater. Sci. **58**(6), 874–985 (2013). https://doi.org/10.1016/j.pmatsci.2013.03.002
44. A. Janotti, C.G. Van de Walle, New insights into the role of native point defects in ZnO. J. Cryst. Growth **287**(1), 58–65 (2006). https://doi.org/10.1016/j.jcrysgro.2005.10.043
45. D.P. Chen, Y. Losovyj, S.E. Skrabalak, n-type doping of visible-light-absorbing (GaN)1 − x(ZnO)x with Aliovalent Sn/Si substitutions. J. Phys. Chem. C **122**(25), 13250–13258 (2018). https://doi.org/10.1021/acs.jpcc.7b08304
46. L. Liu et al., Oxygen vacancies: the origin of n-type conductivity in ZnO. Phys. Rev. B **93**(23), 235305 (2016). https://doi.org/10.1103/PhysRevB.93.235305
47. J.E. Eixenberger et al., Defect engineering of ZnO nanoparticles for bioimaging applications. ACS Appl. Mater. Interfaces **11**(28), 24933–24944 (2019). https://doi.org/10.1021/acsami.9b01582
48. L. Schmidt-Mende, J.L. MacManus-Driscoll, ZnO—nanostructures, defects, and devices. Mater. Today **10**(5), 40–48 (2007). https://doi.org/10.1016/S1369-7021(07)70078-0
49. F.M. Simanjuntak, D. Panda, K.-H. Wei, T.-Y. Tseng, Status and prospects of ZnO-based resistive switching memory devices. Nanosc. Res. Lett. **11**(1), 368 (2016). https://doi.org/10.1186/s11671-016-1570-y
50. W. Liu, W. Li, Z. Hu, Z. Tang, X. Tang, Effect of oxygen defects on ferromagnetic of undoped ZnO. J. Appl. Phys. **110**(1), 013901 (2011). https://doi.org/10.1063/1.3601107
51. G. Xing et al., Correlated d0 ferromagnetism and photoluminescence in undoped ZnO nanowires. Appl. Phys. Lett. **96**(11), 112511 (2010). https://doi.org/10.1063/1.3340930
52. I. Ayoub, V. Kumar, R. Abolhassani, R. Sehgal, V. Sharma, R. Sehgal, H.C. Swart, Y.K. Mishra, Advances in ZnO: manipulation of defects for enhancing their technological potentials. Nanotechnol. Rev. **11**(1), 575–619 (2022). https://doi.org/10.1515/ntrev-2022-0035
53. M.-C. Jun, S.-U. Park, J.-H. Koh, Comparative studies of Al-doped ZnO and Ga-doped ZnO transparent conducting oxide thin films. Nanosc. Res. Lett. **7**(1), 639 (2012). https://doi.org/10.1186/1556-276X-7-639
54. J. Akter, K.P. Sapkota, Md.A. Hanif, Md.A. Islam, H.G. Abbas, J.R. Hahn, Kinetically controlled selective synthesis of Cu2O and CuO nanoparticles toward enhanced degradation of methylene blue using ultraviolet and sun light. Mater. Sci. Semicond. Process. **123**, 105570 (2021). https://doi.org/10.1016/j.mssp.2020.105570
55. N. Dasineh Khiavi, R. Katal, S. Kholghi Eshkalak, S. Masudy-Panah, S. Ramakrishna, H. Jiangyong, Visible light driven heterojunction photocatalyst of CuO–Cu_2O thin films for photocatalytic degradation of organic pollutants. Nanomaterials **9**(7) (2019). https://doi.org/10.3390/nano9071011
56. Y. Du, X. Meng, X. Gao, Variations of the microstructure and the optical and electrical properties with sputtering power for direct-current-magnetron-sputtered indium-doped CuO thin films at room temperature. Thin Solid Films **684**, 53–58 (2019). https://doi.org/10.1016/j.tsf.2019.05.056

57. K. Dulta, G. Koşarsoy Ağçeli, P. Chauhan, R. Jasrotia, P.K. Chauhan, J.O. Ighalo, Multifunctional CuO nanoparticles with enhanced photocatalytic dye degradation and antibacterial activity. Sustain. Environ. Res. **32**(1), 2 (2022). https://doi.org/10.1186/s42834-021-00111-w
58. Synthesis of Cu_2O from CuO thin films: Optical and electrical properties. AIP Adv. **5**(4). https://aip.scitation.org/doi/10.1063/1.4919323
59. M. Nazim, A.A.P. Khan, A.M. Asiri, J.H. Kim, Exploring rapid photocatalytic degradation of organic pollutants with porous CuO nanosheets: synthesis, dye removal, and kinetic studies at room temperature. ACS Omega **6**(4), 2601–2612 (2021). https://doi.org/10.1021/acsomega.0c04747
60. L. Ben Saad, L. Soltane, F. Sediri, Pure and Cu-doped ZnO nanoparticles: hydrothermal synthesis, structural, and optical properties. Russ. J. Phys. Chem. A **93**(13), 2782–2788 (2019). https://doi.org/10.1134/S0036024419130259
61. M. Outokesh, M. Hosseinpour, S.J. Ahmadi, T. Mousavand, S. Sadjadi, W. Soltanian, Hydrothermal synthesis of CuO nanoparticles: study on effects of operational conditions on yield, purity, and size of the nanoparticles. Ind. Eng. Chem. Res. **50**(6), 3540–3554 (2011). https://doi.org/10.1021/ie1017089
62. Identification of 5–7 defects in a copper oxide surface. J. Am. Chem. Soc. https://pubs.acs.org/doi/10.1021/ja204652v
63. A. Živković, N.H. de Leeuw, Exploring the formation of intrinsic *p*-type and *n*-type defects in CuO. Phys. Rev. Mater. **4**(7), 074606 (2020). https://doi.org/10.1103/PhysRevMaterials.4.074606
64. J. Ghijsen et al., Electronic structure of Cu_2O and CuO. Phys. Rev. B **38**(16), 11322–11330 (1988). https://doi.org/10.1103/PhysRevB.38.11322
65. M. Nolan, S.D. Elliott, The p-type conduction mechanism in Cu_2O: a first principles study. Phys. Chem. Chem. Phys. **8**(45), 5350–5358 (2006). https://doi.org/10.1039/B611969G
66. L.V. Devi, T. Selvalakshmi, S. Sellaiyan, A. Uedono, K. Sivaji, S. Sankar, Effect of La doping on the lattice defects and photoluminescence properties of CuO. J. Alloys Compd. **709**, 496–504 (2017). https://doi.org/10.1016/j.jallcom.2017.03.148
67. M. Iqbal, M.A.Z.G. Sial, S. Shabbir, M. Siddiq, A. Iqbal, Effect of Fe doping on the crystallinity of CuO nanotubes and the efficiency of the hybrid solar cells. J. Photochem. Photobiol. Chem. **335**, 112–118 (2017). https://doi.org/10.1016/j.jphotochem.2016.11.020
68. I.A. Zgair, A.H. Omran Alkhayatt, A.A. Muhmood, S.K. Hussain, Investigation of structure, optical and photoluminescence characteristics of Li doped CuO nanostructure thin films synthesized by SILAR method. Optik **191**, 48–54 (2019). https://doi.org/10.1016/j.ijleo.2019.06.008
69. H. Siddiqui, M.R. Parra, M.M. Malik, F.Z. Haque, Structural and optical properties of Li substituted CuO nanoparticles. Opt. Quantum Electron. **50**(6), 260 (2018). https://doi.org/10.1007/s11082-018-1527-8
70. A. Albert manoharan et al., Facile synthesis and characterization of undoped, Mn doped and Nd co-doped CuO nanoparticles for optoelectronic and magnetic applications. J. Mol. Struct. **1171**, 388–395 (2018). https://doi.org/10.1016/j.molstruc.2018.06.018
71. N.M. Basith, J.J. Vijaya, L.J. Kennedy, M. Bououdina, S. Hussain, Optical and magnetic properties of Co-doped CuO flower/plates/particles-like nanostructures. J. Nanosci. Nanotechnol. **14**(3), 2577–2583 (2014). https://doi.org/10.1166/jnn.2014.8514
72. A.F. Wright, J.S. Nelson, Theory of the copper vacancy in cuprous oxide. J. Appl. Phys. **92**(10), 5849–5851 (2002). https://doi.org/10.1063/1.1516620
73. H. Raebiger, S. Lany, A. Zunger, Origins of the p-type nature and cation deficiency in Cu2O and related materials. Phys. Rev. B **76**(4), 045209 (2007). https://doi.org/10.1103/PhysRevB.76.045209
74. D.O. Scanlon, B.J. Morgan, G.W. Watson, A. Walsh, Acceptor Levels in p-Type Cu2O: rationalizing theory and experiment. Phys. Rev. Lett. **103**(9), 096405 (2009). https://doi.org/10.1103/PhysRevLett.103.096405
75. C. Persson, Y.-J. Zhao, S. Lany, A. Zunger, n-type doping of CuInSe2 and CuGaSe2. Phys. Rev. B **72**(3) (2005). https://doi.org/10.1103/PhysRevB.72.035211

76. A. Soon, X.-Y. Cui, B. Delley, S.-H. Wei, C. Stampfl, Native defect-induced multifarious magnetism in nonstoichiometric cuprous oxide: first-principles study of bulk and surface properties of Cu2-δO. Phys. Rev. B **79**(3), 035205 (2009). https://doi.org/10.1103/PhysRevB.79.035205
77. M. Iqbal et al., Synthesis and characterization of transition metals doped CuO nanostructure and their application in hybrid bulk heterojunction solar cells. SN Appl. Sci. **1**(6), 647 (2019). https://doi.org/10.1007/s42452-019-0663-5
78. H. Kidowaki, T. Oku, T. Akiyama, A. Suzuki, B. Jeyadevan, J. Cuya, Fabrication and characterization of CuO-based solar cells. J. Mater. Sci. Res. **1**(1). https://doi.org/10.5539/jmsr.v1n1p138
79. One-dimensional CuO nanowire: synthesis, electrical, and optoelectronic devices application | Nanoscale Research Letters | Full Text. https://nanoscalereslett.springeropen.com/articles/10.1186/1556-276X-9-637
80. A.A. Oliveira, M.I. Valerio-Cuadros, L.F.S. Tupan, F.F. Ivashita, A. Paesano, Size-effect on the optical behaviour of Fe-doped CuO nanoparticles synthesized by a freeze-drying process. Mater. Lett. **229**, 327–330 (2018). https://doi.org/10.1016/j.matlet.2018.07.009
81. NiO nanostructure by RF sputtering for gas sensing applications. Mater. Technol. **35**(1). https://www.tandfonline.com/doi/full/10.1080/10667857.2019.1653595
82. K.O. Ukoba, A.C. Eloka-Eboka, F.L. Inambao, Review of nanostructured NiO thin film deposition using the spray pyrolysis technique. Renew. Sustain. Energy Rev. **82**, 2900–2915 (2018). https://doi.org/10.1016/j.rser.2017.10.041
83. O. Seo et al., Tuning of structural, optical band gap, and electrical properties of room-temperature-grown epitaxial thin films through the Fe2O3:NiO ratio. Sci. Rep. **9**(1) (2019). https://doi.org/10.1038/s41598-019-41049-9
84. S. Park et al., Interaction and ordering of vacancy defects in NiO. Phys. Rev. B **77**(13), 134103 (2008). https://doi.org/10.1103/PhysRevB.77.134103
85. J. Dawson, Y. Guo, J. Robertson, Energetics of intrinsic defects in NiO and the consequences for its resistive random access memory performance. Appl. Phys. Lett. **107**, 122110 (2015). https://doi.org/10.1063/1.4931751
86. N.L. Peterson, C.L. Wiley, Point defects and diffusion in NiO. J. Phys. Chem. Solids **46**(1), 43–52 (1985). https://doi.org/10.1016/0022-3697(85)90194-5
87. C. Alex, G. Shukla, N.S. John, Introduction of surface defects in NiO with effective removal of adsorbed catalyst poisons for improved electrochemical urea oxidation. Electrochim. Acta **385**, 138425 (2021). https://doi.org/10.1016/j.electacta.2021.138425
88. A.C. Gandhi, S.Y. Wu, Strong deep-level-emission photoluminescence in NiO nanoparticles. Nanomaterials **7**(8) (2017). https://doi.org/10.3390/nano7080231
89. L. Kumari, Vertically aligned and interconnected nickel oxide nanowalls fabricated by hydrothermal route. Cryst. Res. Technol. (Wiley Online Library, 2009). https://onlinelibrary.wiley.com/doi/10.1002/crat.200800583
90. Structural, thermal and optical properties of nickel oxide (NiO) nanoparticles synthesized by chemical precipitation method. Scientific.Net. https://www.scientific.net/AMR.1141.65
91. Effect of Mg doping on the improvement of photoluminescence and magnetic properties of NiO nanoparticles. IOPscience. https://iopscience.iop.org/article/10.1088/2632-959X/aba285
92. A.S. Bhatt et al., Optical and electrochemical applications of Li-doped NiO nanostructures synthesized via facile microwave technique. Materials **13**(13) (2020). https://doi.org/10.3390/ma13132961
93. N.S. Gonçalves, J.A. Carvalho, Z.M. Lima, J.M. Sasaki, Size–strain study of NiO nanoparticles by X-ray powder diffraction line broadening. Mater. Lett. **72**, 36–38 (2012). https://doi.org/10.1016/j.matlet.2011.12.046
94. M. Taeño, D. Maestre, J. Ramírez-Castellanos, S. Li, P. S. Lee, A. Cremades, Towards control of the size, composition and surface area of NiO nanostructures by Sn doping. Nanomaterials **11**(2) (2021). https://doi.org/10.3390/nano11020444

95. S.S. Alias, A.B. Ismail, A.A. Mohamad, Effect of pH on ZnO nanoparticle properties synthesized by sol–gel centrifugation. J. Alloys Compd. **499**(2), 231–237 (2010). https://doi.org/10.1016/j.jallcom.2010.03.174
96. Size-dependent bandgap modulation of ZnO nanowires by tensile strain. Nano Lett. https://pubs.acs.org/doi/10.1021/nl301897q
97. The impact of ultraviolet light on the switching characteristics of NiO resistive random-access memory (ReRAM) devices. https://www.spiedigitallibrary.org/conference-proceedings-of-spie/11283/2545485/The-impact-of-ultraviolet-light-on-the-switching-characteristics-of/10.1117/12.2545485.short?SSO=1
98. Y. Kokubun, Electrical conductivity studies in sol–gel-derived Li-doped NiO epitaxial thin films. Phys. Status Solidi (b) (Wiley Online Library). https://onlinelibrary.wiley.com/doi/10.1002/pssb.202000330
99. K. Khun, Development of fast and sensitive ultraviolet photodetector using p-type NiO/n-type TiO2 heterostructures. Phys. Status Solidi (a) (Wiley Online Library). https://onlinelibrary.wiley.com/doi/10.1002/pssa.201330149
100. S. Zhang, G. Liu, W. Qiao, J. Wang, L. Ling, Oxygen vacancies enhance the lithium ion intercalation pseudocapacitive properties of orthorhombic niobium pentoxide. J. Colloid Interface Sci. **562**, 193–203 (2020). https://doi.org/10.1016/j.jcis.2019.12.015
101. Perovskite-based facile NiO/CH3NH3PbI3 heterojunction self-powered broadband photodetector. ACS Appl. Electron. Mater. https://pubs.acs.org/doi/10.1021/acsaelm.1c00679
102. Y. Sun, J. Yang, S. Li, D. Wang, Defect engineering in perovskite oxide thin films. Chem. Commun. **57**(68), 8402–8420 (2021). https://doi.org/10.1039/D1CC02276H

Metal Oxide Nanostructures as an Electron Transport Layer for Dye-Sensitized Solar Cells

Siddan Gouthaman and K. R. Justin Thomas

Abstract The optical features of metal oxide nanoparticles include the capacity to absorb light sources and emit photons. Metal oxide-dye nanocomposites combine these unique properties with processability morphologies and have numerous applications in solar cells, light-emitting diodes, transistors, optoelectronic packaging, and coatings. It contains a diverse spectrum of sophisticated research and novel concepts. Different metal oxides have been used to make other photovoltaic generation devices that can convert solar energy continuously and efficiently. The relevance of MO-NPs as an electron transport layer (ETL) in dye-sensitized solar cells is discussed in this chapter (DSSCs). The fact that these metal oxide materials are easy to get without breakdown and are chemically stable is a good sign. Under normal conditions, they can give you diverse options for designing and making the device. The physical, chemical, and electrical properties of metal oxides make DSSC devices work better and stay stable for longer. Optimizing the preparation parameters for metal oxides is a new, reliable, and all-encompassing way to make metal oxide-based flexible DSSCs that work well. The first section of this chapter provides readers with an introduction to DSSC technology and its workings. The second section discusses potential ETL materials based on metal oxides and how they may be utilized for device efficiency by using novel methodologies, modifying the structure of morphologies, and designing the topology. Then it discusses how ETL based on titanium dioxide (TiO_2) is remarkably interesting and could be changed to show how important TiO_2 and other metals oxidize to DSSCs. Hence, experts are still looking to cut costs by improving operational efficiency, lowering the cost of precursors, and streamlining processes, which results in greener and cleaner technologies. The achievement of DSSCs with an efficiency of 10–24% provides optimism that roll-to-roll devices made of organic and inorganic materials may be manufactured. Here, we examine the current state of these potential approaches and their prospects for the future. In addition, we explore the opportunity of a new alternative to fossil fuels that may be employed economically to augment the existing PV solar cells.

S. Gouthaman · K. R. J. Thomas (✉)
Organic Materials Laboratory, Indian Institute of Technology Roorkee, Roorkee 247667, India
e-mail: krjt@cy.iitr.ac.in

V. Kumar et al. (eds.), *Optical Properties of Metal Oxide Nanostructures*,
Progress in Optical Science and Photonics 26,
https://doi.org/10.1007/978-981-99-5640-1_8

Keywords Metal oxide nanostructures · Semiconductors · Nanocomposite · Electron-transport layer · Dye-sensitized solar cells (DSSCs)

1 Introduction

The global energy crisis and climate change are two of the most important and related scientific problems of our time. Population and economic expansion result in an ever-increasing need for energy. This is especially true for electricity, the most utilized kind of energy. Also, the quantity of greenhouse gases emitting into the air is very worrying. The price of non-renewable fossil fuels is rising, and geopolitical restrictions show how important it is to find other ways to make energy. That indicates that the current energy trends are alarming for the economy and the environment. According to the 2014 energy-technology views of the international energy agency (IEA), the contemporary method of producing electricity might have disastrous consequences if it continues. By 2050, energy demand could rise by 70%, and CO_2 emissions could increase by 60% [1] (Fig. 1). Solar radiation is the most prevalent and environmentally sound renewable energy, capable of providing 1.4×10^5 TW (Terawatt) times the daily energy requirements of the entire planet. Hassanien and colleagues [2] say that only 3.6×10^4 TW of energy can be used now [2]. The volume of solar radiation striking the globe in one hour equals the power utilized by all human and industrial activities in one year [3]. It might help satisfy the world's energy needs.

On the other hand, one of the most significant critical challenges for our civilization is finding techniques to get energy that does not harm the environment and can repeatedly use green protocols. Appropriate and reliable renewable power sources and accompanying energy conversion techniques and systems must be developed to address the issues mentioned. Undoubtedly, there is a growing awareness of the significance of producing solar energy and power conversion technology. According to the IEA, global photovoltaic (PV) capacity has increased by an average of 40% per year since 2000. Solar energy production systems have been growing recently, and it is expected that they will meet 5% of the world's electricity needs by 2030 and 11% by 2050. This will prevent the yearly release of 2.3 gigatonnes (Gt) of carbon dioxide [4].

The transformation of solar energy into electricity is projected to become the primary energy source by 2050 (Fig. 2), providing that efforts to minimize carbon emissions and replace fossil fuels in the energy infrastructure continue. In other words, if we keep progressing toward a carbon-free energy system, solar power will be the primary energy source by 2050. There are three applications for turning light into energy: Photovoltaics (PV) directly turns sunlight into electricity. Concentrating solar power systems (CSP) use the sun's radiation heat energies [5]. Solar thermal collectors (SHC) use the sun's radiation heat energy [6]. There are established and emerging technologies in the portfolio of photovoltaic applications. Edmond Becquerel first documented such an effect in 1839 and utilized a contact with a silver halide as a

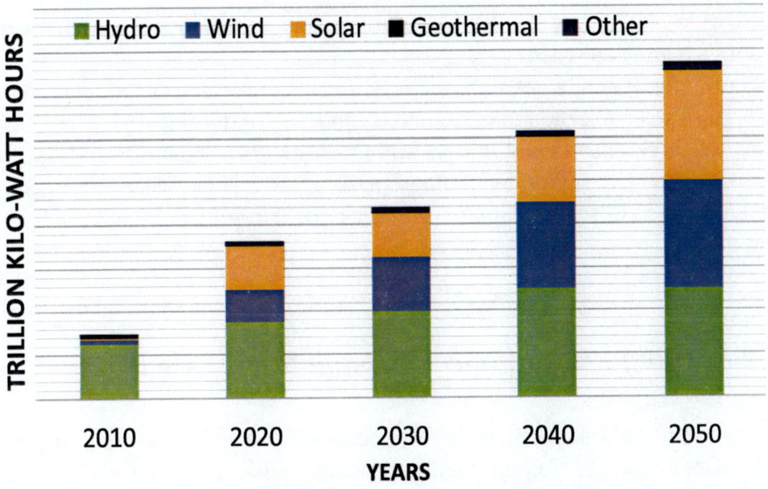

Fig. 1 Global renewable energy generation projection 2010–2050

semiconductor [7]. Then, solid-state photocells using selenium were introduced in 1877 [8].

Numerous novel technologies have been developed to convert renewable incident solar energy to electricity. These technologies are the most passable wind, thermal, and geothermal substitutes. Solar cells, or photovoltaic cells, are the most efficient for converting light energy into electrical potential [9, 10]. Their potential to meet the world's current energy requirements has generated considerable attention. It could help in remote areas where extending the power grid infrastructure is impossible for

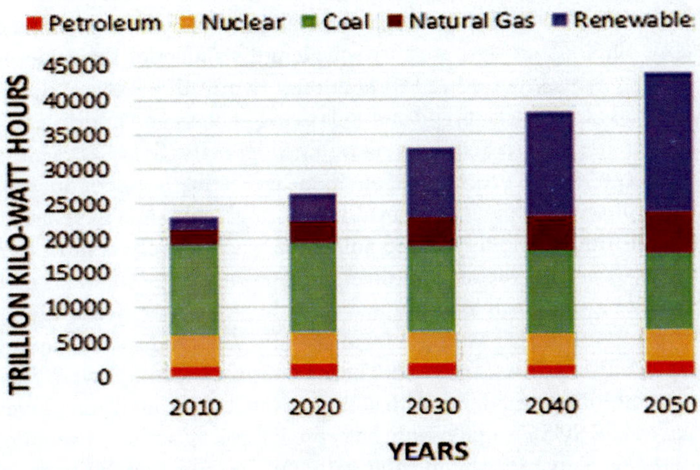

Fig. 2 Global fuel power generation, including renewables 2010–2050

practical or logistical reasons. It is simpler to set up, install, and run without emitting hazardous greenhouse gases [5].

Additionally, PV may decrease the need for fossil fuels. Consequenly, renewable PV energy sources may reduce COx, NOx, and SOx emissions [6]. The DSSCs and their inventors earned several honors and many awards, including the Balzan prize in 2009 and the Millennium technology prize in 2010 [11]. Solar intelligent energy networks can be built using smart and mini-grids, simplifying generation, demand, and supply.

1.1 Emerging Photovoltaics First and Second Generation

All photovoltaic and semiconductor devices work on the same idea as the photoelectric effect. The PV technologies, or solar cells, are usually put into three groups (Table 1), as shown in Fig. 3. PV systems use first-generation solar cells and single-crystalline silicon (Si) and have dominated the market for the last half-century. Based on the Shockley-Queisser limit, these single-junction devices have a photoconversion efficiency limit of 33%, one-third of the 95% Carnot limit for turning sunlight into electricity [12]. The rear and front PV cell contacts are screen-printed. Silicon-based solar cells have the best photoconversion efficiency (PCE) of 25% productivity on a laboratory scale, whereas commercial PV or solar cells have 22%. Still, making this technology is expensive [13], and because it is so rigid, it is a significant threat to Si-PC. Si does not absorb light well (absorption coefficient of 100 cm^{-1}), so it takes several hundred microns to absorb enough light.

Single-crystalline Si-based solar or PV cells have been most prevalent because they have a wide range of spectral absorption, are more efficient, and have mobile carriers with a high charge. Silicon solar cells are easy to maintain and run, but this method requires expensive equipment and installation, which makes it hard to make them cheaply. Photons created at short wavelengths squander their energy as heat, and the energy payback period (EPP) is around two to three years. So, Si-PV cell technology needs new materials and methods to lower costs and improve efficiency so more people can use it [14]. Second-generation solar cells (SGSCs) use less material and cover a large area. Direct bandgap light-harvesting semiconductors have an excellent absorption coefficient and can absorb sunlight with less than one-millimetre thickness. Thin-film technology-based solar cells successively deposit thin films of 1–10 μm thick film in a vacuum on diverse materials (e.g., polymer, metal, glass, etc.) across vast regions with incorporated module production by forming the layers for interconnectivity [15]. Materials from the III–V Periodic table include GaAs, GaAlAs, GaInAsP, InP, InAs, and InSb. These materials and alloys were easy to make, so monolithic multilayer tandems with solar cells that work at different wavelengths could be made. The SGSCs, made from heterojunctions of two other semiconductors, are a popular alternative to silicon solar cells due to their scalability in large-scale production. This combination can use the solar spectrum to make a device that works well and is very efficient. Like first-generation PV or solar cells, their elevated

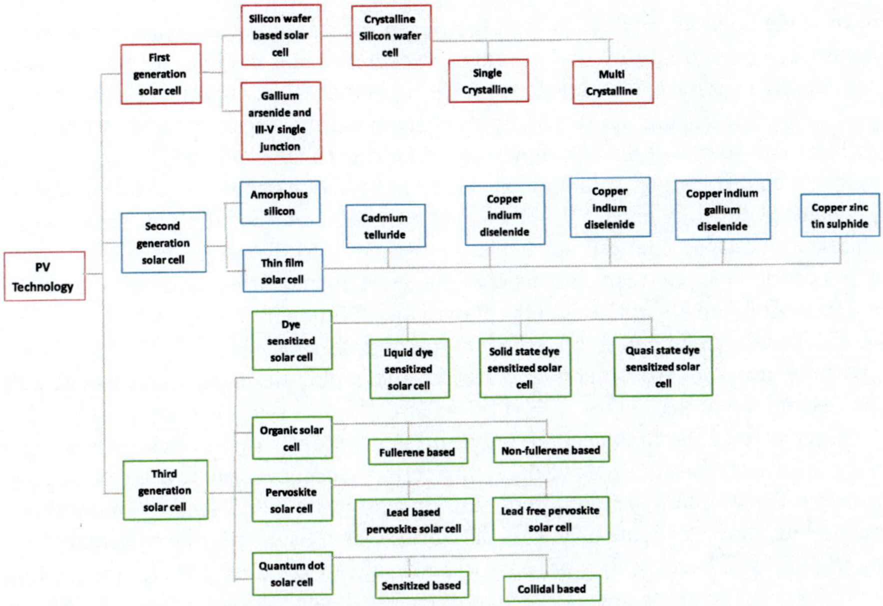

Fig. 3 Different generations of PV technology

temperature and energy remedies augment fabrication or production costs. They are constructed from amorphous silicon, copper indium gallium selenide (CIGS), and cadmium telluride (CdTe). This kind of SGSC thin film is hard to make because the deposition process is complicated; controlling stoichiometry and common structural flaws is hard. SGSCs have poor charge transport as a limiting factor. Their distinct morphology and scare elements (shown in Fig. 3) are other demerit features of these types of cells [16]. These things make it hard to put photovoltaics of the second generation on the market and force researchers to investigate PV technologies of the third generation.

1.2 *Third Generation*

Third-generation solar (or) PV cells are based on novel concepts, including up/down conversion, multiple excitons, multijunction, and hot carrier collection. Researchers have begun working on third-generation or highly emerging photovoltaic-solar cells to address issues with first- and second-generation photovoltaic-solar cells. This was made possible by a lot of research and development in materials science. Four types of new photovoltaics are dye-sensitized solar or PV cells (DSSC), bulk heterojunction solar cells (BHJ), quantum dot solar cells (QDSSC), and perovskite (or) CZTS solar cells, and these are all types of organic thin-film PV cells. This new photovoltaics have many benefits, such as straightforward processes, low-cost (or) cost-effective materials, great portability, and printing solar cells on non-flexible (or) flexible semi-conducting materials [17]. Sensitizer DSSC uses inorganic, organic, and organic dye molecules. Sensitizers' most significant challenge is making technology work better and be more stable while making it easy and cheap to fabricate the device [18]. Due to the tuneable bandgap, inorganic quantum dots provide benefits over dye molecules in quantum dot solar cells. Still, it was not easy difficult quantum dots sensitive and choose electrolytes that did not break them down [15]. Donor materials are conjugated conductive polymers and accept consequently are carbon derivatives like C60 in BHJ. Moisture and ambient light may reduce BHJ's effectiveness.

CZTS solar cells are 12.6% efficient, perovskite solar cells are 25.2% efficient, and both are being studied. In this regard, these polymer-solar photovoltaic cells are less effective than DSSCs. Hybrid solar cells, this type of photovoltaic cell is a mixed state of the first, second, and third generations. It uses each generation of solar cells' best parts to improve operating performance, consistency, and price effectiveness. Perovskite/Si-tandem (monolithic), perovskite/CIGS-tandem (monolithic), perovskite/DSSC, etc., are examples of hybrid solar cells. People are extremely interested in DSSCs because they are easy to make, cheap to make, good for the environment, and can be made in massive quantities. Third-generation PV cells (DSSC) are almost ready for the market and could be built on flexible or rigid semi-conducting substrates [20].

Table 1 Sorted a list of solar cells based on their current efficiency [19]

Type of generation	Type of solar cell	Efficiency (%)
First-generation	Single-crystalline-solar cell	26.1
	Multi-crystalline-solar cell	23.3
	Single-junction gallium-arsenide-solar cell	27.8
Second-generation	Amorphous-silicon-solar cell	14.0
	Cadmium-telluride	22.1
	Copper-indium-gallium di-selenide	23.4
Third-generation	Dye-sensitized-solar cell	12.3
	Organic-solar cell	17.4
	Perovskite-solar cell	25.2
	Quantum-dot-solar cell	16.6
Hybrid-solar cell	Perovskite/Si-tandem (Monolithic)	29.1
	Perovskite/CIGS-tandem (Monolithic)	24.2

1.3 Current Research in DSSCs

The PCE of DSSCs has grown on or after 7–14% [21, 22] (Fig. 4), and DSSCs are the only type of organic molecules-based competent solar cell used in more than one commercial setting, such as a facility in Switzerland that can make 2000 kWh of electricity per year [23]. After DSSCs were found in the late 1960s, scientists thought natural dyes that give off light could generate power in electrochemical cells. In this continuing experiment, the University of California at Berkeley used chlorophyll in spinach (a photosynthetic method) to test this theory. In 1972, the same group developed the first chlorophyll-sensitized zinc oxide (ZnO-electrode)-based solar cell. First, the electrons from excited organic-dye molecules are hooked on large-band gaps in semiconducting substrates, turning light (photons) into electric powers [23, 24]. Numerous studies have been conducted on ZnO single crystals [25]. These DSSCs were not particularly good because the single layer of dye molecules could only absorb up to 1% of the sunlight. So, light harvesting efficiency (LHE) could be raised by making the electrode of fine oxide powder as porous as possible. DSSCs in 1991 showed unexpectedly superior efficiency of around 7%, given that affordable TiO_2 was employed for the bulk of the photovoltaic solar cells. These are also called Gratzel cells and were first made by Brian O'Regan and Michael Gratzel at UC Berkeley in 1988. Scientists at the solar-cell technology at Ecole Polytechnique Fédèrale de Lausanne (EPFL) continued to work on them until 1991 [26]. Brian O'Regan. Michael Gratzel made a 10-μm-thick, high-surface-area, optically high-transparent TiO_2 nanocomposite thin film covered with a charge transfer organic dye with favourable spectrum properties for light harvesting at a high rate of device harvesting (46% of incident-solar-energy flow). Its efficiency at turning photons (light) into electricity (power) was above 80%, and its efficiency at turning incident photons into current (IPCE) was 7.1–7.9% in simulated solar light and 12% in diffuse daylight. With a large-short circuit current density (J_{SC}) of higher than 12 mA cm^{-2}, excellent stability (five million turns without falling apart), and a low price, it could be used in real life. Gratzel et al. developed in 1993 that the efficiency of cells was 9.6%. In 1997, the national renewable energy laboratory (NREL) achieved 10% efficiency [27]. The reactive functional groups such as COOH, PO_3H_2, and $B(OH)_2$ are frequently utilized to make sensitizers that adhere stably to the semiconductor substrate [26]. There are many ways to improve power conversion efficiencies (PCE), including using bright panchromatic dyes or multi-dye analogs to absorb more light, nanostructured metal oxides, and compact blocking layers to make it easier to inject and collect photo generated charge carriers [28]. Because photosensitive dyes make photoelectrons on their own and a semiconductor. Consequently, more efficiency may be attained. The DSSC's photovoltaic electrodes (photoanode and counter-electrode) are made of semiconductor material and are crucial to its work. However, we still do not understand how electrons move from one place to another. It is difficult for DSSCs to convert light energy into electrical energy because of their poor light-harvesting performance, electron-hole recombination, dark current, sluggish electron transfer rate, and high counter-electrode interface resistance. Choosing suitable electrode materials may prevent these issues. The development of DSSCs has moved slowly, but in early 2015, it was said that they were stable for an extended period and had the highest calculated efficiency of 14.3% [29] (Fig. 4). The most widely used conventionally commercialized modules are made of monocrystalline, multi-crystalline, and amorphous crystalline silicon, which account for about 90% of the global market share. Several important things happened simultaneously as the price dropped to $1 per W because silicon PV production increased [30]. Significant research institutes and groups have been set up worldwide throughout the preceding decades, and 181,895 academic papers have been published. IDTechEX thinks that the market for DSSCs used in wireless sensors, portable electronics, automobiles, and mobile devices will reach $130 million by 2023 and $6.9 billion by 2031 for printed and electronic materials [31]. Nevertheless, the low power conversion efficiency (PCE) of DSSC technology still makes it hard for businesses to use it more. To reach the energy (TW) goal, "smart" nanostructured materials must convert solar energy more efficiently and have better electrical, morphological, and optical properties to boost the PCE of DSSC. This may be done by balancing the efficiencies of new materials with their manufacturing costs and choosing improved solar cell technologies. This chapter will discuss how vital metal oxide nanostructures are as electron transport layer (ETL) materials in DSSCs, also called third-generation solar cells. Brief literature reviews, properties (morphologies, optical, electrical, and electronic), and the most recent research advances and trends about the development and comparison of different metal oxide nanoparticles (MO-NPs) and their uses in photovoltaic solar cells cover a wide range of new advancements and studies in DSSCs fields. This work will give researchers a plan for improving material engineering to improve DSSCs.

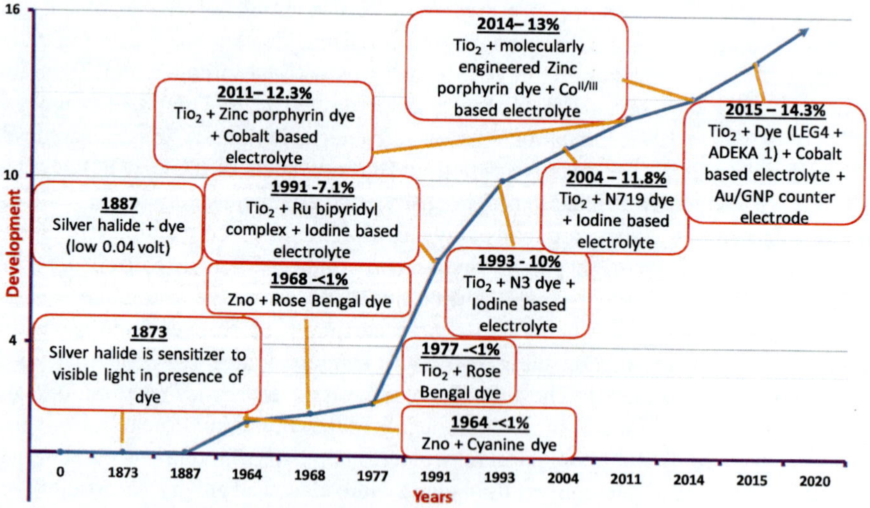

Fig. 4 Chronological development of DSSC

2 Dye-Sensitized Solar Cells (DSSCs)

Compared to silicon-based solar cells, DSSCs have unique topologies and operating procedures, and the following part will describe the construction and operation of DSSCs. Figure 5 illustrates each component's standard structure and placement in DSSCs [32]. A glass substrate coated with platinum is placed in direct contact (face-to-face) with the working electrode to create the counter electrode. The 40 μm gap between these two electrodes is loaded with a conducting substance that serves as an electrolyte, either liquid or solid. The purpose of each part is described below:

(i) In a traditional DSSC substrate, the nanocrystalline layer must be transparent so that the active solar material can soak up sunlight. On top of the transparent conducting oxide, the anode (as semiconductor metal oxide) and cathode (metal as a catalyst) semiconducting organic materials may also be placed (TCO). Since it also works as a current collector, it must conduct electricity well. The conductivity and transmittance of the substrate also affect the overall performance of DSSCs. According to reports, transmittance must be at least 80%.

(ii) A mesoporous **photoanode** in its paste form and a photoanode composed of semiconducting metal oxide offer a high surface area for more dye adsorption. PCE is based on the semiconductor material's shape, size, porosity, conductivity, and crystallinity.

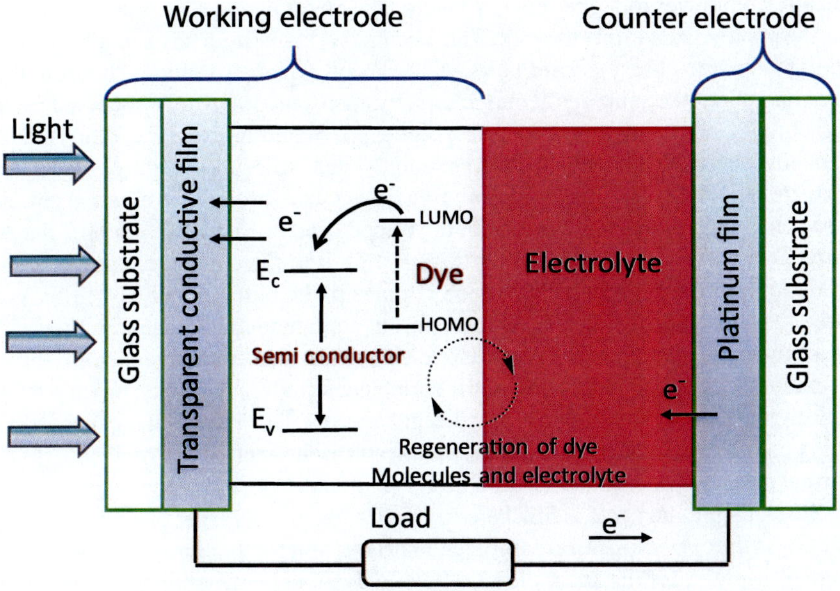

Fig. 5 Components and standard architecture of DSSCs

(iii) The DSSC covalent link is formed between a **dye sensitizer** and a metal oxide semiconductor. This is an essential part of DSSCs, which absorb and energize sunlight. Two kinds of photosensitizers are (a) metal sensitizers that use ruthenium and (b) organic sensitizers that do not use any metal in DSCs, dye molecules that stick to the oxide act as "antennas" that catch photons. So, while DSCs were being made, much research was done on organic dyes to increase the extinction coefficient and broaden the optical absorption spectrum.

(iv) To revive the dye after oxidizing, a salt solution containing an **electrolyte** with a redox couple was injected between the electrodes.

(v) A thin catalyst layer is put over a highly transparent, conductive glass substrate to form the **counter-electrode (CE)**.

2.1 DSSC Charge Transfer Mechanisms and Kinetics

Figure 6 is a clear representation of an energy level diagram that shows how DSSCs work [33]. The following factors must be considered while selecting the contact substances, working electrode (photoanode), dye, and electrolyte for the DSSC to inhibit the reverse transfer process (recombination reaction) and improve performance: The following are the operational phases of DSSCs: (i) excitation; the covalent bonds connect the molecular dye materials to the metal-oxide semiconductors, and dye (D) molecules inside the DSSC absorb the photon and perform an electronic state transition from the ground to the higher-level-excited state (D*). Electrons tend to shift from lower to higher energy levels. The electron and hole must be separated for the photovoltaic effect to work. This occurs (D^+) after light absorbs a light photon and creates free-charge carrier pairs. A chemically distinct semiconductor makes a heterojunction, the same semiconductor substance, and oppositely polarised carriers for a homojunction. (ii) Injection: The highest occupied molecular orbital (HOMO) has a low energy level, while the lowest unoccupied molecular orbital (LUMO) has a high energy level. (iii) Transportation: Electrons move through the external circuit of the semiconductor and reach CE by being trapped and released. Meanwhile, the electrolyte restores the oxidized dye by transporting it to the counter electrode through oxidation-reduction processes in redox couples in the outer circuit. (iv) After going through a thin transparent layer of mesoporous semiconducting material, they gather on a conductive material. The photoanode's electrons subsequently reach CE through an external circuit. (v) Diffusion: At the counter electrode, a reduction process occurs, making the electrolyte fresh. (vi) Dye-regeneration: The process happens when the redox couple reduces the oxidized dye. This returns the excited dye molecules to their normal state and makes them available for photoelectron activation again; thus, the DSSC's operational cycle is finished. In DSSCs, the energy level of each component must be in the correct position or molecular orientation for the charge to travel through them efficiently; for instance, the LUMO of a dye must have a superior energy level than the CB of a semiconductor [34]. Similarly, the HOMO level of the dye must be greater than the redox potential of the electrolytes. Also, the metal oxide semiconductor's CB must be better than the redox potential of the electrolyte material. Figure 7 depicts the kinetics of several approaches (excitation, injection, and regeneration) in DSSCs. All these activities occur on a finite timescale. The arrows show how an electron in a dye moves on or after the HOMO state to the LUMO state. The arrows indicate the dye's HOMO-to-LUMO electron transition, which takes 60 ns takes or 50 fs, or 1.7 picoseconds, for one electron to go from the dye's LUMO to its semiconductor counterpart. Like how long it takes to bring back together injected electrons and oxidized dye, it takes 10 μs to get back together injected electrons and clectrolytes in a semiconductor substrate. In contrast, the regeneration of the sensitizer by electrolytes substant takes only ten nanoseconds. So, it is essential to quickly regenerate the sensitizer and electrolyte to keep the device stable and get a high PCE. However, charge recombination causes kinetic slowdown. Recombination happens at the photoanode electrode's PE/dye and PE/electrolyte interfaces. Lowering it can lower the dark current. Electrolyte-solid contact occurs faster than electron escape, redox, or cation capture [35]. The minimum energy required for silicon to absorb light is 1.1 eV, as is the band gap for the valence-to-conduction. In the visible and near-infrared spectra, photons with wavelengths shorter than 1.13 μm are absorbed by it. Narrower band gap materials absorb more photons and give a larger current.

Recent research has focused on developing and changing the MOS structure, passivating photoanode surfaces, enlightening counter-electrode topology, changing dye molecular engineering, and optimizing and changing electrolytes to increase light absorption, scattering, charge transport, and interface energy by stopping charge-recombination. Researchers have recently made a practical thin tunneling barrier effect for DSSC by putting DNA double-helix fragments at the TiO_2 interface. This effect reduces charge recombination and enhances electron injection. In the following sections, these enhancements are examined in further depth [36]. Transparent conductive oxide (TCO) glass is the most critical aspect of DSSC construction in material

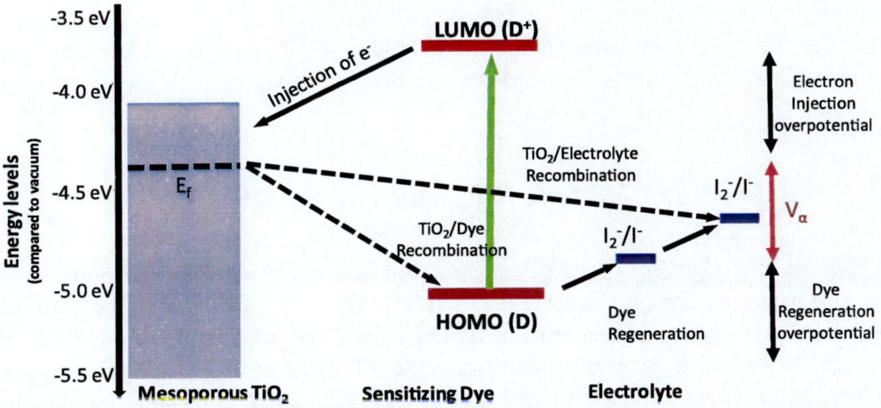

Fig. 6 DSSCs energy level diagram and operating concept

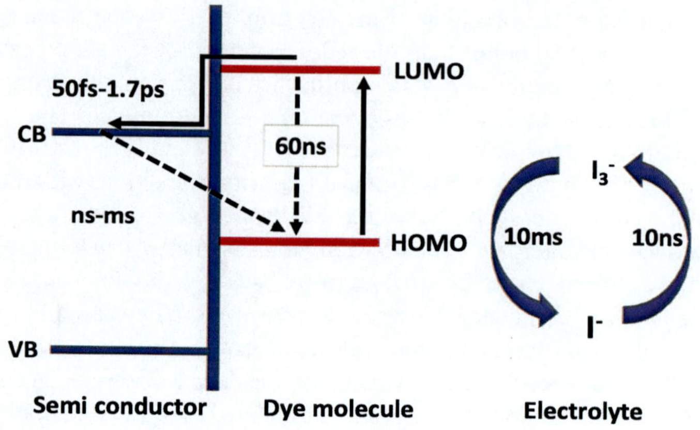

Fig. 7 Electron transport kinetics in the DSSC

selection and substrate (DOS) characteristics. The two most common TCO materials in DSSCs are (i) fluorine-doped tin oxide (FTO), SnO_2: F, and (ii) indium-doped tin-oxide (ITO), SnO_2: In. FTO and ITO have sheet resistances of 18 Ohm/cm^2 and 8.5 O/cm^2, and they let through 75% and 80% of the visible light spectrum, respectively. Due to its temperature resistance and conductivity, FTO is widely used in DSSC manufacture. TCO was created by covering glass substrates with TiO_2 and CTO using magnetron sputtering. The new CTO has an optical transmittance of 90% and a sheet resistance of 15 Ohm/cm^2. Plastic substrates are popular because they are cheap and flexible, and plastic's fragility limits its use. DSSCs use transparent FTO/ITO-dipped polymer films such as polyethylene terephthalate (PET) as glass components. This film is cheap and adaptable. Stainless steel, tungsten, and titanium are alternate substrates. Opaque metal substrates reduce cell efficiency. DSSC substrates should be: DSSC substrates must (a) allow incident light to pass through. (b) a low, continuous sheet resistance that permits electron flow. (c) MOS's good conduction transfers electrons to the external circuit. (d) High-optical transmittance reduces light contemplation and absorption. (e) that strengthens the device.

2.2 *Photoanode Electrode (PE) Material for DSSCs*

The most crucial aspect of DSSC construction is material selection for optimizing its performance. The photoanode also acts as a place to collect electrons, and the whole device's performance depends on the size, shape, topology, and band gap of the PE substance. As a result, it must have the following crucial characteristics and a large surface area for high dye adsorption: The PE's conduction band should make it easy for the dye to send many electrons to the semiconductor. PE should have the right size of pores for dyes to move through them quickly and for them to be resistant to photo corrosion. PE must acquire high photon scattering capacity, large surface roughness, and better electron acceptor capacity. It is required to adhere effectively to conductive substrates. The incremental enhancements to these fundamentals increase charge collection efficiency [37].

3 Metal Oxide Semiconductor (MOS)

Since making MOS assemblies for use in solar cells has become a more recent thrust, understanding the kinetics and dynamics of electron and hole transfer at interfaces is more critical than ever [38]. Research on materials in the modern era has led to a wide range of MOS, including elements in group IV. Electronic device developers can choose between band gaps and carrier mobilities. Other optoelectronic properties are helpful for PV and OLEDS, which are the opposite of photovoltaics. Researchers are developing new materials and planning for new buildings to meet the growing energy needs. The interior surface area grew significantly with the addition of a mesoporous MOS layer. This made it possible to add more dye without changing the way the dye and electrolytes interact. This led to a significant rise in PCE and boosted the DSSC research field. Thus, a large-area photosensitive cell may be created by contacting a MOS with a liquid electrolyte as a cost-effective option for junction formation. Strongly bonded MOS have more significant band gaps, limiting their solar photovoltaic response. Electrolytes are complex on MOS, so materials with a narrow gap that could respond quickly to light over a broad spectrum of wavelengths corrode. In the real world, the only inherently stable photoelectrodes are oxides of reactive metals with a large band gap, such as titanium dioxide (TiO_2). Unfortunately, due to its 3.1 eV band gap, this substance only responds to UV light and is insensitive to visible light. The first devices with single-crystal had limited optical absorption and PV efficiency because a single layer of the sensitizing dye could transfer charge to the substrate. The dye was combined with a nanostructured MOS to absorb light better and move free-carrier charges. Effective DSSCs and MOS structuring are employed on three ascending scales: Mesoporous semiconductor structure, dye molecular engineering, the self-assembly of an adsorbent surface monomolecular dye layer on the MOS surface, and the manufacturing of mesoporous semiconductors. The monomolecular dye layer is more prominent than the substrate by two to three orders of magnitude. The dye's MOS lets it soak up all the light that comes in, which helps it convert solar energy well. The quick injection of electrons into a nanostructured MOS with an excited dye molecule makes this performance possible. High photoanode roughness does not induce charge carrier recombination loss. This allowed the device's opacity and light absorption to compete with solid-state devices. Recombination losses are avoided because the charge carriers are kept apart, and electrons enter an n-type material with most n-type carriers. After crossing the MOS-electrolyte phase barrier, only dye cations or redox electrolytes may absorb back electrons. When electrons and holes in a solid MOS lattice occupy the same region, the rate of this operation is at its slowest. To move electrons and holes between a photon absorber and the external circuit, a wide-range band gap MOS with n-type and p-type contacts is used. Designing the engineering has led to sensitive dyes that work well and stay stable and MOS substrates that require much development (Table 2).

Table 2 Different MOS morphologies and their effectiveness [39]

S. No	Metal oxide semiconductor	Morphology	PCE %	References
1	TiO_2	Nano-particle	10.6	Grätzel et al. [40]
2		Nano-tube	9.1	Lin et al. [41]
3		Nano-rod	9.5	Lee et al. [42]
4		Nano-wire	6.2	Lee et al. [39, 43]
5		Nano-flakes	8.2	Shanmugam et al. [44, 45]
6		Nano-sphere	8.5	Kim et al. [45]
7	ZnO	Nano-particle	6.6	Saito et al. [46]
8		Nano-wire	5.7	Barpuzary et al. [47]
9		Nano-rods	4.7	Rao et al. [48]
10		3D-Sponge	6.7	Sacco et al. [49]
11		Nano-sheet	7.1	Lin et al. [50]
12		Micro-spheres	5.2	Li et al. [51]
13	SnO_2	Nano-crystal	3.2	Birkel et al. [52]
14		Nano-particles	1.7	Lee et al. [53]
15		Nano-rod	1.4	Wijeratne et al. [54]
16		Nano-tubes	1.1	Desai et al. [55]
17		Nano-grains	3.0	Lee et al. [53]
18		Nano-crystals	3.2	Birkel et al. [52]
19		Micro-spheres	2.9	Basu et al. [56]
20		Nano-particles	2.9	Wanninayake et al. [57]
21	Nb_2O_5	Nano-crystal	5.0	Guo et al. [58]
22		Nano-porous	4.1	Ou et al. [59]
23		Nano-tube	3.2	Liu et al. [60]
24	WO_3	Nano-particle	0.8	Zheng et al. [61]
25		Nano-crystals	0.3	Hara et al. [62]
26		Nano-tubes	0.8	Hara et al. [63]
27		Nano-pristine	3.3	Elbohy et al. [64]
28		Nano-rods	1.5	Yong et al. [65]

4 The Nanostructured Semiconductor

In the last 30 years, improvements in nanotechnology and nanoscience have resulted in the invention of several nano-morphology and nano topology, such as nanoparticles, nanowires, nanotubes, nanobelts, and their groups, such as oxide aggregates. The nano-meter-scale dimensions of its fundamental components are a defining characteristic of nanostructures. This phenomenon, known as the "quantum confinement effect," causes the inner surface area of nanostructured materials to be greater than that of bulk materials. Nanostructured metal-oxide nanoparticles have a large surface area that can be used for many distinct functions. They also increase the surface/interface recombination events between the dye and the nanostructures, giving electron scattering a place to start. Nevertheless, the overall performance of the cells is harmed by the disordered nanoscale morphology or shape of dye-sensitized devices. In this method for nanodevices and nanoelectronics, nanostructures are perfect in shape and size, crystallographically perfect, and have no morphological flaws. This is because of their unique properties. Nanostructured metal-oxide semiconductors have gotten much attention. They have been studied for important or complicated uses like electrical, optoelectronic, photovoltaic, photocatalytic, microelectronics, gas sensing, energy conversion, and storage. Their success comes from the fact that their electrical, optical, and charge-transport properties can be fine-tuned over a broad range of values. This lets their functional properties be tailored to applications. The area of the metal oxide's surface, shape, and purity of how well the photoanode electrode (PE) and the highly transparent conventional MOS (TiO_2) film work. Poor light scattering lowers light-harvesting efficiency (LHE) because MOS nanoparticles make it hard for light to spread out, and MOS particles are only a few nanometres in size. Also, the downward mobility of MOS nanoparticles makes it easier for injected electrons, photosensitizers, and electrolytes to recombine their charges. In recent years, being able to shape metal-oxide semiconductor substrates at the nanoscale level has enabled novel approaches to sell them and improved the performance of end-user devices. The Fermi level determines how full the bands are. Titanium dioxide and other nanostructured metal oxides and composite materials (like ZnO, NiO, CeO_2, SnO_2, WO_3, Nb_2O_5, FeO_3, Zn_2SnO_4, $ZnSn_2O_4$, Ag_2O–ZnO, Au–TiO_2, $SrTiO_3$, and $CuGaO_2$) are easy to make and do not cost much to make. Due to their exceptional photo-corrosion resistance and superb electronic characteristics, principal metal-oxides with band gaps larger than 3 eV ($E_g > 3$ eV) are typically used as PE materials. This list shows how different metal oxides can be, and it goes from transition metals to inner-transition metals (lanthanides), from binary to ternary, and from p-type to n-type. Doping a semiconductor with exotic atoms or vacancies alters the electron distribution and band occupancy [66]. Fermi's level is placed just above the valence band in p-type semiconductors and slightly below the conduction band in n-type semiconductors. Transporting charge needs electrons to occupy partially filled orbitals. So, n-type substrate conduct electricity by moving electrons in the conduction band, while p-type materials do the same by moving holes in the valence band. Because of their different chemical potentials, two metal oxide semiconductors and an electrolyte solution flow charges, electrons, and electrolytes until equilibrium is reached. This results in surplus charge regions and MOS; these ideas are crucial for understanding how light quanta are absorbed and how photo-excited charges are made and moved [67]. Semiconductors should "do it all" in direct bandgap solar fuels without degradation or disintegration from light quanta absorption through charge separation, transport, and catalysis. Even though it is important, much trial-and-error searching for single semiconductors for photoelectrochemical applications has failed. When a thick semiconductor absorbs light photons in a heterojunction solar cell, the e^- and h^+ charges may separate inside the semiconductor. Only the selective electrode receives an e- or h+ from the metal oxide. At the interface, there is an electrolyte with a high Helmholtz capacitance. It is easy to determine the EIS data's meaning based on how the metal oxide works electronically. Notable exceptions include studies on other solar cells, where the composition of the metal oxide significantly affects the quantity of the interfacial capacitance at low frequencies. It can also tell essential things about the structure of the bands, the number of states, the number and type of charge carriers, and how well the semiconducting metal-oxide electrode conducts electricity. This segment looked at the key features of metal oxide nanoparticles' size and morphology used to turn solar energy into electricity and the models used to determine how their properties affect how well the photoanode performs so DSSCs function effectively. In both theoretical and practical aspects, electron flow in nanostructured MOS substances or porous materials is an important and useful topic.

MOS nanomaterials are vital to DSSC, DSPEC, and most other solar energy devices to make electricity or chemicals. Most of the research has been done on using different nanostructures in DSCs to make photoelectrode films. This is because MOS nanomaterials-based films can provide an exact surface area. It has been established that a film composed of 15 nm nanoparticles with a thickness of 10 μm has an internal surface area as superior as 780 cm^2 per 1 cm^2 of the geometric (or) morphological surface. Using simple chemical solutions to create nanoparticles is also an explanation. Numerous MOS with broadband gaps needs excellent photodecomposition permanence. Low-light-photon-absorption single-molecule dyes rise; extensive surface areas are required to connect many dye molecules. As we have already said, this can be done with nano-porous architectures, which increase by over 1000-fold the functional oxide surface area on flat, compact films. This porous medium's electron flow influences the performance of DSSC. Recently, Monte Carlo random walk simulations have been used. Two main methods have been used to explain how electrons move through nanostructured media with traps: (i) The multi-trapping model says that charge carriers move from localized situations to the transport level by trapping and releasing electrons. (ii) Hoping model: direct transport of charge carriers between localized states. Javadi and Abdi used a Monte Carlo continuous time random walk simulation to determine how electrons move in the thick layer and columnar structure of porous TiO_2 [68]. Naser. His colleagues made equations starting with how the geometry of nanostructured-morphology solar cells impacts electron flow. They could replace high-tech devices, leading to new and better technology uses [69].

4.1 Binary Metal Oxides

DSCs were made from photoelectrochemical systems produced in the late 1960s. Most active electrodes were constructed of group IV (such as silicon) or III–V semiconducting materials, which had small band gaps and absorbed visible light well. However, it was found that the small band gap of group IV or III–V semiconducting substrate caused a big problem with corrosion on the working electrode when the light was shining on it [70]. Moreover, binary-metal oxides have been observed to predominate as a photoanode in dye-sensitized solar cells. Gratzel's research group reported a development in PCE in 1991 by using mesoporous TiO_2 nanoparticles as a photoanode [26]. It has become the most often used semiconducting metal oxide for DSSCs because it is more stable and works better than ZnO, SnO_2, and other metal oxides. In recent years, exciting trends and strategies have emerged for redesigning traditional TiO_2-based photoelectrodes to improve photovoltaic performance with improved molecular dyes that collect (light) photo-harvesters and redox shuttles. For the first generation of DSSCs, typically 10–20 mm thick layers of TiO_2 were used with traditional ruthenium-based sensitizers to get high short-circuit current densities (J_{SC}). With this thick TiO_2 layer, advanced redox couple-based electrolytes have a high diffusion resistance (RD), which makes it harder to get higher. With this thick layer of TiO_2, advanced redox couple-based electrolytes have an extremely high diffusion resistance (RD). They cannot have J_{SC} values that are much higher than those obtained with iodide or triiodide-based electrolytes. Even though ZnO and SnO_2 are more electrically mobile than TiO_2, and ZnO has a bandgap (Fig. 8] that is the same as TiO_2, TiO_2 is still an excellent choice because it is chemically stable [71]. SnO_2 has a significant band gap energy (3.8 eV), and its CB is about 500 mV more optimistic than TiO_2. Hence, SnO_2 has a higher charge injection rate than ZnO and TiO_2. Despite their benefits, ZnO and SnO_2 have inadequate PCE. Several studies explain the performance differences between ZnO, TiO_2, and SnO_2 solar cells. Still, as light-harvesting technology has improved, dyes with high absorption coefficients have become possible. Due to them, getting the same J_{SC}s by creating sensitive TiO_2 photoelectrodes that are much thinner is now feasible. For this reason, achieving the necessary barriers could be advantageous for the DSSCs. TiO_2 is one of the most discussed nanostructures, undergoing some processes to improve its performance.

In the scientific literature, it is essential to understand heterogeneous electrochemistry using electrochemical-impedance spectroscopy (EIS) for TiO_2, ZnO, and SnO_2 for electron transfer (n-type) and NiO and $CuGaO_2$ for hole moving (p-type). Several times, a specific interface between a semiconducting metal oxide and an electrolyte substrate was studied. In the section on TiO_2, due to current research on PSCs, there is an additional discussion of metal oxide-based semiconductor interactions. Notably, chemical capacitance ($C\mu$) can also be called the metal oxide's density of states (DOS), which tells us how the defect/trap states are distributed at the semiconducting metal oxide interface. ($C\mu$) is also affected by the shape and doping of metal oxides due to changes in surface area, exposure of surface states, conductivity, and the introduction of defects. The density of states in metal oxide

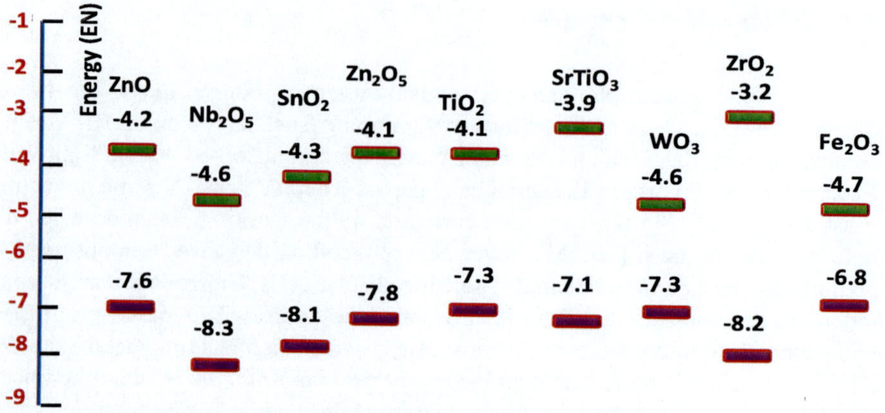

Fig. 8 Band locations of various MOS used in photoanodes

is one of the best ways to use excitonic solar cells, and they are now a pivotal way to make DSSCs work better. Recently, these devices have become popular again, leading to photoconversion efficiencies (PCE) in DSSCs and other unheard solar cells.

4.1.1 Nanostructures Based on TiO_2 Photoanodes

Bottom-up fabrication of the photoanode for thick nanorod morphological array sheets (up to 30 m) was compact rods was hydrothermally etched to create porous rods with a high specific surface. Increased dye loading in the upgraded materials system results in a high PCE (7.91% against 6.5% for the typical P25 NP photoanode), while TiO_2 nanosheets improve electron collecting and specific surface area. In his recent review of this material, Prof. Wang predicted the benefits of the unusual shape of thin (sub-1 nm), fragile TiO_2 sheets. In a composite material, nanosheets and NPs of TiO_2 scatter light, hold dye, and move electrons quickly. The PCE of the improved photoanode was 10.1%, much higher than P25's 7.1%. The geometry of nanotubes is being investigated for DSSCs [72]. Anatase Despite the amorphous nature of anodized nanotubes, TiO_2 is suitable for DSSCs. Air annealing at 350–500 °C changes the material from amorphous to anatase. Most polymeric substrates are destroyed by heat damage. Kapton HN is a polyimide that can be heated in a controlled way without losing its mechanical properties. When TiO_2 nanotubes were put on Kapton HN, they changed from amorphous to anatase and formed flexible solar cells. At the same time, Kapton HN substrates are colored, meaning the solar cell structure must be backlit. The proposed solution's PCE (3.5%) is the same as the PCE (3.58%) of the titanium substrate, but titanium foil is heavier and less flexible than Kapton HN. Therefore, the proposed approach is an excellent, adaptable DSSC invention. It also demonstrates how to get a nanotube surface free of debris by applying alcohol-based ink with a felt-tip pen to the Ti surface before anodization. anodization [73]. For optimal photoanode thickness, a layer 50–90 nm thick improves photoconversion efficiency by 15–20% related to the cell short of the blocking layer (10–15 microns), and a recent article found that a 5 nm weak layer reduces series resistance in the working solar cell, making it more effective. In that article, small photoanodes (2.5 μm) were studied. This was a long way below the length over which electrons could move, so PCE could only be 3.12%. The results cannot be used for thicker photoanodes, where the TiO_2 mesoporous network will have more electron recombination [74]. The light was used to make solar photons stay longer in the photoanode so that electrons could move more quickly in DSSCs. The first photoanode used a transparent mesoporous sheet with tiny NPs (about 20 nm). Unabsorbed light is lost after passing through the sensitive layer. It was suggested to use an impacting PCE, a layer of big particles (200–400 nm) that scatters light inside the photoanode and boost photon absorption. Using opaque coatings to control light-boosted PCE (the "haze idea"), they get 11.1%, where the haze is the ratio of diffuse transmission to total transmission. The DSSC-IPCE increases as the haze in the TiO_2 electrode's near-infrared specific wavelength range worsens [75]. The haze issue is that the large particles, effective at scattering visible light, have a smaller surface area than the nanoparticles used in photoanodes. So, scientists are trying to build structures that scatter visible light while keeping a mesoporous layer's high specific surface area. Caruso and his team developed a mesoporous topology of nanocrystalline with nanocrystalline TiO_2 with a maximum surface area of 108.0 $m^2\ g^{-1}$ and adjustable pore diameters. Anatase TiO_2 nanocrystals made mesoporous beads and were compared to Degussa-P25 with TiO_2 photoanodes for DSSC functionality, and IPCE increased due to the photoanode's high specific surface and light scattering [76]. L. De Marco has suggested a layered system for light harvesting and charge collection that consists of three mesoporous aggregated layers constructed of topology-tailored TiO_2 anatase nanocrystals generated ad-hoc through proper colloidal techniques. The increased effectiveness of this multi-layered DSSC was owing to the increase in Jsc, which was caused by a large amount of dye adsorbed (2.7×10^{-7} mol cm^{-2}, contrasted to 2.0×10^{-7} mol cm^{-2} for the source DSSCs). Additionally, the approach was far more efficient in collecting the amount. This was seen by measuring the length of electron diffusion, which was 30% longer in the multi-layered PV cells than in the reference photoanode [77]. To convince people about DSSCs based on TiO_2, we point out that scientists are trying to advance scientifically relevant stands to shorten the time it takes to make these PV cells. Recently, O'Regan and his colleagues showed that an ultrafast sintering technique that utilizes near-infrared (NIR) light is reliable. In 12.5 s, they were able to make a functional photoanode. Photovoltaic devices fabricated with NIR sintered films can be used the same way as those fabricated with traditional sintered films heated for 1800s. Researchers showed that heating TiO_2 quickly (up to 785 °C) during NIR processing did not significantly change the rutile phase or decrease photocurrent [78] (Table 3)

Table 3 DSSC photovoltaic performance by different morphologies

S. No	Structure of a photoanode based on TiO_2	PCE (%)	References
1	Nano-fibers	10.3	Chuangchote et al. [79]
	Nano-fibers	8.80	Joshi et al. [80]
	Nano-fibers	7.00	Yang et al. [81]
	Nano-fibers	8.50	Hwang et al. [82]
	Nano-fibers	7.53	Krysova et al. [83]
	Nano-fibers	6.26	Cao et al. [84]
2	Nano-wires	9.30	Adachi et al. [85]
	Nano-wires	8.51	Wang et al. [86]
	Nano-wires	3.43	Liu et al. [87]
	Nano-wires	7.34	Lee et al. [88]
	Nano-wires	6.00	Liu et al. [89]
	Nano-wires	5.30	Liu et al. [90]
3	Hollow-spheres	10.34	Koo et al. [91]
	Hollow-spheres/nano-rods	4.57	Dai et al. [92]
	Yolk-shell-micro-spheres	11.0	Li et al. [93]
	Hollow-spheres	5.13	Zhang et al. [94]
	Hollow-spheres/nano-particles	7.50	Sun et al. [95]
	Hollo-spheres doped with Nitrogen	8.08	Cui et al. [96]
4	Nano-tubes (L = 3.6 μm, T = 17 nm, D = 46 nm)	4.70	Paulose et al. [97]
	Nano-tubes (L = 6.0 μm, T = 20 nm, D = 110 nm)	4.24	Paulose et al. [98]
	Nano-tubes (L = 5.7 μm, T = 8 nm, D = 30 nm)	3.00	Zhu et al. [99]
	Nano-tubes (L = 20.0 μm, T = 24 nm, D = 132 nm)	6.89	Shankar et al. [100]
	Nano-tubes (L = 5.0 μm, T = 20 nm, D = 100 nm)	4.80	Bwana et al. [101]

4.1.2 Zinc Oxide Nanostructures (ZnO)

The semiconductor is built on ZnO nanostructures with a 3.37 eV gap and a comprehensive energy band gap. It has been studied for a long time because of its interesting optoelectronic properties and well-known polymorphism. It is hard to explain how this exciting and versatile material is made, its shapes, and what it could be used for. As was already said, the focus is on DSSC anodes with what are said to be new ZnO structures, and ZnO photoanodes for XSCs have come back in the last three years. Several articles aim to improve ZnO-based DSSCs, and hierarchical ZnO structures are commonly used. Particle assemblies are currently a superb option for high-performance ZnO-DSSCs [102]. Another important thing is how well the excited dye sends electrons into the ZnO conduction band. Anta and his team chose this as their top priority, which means that this critical process has been studied, and this observation is actual and needs to be looked at immediately. In 2001, Hagfeldt published a fascinating study about how electrons move into and out of N719-ZnO nanostructures. Previous research has shown varied findings, and Hagfeldt's work is particularly intriguing. The ultrafast electron injection (300 fs) and recombination (injected electrons and oxidized dye) equal to TiO_2 systems were shown using femto-nanosecond spectroscopy. These results indicate that ZnO-based DSSCs with titania-like performance may be optimized for cell shape and manufacturing. The TiO_2-based electrodes' dye absorption is better. However, ZnO's acidic dye-induced environment needs ruthenium compounds from the start, and dye aggregation in ZnO porous films lowers injection efficiency and IPCE by blocking photon absorption by dye linked to the metal oxide. Solar light conversion efficiency needs to be improved through DSSC research, and dye loading time is significant for devices that use the common N710 dye to get the most photocurrent [103]. As photoanodes, most DSSC articles use different shapes of ZnO, and many micro- and nano-ZnO structures have been studied. Anta and his colleagues carefully looked at papers published before 2011, and in 2012 they showed problems with the DSSC ZnO anode. Photochemists found that ZnO could eat dye and get an irreversible charge from the dye molecules if the electrode had an electrochemical gradient [104]. Due to enhanced electron transport through single-crystal electrodes, 1D designs are also prevalent. Due to problems with dye absorption that make light harvesting and photocurrent less effective, 1D ZnO nanostructure photoanodes are only sometimes found in scientific papers. In 2011, Pauporté and his team used electrochemical deposition without a template to make ZnO nanowires and nanoparticles that could be used as anodes in DSSCs with D149 indoline fluorescent dye added to make them sensitive. A hybrid network of ZnO NWs loaded with ZnO NPs has a PCE of 4.08%, and it is exciting to compare the two similar structures in this research. The h2-ZnO photoanode had the same structure as the h1-ZnO photocathode, except there were ZnO NPs between the nanowires. Photoanodes with equal thicknesses behaved differently due to an increase in PCE caused by increases in Jsc (5.4–9.04 mA cm^{-2}) and FF (60–66%). As anticipated, increasing surface area and dye loading increased photogenerated charges, enhancing J_{SC}. Nevertheless, device R shunt modification may improve FF, and nanoparticles increase electron lifetime (20–23 ms), suggesting their significance in charge transport channels. The same research group described a co-sensitization technique for ZnO photoanodes with excellent PCE.

4.1.3 Tin Dioxide Nanostructures (SnO_2)

Tin dioxide (SnO_2) is one of several binary metal oxides that can be used as a semiconducting mesoporous photoanode in DSSCs because it has essential properties. In photovoltaic panels, hierarchical nanomorphology and nanostructures composed of single-crystalline 1D building blocks such as SnO_2 nanowires (NWs), nanorods (NRs), and nanotubes (NTs) are employed to boost current density by providing direct paths for photo-generated electrons [105]. These materials' energy bandgap (3.8 eV) is more significant than TiO_2's (3.2 eV), and SnO_2-coated thin films are typically quite transparent (>80% transmission). The CB of SnO_2 is 0.4 eV lower than TiO_2 and is suitable for low-LUMO sensitizers. The cassiterite rutile phase is the most stable of the SnO_2 polymorphs. Its conducting band edge (CB) is positioned appropriately for PV solar cells, and cassiterite is the sole SnO_2 polymorph utilized in PV-solar cells; thus, we will address it [106]. In this study, the shift in Voc was related to the distinction between the two materials' bulk Fermi values. The best findings were obtained when nanowires (NWs) were coated with nanoparticles (NPs) (PCE = 2.1%). Electrodes comprised of NWs that were branched and interconnected performed better. Voc values for DSSC devices with SnO_2 NPs range from 250 to 400 mV, which makes the devices less efficient. Other shapes for nanostructures have been thought of as a solution. Due to their smaller surface area, NW-based cells showed a higher Voc (0.52 vs. 0.35 V) but a lower Jsc (5.7 vs. 11 mA cm^{-2}) [107]. The nanoflower-morphology established mesoporous layer had an excellent surface area of 40 m^2 g^{-1}, double that of the nanofiber-based sheet (19 m^2 g^{-1}). Nanostructures made ranged from NWs to multiparous nanofibers topology. The most recent electrodes produced had the same surface area as profitable TiO_2 particles P25 (78 and 100 m^2 g^{-1}). These electrodes possessed the greatest J_{SC} reported for SnO_2-based DSSC (18 mA cm^{-2}); however, the PCE (4.3% of the solar cells) remained constrained by a low Voc (0.51 V) owing to the oxide's low CB energy level. J_{SC}'s active surface area and scattering effect increase with the number of shells used (from 12.54 mA/cm^2 with one shell to 17.62 mA/cm^2 with five shells). For both types of cells, the CB edge moved by 200 mV, increasing the open-circuit voltage. V_{OC} for liquid and solid electrolyte DSSCs rises from 0.45 to 0.67 V because electron lifetimes improve by 2–10%.

4.1.4 Niobium Pentoxide Nanostructures (Nb_2O_5)

Nb_2O_5 is an n-type transition-metal oxide semiconductor with a bandgap of 3.2–4.0 eV, and the conduction band edge of this metal oxide is −4 eV higher than TiO_2 [108]. Theoretically, this compound could get a higher Voc and photo conversion efficiency. From a structural perspective, Nb_2O_5 is made of several polymorphs or topologies. Pseudohexagonal H–Nb_2O_5, orthorhombic O–Nb_2O_5, tetragonal T–Nb_2O_5, and monoclinic M–Nb_2O_5 are the most prevalent [109]. The H phase, which is the least stable and efficient, may change into the M phase when heat is applied. Later Nb_2O_5 was used to create DSSC devices, with most published papers ignoring the crystallinity of Nb_2O_5 polymorph topologies. While making Nb_2O_5 films with varying degrees of transparency and roughness, the pH and medium are altered. Using a complex ruthenium dye to sensitize the film, they observed a photovoltaic effect with a Voc of 0.4 V and an IPCE of 40%. These Nb_2O_5 mesoporous thin films were sensitized with a ruthenium-based dye until 1999 when such current density and FF photoconversion performance of the corresponding DSSCs reached 2.2% (Jsc = 4.9 mA cm^{-2}, Voc = 0.65 V, FF = 66%). The performance of Nb_2O_5 electrodes is constrained by the difficulties of generating a high-surface area. Hence, several research institutes have produced unique nano-structural morphologies to improve this quality. Varying the sintering temperature enabled us to modify the effective surface area and pore volume of the hierarchical thin-porous films (19 $m^2\ g^{-1}$) at 400 °C (3.5–16.8 nm, 26 $m^2\ g^{-1}$) and at 500 °C (6 nm, 25 $m^2\ g^{-1}$). Even though the scientists found that these nanobelts work better than the standard Nb_2O_5 NPs they used to compare, the photoconversion efficiency was still only 1.42% [58]. Ou et al. [59] made nano-porous networks with holes that crossed each other. Placing these networks vertically makes electrons live longer and move faster. When these improvements are added to a higher V_{OC}, the photoconversion efficiency rises to 4.10%. The same research team demonstrated that choosing anodization electrolytes might alter the performance of photovoltaics [110]. In 2012, Zhang et al. utilized a material with a significant active surface area of 104 $m^2\ g^{-1}$, comparable to typical TiO_2 nanoparticles. Consequently, the J_{SC} of the PV-solar cells from these fabricated materials achieved 15.0 mA cm^2, and the PCE was near 6.8% [111]. In 2016, highly ordered single-crystalline nanowires W arrays were used to build superstructures that increased the active high-surface area to 121.5 $m^2\ g^{-1}$. A layer of NWs develops over the FTO substrate and functions as a blocking layer. Due to their nanoscale size, NWs can function as a thin light-scattering layer. The photoconversion's effectiveness was lower (6.4%) than in their earlier study [112].

4.1.5 Indium Oxide Nanostructures (In_2O_3)

In_2O_3 is a high bandgap material, having direct and indirect bandgaps of 3.6 and 2.6 eV, respectively. In thin film form, In_2O_3 is transparent in the visible spectrum. This oxide's 80-atom unit cell makes it hard to determine how it conducts electricity. However, the fact that In_2O_3 is hydrophilic makes the solid–liquid interface bigger and makes it a better transport layer. In_2O_3-based DSSCs exhibit worse PCE performance compared to TiO_2-based DSSCs. Sayama et al. [113] made In_2O_3 nanomaterials as nanoparticle morphologies with a 100–500 nm diameter, decreasing the electrode's surface area. The performance was inadequate because of a lower dye loading than other oxides [113]. To determine how injection works in In_2O_3-based photoanodes, Furube et al. and Guo et al. examined the injection and recombination kinetics at the fluorescent dye-metal oxide interface. In_2O_3, unlike TiO_2, has two electron injections, one that happens quickly (<250 fs) and one that happens slowly (8.7 ps). The sp-type of In_2O_3 CBs has a 200-fold lower density of states than the d-type TiO_2 CBs [114]. Mori et al. later showed that the low FF often found with In_2O_3-based photoanodes are caused by current transients that depend on the potential. Also, they showed that indium oxide's electrons live longer, giving it a better CB value than titanium oxide (compared to NHE) [115]. Sharma et al. used electrodeposition, instead of doctor blades or screen printing, to make mesoporous films. For 250 s at 65 °C, indium chloride with a citric acid precursor was deposited on the ITO substrate at three mA cm^{-2}. Air annealing for one h at 300 °C produced a mesh-like nanosphere film of 330 nm diameter of 20–30 nm aggregated spheres.

Hierarchically built nanostructure-based DSSCs have a 0.5% efficiency because the FF and JSC are minimal (40% and 0.74 mA cm^{-2}, respectively) [116]. The higher CB of In_2O_3 compared to TiO_2 (compared to NHE) lowers Voc, so Kong et al. made lanthanide-embedded indium-oxide to fix this problem. The inverse opal structural morphologies of the mesoporous thin film made it better at absorbing light, and structures doped with thulium and ytterbium had Voc rates of 0.42 and 0.46 V, respectively, compared to 0.35 V for undoped In_2O_3. This study has the best power conversion efficiency for indium oxide photoanodes at 0.96 [117].

4.1.6 Tungsten Oxide Nanostructures (WO_3)

The bandgap of tungsten (VI) oxide (WO_3) is between 2.5 and −2.8 eV at room temperature (RT), and it is a semiconducting transition-metal oxide with a bandgap that can be tuned. They can make these metal oxide nanoparticles using solution-based colloidal, electrochemical, or chemical vapor deposition methods. Hydrothermal growth in alkaline metal sulfates makes oriented WO_3 NWs like other metal oxides, and in 1998, DSSC photoanodes made from WO_3 became customary [118]. But before 2010, there was no record of any new work, probably because it needed to be more remarkably effective. Zheng et al. found that the WO_3 surface's acidity hampers dye grafting. Photoanodes were made from WO_3 NPs with orthorhombic and monoclinic phases. A ruthenium-based dye was used to make the photoanodes sensitized, and 12 μm is the best thickness for solar cells with a PCE of 0.75%. WO_3's higher CB edge level (vs. NHE) than TiO_2 reduced PV-solar cell V_{oc} due to poor dye coverage on the J_{SC} and FF also affects recombination. The film was treated with $TiCl_4$, such as TiO_2, SnO_2, or ZnO, to fix this problem. V_{OC} and J_{SC} increased, resulting in a 1.46% productivity boost [61]. Yong et al. also used spherical particles of WO_3 to form nanorods after 30 min of heat treatment in the air; $W_{18}O_{49}$ NRs yielded monoclinic phase WO_3 NRs at 500 °C. Despite the oxide nanostructure modification, NPs performed similarly [65]. Rashad et al. recently used WO_3 hierarchical nanostructures to increase DSSC performance. Hydrothermal synthesis produced a pure orthorhombic-phase flower-like structure. A wider bandgap increases Voc to 0.65 V without annealing, which improves performance. This nanostructure reduces recombination with the redox pair in the electrolyte, which improves FF by up to 70% for WO_3 that has been annealed at 300 °C. The best WO_3-based DSSC photoconversion efficiency was 3.2% [119].

5 Ternary Oxides

As was said in the previous paragraphs, most research on replacing TiO_2 in DSSCs has been done on binary oxides. On the other hand, multi-cation oxides have yet to be studied as much, but this approach has become more popular in the last few decades [120]. Compared with ordinary binary oxides, multi-cation oxides give additional opportunities to modify the materials' chemical and physical characteristics by varying their compositions. $SrTiO_3$ and Zn_2SnO_4 are the most investigated ternary oxides for DSSCs. Changing the relative element ratio is one of their best features because it lets the work function, bandgap energy, and electrical resistivity vary. DSSC applications of multi-cation- oxides are intriguing, given their abundant availability [121].

5.1 Zinc Tin Oxide Nanostructures ($ZnSnO_3$)

Zinc stannate has two oxides, $ZnSnO_3$ and Zn_2SnO_4, with different Zn-to-Sn ratios and crystallographic structures [122]. The crystal structure of Zn_2SnO_4 is cubic spinel, while that of $ZnSnO_3$ is orthorhombic and unstable at high temperatures. The material studied the most is Zn_2SnO_4, which has a great bandgap of 3.3–3.8 eV and a straight forbidden transition [123]. DSSCs may benefit from this metal oxide material in several ways: first, its CB energy level position is equal to −4.1 eV, which also is further optimistic than TiO_2 anatase; furthermore, zinc-tin oxide has much higher conduction capabilities than TiO_2, with expected electron mobility of 10–15 $cm^2\ V^{-1}\ s^{-1}$ [124]. Another essential benefit of this oxide is that it is more durable and has better corrosion resistance than ZnO because it is stable in a more comprehensive pH range, from 3.15 to 14. Finally, various techniques can quickly make this oxide, including spray pyrolysis, sputtering, and hydrothermal synthesis. Reports say that NPs, NWs, and nanobelts, among other shapes, could be used in photovoltaics [125]. In the first work, since the Hagfeldt group showed their first work in 2007 on Zn_2SnO_4 as a DSSC photoanode, N719 was used to make hydrothermal zinc stannate NPs with a 27 nm diameter that was sensitized. Even though Zn_2SnO_4's higher positive CB edge level gave it a better Voc, the PV-solar cells' PCE was near 1%, drastically less than TiO_2 NPs'. Both oxides behaved differently due to reduced injection efficiency and current density. Several teams investigated the effect of Zn_2SnO_4 NP morphology on electrode preparation in 2012. Chen and his team made mesoporous thin films with NWs instead of spherical NPs. The NWs were made through chemical vapor deposition [126]. Finding the right conditions for making Zn_2SnO_4 nanostructures so that the SnO_2 phase does not form hard. After seeing the best reagent ratio, growth temperature, and oxygen flow rate, periodic rhombohedral Zn_2SnO_4 NWs were made and sensitized with N719 to make PV-solar cells with a PCE of 2.8%, noting that the Voc is high at 0.74 V, NW-based Zn_2SnO_4 photoanodes have rapid direct transport mechanisms in the crystalline morphologies and a delayed percolation transport process that involves NW active-surface states. This study found that NW transport properties could be improved by passivation [127]. Nanostructure engineering and hierarchical Zn_2SnO_4 electrodes have kept pace since 2012, and electrodes are made using electro-spinning or template-assisted hydrothermal synthesis. Choi et al. electro-spun $Zn(OAc)_2$ and $Sn(OAc)_4$ as building scaffolds in PVAc to create a Zn_2SnO_4 photoanode and improve fiber adhesion to FTO. 450 °C calcination produces amorphous and porous nanofibers with NPs of less than

5 nm. The nanofiber network had 124 m^2 g^{-1}, like TiO_2 NP electrodes, and DSSCs with only organic dyes on nanofiber-based electrodes had 3.7% PCE [128]. Wang et al. hydrothermally synthesized hierarchical Zn_2SnO_4 NSs with NH_4F to increase electrode-specific area, and NSs-based working electrodes had 108.1 m^2 g^{-1}. These electrodes surpass the reference system, which contains these NPs, according to electrochemical and optical measurements regarding electron transport rate, rate of recombination, and light scattering characteristics. These parameters increased PCE by 4.8 and 5.4% after $TiCl_4$ curing. The same research group developed the template-assisted formulation of hierarchical macro-porous nanostructures made of Zn_2SnO_4 NPs to make electrodes with precise surfaces near 100 m^2 g^{-1}. By adjusting the size of the electrode particle, PCE may reach 6.1% [129]. In 2015, hierarchically structured Zn_2SnO_4 NPs-in-beads were found, and 20 nm NPs were embedded in submicron-sized beads by electro-spraying a colloidal suspension of amphiphilic Zn_2SnO_4 NPs. The Zn_2SnO_4 beads' exceptional nano-structural-topology improved dye adsorption when sensitized with organic dyes. Organic dyes improved charge recombination lifetime, molar absorbability, light scattering from electrodes, electrolyte penetration, and molar absorbability. Optimized DSSCs had a 6.3% PCE, a Voc of 0.71 V, a Jsc of 12.2 mA/cm^2, and an FF of 0.72; this is the ternary metal oxide electrodes' highest PCE [130].

5.2 *Strontium Titanium Oxide Nanostructures ($SrTiO_3$)*

Strontium titanate has a more noticeable conduction band edge than TiO_2, like zinc-stannate. In 1999, Burnside et al. reported the initial DSSC use of mesoporous $SrTiO_3$ [131]. Nano-crystalline $SrTiO_3$ with perovskite structural morphologies are produced by treating nano-crystalline TiO_2 and strontium hydroxide with water at elevated temperatures. This work's photoanode has a surface area of 85 m^2 g^{-1}, with a bi-modal dispersion of tiny sphere-shaped particles (10–15 nm) and larger cubic-three-dimensional particles (40–60 nm). The electrode can absorb dye and scatter light well with this distribution. If their Voc was more significant than ordered mesoporous TiO_2-constructed DSSC (0.79 and 0.69 V), their efficacy was restricted to 1.8% (against 6.0% for the reference devices) caused by the minimal current and dye absorption at the electrode. The longer Ti–Ti gap in $SrTiO_3$ compared to TiO_2 may make it harder for the dye to stick to the surface [132]. They got 620 nm spherical particles with sponge-like mesoporous shapes with 42 nm pores, and photoconversion remained low at 0.51% due to cell efficiency. Hao's group doped $SrTiO_3$ with holmium or samarium cations to create a cascade-free charge carriers' injection. Despite the improvement, photoconversion efficiency was still low, and DSSCs needed a higher CB to increase Voc. Even though N719-sensitized $SrTiO_3$ photoanodes show injection from higher states, a dye's LUMO level might not be enough to send enough energy [133]. Yang and his team changed the number of electrolytes to change $SrTiO_3$'s flat-band properties. The flat band potential increased when tetra-butylammonium perchloride and acetylacetone were added to the propionitrile-based electrolyte. This increased the photocurrent by making it easier for electrons to move from the dye (N3) to the $SrTiO_3$, and Voc was also affected [134]. Although this method improves performance, it loses the main benefit of perovskite-type polycrystalline $SrTiO_3$, $CaTiO_3$, and $BaTiO_3$, which are ternary oxides based on semiconductors [135]. DSSCs with $SrTiO_3$ photoanodes have a more excellent Voc than DSSCs with TiO_2 photoanodes Voc. Nevertheless, photocurrents are reduced due to insufficient dye loading and an energy mismatch at the Fermi level of the photoanode. Due to the inappropriate NP shape, the same group tried $MgTiO_3$, $MgTi_2O_5$, and Mg_2TiO_4; these oxides performed less than 1% [136].

5.3 *Barium Stannate Oxide Nanostructures ($BaSnO_3$)*

The scientists reported a 3.1 eV gap comparable to TiO_2 but with a conduction band edge (CB) that moved by 0.9 eV and shifted a more positive value. Thus, the irregular distribution of particle sizes resulted in low Jsc and FF, but the position of the CB edge influenced Voc., the photoconversion effectiveness was 1.42%. Guo and his colleagues created barium stannate ($BaSnO_3$)'s perovskite structure using hydrothermal, co-precipitation, and solid-state techniques. During manufacture, the size and surface area of the particles fluctuate, resulting in varying quantities of dye. The photovoltaic characteristics of PV-DSSCs with a co-precipitated $BaSnO_3$ photoanode (with the greatest dye absorption) achieved 1.1% of PCE [137]. Li and his colleagues started with polycrystalline $BaSnO_3$ and used a solid-state reaction to make porous $BaSnO_3$ hollow structures with a $BaCO_3$ core and a SnO_2 shell [138]. Unfortunately, DSSCs with open designs were inefficient, with a PCE of 0.12%. Using $BaSnO_3$ electrodes, Shin et al. considerably increased the effectiveness of photoconversion by 4.5%. $TiCl_4$ curing, known to reduce recombination and increase charge absorption, also made photoconversion 6.2% more efficient [139].

6 Composites

Nanocomposites made from metal oxide semiconductors have high specific surface areas of ordered mesoporous nanoparticle thin films, the fast electron transport of crystalline nanowires, and their ability to scatter light. At all photoanode thicknesses, NWs outperform NPs owing to an optimal ratio [140] The finest NP-only photoanode had a PCE of 6.8%, while the most fantastic device got a PCE of 8.6%. The optimal device employs a 13 μm photoanode, whereas the NP film employs a ten μm photoanode. As expected, single-crystal nanowires make it easier for electrons to move through a composite network by making it longer for electrons to move through the web. By putting together NPs and NWs that were the same as the photoanode, other research groups were able to confirm the results. Using nanocomposites to alter the band arrangement inside the photoelectrode to increase its charge-collecting characteristics and prevent recombination is an inventive method for enhancing photoanode performance. Although pure TiO_2 has certain favorable characteristics for the ideal PE, the efficiency of DSSCs, including TiO_2, is inadequate for high-performance operations. In contrast, synthesizing diverse TiO_2 morphologies needs a complicated and intricate procedure. Low and Lai [139] found that the length, width, and thickness of nanotubes and nanorods depended on how the reaction went. So, nanocomposites made of a carbon-based substrate and TiO_2 could create highly efficient DSSCs without complicated manufacturing processes [141]. Nanoparticles can be shaped like rods or wires but have a "core-shell" structure. Recent research by Karlsson [140] shows that electrons can be kept in the SnO_2 nucleus of a core-shell SnO_2–TiO_2 nanoparticle configuration. The electron confinement showed that the PV-solar cell could keep its photovoltage for more than 500 s, even in complete darkness. The outcome was a rise in the short-circuit current, electron lifetime, and free-circuit ability. This increased the best device's ability to convert light into electricity from 0.7 to 3.5% [142]. The ability to quickly fabricate parallax-oriented ZnO microrod morphologies and distribute TiO_2 nanoparticle topologies within the low-density array led to a composite thin film in which the TiO_2 nanoparticles may behave as a concretely large, highly surface porous film. On the other hand, ZnO micro-rods may collect electrons to decrease charge recombination rapidly. Electrochemical impedance spectra revealed increased electron lifetime, effective diffusion length, and electron collection efficiency. The hybrid system minimized charge recombination compared to the typical mesoporous TiO_2 nanoparticle layer. Due to this, photoconversion performance might increase from 6.6 to 7.8%. This demonstrates that rapid mobility in ZnO crystalline forms may enhance PCE [143]. To discover what makes ZnO and TiO_2 different in DSSCs, Gratzel and his colleagues looked at how electrons moved through thin ZnO and TiO_2 made by atomic layer implantation on a porous, hierarchical insulating framework. Doing so addressed the form differences that made comparing ZnO and TiO_2 photoanodes difficult in prior studies. Under ideal circumstances, ZnO and TiO_2 have similar photovoltaic (PV) efficiencies. However, their capacity to transfer electrons differs significantly. TiO_2 has a lower recombination process, a larger dye loading, and the ability to inject electrons rapidly than ZnO [144]. Recent research has found an elegant way to turn rare earth metal submicron particles into multi-shell structures. The TiO_2 nanoparticles of size 20 nm absorb standard N719 dye. The upconverting core changes near-infrared light into visible light, which lets it absorb the N719 dye's upconverted radiation. As a passivating film, the SiO_2 layer stops the rare-earth crystal from being able to trap electrons. This solves the problem of making the core and TiO_2 shell work together electrically while keeping the rare earth metal's ability to change energy. It was possible to get a PCE as superior as 8.65%, which is better than the PCE of the reference cell, which was only 7.80% without the upconverting multishell structures [145]. To get the most out of the effect, the position of where the down-converter emits and where the light harvester takes in the light must be well matched. People have thought of many ways to set up the cell, like putting the front surface layer of the PV cell or putting nano-structures that change the charge in the opposite direction inside the TiO_2 anode. Before the UV rays are taken away from the front glass, the main benefit of the front film is that it changes them into less harmful forms. Hosseini et al. [144] say that the biggest problem is that the light produced in an isotropic way spreads out. This light can be collected better if the down-converting particles are close to the dye in the photoanode. In all designs, the IPCE was improved by the down-converting layer, which can take in transmitted light and send it out at around 500 nm, close to where the direct dye in the DSSC can absorb the lightest [146].

7 P-Type Semiconductors

Most perovskite and sensitized solar cells have electrodes made of TiO_2 and other n-type electron acceptors. The second generation of these PV cells is built on an inverted design that uses p-type materials as the framework for moving holes. For tandem-sensitized PV cells to work, the way the photo absorber sends holes into the PV cell must be fundamentally different. The first instance of a functioning p-type cell was described in 1999, and it only functioned 0.0076% of the time. Since then, much progress has been made, but p-type DSSCs have received far less attention, and their efficiencies continue to lag those of typical n-type PV cells. The fact that p-type materials may perform better is one of their biggest problems. Compared to standard n-type oxides like TiO_2 and ZnO, substances with p-type conductivity, a broad gap, high conductivity, and resistance to photo-corrosion are considered rare. NiO is made in hundreds of tons annually, and the transition metal oxide has been studied the most. It is the most used and investigated of these compounds. Most of the nickel oxide is used to make nickel salts. Nickel oxide can also be used as a flexible hydrogenation catalyst. Electrochromic devices, memory, and, more recently, supercapacitors are further utilized for NiO [147]. Nickel oxide is often nonstoichiometric, just like other binary metal oxides. Stoichiometric NiO is green, but nonstoichiometric NiO is usually black in color. Its electrical characteristics still need to be determined since their levels can change depending on how it was made and measured. Consequently, the high bandgap of NiO is 3.4–4.3 eV (with 3.6 eV presenting the most typical value), but the optimum of its valence band is between 4.2 and 5.8 eV. The conductivity of NiO ranges between 10^{-12} and 10^{-15} S cm^{-1}, and its Fermi level is often believed to be 0.4 eV or above the valence band [148]. $Ni(OH)_2$ was synthesized using nickel salts and additives as precursors. $Ni(OH)_2$ films were created in one or more steps using hydrothermal and solvothermal techniques. Organic (co-polymers) or inorganic (granules) templates may improve film porosity and homogeneity. Lastly, mesoporous films can be made by making NiO NPs outside the film and then spin-coating or screen-printing them onto the film. This is followed by a sintering phase, like how TiO_2 is usually deposited. NiO's structure, chemical state, and doping fluctuate throughout manufacturing, which affects its optoelectronic capabilities [149]. Due to the variable quantities of voltage in the gap and the metallic NiO, the hue and opacity of the NiO-coated film vary based on the manufacturing process. Traditionally, the first experimental studies employing p-type metal oxides focused on DSSCs. NiO is a common material for sensitive

PV cells, but it has two major problems: a high rate of geminate recombination and limited hole mobility. When a charge is injected from a dye's highest occupied molecular orbital (HOMO) into the valence band of nickel oxide, the injected hole may mix with excited electrons in the reduced dye's lowest unoccupied molecular orbital (LUMO). NiO's low hole mobility worsens recombination by sticking injected holes to the oxide interface. This makes it more likely that fewer dye molecules will interact with them on their way to the collecting electrode. Most research has been done to find ways to fix these weak spots and make new, healthy cells. Standard techniques to reduce the negative impacts of recombination include.

- While spatially separating the excited dye's electron and hole,
- Adjusting the redox electrolyte.
- Accelerating the injected hole's delivery to the collecting electrode by increasing the NiO's mobility.

So, cobalt poly-pyridyl complexes are good alternatives to redox electrolytes because they absorb less light and have a much higher redox potential than iodide pairs. NiO electrodes and similar complexes were used to raise Voc to 0.35 V. Other electrolytes raised Voc even more. Diaminoethane-cobalt organometallic complexes in NiO-based p-DSSCs produced a remarkable 0.71 V potential. The phosphonic and carboxylic acid attachment structural characteristics give the highest open-circuit voltage for dyes bound to mesoporous NiO through monodentate binding mode. This means that new ways to make dyes bind in this way could improve cell performance [150]. As previously stated, NiO is an electrochromic substance that absorbs photons without producing electricity. This lowers the sensitized electrode's light-gathering efficiency and the solar cell's performance. (i) change the nanostructure of mesoporous NiO to make it less dense and better at absorbing light, or (ii) make dyes with a higher absorption coefficient [151].

8 Future Trends in Technological Development

Since the cost of making DSSCs is about the same as making other thin-film photovoltaic technology, they are almost ready to be sold to the public. In 2014, the international demand for DSSC technology was estimated at around $49.6 million, and between 2015 and 2022, it is anticipated to rise at a CAGR of more than 12 percent Module/PV cell production prices and related economics influence the DSSC market [152]. Kalowekamo et al. looked at how much it costs to make DSSC and compared it to how much it costs to make the two previous versions of solar cells before it [153]. DSSC is anticipated to cost between $0.50 and $1.00/W, with a performance of 5–15% [18]. The DSSC is more cost-effective than photovoltaic organic solar cells, Si solar cells, and perovskite PV solar cells. Australia, Israel, Japan, Sweden, Switzerland, the United Kingdom, and the United States are among the countries that have worked on this solar technology. These brands include Dynamo, Dyesol, G24 Power, Konarka, SHARP, Solaronix, and 3G Solar. The first time DSSC

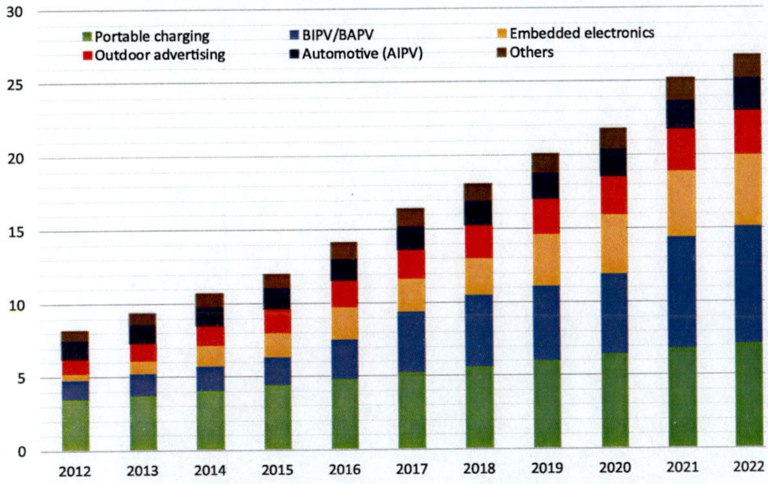

Fig. 9 The prevalence rate and advancement of DSSC technology

devices were used for business, they were attached to backpacks as portable solar chargers. In 2014, research on the market showed that portable charging devices were the most popular type of application. They made up 33.0% of the industry's revenue share and are expected to grow. Campillo and Foster [154] say that it gives off the most incredible energy in light indoor settings, which makes it suitable for portable devices [154]. The recent invention of a solar-powered, battery-free portable keyboard reduces expenses and time. But the main goal of this technology will be to build PV into building systems (also called building-integrated PV or construction-applied PV). The value of building-integrated PV and building-applied PV will be more than $30 million [155]. It is projected that the need for energy consumption in residential areas, such as solar panels on roofs and water heating, would boost the expansion of the DSSC business [155]. DSSC technology is anticipated to expand in building-integrated or architecture photovoltaics, automotive, portable, or compact charging, embedded electronics, etc (Fig. 9). Figure 9 shows that DSSC has been chosen as one of the most useful applications for the next few years, even though it still needs to be developed for industrial or commercial use [155, 156].

9 Conclusions

PV devices utilize solar energy for electricity with no negative environmental consequences. Many studies on DSSC were conducted in the 1990s because changing power is scientifically interesting and important in science and engineering. Even

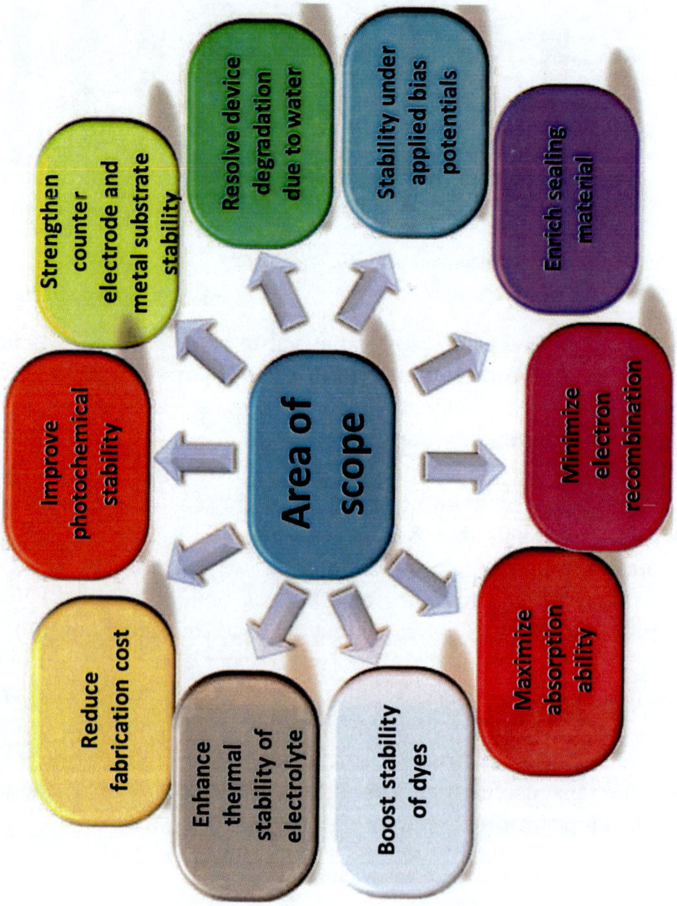

Fig. 10 Future perspective for efficient DSSC

though it does not work as well as other solar cells, DSSC is a desirable alternative because it can adapt to different lighting conditions, be used with a flexible substrate, and come in assorted colors. In DSSCs, n-type semiconductors are employed as photoanodes or working electrodes, while p-type semiconductors are utilized as photoactive cathodes or counter electrodes, according to this chapter. In n-type semiconductors, improving particle size improves performance significantly. In their ground-breaking work on mesoscopic PV solar cells in 1991, O'Regan and Gratzel used different dyes, electrolytes, and device topologies that made the cells much more efficient and stable. Sensitized cells are nearing commercialization and are a viable strategy for constructing PV modules (BIPV), and TiO_2 is the most used and researched electrode for this sort of cell. In addition, other wide-band-gap metal oxide semiconductors offer inherent benefits.

Due to how easy and flexible it is to make photoanodes, using them can make a substantial difference in how well and stable mesoscopic solar cells work. This is due to the flexibility of materials' components or chemical compositions and optoelectronic properties. This chapter looked at the different morphologies of binary and ternary metal oxide nanomaterials that could be used as photoanodes in DSSCs and, by extension, PV solar cells. According to reports, semiconductors can make effective photoanodes due to their band gap, morphology, composition, and doping. Several research institutes and groups worked together to explain nanostructures and their shapes (0D, 1D, 2D, and 3D of ordered mesoporous nanomaterials) and how doping affects the performance of DSSCs (mostly TiO_2 and ZnO). The development of manufacturing and nano structuration techniques to regulate their composition and shape is a rich research topic with the potential to advance the efficiency, processability, and cost of solar energy conversion systems. Also, research on metal oxides could help store energy, make sensors, and produce highly sustainable microelectronic systems. Many tasks need to require more concentration and effort (Fig. 10). Other methods, such as core-shell layer-by-layer construction in MOS, up-and-down switchover materials, thin-layer passivation, and large surface area manipulation, are diverse ways to increase absorption capacity and slow recombination. New materials could quickly soak up near-infrared light, improve photothermal stability, and slow down dye decor. Using stabilizing chemicals, they investigate and use less volatile, quasi-state, and solid-state electrolytes. As DSSC technology has advanced, their study on interfacial contact has received less attention. As a result, interfacial interactions may improve the performance of DSSC. Some DSSC manufacturing techniques, such as screen printing, electrophoretic deposition, and roll-to-roll forming, might be expedited to manufacture more items. For optimal encapsulation, it may be necessary to investigate more effective materials and techniques.

Acknowledgements Siddan Gouthaman (SG) and K. R. Justin Thomas (KRJT) thank the Indian Institute of Technology Roorkee (IITR), for providing funding through an Institute Post-doctoral fellowship (IPDF) to SG and infrastructure facilities.

References

1. B. Diczfalusy, What does a post-peak carbon economy look like? in *Carbon Governance, Climate Change, and Business Transformation* (Routledge, 2014), pp. 12–26
2. R.H.E. Hassanien, M. Li, F. Yin, The integration of semi-transparent photovoltaics on greenhouse roof for energy and plant production. Renew. Energy **121**, 377–388 (2018)
3. N.S. Lewis, D.G. Nocera, Powering the planet: chemical challenges in solar energy utilization. Proc. Natl. Acad. Sci. **103**, 15729–15735 (2006)
4. P. Frankl, S. Nowak, M. Gutschner et al., Technology roadmap: solar photovoltaic energy. Int. Energy Assoc. (2010)
5. M.-E. Ragoussi, T. Torres, New generation solar cells: concepts, trends, and perspectives. Chem. Commun. **51**, 3957–3972 (2015)
6. H. Wu, G. Xie, Z. Jie et al., Research progress about chemical energy storage of solar energy. IOP Conf. Ser.: Earth Environ. Sci. (IOP Publishing, 2018), p. 052070
7. M.E. Becquerel, Mémoire sur les effets électriques produits sous l'influence des rayons solaires. C R Hebd Seances Acad. Sci. **9**, 561–567 (1839)
8. W.G. Adams, R.E. Day, V. The action of light on selenium. Proc. R. Soc. Lond. **25**, 113–117 (1877)
9. M. Hosenuzzaman, N.A. Rahim, J. Selvaraj et al., Global prospects, progress, policies, and environmental impact of solar photovoltaic power generation. Renew. Sustain. Energy Rev. **41**, 284–297 (2015)
10. A. Shahsavari, M. Akbari, Solar energy potential in developing countries for reducing energy-related emissions. Renew. Sustain. Energy Rev. **90**, 275–291 (2018)
11. J. Zhao, A. Wang, M.A. Green, 24.5% efficiency PERT silicon solar cells on SEH MCZ substrates and cell performance on other SEH CZ and FZ substrates. Sol. Energy Mater. Sol. Cells **66**, 27–36 (2001)
12. A. Agrawal, S.A. Siddiqui, A. Soni, G.D. Sharma, Recent development in perovskite solar cell based on planar structures, in *Intelligent Computing Techniques for Smart Energy Systems: Proceedings of ICTSES 2018* (Springer, (2020), pp. 1039–1046
13. T.D. Lee, A.U. Ebong, A review of thin film solar cell technologies and challenges. Renew. Sustain. Energy Rev. **70**, 1286–1297 (2017)
14. N. Chadwick, D.K. Kumar, A. Ivaturi et al., Isomer dependence of efficiency and charge recombination in dye-sensitized solar cells using Ru complex dyes bearing halogen substituents. Eur. J. Inorg. Chem. **2015**, 4878–4884 (2015)
15. E. Akman, Y. Altintas, M. Gulen et al., Improving performance and stability in the quantum dot-sensitized solar cell through single-layer graphene/Cu_2S nanocomposite counter electrode. Renew. Energy **145**, 2192–2200 (2020)
16. C.-P. Lee, C.-T. Li, K.-C. Ho, Use of organic materials in dye-sensitized solar cells. Mater. Today **20**, 267–283 (2017)
17. A. Fakharuddin, R. Jose, T.M. Brown et al., A perspective on producing dye-sensitized solar modules. Energy Environ. Sci. **7**, 3952–3981 (2014)
18. J. Yan, B.R. Saunders, Third-generation solar cells: a review and comparison of polymer: fullerene, hybrid polymer, and perovskite solar cells. RSC Adv. **4**, 43286–43314 (2014)
19. E.A. Gibson, *Solar Energy Capture Materials* (Royal Society of Chemistry, 2019)
20. W. Cole, B. Frew, P. Gagnon et al., Envisioning a low-cost solar future: exploring the potential impact of achieving the SunShot 2030 targets for photovoltaics. Energy **155**, 690–704 (2018)
21. M. Khan, M.A. Iqbal, M. Malik et al., Improving the efficiency of dye-sensitized solar cells based on rare-earth metal modified bismuth ferrites. Sci. Rep. **13**, 3123 (2023)
22. M.A. Green, *High-Efficiency Silicon Solar Cells*, pp. 1–240 (1987)
23. H. Tributsch, M. Calvin, Electrochemistry of excited molecules: photo-electrochemical reactions of chlorophylls. Photochem. Photobiol. **14**, 95–112 (1971)
24. H. Tsubomura, M. Matsumura, Y. Nomura, T. Amamiya, Dye-sensitized zinc oxide: aqueous electrolyte: platinum photocell. Nature **261**, 402–403 (1976)

25. B. O'regan, M. Grätzel, A low-cost, high-efficiency solar cell based on dye-sensitized colloidal TiO2 films. Nature **353**, 737–740 (1991)
26. M.K. Nazeeruddin, E. Baranoff, M. Grätzel, Dye-sensitized solar cells: a brief overview. Sol. Energy **85**, 1172–1178 (2011)
27. C. Frei, R. Whitney, H.W. Schiffer et al., World energy scenarios: composing energy futures to 2050. Conseil Francais de l'energie (2013)
28. H.E. Murdock, D. Gibb, T. André et al., Renewables 2021-global status report (2021)
29. A. Jäger-Waldau, *PV Status Report 2019* (Publications Office of the European Union, Luxembourg, 2019), pp.7–94
30. J.B. Baxter, Commercialization of dye-sensitized solar cells: present status and future research need to improve efficiency, stability, and manufacturing. J. Vac. Sci. Technol. A: Vac. Surf. Films **30**, 020801 (2012)
31. X. Yang, M. Yanagida, L. Han, Reliable evaluation of dye-sensitized solar cells. Energy Environ. Sci. **6**, 54–66 (2013)
32. S. Holmberg, A. Perebikovsky, L. Kulinsky, M. Madou, 3-D micro and nano technologies for improvements in electrochemical power devices. Micromachines (Basel) **5**, 171–203 (2014)
33. W.A. Dhafina, H. Salleh, M.Z. Daud, The application of sensitizers from red frangipani flowers and star gooseberry leaves in dye-sensitized solar cells. J. Phys.: Conf. Ser. (IOP Publishing, 2018), p. 012016
34. K. Kakiage, Y. Aoyama, T. Yano et al., Highly efficient dye-sensitized solar cells with collaborative sensitization by silyl-anchor and carboxy-anchor dyes. Chem. Commun. **51**, 15894–15897 (2015)
35. A. Hagfeldt, M. Grätzel, Molecular photovoltaics. Acc. Chem. Res. **33**, 269–277 (2000)
36. M.S. Ahmad, A.K. Pandey, N. Abd Rahim, Advancements in developing TiO_2 photoanodes and its fabrication methods for dye-sensitized solar cell (DSSC) applications. A review. Renew. Sustain. Energy Rev. **77**, 89–108 (2017)
37. B. Chen, Z. Yang, Q. Jia et al., Emerging applications of metal-organic frameworks and derivatives in solar cells: recent advances and challenges. Mater. Sci. Eng. R. Rep. **152**, 100714 (2023)
38. D. Lee, Y. Rho, F.I. Allen et al., Synthesis of hierarchical TiO_2 nanowires with densely-packed and omnidirectional branches. Nanoscale **5**, 11147–11152 (2013)
39. L. Vesce, Novel materials and processes for photovoltaic technology. Energies (Basel) **16**, 425 (2023)
40. C.-J. Lin, W.-Y. Yu, S.-H. Chien, Transparent electrodes of ordered opened-end TiO_2-nanotube arrays for highly efficient dye-sensitized solar cells. J. Mater. Chem. **20**, 1073–1077 (2010)
41. B.H. Lee, M.Y. Song, S.-Y. Jang et al., Charge transport characteristics of high efficiency dye-sensitized solar cells based on electrospun TiO_2 nanorod photoelectrodes. J. Phys. Chem. C **113**, 21453–21457 (2009)
42. E.C. Okpara, O.C. Olatunde, O.B. Wojuola, D.C. Onwudiwe, Applications of transition metal oxides and chalcogenides and their composites in water treatment: a review. Environ. Adv. **100341** (2023)
43. M. Shanmugam, R. Jacobs-Gedrim, C. Durcan, B. Yu, 2D layered insulator hexagonal boron nitride enabled surface passivation in dye-sensitized solar cells. Nanoscale **5**, 11275–11282 (2013)
44. Y.J. Kim, M.H. Lee, H.J. Kim et al., Formation of highly efficient dye-sensitized solar cells by hierarchical pore generation with nanoporous TiO_2 spheres. Adv. Mater. **21**, 3668–3673 (2009)
45. M. Saito, S. Fujihara, Large photocurrent generation in dye-sensitized ZnO solar cells. Energy Environ. Sci. **1**, 280–283 (2008)
46. D. Barpuzary, A.S. Patra, J.V. Vaghasiya et al., Highly efficient one-dimensional ZnO nanowire-based dye-sensitized solar cell using a metal-free, D – π – A-type, carbazole derivative with more than 5% power conversion. ACS Appl. Mater. Interfaces **6**, 12629–12639 (2014)

47. A.R. Rao, V. Dutta, Achievement of 4.7% conversion efficiency in ZnO dye-sensitized solar cells fabricated by spray deposition using hydrothermally synthesized nanoparticles. Nanotechnology **19**, 445712 (2008)
48. A. Sacco, A. Lamberti, R. Gazia et al., High efficiency dye-sensitized solar cells exploiting sponge-like ZnO nanostructures. Phys. Chem. Chem. Phys. **14**, 16203–16208 (2012)
49. C.-Y. Lin, Y.-H. Lai, H.-W. Chen et al., Highly efficient dye-sensitized solar cell with a ZnO nanosheet-based photoanode. Energy Environ. Sci. **4**, 3448–3455 (2011)
50. Z. Li, Y. Zhou, G. Xue et al., Fabrication of hierarchically assembled microspheres consisting of nanoporous ZnO nanosheets for high-efficiency dye-sensitized solar cells. J. Mater. Chem. **22**, 14341–14345 (2012)
51. A. Birkel, Y.-G. Lee, D. Koll et al., Highly efficient and stable dye-sensitized solar cells based on SnO_2 nanocrystals prepared by microwave-assisted synthesis. Energy Environ. Sci. **5**, 5392–5400 (2012)
52. J.-H. Lee, N.-G. Park, Y.-J. Shin, Nano-grain SnO_2 electrodes for high conversion efficiency SnO2–DSSC. Sol. Energy Mater. Sol. Cells **95**, 179–183 (2011)
53. K. Wijeratne, J. Akilavasan, M. Thelakkat, J. Bandara, Enhancing the solar cell efficiency through pristine 1-dimensional SnO_2 nanostructures: comparison of charge transport and carrier lifetime of SnO_2 particles vs. nanorods. Electrochim. Acta **72**, 192–198 (2012)
54. U.V. Desai, C. Xu, J. Wu, D. Gao, Hybrid TiO_2–SnO_2 nanotube arrays for dye-sensitized solar cells. J. Phys. Chem. C **117**, 3232–3239 (2013)
55. K. Basu, D. Benetti, H. Zhao et al., Enhanced photovoltaic properties in dye-sensitized solar cells by surface treatment of SnO_2 photoanodes. Sci. Rep. **6**, 1–10 (2016)
56. W. Wanninayake, K. Premaratne, R.M.G. Rajapakse, High efficient dye-sensitized solar cells based on synthesized SnO_2 nanoparticles. J. Nanomater. (2016)
57. P. Guo, M.A. Aegerter, RU (II) sensitized Nb_2O_5 solar cell made by the sol-gel process. Thin Solid Films **351**, 290–294 (1999)
58. J.Z. Ou, R.A. Rani, M.-H. Ham et al., Elevated temperature anodized Nb_2O_5: a photoanode material with exceptionally large photoconversion efficiencies. ACS Nano **6**, 4045–4053 (2012)
59. X. Liu, R. Yuan, Y. Liu et al., Niobium pentoxide nanotube powder for efficient dye-sensitized solar cells. New J. Chem. **40**, 6276–6280 (2016)
60. H. Zheng, Y. Tachibana, K. Kalantar-Zadeh, Dye-sensitized solar cells based on WO_3. Langmuir **26**, 19148–19152 (2010)
61. K. Hara, Z.-G. Zhao, Y. Cui et al., Nanocrystalline electrodes based on nanoporous-walled WO_3 nanotubes for organic-dye-sensitized solar cells. Langmuir **27**, 12730–12736 (2011)
62. S. Gholamrezaei, M. Salavati Niasari, M. Dadkhah, B. Sarkhosh, New modified sol–gel method for preparation $SrTiO_3$ nanostructures and their application in dye-sensitized solar cells. J. Mater. Sci.: Mater. Electron. **27**, 118–125 (2016)
63. H. Elbohy, K.M. Reza, S. Abdulkarim, Q. Qiao, Creation of oxygen vacancies to activate WO_3 for higher efficiency dye-sensitized solar cells. Sustain. Energy Fuels **2**, 403–412 (2018)
64. S.-M. Yong, T. Nikolay, B.T. Ahn, D.K. Kim, One-dimensional WO_3 nanorods as photoelectrodes for dye-sensitized solar cells. J. Alloys Compd. **547**, 113–117 (2013)
65. J. Highfield, Advances and recent trends in the heterogeneous photo (electro)-catalysis for solar fuels and chemicals. Molecules **20**, 6739–6793 (2015)
66. J. Gong, K. Sumathy, Q. Qiao, Z. Zhou, Review on dye-sensitized solar cells (DSSCs): advanced techniques and research trends. Renew. Sustain. Energy Rev. **68**, 234–246 (2017)
67. N. Abdi, Y. Abdi, Z. Alemipour, Effects of morphology and thickness of Al_2O_3 scaffold on charge transport in Perovskite-based solar cells. Sol. Energy **153**, 379–382 (2017)
68. N. Abdi, Y. Abdi, Z. Alemipour, E. NedaaeeOskoee, Chemical diffusion coefficient in dye-sensitized solar cells as a function of porosity and surface roughness. Sol. Energy **135**, 506–511 (2016)
69. F. Cao, G. Oskam, P.C. Searson, A solid-state, dye-sensitized photoelectrochemical cell. J. Phys. Chem. **99**, 17071–17073 (1995)

70. F. Babar, U. Mehmood, H. Asghar et al., Nanostructured photoanode materials and their deposition methods for efficient and economical third generation dye-sensitized solar cells: a comprehensive review. Renew. Sustain. Energy Rev. **129**, 109919 (2020)
71. L. Wang, T. Sasaki, Titanium oxide nanosheets: graphene analogs with versatile functionalities. Chem. Rev. **114**, 9455–9486 (2014)
72. A. Vomiero, V. Galstyan, A. Braga et al., Flexible dye-sensitized solar cells using TiO_2 nanotubes. Energy Environ. Sci. **4**, 3408–3413 (2011)
73. A. Braga, C. Baratto, P. Colombi et al., An ultrathin TiO_2 blocking layer on Cd stannate as highly efficient front contact for dye-sensitized solar cells. Phys. Chem. Chem. Phys. **15**, 16812–16818 (2013)
74. Y. Chiba, A. Islam, Y. Watanabe et al., Dye-sensitized solar cells with a conversion efficiency of 11.1%. Jpn. J. Appl. Phys. **45**, L638 (2006)
75. D. Chen, F. Huang, Y. Cheng, R.A. Caruso, Mesoporous anatase TiO_2 beads with high surface areas and controllable pore sizes: a superior candidate for high-performance dye-sensitized solar cells. Adv. Mater. **21**, 2206–2210 (2009)
76. V. Roiati, S. Colella, G. Lerario et al., Investigating charge dynamics in halide perovskite-sensitized mesostructured solar cells. Energy Environ. Sci. **7**, 1889–1894 (2014)
77. M.J. Carnie, C. Charbonneau, P.R.F. Barnes et al., Ultra-fast sintered TiO_2 films in dye-sensitized solar cells: phase variation, electron transport, and recombination. J. Mater. Chem. A Mater. **1**, 2225–2230 (2013)
78. S. Chuangchote, T. Sagawa, S. Yoshikawa, Efficient dye-sensitized solar cells using electrospun TiO_2 nanofibers as a light-harvesting layer. Appl. Phys. Lett **93**, 266 (2008)
79. P. Joshi, L. Zhang, D. Davoux et al., Composite of TiO_2 nanofibers and nanoparticles for dye-sensitized solar cells with significantly improved efficiency. Energy Environ. Sci. **3**, 1507–1510 (2010)
80. L. Yang, W.W.-F. Leung, Improvement of dye-sensitized solar cells with nanofiber-based anode. In: *ASME International Mechanical Engineering Congress and Exposition* (2011), pp 1751–1752
81. S.H. Hwang, C. Kim, H. Song et al., Designed architecture of multiscale porous TiO_2 nanofibers for dye-sensitized solar cells photoanode. ACS Appl. Mater. Interfaces **4**, 5287–5292 (2012)
82. H. Krysova, A. Zukal, J. Trckova-Barakova et al., The application of electrospun titania nanofibers in dye-sensitized solar cells. Chimia (Aarau) **67**, 149 (2013)
83. Y. Cao, Y.-J. Dong, H.-L. Feng et al., Electrospun TiO_2 nanofiber-based hierarchical photoanode for efficient dye-sensitized solar cells. Electrochim. Acta **189**, 259–264 (2016)
84. M. Adachi, Y. Murata, J. Takao et al., Highly efficient dye-sensitized solar cells with a titania thin-film electrode composed of a network structure of single-crystal-like TiO_2 nanowires made by the "oriented attachment" mechanism. J. Am. Chem. Soc. **126**, 14943–14949 (2004)
85. J. Wang, D. Bai, Network TiO_2 nanowires for dye-sensitized solar cells. J. Electrochem. Soc. **161**, H265 (2014)
86. W. Liu, H. Lu, M. Zhang, M. Guo, Controllable preparation of TiO_2 nanowire arrays on titanium mesh for flexible dye-sensitized solar cells. Appl. Surf. Sci. **347**, 214–223 (2015)
87. K.-W. Lee, M. Kim, J.-M. Kim et al., Enhanced photovoltaic performance of back-illuminated dye-sensitized solar cell based on TiO_2 nanoparticle/nanowire composite film in cobalt redox system. J. Alloys Compd. **656**, 568–572 (2016)
88. Y.-Y. Liu, Y.-L. Wang, X.-Y. Ye et al., Ultra-long hierarchical bud-like branched TiO_2 nanowire arrays for dye-sensitized solar cells. Thin Solid Films **640**, 14–19 (2017)
89. Y.-Y. Liu, X.-Y. Ye, Q.-Q. An et al., A novel synthesis of the bottom-straight and top-bent dual TiO_2 nanowires for dye-sensitized solar cells. Adv. Powder Technol. **29**, 1455–1462 (2018)
90. H. Koo, Y.J. Kim, Y.H. Lee et al., Nano-embossed hollow spherical TiO_2 as bifunctional material for high-efficiency dye-sensitized solar cells. Adv. Mater. **20**, 195–199 (2008)
91. G. Dai, L. Zhao, J. Li et al., A novel photoanode architecture of dye-sensitized solar cells based on TiO_2 hollow sphere/nanorod array double-layer film. J. Colloid Interface Sci. **365**, 46–52 (2012)

92. Z.-Q. Li, W.-C. Chen, F.-L. Guo et al., Mesoporous TiO_2 yolk-shell microspheres for dye-sensitized solar cells with a high efficiency exceeding 11%. Sci. Rep. **5**, 1–8 (2015)
93. P. Zhang, Z. Hu, Y. Wang et al., A bi-layer composite film based on TiO_2 hollow spheres, P25, and multi-walled carbon nanotubes for efficient photoanode of the dye-sensitized solar cell. Nanomicro. Lett. **8**, 232–239 (2016)
94. J. Sun, H. Guo, L. Zhao et al., Co-sensitized efficient dye-sensitized solar cells with TiO_2 hollow sphere/nanoparticle double-layer film electrodes by Bi_2S_3 quantum dots and N719. Int. J. Electrochem. Sci. **12**, 7941–7955 (2017)
95. Z. Cui, K. Zhang, G. Xing et al., Multi-functional 3D N-doped TiO_2 microspheres used as scattering layers for dye-sensitized solar cells. Front. Chem. Sci. Eng. **11**, 395–404 (2017)
96. M. Paulose, K. Shankar, O.K. Varghese et al., Application of highly-ordered TiO_2 nanotube-arrays in heterojunction dye-sensitized solar cells. J. Phys. D Appl. Phys. **39**, 2498 (2006)
97. M. Paulose, K. Shankar, O.K. Varghese et al., Backside illuminated dye-sensitized solar cells based on titania nanotube array electrodes. Nanotechnology **17**, 1446 (2006)
98. K. Zhu, T.B. Vinzant, N.R. Neale, A.J. Frank, Removing structural disorder from oriented TiO_2 nanotube arrays: reducing the dimensionality of transport and recombination in dye-sensitized solar cells. Nano Lett. **7**, 3739–3746 (2007)
99. K. Shankar, G.K. Mor, H.E. Prakasam et al., Highly-ordered TiO_2 nanotube arrays up to 220 μm in length: use in water photoelectrolysis and dye-sensitized solar cells. Nanotechnology **18**, 065707 (2007)
100. N.N. Bwana, Improved short-circuit photocurrent densities in dye-sensitized solar cells based on ordered arrays of titania nanotubule electrodes. Curr. Appl. Phys. **9**, 104–107 (2009)
101. M. Ali, M. Winterer, ZnO nanocrystals: surprisingly 'alive.' Chem. Mater. **22**, 85–91 (2010)
102. J.A. Anta, E. Guillén, R. Tena-Zaera, ZnO-based dye-sensitized solar cells. J. Phys. Chem. C **116**, 11413–11425 (2012)
103. T.P. Chou, Q. Zhang, G. Cao, Effects of dye loading conditions on the energy conversion efficiency of ZnO and TiO_2 dye-sensitized solar cells. J. Phys. Chem. C **111**, 18804–18811 (2007)
104. H.J. Snaith, C. Ducati, SnO_2-based dye-sensitized hybrid solar cells exhibiting near unity absorbed photon-to-electron conversion efficiency. Nano Lett. **10**, 1259–1265 (2010)
105. S. Ferrere, A. Zaban, B.A. Gregg, Dye sensitization of nanocrystalline tin oxide by perylene derivatives. J. Phys. Chem. B **101**, 4490–4493 (1997)
106. E.N. Kumar, R. Jose, P.S. Archana et al., High-performance dye-sensitized solar cells with record open circuit voltage using tin oxide nanoflowers developed by electrospinning. Energy Environ. Sci. **5**, 5401–5407 (2012)
107. A. Le Viet, R. Jose, M.V. Reddy et al., Nb_2O_5 photoelectrodes for dye-sensitized solar cells: choice of the polymorph. J. Phys. Chem. C **114**, 21795–21800 (2010)
108. L. Hu, M. Wolf, M. Grätzel, Z. Jiang, Synthesis and properties of niobium films derived from niobium pentaethoxide. J. Solgel Sci. Technol. **5**, 219–226 (1995)
109. M. Wei, Z. Qi, M. Ichihara, H. Zhou, Synthesis of single-crystal niobium pentoxide nanobelts. Acta Mater. **56**, 2488–2494 (2008)
110. R.A. Rani, A.S. Zoolfakar, J.Z. Ou et al., Reduced impurity-driven defect states in anodized nanoporous Nb_2O_5: the possibility of improving the performance of photoanodes. Chem. Commun. **49**, 6349–6351 (2013)
111. H. Zhang, Y. Wang, D. Yang et al., Directly hydrothermal growth of single crystal Nb_3O_7 (OH) nanorod film for high performance dye-sensitized solar cells. Adv. Mater. **24**, 1598–1603 (2012)
112. H. Zhang, Y. Wang, P. Liu et al., Highly ordered single crystalline nanowire array assembled three-dimensional $Nb_3O_7(OH)$ and Nb_2O_5 superstructures for energy storage and conversion applications. ACS Nano **10**, 507–514 (2016)
113. K. Sayama, H. Sugihara, H. Arakawa, Photoelectrochemical properties of a porous Nb_2O_5 electrode sensitized by a ruthenium dye. Chem. Mater. **10**, 3825–3832 (1998)
114. J. Guo, D. Stockwell, X. Ai et al., Electron-transfer dynamics from Ru polypyridyl complexes to In_2O_3 nanocrystalline thin films. J. Phys. Chem. B **110**, 5238–5244 (2006)

115. S. Mori, A. Asano, Light intensity independent electron transport and slow charge recombination in dye-sensitized In_2O_3 solar cells: in contrast to the case of TiO_2. J. Phys. Chem. C **114**, 13113–13117 (2010)
116. R. Sharma, R.S. Mane, S.-K. Min, S.-H. Han, Optimization of growth of In_2O_3 nano-spheres thin films by electrodeposition for dye-sensitized solar cells. J. Alloys Compd. **479**, 840–843 (2009)
117. A. Manimekalai, R. Parimaladevi, M. Umadevi, Enriched In_2O_3/WO_3 nanocomposites as photoanode for a dye-sensitized solar cell in photo powered energy system. Mater. Lett. **319**, 132285 (2022)
118. M.M. Rashad, A.E. Shalan, Hydrothermal synthesis of hierarchical WO_3 nanostructures for dye-sensitized solar cells. Appl. Phys. A **116**, 781–788 (2014)
119. G. Richhariya, A. Kumar, A.K. Shukla et al., Effect of different counter electrodes on power conversion efficiency of DSSCs. J. Electron. Mater. **52**, 60–71 (2023)
120. M. Miyauchi, Z. Liu, Z.-G. Zhao et al., Single crystalline zinc stannate nanoparticles for efficient photo-electrochemical devices. Chem. Commun. **46**, 1529–1531 (2010)
121. M.A. Alpuche-Aviles, Y. Wu, Photoelectrochemical study of the band structure of Zn_2SnO_4 prepared by the hydrothermal method. J. Am. Chem. Soc. **131**, 3216–3224 (2009)
122. S.S. Mali, C. Su Shim, C. Kook Hong, Highly porous Zinc Stannate (Zn_2SnO_4) nanofibers scaffold photoelectrodes for efficient methyl ammonium halide perovskite solar cells. Sci. Rep. **5**, 11424 (2015)
123. T.J. Coutts, D.L. Young, X. Li et al., Search for improved transparent conducting oxides: a fundamental investigation of CdO, Cd_2SnO_4, and Zn_2SnO_4. J. Vac. Sci. Technol. A: Vac. Surf. Films **18**, 2646–2660 (2000)
124. J. Chen, L. Lu, W. Wang, Zn_2SnO_4 nanowires as photoanode for dye-sensitized solar cells and the improvement on open-circuit voltage. J. Phys. Chem. C **116**, 10841–10847 (2012)
125. J. Chen, W. Wang, Carrier transport processes in dye-sensitized solar cells based on Zn_2SnO_4 nanostructures studied by intensity modulated photocurrent/photovoltage spectroscopy. Appl. Phys. Lett. **102**, 213904 (2013)
126. S. Choi, D. Hwang, D. Kim et al., Amorphous Zinc Stannate (Zn_2SnO_4) nanofibers networks as photoelectrodes for organic dye-sensitized solar cells. Adv. Funct. Mater. **23**, 3146–3155 (2013)
127. Y.-F. Wang, K.-N. Li, Y.-F. Xu et al., Hierarchical Zn_2SnO_4 nanosheets of nanoparticles for efficient dye-sensitized solar cells. Nano Energy **2**, 1287–1293 (2013)
128. D. Hwang, J.-S. Jin, H. Lee et al., Hierarchically structured Zn_2SnO_4 nanobeads for high-efficiency dye-sensitized solar cells. Sci. Rep. **4**, 7353 (2014)
129. S. Burnside, J.-E. Moser, K. Brooks et al., Nanocrystalline mesoporous strontium titanate as photoelectrode material for photosensitized solar devices: increasing photovoltage through flat-band potential engineering. J. Phys. Chem. B **103**, 9328–9332 (1999)
130. P. Jayabal, V. Sasirekha, J. Mayandi et al., A facile hydrothermal synthesis of $SrTiO_3$ for dye-sensitized solar cell application. J. Alloys Compd. **586**, 456–461 (2014)
131. Y. Li, W. Guo, H. Hao et al., Enhancing the photoelectrical performance of dye-sensitized solar cell by doping SrTiO3: Sm3+@ SiO_2 core–shell nanoparticles in the photoanode. Electrochim. Acta **173**, 656–664 (2015)
132. S. Yang, H. Kou, J. Wang et al., Tunability of the band energetics of nanostructured $SrTiO_3$ electrodes for dye-sensitized solar cells. J. Phys. Chem. C **114**, 4245–4249 (2010)
133. Y. Okamoto, Y. Suzuki, Perovskite-type $SrTiO_3$, $CaTiO_3$, and $BaTiO_3$ porous film electrodes for dye-sensitized solar cells. J. Ceram. Soc. Jpn. **122**, 728–731 (2014)
134. N. Ishiia, Y. Okamoto, B.Y. Suzuki, Semiconductor $MgTiO_3$, $MgTi_2O_5$ and Mg2TiO4 double-oxide electrodes for dye-sensitized solar cells. Int. Lett. Chem. Phys. Astron. **46**, 9–15 (2015)
135. G. Natu, Y. Wu, Photoelectrochemical study of the ilmenite polymorph of $CdSnO_3$ and its photoanodic application in dye-sensitized solar cells. J. Phys. Chem. C **114**, 6802–6807 (2010)
136. F. Guo, G. Li, W. Zhang, Barium staminate as semiconductor working electrodes for dye-sensitized solar cells. Int. J. Photoenergy (2010)

137. B. Li, Y. Tang, L. Luo et al., Fabrication of porous $BaSnO_3$ hollow architectures using BaCO3@ SnO2 core-shell nanorods as precursors. Appl. Surf. Sci. **257**, 197–202 (2010)
138. S.S. Shin, J.S. Kim, J.H. Suk et al., Improved quantum efficiency of highly efficient perovskite $BaSnO_3$-based dye-sensitized solar cells. ACS Nano **7**, 1027–1035 (2013)
139. F.W. Low, C.W. Lai, Recent developments of graphene-TiO_2 composite nanomaterials as efficient photoelectrodes in dye-sensitized solar cells: a review. Renew. Sustain. Energy Rev. **82**, 103–125 (2018)
140. M. Karlsson, I. Jõgi, S.K. Eriksson et al., Dye-sensitized solar cells employing a SnO_2–TiO_2 core-shell structure made by atomic layer deposition. Chimia (Aarau) **67**, 142 (2013)
141. W. Chen, Y. Wu, Y. Yue et al., (2015) Efficient and stable large-area perovskite solar cells with inorganic charge extraction layers. Science **350**, 944–948 (1979)
142. A.K. Chandiran, M. Abdi-Jalebi, M.K. Nazeeruddin, M. Grätzel, Analysis of electron transfer properties of ZnO and TiO_2 photoanodes for dye-sensitized solar cells. ACS Nano **8**, 2261–2268 (2014)
143. L. Liang, Y. Liu, C. Bu et al., Highly uniform, bifunctional core/double-shell-structured β-$NaYF_4$: Er^{3+}, Yb^{3+}@ SiO_2@TiO_2 hexagonal sub-micro prisms for high-performance dye-sensitized solar cells. Adv. Mater. **25**, 2174–2180 (2013)
144. Z. Hosseini, W.-K. Huang, C.-M. Tsai et al., Enhanced light harvesting with a reflective luminescent down-shifting layer for dye-sensitized solar cells. ACS Appl. Mater. Interfaces **5**, 5397–5402 (2013)
145. J. He, H. Lindström, A. Hagfeldt, S.-E. Lindquist, Dye-sensitized nanostructured p-type nickel oxide film as a photocathode for a solar cell. J. Phys. Chem. B **103**, 8940–8943 (1999)
146. M.D. Irwin, D.B. Buchholz, A.W. Hains et al., p-type semiconducting nickel oxide as an efficiency-enhancing anode interfacial layer in polymer bulk-heterojunction solar cells. Proc. Natl. Acad. Sci. **105**, 2783–2787 (2008)
147. D. Dini, Y. Halpin, J.G. Vos, E.A. Gibson, The influence of the preparation method of NiOx photocathodes on the efficiency of p-type dye-sensitized solar cells. Coord. Chem. Rev. **304**, 179–201 (2015)
148. S. Powar, T. Daeneke, M.T. Ma et al., Highly efficient p-type dye-sensitized solar cells based on tris (1, 2-diaminoethane) cobalt (II)/(III) electrolytes. Angew. Chem. Int. Ed. **52**, 602–605 (2013)
149. T. Daeneke, Z. Yu, G.P. Lee et al., Dominating energy losses in NiO p-type dye-sensitized solar cells. Adv. Energy Mater. **5**, 1401387 (2015)
150. A. Aslam, U. Mehmood, M.H. Arshad et al., Dye-sensitized solar cells (DSSCs) as a potential photovoltaic technology for the self-powered internet of things (IoTs) applications. Sol. Energy **207**, 874–892 (2020)
151. J.E. Trancik, Photovoltaics—A Niche market distraction or a global energy solution? Geo Public Pol'y Rev. **11**, 69–71 (2006)
152. J. Kalowekamo, E. Baker, Estimating the manufacturing cost of purely organic solar cells. Sol. Energy **83**, 1224–1231 (2009)
153. J. Campillo, S. Foster, Global solar photovoltaic industry analysis with focus on the Chinese market (2008)
154. J.W. Matiko, N.J. Grabham, S.P. Beeby, M.J. Tudor, Review of the energy harvesting application in buildings. Meas. Sci. Technol. **25**, 012002 (2013)
155. T.D. Nielsen, C. Cruickshank, S. Foged et al., Business, market, and intellectual property analysis of polymer solar cells. Sol. Energy Mater. Sol. Cells **94**, 1553–1571 (2010)
156. N.A. Karim, U. Mehmood, H.F. Zahid, T. Asif, Nanostructured photoanode and counter electrode materials for efficient dye-sensitized solar cells (DSSCs). Sol. Energy **185**, 165–188 (2019)

Engineering of the Metal Oxides Nanostructures for Solar Energy Harvesting

Lokanath Mohapatra and Ajay Kumar Kushwaha

Abstract Semiconducting metal oxides have been utilized in various roles such as electron transport layer, hole transport layer, and active layer in solar cell. The capability of photo charge generation, separation, and recombination is governed by the quality of the metal oxide layer and eventually affect the solar cell efficiency. It is suggested that the materials used for charge transfer should have resistant to light, moisture, and heat to maintain long-term stability. Metal oxides are ideally designed charge transport materials due to its several favorable properties. Considering the technological significance of metal-oxides, various materials engineering concepts have been developed to improve the quality of metal oxide layers and the performance of solar cell. This chapter discusses application of metal oxides in solar cell in different roles. The important materials engineering concepts (improvement in crystalline properties, size morphological, doping, defect and formation of heterostructures) towards improvement in solar energy harvesting are also presented.

Keywords Solar cells · Photovoltaics · Photoconductivity · Efficiency · Defects · Surface properties · Electron transport · Nanomaterials · Thin films · Nanostructures

1 Introduction

The increasing population and energy demand accelerates the use of fossil fuels. This use of fossil fuel becoming the major source of carbon and greenhouse emission which causes several environmental challenges such as global warming, climate

L. Mohapatra · A. K. Kushwaha (✉)
Department of Metallurgical Engineering and Materials Science, Indian Institute of Technology Indore, Simrol, Indore, M.P. 453552, India
e-mail: akk@iiti.ac.in

A. K. Kushwaha
Centre for Advanced Electronics, Indian Institute of Technology Indore, Simrol, Indore, M.P. 453552, India

V. Kumar et al. (eds.), *Optical Properties of Metal Oxide Nanostructures*, Progress in Optical Science and Photonics 26,
https://doi.org/10.1007/978-981-99-5640-1_9

change. It is very necessary to use the green energy sources to overcome these challenges. The solar energy is the limitless green energy source. Solar energy can be utilized by converting it into electrical energy by using solar cell and in current time this is the most effective way to use the solar energy. Solar energy harvesting is a sustainable method to produce renewable energy. Within an hour, the energy received on the earth from the sun is more than the entire energy used by the planet in a year. Solar cells convert the solar energy into electricity which is more acceptable form of energy that can be used for a variety of purposes in the modern world [1]. Although there are several technologies for harvesting of solar energy, however, solar cell has proven its excellency in this domain.

The fabrication of solar cell required a variety of materials. Among them, metal oxide semiconductors stand out as a class of relatively affordable, eco-friendly materials with a wide variety of characteristics suitable for solar cell application. Numerous techniques have been demonstrated to control and modify the characteristics of metal oxide semiconductors. There are various solar cell technologies, organic solar cells (OPVs), perovskite solar cells, dye-sensitized solar cells (DSSCs) and quantum dot solar cells utilizing metal oxides. In dye-sensitized solar cells, metal oxides act as an active material to load dyes while in perovskite solar cells metal oxide act as an active layer as well as charge transporting layer. The active layer facilitates the electrons and holes generation where the charge transport layer allows one type of charge carrier to move and bloke the other type of charge carrier [2]. The material suitability as an electron or hole transport layer depends on the band alignment of metal oxide and the types of charge carriers present. Metal oxides must have lower valence band maximums and conduction band minimums than the absorber material to be employed as electron transport layer. The valence band maximums and conduction band minimums should be higher than the absorber material for the hole transport layer [3]. The efficiency of the solar cell changes depending on these band positions and the misalignment in the band position leads to the to charge carrier recombination in the charge transport layer.

To achieve the desire properties such as higher charge carrier mobility, proper band position, efficient charge carrier separation, lower inter facial trap state, the tuning of metal oxides properties is an important task. The engineering of the metal oxide nanostructures/thin film is crucial toward controlling these properties to develop more efficient solar energy harvesting devices. This chapter discuss on the role of metal oxides in solar cell and various strategies to control the properties of metal oxides nanostructures to improve of the solar light harvesting performance.

2 Metal Oxides in Solar Cell

2.1 Metal Oxides as an Electron Transporting Layer

There are several features on semiconducting metal-oxides that offers the suitability for electron transport in solar cell (Fig. 1a). The electron transport layer (ETL) act as channel for electro flow by decreasing the potential energy barrier. This layer collect the electron generated in the active layer and transfer them to the electrode before recombination. The electron transport layer work as a link between active materials and electrode to transport the electron and blocks the transfer of the hole (Fig. 1b, c). There are several essential properties are required in electron transport layers such as (i) high transparency to visible light so that energy photon can reach to active layer, (ii) favorable energy level alignment for electron movement from active materials to electrode and (iii) high electrical conductivity with high electron mobility to improve charge separation capability [4–6]. Following oxide materials are commonly investigated as electron transport layers in solar cells, which includes TiO_2, ZnO, WO_3, Zn_2SnO_4, SiO_2, CeO_2, Nb_2O_5, $SrSnO_3$, metal oxides binary and ternary etc. [7, 8]. TiO_2 is the most comprehensively investigated metal oxide semiconductor for solar cells due to suitable band edge positions for generation and transportation of charge carriers. However, high number of trap states and lower electron mobility are major issues with TiO_2 material. The TiO_2 also requires high temperature heat treatment to achieve the crystallization [9].

Zinc oxide (ZnO) is another semiconductor metal oxide with has shown favorable properties as electron transport materials. The ZnO films deposited using wet-chemical approach at low temperature offers surface with nanostructured morphology for solar light harvesting applications [10, 11]. Tin oxide (SnO_2) is another potential electron transport layers used in solar cell due to high transparency in the visible and near infrared regions. The low temperature method has been proposed to develop the SnO_2 layer which has shown PCE of 13.0% with highly stable performance. The SnO_2 has lesser oxygen vacancies hence more stable than TiO_2 and resulted much lower degradation of perovskite solar cell [12]. Zinc tin oxide (Zn_2SnO_4) is also demonstrated as transparent conducting layer for solar cell applications. Zn_2SnO_4 has a wide bandgap approximate 3.8 eV with reasonably good mobility 10–30 cm^2Vs^{-1} [13, 14].

Apart from these a various other metal oxides are investigated for efficient electron transport layer. Binary ZnO is also a promising alternative for the TiO_2. As ZnO has higher electron mobility and the outstanding transmittance of the visible light [15]. Also, ZnO can be crystalized at lower heating temperature which will help to reduce the cost of the solar cell. Various morphology such as nanorod, nanosheets nanoparticle, nanoplate of ZnO was being investigated. But ZnO still have challenges like poor chemical stability and charge recombination at interface limits it potential utilization in perovskite solar cell [16].

The nanostructuring of electron transfer layer helpful to enhancing the efficiency of the solar cell. So far, various morphology with different composition of

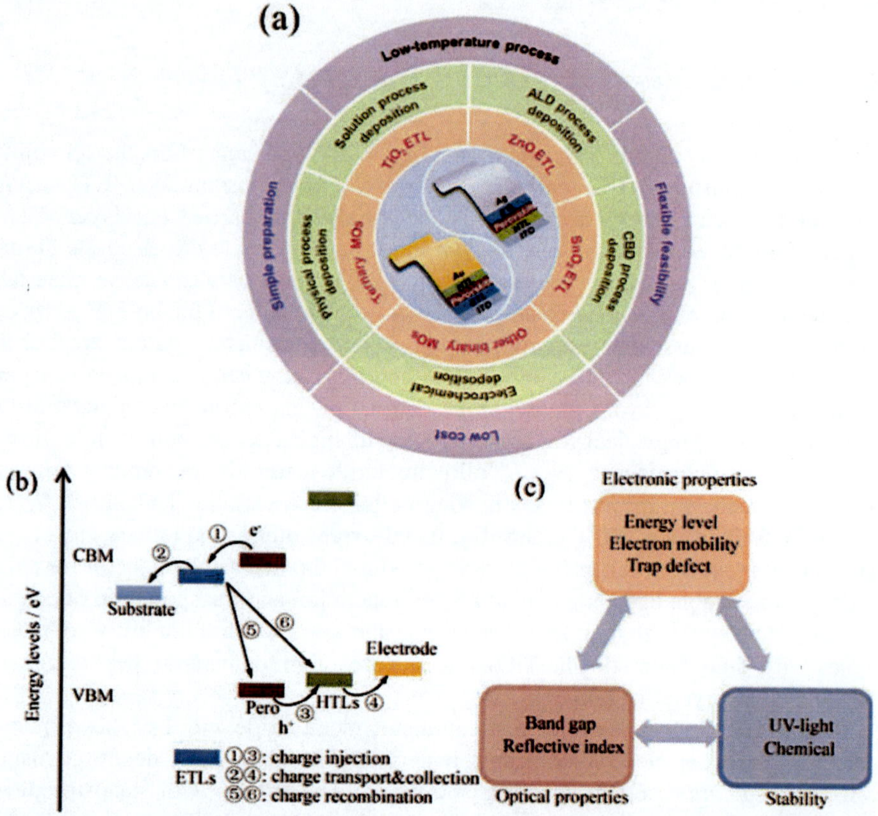

Fig. 1 Schematic digram of **a** metal oxide electrontransfer layer fabrcaition and properties, [Li et al., *Rare Metals* 40, 2730 (2021)] [8], **b** charge transfer process in typical solar-cell based on the metaloxide electrontransfer layer, **c** facters effect the metaloxide electrontransfer layere [Li et al., *Rare Metals* 40, 2730 (2021)] [8]

metal oxides has being synthesized and utilized to improve the electron transfer. In perovskite solar cell in was observed that the ZnO nanorod provide the direct path for the charge transfer and improve the efficiency of the solar cell [17]. Apart from this, the rough surface of ZnO improves the loading of the perovskite and improves the efficiency of the device [18]. Elemental doping is also an alternative way to tune the electronic properties of metal oxides. The doping will help to adjust energy level alignment and improved electron mobility to improve the efficiency which is disused in upcoming section [19]. Surface modification is another way to engineering the metal oxides for the electron transfer layer. The Surface modification can be achieved via coating self-assembled monolayer (SAM). The functional hydrophobicity and high electric dipole of SAM will help to accelerates the charge extraction from the perovskite layer in the perovskite solar cell [20]. Additionally, SAM also passivate the various defect sites (oxygen-vacancy defects, surface defects, and ionic defects) of perovskite reduce the charge linkage and defect site recombination and enhance the efficiency of the device [21].

The performance of Electron transport layer can be improved by the Nanostructuring, Element doping, Surface modification and Multi-layer ETLs. If we discuss about the nanostructuring so far various nanostructure such as nanodisk, nanorod, nanoplate, nanosheets of metal oxides is being used for the ETLs. For example, Mahmood et al. synthesized ZnO nanodisks and nanowells using a low-temperature hydrothermal technique and in comparison to ZnO nanodisks the nanowell device shown the maximum PCE of 16.65% [22]. Moreover, studies on various metal oxide nanostructures have been conducted. To act as the ETL, Fe_2O_3 nano islands were also in situ deposited on the compact Fe_2O_3 layers by Luo et al. Perovskite solar cell based on Fe_2O_3 nanoislands displayed exceptional PCE of 18.2%, thanks to their acceptable light transmission, improved charge extraction, and reduced charge recombination [23]. The best PCE, measuring 20.22% with a Voc of 1.19 V, was achieved by the device using Nb_2O_5 nanoparticles [24]. The Nb_2O_5 ETL energy bands are well aligned with perovskite, which enhances charge extraction at the interface and lowers the energy barrier for electron injection. Liu et al. synthesized Zn_2SnO_4 nanoparticles by the Hydrothermal process and the PCE of 17.7% was achieved in perovskite solar cell [25]. The studies show that Zn_2SnO_4 emphasize the electron transfer and injecting between ELT and Perovskite layer. Also the rescued Zn_2SnO_4 nanorod are deposited on FTO substrate using the spray pyrolysis technique by Tavakoli et al. device based on Zn_2SnO_4 nanorod shown a PCE of 18.24% [26]. metal oxides with Core–shell structure has also been explored widely as ETLs. Core-shell ZnO@SnO_2 nano particle synthesized by Li et al. in which SnO_2 shell cover the core ZnO. The Perovskite solar cell with CPRE cell ETL displayed a PCE of 14.35%. This high PCE is observed due to higher electron mobility of SnO_2 compared to pristine one and well-matched energy level alignment of ZnO and SnO_2 for efficient electron injection [27]. Zhong et al. deposited ZnO@TiO_2 nanorod and it was observed that the ZnO@TiO_2 has PCE of 8.14 and bare ZnO has the PEC of 5.41%. Improved ETL/perovskite interface contact and reduced charge recombination is the main reason for this 50.46% PCE improvement [28].

Elemental doping is also proposed to change the band alignment and improve the charge carrier mobility for the efficient ELT layer. Tseng et al. have reported Al doped ZnO using RF sputtering technique, doping of Al improved the photoconversion efficiency of a perovskite solar cell from 15.1 to 17.6%. This enhancement in the PCE is obtained due to improve electrical conductivity and better band alignment of Al doped ZnO film with the perovskite layer [29]. Zheng et al. studied the effect of I doping on the ZnO nanopillar for formation of ETL which rendered more than 7% increase in the PCE of the perovskite solar cell. The presence of I regulate the crystal growth of ZnO which helps to form the compact nanopillar structure and reduced aspect-ratio [30]. The Li doping in ZnO also improved the solar cell efficiency. The Cs doped ZnO has resulted an improved PCE by 4.7% due to the reduction of ZnO trap density [31]. Azmi et al. have synthesized pure ZnO, Li-doped ZnO, Na-doped ZnO, K-doped ZnO as ETL for perovskite solar cell which has shown the PCE of 16.10, 17.80, 18.90, 19.90% and this improvement in the PCE due to the decrease in surface defect concentration and surface hydroxyl group. Potassium (K) doping on ZnO significantly increase the fermi energy level, leading to an improved internal electric field, which was conducive to charge extraction in the devices [32]. α-Fe_2O_3 was used as ETL and effect of doping in the ETL properties was reported by Guo et al., they synthesized Ni-doped α-Fe_2O_3 for the perovskite solar cell. The improved electron mobility and downshift lower conduction band after Ni doping in α-Fe_2O_3 enhance the electron transfer between ETL and perovskite [33]. Xiaoqin et al. reported Zn doping in Nb_2O_5 by solution-combustion method. The best efficiency of 17.7% was attained by PSCs based on the Zn–Nb_2O_5 ETL [34].

2.2 *Metal Oxides as Hole Transporting Layer*

Several metal oxides have been proposed as excellent hole transport layer (HTL) materials for solar cell. The hole transport layer is mainly used in organic, quantum dots, dye sensitized, perovskite solar cells. The hole transport materials should have, (i) energy level such as, it forms an ohmic contact with active layer to easily extract the hole. (ii) p-type semiconducting nature, lowest electrons mobility. (iii) high hole conductivity to improve the charge separation capability. (iv) transparent to visible light for effective photon transmission to active layer [35]. It is also required to have facile materials processing at low-temperature with large-area deposition possibility for solar cells. The metal oxides such as MoO_x, WO_3, and V_2O_5 have high work function but are n-type semiconductor so unable to block the electron transport leading to higher recombination. Therefore, mostly p-type metal oxides such as NiO_x, $Ni_xCo_{3-x}O_4$, and CuO_x, WO_3, V_2O_5, MoO_x are examined (Fig. 2) [36].

Earth abundant copper-based oxides (CuO_x) are demonstrated as one of the candidates for hole transport layer in solar cell devices. An in situ low-temperature method for synthesis of cuprous oxide (Cu_2O) and copper oxide (CuO) is reported by the

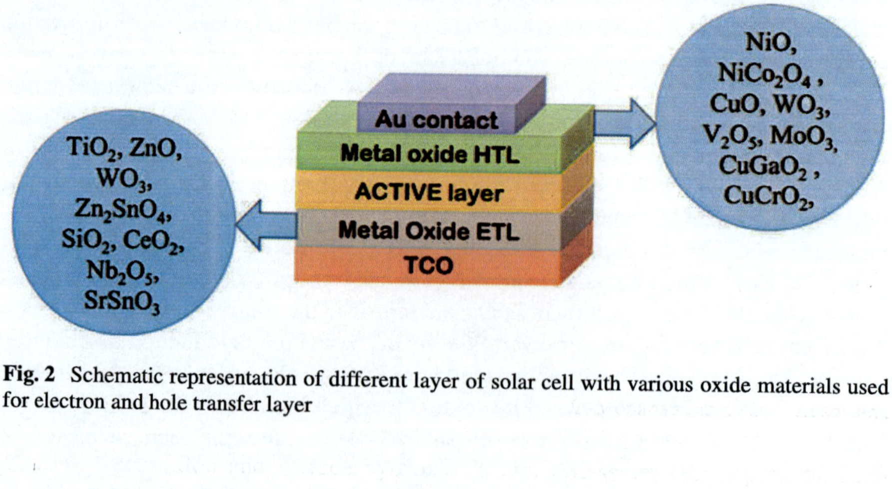

Fig. 2 Schematic representation of different layer of solar cell with various oxide materials used for electron and hole transfer layer

Ding et al., demonstrated as hole transport layer with 13.35% and 12.16% efficiencies of solar cell, respectively. The solution deposited coper oxide film has shown uniform and stable surface with optically transparent for the visible light [37]. Nickel oxide (NiO_x) is another candidate which has demonstrated good potential for hole transport layer due to superior thermal and chemical stability. It has been reported that NiO_x based solar cell has shown improved charge extraction and larger charge career lifetime. Molybdenum oxide (MoO_x) thin films have also been used as either an interlayer or buffer to improve hole injection and hole extraction capability. It is reported that the combination of MoO_x/Al in perovskite solar cell as HTL has shown better results in comparison to only silver-based layer [38].

Another P-type Inorganic semiconductors NiO_x also have been widely studied for the HTL in organic and dye sensitized solar cell devices. The transparency, tunable work function, better hole mobility and stability of NiO_x make it an efficient material for HTL [39]. The NiO_x has potential to replace the (PEDOT: PSS). The replacement of PEDOT:PSS with NiO_x in polymer bulk heterojunction solar cells improved the efficiency of the device [40]. NiO_x also has been utilized in the in the Perovskite solar cell. Jeng et al. for the first time utilized the NiO_x in the hybrid perovskite solar cell and able to achieve the efficiency of 7.8% due to the batter band alignment of the interlayer of NiO_x with the valence band of $CH_3NH_3PbI_3$ perovskite [41]. Later on NiO_x synthesized by the different method sputtering, electrodeposition, sol–gel process, and pulsed laser deposition, are utilized for the $CH_3NH_3PbI_3$ perovskite to improve the efficiency of the device. NiO_x thin film developed by Low-temperature sputtering techniqu has shown enhanced photoconversion efficiency of 11.6% is reported by Wang et al. [42]. Park et al. able to achieve the efficiency of 17.3% by using the pulse laser deposited NiO_x nanostructure with (111) orientation as a HTL in perovskite solar cell [43]. The NiO_x has lower conductivity which limits and hinder the further improvement in the efficiency. Low conductivity of HTL reduces hole extraction and increases recombination of charge carriers. Doping is an efficient way to improve the conductivity. Report shows that the Cu and Cs doped NiO_x has shown the perovskite solar cell efficiency of 20.26% and 19.35% respectively. The alternation of Ni ion with the Cu and Cs helps to improve the hole mobility, carrier concentration, hole extraction and lower the recombination of carrier [44, 45].

Nanostructuring is an efficient way to enhance the Solar cell performance by improving the HTL efficiency. Kwon et al. synthesized precrystallized NiO nanoparticles by the simple solution process to use in HTL for the Perovskite solar cell. The solar cell with 45 nm tick NiO HTL has shown the Higher Photo conversion efficiency of 15.4% [46]. Whereas, Yin et al. studied Nanoforest Architecture of NiO was grown by Hydrothermal method for HTL in the Perovskite solar cell. The device has the optimal PCE of 18.77%. mesoporous structure of Nanoforest Architecture has provided a highly conductive pathway for rapid hole extraction and reduces charge recombination at perovskite and HTL interface [47]. Furthermore, by adjusting the length and branch growth of the NiO nanoforest improved photo capturing could be accomplished. Mesoporous NiO nanosheet deposited by the hydrothermal method is also investigated for HTL and 12.19% PCE was observed with an improved stability in the perovskite solar cell [48].

Doping in metal oxides has shown increased electrical conductivity. The dopant ions typically featured distinctive structural and electrical features that easily merged into HTL. The doping in oxide HTL is regard as a crucial approach to augment the solar cell performance. Copper is one of the most common dopant elements of NiO_x HTL. The Cu can easily occupy the Ni site and by including the proper amount of Cu, the crystallinity, transmittance, and conductivity of NiO_x can be altered. The addition of Cu as a dopant change the native defect which is responsible for the increase in conductivity or hole density in NiO_x [49]. Jae et al. fabricated Cu-doped NiO_x films by the sol-gel method and utilized as HTL in the Perovskite solar cell and the highest PCE of 15.52% is achieved [50]. The monovalent alkali metals lithium (Li^+) has ionic radius (0.76), which is closer to Ni^{2+} (0.69) due to which lithium (Li^+) is used as dopants for NiO_x in order to boost the electrical conductivity. The addition of Li^+ improved the concentration of Ni^{3+} and decrease in resistivity of NiO_x at the same time. Qiu et al. used pulsed laser deposition (PLD) technique to deposit Li doped NiO for the HTL. After Li-doping the perovskite solar cell with structure of ITO/Li:NiO/$MAPbI_3$–xClx/PCBM/Ag has shown highest PCE of 15.51% with enhanced stability [51]. Besides, Park et al. also deposited Li in NiO by spin-coating process for HTL and PCE was improved from 12.15 to 14.66% [52]. Hu et al. have reported the Co-doped NiO_x as a HTL in the perovskite solar cell [53]. Djurišic et al. investigated the Cesium (Cs+) doped NiO as HTL for the perovskite solar cell. The best efficiency of Cs doped NiO_x devices was 19.35%. The enhanced performance in devices with Cs doped NiO_x is endorsed to a important progress in the hole extraction and optimized band level compared to undoped NiO_x [44]. Wei et al. reported Ag-doped NiO_x films as a HTL and fabricated ITO/Ag:NiO_x/$MAPbI_3$/PCBM/BCP/Ag structure which has PCE of 16.86% [54]. Similarly, the doping of magnesium Zinc iron on NiO_x is reported to use as HTL. Liu et al. synthesized doubles element doped NiO_x films with Li, Cu dopant as a HTL by solution process method. The device prepared by using Li, Cu co-doped NiO_x HTL has shown the maximum PCE of 14.53% [55]. Qin et al. has deposited the Cu doped CrO_x by the RF sputtering technique to use in the HTL which improve the PCE from 9.27 to 10.99%. This enhancement in the device performance is observed due to efficient hole injection with lower charge recombination from perovskite layer to HTL [56].

2.3 *Metal Oxides as Active Layer*

The semiconducting metal oxides are also utilized as light harvesting materials in excitonic solar cells. The few examples are cited below. Dye-sensitized solar cell (DSSC) is a photo electrochemical cell where porous metal oxide electrode with dye molecules is serving as the active layer for harvesting of solar photon. The detailed working principle of DSSC is shown in Fig. 3. In DSSC photo excited electron of dye goes into the conduction band of semiconductor material. Then the oxidized dye reduces by the electron given by the electrolyte. The injected electron travels through

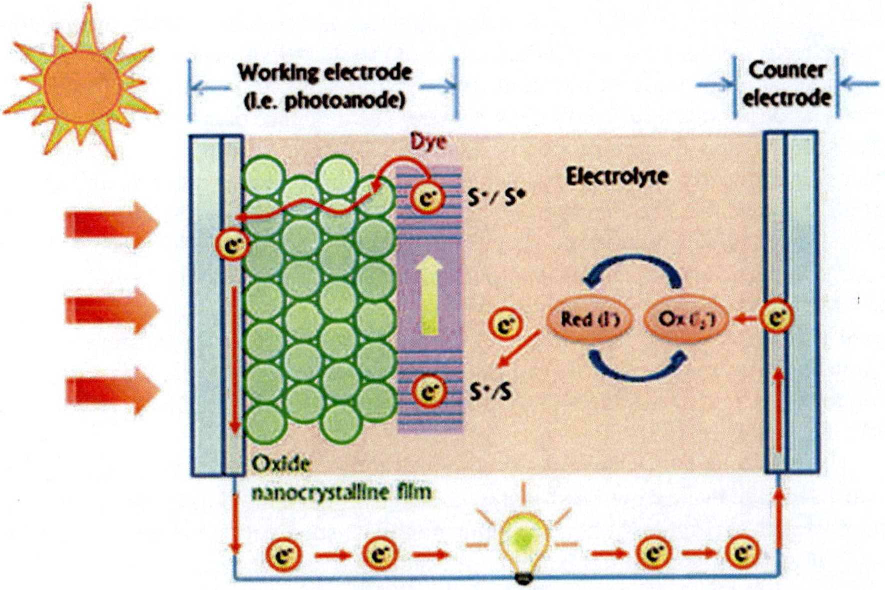

Fig. 3 Structure and operating mechanism of a dye-sensitized solar cell [Zhang et al., *Physical Chemistry Chemical Physics*, 14, 14982, (2012)] [60]

the semiconductor material to the back contact and travel to external circuit. This electron goes to the electrolyte by the counter electrode form the external circuit.

Photoanode is most important part of the DSSC. The selection of the photoanode is very crucial for the Efficient DSSC so the material properties, morphology needs to be select carefully. The key characters for the efficient photoanode are mentioned bellow:

1. Large surface area to enhance the dye absorption ability.
2. Photoanode should have high electron mobility for efficient extraction of electron form dye to external circuit.
3. For diffusion of dye and electrolyte the photoanode should have optimized Pore size.
4. Photoanode needs to be corrosion resistance.
5. For dye to function effectively, the photoanode needs to be able to absorb and scatter sunlight effectively.

Various n-type metal oxides such as TiO_2, ZnO, Nb_2O_5, $SrTiO_3$, and SnO_2 and composites have been investigated for DSSC as photoelectrode materials. The wide bandgap of metal oxide semiconductors (Eg-3 eV) offers suitability for charge transfer. The metal oxide only harvests the UV light while dye molecule harvests the visible light of the solar energy. To improve the performance of DSSC, high amount of dye loading is required which can be achieved by developing high surface metal oxide layer [57–60]. Currently the DSSC also facing various kind of problems such as light scattering capability, charge recombination, interfacial contact, dye pickup, cost effective production of photoanode and Dye/electrolyte sealing which limits it performance and commercialization. To overcome this problem, the engineering of metal oxide photoanode is important. If we see the influence of morphology different Morphology of TiO_2 photoanode is investigated by the scientist. The Mesoporous photoanode mostly have the improving surface area and light scattering ability Cao et al. reported that the mesoporous TiO_2 based DSSC of give an improved light absorption and have best Photo conversion efficiency of 10.5%. Whereas the lower charge conductivity of the mesoporous structure can be overcome by the nanorod structure [61]. Adachi et al. studied the anatase TiO_2 nanowires for the DSSC and able archive the PCE of 9.3% [62]. The low active surface area of the nanowires also limits the efficiency of DSSC. Hollow sphere structure of TiO_2 are also studied to improve the light scattering and the Hollow sphere TiO_2 based DSSC revealed an efficiency of 8.3% [63].

The mechanism of quantum dots solar cell (QDSC) is similar to DSSCs, herein in place of dye the narrow-band-gap semiconductor quantum dots (QDs) are used as photosensitizer. Figure 4 shows the Quantum dots solar cell structure and detailed working principle in which metal oxide semiconductor work as photoanode and quantum dots as sensitizer [64]. Upon light irradiation The QD absorbs the light and generates exciton and the electron transfer to the CB of QDs then it jumps to mental oxide to anode to the external circuit. Meanwhile, the oxidized QDs are regenerated by reduced species in the electrolyte, while the oxidized species reduced by the electrons from the external circuit under the catalysis of cathode.

Mostly narrow band gap QDs are preferred for QDSC. The material should have the following key features for efficient QDSC.

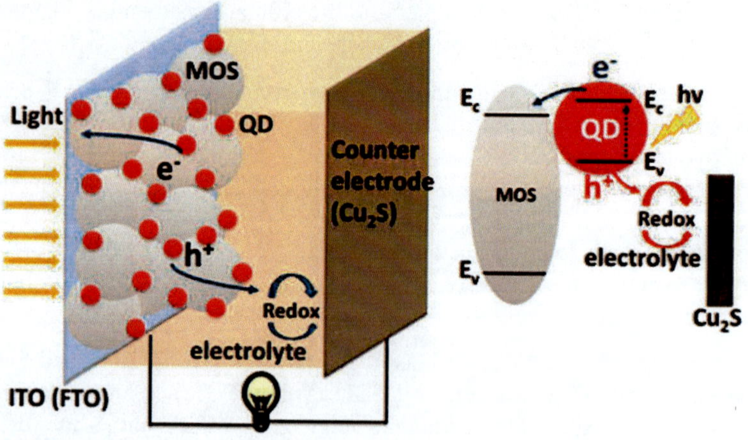

Fig. 4 Structure and operating mechanism of a quantum dots solar cell [Cao et al., *J Phys Chem Lett*, 6, 1869, (2015)] [64]

1. Suitable band alignment between the QD and the ETL is quite essential for the efficient charge transfer and extraction.
2. Good electron mobility of the material is also required for faster extraction of electron from the QDs to substrate.
3. High surface area of the metal oxide is required to get appropriate quantity QD loading for efficient photo absorption.
4. Simple preparation method high chemical stability, and low toxicity are also necessary for the commercialization.

TiO_2 and ZnO are the most widely studied material for the QDSC. TiO_2 nanostructure is most suitable for QDSC due to its distinguishing advantages, including noble chemical stability, less costly, and nontoxicity. TiO_2 nanoparticle with mesoporous structure are the most frequently used material in QDSCs due to the large surface area. Fan Xu et al. prepared Hybrid TiO_2 hollow sphere-nanowire for the QDSC and able to achieve the PCE of 6.01%. Hybrid structure of photoanode has a significantly large specific surface area along with keeping sufficient room for effective electrolyte penetration. Besides, the improved light scattering capability leads to an improvement in utilization of light efficiency and due to which improved PCE is observed [65]. ZnO has the High electron mobility and suitable band alignment took the attention of researcher. Tian et al. synthesized a bilayer photoelectrode consist of ZnO nanoparticle film and ZnO microsphere (MS) by solution process followed by surface modification for QDSC. The QDSC with ZnO NPs/MSs has a PCE of 5.01% were as the ZnO NP QDSC has PCE of 1.66%. Using MSs as the top layer, it was possible to get more effective light scattering to the improved photo current [66]. Seol et al. grown ZnO nanowire using the SILAR method for QDSC then get the maximum PCE 4.15% [67]. Apart from these two material SnO_2, ZrO_2, Zn_2SnO_4, NiO are also investigate for the QDSC. The PCE of QDSC with this material are still lag far behind.

3 Engineering of Metal Oxide Nanostructures to Improve the Solar Light Harvesting

Several strategies have been reported to control the properties of metal oxide nanostructures to make them more appropriate for solar light harvesting applications. The tuning of morphology, size, shape, development of heterostructures, doping, defect control surface modification and other process are effectively introduced to improve the charge extraction and mobility, chemical stability, defect control, efficiency of the device as shown in Fig. 5.

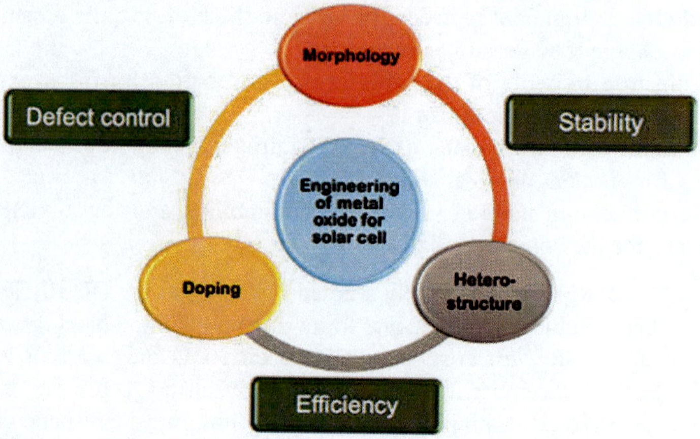

Fig. 5 Schematic representation of deferent engineering process to control the properties of metal oxide nanostructures

3.1 Morphologically Tailored Metal Oxide Nanostructures in Solar Cell Applications

One factor influencing the photo carrier dynamics of the active layers and consequently the solar cell performance that is the morphological arrangement of the donor–acceptor blends. It is recognized that controlling morphology is a key strategy for enhancing phase separation and self-organization. It is known that as materials reduce from bulk to nano dimensions, their surface area increases considerably. Therefore, materials in their nano form have a benefit over their bulk counterparts in terms of demonstrating interesting physical and chemical characteristics. The surface activated phenomena are further modified by the nanoscale material shape. The metal oxides properties can be tuned as per requirement by developing different morphologies in nano sized particles. The efficiency of metal oxides as photo-anodes is largely dependent on their surface area.

It has been reported that different morphologies of TiO_2 nanostructure offer different efficiency of photoconversion in DSSC solar cell. Nanowires morphology possess high surface area and light scattering capability leading to improve light harvesting capabilities. The TiO_2 nanowire deposited by spin coating with regular TiO_2 nanowires and corn like TiO_2 nano has around 1% difference in photoconversion efficiency [68]. The ultra-long anatase TiO_2 nanowire prepared by hydrothermal technique arrays even improved the performance more than 2% and reach to an efficiency of 9.40% in the DSSC application [69]. Also the ultra-long TiO_2 nanowire has shown the photoconversion efficiency of 8.05% shown in Fig. 6c [70]. The nanospindle morphology of the TiO_2 synthesized by hydrothermal technique has shown 8.43% the efficient DSSC due to its improved dye loading light scattering ability [71]. The TiO_2 hallow sphere photoanodes has shown 8.30% photocurrent conversion efficiency [63]. A planer morphology of hexagonal TiO_2 microplates rendered high reflectance of visible light [72]. Along with different morphology such as omni-directional, densely packed, and 3D hierarchically branched TiO_2 nano wires shown in Fig. 6 [73], the rice grain morphology in solar cell resulted efficiency of 5.10, 9.21% [74, 75], star-like TiO_2 microstructures has shown the efficiency of 9.56% [75], hierarchical coronary TiO_2 nanorod arrays has shown the efficiency of 6.54% [76] and TiO_2 nanoparticle/nanofiber shows the efficiency of 10.30% [77]. These reports indicate that the morphology of the TiO_2 nanostructures significantly changes the solar light harvesting and accordingly variations are notices in the photocurrent conversion efficiency (Fig. 6).

Zinc oxide nanostructures include nanowires, nanorods, tetrapods, nanobelts, nanoflowers, nanoparticles, etc. can be achieved by different physical and chemical synthesis technique (Fig. 7). The ZnO nanowire of 13 μm with regular wire and nanoforest morphology having the photoconversion efficiency of 0.75 and 2.51% for DSSC due to the increasing degree of hierarchy and surface area facilities more

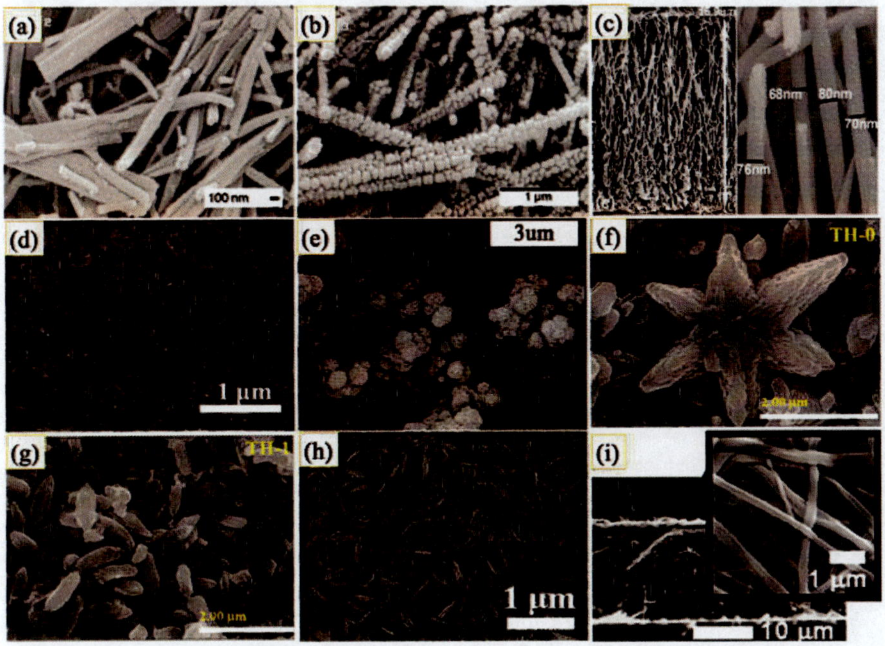

Fig. 6 SEM Images of TiO_2 with the different morphologies. **a** TiO_2 nanorod. **b** Corn-like TiO_2 [Fray et al., *Electrochim Acta*, 90, 302 920130] [68], **c** ultra-long anatase TiO_2 nanowire [Que, Lanfang, et al., Journal of Power Sources 266 (2014)] [70]. **d** TiO_2 nanospindle [Ma, Chao, et al., Colloids and Surfaces A: Physicochemical and Engineering Aspects 538 (2018)] [71]. **e** TiO_2 hollow spheres [Dadgoster et al., *ACS Appl Mater Interfaces*, 4, 2964 (2012)] [63, 71]. **f** star-like TiO_2 microstructures, **g** rice-shaped TiO_2 particles of ∼1 μm [Lekphet, Woranan, et al., Applied Surface Science 382 (2016)] [75]. **h** hierarchical coronary TiO_2 nanorod arrays [Wang, Yinglin, et al., Solar Energy 180 (2019)] [76]. **i** TiO_2 nanoparticle/nanofiber [Chaungchote et al., *Appl Phys Lett*, 93, 033310, (2008)] [77]

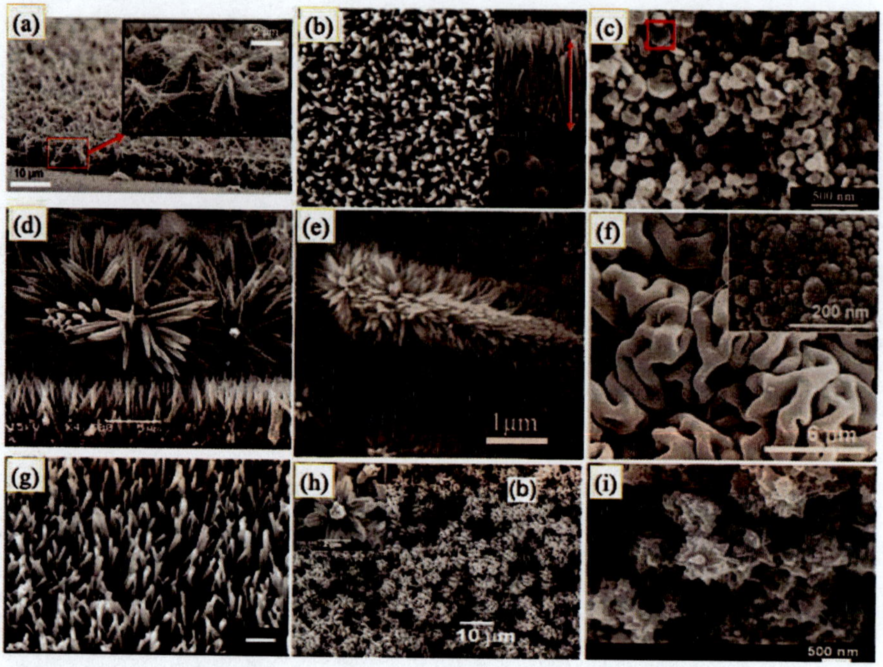

Fig. 7 SEM Images of ZnO with the different morphology. **a** ZnO nanoforest [Ko et al., *Nano Lett*, 11, 666 (2011)] [78]. **b** ZnO nanorod [Urgessa, Z. N., et al., Journal of Alloys and Compounds 798 (2019)] [80]. **c** Hierarchically Structured ZnO Nanorods [Lin et al., *Electrochimica Acta*, 88, 421 (2013)] [81]. **d** Bilayer ZnO nanostructure [Chen, J., et al., Applied Surface Science 255.17 (2009)] [83]. **e** ZnO Nanotips on A Rough Metal Anode [Yang et al., *The Journal of Physical Chemistry C*, 113, 20521 (2009)] [84], **f** network structure of electrospun ZnO nanofiber mats [Kim et al., *Appl Phys Lett*, 91, 163109, (2007)] [85], **g** nanorods array [Fan et al., *The Journal of Physical Chemistry C*, 117, 16349 (2013)] [86]. **h, i** ZnO-nanoflower [*Appl Phys Lett*, 90, 263501 (2007)] [Umar et al., *Materials Research Bulletin*, 47, 2407, (2012)] [87, 88]

dye loading and enhanced photon absorption [78]. The photo current conversion efficiency of 0.22% was achieved for the 2 μm long ZnO nanorod after optimizing the rod length by hydrothermal growth [79] also report shows that the ZnO nanoroad with the length of 1.5 μm has shown the efficiency of 0.8% for the DSSC [80]. ZnO nanorod embedded over ZnO film results in a 2.17% photo-conversion efficiency [81]. By reducing the electron-collection distance, the ZnO nanorods offered a 1D electron-transfer channel and improved charge collection efficiency. Also, hierarchical ZnO nanorods achieved a significant photo-conversion efficiency of 4.13% [82]. The Bilayer ZnO nanostructure synthesized by chemical bath has shown the solar conversion efficiency of 0.45% for QD solar cell [83]. reports show that the ZnO nanotips on rough Zn microtip foil shows a photoconversion efficiency of 1.4% due to the faster charge conduction [84]. The nanofiber mats morphology shown a DSSC efficiency of 1.38% [85] and long ZnO nano wire shown in Fig. 7 has the efficiency of 1.63% [86]. The ZnO flower has the 1.9 and 1.38% photoconversion efficiency in DSSC [87, 88]. Based on the morphological changes the photocurrent conversion efficiency gets changed in ZnO nanostructures.

3.2 *Doped and Defects-Controlled Metal Oxide Nanostructures for Solar Cell Applications*

The bandgap of metal oxides can be modified by doping with the appropriate cation or anion, which modify their electrical characteristics. The altering the materials growth process it has been observed that the band gaps of semiconductor metal oxides get altered [89]. Further the bandgap of metal oxides nanostructures can be modified by altering the dimensions, shape, alloying, doping, defect control. Cations can be added to metal oxide (photo-anode materials) as dopants, which increases the dipole moment and alters the interface's energetics for electron transfer [90]. Strontium doped with TiO_2 causes a 0.6 eV shift in the conduction band's negative direction, or from E_{CB} 4.3 to E_{CB} 3.7 eV [91]. The efficiency of electron-hole separation at the interface is improved by the shift of the conduction band in the negative direction, which also lowers the rate of electron/electrolyte recombination and enhances photocurrent. Efficiency of DSSCs is greatly influenced by the injection and movement of photo-excited electrons inside the cell. A more effective DSSC must have a greater open circuit voltage because of the lower recombination rate [90].

As previously indicated, the TiO_2 and ZnO commonly used as photoanodes only absorb UV light due to their bandgap being in the 3.2–3.3 eV region. The goal of the researchers is to change their optical absorption such that it absorbs more light in the visible range. Different cations and anions are utilized as dopants for TiO_2 to change their photo-conversion efficiencies. TiO_2 is reported with various cationic dopants and their corresponding photoconversion efficiencies (e.g. W (η ~ 9.10%) [92], Nb (η ~ 7.8%) [93], Ce (η ~ 7.65%) [94], Sc (η ~ 9.6%) [95], Cr (η ~ 8.4%) [96], Sn (η ~ 8.31%) [97], Sb (η ~ 8.13%) [98], V (η ~ 7.8%), Nb (η ~ 8.33%), and Ta (η ~ 8.18%) [99]. Similarly, the doping of ZnO with different cations Mg, Al, K and Ga are showing the photoconversion efficiency of 4.11, 0.96, 0.01, and 4.02% respectively [100–103]. The effect of doping amount and type of dopants on photocurrent conversion can be seen in Fig. 8.

In addition to cation doping, various investigators have reported the anion doping into TiO_2. There is evidence that TiO_2 absorbs visible light more effectively when various anions, including N, B, C, S, and F, are present. Where N- doping improved the efficiency from 8.9 to 10.1% [104]. Similarly B, C, S, and F doping improved the efficiency of TiO_2 DSSC from 3.02, 1.43, 5.56, 4.76 to 3.44, 3.39, 6.91, 5.24 respectively [105–108]. The redshift of the absorption spectra in anion doped TiO_2 is typically caused by the creation of a non-oxide phase or a new defect level that is slightly above the valence band. For ZnO, a significant and consistent improvement in efficiency occurs after doping, from 2.30%, 5.08%, 5.60% to 4.50%, 6.25%, and

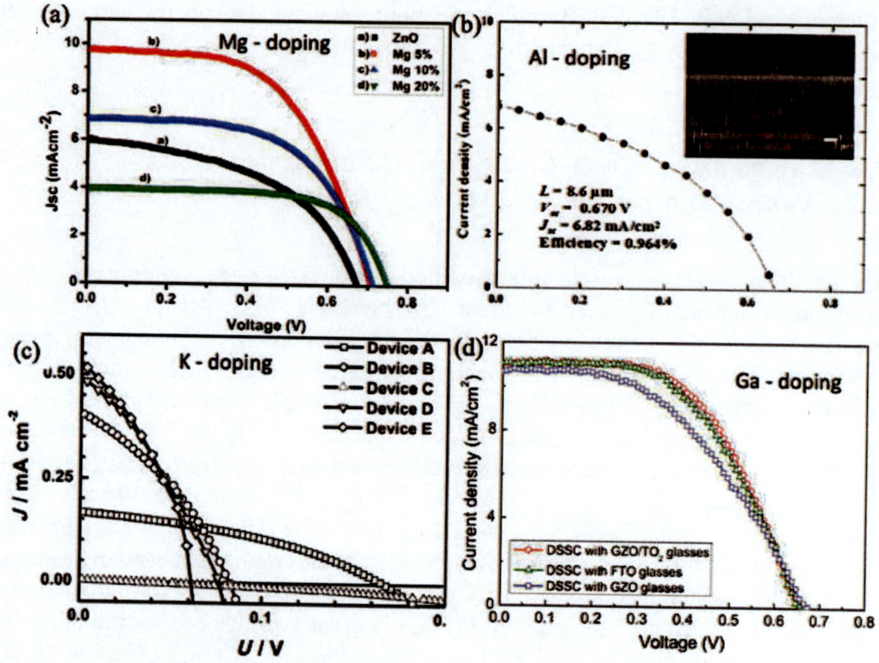

Fig. 8 I–V characteristics of different solar cell after doping of cation in ZnO. **a** Mg-doped [Raj et al., *The Journal of Physical Chemistry C*, 117, 2600 (2013)] [100]; **b** Al-doped [Tao et al., *J Power Sources*, 214, 159 (2012)] [101]; **c** K-doped [Bai et al., *ChemSusChem*, 6, 622 (2013)] [102]; **d** Ga-doped [Kim et al., *Nanoscale Res Lett*, 7, 11, (2012)] [103]

6.10%, for I, N, and B dopant, respectively [109–111]. There have also been reported efforts based on F and B doping in ZnO.

3.3 *Surface Modification of Metal Oxide by Self-assembled Monolayers*

Self-assembled monolayers (SAM) are the order arrangement of the organic molecules on the surface. Molecule and ligands of SAMs are containing some functional group to have high affinity towards the surface of the material and are mostly grown on to the surface by impulsive adsorption of organic molecule on to the surface. The chemical composition and properties of the substrate plays a very significant role in the deposition of SAMs with regulated intermolecular interactions and layer orientation. The SAM are also classified into amorphous and crystalline depending upon the orientation and packing concentration of SAMs. The SAMs is an effective way to engineer interface. SAMs usually consist of three groups: an anchoring group that chemically bonds molecules to surfaces, a linkage that defines how the molecules are arranged, and a functional group that affects surface and interface properties [112].

SAMs are most commonly used in the Perovskite solar cell to tune the interface between perovskite and HTL or ETL. SAMs reduce the defects on the metal oxide surface to get enhance charge extraction and transfer by the HTL and ETL. SAMs also proves a hydrophobicity to the Perovskite layer to improve the stability. The HTL and ETL used in the perovskite solar cell are mostly metal oxides (TiO_2, ZnO, NiO, CuO, etc.). The surface of the oxides is mostly having hydroxyl groups which react with sulfonic, carboxyl, phosphonic, or silane anchoring groups of the SAMs. Afterwards, a dipole moment is get generated at the interface by the anchoring groups which change the band alignment. Also, the defect concentration at the surface got reduced which reduce the charge recombination at the surface and enhance the charge extraction [113].

In ETL Surface modification of metal oxide is achieved by the using self-assembled monolayer (SAM). Which will help to improve interaction of metal oxide and perovskite layer and charge transfer between them. Specially the electric dipole effect has been widely adopted for the SAM to enhance the solar cell efficiency. The high electron dipole and hydrophobic functional group of SAM effects the efficiency of the solar cell. Phosphonic, silane, sulfonic, and carboxyl anchoring groups of the SAM has the great affinity and bind well to metal oxide ETL [114]. Zhang et al. used ionic liquid 1-ethyl-3-methylimidazolium hexafluorophosphate ([EMIM]PF_6) as an inter-layer between the ZnO ETL and $MAPbI_3$ layer. Device with FTO/ZnO/[EMIM]PF_6/$MAPbI_3$/spiro-OMeTAD/Au structure has shown the PCE of 13.50% and device without ionic liquid layer is only having PCE of 10.08%. The enhanced electron extinction is achieved due to suitable band alignment of SAM and also improved long term stability is achieved [115]. Han et al. modified ZnO with Bifunctional 4-picolinic acid interface linker for a higher crystallinity perovskite layer. The deposition of Bifunctional 4-picolinic acid SAM improve the interface transfer property and device exhibits the highest PCE of 18.90% [116]. Liu et al. used SAM to passivate the surface defect of ZnO. 3-(aminopropyl)triethoxysilane [APTES] SAM is used to reduce the oxygen defect site by forming strong O–Si bonding. And the SAM modified ZnO based Perovskite solar cell have a PCE of 18.34% [117].

Surface modification of HTL by the Self assembled monolayer can reduce the defect sites at the surface which enhance the charge charier recombination at the interface during the charge transfer from the active layer to HTL. Wang et al. discussed the effect of SAM on the NiO_x HTL. In this article he engineered the NiO_x surface by a series of benzoic acid self-assembled and used in the perovskite solar cell as a HTL. It was observed that among all 4-bromobenzoic acid SAM effectively done the surface passivation resulting in more stable perovskite solar cells with improved PCE of 18.4%. This SAM layer reduces the charge carrier recombination, reduce the charge transfer resistance between NiO_x NPs and perovskite, and alters the HTL surface wettability, therefore improving the perovskite crystallization with higher stability is obtained [118]. Alghamdi et al. used a 2-(3,6-dimethoxy-9H-carbazol-9-yl)ethyl]phosphonic acid (MeO-2PACz) self-assembled monolayer to modify the surface of NiO_x deposited by sputtering technique for the HTL of the Perovskite solar cell. The overall efficiency of the device was enhanced from 11.9 to 17.2% after the surface modification. This enhancement is noticed due to reduction in charge recombination at the interface, and passivation of the defects site in NiO_x surfaces [119]. Singh et al. reported that the surface modification of NiO_x by 4-cyanophenylphosphonic acid (CNPPA) SAM with electron-withdrawing group can improve the PCE of perovskite solar cell form 17.02 to 18.45% [120].

3.4 Development of Metal Oxide Heterostructures for Solar Cell Applications

The prime objective for the formation of heterojunction structures to enhance the solar light harvesting capability and improve the charge separation [121, 122]. Most metal oxides are used with other active materials to develop heterojunction solar cell. The heterojunction of different semiconducting material is made to develop hybrid solar cell. Most commonly heterojunction solar cell is of two type such as inorganic and organic/inorganic hybrid solar cell. In Inorganic solar cell the heterojunction is made using different semiconducting material whereas hybrid solar cell are the combine two components to convert sunlight into electrical charge.

3.4.1 Inorganic Heterojunction Solar Cell

ZnO is considered as one of the most widely used material while constructing various inorganic heterojunction solar cell, involving thin film solar cells, quantum dot solar cells (QDSC), and various multipurpose excitonic solar cells (e.g., in the active layer, used as hole transport and electron transport layer in n-type semiconductor). During 1970s, Kazmerski et al. constructed a fully inorganic solar cell using the ZnO layer along with ZnO:Al (n-type) in CdS: Cu(InGa)Se_2 thin film solar cell [123]. In between active layer and the cathode, a buffer layer made of ZnO was utilized, and the ZnO:Al mixture worked as a top contact. As time goes on, ZnO became widely used in bulk heterojunction solar cell as n-type semiconductor component, due to its high charge carrier mobility, and the wide band gap of 3.37 eV at room temperature [28].

The most prevalent heterojunction inorganic solar cell is made of a mixture of lead chalcogenides (e.g., PbSe, PbS, PbTe) and zinc oxide. Over the time, PbX has become a great material for photovoltaics due to its high dielectric constant and hence, there are huge Bhor radii, which produce a noticeable quantum confinement effect [124]. The energy level of PbX is advantageous when combined with ZnO because the LUMO of PbX can be modified to reduce the energy gap with the conduction band of ZnO. Leschikes at el. reported a heterojunction solar cell developed on a planar heterojunction between a thin ZnO sheet and a PbSe nanocrystal. The heterojunction solar cell using ZnO thin film achieved greater photocurrent density and open circuit potential values, with an overall PCE of 1.6%, as compared to Schottky solar cell manufactured using comparable PbSe NCs. To increase the efficiency, they replaced the planar films with nanostructured film which increased the interfacial area of the film and hence it resulted in massive improvements nearly doubling the efficiency to that of planar ZnO cells (n = 2.94%) [125].

Apart from lead based solar cells, copper-based p-type semiconductors with combination of n-type ZnO were started to be used more often as they offer a low cost and non-toxic material with high potential solar cells, which holds advantage over Pb and Cd based solar cell with can be toxic. In 2018, Zang et al. fabricated a micrometer grain sized CuO2/ZnO solar cell with a PCE of 3.17%, however this is still low as compared to the theoretical PCE of 20% [126]. While nanoparticles offer a greater surface area, still the nonuniform distribution is causing high recombination losses, hence they came up with the idea of nanowires or nanotubes which provides high interface as well as smooth pathway for transportation of electrons. ZnO nanowires (NWs):CdS was inserted into CIGS solar cells in place CuO_2 to enhance the performance of the solar cell. In the PbS QDs solar cells, ZnO nanowires were used causing in a solar cell with photocurrent of over 20 mA/cm^2 and PCE ranging up to 4.3% [127].

The modification of the ZnO surface by TiO_2 nanoscale coating reduce the charge carrier recombination and enhance the efficiency of the perovskite solar cell. In inorganic bulk heterojunction solar cell ZnO and TiO_2 are used with different p-type lead chalcogenide (PbS, PbSe) [124], p-type semiconductors (Cu_2O [128], Cu_2ZnSnS_4 (CZTS) [129], Cu(In,Ga)Se_2 [130]) and several others are often being investigated. Improve the light collection capacity of QDSCs is possible by depositing various quantum dots (QDs) (Al_2O_3, MgO, TiO_2, SnO_2 etc.) on ZnO NRs and taking advantage in terms of high extinction coefficient, adjustable bandgap, long lifetime, high stability, and simple synthesis. Additionally, ZnO NRs arrays can offer direct electron transport routes and prevent recombination at grain boundaries, enhancing the efficiency of carrier collection [131].

3.4.2 Hybrid Organic/Inorganic Heterojunction Solar Cells

Hybrid solar cell is trusted for mechanical flexibility and low price. Hybrid solar cell having two components to convert the sunlight into electrical energy: (i) An organic semiconductor composed of a conjugated polymer that can act active photo absorbent material and electron donor and (ii) an inorganic semiconductor of a oxide and chalcogenide acting as the electron acceptor [132]. ZnO is used to substitute the organic material in fully organic solar cell to improve the electron mobility in organic hybrid solar cell. Energy production mechanism of the hybrid solar cell is similar to the organic solar cell, where light illumination produces excitons generation which diffuse donor/acceptor edge within a definite diffused length. The difference in donor and acceptor between the HOMO and LUMO creates a driving force that allows excitons to be split into two independent free carriers by charge transfer, by defeating the excitons binding energy. The positive moves towards cathode and

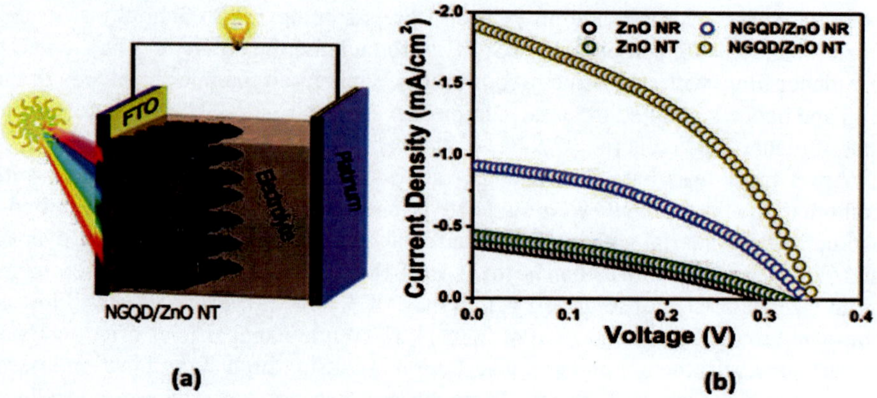

Fig. 9 **a** Schematic diagram of a typical QDSSC device and **b** I–V characteristics of ZnO and ZnO/N-doped graphene quantum dots heterostructure [136]

negative charges go to anode by constant charge separation by donor and acceptor materials [133].

In 2004 first study of ZnO in heterojunction solar cell was reported. Nanocrystalline ZnO of diameter ~5 nm was prepared by hydrolysis to fabricate a bulk heterojunction hybrid solar cell device along with organic p-type semiconductor MDMO-PPV. The forward current density for organic/inorganic hybrid solar cells was substantially greater in ZnO:MDMO-PPV than in pristine MDMO-PPV in the same configuration, showing that the presence of ZnO nano structure does indeed offer a continuous channel for electron transport [134]. Surface modified or core-shell structure are widely used in the development of metal oxide heterostructures [135]. The photoconversion efficiency of 1.6% is achieved for nanocrystalline ZnO/MDMO:PPV hybrid solar cell with 0.71 sun illumination [134]. The photogenerated carriers in quantum dot sensitized solar cells are only separated in a small region close to the interface, making planar solar cells ineffective for absorbing light and collecting photogenerated carriers. ZnO heterostructure developed using the nitrogen-doped graphene QDs, has resulted improvement in photo-current conversion efficiency as shown in Fig. 9 [136].

4 Conclusion

Metal oxide nanostructures have drawn significant attention towards optoelectronic application. The harvesting of solar energy either using solar cell or other electrochemical devices were demonstrated by applying variety of metal oxide nanostructures. It is very interesting to understand the properties of the same materials when it changes the morphology, surface, and size. It has been noticed that material properties are drastically altered when achieves nano dimension. The control over surface defects leads to effective change on photon absorption and emission capabilities. The bandgap can be effectively tuned as per requirement by doping, surface alloying, surface modification and heterostructure. All these possibilities have been explored with variety of metal oxide nanostructures to develop the efficient photoelectrode for solar energy harvesting. The engineering of metal oxide nanostructures has shown significant control over the performance of solar cell. Numerous efforts have been reports and many more are underway to further improve the properties of metal oxide nanostructures to make them useful for not only in optoelectronic but several other leading applications.

References

1. S. Sundaram, D. Benson, T.K. Mallick, Overview of the PV industry and different technologies, in *Solar Photovoltaic Technology Production* (Elsevier, 2016), pp. 7–22
2. N.K. Elumalai, C. Vijila, R. Jose et al., Metal oxide semiconducting interfacial layers for photovoltaic and photocatalytic applications. Mater. Renew. Sustain. Energy **4**, 11 (2015)
3. V.C. Lokhande, C.H. Kim, A.C. Lokhande et al., Metal oxides for perovskite solar cells, in *Chemically Deposited Nanocrystalline Metal Oxide Thin Films* (Springer International Publishing, Cham, 2021), pp.197–233
4. S. Akin, Boosting the efficiency and stability of perovskite solar cells through facile molecular engineering approaches **199**, 136–142 (2020)
5. A.E. Shalan, W. Sharmoukh, A.N. Elshazly et al., Dopant-free hole-transporting polymers for efficient, stable, and hysteresis-less perovskite solar cells **26**, e00226 (2020)
6. A.H. Zaki, A.E. Shalan, A. El-Shafeay et al., Acceleration of ammonium phosphate hydrolysis using TiO_2 microspheres as a catalyst for hydrogen production. Nanosc. Adv. **2**, 2080–2086 (2020)
7. K. Valadi, S. Gharibi, R. Taheri-Ledari et al., Metal oxide electron transport materials for perovskite solar cells: a review (2021)
8. J.X. Song, X.X. Yin, Z.F. Li et al., Low-temperature-processed metal oxide electron transport layers for efficient planar perovskite solar cells (2021)
9. Grätzel M (2003) Dye-sensitized solar cells **4**, 145–153
10. H. Zhou, Y. Shi, K. Wang et al., Low-temperature processed and carbon-based $ZnO/CH_3NH_3PbI_3/C$ planar heterojunction perovskite solar cells **119**, 4600–4605 (2015)
11. T. Singh, J. Singh, T. Miyasaka, Role of metal oxide electron-transport layer modification on the stability of high performing perovskite solar cells (2016)
12. J. Song, E. Zheng, J. Bian et al., Low-temperature SnO_2-based electron selective contact for efficient and stable perovskite solar cells. J. Mater. Chem. A Mater. **3**, 10837–10844 (2015)
13. F. Sadegh, S. Akin, M. Moghadam et al., Highly efficient, stable and hysteresis-less planar perovskite solar cell based on chemical bath treated Zn_2SnO_4 electron transport layer. Nano Energy **75**, 105038 (2020)
14. L. Mohapatra, M. Nagaraju, S. Suman et al., Enhancement in photocurrent conversion efficiency via recrystallization of zinc tin hydroxide nanostructures. J. Alloys Compd. **928**, 167127 (2022)
15. Y. Wang, M. Zhong, W. Wang et al., Effects of ZnSe modification on the perovskite films and perovskite solar cells based on ZnO nanorod arrays. Appl. Surf. Sci. **495**, 143552 (2019)
16. Ü. Özgür, A. YaI, C. Liu et al., A comprehensive review of ZnO materials and devices. J. Appl. Phys. **98**, 041301 (2005)
17. Y. Xu, P.D. Lund, H. Wang et al., Performance improvement of perovskite solar cells based on PCBM-modified ZnO-nanorod arrays. IEEE J. Photovolt. **6**, 1530–1536 (2016)

18. Z. Li, J. Zhang, Y. Xu et al., Improving photovoltaic performance of perovskite solar cells: the interfacial modification role of aluminum chloride and ammonia on ZnO nanorods **10**, 1750017 (2017)
19. F. Khan, J. Hyun Kim, Enhanced charge-transportation properties of low-temperature processed Al-doped ZnO and its impact on PV cell parameters of organic-inorganic perovskite solar cells. Solid State Electron. **164**, 107714 (2020)
20. Y. Ogomi, A. Morita, S. Tsukamoto et al., CH3 NH3 SnxPb (1 − x) I 3 perovskite solar cells covering up to 1060 nm. J. Phys. Chem. Lett. **5**, 1004–1011 (2014)
21. K. Wojciechowski, S.D. Stranks, A. Abate et al., Heterojunction modification for highly efficient organic-inorganic perovskite solar cells. ACS Nano **8**, 12701–12709 (2014)
22. K. Mahmood, A. Khalid, M. Hameed et al., Thin films of nanostructured ZnO used as an electron-transporting material for improved performance of perovskite solar cells **124**, 824 (2018)
23. Q. Luo, H. Chen, Y. Lin et al., Discrete Iron(III) oxide nanoislands for efficient and photostable perovskite solar cells. Adv. Funct. Mater. **27**, 1702090 (2017)
24. Z. Wang, J. Lou, X. Zheng et al., Solution processed Nb_2O_5 electrodes for high efficient ultraviolet light stable planar perovskite solar cells. ACS Sustain. Chem. Eng. **7**, 7421–7429 (2019)
25. X. Liu, C.-C. Chueh, Z. Zhu et al., Highly crystalline Zn_2SnO_4 nanoparticles as efficient electron-transporting layers toward stable inverted and flexible conventional perovskite solar cells. J. Mater. Chem. A Mater. **4**, 15294–15301 (2016)
26. M.M. Tavakoli, D. Prochowicz, P. Yadav et al., Zinc stannate nanorod as an electron transporting layer for highly efficient and hysteresis-less perovskite solar cells (2018)
27. Z. Li, R. Wang, J. Xue et al., Core-shell ZnO@SnO_2 nanoparticles for efficient inorganic perovskite solar cells. J. Am. Chem. Soc. **141**, 17610–17616 (2019)
28. M. Zhong, L. Chai, Y. Wang, Core-shell structure of ZnO@TiO_2 nanorod arrays as electron transport layer for perovskite solar cell with enhanced efficiency and stability. Appl. Surf. Sci. **464**, 301–310 (2019)
29. Z.-L. Tseng, C.-H. Chiang, S.-H. Chang et al., Surface engineering of ZnO electron transporting layer via Al doping for high efficiency planar perovskite solar cells. Nano Energy **28**, 311–318 (2016)
30. Y.-Z. Zheng, E.-F. Zhao, F.-L. Meng et al., Iodine-doped ZnO nanopillar arrays for perovskite solar cells with high efficiency up to 18.24%. J. Mater. Chem. A Mater. **5**, 12416–12425 (2017)
31. Q. An, P. Fassl, Y.J. Hofstetter et al., High performance planar perovskite solar cells by ZnO electron transport layer engineering. Nano Energy **39**, 400–408 (2017)
32. R. Azmi, S. Hwang, W. Yin et al., High efficiency low-temperature processed perovskite solar cells integrated with alkali metal doped ZnO electron transport layers. ACS Energy Lett. **3**, 1241–1246 (2018)
33. Y. Guo, T. Liu, N. Wang et al., Ni-doped α-Fe 2 O 3 as electron transporting material for planar heterojunction perovskite solar cells with improved efficiency, reduced hysteresis and ultraviolet stability. Nano Energy **38**, 193–200 (2017)
34. X. Ye, H. Ling, R. Zhang et al., Low-temperature solution-combustion-processed Zn-Doped Nb_2O_5 as an electron transport layer for efficient and stable perovskite solar cells. J. Power Sour. **448**, 227419 (2020)
35. J. Wang, Y.J. Lee, J.W.P. Hsu, Sub-10 nm copper chromium oxide nanocrystals as a solution processed p-type hole transport layer for organic photovoltaics. J. Mater. Chem. C Mater. **4**, 3607–3613 (2016)
36. J. Wang, V. Ibarra, D. Barrera et al., Solution synthesized p-type copper gallium oxide nanoplates as hole transport layer for organic photovoltaic devices **6**, 1071–1075 (2015)
37. C. Zuo, L. Ding, Solution-processed Cu_2O and CuO as hole transport materials for efficient perovskite solar cells **11**, 5528–5532 (2015)
38. R. Singh, P.K. Singh, B. Bhattacharya et al., Review of current progress in inorganic hole-transport materials for perovskite solar cells (2019)

39. A. Garcia, G.C. Welch, E.L. Ratcliff et al., Improvement of interfacial contacts for new small-molecule bulk-heterojunction organic photovoltaics **24**, 5368–5373 (2012)
40. M.D. Irwin, D.B. Buchholz, A.W. Hains et al., *p*-type semiconducting nickel oxide as an efficiency-enhancing anode interfacial layer in polymer bulk-heterojunction solar cells **105**, 2783–2787 (2008)
41. J.-Y. Jeng, K.-C. Chen, T.-Y. Chiang et al., Nickel oxide electrode interlayer in $CH_3NH_3PbI_3$ perovskite/PCBM planar-heterojunction hybrid solar cells **26**, 4107–4113 (2014)
42. K.-C. Wang, P.-S. Shen, M.-H. Li et al., Low-temperature sputtered nickel oxide compact thin film as effective electron blocking layer for mesoscopic NiO/$CH_3NH_3PbI_3$ perovskite heterojunction solar cells. ACS Appl. Mater. Interfaces **6**, 11851–11858 (2014)
43. J.H. Park, J. Seo, S. Park et al., Efficient $CH_3NH_3PbI_3$ perovskite solar cells employing nanostructured p-type NiO electrode formed by a pulsed laser deposition **27**, 4013–4019 (2015)
44. W. Chen, F.-Z. Liu, X.-Y. Feng et al., Cesium doped NiO_x as an efficient hole extraction layer for inverted planar perovskite solar cells. Adv. Energy Mater. **7**, 1700722 (2017)
45. W. Chen, Y. Wu, J. Fan et al., Understanding the doping effect on NiO: toward high-performance inverted perovskite solar cells. Adv. Energy Mater. **8**, 1703519 (2018)
46. U. Kwon, B.-G. Kim, D.C. Nguyen et al., Solution-processible crystalline NiO nanoparticles for high-performance planar perovskite photovoltaic cells. Sci. Rep. **6**, 30759 (2016)
47. X. Yin, J. Zhai, L. Song et al., Novel NiO nanoforest architecture for efficient inverted mesoporous perovskite solar cells. ACS Appl. Mater. Interfaces **11**, 44308–44314 (2019)
48. X. Yin, J. Zhai, T. Wang et al., Mesoporous NiO nanosheet network as efficient hole transporting layer for stable inverted perovskite solar cells. Mater. Lett. **231**, 101–104 (2018)
49. L. Zhao, G. Su, W. Liu et al., Optical and electrochemical properties of Cu-doped NiO films prepared by electrochemical deposition. Appl. Surf. Sci. **257**, 3974–3979 (2011)
50. J.W. Jung, C.-C. Chueh, A.K.-Y. Jen, A low-temperature, solution-processable, Cu-doped nickel oxide hole-transporting layer via the combustion method for high-performance thin-film perovskite. Sol. Cells **27**, 7874–7880 (2015)
51. Z. Qiu, H. Gong, G. Zheng et al., Enhanced physical properties of pulsed laser deposited NiO films via annealing and lithium doping for improving perovskite solar cell efficiency. J. Mater. Chem. C Mater. **5**, 7084–7094 (2017)
52. M.-A. Park, I.J. Park, S. Park et al., Enhanced electrical properties of Li-doped Ni*Ox* hole extraction layer in p – i — n type perovskite solar cells **18**, S55–S59 (2018)
53. Y. Xie, K. Lu, J. Duan et al., Enhancing photovoltaic performance of inverted planar perovskite solar cells by cobalt-doped nickel oxide hole transport layer. ACS Appl. Mater. Interfaces **10**, 14153–14159 (2018)
54. Y. Wei, K. Yao, X. Wang et al., Improving the efficiency and environmental stability of inverted planar perovskite solar cells via silver-doped nickel oxide hole-transporting layer. Appl. Surf. Sci. **427**, 782–790 (2018)
55. M.-H. Liu, Z.-J. Zhou, P.-P. Zhang et al., p-type Li, Cu-codoped NiOx hole-transporting layer for efficient planar perovskite solar cells. Opt. Express **24**, A1349 (2016)
56. P.-L. Qin, H.-W. Lei, X.-L. Zheng et al., Copper-doped chromium oxide hole-transporting layer for perovskite solar cells: interface engineering and performance improvement. Adv. Mater. Interfaces **3**, 1500799 (2016)
57. M. Grätzel, Solar energy conversion by dye-sensitized photovoltaic cells. Inorg. Chem. **44**, 6841–6851 (2005)
58. R. Jose, V. Thavasi, S. Ramakrishna, Metal oxides for dye-sensitized. Sol. Cells **92**, 289–301 (2009)
59. J. Gong, K. Sumathy, Q. Qiao et al., Review on dye-sensitized solar cells (DSSCs): advanced techniques and research trends **68**, 234–246 (2017)
60. Q. Zhang, D. Myers, J. Lan et al., Applications of light scattering in dye-sensitized solar cells **14**, 14982 (2012)
61. K. Cao, J. Lu, H. Li et al., Efficient dye-sensitized solar cells using mesoporous submicrometer TiO_2 beads. RSC Adv. **5**, 62630–62637 (2015)

62. M. Adachi, Y. Murata, J. Takao et al., Highly efficient dye-sensitized solar cells with a titania thin-film electrode composed of a network structure of single-crystal-like TiO_2 nanowires made by the "oriented attachment" mechanism. J. Am. Chem. Soc. **126**, 14943–14949 (2004)
63. S. Dadgostar, F. Tajabadi, N. Taghavinia, Mesoporous submicrometer TiO_2 hollow spheres as scatterers in dye-sensitized solar cells. ACS Appl. Mater. Interfaces **4**, 2964–2968 (2012)
64. J. Tian, G. Cao, Control of nanostructures and interfaces of metal oxide semiconductors for quantum-dots-sensitized solar cells. J. Phys. Chem. Lett. **6**, 1859–1869 (2015)
65. Y.-F. Xu, W.-Q. Wu, H.-S. Rao et al., CdS/CdSe co-sensitized TiO2 nanowire-coated hollow spheres exceeding 6% photovoltaic performance. Nano Energy **11**, 621–630 (2015)
66. J. Tian, L. Lv, X. Wang et al., Microsphere light-scattering layer assembled by ZnO nanosheets for the construction of high efficiency (>5%) quantum dots sensitized solar cells **118**, 16611–16617 (2014)
67. M. Seol, H. Kim, Y. Tak et al., Novel nanowire array based highly efficient quantum dot sensitized solar cell **46**, 5521 (2010)
68. A.M. Bakhshayesh, M.R. Mohammadi, H. Dadar et al., Improved efficiency of dye-sensitized solar cells aided by corn-like TiO2 nanowires as the light scattering layer. Electrochim Acta **90**, 302–308 (2013)
69. W.-Q. Wu, Y.-F. Xu, C.-Y. Su et al., Ultra-long anatase TiO_2 nanowire arrays with multi-layered configuration on FTO glass for high-efficiency dye-sensitized solar cells **7**, 644–649 (2014)
70. L. Que, Z. Lan, W. Wu et al., High-efficiency dye-sensitized solar cells based on ultra-long single crystalline titanium dioxide nanowires. J. Power Sour. **266**, 440–447 (2014)
71. C. Ma, L. Wang, Z. Guo et al., Monodisperse TiO_2 microspheres assembled by porous spindles for high performance dye-sensitized solar cells. Colloids Surf. A Physicochem. Eng. Asp. **538**, 94–99 (2018)
72. W. Peng, L. Han, Hexagonal TiO2 microplates with superior light scattering for dye-sensitized solar cells. J. Mater. Chem. **22**, 20773 (2012)
73. D. Lee, Y. Rho, F.I. Allen et al., Synthesis of hierarchical TiO2 nanowires with densely-packed and omnidirectional branches. Nanoscale **5**, 11147 (2013)
74. A.S. Nair, Y. Shengyuan, Z. Peining et al., Rice grain-shaped TiO_2 mesostructures by electrospinning for dye-sensitized solar cells **46**, 7421 (2010)
75. W. Lekphet, T.-C. Ke, C. Su et al., Morphology control studies of TiO_2 microstructures via surfactant-assisted hydrothermal process for dye-sensitized solar cell applications. Appl. Surf. Sci. **382**, 15–26 (2016)
76. Y. Wang, P. Cheng, C. Feng et al., High performance flexible dye-sensitized solar cells base on multiple functional optimizations **180**, 423–428 (2019)
77. S. Chuangchote, T. Sagawa, S. Yoshikawa, Efficient dye-sensitized solar cells using electrospun TiO_2 nanofibers as a light harvesting layer. Appl. Phys. Lett. **93**, 033310 (2008)
78. S.H. Ko, D. Lee, H.W. Kang et al., Nanoforest of hydrothermally grown hierarchical ZnO nanowires for a high efficiency dye-sensitized solar cell. Nano Lett. **11**, 666–671 (2011)
79. W.K. Tan, Z. Lockman, K. Abdul Razak et al., Enhanced dye-sensitized solar cells performance of ZnO nanorod arrays grown by low-temperature hydrothermal reaction. Int. J. Energy Res. n/a-n/a (2013)
80. Z.N. Urgessa, R. Ruess, S.R. Tankio Djiokap et al., Effect of morphology and surface treatment on the performance of ZnO nanorod-based dye-sensitized solar cells. J. Alloys Compd. **798**, 249–256 (2019)
81. L.-Y. Lin, M.-H. Yeh, C.-P. Lee et al., Flexible dye-sensitized solar cells with one-dimensional ZnO nanorods as electron collection centers in photoanodes. Electrochim. Acta **88**, 421–428 (2013)
82. W. Peng, L. Han, Z. Wang, Hierarchically structured ZnO nanorods as an efficient photoanode for dye-sensitized solar cells **20**, 8483–8487 (2014)
83. J. Chen, L. Chen, J.L. Song et al., Bilayer ZnO nanostructure fabricated by chemical bath and its application in quantum dot sensitized solar cell. Appl. Surf. Sci. **255**, 7508–7511 (2009)

84. Z. Yang, T. Xu, Y. Ito et al., Enhanced electron transport in dye-sensitized solar cells using short ZnO nanotips on a rough metal anode **113**, 20521–20526 (2009)
85. I.-D. Kim, J.-M. Hong, B.H. Lee et al., Dye-sensitized solar cells using network structure of electrospun ZnO nanofiber mats. Appl. Phys. Lett. **91**, 163109 (2007)
86. J. Fan, Y. Hao, C. Munuera et al., Influence of the annealing atmosphere on the performance of ZnO nanowire dye-sensitized solar cells **117**, 16349–16356 (2013)
87. C.Y. Jiang, X.W. Sun, G.Q. Lo et al., Improved dye-sensitized solar cells with a ZnO-nanoflower photoanode. Appl. Phys. Lett. **90**, 263501 (2007)
88. A. Umar, M.S. Akhtar, A. Al-Hajry et al., Hydrothermally grown ZnO nanoflowers for environmental remediation and clean energy applications. Mater. Res. Bull. **47**, 2407–2414 (2012)
89. F. Capasso, Band-gap engineering: from physics and materials to new semiconductor devices. Science **235**, 172–176 (1979)
90. D.F. Watson, G.J. Meyer, Cation effects in nanocrystalline solar cells. Coord. Chem. Rev. **248**, 1391–1406 (2004)
91. Y. Diamant, S.G. Chen, O. Melamed et al., Core-shell nanoporous electrode for dye sensitized solar cells: the effect of the $SrTiO_3$ shell on the electronic properties of the TiO_2 core. J. Phys. Chem. B **107**, 1977–1981 (2003)
92. X. Zhang, F. Liu, Q.-L. Huang et al., Dye-sensitized W-doped TiO_2 solar cells with a tunable conduction band and suppressed charge recombination **115**, 12665–12671 (2011)
93. X. Lü, X. Mou, J. Wu et al., Improved-performance dye-sensitized solar cells using Nb-doped TiO_2 electrodes: efficient electron injection and transfer. Adv. Funct. Mater. **20**, 509–515 (2010)
94. J. Zhang, W. Peng, Z. Chen et al., Effect of cerium doping in the TiO_2 photoanode on the electron transport of dye-sensitized solar cells **116**, 19182–19190 (2012)
95. A. Latini, C. Cavallo, F.K. Aldibaja et al., Efficiency improvement of DSSC photoanode by scandium doping of mesoporous titania beads **117**, 25276–25289 (2013)
96. C. Kim, K.-S. Kim, H.Y. Kim et al., Modification of a TiO2 photoanode by using Cr-doped TiO2 with an influence on the photovoltaic efficiency of a dye-sensitized solar cell. J. Mater. Chem. **18**, 5809 (2008)
97. Y. Duan, N. Fu, Q. Liu et al., Sn-doped TiO_2 photoanode for dye-sensitized solar cells **116**, 8888–8893 (2012)
98. M. Wang, S. Bai, A. Chen et al., Improved photovoltaic performance of dye-sensitized solar cells by Sb-doped TiO2 photoanode. Electrochim. Acta **77**, 54–59 (2012)
99. J. Liu, Y. Duan, X. Zhou et al., Influence of VB group doped TiO2 on photovoltaic performance of dye-sensitized solar cells. Appl. Surf. Sci. **277**, 231–236 (2013)
100. C.J. Raj, K. Prabakar, S.N. Karthick et al., Banyan root structured Mg-doped ZnO photoanode dye-sensitized solar cells **117**, 2600–2607 (2013)
101. R. Tao, T. Tomita, R.A. Wong et al., Electrochemical and structural analysis of Al-doped ZnO nanorod arrays in dye-sensitized solar cells. J. Power Sour. **214**, 159–165 (2012)
102. J. Bai, X. Xu, L. Xu et al., Potassium-doped Zinc Oxide as photocathode material in dye-sensitized solar cells. Chemsuschem **6**, 622–629 (2013)
103. J.-H. Kim, K.-J. Lee, J.-H. Roh et al., Ga-doped ZnO transparent electrodes with TiO2 blocking layer/nanoparticles for dye-sensitized solar cells. Nanosc. Res. Lett. **7**, 11 (2012)
104. W. Guo, L. Wu, Z. Chen et al., Highly efficient dye-sensitized solar cells based on nitrogen-doped titania with excellent stability. J. Photochem. Photobiol. A Chem. **219**, 180–187 (2011)
105. A. Subramanian, H.-W. Wang, Effects of boron doping in TiO2 nanotubes and the performance of dye-sensitized solar cells. Appl. Surf. Sci. **258**, 6479–6484 (2012)
106. N.T. Hieu, S.J. Baik, O.H. Chung et al., Fabrication and characterization of electrospun carbon nanotubes/titanium dioxide nanofibers used in anodes of dye-sensitized solar cells. Synth. Met. **193**, 125–131 (2014)
107. Q. Sun, J. Zhang, P. Wang et al., Sulfur-doped TiO_2 nanocrystalline photoanodes for dye-sensitized solar cells **4**, 023104 (2012)

108. S.I. Noh, K.-N. Bae, H.-J. Ahn et al., Improved efficiency of dye-sensitized solar cells through fluorine-doped TiO2 blocking layer. Ceram. Int. **39**, 8097–8101 (2013)
109. H. Tian, L. Hu, C. Zhang et al., Enhanced photovoltaic performance of dye-sensitized solar cells using a highly crystallized mesoporous TiO_2 electrode modified by boron doping **21**, 863–868 (2011)
110. Y. Xie, N. Huang, Y. Liu et al., Photoelectrodes modification by N doping for dye-sensitized solar cells. Electrochim. Acta **93**, 202–206 (2013)
111. Y.-Z. Zheng, X. Tao, Q. Hou et al., Iodine-doped ZnO nanocrystalline aggregates for improved dye-sensitized solar cells **23**, 3–5 (2011)
112. A. Ulman, Formation and structure of self-assembled monolayers. Chem. Rev. **96**, 1533–1554 (1996)
113. Z. Li, J. Guo, Z. Li et al., Incorporating self-assembled silane-crosslinked carbon dots into perovskite solar cells to improve efficiency and stability. J. Mater. Chem. A Mater. **8**, 5629–5637 (2020)
114. F. Ambrosio, N. Martsinovich, A. Troisi, What is the best anchoring group for a dye in a dye-sensitized solar cell? J. Phys. Chem. Lett. **3**, 1531–1535 (2012)
115. W. Zhang, Z. Ren, Y. Guo et al., Improved the long-term air stability of ZnO-based perovskite solar cells prepared under ambient conditions via surface modification of the electron transport layer using an ionic liquid. Electrochim. Acta **268**, 539–545 (2018)
116. F. Han, G. Hao, Z. Wan et al., Bifunctional electron transporting layer/perovskite interface linker for highly efficient perovskite solar cells. Electrochim. Acta **296**, 75–81 (2019)
117. D. Liu, Y. Wang, Z. She et al., Suppressed decomposition of perovskite film on ZnO via a self-assembly monolayer of methoxysilane **2**, 1800240 (2018)
118. Q. Wang, C. Chueh, T. Zhao et al., Effects of self-assembled monolayer modification of nickel oxide nanoparticles layer on the performance and application of inverted perovskite solar cells. Chemsuschem **10**, 3794–3803 (2017)
119. A.R.M. Alghamdi, M. Yanagida, Y. Shirai et al., Surface passivation of sputtered NiO_x using a SAM interface layer to enhance the performance of perovskite solar cells. ACS Omega **7**, 12147–12157 (2022)
120. N. Singh, Y. Tao, Effect of surface modification of nickel oxide hole-transport layer via self-assembled monolayers in perovskite solar cells **2**, 2390–2399 (2021)
121. H.S. Jung, J.-K. Lee, M. Nastasi et al., Preparation of nanoporous MgO-coated TiO_2 nanoparticles and their application to the electrode of dye-sensitized solar cells **21**, 10332–10335 (2005)
122. S.S. Kanmani, K. Ramachandran, Synthesis and characterization of TiO_2/ZnO core/shell nanomaterials for solar cell applications. Renew. Energy **43**, 149–156 (2012)
123. L.L. Kazmerski, F.R. White, G.K. Morgan, Thin-film $CuInSe_2$/CdS heterojunction solar cells. Appl. Phys. Lett. **29**, 268–270 (1976)
124. C.K. Miskin, S.D. Deshmukh, V. Vasiraju et al., Lead chalcogenide nanoparticles and their size-controlled self-assemblies for thermoelectric and photovoltaic applications. ACS Appl. Nano Mater. **2**, 1242–1252 (2019)
125. K.S. Leschkies, T.J. Beatty, M.S. Kang et al., Solar cells based on junctions between colloidal PbSe nanocrystals and thin ZnO films. ACS Nano **3**, 3638–3648 (2009)
126. Z. Zang, Efficiency enhancement of ZnO/Cu_2O solar cells with well oriented and micrometer grain sized Cu_2O films. Appl. Phys. Lett. **112**, 042106 (2018)
127. J. Jean, S. Chang, P.R. Brown et al., ZnO nanowire arrays for enhanced photocurrent in PbS quantum dot solar cells **25**, 2790–2796 (2013)
128. N.M. Rosas-Laverde, A. Pruna, J. Cembrero et al., Performance of graphene oxide-modified electrodeposited ZnO/Cu_2O heterojunction solar cells **58**, 263–273 (2019)
129. R.N. Gayen, T. Chakrabarti, Effect of series and shunt resistance on the photovoltaic properties of solution-processed zinc oxide nanowire based CZTS solar cell in superstrate configuration. Mater. Sci. Semicond. Process. **100**, 1–7 (2019)
130. S. Qiao, J. Liu, G. Fu et al., ZnO nanowire based CIGS solar cell and its efficiency enhancement by the piezo-phototronic effect. Nano Energy **49**, 508–514 (2018)

131. B. Boro, B. Gogoi, B.M. Rajbongshi et al., Nano-structured TiO2/ZnO nanocomposite for dye-sensitized solar cells application: a review **81**, 2264–2270 (2018)
132. M. Halim, Harnessing sun's energy with quantum dots based next generation solar cell **3**, 22–47(2012)
133. J. Huang, Z. Yin, Q. Zheng, Applications of ZnO in organic and hybrid solar cells. Energy Environ. Sci. **4**, 3861 (2011)
134. W.J.E. Beek, M.M. Wienk, R.A.J. Janssen, Efficient hybrid solar cells from Zinc Oxide nanoparticles and a conjugated polymer **16**, 1009–1013 (2004)
135. F. Qiao, K. Sun, H. Chu et al., Design strategies of ZnO heterojunction arrays towards effective photovoltaic applications **1**, 20210008 (2022)
136. T. Majumder, S. Dhar, P. Chakraborty et al., Advantages of ZnO nanotaper photoanodes in photoelectrochemical cells and graphene quantum dot sensitized solar cell applications **813**, 92–101 (2018)

Metal Oxides for Future Electrochemical Energy Storage Devices: Batteries and Supercapacitors

Chetna Madan, Sonu Kumari, and Aditi Halder

Abstract The focus in the energy supply has been shifted towards renewable energy resources and thus storing energy derived from such intermittent sources needs rigorous attention. The success of renewable energy usage is largely dependent upon energy storage devices. Our energy-intensive society measures its progress in terms of per capita energy consumption, with a higher figure indicating more development. To keep abreast with the anticipated rise in energy requirements, the surplus energy must be stored in environmentally and economically sustainable technology. Electrochemical energy storage devices, considered to be the future of energy storage, make use of chemical reactions to reversibly store energy as electric charge. Battery energy storage systems (BESS) store the charge from an electrochemical redox reaction thereby contributing to a profound energy storage capacity. Supercapacitors, on the other hand, store the charge electrostatically thus being rapid, recurrent, and immediate in energy deliverance. This chapter elaborates on the fundamental operating mechanism of these energy storage devices. The emergence of both these technologies as viable future energy storage systems depends on the kinetics of electrode reactions. Electrode materials are selected based on their performance, abundance, and operational safety. Of the wide palate of materials to opt from, transition metal oxides serve as an outstanding candidate along with other desirable properties like variable electronic configuration, ease of opting for different morphology, porosity, and size, synergistic interactions among different metals, chelation, and multiple stable oxidation states. A rigorous correlation between the structure, properties, and performance of numerous transition metal oxides for storage energy is summarised here.

C. Madan · S. Kumari · A. Halder (✉)
School of Chemical Sciences, Indian Institute of Technology Mandi, Mandi, HP 175005, India
e-mail: aditi@iitmandi.ac.in

V. Kumar et al. (eds.), *Optical Properties of Metal Oxide Nanostructures*,
Progress in Optical Science and Photonics 26,
https://doi.org/10.1007/978-981-99-5640-1_10

1 Introduction

The alarming concerns about environmental pollution, global warming, and clean energy security associated with the injudicious utilization of the limited fossil fuel reserves have challenged all the nations of the world to collaboratively propose adaptable transition schemes. The sustainable development goals laid by the UN encourage global participation and facilitation to address the urgent concerns for a harmonious and symbiotic life on the planet. The head-strong challenges that require immediate and effective actions comprise curtailing carbon emissions, conversion of carbon dioxide to value-added commercial products, striving to prevent an average global temperature rise and ill effects of climate change on the fragile ecosystem, and finding cleaner and more sustainable alternative fuels to propel the energy demands [1]. Transitioning to renewable energy is going to be a key contributor as the appropriate action steps are urgently needed to avoid severe ecological and climate disasters.

Combustion of fossil fuels to meet our energy requirements and depleting forest reserves have led to catastrophic levels of pollutants in the environment that costs the lives of nearly 9 million people every year. To limit the rise in global average temperatures [2] from anthropogenic activities to only 1.5°, all the nations have agreed in the **2021 United Nations Climate Change Conference** (COP26) to decarbonize their energy sector, achieving net-zero carbon emissions by mid-century and promote usages of renewable energy through various economic, social, technical, and tariff policies.

Renewable energy comprises the energy procured from naturally replenishable sources, that have no harmful emissions or environmental and climatic effects. The common sources of renewable energy include offshore and onshore wind turbines, concentrated solar power plants, solar photovoltaics, small and large hydropower plants, and geothermal plants, collectively known as wind-water-solar (WWS) technologies. However, even though the advantages of using renewable energy to meet electricity demands seem enticing, they fall behind owing to their inconsistency and variable availability. One of the solutions to promote renewable energy could be improving their predictability by better forecasting. Another and more reliable way to incorporate renewable energy in the electricity sector is to complement it with energy storage systems to compensate for the intermittency.

Energy storage systems (ESS) [3] are indispensable for the growth of the renewable energy sector. According to research projections by a Precedence study [4], the global market for ESS is going to cross the figure of 435 billion USD by the end of this decade. Storing the surplus renewable energy when it is readily available or cheap and using it later to manage the peak demand helps restore the lag created by its intermittent nature. According to the Indian Power System Operation Corporation (POSOCO)-National Load Dispatch Centre, the demand for energy increases by roughly 20% every fiscal year summer. Achieving the mammoth task of integrating 500 GW of renewable energy in India's energy sector by 2030 has already led to various incentives and obligations, thus expanding the scope for future energy storage technologies [5]. The expedited interest in e-vehicles and hybrid vehicles by governments across the globe to cut down their carbon dioxide emissions also creates new horizons for the growth of renewable energy complemented by ESS technologies. Energy storage systems [6] that are widely being explored for assisting renewable energy adoption include pumped hydro energy storage (PHES) and compressed air energy storage (CAES); based on potential energy storage, flywheels; based on kinetic energy storage, supercapacitors, and batteries; based on electrical energy storage. Owing to a large installation capital, site-specific procurement, re-allocation challenges, and serious environmental implications due to a large set-up requirement pumped hydro storage although attractive for large-scale grid stabilization, face repercussions. Flywheel energy storage technology suffers poor energy density and critical self-discharge and standby losses thus limiting its long-term energy storage.

Electrochemical energy storage (EES) devices constitute storing of energy as electrical charges mediated via chemical reactions. Battery technology uses the stored chemical potential of a redox reaction occurring at its electrodes and converts it into electrical energy when needed. The terminals of a battery, namely the cathode and anode are separated by ionically conducting electrolytes. Depending on the nature of the materials employed as the electrodes, the potential a battery can deliver, its cyclability, and its lifetime is determined. Hence, the selection of electrode materials is the key criterion in governing the overall performance of a battery. Also, the battery energy storage systems (BESS) provide the freedom of modulating the capacity by giving the option of stacking as per the requirement, thus being compatible with the operation site and facilitating maintenance. Of the numerous battery chemistries being researched and evaluated at different readiness levels, many are being successfully transferred for commercial applications.

In comparison to a conventional capacitor, a supercapacitor has a substantially larger capacitance. Supercapacitors can either store electrical energy by electrostatic separation of charges on the electrodes across a dielectric medium or through the Faradaic process. The formation of an electrochemical double layer of separated charges at the electrode–electrolyte interface decides the electrochemical storage of energy. Hence, supercapacitors are very quick in delivering high-power electricity for a brief time, giving them a high-power density. The well-known Ragone plot in Fig. 1 organizes the electrical energy storage devices in a comparative layout in terms of their energy density versus power density. A high-energy device being able to store a larger amount of energy per unit mass of the active material is ideal for a steady withdrawal of energy over a defined period, for example, batteries. On the other hand, a high-power density device such as a supercapacitor would be preferred in applications where an instant inflow of a large amount of energy is needed at once to support an energy-intensive application.

The evaluation of the performance of an EES device depends on the parameters like specific capacity, energy, power density, cyclic life, and durability, for which the materials used as the electrodes and electrolytes have a critical role. Hence, a thorough evaluation of the materials to be employed for various applications in electrical energy storage devices is significant to enhance their performance, lifespan, and safety.

Metal oxides have been a key player in the progression of energy storage technologies (ESTs). Due to their versatile properties such as multiple oxidized/reduced

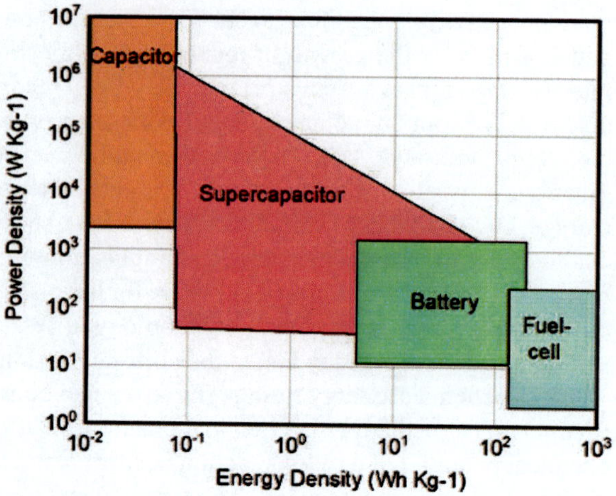

Fig. 1 Ragone Plot for various energy storage/conversion devices showing approximate area according to their energy densities versus power densities

states, band gap lying in the semiconducting region, optimum binding energies with the intermediates involved in various electrochemical redox reactions, layered crystallographic structure, large surface area, and porous morphology, and feasible introduction of numerous types of defects, metal oxides give enormous grounds to explore. This chapter is dedicated to compiling the resourcefulness of metal oxides in different electrochemical energy storage applications.

2 Supercapacitors as Electrochemical Energy Storage Systems

2.1 Introduction and Basic Principles

It is desirable to have an electrochemical system that can store energy and at the same time deliver considerable energy density and significant power density on top of prolonged recycling duration. Battery-capacitor combination devices are intriguing due to their potential use in future electric vehicles, smart electric grids, and even tiny electronic/optoelectronic devices, among other things. High performance, low cost, safety, and environmental responsibility need proper design and production of battery-capacitor combo devices. A supercapacitor is a type of capacitor that bridges the gap between a regular capacitor and a battery. An electrochemical device that stores charges is titled as a capacitor. By the electrostatic field created between the metal plates in capacitors, direct electrochemical energy storage is viable. The type of dielectric and electrode material used in a capacitor defines its application. The dielectric substance is responsible for storing the majority of the energy. Capacitance is the terminology for the measurement of charge storage at a given voltage and can be denoted by the formula below in Eqs. (1) and (2):

$$\textbf{Capacitance} = \textbf{Q/V} \tag{1}$$

$$\textbf{Capacitance} = \boldsymbol{\varepsilon}\textbf{A}\boldsymbol{d} \tag{2}$$

Q Total charge stored on conducting plate
V Applied voltage to the capacitor
ε Electrostatic constant or permittivity of the dielectric medium
A Area of one conducting plate
d Distance between two conducting plates.

Ceramic, polymer films, or aluminum oxide are commonly employed as dielectric media in capacitors to separate conducting plates, which are where we store our charge. Activated carbon is utilized as a physical barrier between the plates or electrodes; permitting the establishment of a double electric field when such material is electrically charged, which functions as a dielectric. Supercapacitors have unique benefits over capacitors, such as greater energy storage and higher energy density. When compared to batteries, supercapacitors feature lower internal resistance, allowing for higher power density and lower equivalent series resistance (ESR).

2.2 The Evolution of the Supercapacitor Over Time

It took 150 years for a concept conceived in the 1800s to become a reality and another 20 years before commercial availability. Throughout the last decade, research toward clean and renewable energy sources have exploded due to the dramatic advancement in the globe's energy requirements and the associated issues of traditional sources of energy on the environment. Because new energy forms are intermittent or regionally constrained, better energy storage systems, like supercapacitors, are urgently required for successful storage. Nanotechnology has created novel materials and structures for effective energy storage, which has opened up new frontiers.

2.3 Classification of Supercapacitors

Based on energy storage and release mechanisms, the classification of supercapacitors leads to three such types; EDLC (electrochemical Double Layer Capacitor), pseudo-capacitor, and hybrid supercapacitors. And since the charge separation occurs when an electrode and an electrolyte come into contact electrochemically, EDLC stores energy at the nanoscale. Non-faradaic redox is involved, which means there are no chemical oxidation-reduction reactions or electron transfer processes taking place. Because only physical charge transfer occurs, cycle life is increased, resulting in a reasonably long cycle life. Pseudo capacitors, on the other hand, uses metal oxides/metal-doped carbons/conductive polymers; all are employed as high-energy electrode materials to bring out faradaic redox processes. Compared to EDLC, it has a higher energy density but shorter life cycles. Because it is made up of EDLC and pseudo-capacitor components, hybrid SC has properties of both. Asymmetric, composite, and hybrid SC are the three types of hybrid SC. In composite electrodes, either metal oxides or conducting polymers are combined with carbon-based materials to construct an electrode that possesses the ability to hold the charge in both ways physically as well as chemically. With its large surface area, carbon promotes contact between both the material and the electrolyte. Increased capacitance through composite electrodes is caused by the faradic reaction and pseudocapacitive material. At the current time, two prevalent types of composites are binary; two different electrode materials, and ternary in which three different electrode materials are used in making electrodes. Asymmetric hybrids are comprised of EDLC; non-faradaic and pseudo capacitors; faradaic electrodes. Conducting polymer/metal oxide; as positive electrode while carbon-based electrode as a negative counterpart. A battery-type hybrid, like asymmetric hybrids, combines two different electrodes, however at this moment one is a supercapacitor-electrode and another is a battery-electrode. Hence, a single cell has characteristic features of both batteries and supercapacitors. The different categories of supercapacitors are depicted in Fig. 2.

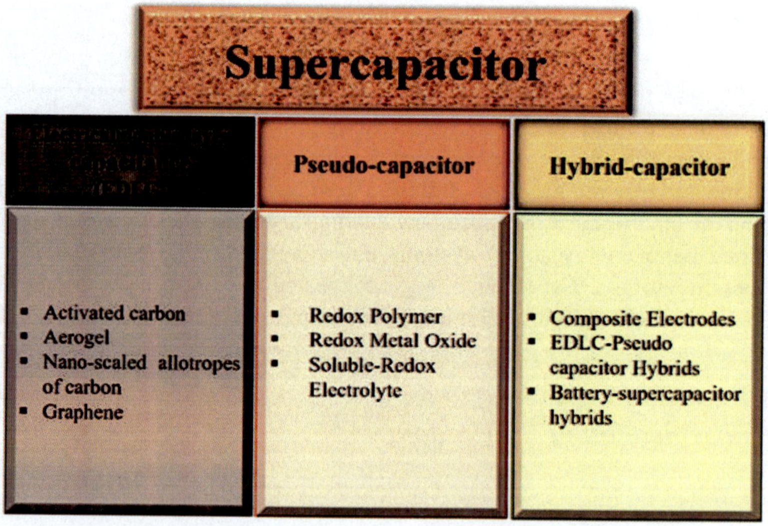

Fig. 2 Depending on construction and operation, classification of supercapacitors

2.4 What Makes Transition Metal Oxide a Potential Electrode Material for Supercapacitors?

Three types of materials could be identified as supercapacitor electrodes: transition metal oxides, conductive polymers, and materials made of carbon. Widely, carbonous materials are utilized for the making of electrodes in EDLCs. Their character traits include high specific surface area and controllable pore-size distribution as well as superior electrical conductivity [7]. Carbon-based materials have high power densities but a lower value of energy densities; resulting from their charge storage method which limits the overall performance of EDLCs [8, 9]. Graphene, carbon nanotubes, and carbon nanofibers just are a couple of the carbonous materials which have been studied as potential electrode materials. The lower specific capacitance of carbon-based materials, however, restricts the ability of EDLCs [10], and their high cost limits their use [11]. Conductive polymers have strong conductivity and are pseudocapacitive, but their stability is low, making them easily detachable from the substrate [12, 13]. For instance, polyaniline has a slow charge and discharge rate and poor charge/discharge process stability [14]. TMOs have superior chemical stability over conductive polymers and higher specific capacitance (100–2000 F/g), energy density, and energy density compared to carbon materials [15–17].

2.5 Concentrating Supercapacitor Energy Density

When we compare supercapacitors' energy densities with batteries; they aren't all that high (20 vs. 30–200 Wh/kg), and striving to improve supercapacitors' energy density remains a challenging and time-consuming research project. Advanced manufacturing methods and technologies can boost the storage capacity of supercapacitors, but it is crucial and difficult too; to find new electrolytes or electrode-active materials that offer better corresponding electrochemical performance over the long term. The lower energy density of supercapacitors resulting their larger size and consequently portability issues arose. In essence, expanding the accessible surface area of active materials in double-layer capacitors; and also, by widening the operational voltage window can improve the energy densities of supercapacitors. The development of novel materials with enormous surface areas and organic electrolytes which suitably can sustain a wide-ranging voltage window is the theme of an escalating volume of research. If these issues are adequately resolved, supercapacitor energy densities could be nearly comparable to batteries.

2.5.1 High-Specific Area

Due to greater interactions between electrolytes and active electrode materials, electrode materials having a wide surface area can improve electrochemical performance.

Increased interactions also lead to an increase in faradaic reactions at the interfaces, which increases the contribution of the pseudocapacitive component to the total capacitance value. Expanding active material surface area through porosity could also strengthen many critical SC variables. This is because as porousness (permeability) of active material increases the diffusibility of electrons and ions through into electrodes, which ultimately lowers internal resistance and controls changes in volume during charge–discharge cycles; further, the cycling life would be improved. Several different dimensions and morphologies of metallic oxide materials, which would include nanoparticles, nanorods, nanoflower, nanowires, nanosheets, and nanotubes, could be developed by using various techniques. The structural, physical, and chemical features of each morphology are distinctive. Due to its superior electrochemical performance, eco-friendliness, and inexpensive price, like nickel oxide seems to be a viable electrode material for pseudocapacitors. As opposed to their bulk counterparts, nanostructured materials have distinctive and interesting features that are well recognized. Therefore, supercapacitor electrode materials having nanoscale crystalline particle size will have large specific surface areas, which produce high electro-chemical activities plus superior capacitive performance. According to Zheng et al. [18], post-heating at 300 °C was employed to produce nanoflake-like NiO using the hydrothermal synthesis method. The morphology of NiO particles shows distributed sandwich and shaggy cauliflower-shaped micaceous nanoflakes as displayed in Fig. 3a. The NiO nanoflakes have a thickness of 20 nm and a width of 50–80 nm. In an alkaline electrolyte solution, the supercapacitor properties of the NiO sample were investigated. The NiO produced through heat treatment after the hydrothermal process at 140 °C displays good capacitive performance. The value of specific capacitance for prepared NiO was about 137.7 F/g @ current density of 0.2 A/g, which is about four times greater than NiO synthesized by directly calcining $Ni(NO_3)_2$. $6H_2O$ at 550 °C temperature [19]. The cauliflower-like nanoflake structure within NiO increases the effective surface area facilitating faradaic redox reactions and simultaneously improving pseudo-capacitance as well as tremendous cyclability and retained capacitance up to 91.6% after 1000 cycles of charging and discharging. According to an electrochemical impedance spectroscopy investigation, it has low internal resistance is 0.2 V, lower than the NiO electrode made via molten salt synthesis [20], indicating the main reason was mass transfer constraint. As depicted in Fig. 4a, the NiO samples' curve shapes show that their capacitance properties are substantially different from those of electric double-layer capacitance where the form is typically extremely similar to an ideal rectangular shape. It suggests that the pseudocapacitive capacitance, which would be based on the redox mechanism, is the primary cause of the capacity of NiO electrodes and other metal oxides. Due to sample 1's shaggy surface and large specific surface area for quick, reversible Faradaic reactions, the redox current intensity provided by sample 1 is noticeably higher than that of sample 2.

Besides, it is known that this surface Faradaic redox reaction [Ni^{2+}/Ni^{3+}], takes place in an alkaline solution at the surface of NiO as indicated below in Eq. (3) [23]

$$[NiO] + (OH)^{-} \leftrightharpoons [NiOOH] + (e)^{-} \tag{3}$$

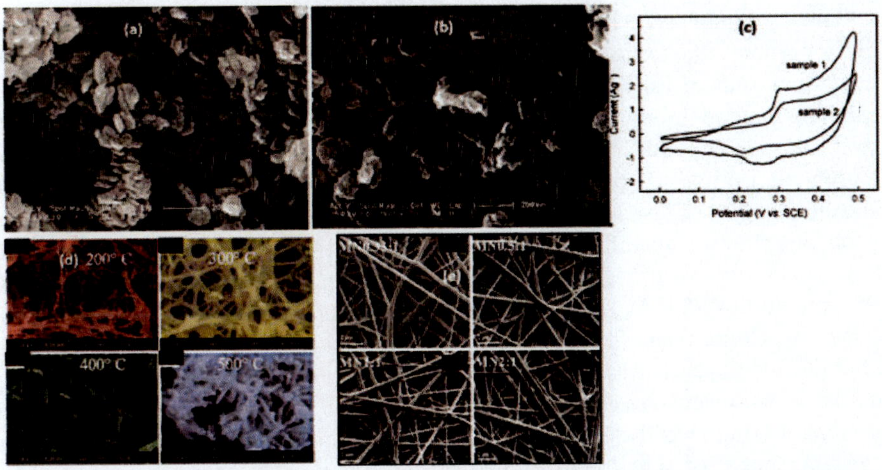

Fig. 3 SEM images showing the nanoflake-like morphology of NiO, **a** sample 1 and **b** sample 2, **c** CV for NiO electrodes (samples 1,2) sample 1 has a greater surface area accessible to faradaic redox reactions than sample 2, it exhibits a greater current density [14]. Reproduced from Ref. [18] with permission from Elsevier, Copyright 2009. **d** SEM images showing samples of IrO_2 undergo annealing at different temperatures; as-deposited IrO2 annealed @ 200 °C; forming nanofibers IrO_2 having small buds; indicating the presence of polymeric content, at 300 °C showing reduced buds probably because of evaporation of polymer content, for 400 °C, buds get completely disappeared, and at 500 °C, quality of IrO_2 nanofibers get damaged and aggregation occurs. Reproduced from Ref. [21] with permission from Elsevier, Copyright 2021. **e** SEM image of as-spun nanofibers with a diameter in the range of 200 to 300 nm. Reproduced from Ref. [22] with permission from Elsevier, Copyright 2015

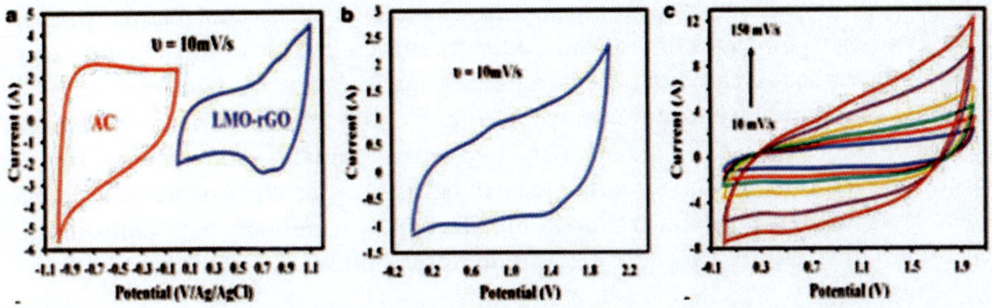

Fig. 4 **a** CV scans for the AC and LMO-rGO electrodes in 0.5 M Li_2SO_4; in aqueous solution @ 10 mV/s scan rate. The electrodes with AC and LMO-rGO have potential windows of −1.0 to 0 V and 0 to 1.1 V, respectively. **b** The CV curve of an asymmetric device (AC/LMO-rGO) with an increased working voltage window of up to 2 V. **c** CVs of the asymmetric device at various scan rates from 10 to 150 mV/s; the shape of the CV curves is not significantly changed as the scan rate increases from 10 to 150 mV/s, demonstrating the AC//LMO-rGO device's good rate capability as well as reversibility. Reproduced from Ref. [34] with permission from Springer Nature, Copyright 2017

This also indicates that the electrochemical redox process for NiO in a supercapacitor is not a simple adsorption/desorption process of OH ions. In another research work, nanorods of RuO_2 were deposited over carbon nanofibers using the electrospinning method through precipitation and recrystallization methods. This results in an increase in large surface area with electrochemically active functional sites at the mesopores leading to better performance. RuO_2-carbon nanofibers that have been annealed at 220 °C used as electrode material with KOH aqueous electrolyte have good capacitance capabilities (188 F/g at value 1 mA/cm^2 current density), as well as with improved energy density valued (22–15) Wh/kg and power density from 400 W/kg to 4000 W/kg. At 1 mA/cm^2 discharge current density, the capacitance value for RuO_2. Carbon nanofibers maintained approximately 93% of their initial value; after the 3000 cycles. To achieve a lower value for internal resistance and promotion of ion-diffusion and charge-storage in the majority of amorphous materials, a bigger mesopore volume with low-dimensional RuO_2 nano-rods at carbon nanofibers is advantageous [24]. The electrospinning method is used to generate iridium oxide (IrO_2) nanofibers [21], and the annealing temperature is optimized to produce high-quality IrO_2 nanofibers. The four different annealing temperatures 200, 300, 400, and 500 °C resulted in a hollow nanofibrous shape with about 45 nm diameter on average as shown in their SEM image in Fig. 3d. With a 2.1 nm average pore size; the surface area measured was 80 m^2/g for a sample that was annealed at 400 °C, so it can be deduced that the nanofibers that were created had higher porosity. The sample also had a maximum specific capacitance of 705 F/g at 1 mA/cm^2, which was caused by the fully developed morphology. Owing to the such hollowness of nanofibers, which contributes to their higher surface area, electrolyte's permeability may be improved as well as the likelihood of efficient ion and electron transport in the electrode, which would result in good electrochemical performance [21]. The contact area between both electrolyte ions and that of the electrode surface could be enhanced by using 1D nanofibers to increase both the surface area and ion transport pathways [25]. The supercapacitors' ability to store charge is significantly influenced by the porosity of such nanofibers [26, 27]. Due to their large surface area, mesopores can improve electrolyte ion transit, while micropores can help boost the electrochemical double-layer capacitance. As an illustration in Fig. 3e, MnO_x nanofibers with pores that are 10–30 nm in size offer excellent electrolyte diffusion [22].

2.5.2 Boosting the Working Potential

The challenge for the field of SCs is to create devices having high energy density but not at the cost of their high-power densities or long-cycle lives.

$$E = 1/2\left(CV^2\right) \quad (4)$$

E = Energy Density, C = Capacitance, and V = Operating Potential Window.

Equation 4 conclusion shows that; one option to boost an SC's energy density is to do so by developing more effective electrode materials. The specific capacitance of a cell can be enhanced by creating high surface area, high conductivity, and porous, and lightweight electrode materials; which will increase the system's overall capacitance. Increasing the electrolyte's operating potential window is another way to boost energy density, provided the electrode materials stay stable well within the working potential range. It is more effective to increase the working cell voltage than the capacitance of such active materials because the energy density is proportional to the square of the operating potential window. Despite having large working potential windows, ionic liquids lie from 3.5 to 4.0 V; and range from 2.5 to 2.7 V in the case of organic electrolytes, have some drawbacks, such as limited ionic-conductivity, higher cost, flammability, high viscosity, corrosiveness, sensitive for humidity, safety, and disposal risks [28, 29]. The creation of a green storage system depending on neutral aqueous electrolytes is thus urgently required on a global scale to establish a secure and consistent energy supply. Aqueous electrolytes' operating cell voltage must be increased above their thermodynamic limit of 1.3 V to meet the continuously rising demand for green and sustainable energy. This requires the development of improved electrode material. When it comes to aqueous electrolytes, this thermodynamically controlled operating cell voltage can be increased by switching from such a symmetric to an asymmetric configuration. The typical operating potential window for a symmetric supercapacitor in an aqueous electrolyte is less than 1 V. Asymmetric supercapacitors may often fully utilize their two distinct electrodes and would display two distinct operational potential-window in that same electrolyte only. Due to that expansion of the working voltage window, the device's energy density may also be greatly enhanced. The electrochemical oxidation/reduction potential and the active material work function are intimately related, therefore the asymmetric supercapacitor's operation voltage can be raised by simply selecting metallic oxide active material with a significant difference in their work function [30]. To increase the operational potential window of a cell, an asymmetric arrangement makes use of the two electrodes' distinct potential windows [31]. Weijie Liu et al. [32] due to the large work function difference of two electrode materials (MnO_2 and V_2O_5), the device fabricated worked at a wider operational potential window of 2.0 V and excellent electrochemical performances owning due to a larger work function difference of electrodic materials MnO_2 and V_2O_5. A higher areal capacitance of 0.613 F/cm^2@ 1.5 mA/cm^2 discharge current density was exhibited, and the energy density of 0.340 mWh/cm^2 at 1.5 mW/cm^2 power density was obtained. This device showed good flexibility, that it could withstand the bend testing and drive LED even under severe bending positions. The device was also found quite stable even after 5000 cycles at such a higher current density of 30 mA/cm^2. The device can function over a wide potential range from 0 to 2.0 V and could maintain a similar shape in the potential range of 0–0.8 V. In general, supercapacitors with a large potential window have high power densities. In comparison to conventional symmetric supercapacitors, asymmetrical supercapacitors have this significant benefit and are one of the key components in meeting application demand. CV curves exhibit good symmetry and nearly rectangular forms; at 100 mV/s comparatively higher one, indicating that the device also exhibits strong capacitive performance. With a significantly wider operating potential window from −0.8 to 0.8 V; versus SCE; 0.5 M Na_2SO_4 aqueous solution as an electrolyte, Song Z. et al. synthesized Fe_2O_3/GA composite using the hydrothermal method, the yielded specific-capacitance valued 81.3 F/g @ 1 A/g [33]. The asymmetric AC/LMO-rGO aqueous supercapacitor was created by Azari et al. [34] as shown by the CV scans in Fig. 4a–c and has a 2.0 V extended potential window. Its specific energy and power are comparable to those of commercial lead-acid batteries. The exceptional super-capability of LMO-rGO nanocomposite is because of their synergistic effect involving graphene & $LiMn_2O_4$ (LMO). The capacitance value of LMO (positive electrode) raised from 162.5 to 268.75 F/g @ 1 A/g on nano-compositing.

The nanocomposite's superior architecture, high specific surface area, outstanding electrical conductivity, and improved ion transport routes, which allow for quick electron & lithium-ion transfer well within nanocomposite, are what account for the device's increase in capacitance. Combination with an activated carbon (AC) electrode; the device exhibits a higher hydrogen evolution overpotential and with the outstanding capacitive properties of the LMO-rGO nanocomposite plus high oxygen evolution overpotential effectively increases the operating voltage range of the device to 2.0 V. With just a 5% drop of capacitance across 6000 cycles at 0.74 A/g current density value, these nanocomposite demonstrating good cycling stability as well. The findings demonstrate that the synthesized nanocomposite is a suitable active component for the production of asymmetric supercapacitors exhibiting high energy & high-power densities, with potential applications in high-energy devices found in the real world.

2.6 Improvements in Cyclic Stability

Cyclic stability is without a doubt the most crucial factor for the practical form of supercapacitors. Cyclic-stability of a supercapacitor is defined as the rate of capacitance retention of such electrode materials/devices given standard and realistic test circumstances, following a specific amount of working time or a specific number of charging and discharging cycles. Although other components like the electrolyte, and separator are also significant, the electrode materials serve as the primary component that determines the cyclic stability of supercapacitors. The selection of materials and their energy storage mechanism is directly impacting the cyclability of active materials. Cyclic stability of electrode materials exhibiting EDLCs behavior (such as carbon electrode materials) often falls under type I, which is primarily because the electrode material does not experience chemical charge transfer reactions and phase changes throughout the charge and discharge process. In other words, charge storage is accomplished via the physical process involving reversible ion adsorption and desorption just on the electrode/electrolyte interface. As a result, during extended cycles, the capacitance rarely diminishes. Battery-type and pseudocapacitance cyclic stability Supercapacitor electrode materials, like metal nitrides, metal oxides/hydroxides, and metal carbides, primarily fall under type II, which is due to the material's self-activation, an obvious increase in specific capacitance relative to initial specific capacitance, and constant retention rate of greater than 100%. The capacitance-retention rate falls to its original capacitance or is significantly less than 100% after the self-activation stage. During the charge/discharge process, the conductive-polymer electrode material readily triggers the collapsing of the material structure, resulting in a capacitance retention rate that is less than 100%; these are categorized as type III materials.

The test conditions for stability assessment should be given careful consideration, together with other performance indicators such as mass load, electrolyte composition, potential window, scanning rate, test temperature, and others as enlisted in Fig. 5. The test parameters, including the potential window and current density/scanning rate, have a significant impact on the cyclic stability of electrode materials. For instance, a material used for electrodes exhibits higher specific capacitance at low current densities while exhibiting extremely lower values at high current densities. The electrode material undergoes its electrochemical reaction more effectively at low current densities and scanning rates, and as a result, the electrode material exhibits a higher value of specific capacitance. Metal oxides as electrode materials are debatable because they exhibit both pseudocapacitive as well as battery-type characteristics during the charge/discharge process. The cyclic stability of metal oxides varies based on the type of material. To take MnO_2 as a reference, it is well known that the pseudo-capacitance, that is a reversible redox transition induced by either proton exchange or cation exchange or both in the electrolytes, and Mn(III)/Mn(II), Mn(IV)/Mn(III), and Mn(VI)/Mn(IV) caused by the transition accounts for the majority of such capacitance of manganese oxide. Equation (5) describes the unique mechanism [28, 35].

$$\mathrm{MnO(OC)}\beta + \delta \mathrm{C} + \delta \mathrm{e} \leftrightharpoons \mathrm{MnO}\alpha - \beta\mathrm{(OC)}\beta + \delta \quad (5)$$

Alkali metal cations (Li+, Na+, and K+) and protons are represented by the symbol C+ in the electrolyte, while MnO(OC)β & MnOα − β(OC)β + δ MnO(OC) designate $MnO_2.nH_2O$ in High and low oxidation states respectively. We can see from Eq. (5) that the redox process involves both cations and electrons. As a result, for MnO_2 with weak conductivity, the redox reaction rate is significantly smaller than that of the electron transport rate whenever the current density or scanning rate is high. As a result, there are fewer materials available to supply this capacitance, which causes a quick decline to a value that is less than the capacitance. As described in the study of Bag et al., if the current density or scanning rate is employed to evaluate the cyclic stability, the phenomena that the capacitance value is very little and the cyclability is very superior would exist [36]. However, if the cycle performance is examined with a lower current density/scanning rate, according to the findings of Xia and colleagues [37], it shows far worse stability for MnO_2 as compared to Bag's and Huang's results. Oxides; other than MnO_2, similarly experiencing comparable circumstances. The Ni–Co–O@CFP exhibits excellent electrochemical performance, however, it may be much better if it showed a rate capability of 20 A/g before being tested for cyclic

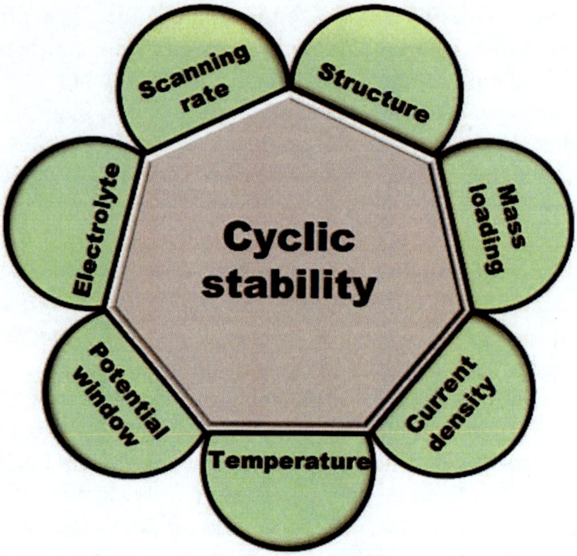

Fig. 5 Variables that affect the supercapacitors cyclic stability

stability, or tested for cyclic stability at 10 A/g. As metal oxides generally have poor conductivity, this phenomenon can also be seen in widely used metal oxides, such as V_2O_5, Fe_2O_3, and Fe_3O_4. Because of this, the main goal of describing the cyclic stability of pseudocapacitive metal oxides is to make sure that the current density/scanning rate employed can essentially trigger the redox reaction of all electrode materials. Supercapacitors' cycle life is influenced by a variety of factors. Due to this, many scientists are working to enhance the cyclic stability of electrode materials by modifying the materials' composition, microstructure, and preparation of composite materials. Creating heterojunction nanocomposites of the electrode material can increase the cyclic stability of the electrode material. It is possible to create V_2O_5@PPy nanocomposites with outstanding cycle stability by utilizing the synergies between V_2O_5 and PPy. Only 17.5% of the capacitance is attenuated [38] after 2000 cycles (Fig. 6a). In one more recent research, VN@VO_2 is created by depositing a single oxide-layer (VO_2) just on the surface of VN, which exhibited 1000 cycle of charge and discharge @ 1 A/g current density [39], with an 85% capacitance retention rate. Because microstructure and morphology directly affect the electrolyte ions' adsorption on the electrode material (primarily in electric DL supercapacitors) and diffusion, they became crucial for determining its cyclic stability (mainly in the case of pseudocapacitive supercapacitors). Generally speaking, innovative microstructure and shape could create favorable conditions that help enhance quick electrochemical processes, such as high specific surface area, optimized pore size distribution, and so forth. Numerous studies have shown that creating specific microscopic morphologies of the electrode material, such as spherical nanoparticles, hollow structures, porous structures, layered structures, etc., has improved the cyclic stability of supercapacitors. One of the fundamental elements that govern the physical/chemical/mechanical properties of solid metals is crystal structure. As a result, altering the crystal structure is observed to play a key role in improving cyclic stability. Due to the varying crystal structures of MnO_2, they have tunnels or intermediate layers with gaps of different sizes. The intercalation/deintercalation of protons/cations in MnO_2 is what gives MnO_2 electrode materials their capacitance characteristics. The only crystal structures that can display great cyclic stability are those that have enough space to incorporate these ions. Because the small tunnel in β-MnO_2 cannot accommodate cations throughout charging-discharging cycles, there is almost no specific capacitance, making it not a good material for capacitor applications. Despite having strong cyclability, it has no significance. The specific capacitance is still quite big even though the cyclability of α-MnO_2 and δ-MnO_2 has reduced as depicted in Fig. 6b–d. They are excellent potential supercapacitor electrode materials because of their huge interlayer spacing and big tunnel size [40]. The strategy of coating and encapsulating has a primary purpose that, in addition to employing the coating to shield the structure of the electrode materials to increase their cycle stability, is that the synergy between two materials will indeed help to the ultra-long cyclability.

The enhancement of the cyclic durability of electrode materials can be accomplished by adding "impurity atoms, ions, or nanoparticles" in addition to controlling the microstructure and morphology. The following factors play a major role in how these impurity materials affect cyclic stability. (i) The bulk material may structurally deform due to the presence of impurity materials, and/or charge density may change. (ii) Impurity substances operate as "catalysts" in the electrode process, efficiently accelerating it and reducing the dissolution of electrode components in the electrolyte, enhancing cyclic stability. Introduction of gold nanoparticles done to nano Co_3O_4 bulk (AuNP/nano-Co_3O_4) to lower the internal resistance of charge transfer. The specific capacitance of the gold nanoparticle-decorated nano-Co_3O_4 is higher (681 F/g) than that of the pure Co_3O_4 (368 F/g), as well as the capacitive retention rate of AuNP/nano Co_3O_4 & nano Co_3O_4 was 83.1% and 79.8%, respectively, after 13000 cycles. This is a result of the addition of Au nanoparticles, which increased the conductivity of the active material; also decreasing its charge's diffusion length during charging and discharging, further boosting the cycle stability [41]. In recent years, combining two metal oxides to create composite nanostructures has emerged as a new research avenue and a hotspot in the field of supercapacitors. Nickel, cobalt, and manganese-based binary oxides are the most prevalent binary oxide-based electrode material. It has been suggested that adding cobalt compounds to nickel oxide (nickel hydroxide) can enhance the material's electronic conductivity. The material's electrochemically active sites are improved by the addition of nickel and cobalt oxide (cobalt-hydroxide) [42]. Because of its benefits of having higher electron conductivity, low diffusion barrier to protons/ions, and simple electrolyte penetration, $NiCo_2O_4$ (NCO) has emerged as a potential hybrid supercapacitor electrode material. However, NCO electrode materials have always had issues with cyclic robustness and multiplier characteristics, and numerous researchers have been striving to fix these issues. Mesoporous NCO nanowires synthesized on carbon fiber textiles, according to Zhang et al. [43] showed high capacitance, great multiplier performance, and good cyclic stability as the flexible electrodes in supercapacitors.

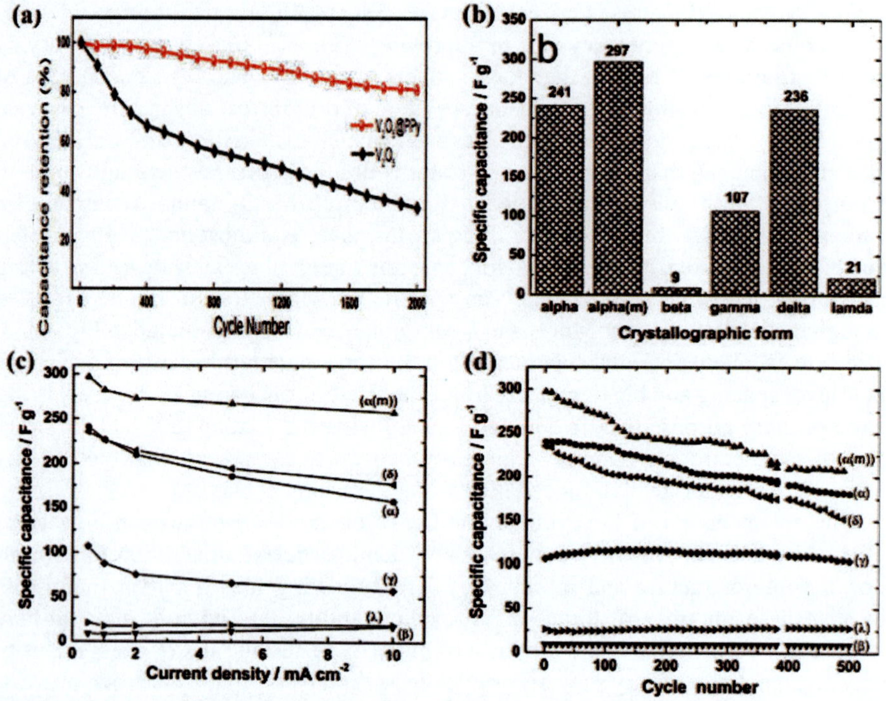

Fig. 6 **a** Comparison of cycling performance of V_2O_5@PPy, a synergic effect between V_2O_5 and PPy providing higher cyclability than pure V_2O_5 electrodes. Reproduced from Ref. [38] with permission from American Chemical Society, Copyright 2018. **b** Specific capacitance values of different crystallographic forms MnO_2, **c** Variation in specific capacitance value observed at various current densities, **d** Comparison of cyclic stability of different forms of MnO_2. Reproduced from Ref. [40] with permission from American Chemical Society, Copyright 2008

3 Batteries as Electrochemical Energy Storage Systems

3.1 Introduction and Basic Principle

Batteries store electric energy from the electrochemical reactions taking place at their two electrodes [44]. The transfer of electrons takes place through the external circuit powering the load. The materials ideal for the anode (which is the negative electrode) should release electrons for the load, get oxidized readily, and be capable of continuous reversible ion exchange with the electrolyte, while a cathode (positive electrode) material should accept electrons from the load and get reduced favorably and facilitate mercurial intercalation of ions [45]. The potential of a battery is ascertained by calculating the difference between the chemical potential of the cathode and anode and is given by the following Eq. (6):

$$E^{\circ}_{cell} = E^{\circ}_{cathode} - E^{\circ}_{anode} \tag{6}$$

From Gibb's free energy change between the reactants and products and the faradic capacity of charge transferred, the value of cell potential can be determined from Eq. (7):

$$E^{\circ} = -\Delta G/nF \tag{7}$$

The electrolyte becomes the medium of reversible ion transfer and should have a high transference number.

A major difference between a capacitor and a battery in storing electrical energy lies in the way the ions are arranged inside them. The capacitors store electricity by the charged ions residing at the electrode–electrolyte interface (electrochemical double layer), whereas the batteries store electrical energy as the charged ions reside inside the whole electrode itself (Faradic reactions). This gives batteries a higher capacity than capacitors, while capacitors exhibit faster charge–discharge because of the absence of hindered kinetics from Faradic reactions. A rechargeable battery is also known as a secondary battery and functions on the electrochemical reactions taking place at the electrodes and electrolyte interface. Depending on the chemistry of the electrode–electrolyte interface, and the size of the battery, its voltage, capacity, and energy density may vary [46].

3.2 Historical Overview

The battery is an age-old technology [47] invented to power electrical needs and is still undergoing new research every day. Dating back to the 19th century, Alessandro Volta's observation regarding the conduction of electrical charge by fluids (electrolytes) lead to the inception of stacked voltaic cells where silver and zinc metal separated by electrolyte-soaked paper were used. It became evident that higher voltage could be achieved by a combination of metals having a greater difference in their redox potentials. This ignited the interest of scientists worldwide and several new combinations began to be explored to acquire electricity from the combination of chemicals. Leclanchè in 1866, developed a battery where zinc anode and manganese-dioxide cathode were used along with ammonium chloride electrolyte. The first rechargeable lead-acid battery system, used to date, was invented by the French physician Plantè in 1859. In 1899, the invention of the Ni–Cd cell by Swedish scientist Waldmar Jungner made use of nickel and cadmium at the two terminals using KOH as the electrolyte. The developments in battery chemistry kept the scenario competitive as new systems having better voltage and discharge properties kept emerging with time as depicted by the timeline in Fig. 7. The primary alkaline battery invented by Canadian scientist Lew Urr evolved in 1959. The breakthrough battery technology involving lithium metal on the anode changed the world. Lithium-ion battery technology proposed by Whittingham in 1974 used TiS_2 as the cathode,

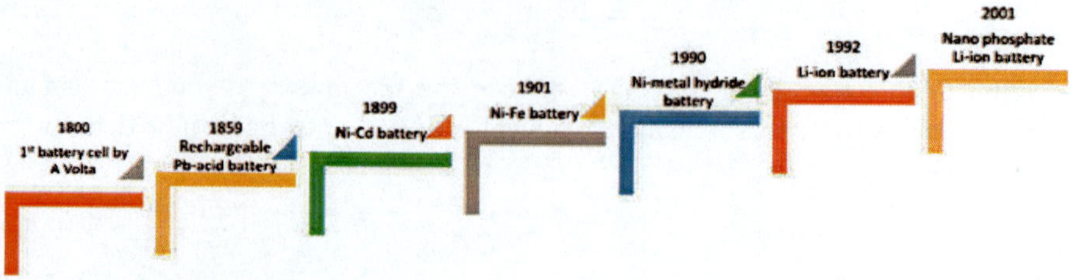

Fig. 7 Timeline of battery development

while Goodenough et al. used $LiCoO_2$ as the cathode and graphite as the anode. The battery market is largely dominated by lithium-based batteries these days whether in portable consumer electronics, electric vehicles, or grid ancillary services. However, more innovative and environmentally friendly battery technologies (as discussed in a further section) are being explored because of the dynamic interest in shifting our energy sector to non-fossil fuel sources.

3.3 *Need for Battery Energy Storage*

There is a global surge in shifting the source away from fossil fuels to meet our energy needs. The rise in average earth temperature, increasing ocean acidification, abrupt changes in weather patterns, and frequent floods and forest fires have all pushed the agenda of adopting clean and sustainably-sourced energy. The thriving interest in harnessing renewable solar and wind energy has caused the decentralization of the energy source. The discontinuity in renewable energy generation arising due to diurnal and seasonal fluctuations renders it unreliable and unsuitable for the power grid. However, this intermittency must be balanced with adequate energy storage systems where battery energy storage gains a huge credit. Battery energy storage systems (BESS) like lithium-ion batteries, and lead-acid batteries attached to renewable sources of energy store the surplus energy and can either be utilized in the peak hours of demand or when the prices of electricity are higher, it can sell energy or feed into the grid. For providing ancillary services to the grid [48] when an event of destabilization or fluctuation occurs BESS plays a moderator role. Also largely sourced energy from renewable sources when fed into the grid may suffer from more fluctuations owing to weather disturbances, and must be stabilized by employing BESS [49] at the generator and distributors end to maintain the grid. For a greater adoption of e-vehicles, the automobile industry demands reliable and cost-effective batteries that can compete with the internal combustion engine's performance in terms of cost, maintenance, refilling or recharging time, and lifetime, and poses no serious environmental implications. Therefore, the coming decade sees a huge rise in the development and demand for BESS [50].

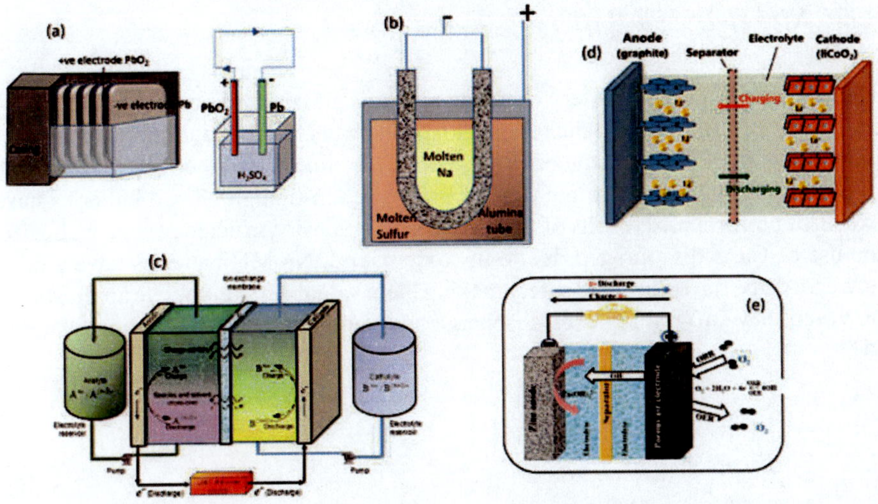

Fig. 8 Schematic illustration of different predominant battery energy storage systems: **a** Lead-acid battery, **b** Sodium-sulfur battery, **c** redox-flow battery. Reproduced from Ref. [56] with permission from Springer Nature, Copyright 2011, **d** Li-ion battery. Reproduced from Ref. [57] with permission from Elsevier, Copyright 2020, and **e** zinc-air battery. Reproduced from Ref. [58] with permission from Elsevier, Copyright 2021

3.4 Types of Battery-Chemistry

Battery systems can be classified based on the different materials used on the electrodes and electrolytes, and various battery chemistries as summarized in Fig. 8. Thus, different systems exhibit varying energy densities, specific capacities, operating voltage, cyclability, charge–discharge ability, lifespan, and cost and therefore are suited for different applications.

3.5 Lead-Acid Battery

This battery system is one of the oldest functional battery technologies and uses spongy lead as the anode, lead dioxide serves as the cathode, and dilute sulfuric acid functions as the electrolyte. They have a 70–80% efficiency and only 30 Wh/kg of energy density and are widely used in automobiles, and DC back-ups. During discharge, there is a production of lead sulfate which grow larger and is difficult to reverse on over-discharging. Also, the loss of water due to the generation of hydrogen on the cathode requires periodic replenishment. These batteries require more maintenance and the use of lead, and corrosive acid makes these batteries less attractive from an environmental point of view. A schematic of lead-acid batteries is shown in Fig. 8a.

3.6 Nickel-Based Battery

There are several nickel battery chemistries in this class such as nickel-zinc, nickel–cadmium, nickel–iron, and nickel-metal hydride where nickel hydroxide comprises the positive electrode and aqueous potassium hydroxide is the electrolyte. The anode materials vary according to the type of battery. Even though Ni–Cd batteries have excellent performance results of their high energy density, efficiency, and cyclic life, the use of Cd is discouraged due to its toxic nature. Ni–MH batteries have a good power density, durability, and safety making them valuable for e-vehicle applications, however, they suffer a high self-discharge and hydrogen evolution during charging [51].

3.7 Sodium-Sulfur Battery

This system has a very high energy density (120 Wh/kg) and power density (120 W/kg), and also has an optimum cyclic life and low self-discharge. The anode contains molten sodium and the cathode contains molten sulfur separated by an alumina ceramic electrolyte [52] as illustrated by the diagrammatic representation in Fig. 8b. This battery wields the potential to be used in mini-grid applications, however, is inhibited due to its high cost.

3.8 Flow Battery

This is a relatively new and promising technology that stores the electrolyte, known as the anolyte and catholyte in separate tanks. The electrolytes are released in the reaction tank by the controlling pumps as per the operation demand. This battery is unique as it utilizes redox reactions occurring between the two electrolytes rather than commonly occurring reactions between electrode and electrolyte and gives high energy and power density, cyclability, longer lifetime, low temperature, and maintenance-free operation [53]. Vanadium redox flow batteries and zinc-bromide batteries are the best examples of emerging flow battery technology. Figure 8c illustrates a prototype typical flow battery cell configuration with all the mentioned components.

3.9 Lithium Battery

This cutting-edge technology shook the world and is the most widely adopted battery in portable electronics and electric vehicles due to its high energy density, cyclic life, and low self-discharge [54]. The reversible exchange of Li ions between the graphite-based-Li intercalated anode and layered oxide cathode across the Li-containing organic electrolyte, as shown in Fig. 8d gives the battery a high operating voltage. The high cost of lithium and the requirement of a battery management system to tackle internal heat up and failure, over-charging, or discharging are some issues that require immediate attention.

3.10 Metal-Air Battery

Rechargeable metal-air batteries [55] as shown in Fig. 8e employ metals like aluminum, iron, zinc, magnesium, etc., and are recently gaining popularity because of their appreciable energy density, low cost, and non-toxicity. Metal-air batteries, a type of flow battery make use of metal in the form of sheets or nano-powder coating at the anode while oxygen from ambient air is reduced by a suitable electrocatalyst during discharging at the cathode. The electrolyte used is typically a highly-concentrated alkali in aqueous form or an anion exchange membrane in a solid state. Such batteries have a high utility in a diverse array of applications, from grid storage to portable electronics.

3.11 Metal Oxides for Battery Energy Storage

The abundance and properties such as high mechanical and chemical stability, and tuneable combinations of electronic arrangement, make metal oxides attractive candidates for a multitude of electrochemical reactions [59]. The processing and fabrication of metal oxides are usually facile and require minimal use of harsh chemicals and stringent reaction conditions [60]. Metal oxides contribute to heterogeneous catalysis [61] due to their electrical and ionic conductivity, large surface area, layered structure, and active site participation. The electronic behavior, adsorption energy, and catalytic activity of the metal oxides can be tuned easily by simple strategies like the introduction of dopants or lattice defects. In their study, Sun et al. [62] proposed that substitutional doping of copper in IrO_2 ($Cu_{0.3}Ir_{0.7}O_\delta$) leads to an improved oxygen evolution catalysis owing to the increased Jahn-Teller effect in CuO_6 octahedra and partial oxygen defects in the lattice of IrO_6 octahedra, thus breaking the degeneracy and manipulating the electronic occupancy of t_{2g} and e_g orbitals and by affecting the intermediate adsorption energies too. Huang et al. [63] reported that with increasing entropy, the electronic states of a multi-metal oxide get tuned and lead to superior OER activity as a result of modified adsorption energy of the intermediates on the diverse active metal binding sites. Conclusively, the adsorption energies of metal oxides with the involved oxygenated intermediates (O*, *OO, *OH) are governed by the optimum coupling strength [64, 65] between the valence states of the intermediate with the d-band states of the transition metal and are significant for designing the electrocatalyst. Such structural and electronic features of the metal oxide make them enticing as electrocatalysts in electrochemical applications because of the facilitated intermediate adsorption, enhanced charge transfer dynamics [66, 67], and convenient desorption of the product. Thereby improving the overall kinetics of an otherwise hindered and slow electrochemical reaction.

Metal oxides prove their relevance as electrocatalysts owing to their numerous properties discussed below that make their application in energy storage applications promising. The versatile properties of metal oxide make them eligible to be utilized as the cathode, anode, or electrolyte for certain types of BESS [68]. This section discusses the most enticing qualities of metal oxides that make them suitable for different roles of different types of batteries.

3.11.1 Metal-Oxides Have Optimal Intermediate Binding Energy

In any catalysis reaction, the binding of the reactant or intermediates with the catalyst surface site is a crucial step in determining the overall dynamics of the catalysis. Therefore, the binding energy of the intermediates at the catalyst surface plays a crucial determinant of the activity by contributing to the required energy for catalyzing a target reaction. According to the famous Sabatier's principle [69], the binding energy between the surface of the catalyst and the reacting intermediate must not be too weak to permit complete adsorption. Also, it must not be too strong to prevent the diffusion of the formed product away and regenerate the catalyst surface again for more catalytic cycles and have a higher turnover number. Plots correlating the intrinsic adsorption and electronic behavior to the activity of a catalyst are termed Volcano Plots. The volcano plots are useful to identify the nature of the binding energy of a particular metal oxide with the reaction intermediates, usually the most suitable ones being near the top of the peak [70]. Figure 9a illustrates a Volcano plot between the adsorption-free energies difference versus the overpotential for OER of different types of metal oxide systems. Metal oxides that are not easily oxidized are not very active towards OER because the intermediates adsorb weakly. Likewise, the metal oxides which get easily oxidized are also not very active catalytically because of the strong adsorption of the intermediates [71]. These curves indicate a higher activity for oxides with less stability. The uncoordinated surface sites at metal oxides usually behave as the key descriptor in the activity of OER. A highly unsaturated site gives better catalytic activity for weak binding oxides (left side) while for strong binding oxides, the activity decreases (right), as shown in Fig. 9b. In reversible air-breathing batteries like lithium-air [72], zinc-air [73], and other metal-air systems [74, 75], the use of a metal oxide-based catalyst to catalyze the sluggish kinetics of oxygen redox reactions has been well evaluated using various bifunctional catalysts. Such a bifunctional metal-oxide electrocatalyst has the optimal binding energy with the intermediates at its active sites [71, 76, 77] to facilitate the fundamentally inverse oxygen evolution reaction (OER) as well as oxygen reduction reaction (ORR) at the cathode during charging and discharging respectively [78–80].

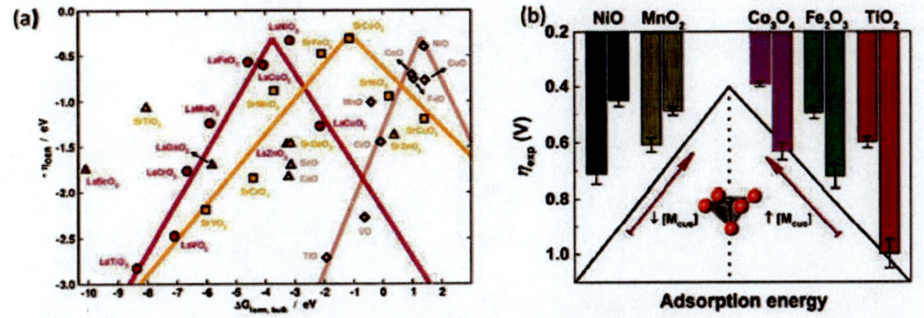

Fig. 9 **a** Volcano plot representing the binding energy relation with the corresponding over potential of different metal oxides as a crucial parameter in designing a metal oxide electrocatalyst. **b** The relation of uncoordinated sites on metal oxide surface with the activity in terms of overpotential requirements. Reproduced from Ref. [70] with permission from Springer Nature, Copyright 2022

3.11.2 Metal-Oxide with a Layered Structure as Host

Some metal oxides with high oxidization states of metals tend to exist in the layered structure by sharing the edge or vertex in an octahedral geometry. For example, various redox-active oxides like delta-MnO_2, orthorhombic V_2O_5, monoclinic $WO_3{\cdot}2H_2O$, etc., exhibit layered structures. They usually have a wide interlayer spacing, thus allowing the intercalation of various guest ions such as cations, anions, or water molecules [81]. Battery chemistries like sodium-ion [82] and lithium-ion [83] require host materials at the electrodes to be a layered structure that can reversibly intercalate sodium and lithium ions respectively. The layered structure of metal oxides with the general formula $LiMO_2$ and Na_xMO_2 where M denotes Mn, Co, Fe Ni, Ti, Cr, V, and their different stoichiometric combinations allow the facile intercalation and extraction of ions thus working efficiently as the cathode of such metal-ion batteries [84]. The Na_xMO_2 comprises sheets of MO_6 octahedra having a 2D transport channel for facilitated Na^+ ion interchange. The commonly occurring phases P_2 and O_3, are based on the Na^+ ion environment and the number of oxygen atoms stacked in sequence [85] as shown in Fig. 10a, b. If the Na^+ ion is at trigonal prismatic sites then it's denoted by P, while for octahedral sites it is denoted by O and the number represents the repetition of transition metal layers within the unit cell. Advanced studies have also demonstrated the applicability of the formation of binary or tertiary Na_xMO_2 compounds such as $Na_{0.3}Fe_{0.5}Co_{0.5}O_2$, P_2–$Na_{2/3}Ni_{1/3}Mn_{2/3}O_2$ have been found to exhibit better capacity retention and performance. Similarly for a lithium-ion battery, the layered metal oxides like $LiMO_2$ where M could be Co, Mn, or Ni, the metal ions occupy octahedral sites between the layers of cubic close-packed oxygen ions with interstitial layers of Li. At the time of charging, Li ions get extracted from the layered $LiMO_2$, resulting in non-stoichiometric $Li_{1-x}MO_2$ where the charge is compensated by the variable oxidation state of the transition metal, usually within the potential range of 3.5–4 V versus Li^+/Li. The cathode of the three metals (Ni, Co, Mn) in a stoichiometric composition of Li $Ni_{1/3}Co_{1/3}Mn_{1/3}O_2$

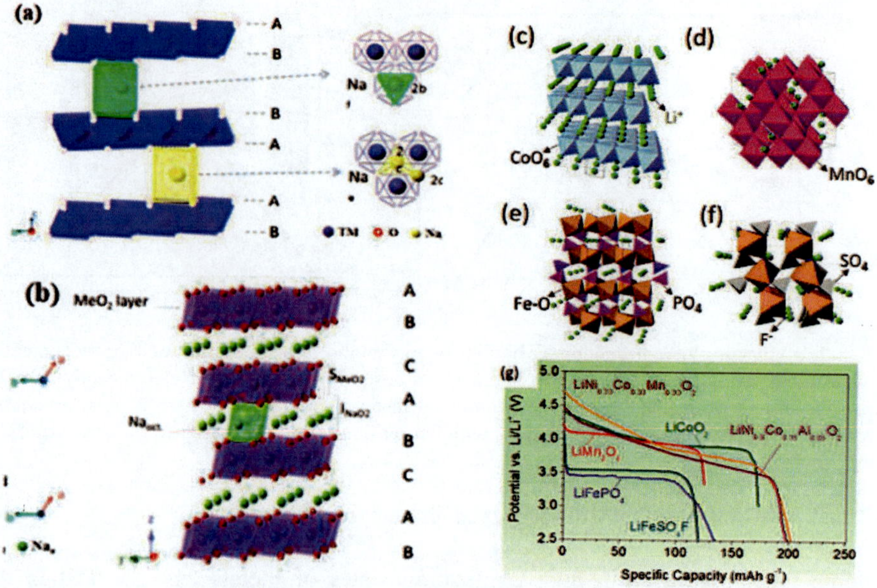

Fig. 10 Structural representation of P_2 and O_3-type layered metal oxides used as the cathode of sodium-ion batteries where **a** denotes P_2-type layered structure while, **b** represents O_3-type layered structure. Reproduced from ref. [85] with permission from Elsevier, Copyright 2016. **c–f** Structural schematic diagrams of intercalation cathodes of lithium-ion batteries corresponding to layered ($LiCoO_2$), spinel ($LiMn_2O_4$), olivine ($LiFePO_4$), and tavorite ($LiFeSO_4F$) respectively. **g** The discharge profiles of different metal oxides. Reproduced from Ref. [88] with permission from Elsevier, Copyright 2015

has found commercial applications due to its high capacity [86]. Tin oxide (SnO_2) is prominently used as the anode material for sodium-ion and lithium-ion batteries [87]. Several different types of metal oxides, spinels, layered, olivine, etc., can serve as the host cathode for lithium-ion batteries as shown in Fig. 10c–f. This intercalation-type cathode [88] wields a specific capacity of 100–200 mAh/g and an average voltage between 3 and 5 V versus Li/Li+, Fig. 10g. The layered structure of metal oxides capable of intercalating ions also emerged as an electrocatalyst for metal-air battery operations. As reported by Mathur et al. the doping of cobalt in 2D δ-MnO_2 caused an electronic rearrangement and led to superior bifunctional oxygen evolution and reduction reactivity utilized for zinc-air battery [89].

3.11.3 Metal-Oxide Exhibiting Variable Oxidation States

The ability to exist in multiple oxidation states attributed to the variable electronic arrangement of the d-band bestows transition metal oxide a high intrinsic catalytic activity. Different features of metal oxides arise from the metal existing in variable oxidation states. For instance, manganese exhibiting oxidation states from +2 to +

7 exists as different oxides like MnO [90], Mn_3O_4 [91], Mn_2O_3 [92], and MnO_2 [93] that have their peculiar catalytic activity owing to a difference in the electronic arrangement in d-bands, phase, morphology, crystal structure, and porosity. As shown in Fig. 11a, in a study, Yang et al. synthesized an epitaxial layer of Mn_3O_4 on MnO by the oxidation of the latter which resulted in improved porosity and charge storage thus a better capacitance [90]. Transition metal oxides can also perform as bifunctional catalysts due to different phases of oxides co-existing in a catalyst system as heterogeneous interfaces as the metal can adopt variable oxidation states. This is proposed in our previous study [94] wherein different oxides of iron, FeO, Fe_2O_3, and Fe_3O_4, co-existing in a catalyst as heterogeneous interfaces were responsible for bifunctional oxygen electrocatalytic activity. The bifunctional oxygen activity of a perovskite oxide was demonstrated to be enhanced by creating oxygen vacancy as proved by Lee et al. [95] With the introduction of oxygen vacancy and the partial reduction of cobalt +4 into +3 oxidation state, there occurs a rise in d band center, thus decreasing band gap between metal d band and oxygen p band centers and favoring charge transfer dynamics.

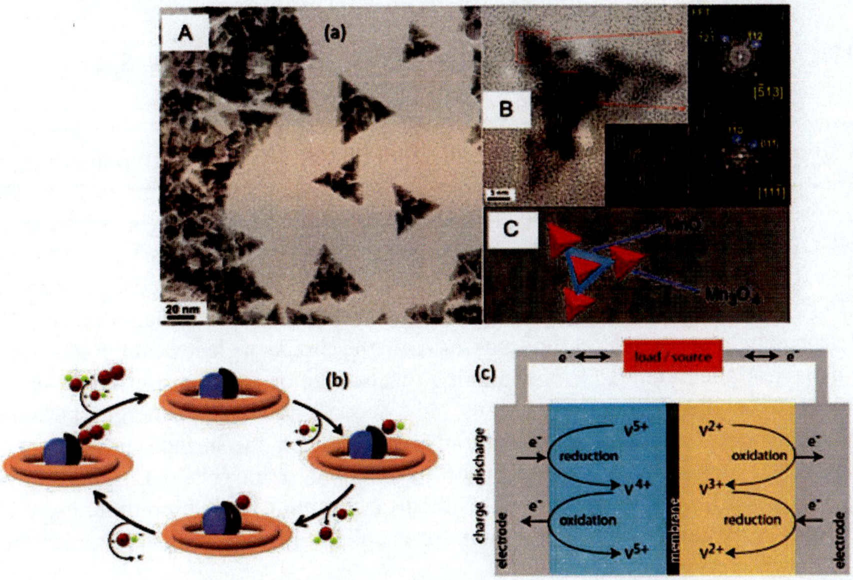

Fig. 11 **a** Different oxidation states exhibiting manganese oxides forming a composite catalyst for energy storage application. Reproduced from Ref. [90] with permission from Elsevier, Copyright 2015. **b** The surface reconstruction of nickel nanoparticles in an acidic environment, giving rise to the catalytic site, a heterojunction between oxidized nickel and metallic nickel. Reproduced from Ref. [96] with permission from Royal Society of Chemistry, Copyright 2020. **c** Vanadium exhibits four different oxidation states that make its use feasible in a redox flow battery with vanadium oxide ions VO^{2+} and VO_2^+ showing V^{+4} and V^{+5}. Reproduced from Ref. [99] with permission from IntechOpen, Copyright 2010

It is a well-observed fact that metal nanoparticles have the tendency of forming oxides/oxyhydroxides by getting their outer surface oxidized during certain experimental or operational conditions. The emerging heterojunction between different oxidation states of the transition metal creates a difference in the energy band and thus affects the electron transfer properties. The phenomenon of surface reconstruction generates the highly oxidized catalytically active site in the metal oxide catalysts The synergistic interaction at the heterojunction of metal in a lower oxidation state with its higher oxidation state, is identified to be the zone of activity for many electrocatalytic applications. As shown by our study, the layer of nickel oxide formed on nickel nanoparticles during acid leaching synthesis results in enhanced catalytic activity towards oxygen evolution [96–98], Fig. 11b.

Another, energy storage application, which is the redox flow battery, Fig. 11c, is in principle based on the four different oxides of vanadium in +2, +3, +4, and +5 oxidation states as already described earlier. The reversible oxidation and reduction of vanadium oxide ions VO^{2+} and VO_2^+ where vanadium exists in +4 and +5 oxidation states [45] and V^{2+} and V^{3+} proves the utility of different metal oxides in different oxidation states towards the energy storage applications.

3.11.4 Crystal Structures: Polymorphs & Mixed Metal Oxides: Spinel, and Perovskite

Metal oxides have the unique feature to acquire different crystallographic arrangements owing to the variety of oxidation states they can adopt, atomic radii, and defect sites the metal ion may occupy. Spinel metal oxides with a general formula $A_xB_yO_4$ where $x + y = 3$ and A is a divalent cation (Ni^{2+}, Co^{2+}, Fe^{2+}, Mn^{2+}, etc.) while B is a trivalent cation (Fe^{3+}, Al^{3+}, Mn^{3+}, and Co^{3+}) [100] occupy the tetrahedral and octahedral sites respectively in the lattice and have remarkable oxygen electrocatalytic performance, thus making them the choice of electrocatalyst for zinc-air battery application [101]. The hopping mechanism of electron transfer in spinel oxides [102] makes conductivity very efficient due to low activation energies between different oxidation states cations. In a study by Zhao et al., the surface engineering of active metal sites in spinel nanosheets of nickel cobalt oxide ($NiCo_2O_4$) by thermal annealing at different temperatures resulted in optimum surface configuration, as shown in Fig. 12a, b which is desirable for excellent bifunctional catalytic activity in zinc-air batteries [80].

The Perovskite-type metal oxides [103, 104] have the general formula ABO_3 where divalent A cation belongs to a rare-earth metal class (La^{2+}, Bi^{2+}, etc.) or alkaline-earth metal class (Ca^{2+}, Ba^{2+}, Sr^{2+}etc.), the B^{4+} cation belongs to transition metal class (Ti, W, Co, Mn, Fe, Ni, Cr, etc.) [105]. These perovskite oxides can opt for a mechanism known as the lattice oxygen mediated (LOM) mechanism where they make use of the lattice oxygen for the catalysis of oxygen evolution reactions [106]. This occurs when the oxygen p band is higher in energy than the metal d band, the electron transfer occurs from the p band to the d band, suggesting a higher

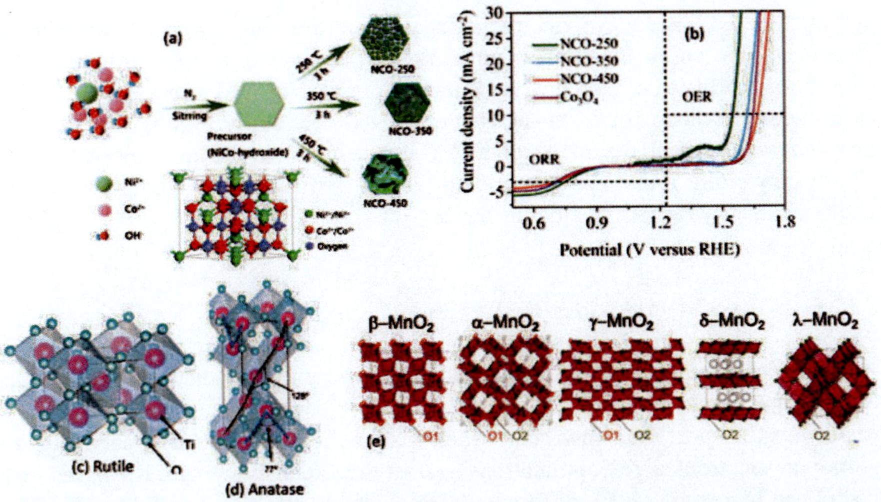

Fig. 12 **a** Spinel oxide $NiCoO_4$ prepared in different sintering conditions having different oxygen bifunctional activity shown in (**b**). Reproduced from Ref. [80] with permission from American Chemical Society, Copyright 2019. The most predominant and stable polymorphs of TiO_2 rutile (**c**) and anatase **d** have different catalytic activity derived from a difference in the crystal structure. Reproduced from Ref. [112] with permission from American Chemical Society, Copyright 2022. **e** Different polymorphs of MnO_2 arise from the various bonding patterns and exhibit unique structural, electronic, and catalytic features. Reproduced from Ref. [113] with permission from Elsevier, Copyright 2022

covalency [107]. This mechanism is energetically less demanding due to the non-existing scaling relation between the catalyst surface and intermediates M-OH and M-OOH [108, 109].

Mixed metal oxides (A_x B_{1-x} $(OH)_y$) have a synergistic charge transfer between two or more metallic sites that influences the local electron density and governs the interaction with the catalytic intermediates. Ni–Fe oxide, regarded as one of the most suitable OER catalysts [110], has a synergistic interaction between the conductive nickel oxide and the site of catalytic activity FeOOH. Similarly, for Ni–Fe–Co mixed metal oxide [111], it is proved by X-ray absorption spectroscopy (XAS) that Co facilitates the formation of conductive NiOOH from $Ni(OH)_2$ which activates Fe sites by shrinking the Fe–OH/OOH bonds and reducing the intermediate adsorption energy.

Certain metal oxides like TiO_2, MnO_2, etc., may also exist as different stable polymorphs depending on the synthesis and persisting temperature conditions. These polymorphs differ from each other by distinct crystallographic features. The rutile and anatase polymorphs of TiO_2 as shown in Fig. 12c, d is different based on the voids titanium occupies and the sharing of edges in TiO_6 octahedra [112].

MnO_2 exists in various structural polymorphs namely Hollandite-α, Pyrolusite-β, Intergrowth-γ, Birnessite-δ, and defective spinel-λ structure as illustrated in Fig. 12e owing to the myriad of bonding patterns between Mn^{+4} and O bipyramids. These polymorphs exhibit diverse electrochemical properties and suitability for different electrocatalytic applications. The spinel-type MnO_2 has a layered structure [113] with the capability to act as a host for intercalating Li, Na, or Mg thus suggesting it is an attractive cathode for metal-ion batteries. Another study highlights the synergism between composite made of alpha and delta MnO_2. The composite being photocatalytically and electrochemically active performs exceptionally well in a photochemically recharged zinc-air battery [114].

3.11.5 Mixed Metal Oxides/Hydroxides

Metal oxides may get hydrolyzed in aqueous conditions to form metal hydroxides that exhibit large spaces between the interlayered structure, good electrochemical activity, easier ion transport, and low cost which makes them an attractive candidate for numerous electrocatalytic applications against the expensive Pt and Ru-based catalyst [115]. In a study [116], different electrochemical routes to synthesize $Ni(OH)_2$ nanoparticles supported on carbon nanotubes were explored to conclude that the disordered δ-$Ni(OH)_2$ formed by direct approach of coprecipitation performed better towards methanol/ethanol oxidation for direct alcohol-based fuel cells as compared to the more ordered and thermodynamically stable β-$Ni(OH)_2$ synthesized indirectly via electrodeposition and potential cycling in alkaline conditions. Metal hydroxides act as a precursor to the oxidized and catalytically active MOOH species. Another class of metal hydroxides is the layered double hydroxides (LDHs) formed as a result of coprecipitation of M^{3+} ions and divalent hydroxides [117], having the general formula: $M^{II}{}_{1-x}N^{III}{}_x\ (OH)_2\ A^{n-}_{x/n}\cdot yH_2O$, where M^{II} could be Ni^{2+}, Co^{2+}, Mg^{2+} etc., N^{III} could be Fe^{3+}, Al^{3+}, Mn^{3+} etc., A denotes a suitable anion like carbonate, nitrate, halide, chalcogenide, etc. The layered structure of these mixed metal hydroxides intercalates anions and water molecules in between the layers to balance charge and also empowers LDHs with exceptional catalytic activity. The unique structural features of LDH present many tuneable properties as per application [118]. The anions between the layers can be exchanged to modify the interlayer spacing. This can also be used to exfoliate the LDH via sonication or dispersion into unilamellar sheets with almost molecular-level thickness. Also, this approach can be used to create super-lattice composites and heterojunctions which create a specific electronic arrangement. The morphology and crystal structure of LDHs can be optimized to regulate the exposed surface, interface, and thus catalytic performance. Doping of heteroatoms and defect engineering in the LDH also contributes to optimizing the local electronic arrangement and catalytic behavior. In a work by Ma et al. [119], a bifunctional electrocatalyst was synthesized by exfoliating a Ni–Fe LDH and creating its superlattice nanocomposite with graphene, as shown in Fig. 13a. Figure 13b represents the formation of layer-by-layer self-assembly of Ni–Fe LDH and Ni–Al LDH with MoS_2 [120] as confirmed by HR-TEM imaging in Fig. 13c, resulted in strong synergistic electronic interactions between them and enhanced the

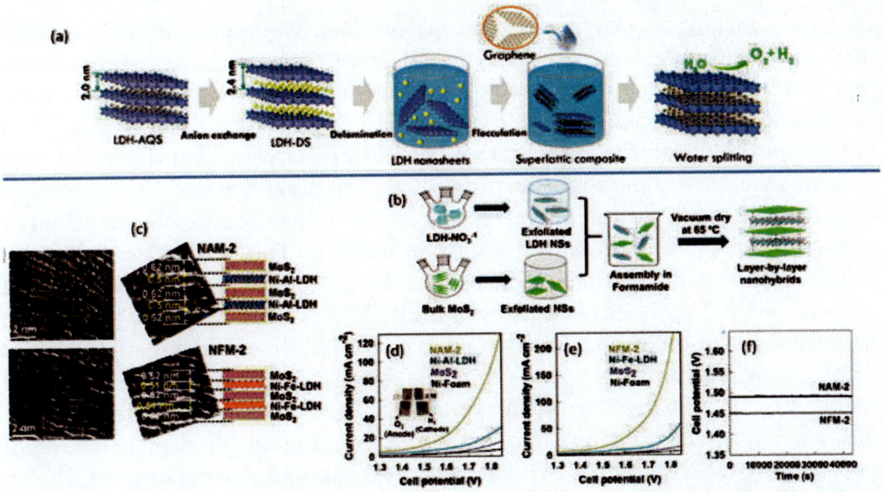

Fig. 13 **a** A schematic representation of forming a superlattice composite between molecular level thick LDH nanosheets with graphene oxide nanosheets to produce a bifunctional electrocatalyst for water electrolysis. Reproduced from Ref. [119] with permission from American Chemical Society, Copyright 2015. **b** The formation of layer-by-layer self-assembly of Ni–Fe LDH and Ni–Al LDH with MoS_2 and its confirmation given by the lattice fringes obtained from the HR-TEM (**c**). The bifunctional performance of such fabricated electrocatalysts for overall water splitting is given for **d** Ni–Al LDH with MoS_2. **e** Ni–Fe LDH with MoS_2 and **f** their electrochemical stability study at a fixed current density. Reproduced from Ref. [120] with permission from American Chemical Society, Copyright 2018

intermediate adsorption for OER and HER. The exceptional water splitting performance Fig. 13d–f of these electrocatalysts has created a result of interfacial electron coupling through the hetero-layers of the nanohybrid.

3.11.6 Defects in Metal Oxide

Defect engineering is a crucial strategy for impacting the electronic behavior of a metal oxide. The generation of defects also manipulates the intermediate adsorption energy for electrocatalytic interface reactions. Strategic incorporation of defects has been found to activate the metal oxide toward advanced performance in lithium-ion batteries [121, 122]. The nascent low-coordinated or free sites evolved due to defect creation participation as the active site thus increasing the intrinsic catalytic activity. Metal oxides have several structural features like layered structure to intercalate anions, variable oxidation states to balance the charge in the case of a metal vacancy defect, or substitution of cation, which stabilizes them after the defect generation [123]. In perovskite oxide materials the partial substitution of the A site or B site generates significant defects that influence the activity of the catalyst. The creation of oxygen vacancy has been many times found to alter the electronic properties of a compound making it catalytically more active [124]. Various studies have focussed their attention on evolving the catalytic properties of metal oxides using different defect-creation strategies like atomic vacancies, creating hetero-interfaces, substitutional doping, etc. [125, 126]. It was found that the creation of a ternary form $NiCo_2O_4$ from the octahedral spinel structured Co_3O_4 by the inclusion of Ni atoms improved the activity and corrosion resistance [127]. In another study, oxygen defects created in $NiCo_2O_4$ play an important role in expediting the bifunctional activity of oxygen electrocatalysis employed in zinc-air batteries. This occurs due to the favorable adsorption of oxygen intermediates from the oxygen defect sites which are easily fabricated in $NiCo_2O_4$ nanosheets by annealing and etching. Figure 14a shows the uniform distribution of Ni, Co, and O in the HR-TEM elemental mapping despite the structural defects arising from oxygen deficiency. The same is confirmed by O1s spectra from XPS in Fig. 14b where an additional peak deconvoluted as a defect site of lower oxygen coordination appears at 530.5 eV. Figure 14c depicts the bifunctional activity of OER and OER on the $NiCo_2O_4$ nanosheet electrocatalyst with and without the defects, clearly indicating the improvement in performance with defects [128]. Doping of heteroatom with different oxidation states and atomic radius brings defects and lattice strain in the host metal oxide structure. It can also modulate the electron occupancy as happened when Rao et al. doped vanadium in spinel Co_3O_4. The occupancy of e_g orbital changed from zero to 1.01and the spin-modified catalyst was found to have improved bifunctional activity towards OER and ORR and performed exceptionally well in zinc-air battery [129]. Figure 14d shows the TEM elemental mapping of vanadium-doped Co_3O_4 nanorod with uniformly doped vanadium. The test set-up for an aqueous zinc-air battery which was used in this study is demonstrated in Fig. 14e, while the comparative polarization curve and power density plot is shown in Fig. 14f. Certainly, the performance of V-Co_3O_4 is enhanced by doping vanadium. This is also confirmed in Fig. 14g by its high specific capacity at a 5 mA cm^{-2} current density.

3.11.7 Activation of Catalytically Active Facet/Site of Metal Oxide: Surface Reconstruction

With the help of in-situ characterization techniques [130] it has been identified that the reaction conditions subjected are responsible for activating the catalytically active site or phase or emergence of other active facets by oxidation, surface restructuring, or other processes. It is a well-understood phenomenon that certain metal compounds such as metal nitrides, phosphides, sulfides, etc. when immersed in an alkaline medium and subjected to anodic conditions, evolve into their corresponding oxides and oxyhydroxide species which are believed to be the actual active site for the ongoing electrocatalytic reaction. This is especially observed for nickel, cobalt, and iron-based catalysts to convert into their respective oxides for catalyzing oxygen evolution reactions during the performance of a battery reinforcing the superior catalytic behavior of metal oxides as electrocatalysts. Similarly, certain high-energy facets of metal oxide are shown to have better catalytic activity than others.

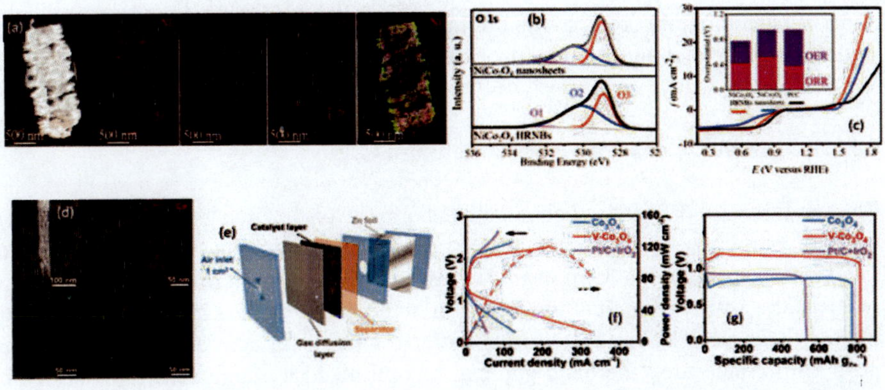

Fig. 14 **a** TEM elemental mapping of oxygen defect containing $NiCo_2O_4$ nanosheets showing the uniform distribution of each element. **b** High-resolution O 1 s spectra of this catalyst indicate the presence of oxygen defect. **c** Comparison of bifunctional OER and ORR activity of the catalyst with undefective and standard Pt/C catalyst. The minimum difference in the potential values indicates better bifunctional catalytic performance. Reproduced from Ref. [128] with permission from American Chemical Society, Copyright 2022. **d** TEM elemental mapping of V-Co_3O_4 confirms the uniform doping of vanadium, **e** the diagram of the setup used for the aqueous zinc-air battery performance evaluation of V-Co_3O_4. **f** Galvanostatic polarization curves and power density curves of the catalyst in comparison to the standard catalyst. **g** Specific capacity of V-Co_3O_4 proves its enhanced zinc-air battery performance. Reproduced from Ref. [129] with permission from American Chemical Society, Copyright 2021

For instance, the anatase crystal of TiO_2 exists most predominantly in {101} facet however the reconstruction into a more reactive {001} facet is more electrochemically active [131]. In a study by Nien-Chu Lai et al., they demonstrated that the growth of the high-energy facet of Cr_2O_3 (110) exhibits better bonding with intermediates owing to the presence of uncoordinated sites and thus gives a higher OER catalytic activity in a rechargeable Li–O_2 battery [132]. Surface atomic structure engineering on one-dimensional cobalt oxide CoO nanorods having oxygen vacancies on the oxygen-terminated {111} nano facets exhibit enhanced electron transfer and intermediate binding energies, thus giving a bifunctional oxygen electrocatalytic activity [133]. Acid leaching is also known as a method to modify the surface and restructures the active site. For, a nickel-nickel oxide core-shell nanoparticle encapsulated within graphitic rings, acid leaching favors the emergence of the active NiO {110} facet and thereby improves the OER activity [96]. One more advantage of using metal oxides as the electrocatalyst is that it offers a variety of active sites differing in performance or catalytic nature. Sometimes the surface facet is less active than the edges, as discovered in a theoretical study by Yi Cui's group. The stable (0001) surface of lithium cobalt oxide (LCO) is less active as compared to the nonpolar (1120) and polar (0112) surfaces for the OER activity. Further acid etching on LCO creates more active edge sites and improves the catalytic performance [134]. Surface engineering of Co_3O_4 into different morphologies revealed that polyhedron facet {112} performs better as OER and ORR catalyst compared to the nanocube and nano truncated octahedron having dominant {001} and {111} facets, thus being an outstanding performing catalyst in zinc-air battery [135].

3.11.8 Metal Oxides with a Suitable Band Gap

The fundamental difference between the energy levels valence band and conduction band of particular metal oxides gives it unique semi-conducting properties. The ideal band gap spanning from near UV-Vis-near IR of the electromagnetic spectrum receives the ability of optical transmission in metal oxides making them suitable for photovoltaic, photochemical, and photoelectrochemical catalytic applications. The properties concerning the band gap can be optimized by engineering the structure and monitoring defects. Transparent conducting oxides (TCO) like tin-doped indium oxide (ITO, Sn–In_2O_3) have a high photon transmitting capability in the visible range and good conductivity. These semiconductors have a wide bandgap and a high concentration of free electrons or holes in their conduction or valence bands and are dominantly n-type. The combined effect of the layered structure and optimum band gap gives a diverse utility to metal oxides like MoO_3, TiO_2, and MnO_2 in various energy storage applications [136]. Transition metal oxides are also observed to show quantum transport behavior modified by substitutional doping with another element [137, 138]. Modifications in the band gap can be performed by doping hetero atom and creating oxygen vacancy and other structural defects to enhance visible light activity [139].

The emerging technology of photoelectrochemically chargeable zinc-air battery comprises a zinc electrode and a photon-sensitized metal oxide-based catalyst. The charging of such batteries takes place when placed in ambient sunlight, along with necessary external biasing [140]. Some heterojunctions formed by metal oxides like Fe_2O_3 (band gap 1.9–2.2 eV) combined with metal oxyhydroxides like NiOOH exhibit excellent oxygen evolution reactions for lithium-air batteries [141]. The heterojunction of TiO_2 and V_2O_5 behaves as optimum photoelectrode for lithium-ion battery [142]. The oxide V_2O_5 can act as an excellent photoelectrode itself owing to its bandgap thus being ambient for zinc-ion and lithium-ion battery purposes [143, 144]. Similarly, the composite of α- and δ-MnO_2 having unique crystallographic orientation and phases was shown to be highly responsive towards visible-light-driven charging/discharging of zinc-air battery [114]. The photoelectrode designed using α-Fe_2O_3 requires a reduced charging potential by approx. 0.5–0.8 V under the irradiation of sunlight [143]. Another group developed a $BiVO_4$ and V_2O_5 heterojunction photoelectrode that reduced the charging voltage to 1.2 V under visible-light illumination [144].

4 Conclusion

This chapter summarises the distinctive features of metal oxide that make them an attractive candidate for different roles in energy storage devices. Here the two energy storage technologies discussed at length include supercapacitors and different battery chemistries. The layered structure of metal oxides allows it to become a host for the reversible intercalation of ions in lithium and sodium-ion batteries. The variety of structural polymorphs that metal oxides can exhibit and the specific catalytic activity of different facets of a crystal structure further justify the suitability of metal oxides in different energy storage applications. With the possibility to survive in different oxidation states, the transition metal oxides can give rise to simultaneous oxides at heterogeneous interfaces, stabilize doping and vacancy defects, and have an instinctive conversion to the catalytically active metastable states. While the optimum bandgap makes them ideal semiconductors for photon-enhanced charge transfer performance. Metal oxides offer innumerable structural and electronic features with convenient tunability which enables their widespread acceptance for feasible energy storage.

References

1. R.G. Gusmão Caiado, W. Leal Filho, O.L.G. Quelhas, D. Luiz de Mattos Nascimento, L.V. Ávila, A literature-based review on potentials and constraints in the implementation of the sustainable development goals. J. Clean. Prod. **198**, 1276–1288 (2018)
2. A. Wiltshire, D. Bernie, L. Gohar, J. Lowe, C. Mathison, C. Smith, Post COP26: does the 1.5 °C climate target remain alive?. Weather **77**, 412–417 (2022)
3. D.O. Akinyele, R.K. Rayudu, Review of energy storage technologies for sustainable power networks. Sustain. Energy Technol. Assess. **8**, 74–91 (2014)
4. Precedence Research, Energy Storage Systems Market (By Technology: Compressed Air, Pumped Hydro Storage, Lithium Ion, Sodium Sulphur, Lead Acid, Redox flow, Nickel Cadmium, Flywheel, By Application, Transportation, Grid Management; By End User: Residential, Non-Residential, U, 2022
5. IEA, *India Energy Outlook 2021* (2021)
6. M. Aneke, M. Wang, Energy storage technologies and real life applications—A state of the art review. Appl. Energy **179**, 350–377 (2016)
7. R. Liang, Y. Du, P. Xiao, J. Cheng, S. Yuan, Y. Chen, J. Yuan, J. Chen, Transition metal oxide electrode materials for supercapacitors: a review of recent developments. Nanomaterials **11**, 1248 (2021)
8. Y. Zhang, L. Zhang, L. Cheng, D. Yang, L. Wan, C. Du, J. Chen, M. Xie, Synthesis of faradaic-active N, O-doped carbon nanosheets from m-trihydroxybenzene and piperazine for high-performance supercapacitor. Appl. Surf. Sci. **538**, 148040 (2021)
9. P. Avasthi, N. Arya, M. Singh, V. Balakrishnan, Fabrication of iron oxide-CNT based flexible asymmetric solid state supercapacitor device with high cyclic stability. Nanotechnology **31**, 435402 (2020)
10. G. Yang, S.-J. Park, MnO2 and biomass-derived 3D porous carbon composites electrodes for high performance supercapacitor applications. J. Alloys Compd. **741**, 360 367 (2018)
11. L. Guan, L. Pan, T. Peng, C. Gao, W. Zhao, Z. Yang, H. Hu, M. Wu, Synthesis of biomass-derived nitrogen-doped porous carbon nanosheests for high-performance supercapacitors. ACS Sustain. Chem. Eng. **7**, 8405–8412 (2019)

12. N. Wang, P. Zhao, K. Liang, M. Yao, Y. Yang, W. Hu, CVD-grown polypyrrole nanofilms on highly mesoporous structure MnO2 for high performance asymmetric supercapacitors. Chem. Eng. J. **307**, 105–112 (2017)
13. H. Chen, W. Li, M. He, X. Chang, X. Zheng, Z. Ren, Vertically oriented carbon nanotube as a stable frame to support the Co0.85Se nanoparticles for high performance supercapacitor electrode. J. Alloys Compd. **855**, 157506 (2021)
14. M.A. Arvizu, F.J. González, A. Romero-Galarza, F.J. Rodríguez-Varela, C.R. Garcia, M.A. Garcia-Lobato, Symmetric supercapacitors of PANI coated RuO2/TiO2 macroporous structures prepared by electrostatic spray deposition. J. Electrochem. Soc. **169**, 020564 (2022)
15. G. Wang, Z. Jin, Q. Guo, Ordered self-supporting NiV LDHs@P-Nickel foam nano-array as high-performance supercapacitor electrode. J. Colloid Interface Sci. **583**, 1–12 (2021)
16. D. Wu, H. Han, X. Hong, S. Tao, S. Xu, B. Qian, L. Wang, X. Chen, P.K. Chu, A novel self-branching MnCo2O4/ nanographene hybrid composites on macroporous electrically conductive network as bifunctional electrodes for boosting miniature supercapacitors and sodium ion batteries. J. Alloys Compd. **846**, 155720 (2020)
17. P. Saren, A. De Adhikari, S. Khan, G.C. Nayak, Self-assembled GNS wrapped flower-like MnCo2O4 nanostructures for supercapacitor application. J. Solid State Chem. **271**, 282–291 (2019)
18. Y. Zheng, H. Ding, M. Zhang, Preparation and electrochemical properties of nickel oxide as a supercapacitor electrode material. Mater. Res. Bull. **44**, 403–407 (2009)
19. Y. Wang, Y. Xia, Electrochemical capacitance characterization of NiO with ordered mesoporous structure synthesized by template SBA-15. Electrochim. Acta **51**, 3223–3227 (2006)
20. Y.-Z. Zheng, M.-L. Zhang, Preparation and electrochemical properties of nickel oxide by molten-salt synthesis. Mater. Lett. **61**, 3967–3969 (2007)
21. S.A. Beknalkar, A.M. Teli, N.S. Harale, D.S. Patil, S.A. Pawar, J.C. Shin, P.S. Patil, Fabrication of high energy density supercapacitor device based on hollow iridium oxide nanofibers by single nozzle electrospinning. Appl. Surf. Sci. **546**, 149102 (2021)
22. J. Bhagwan, A. Sahoo, K.L. Yadav, Y. Sharma, Porous, one dimensional and high aspect ratio Mn3O4 nanofibers: fabrication and optimization for enhanced supercapacitive properties. Electrochim. Acta **174**, 992–1001 (2015)
23. W. Xing, F. Li, Z. Yan, G.Q. Lu, Synthesis and electrochemical properties of mesoporous nickel oxide. J. Power Sour. **134**, 324–330 (2004)
24. S. Jeon, J.H. Jeong, H. Yoo, H.K. Yu, B.-H. Kim, M.H. Kim, RuO2 nanorods on electrospun carbon nanofibers for supercapacitors. ACS Appl. Nano Mater. **3**, 3847–3858 (2020)
25. E. Lee, T. Lee, B.-S. Kim, Electrospun nanofiber of hybrid manganese oxides for supercapacitor: relevance to mixed inorganic interfaces. J. Power Sour. **255**, 335–340 (2014)
26. J. Zhao, C. Lai, Y. Dai, J. Xie, Pore structure control of mesoporous carbon as supercapacitor material. Mater. Lett. **61**, 4639–4642 (2007)
27. J. Piao, D. Bin, S. Duan, X. Lin, D. Zhang, A. Cao, A facile template free synthesis of porous carbon nanospheres with high capacitive performance. Sci. China Chem. **61**, 538–544 (2018)
28. G. Wang, L. Zhang, J. Zhang, A review of electrode materials for electrochemical supercapacitors. Chem. Soc. Rev. **41**, 797–828 (2012)
29. J. Yan, Q. Wang, T. Wei, Z. Fan, Recent advances in design and fabrication of electrochemical supercapacitors with high energy densities. Adv. Energy Mater. **4**, 1300816 (2014)
30. J. Chang, M. Jin, F. Yao, T.H. Kim, V.T. Le, H. Yue, F. Gunes, B. Li, A. Ghosh, S. Xie, Y.H. Lee, Asymmetric supercapacitors based on graphene/MnO 2 nanospheres and graphene/MoO3 nanosheets with high energy density. Adv. Funct. Mater. **23**, 5074–5083 (2013)
31. F. Bonaccorso, L. Colombo, G. Yu, M. Stoller, V. Tozzini, A.C. Ferrari, R.S. Ruoff, V. Pellegrini, Graphene, related two-dimensional crystals, and hybrid systems for energy conversion and storage. Science (80). https://doi.org/10.1126/science.1246501
32. W. Liu, N. Liu, Y. Shi, Y. Chen, C. Yang, J. Tao, S. Wang, Y. Wang, J. Su, L. Li, Y. Gao, A wire-shaped flexible asymmetric supercapacitor based on carbon fiber coated with a metal oxide and a polymer. J. Mater. Chem. A **3**, 13461–13467 (2015)

33. Z. Song, W. Liu, P. Xiao, Z. Zhao, G. Liu, J. Qiu, Nano-iron oxide (Fe2O3)/three-dimensional graphene aerogel composite as supercapacitor electrode materials with extremely wide working potential window. Mater. Lett. **145**, 44–47 (2015)
34. S.R. Azari, M.S. Rahmanifar, M.F. El-Kady, A. Noori, M.F. Mousavi, R.B. Kaner, A wide potential window aqueous supercapacitor based on LiMn2O4–rGO nanocomposite. J. Iran. Chem. Soc. **14**, 2579–2590 (2017)
35. C. Xu, F. Kang, B. Li, H. Du, Recent progress on manganese dioxide based supercapacitors. J. Mater. Res. **25**, 1421–1432 (2010)
36. S. Bag, C.R. Raj, Hierarchical three-dimensional mesoporous MnO2 nanostructures for high performance aqueous asymmetric supercapacitors. J. Mater. Chem. A **4**, 587–595 (2016)
37. H. Xia, J. Feng, H. Wang, M.O. Lai, L. Lu, MnO2 nanotube and nanowire arrays by electrochemical deposition for supercapacitors. J. Power Sour. **195**, 4410–4413 (2010)
38. J.-G. Wang, H. Liu, H. Liu, W. Hua, M. Shao, Interfacial constructing flexible V2O5@Polypyrrole Core-Shell nanowire membrane with superior supercapacitive performance. ACS Appl. Mater. Interfaces **10**, 18816–18823 (2018)
39. Y. Liu, L. Liu, L. Kang, F. Ran, Energy storage mechanism of vanadium nitride via intercalating different atomic radius for expanding interplanar spacing. Energy Environ. Mater. **5**, 565–571 (2022)
40. S. Devaraj, N. Munichandraiah, Effect of crystallographic structure of MnO2 on its electrochemical capacitance properties. J. Phys. Chem. C **112**, 4406–4417 (2008)
41. Y. Tan, Y. Liu, L. Kong, L. Kang, F. Ran, Supercapacitor electrode of nano-Co3O4 decorated with gold nanoparticles via in-situ reduction method. J. Power Sour. **363**, 1–8 (2017)
42. J.-H. Zhong, A.-L. Wang, G.-R. Li, J.-W. Wang, Y.-N. Ou, Y.-X. Tong, Co3O4/Ni(OH)2 composite mesoporous nanosheet networks as a promising electrode for supercapacitor applications. J. Mater. Chem. **22**, 5656 (2012)
43. L. Shen, Q. Che, H. Li, X. Zhang, Mesoporous NiCo2O4 nanowire arrays grown on carbon textiles as binder-free flexible electrodes for energy storage. Adv. Funct. Mater. **24**, 2630–2637 (2014)
44. J.O.B. Dr. Ing. Claus Daniel, *Handbook of Battery Materials* (Wiley, 2011)
45. X. Yang, A.L. Rogach, Electrochemical techniques in battery research: a tutorial for nonelectrochemists. Adv. Energy Mater. **9**, 1–10 (2019)
46. B. Viswanathan, B. Viswanathan, *Chapter 12—Batteries* (2017)
47. BOBBY, History of batteries: a timeline. https://www.upsbatterycenter.com/blog/history-batteries-timeline/
48. X. Fan, B. Liu, J. Liu, J. Ding, X. Han, Y. Deng, X. Lv, Y. Xie, B. Chen, W. Hu, C. Zhong, Trans. Tianjin Univ. **26**, 92–103 (2020)
49. K.C. Divya, J. Østergaard, Battery energy storage technology for power systems-an overview. Electr. Power Syst. Res. **79**, 511–520 (2009)
50. O. Schmidt, A. Hawkes, A. Gambhir, I. Staffell, The future cost of electrical energy storage based on experience rates. Nat. Energy **2**, 1–8 (2017)
51. A.K. Shukla, S. Venugopalan, B. Hariprakash, Nickel-based rechargeable batteries. J. Power Sour. **100**, 125–148 (2001)
52. Z. Wen, J. Cao, Z. Gu, X. Xu, F. Zhang, Z. Lin, Research on sodium sulfur battery for energy storage. Solid State Ionics **179**, 1697–1701 (2008)
53. H. Zhang, W. Lu, X. Li, Progress and perspectives of flow battery technologies. Electrochem. Energy Rev. **2**, 492–506 (2019)
54. A. Manthiram, X. Yu, S. Wang, Lithium battery chemistries enabled by solid-state electrolytes. Nat. Rev. Mater. **2**, 1–16 (2017)
55. N. Wang, H. Wan, J. Duan, X. Wang, L. Tao, J. Zhang, H. Wang, A review of zinc-based battery from alkaline to acid. Mater. Today Adv. **11**, 100149 (2021)
56. A.Z. Weber, M.M. Mench, J.P. Meyers, P.N. Ross, J.T. Gostick, Q. Liu, Redox flow batteries: a review. J. Appl. Electrochem. **41**, 1137–1164 (2011)
57. H. Xu, Z. Sun, J. Chen, in *Emerging 2D Materials and Devices for the Internet of Things* (Elsevier, 2020), pp. 139–164

58. A.K. Worku, D.W. Ayele, N.G. Habtu, Recent advances and future perspectives in engineering of bifunctional electrocatalysts for rechargeable zinc–air batteries. Mater. Today Adv. **9**, 100116 (2021)
59. S. Yan, K.P. Abhilash, L. Tang, M. Yang, Y. Ma, Q. Xia, Q. Guo, H. Xia, Research advances of amorphous metal oxides in electrochemical energy storage and conversion. Small **15**, 1–30 (2019)
60. A.A. Yaqoob, A. Ahmad, M.N.M. Ibrahim, R.R. Karri, M. Rashid, Z. Ahamd, in *Sustainable Nanotechnology for Environmental Remediation* (Elsevier, 2022), pp. 611–635
61. A. Huang, Y. He, Y. Zhou, Y. Zhou, Y. Yang, J. Zhang, L. Luo, Q. Mao, D. Hou, J. Yang, A review of recent applications of porous metals and metal oxide in energy storage, sensing and catalysis. J. Mater. Sci. **54**, 949–973 (2019)
62. W. Sun, Y. Song, X.-Q. Gong, L. Cao, J. Yang, An efficiently tuned d-orbital occupation of IrO2 by doping with Cu for enhancing the oxygen evolution reaction activity. Chem. Sci. **6**, 4993–4999 (2015)
63. W. Huang, J. Zhang, D. Liu, W. Xu, Y. Wang, J. Yao, H.T. Tan, K.N. Dinh, C. Wu, M. Kuang, W. Fang, R. Dangol, L. Song, K. Zhou, C. Liu, J.W. Xu, B. Liu, Q. Yan, Tuning the electronic structures of multimetal oxide nanoplates to realize favorable adsorption energies of oxygenated intermediates. ACS Nano **14**, 17640–17651 (2020)
64. E.M. Fernndez, P.G. Moses, A. Toftelund, H.A. Hansen, J.I. Martínez, F. Abild-Pedersen, J. Kleis, B. Hinnemann, J. Rossmeisl, T. Bligaard, J.K. Nørskov, Scaling relationships for adsorption energies on transition metal oxide, sulfide, and nitride surfaces. Angew. Chem. Int. Ed. **47**, 4683–4686 (2008)
65. Z. Bian, T. Tachikawa, P. Zhang, M. Fujitsuka, T. Majima, A nanocomposite superstructure of metal oxides with effective charge transfer interfaces. Nat. Commun. **5**, 1–9 (2014)
66. M.K. Bates, Q. Jia, H. Doan, W. Liang, S. Mukerjee, Charge-transfer effects in Ni–Fe and Ni–Fe–Co mixed-metal oxides for the alkaline oxygen evolution reaction. ACS Catal. **6**, 155–161 (2016)
67. F. Wang, Y. Zhang, N. Yu, L. Fu, Y. Zhu, Y. Wu, T. van Ree, in *Metal Oxides in Energy Technologies* (Elsevier, 2018), pp. 127–167
68. H. Ooka, J. Huang, K.S. Exner, The sabatier principle in electrocatalysis: basics, limitations, and extensions. Front. Energy Res. **9**, 1–20 (2021)
69. J. Li, Oxygen evolution reaction in energy conversion and storage: design strategies under and beyond the energy scaling relationship. Nano-Micro Lett. **14**, 112 (2022)
70. Y. Jiao, Y. Zheng, M. Jaroniec, S.Z. Qiao, Design of electrocatalysts for oxygen- and hydrogen-involving energy conversion reactions. Chem. Soc. Rev. **44**, 2060–2086 (2015)
71. M.A. Rahman, X. Wang, C. Wen, A review of high energy density lithium-air battery technology. J. Appl. Electrochem. **44**, 5–22 (2014)
72. G.M. Tomboc, P. Yu, T. Kwon, K. Lee, J. Li, Ideal design of air electrode—A step closer toward robust rechargeable Zn–air battery. APL Mater. **8**, 050905 (2020)
73. J. Pan, X.L. Tian, S. Zaman, Z. Dong, H. Liu, H.S. Park, B.Y. Xia, Batter. Supercaps **2**, 336–347 (2019)
74. Q. Liu, Z. Pan, E. Wang, L. An, G. Sun, Aqueous metal-air batteries: fundamentals and applications. Energy Storage Mater. **27**, 478–505 (2020)
75. Z.W. Seh, J. Kibsgaard, C.F. Dickens, I. Chorkendorff, J.K. Nørskov, T.F. Jaramillo, Combining theory and experiment in electrocatalysis: insights into materials design. Science (80). https://doi.org/10.1126/science.aad4998
76. C. Wang, Y. Yu, J. Niu, Y. Liu, D. Bridges, X. Liu, J. Pooran, Y. Zhang, A. Hu, Recent progress of metal-air batteries—A mini review. Appl. Sci. **9**, 2787 (2019)
77. M.M. Montemore, M.A. van Spronsen, R.J. Madix, C.M. Friend, O2 activation by metal surfaces: implications for bonding and reactivity on heterogeneous catalysts. Chem. Rev. **118**, 2816–2862 (2018)
78. S. Park, Y. Shao, J. Liu, Y. Wang, Oxygen electrocatalysts for water electrolyzers and reversible fuel cells: status and perspective. Energy Environ. Sci. **5**, 9331 (2012)

79. J. Zhao, Y. He, Z. Chen, X. Zheng, X. Han, D. Rao, C. Zhong, W. Hu, Y. Deng, Engineering the surface metal active sites of nickel cobalt oxide nanoplates toward enhanced oxygen electrocatalysis for Zn–air battery. ACS Appl. Mater. Interfaces **11**, 4915–4921 (2019)
80. V. Augustyn, Tuning the interlayer of transition metal oxides for electrochemical energy storage. J. Mater. Res. **32**, 2–15 (2017)
81. N. Yabuuchi, K. Kubota, M. Dahbi, S. Komaba, Research development on sodium-ion batteries. Chem. Rev. **114**, 11636–11682 (2014)
82. T. Kim, W. Song, D.-Y. Son, L.K. Ono, Y. Qi, Lithium-ion batteries: outlook on present, future, and hybridized technologies. J. Mater. Chem. A **7**, 2942–2964 (2019)
83. Q. Liu, Z. Hu, M. Chen, C. Zou, H. Jin, S. Wang, S.L. Chou, S.X. Dou, Recent progress of layered transition metal oxide cathodes for sodium-ion batteries. Small **15**, 1–24 (2019)
84. H. Su, S. Jaffer, H. Yu, Transition metal oxides for sodium-ion batteries. Energy Storage Mater. **5**, 116–131 (2016)
85. S. Kumar, S. Saralch, U. Jabeen and D. Pathak, Metal oxides for energy applications (2020)
86. L.P. Wang, (2017)
87. N. Nitta, F. Wu, J.T. Lee, G. Yushin, Li-ion battery materials: present and future. Mater. Today **18**, 252–264 (2015)
88. A. Mathur, R. Kaushik, A. Halder, Photoenhanced performance of Cobalt-intercalated 2-D manganese oxide sheets for rechargeable zinc–air batteries. Mater. Today Energy **19**, 100612 (2021)
89. J. Yang, X. Yang, Y.L. Zhong, J.Y. Ying, Porous MnO/Mn3O4 nanocomposites for electrochemical energy storage. Nano Energy **13**, 702–708 (2015)
90. P.J. Rosaiah, D.P. Zhu, H.O.M. Shaik, Y. Qiu, L. Zhao, Reduced graphene oxide/Mn3O4 nanocomposite electrodes with enhanced electrochemical performance for energy storage applications. J. Electroanal. Chem. **794**, 78–85 (2017)
91. A.J. Carrillo, D.P. Serrano, P. Pizarro, J.M. Coronado, Improving the thermochemical energy storage performance of the Mn2O3/Mn3O4 redox couple by the incorporation of iron. Chemsuschem **8**, 1947–1954 (2015)
92. N.H. Kwon, K.-G. Lee, H.K. Kim, S.-J. Hwang, MnO2-based nanostructured materials for various energy applications. Mater. Chem. Front. **5**, 3549–3575 (2021)
93. C. Madan, A. Mathur, A. Halder, Facile generation of a stable Bi-functional mixed phase Fe3O4/Fe-N4 electrocatalyst for rechargeable zinc-air battery. J. Electrochem. Soc. **169**, 020516 (2022)
94. H. Lee, O. Gwon, K. Choi, L. Zhang, J. Zhou, J. Park, J.-W. Yoo, J.-Q. Wang, J.H. Lee, G. Kim, Enhancing bifunctional electrocatalytic activities via metal d-band center lift induced by oxygen vacancy on the subsurface of perovskites. ACS Catal. **10**, 4664–4670 (2020)
95. C. Madan, C.S. Tiwary, A. Halder, Enhancing the oxygen evolution activity of nitrogen-doped graphitic carbon shell-embedded nickel/nickel oxide nanoparticles by surface dissolution. Mater. Chem. Front. **4**, 3267–3279 (2020)
96. A. Aijaz, J. Masa, C. Rösler, W. Xia, P. Weide, A.J.R. Botz, R.A. Fischer, W. Schuhmann, M. Muhler, Co@Co3O4 encapsulated in carbon nanotube-grafted nitrogen-doped carbon polyhedra as an advanced bifunctional oxygen electrode. Angew. Chemie Int. Ed. **55**, 4087–4091 (2016)
97. Y. Jiang, Y.-P. Deng, J. Fu, D.U. Lee, R. Liang, Z.P. Cano, Y. Liu, Z. Bai, S. Hwang, L. Yang, D. Su, W. Chu, Z. Chen, Interpenetrating triphase cobalt-based nanocomposites as efficient bifunctional oxygen electrocatalysts for long-lasting rechargeable Zn-air batteries. Adv. Energy Mater. **8**, 1702900 (2018)
98. C. Blanc, A. Rufer, in *Paths to Sustainable Energy* (InTech, 2010)
99. Y.-T. Lu, Y.-J. Chien, C.-F. Liu, T.-H. You, C.-C. Hu, Active site-engineered bifunctional electrocatalysts of ternary spinel oxides, M0.1Ni0.9Co2O4 (M: Mn, Fe, Cu, Zn) for the air electrode of rechargeable zinc–air batteries. J. Mater. Chem. A **5**, 21016–21026 (2017)
100. M.-S. Park, J. Kim, K.J. Kim, J.-W. Lee, J.H. Kim, Y. Yamauchi, Porous nanoarchitectures of spinel-type transition metal oxides for electrochemical energy storage systems. Phys. Chem. Chem. Phys. **17**, 30963–30977 (2015)

101. S. Li, X. Zhou, G. Fang, G. Xie, X. Liu, X. Lin, H.-J. Qiu, Multicomponent spinel metal oxide nanocomposites as high-performance bifunctional catalysts in Zn–air batteries. ACS Appl. Energy Mater. **3**, 7710–7718 (2020)
102. Y. Dai, J. Yu, C. Cheng, P. Tan, M. Ni, Mini-review of perovskite oxides as oxygen electrocatalysts for rechargeable zinc–air batteries. Chem. Eng. J. **397**, 125516 (2020)
103. S. Gupta, W. Kellogg, H. Xu, X. Liu, J. Cho, G. Wu, Bifunctional perovskite oxide catalysts for oxygen reduction and evolution in alkaline media. Chem. Asian J. **11**, 10–21 (2016)
104. T. Takeguchi, T. Yamanaka, H. Takahashi, H. Watanabe, T. Kuroki, H. Nakanishi, Y. Orikasa, Y. Uchimoto, H. Takano, N. Ohguri, M. Matsuda, T. Murota, K. Uosaki, W. Ueda, Layered perovskite oxide: a reversible air electrode for oxygen evolution/reduction in rechargeable metal-air batteries. J. Am. Chem. Soc. **135**, 11125–11130 (2013)
105. J.S. Yoo, X. Rong, Y. Liu, A.M. Kolpak, Role of lattice oxygen participation in understanding trends in the oxygen evolution reaction on perovskites. ACS Catal. **8**, 4628–4636 (2018)
106. J. Suntivich, K.J. May, H.A. Gasteiger, J.B. Goodenough, Y. Shao-Horn, A perovskite oxide optimized for oxygen evolution catalysis from molecular orbital principles. Science (80) **334**, 1383–1385 (2011)
107. A. Grimaud, W.T. Hong, Y. Shao-Horn, J.-M. Tarascon, Anionic redox processes for electrochemical devices. Nat. Mater. **15**, 121–126 (2016)
108. J. Song, C. Wei, Z.-F. Huang, C. Liu, L. Zeng, X. Wang, Z.J. Xu, A review on fundamentals for designing oxygen evolution electrocatalysts. Chem. Soc. Rev. **49**, 2196–2214 (2020)
109. D. Friebel, M.W. Louie, M. Bajdich, K.E. Sanwald, Y. Cai, A.M. Wise, M.-J. Cheng, D. Sokaras, T.-C. Weng, R. Alonso-Mori, R.C. Davis, J.R. Bargar, J.K. Nørskov, A. Nilsson, A.T. Bell, Identification of highly active Fe sites in (Ni, Fe)OOH for electrocatalytic water splitting. J. Am. Chem. Soc. **137**, 1305–1313 (2015)
110. Z. Hiroi, Inorganic structural chemistry of titanium dioxide polymorphs. Inorg. Chem. **61**, 8393–8401 (2022)
111. T. Hatakeyama, N.L. Okamoto, T. Ichitsubo, Thermal stability of MnO2 polymorphs. J. Solid State Chem. **305**, 122683 (2022)
112. A. Mathur, R. Kaushik, A. Halder, Visible-light-driven photo-enhanced zinc–air batteries using synergistic effect of different types of MnO2 nanostructures. Catal. Sci. Technol. **10**, 7352–7364 (2020)
113. Z. Zhang, X. Li, C. Zhong, N. Zhao, Y. Deng, X. Han, W. Hu, Spontaneous synthesis of silver-nanoparticle-decorated transition-metal hydroxides for enhanced oxygen evolution reaction. Angew. Chem. Int. Ed. **59**, 7245–7250 (2020)
114. D. Liu, R.A. Lazenby, J. Sloan, M. Vidotti, P.R. Unwin, J.V. Macpherson, Electrodeposition of nickel hydroxide nanoparticles on carbon nanotube electrodes: correlation of particle crystallography with electrocatalytic properties. J. Phys. Chem. C**120**, 16059–16068 (2016)
115. G.H.A. Therese, P.V. Kamath, Electrochemical synthesis of metal oxides and hydroxides. Chem. Mater. **12**, 1195–1204 (2000)
116. G. Chen, H. Wan, W. Ma, N. Zhang, Y. Cao, X. Liu, J. Wang, R. Ma, Layered metal hydroxides and their derivatives: controllable synthesis, chemical exfoliation, and electrocatalytic applications. Adv. Energy Mater. **10**, 1902535 (2020)
117. W. Ma, R. Ma, C. Wang, J. Liang, X. Liu, K. Zhou, T. Sasaki, A superlattice of alternately stacked Ni–Fe hydroxide nanosheets and graphene for efficient splitting of water. ACS Nano **9**, 1977–1984 (2015)
118. M.S. Islam, M. Kim, X. Jin, S.M. Oh, N.-S. Lee, H. Kim, S.-J. Hwang, Bifunctional 2D superlattice electrocatalysts of layered double hydroxide-transition metal dichalcogenide active for overall water splitting. ACS Energy Lett. **3**, 952–960 (2018)
119. G. Qian, F. Monaco, D. Meng, S.-J. Lee, G. Zan, J. Li, D. Karpov, S. Gul, D. Vine, B. Stripe, J. Zhang, J.-S. Lee, Z.-F. Ma, W. Yun, P. Pianetta, X. Yu, L. Li, P. Cloeten, Y. Liu, The role of structural defects in commercial lithium-ion batteries. Cell Rep. Phys. Sci. **2**, 100554 (2021)
120. C. Dong, W. Dong, X. Lin, Y. Zhao, R. Li, F. Huang, Recent progress and perspectives of defective oxide anode materials for advanced lithium ion battery. EnergyChem **2**, 100045 (2020)

121. Z. Lei, Y. Tan, Z. Zhang, W. Wu, N. Cheng, R. Chen, S. Mu, X. Sun, Defects enriched hollow porous Co–N-doped carbons embedded with ultrafine CoFe/Co nanoparticles as bifunctional oxygen electrocatalyst for rechargeable flexible solid zinc-air batteries. Nano Res. **14**, 868–878 (2021)
122. H.-F. Wang, C. Tang, B.-Q. Li, Q. Zhang, A review of anion-regulated multi-anion transition metal compounds for oxygen evolution electrocatalysis. Inorg. Chem. Front. **5**, 521–534 (2018)
123. D.U. Lee, P. Xu, Z.P. Cano, A.G. Kashkooli, M.G. Park, Z. Chen, Recent progress and perspectives on bi-functional oxygen electrocatalysts for advanced rechargeable metal–air batteries. J. Mater. Chem. A **4**, 7107–7134 (2016)
124. M. Rana, S. Mondal, L. Sahoo, K. Chatterjee, P.E. Karthik, U.K. Gautam, Emerging materials in heterogeneous electrocatalysis involving oxygen for energy harvesting. ACS Appl. Mater. Interfaces **10**, 33737–33767 (2018)
125. C. Jin, F. Lu, X. Cao, Z. Yang, R. Yang, Facile synthesis and excellent electrochemical properties of NiCo2O4 spinel nanowire arrays as a bifunctional catalyst for the oxygen reduction and evolution reaction. J. Mater. Chem. A **1**, 12170 (2013)
126. K.-L. Bao, J.-Y. Xu, N.-F. Yu, J.-B. Kuang, Z.-T. Yang, H. Chen, J.-L. Ye, Y.-P. Wu, Oxygen defect-enriched hierarchical NiCo2O4 hollow rectangular nanobars with enhanced bifunctional oxygen electrocatalysis for efficient rechargeable zinc-air batteries. Energy Fuels **36**, 6542–6551 (2022)
127. Y. Rao, S. Chen, Q. Yue, Y. Kang, Optimizing the spin states of mesoporous Co3O4 nanorods through vanadium doping for long-lasting and flexible rechargeable Zn–air batteries. ACS Catal. **11**, 8097–8103 (2021)
128. S. Chen, L. Ma, Z. Huang, G. Liang, C. Zhi, In situ/operando analysis of surface reconstruction of transition metal-based oxygen evolution electrocatalysts. Cell Rep. Phys. Sci. **3**, 100729 (2022)
129. H.G. Yang, C.H. Sun, S.Z. Qiao, J. Zou, G. Liu, S.C. Smith, H.M. Cheng, G.Q. Lu, Anatase TiO2 single crystals with a large percentage of reactive facets. Nature **453**, 638–641 (2008)
130. N.-C. Lai, G. Cong, Z. Liang, Y.-C. Lu, A highly active oxygen evolution catalyst for lithium-oxygen batteries enabled by high-surface-energy facets. Joule **2**, 1511–1521 (2018)
131. T. Ling, D.-Y. Yan, Y. Jiao, H. Wang, Y. Zheng, X. Zheng, J. Mao, X.-W. Du, Z. Hu, M. Jaroniec, S.-Z. Qiao, Engineering surface atomic structure of single-crystal cobalt (II) oxide nanorods for superior electrocatalysis. Nat. Commun. **7**, 12876 (2016)
132. Z. Lu, G. Chen, Y. Li, H. Wang, J. Xie, L. Liao, C. Liu, Y. Liu, T. Wu, Y. Li, A.C. Luntz, M. Bajdich, Y. Cui, Identifying the active surfaces of electrochemically tuned LiCoO2 for oxygen evolution reaction. J. Am. Chem. Soc. **139**, 6270–6276 (2017)
133. X. Han, G. He, Y. He, J. Zhang, X. Zheng, L. Li, C. Zhong, W. Hu, Y. Deng, T. Ma, Engineering catalytic active sites on cobalt oxide surface for enhanced oxygen electrocatalysis. Adv. Energy Mater. **8**, 1702222 (2018)
134. Y. Guo, L. Ma, K. Mao, M. Ju, Y. Bai, J. Zhao, X.C. Zeng, Eighteen functional monolayer metal oxides: wide bandgap semiconductors with superior oxidation resistance and ultrahigh carrier mobility. Nanosc. Horiz. **4**, 592–600 (2019)
135. W.S. Choi, M.F. Chisholm, D.J. Singh, T. Choi, G.E. Jellison, H.N. Lee, Wide bandgap tunability in complex transition metal oxides by site-specific substitution. Nat. Commun. **3**, 689 (2012)
136. S.A. Razek, M.R. Popeil, L. Wangoh, J. Rana, N. Suwandaratne, J.L. Andrews, D.F. Watson, S. Banerjee, L.F.J. Piper, Designing catalysts for water splitting based on electronic structure considerations. Electron. Struct. **2**, 023001 (2020)
137. F. Li, J. Ran, M. Jaroniec, S.Z. Qiao, Solution combustion synthesis of metal oxide nanomaterials for energy storage and conversion. Nanoscale **7**, 17590–17610 (2015)
138. T.S. Andrade, V. Dracopoulos, M.C. Pereira, P. Lianos, Unmediated photoelectrochemical charging of a Zn-air battery: the realization of the photoelectrochemical battery. J. Electroanal. Chem. **878**, 114709 (2020)

139. H. Gong, H. Xue, B. Gao, Y. Li, X. Yu, X. Fan, S. Zhang, T. Wang, J. He, Solar-enhanced hybrid lithium–oxygen batteries with a low voltage and superior long-life stability. Chem. Commun. **56**, 13642–13645 (2020)
140. C.-T. Wang, H.-H. Huang, Photo-chargeable titanium/vanadium oxide composites. J. Non. Cryst. Solids **354**, 3336–3342 (2008)
141. B.D. Boruah, B. Wen, M. De Volder, Light rechargeable lithium-ion batteries using V2O5 cathodes. Nano Lett. **21**, 3527–3532 (2021)
142. B.D. Boruah, A. Mathieson, B. Wen, S. Feldmann, W.M. Dose, M. De Volder, Photo-rechargeable zinc-ion batteries. Energy Environ. Sci. **13**, 2414–2421 (2020)
143. X. Liu, Y. Yuan, J. Liu, B. Liu, X. Chen, J. Ding, X. Han, Y. Deng, C. Zhong, W. Hu, Utilizing solar energy to improve the oxygen evolution reaction kinetics in zinc–air battery. Nat. Commun. **10**, 4767 (2019)
144. T.S. Andrade, I.C. Sena, L.C.A. de Oliveira, P. Lianos, M.C. Pereira, Decreasing the charging voltage of a zinc-air battery using a bifunctional W:BiVO4/V2O5 photoelectrode and sulfite as a sacrificial agent. Mater. Today Commun. **28**, 102546 (2021)

Graphene/Metal Oxide-Based Nanocomposite for Electrochemical Sensors

Ahmet Aykaç and İzel Ok

Abstract In recent years, nanotechnological developments have brought revolutionary changes and developments to our lives. Among them, the preparation and applications of new generation nanocomposites have been one of the most interesting topics. In this context, nanocomposites prepared from graphene and its derivatives and metal oxides have received great attention. Graphene is an allotrope of carbon consisting of a single layer of atoms in a two-dimensional (2D) honeycomb lattice nanostructure. The Nobel Prize in Physics 2010 was jointly awarded to Andre Geim and Konstantin Novoselov for groundbreaking experiments on the 2D material graphene. Since then, graphene has been widely used in many fields such as solar cells, batteries, biosensors, filtration systems, and drug delivery systems due to its unique properties, including excellent lightness, durability, electrochemical, photochemical, thermal and electrical conductivity. The use of graphene with many materials, such as glass, metals, and metal oxides (MO), metal oxide nanostructures (MONS) either enhances the existing superior properties or adds new ones. In particular, there has been a significant increase in the number of studies in graphene/MONS-based nanocomposites in recent years. Nanocomposites of graphene and MONS have more advanced properties than MONS and graphene matrices, because nanocomposites are developed by combining the superior properties of these two different structures, and some missing properties can be eliminated and/or improved. This chapter covers recent trends in the synthesis of graphene/MONS-based nanocomposites and their applications for electrochemical sensing platforms. First, a brief introduction on the properties, synthesis strategies, and applications of graphene is given. Next, a detailed review of the fabrication and sensor applications of graphene/MONS-based nanocomposites is presented. Finally, some

A. Aykaç (✉)
Department of Engineering Sciences, Faculty of Engineering and Architecture, Izmir Katip Çelebi University, 35620 Izmir, Turkey
e-mail: ahmet.aykac@ikcu.edu.tr

A. Aykaç · İ. Ok
Graduate School of Natural and Applied Sciences, Department of Nanoscience and Nanotechnology, Izmir Katip Çelebi University, 35620 Izmir, Turkey

V. Kumar et al. (eds.), *Optical Properties of Metal Oxide Nanostructures*, Progress in Optical Science and Photonics 26,
https://doi.org/10.1007/978-981-99-5640-1_11

outstanding research on the use of carbonaceous/MONS-based nanocomposites in electrochemical sensor platforms is presented.

Keywords Graphene · Carbonaceous materials · Metal oxides · Nanocomposites · Electrochemical performance · Electrochemical sensors

1 Introduction

Carbon (C) is one of the most well-known elements since ancient times, that belonging to 4A group. C is the fourth most abundant element in nature after hydrogen (H_2), helium (He), and oxygen (O_2) and it is the second most abundant element in the human body after O_2 by weight. The four electrons in the valence orbital of C allow forming covalent chemical bonds. C can form not only organic compounds but also inorganic compounds as well. Atoms of C can be bonded together in various ways to form several C allotropes such as diamond, fullerenes, amorphous C, graphite, graphene, etc. [1].

Recently, carbonaceous-based materials have risen to prominence with their structural and electronic properties. Among these materials, especially graphene, is one of the most studied materials in fields such as physics, nanoscience and nanotechnology, chemistry, and material science. Graphene is a single layer, two-dimensional (2D), sp^2-hybridized C atoms with a honeycomb structure that is one of the most well-known allotropes, along with carbon nanotubes (CNTs), fullerenes, graphite, and diamond as can be seen in Fig. 1 [2]. Although graphene is defined as a single layer 2D material, considering its morphological characteristics, it can be classified into two layers (bi-layer graphene) (BG), few-layer graphene (number of layers $\leq$ 10) (FG) and graphite nano- and micro-platelets. Graphene and its derivatives are distinguished from other carbonaceous-based materials by its extraordinary optical, magnetic, mechanical, electrical and thermal properties, as well as its enormous special surface area [3]. Depending on the application area, graphene can be classified into graphite oxide, graphene oxide (GO), reduced graphene oxide (rGO), graphene nanoplatelets (GNPs), graphene nanosheets (GNSs), graphene quantum dots (GQDs), and graphene nanoribbons (GNRs) [4]. Detailed information on the properties of graphene and its derivatives will be detailed in Sect. 2.1. Graphene and its derivatives are synthesized by various methods.

These methods differ according to the physical and chemical synthesis conditions, the types of auxiliary materials used, the synthesis time, costs and the quality of the synthesized product. The most commonly used methods in the synthesis of graphene and its derivatives are mechanical and chemical exfoliation, chemical synthesis, pyrolysis, epitaxial growth, and chemical vapor deposition (CVD). Characterization processes play key roles to determine the quality of the sample by evaluating the number of layers, defects and morphology. Raman spectroscopy, optical microscopy, atomic force spectroscopy (AFM), scanning electron microscopy (SEM), transmission electron microscopy (TEM), and X-ray diffraction (XRD) are

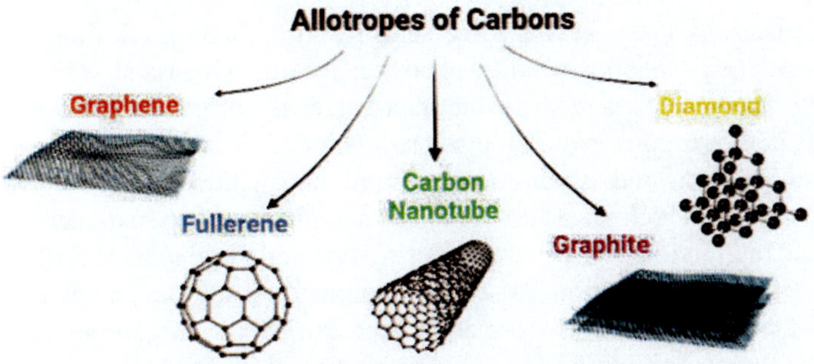

Fig. 1 Allotropes of carbons as graphene, fullerene, carbon nanotubes, graphite, and diamond

commonly performed methods for the characterization of graphene and its derivatives in terms of morphology, quality, and structure [5]. Raman spectroscopy is the most preferred characterization method for the identification of graphene and its derivatives. In this characterization method, characteristic peaks defining graphene are accepted as D band (~1360 cm^{-1}), G band (~1580 cm^{-1}), and 2D band (~2700 cm^{-1}) [5]. Synthesis methods and types of characterization of graphene and its derivatives will be discussed under Sect. 2.2.

Apart from graphene and its derivatives from carbonaceous-based materials, various nanocomposite materials are obtained with other carbonaceous-based materials. The obtained nanocomposites are used in fields such as materials science, medicine, biomedical engineering, physics, environment, electrical energy, sensors, etc. due to their superior properties. Particularly, nanocomposites formed with different metal nanoparticles (NP), metal oxide nanostructures (MONS) and some polymers have attracted great attention. Recently, carbonaceous/MONS-based nanocomposite materials have been preferred especially in the production of nanobiosensor and electrochemical sensor electrodes.

MONS present good catalytic activity, switchable magnetic, conductive, chemical, and electronic properties, large surface area, and ion/electron transport path shortening capabilities [6]. Although they present extraordinary features compare to their bulk materials, there are some difficulties in the control of the surface area, pore distribution, and porosity that limited their usage in applications [7]. Therefore, nanocomposites generated with MONS and carbonaceous materials with superior electronic and mechanical properties including graphene and its derivatives, activated carbon (AC), carbon nanospheres, mesoporous carbon, CNTs, carbon nanofibers (CNF), and hierarchical porous graphite carbon (HPGC) are preferred in many applications, especially as electrode in electrochemical sensor studies [8]. Section 3.1 includes a detailed review of the types, and properties of carbonaceous/MONS-based nanocomposite materials.

Sensor applications, particularly electrochemical sensor studies, are one of the most common applications for carbonaceous/MONS-based nanocomposite. Sensors are devices that detect changes in a sample and produce meaningful output. If a chemical reaction occurs between the sensor and the sample during this recognition process, such sensors are called chemical sensors. Chemical sensors consist of three main components: a receptor that recognizes the target compound (analyte), a transducer in the receptor part that converts chemical information into a measurable signal, and an electronic system usually combined with a transducer [9]. Electrochemical sensors, which are a sub-branch of chemical sensors, can measure by electrochemical interaction between the target analyte and the electrode [10]. Section 4.2 contains detailed explanations of electrochemical sensors and their designs. The choice of electrode type and electrolyte used during the measurement of electrochemical sensors, chemical and physical properties of electrode surfaces, and applied potential affect the operating performance of the sensor [11]. For example, the surface area, electrical conductivity, catalytic performance, selectivity or sensitivity of the electrodes can be increased by modifications on the electrode [12]. In this context, due to the synergistic effects such as increased conductivity, fast electron transfer, high catalytic activity, large surface area, homogeneous distribution, sensitivity, and selectivity between carbonaceous/MONS-based nanocomposites might improve the performance of electrodes in electroanalytical applications. Application areas of graphene-based materials are discussed in Sect. 2.3, additionally, some applications of carbonaceous/MONS-based nanocomposites are given in Sect. 3.3, and lastly, electrochemical sensor applications of carbonaceous/MONS-based nanocomposites are detailed under Sect. 4.3.

2 Graphene and Its Derivatives

The graphene journey, which is thought to have started with the discovery of pencil by Italian scientists in the 1560 s, gained momentum with the discovery of graphene by Geim and Novoselov [13] in 2004. Graphene, consisting of a honeycomb monolayer, 2D, sp^2-hybridized C atoms, is considered the "mother of all graphite forms" [2]. On the other hand, although graphene is defined as a single-layer, 2D material, considering its morphological properties, it can be classified as two-layer graphene (bilayer graphene) (BG), several-layer graphene (layer number $\leq$ 10) (FG) and graphite [3]. Studies are continuing on obtaining different graphene derivatives such as graphite oxide, GO, rGO, GNPs, GNSs, GQDs, and GNRs as illustrated in Fig. 2 with various modifications in order to improve the above-mentioned properties of graphene.

GO, one of the most important derivatives of graphene has a 2D structure like graphene and exhibits similar properties. Moreover, it is also covalently functionalized with many O_2-containing functional groups such as epoxy, hydroxyl, carbonyl, and carboxyl groups at the edges and basal planes of the sheet [14–16]. The presence of these functional groups enables GO to be easily processed in dispersions, have hydrophilic qualities, and exhibit highly effective mechanical, colloidal or optical properties. GO has extraordinary stability and it readily disperses in water owing to the ionized O_2-containing functional groups [17]. Different reduction techniques can be used to convert GO to rGO. rGO is usually considered a bridge between graphene

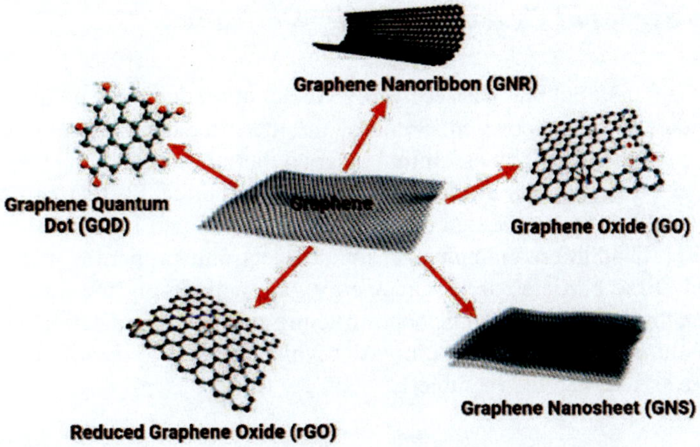

Fig. 2 Presentation of most popular graphene derivatives

and GO. rGO exhibits graphene-like properties including reasonably strong conductivity and it is simple to create needed quantities of rGO from affordable GO thanks to its conversion methods [18]. GNPs are a form of carbon that ranges in thickness from 0.34 to 100 nm and it is composed of graphite under certain conditions. GNPs exhibit attractive properties in terms of lightweight, high aspect ratio, electrical and thermal conductivity, mechanical strength, low cost, and planar structure [19, 20]. Nanocomposites can be made by combining GNPs with glass fibers, polymers, or another matrix and GNPs can improve the mechanical properties of diverse nanocomposites due to the strong interfacial interaction of nanoplates with matrices [21] GNS is a carbonaceous-based material with a 2D sheet structure consisting of a few layers sp^2-hybridized carbon lattices, exhibiting high electrical conductivity, outstanding catalytic properties, large surface area, and corrosion resistance [22]. GQDs are usually derived from graphene and/or GO. In general, GQDs have properties such as low toxicity, highly soluble in a variety of solvents, large surface area, stable photoluminescence, adjustable physical properties, and chemical stability [23–25]. Another type of graphene is GNRs, which are a semi-1D form of graphene with an ultrathin width. GNRs are made by unzipping or exfoliating the walls of multi-walled carbon nanotubes (MWCNTs) [26]. Due to its non-traditional metal-free magnetic configuration, zigzag graphene nanoribbons (ZGNRs) stand out among the large spectrum of atomically accurate graphene edges created via surface synthesis. Regardless of width, ZGNRs have magnetic moments linked ferromagnetically along the edge and antiferromagnetically linked across it [27].

2.1 Properties of Graphene and Its Derivatives

Graphene is a material that has received great attention due to its unique properties since the day it was discovered. Some of the these remarkable features are good electrical conductivity, lightness, thinness, good thermal conductivity (5000 W/mK) [28] high electron mobility at room temperature, hall effect [29], large surface area (2630 m^2/g) [30], high modulus of elasticity (21 TPa) and intrinsic strength (130 GPa) [4, 31]. In addition, graphene has unique electronic properties such as linearly distributed Dirac particles, transport energy gap, and simply the light absorption coefficient, thanks to its electronic band structure and sp^2-hybridization, as indicated in the literature [2, 32]. The main properties which are used to identify graphene and its derivatives summarized in Table 1.

2.2 Synthesis and Characterizations

Graphene was isolated from graphite by using the exfoliation method by Geim and Novoselov in 2004, and its widespread use began exactly after this date. However, several layers of graphite were synthesized on a single crystal platinum surface by decomposition methods by Lang et al. in 1975. Since the characterization techniques at that time were not sufficient, this synthesis could not be defined as graphene synthesis [33]. In another study, Lu et al. [34] produced monoatomic monolayer graphene from several layers by mechanical friction of highly ordered pyrolytic graphite (HOPG) with AFM tips [34].

Although the synthesis of graphene and its derivatives has been described many times in the literature, studies on cost-effective, environmentally-friendly, fast, and practical synthesis methods are still under investigation. Synthesis of the graphene and its derivatives can be divided into two main categories; top-down approach and bottom-up approach as pointed out in Fig. 3. The top-down approach has a design that can also be described as reverse engineering and is based on obtaining small materials from bulk materials. The synthesis of the graphene and its derivatives with top-down approach include mechanical exfoliation, chemical exfoliation, and chemical synthesis. The bottom-up approach is also known as the self-assembly method. It is the creation of larger structures by combining simple and small units in an appropriate way. The bottom-up method includes pyrolysis, epitaxial growth, and chemical vapor deposition (CVD) which is the most popular method for producing large-scale graphene. If the process is carried out in a resistive heating furnace, it is known as thermal CVD; if the process includes plasma assisted growth, it is known as plasma enhanced CVD (PECVD) [32]. The CVD method and other commonly used synthesis methods are discussed in detail below with their advantages and disadvantages.

Table 1 The most common properties of graphene and its derivatives

Properties	Graphene	Graphene oxide (GO)	Reduced graphene oxide (rGO)	Graphene nanoplateles (GNPs)	Graphene nanosheets (GNS)	Graphene quantum dots (GQDs)	Graphene nanoribbons (GNRs)
Electrical	✓	✓	✓	✓	✓	✓	✓
Conductivity	✓	✓	✓	✓	✓	✓	✓
Large surface area	✓	✓	✓	✓	✓	✓	✓
Mechanical Strength	✓	✓	✓	✓	✓	✓	✓
Hydrophilicity	X	✓	✓	X	X	✓	X
Catalytic activity	✓	✓	✓	✓	✓	✓	✓
Photolum in escence	V	✓	✓	X	X	✓	X

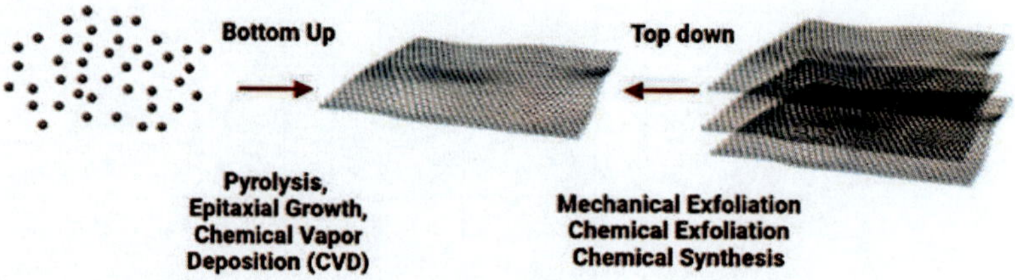

Fig. 3 Basic illustration of bottom-up and top-down synthesis of graphene

2.2.1 Mechanical and Chemical Exfoliation

Mechanical exfoliation is the first recognized and simplest method for graphene synthesis. In this technique, a piece of graphite is subjected to repeated strip exfoliation and then transferred to a substrate. Graphene sheets of varying thickness can be obtained by mechanical exfoliation or by peeling layers from graphite materials such as highly ordered pyrolytic graphite (HOPG), single-crystalline graphite, or natural graphite [30, 32]. In this method, the adhesive tape is pressed onto a HOPG surface so that the top few layers are attached to the tape, then the tape containing crystals of the layered material is pressed to a preferred surface and after peeling, the bottom layer is left on the substrate [35]. Optical microscopy, Raman spectroscopy, and AFM are commonly used to characterize graphene flakes generated using this technique. It is feasible to produce high-quality crystals using this procedure, one of the disadvantages of this technique is only suitable for laboratory experimentation and prototyping so it is not suitable for large-scale production [30, 32].

In addition to mechanical exfoliation, another commonly preferred method for graphene synthesis is chemical exfoliation. In this method, the Van der Waals interactions between the graphite layers are reduced through oxidation or with solvent treatments, and then the layers are separated by sonication resulting in obtaining graphene. Alkali metals such as potassium (K), cesium (Cs) and sodium–potassium alloy (NaK), hydrazine hydrate (N_2H_4), methanesulfonic acid (CH_4O_3S), and ionic solvents are the most commonly used reducing solvents to weaken the bonds between the layers [36]. The schematic illustration of the mechanical and chemical exfoliation methods are given the Fig. 4.

2.2.2 Chemical Synthesis

Chemical synthesis is another approach for the production of graphene by using chemical reduction of GO and sonication. In chemical reduction of GO, phenyl hydrazine ($C_6H_8N_2$), hydroxylamine (H_3NO), glucose ($C_6H_{12}O_6$), ascorbic acid ($C_6H_8O_6$), hydroquinone ($C_6H_6O_2$), sodium borohydride ($NaBH_4$), N_2H_4 alkaline solutions and pyrrole are used as reducing agents [32]. Production of GO, can be carried out by oxidation of graphite using oxidants including concentrated sulfuric acid (H_2SO_4), nitric acid (HNO_3) and potassium permanganate ($KMnO_4$), Brodie

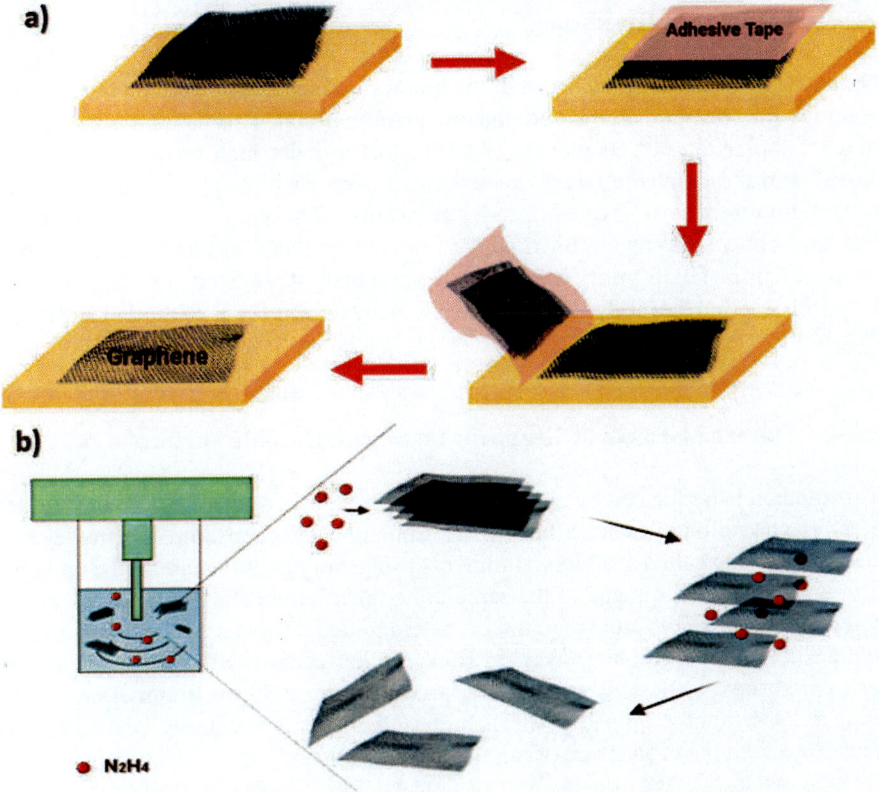

Fig. 4 The schematic representation of the mechanical and chemical exfoliation methods. **a** mechanical exfoliation, and **b** chemical exfoliation

method [37], Staudenmaier method [38], and Hummers method [39]. As graphite is converted into GO, the interlayer spacing increases due to oxidation using reducing agents and sonication, resulting in different layer suspensions. Brodie et al. demonstrated the synthesis of GO by adding potassium chlorate ($KClO_3$) to a graphite slurry in the presence of HNO_3. However, the process had some drawbacks including several steps, consumption of time, and production of harmful chemical vapors. Alternatively, Staudenmaier et al. succeeded in producing highly oxidized GO in a one-step process by incorporating concentrated H_2SO_4 into the process. Furthermore, Hummers et al. created a new method used today. In this method, as an acid instead of using the HNO_3 the reaction is performed in the presence of sodium nitrite ($NaNO_3$), concentrated H_2SO_4 and $KMnO_4$ [40]. Throughout the time, Marcano et al. modified further Hummer's method resulting in obtaining the method which is the most widely used nowadays. This method employed a larger amount of the $KMnO_4$, the exclusion of $NaNO_3$, and a 9:1 reaction mixture of H_2SO_4/H_3PO_4. As a result of the enhanced approach, a better oxidation efficiency and a higher amount of hydrophilic oxidized graphene were obtained [41, 42].

2.2.3 Pyrolysis of Graphene

Graphene pyrolysis method is one of the bottom-up methods performed by the interaction of the solvent material with the precursor material. This method is performed in a 1:1 molar ratio of ethanol (C_2H_5OH) and Na under high pressure in a sealed vessel, and the graphene layers are separated from each other using the pyrolysis of sodium ethoxide (C_2H_5ONa) under sonication. The graphene sheets produced can have dimensions up to 10 μm. It is an advantage that it has low cost and allows the production of high-purity functionalized graphene at relatively low temperatures, while the fact that the produced graphene has many defects is one of the disadvantages [32, 43].

2.2.4 Epitaxial Growth of Graphene on Silicon Carbide Surface

This method is performed by epitaxial thermal synthesis of graphene on the surface of single-crystalline silicon carbide (SiC). Epitaxy is the deposition of a single crystalline film, also called an epitaxial film, on a single crystal substrate. If the epitaxial film and substrate are made of the same material, a homoepitaxial layer is formed, and if it consists of different materials, a heteroepitaxial layer is formed. In graphene synthesis, a single graphite layer or graphene heteroepitaxial layer is formed on the SiC surface [40]. This method was first used in 1975 by Bommel et al. The epitaxial growth of graphene on SiC is preferred due to its scalability, best electronic properties, and the production of high-quality graphene.

2.2.5 Chemical Vapor Deposition

The chemical vapor deposition (CVD) method is one of the most preferred methods for preparing high-quality, single-layer graphene. In this method, large-area samples can be produced by exposing the metal to various hydrocarbon precursors at high temperatures [30]. Critical factors such as catalyst, precursor, flow rate, temperature, pressure, and time play crucial role in the production of graphene by the CVD method. During the CVD process, graphene growth on metal can be summarized in four steps. These; (i) precursor gas is adsorbed and catalytically decomposed, (ii) decomposed carbon species diffuse and dissolve on the surface of the bulk metal, (iii) these carbon atoms are separated on the metal surface, (iv) surface nucleation and growth produce graphene [44].

The CVD method is classified into several type depending of the precursor, desired product thickness, and structure. Among them, thermal CVD and plasma-enhanced CVD (PECVD) are the most prevalent types of CVD to produce graphene. Thermal CVD technique for graphene production is the process of flowing different hydrocarbons such as methane (CH_4), ethylene (C_2H_4), acetylene (C_2H_2), and benzene (C_6H_6) gaseous or hydrogen (H) and argon (Ar) as precursor into a quartz tube containing substrates such as nickel (Ni), copper (Cu), palladium (Pd), iridium (Ir),

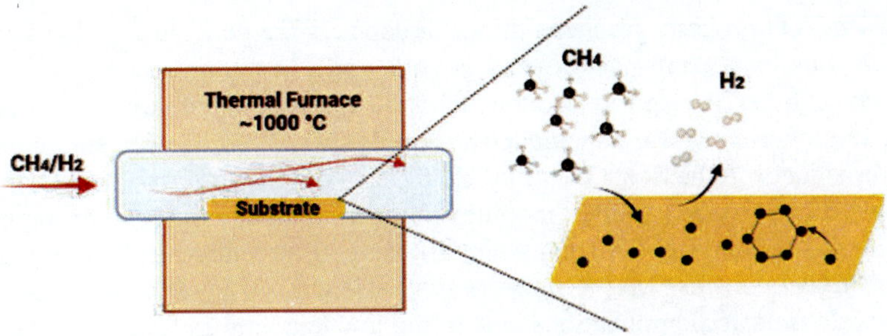

Fig. 5 A representative figure of CVD mechanism for production of graphene

and ruthenium (Ru) at certain rates in a furnace at high temperatures as represented in Fig. 5. Among these substrates, Cu and Ni are the most preferred metals [30, 32].

Cu is one of the most commonly used materials in the thermal CVD amplification technique. Li et al. used thermal catalytic decomposition of CH_4 on a metal surface using polycrystalline Cu foils to generate the first CVD growth of uniform and large-area graphene [45]. The reason why Cu substrate is preferred in CVD process is that Cu allows for the synthesis of single-layer graphene, while graphene synthesized on other metals has an inhomogeneous and multi-layered structure. During synthesis processes, it is possible to obtain high quality graphene films by controlling the nucleation of the carbon atom and the growth of the graphene nucleus by changes in production parameters such as temperature, pressure, gas flow etc. [46]. However, since the growth of graphene on Cu is limited to the surface, the smoothness of the Cu surface plays an important role in covering the entire surface of the substrate [47]. Cu surface can be chemically and mechanically polished to reduce surface roughness and contamination [48]. Since Cu has some disadvantages such as surface roughness and sublimation, Ni substrate is used as another alternative in the synthesis of graphene by CVD method. In these processes, Ni substrates such as Ni foil, polycrystalline Ni thin film and Ni foam are used as an alternative substrate for the growth of graphene and its derivatives. In this approach, different cooling rates affect the separation behavior, thickness and quality of graphene films [32, 49]. For example, the homogeneity of graphene layers is affected by the Ni orientation used as the substrate. When a Ni substrate is exposed to high temperature hydrocarbon gases, the hydrocarbon gases decompose and the released C atoms are incorporated into the Ni stack through catalytic boosting. Since the solubility of C decreases with cooling, excess C atoms diffuse onto the surface of the Ni substrate and start to form graphene layers [50]. The problem encountered in Ni synthesis is that the produced graphene usually contains many wrinkles and folds and is time-consuming [32]. The differences between the synthesis methods using these two different substrates can be thought of as the fact that C has a higher solubility in Ni (>0.1% atomic). Due to the high solubility of carbon in Ni and graphene layers (single and multi-layers) growth is limited to the micron's region. Cu has a lower C solubility (0.001% atomic), so that Cu is the best substrate for the growth of mainly monolayer [46]. On the other hand, in the PECVD process, graphene thin film deposits are generated on the surface of the substrate as a result of the reaction with gases in the presence of plasma in a vacuum chamber. As a plasma source, the PECVD system employs radio frequency (RF), microwave (MW), and inductive coupling that electric currents created by electromagnetic induction [32, 40]. Titanium (Ti), silicon dioxide (SiO_2), silicon (Si), aluminum oxide (Al_2O_3), niobium (Nb), molybdenum (Mo), chromium (Cr), zirconium (Zr), tungsten (W), tantalum (Ta), and 304 stainless steel are all employed in the PECVD process [51]. Graphene synthesis with PECVD is more suitable for large-scale industrial applications, due to the low temperature and quick deposition time. In addition, the technique allows the formation of catalyst-free graphene [52].

The morphology, crystallinity, shape, and size of the resulting product may be controlled by adjusting the CVD process parameters [43, 53]. CVD method allows obtaining high quality and yield products for large-scale practical applications. However, the CVD process has some drawbacks such as complexity, difficulties in transferring graphene between substrates, and high energy requirement for a particular method [54]. Synthesized graphene and its derivatives using the aforementioned methods can be thoroughly characterized using the numerous characterization techniques available thanks to today's advanced technology. Raman spectroscopy, optical microscopy, AFM, SEM, TEM, HRTEM, and XRD are among the most preferred microscopic and spectroscopic characterization techniques to determine the morphology, structure, quality, and layer numbers of graphene and its derivatives.

Raman spectroscopy: Raman spectroscopy is a frequently used method for characterization of graphene and its derivatives. The measurement is based on sending a monochromatic light source (laser) to the sample and detecting the change in energy of the light. The interactions between the vibrational energies of the molecules in the sample and the first electromagnetic wave cause changes in the energy level, and the materials are characterized according to these molecular vibrations [55]. The characteristic peaks of graphene are referred to as the D band (~1360 cm^{-1}), G band (~1580 cm^{-1}), and 2D band (~2700 cm^{-1}) [5].

Optical Microscopy: This technique is a practical, fast and cheaper method compared to other characterization techniques. In particular, it provides basic information about graphene whose surface has been functionalized by using the contrast difference between the graphene layers and the substrate. In addition, the size and shape of the graphene structure might able to be observed by using this technique [40].

Scanning electron microscopy (SEM): SEM is a high-magnification method used to examine the shape and surface morphology of graphene [56]. As in optical microscopy, a contrast difference is used between the graphene layers and the substrate, and an image is created by focusing the intense electron beam on the sample [40]. SEM also may provide information about whether the substrates used for graphene and its derivatives have been removed or not.

Atomic force spectroscopy (AFM): This method is used to characterize the topography of graphene in terms of the number of layers by scanning it with a probe. The working principle of AFM is to detect the surface by scanning line by line on the surface using a sharp tip of a micro-machined probe.

Transmission electron microscopy (TEM): TEM is used to analyze the properties of graphene such as atomic structure, gap defects, grain boundaries, and layer stacking. The working principle of TEM is that when an electron beam passes through a sample, an image is formed as a result of the interaction of electrons with the material. HRTEM, a sub-technique of TEM, provides extra information about the material such as the thickness, morphology and lattice arrangement of the material [57].

X-Ray Diffraction (XRD): The XRD technique provides information about the crystal structure of materials, the crystal or grain size of metal matrices, and the chemical composition of materials [55]. The XRD model of graphene shows three significant diffraction peaks at 26.5°, 54.6°, and 87° attributed to the (002), (006), and (006) indices of graphite. The peak at 26.4° is evidence of the high crystallinity of several-layer graphene [5, 58].

2.3 Applications of Graphene-Based Nanomaterials

New features have been added to the existing superior properties of graphene as a consequence, graphene derivatives more actively been used in many fields such as energy storage and conversion, sensors, catalysis chemistry, biomedical applications, etc. Some of the applications of graphene and its derivatives are given in light of scientific aspects under this section.

Lithium-ion batteries (LIBs), super capacitors, solar cells, photocatalytic activity, environmental remediation, electrochemical sensors are among the areas where graphene and its derivatives are widely used. The increasing world population has brought with it many environmental problems. At the beginning of these problems, global warming, decrease in underground water resources, unsustainable human-induced pollution in lakes, seas and oceans, and the increase in the need for energy have become extremely critical. Scientists continue working intensively to find the solution for environmental problems. At this point, environmentalist approaches have come to the fore and the orientation to alternative energy sources (solar cells, wind energy etc.) has increased. One of the most fundamental problems of developed alternative energy sources is the problem of energy storage. In order to solve this problem, energy storage and conversion devices with high energy and power density are being developed.

LIBs are widely used in energy storage and portable electronic device applications, and their performance is greatly influenced by the physical and chemical properties of the cathode and anode materials. Graphene is one of the most preferred materials for improving the performance of LIBs due to its large specific surface areas, high conductivity and suitable pore sizes [59]. In a study, sulphur-doped graphene nanosheets (S-GNS) were investigated for their use as anode material in LIBs. According to the study, as the electrical conductivity increase, the speed of S-GNS increased compared to GNS. The initial discharge and charge profiles of S-GNS have large specific capacities of ~1700 and 870 mA h g^{-1} at current density of 374 mA g^{-1} (1C) respectively. Also, the reversible capacity of S-GNS is two and three times that of GNS at current densities of 372 mA g^{-1} (1C) and 11,160 mA g^{-1} (30C), respectively. Furthermore, even after 500 cycles at a high current density of 1488 mA g^{-1} (4C), the S-GNS maintained good cyclic stability [60]. Supercapacitors are other energy storage devices with high power density and short charge/discharge times. Supercapacitor electrodes consists of two units as support and active units, and the support unit is responsible for ion transfer in the electrolyte. Therefore, the support unit materials should have a large surface area as well as electrical conductivity, such as graphene and its derivatives [61]. Cao et al. investigated the effect of oxygen functional groups on graphene and different annealing temperatures used during graphene synthesis on capacitive performance. In the study, the specific capacitances of graphene at different annealing temperatures such as 200 °C, 400 °C and 800 °C are 201, 153 and 34 Fg^{-1}, respectively. In addition, the study pointed out that the specific capacitance decreases when the oxygenated species decreases, and that the oxygen functional groups affect the electrochemical activity [62]. Solar cells are electrical devices that use the photovoltaic effect to convert sunlight energy directly into electricity. Solar cells are divided into two as dye-sensitized solar cells (DSSC), and Perovskite solar cells (PSCs) [63]. The electrodes used in the design of solar cell batteries should have high electrical conductivity and a large surface area. Therefore, graphene-based materials are good candidate for solar cells applications. One study evaluated the effect of graphene thickness on perovskite/graphene solar cells due to changes in the number of graphene layers. According to the study, the graphene/peroxide solar cell consisting of seven layers of graphene showed the best performance with a conversion efficiency of 9.4% [64].

In terms of photocatalytic applications, efforts to prevent environmental pollution by minimizing the damage caused by dyestuffs, heavy, and toxic chemicals are continuing rapidly. One of these studies was performed by using boron-doped graphene nanostrips (BGNRs) produced by Xing et al. for the photodegradation of Rhodamine B. B-GNRs with high conductivity were effective in the degradation of Rhodamine B with their photocatalytic activities [65].

Graphene-based materials are also widely used in sensor technologies, apart from supercapacitor applications, photocatalytic activity, and new generation biomedical materials. These sensors are widely used in the detection of analytes such as glucose, cholesterol, dopamine, and uric acid, in the detection of heavy metals, and in the detection of preservatives/additives in foods. In addition, with the rapid development of modern industry, the effect of toxic heavy metal ions such as Pb^{2+} and Cd^{2+} heavy metal ions on environmental pollution are increasing. These ions are absorbed by simple microorganisms and join the food chain, accumulate over time in the metabolic cycle, and become harmful in more advanced organisms. Therefore, the detection of these heavy metals is crucial in terms of environment and water remediation [66]. In a study, a polyglycine modified graphene paste electrode (PGMGPE) has been developed for the detection of Hg (II) and Pb (II) ions, which are heavy metal ions. The PGMGPE electrode showed good sensitivity, selectivity, as well as stability retaining 85% of the initial current even after 25 cycles. Moreover, the electrode showed a relative standard deviation of 2.85% for the analyte considered a reasonable reproducibility [67]. In another heavy metal detection study, directly thiolated graphene (G-SH) thin film was formed on a glassy carbon electrode (GCE) and used for the detection of Hg (II) ion. The study indicated that both fast and sensitive measurements were achieved by increasing the adhesion surface of graphene, and the G-SH concentration on the electrode and sample pH were playing a role in the performance of the electrode [68]. Graphene derivatives modified electrochemical sensors are not only used for the detection of environmentally hazardous chemicals but also have been designed for the determination of different types of drugs/ingredients and food additives. In that aspect, a study presented the detection of acetaminophen (AP) by using phosphorus-doped graphene (P-rGO) based GCE (P-rGO/GCE). Owing to the conductive and electrochemical properties of graphene, the electrode exhibited an excellent electrocatalytic activity for the AP and a detection limit of 0.36 μM (S/N = 3) in the range of 1.5–120 μM [69]. Moreover, the GO-based electrochemical sensor developed by Qian et al. exhibited a wide linear range, selective and sensitive performance in naproxen detection. This is because the oxygen content of GO facilitates the catalytic oxidation of naproxen [70]. Oghli et al. synthesized a polyoxometalate/rGO modified pencil graphite electrode (PGE) for trace detection of paroxetine[71]. Another nano platform for the drug detection have been designed based on decoration of Au nanoparticles on rGO for the detection of nitrazepam, and the prepared sensor exhibited wide linear range (0.5 μM–400 μM), low LOD, very good sensitivity and high stability [72]. A study was conducted on the detection of hydrazine as food additive. In this study, laser-induced flexible graphene was used as the sensing electrode. These graphene electrodes were produced by laser processing of polyimide films, and they concluded that the sensor presented good repeatability (500 cycles), sensitivity (up to 115% at 0.5 mM hydrazine), and selectivity (limit of detection (LOD) of 70 μM) for hydrazine detection [73].

3 Carbonaceous/Metal Oxides-Based Nanocomposites

Composite material is known as the combination of two or more materials with different physical and chemical properties. Composite material exhibits an improved performance with new features by incorporating the properties of all raw materials from which it is obtained. For example, composite materials become stronger, lighter, or electrically resistant. Composites are classified according to the matrix type, size, and reinforcement elements used. Different composites can be obtained by using various reinforcement elements in different matrix types. Materials such as fiber reinforcement, sheet reinforcement, and particle reinforcement can be used, and polymeric, ceramic, or metallic composites can be produced depending on the matrix used. If one of the materials used in composites is one-dimensional (1D), two-dimensional (2D), or three-dimensional (3D) and smaller than 100 nm, these composites are defined as nanocomposites. Nanocomposites can be classified into three main classes; metal matrix nanocomposites, ceramic matrix nanocomposites, and polymer matrix nanocomposites. Especially, during the last two decades, composite materials functionalized with various nanostructures for the preparation of nanocomposite materials with superior properties have been one of the most studied subjects. Metallic nanostructures (spherical, cubic, ribbon-like, flower-like, rod, wire, etc.), polymeric nanomaterials, hybrid nanoclusters, carbonaceous-based nanostructures, and metal oxide-based nanoparticles (MONP) are among the most widely used nanostructures for the production of nanocomposites.

MO are compounds formed by the reaction of metals with oxygen. MO nanostructures (MONS) is formed by synthesizing MO compounds at the nanoscale. MONS have hierarchical porous structures, a large surface area, ion/electron transport path shortening abilities, highly conductive networks, sensitivity to surface modifications by adding hydroxyl groups (−OH), and the ability to form nanocomposites by combining with various materials. Furthermore, these MONS present superparamagnetic activity, good conductivity, sensitivity, and high catalytic activity. As a result, copper oxide (CuO), zinc oxide (ZnO), Al_2O_3, magnesium oxide (MgO), zirconium dioxide (ZrO_2), titanium dioxide (TiO_2), cerium oxide (CeO_2), etc. MONS are used in various fields such as material science, chemistry, medicine, pharmaceutical science, sensor technology, optics, electronics, and energy [6, 74].

Despite the many advantages mentioned above, MONS can be inadequate in applications. The main reason for this is that it can be difficult to control the surface area, pore distribution, and porosity in MONS, which can affect their performance in applications [7]. Therefore, research to obtain nanocomposites with more advanced properties by combining MONS with other materials has gained popularity in recent years. In this context, graphene and its derivatives, AC, carbon nanospheres, mesoporous carbon, CNTs, CNFs, and HPGC have a synergistic effect with MONS since they have strong corrosion resistance, long cycle life, good electrical conductivity, and chemical stability. Especially in sensor applications, there are many studies in which nanocomposite structures of MONP and carbonaceous-based materials are used for the electrode systems in the sensor [7, 8, 75]. In the developed nanocomposite electrodes, carbonaceous nanostructures provide several advantages to MONS such as preventing the agglomeration of MONS, providing physical support to MONS, and facilitate the ion/electron transfer as illustrated in Fig. 6. The composition, microstructure, electrode porosity, pore size distribution, electrical conductivity, specific surface area, and physical properties of carbonaceous/MONS-based nanocomposites influence the electrode performance resulting in the performance of the application [7].

Nanocomposites created using graphene have excellent catalytic, magnetic and optoelectronic capabilities due to the electrical conductivity provided by the unique sp^2 hybridization of C bonds in graphene [76]. Besides graphene, CNTs are among the most frequently utilized carbonaceous materials, and their nanocomposite structure with MONS is of interest [77]. Another carbonaceous material is 3D porous carbons, which can be classified as macroporous, mesoporous and microporous. These structures are used as a matrix for the accommodation of active particles, owing to their properties such as adjustable pore size, high pore volume and surface area [78, 79].

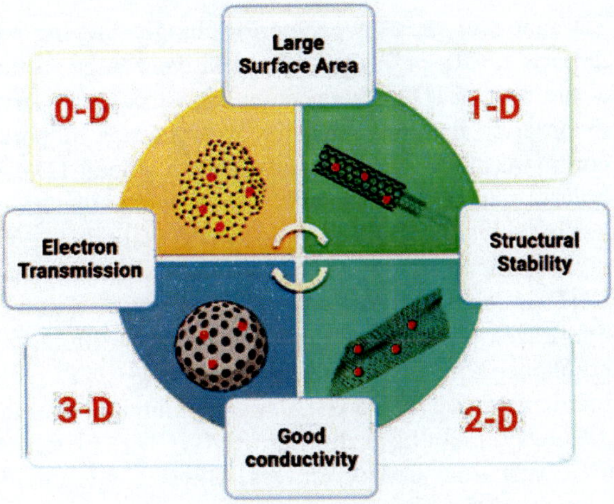

Fig. 6 Illustrative representation of different dimension carbonaceous-based nanomaterial contributions on their nanocomposites

3.1 Classification and Properties of Carbonaceous/Metal Oxides-Based Nanocomposites

Since carbonaceous-based materials belong to a large family, it would be useful and practical to classify these materials according to their dimension and to examine the final carbonaceous/MO-based nanocomposite materials obtained. In this content, the nanocomposites of ZnO, TiO_2, CuO etc. MONS with 0D, 1D, 2D and 3D carbonaceous-based materials are discussed.

Zero dimensional (OD) carbonaceous/MONS-based nano*composite*: Activated carbon (AC), one of the 0D carbons, is a material produced by the carbonization process, usually from carbon-rich sources such as wood, lignite and coconut shells. AC is employed in adsorption, catalysis, energy storage and supercapacitor applications owing to its porous structures, electronic conductivity, large surface area and highly reactive surface function [80–82]. Furthermore, mesoporous carbon is also preferred in the development of nanocomposite electrodes due to its large specific surface area and adjustable pore size. These carbons have a hierarchical pore structure that includes macropores (>50 nm), mesopores (2–50 nm), and nanopores (<2 nm) [83]. Cell performance is dependent on ion transport in the electrolyte and access of ions at the electrode. In this transport, nanopores have a minor role while mesopores facilitate the mass transfer of ions on the electrode surface [7, 84]. Carbon nanospheres are another type of 0D carbonaceous nanomaterial, and the expanded hollow carbon spheres are high graphite, which provides good electronic conductivity. 1D dimensional carbonaceous/MONS-based nanocomposites are more preferred than 0D derivatives.

One dimensional (1D) carbonaceous/MONS-based nano*composite*: In contrast to 0D carbonaceous based materials, 1D carbonaceous nanostructures facilitate ion/electron transport. Due to the reduced contact resistance between neighboring NPs,

1D nanostructures have superior charge carrying abilities than their 0D NP counterparts. CNTs or CNFs with a relatively large surface area, that are widely used as a member of 1D carbonaceous nanostructures, serve as a scaffold for the MONS in order to increase mechanical properties of pure MONS [7]. CNTs, the most common carbonaceous-based material among 1D carbonaceous nanostructures, are hollow tubular materials with nano-scale diameters and lengths. CNTs are classified as single-walled (SWCNTs) or multi-walled (MWCNTs) based on the number of layers formed by cylindrically rolling graphene sheets. The chemical bonding of CNTs is composed entirely of sp^2 hybridized bonds which bonds stable than the sp^3 hybridized bonds that found in diamond. Similarly, sp^2 hybridized bonds exist in graphene. This sp^2 hybridization provide CNTs with the unique strength and electrical properties [85]. SWCNTs can be metallic or semiconductor whereas MWCNTs are always metallic each layer in MWCNTs may have different properties and there are connections between layers. The size, alignment and number of CNTwalls can vary depending on the rolling process, which affects the electrical, mechanical and thermal properties of the materials [85, 86]. Therefore, the method to be followed in CNTs synthesis and the final product to be obtained at the end of the synthesis are determined according to the application areas and objective of the usage. For example, metallic SWCNTs can be used as interconnect materials, whereas semiconductor SWCNTs are preferred to produce nanoelectronic devices [77]. CNTs and/or their nanocomposites with MONS are widely used in, biomedical applications, catalytic applications, electronics, energy storage, and sensors as electrodes due to their unique properties such as switchable conductivity, electronics, optoelectronics, porous texture, and flexibility. The reasons for the usage of CNTs as electrodes are their mesoporous structure, accelerate charge transfer during the oxidation and reduction of reagents and accelerate the diffusion of components in the environment [87]. There are two different approaches during the synthesis of CNTs with MONS. These are; (i) in-situ approach in which simultaneous growth of CNTs and MONS is achieved; and (ii) ex-situ or post-growth decoration approach performed in a separate step following synthesis of the CNTs. In the in-situ approach, one of the hybrid components is produced in the presence of the other, or both components are synthesized simultaneously. As a result, during manufacturing process, one component can affect the size, crystal structure or morphology of the other, providing the hybrid material to acquire new properties. On the other hand, in the ex-situ approach, the two components are produced separately in the desired size, crystal structure and morphology. These materials are then modified with appropriate functional groups or binding molecules and joined via such as covalent bonding, π–π stacking, hydrogen bonds, electrostatic forces, and hydrophobic interaction [77, 88]. The advantages of CNT/MONS-based nanocomposites, which demonstrate more advanced properties in the form of nanocomposite structures, make them more attractive in applications. Because in CNT/MONS-based nanocomposite structures, (i) low electrical conductivity of MONS can be compensated by CNTs providing a facilitated electron transport channel, (ii) the large specific surface area of CNTs can increase the contact area between electrolyte and electrode, (iii) CNTs with improved mechanical toughness than pure MONS can serve as a carbon scaffold for MONS. Some of the examples of studied CNTs/MONS-based nanocomposites are ZnO/CNTs, nickel oxide/CNTs (NiO/CNTs), vanadium (V) oxide/MWCNTs (V_2O_5/MWCNTs), manganese dioxide/CNTs (MnO_2/CNTs). Detailed examples of CNTs/MONS-based nanocomposite applications are given in the following sections [79].

In addition to CNTs, CNFs are also recognized as another type of 1D carbonaceous nanomaterial. These materials are sp^2 hybridized linear filaments and typical aspect ratios higher than 100:1. CNFs have a more complex structure than CNTs because unlike CNTs, these materials consist of an arrangement of graphene structures found as stacked cones. The orientation of these carbon layers in their structures affects the mechanical properties of CNFs [89, 90]. CNFs can be synthesized by the appropriate combination of electrospinning of organic polymers and thermal post-treatment. In the electrospinning process, electrospun polymer nanofibers are produced first, followed by stabilization and carbonization. Finally, thermal after treatment is carried out in an atmosphere containing inert Ar or nitrogen (N) gas [91]. They are frequently preferred in fields such as biomedical, sensor, textile, and electrical applications due to their excellent mechanical, physical and chemical properties including good thermal conductivity, corrosion resistance, high strength, large specific surface area, chemical stability, electrical conductivity, porosity, light weight, surface roughness, and tendency to surface modifications [90, 92].

Two dimensional (2D) carbonaceous/MONS-based nano*composite*: Due to their high electrical conductivity, large surface area, and relatively facile functionalization features, 2D carbonaceous based nanomaterials such as graphene, GO and rGO are suitable for modification with MONS and to be obtained in nanocomposite forms [93]. These nanocomposites are used as electrodes in electrochemical sensors, catalytic and photocatalytic applications as well as in the field of energy storage [94–96]. The most common MONS used for preparation of nanocomposite with 2D carbonaceous-based materials include ZnO, MnO_2, CuO, CeO_2, TiO_2, ZrO_2, NiO, molybdenum trioxide (MoO_3), trimanganese tetraoxide (Mn_3O_4), palladium dioxide (PdO_2), ruthenium (IV) oxide (RuO_2) cobalt oxide (Co_3O_4), iron(III) oxide (Fe_2O_3), and iron(II,III) oxide (Fe_3O_4) [97].

Three dimensional (3D) carbonaceous/MONS-based nano*composite*: Another type of carbonaceous-based material is 3D porous carbon structures such as diamond, GO, graphite, etc. These structures have larger surface areas due to their pores which provides to increase the efficiency of the electrode by loading as much active material as possible onto the electrode. Furthermore, in order to further improve the efficiency of the electrode, these carbon structures are used in nanocomposite with MONS, creating a synergetic effect with the properties of MONS [5, 78, 79].

3.2 *Preparation of Graphene/Metal Oxide-Based Nanocomposites*

Since graphene-based materials and their combinations with MONS are of great interest any different synthesis methods are developed depending on the application area of these materials or the type and quality of the final product to be obtained. The most widely used methods for forming nanocomposites of graphene-based materials with MONS are hydrothermal/solvothermal synthesis, chemical reduction, microwave (MW) assisted synthetic approaches, and electrochemical synthetic approaches. During the synthesis of graphene and its derivatives/MONS-based nanocomposites, the graphene, GO or rGO surfaces are modified with hydrophilic and oxygen-containing functional groups such as epoxides, hydroxides, and carboxylic groups to promote them to be more active in chemical reactions with metal ions. Afterwards, the nanocomposite material formed by final annealing is stabilized to obtain a sandwich-like structure [98].

3.2.1 Solvothermal and Hydrothermal Synthetic Approaches

Solvothermal and hydrothermal synthetic approaches are the most common methods used for the development of graphene-based nanocomposites. These reactions occur at high temperatures, typically between 100 and 200 °C, followed by purification (usually with water) and drying to obtain the final product [99]. In that aspect, Song et al. synthesized CuO/GO nanocomposite by hydrothermal synthesis that the reaction carried out at different temperatures (120, 150, and 180 °C) for 10 h in order to modify and improve the properties of the GCE used for non-enzymatic glucose detection. According to the study, the efficiency of the reaction carried out at 120 °C was higher, and GO prevents aggregation of CuO NPs and improves their sensing properties due to its electroactive surface area. In addition, the performance improved GCE had a detection limit of 0.69 μM (S/N = 3) in glucose detection, a sensitivity of 262.52 $\mu A\ mM^{-1}\ cm^{-2}$, and a linear range from 2.79 μM to 2.03 mM under a working potential of +0.7. All this means that the sensor using the modified electrode has high stability, good repeatability, and excellent selectivity [100]. The strong electrolytic water used in the hydrothermal synthesis reaction has a high diffusion coefficient and dielectric constant. This facilitates the removal of oxygen-containing groups by dehydration. The method is environmentally friendly and has a simple installation. In addition, the degree of reduction desired in the reaction and the properties of the hybrid material can be controlled by adjusting the temperature and pressure of the reaction. On the other hand, a possible disadvantage of the method is the significant energy consumption [99].

3.2.2 Chemical Reduction Approach

The chemical reduction method is a method that uses different chemicals such as sodium hydroxide (NaOH), ammonium hydroxide (NH_4OH), N_2H_4, and diethylene glycol (DEG) as reducing agents [101]. In this method, the reduction degree and other properties can be adjusted by using different reducing agents. Furthermore, this method does not require high energy consumption. One of the disadvantages of this method is that in some cases the resulting product cannot be easily purified from reducing agents [99]. A study that performed by using chemical reduction to produce europium oxide nanoparticles (EuNs) on rGO and obtaining EuN/rGO nanocomposite structure carried out by Naderi et al. During the synthesis, they gradually added a 32% NH_4OH aqueous solution to a solution of $Eu(NO_3)_3 \cdot 6H_2O$ in an ultrasonic bath, then to this solution GO suspension was added under sonication and allowed the NPs to bind to the GO layers. Then GO was converted to rGO by adding N_2H_4 and as a result they obtained the EuN–rGO nanocomposite structure [101].

3.2.3 Microwave-Assisted Synthesis Approach

MW-assisted synthesis method is another method used to produce graphene/MONS-based nanocomposite structures with homogeneous distribution, and high purity. MW is electromagnetic radiation with a frequency range of 0.3–300 GHz and a wavelength range of 1 mm–1 m. Because MWs travel at the speed of light, breaking a chemical bond requires less energy. As a result, MWs have no effect on the framework of the chemical molecule [102]. In contrary to the traditional heating method, in MW heating, materials are heated directly through dielectric loss, resulting in rapid heating. Moreover, due to the different dielectric constants of the reactants and the solvent, MW can be used to achieve selective dielectric heating. In this type of heating, a significant increase in the transfer of energy directly to the reactants is achieved, resulting in a sudden increase in the internal temperature. As a result, MW can facilitate the nucleation of NPs and shorten the synthesis time [76, 103]. The reaction time is greatly reduced with this method due to the lower activation energy barrier. Moreover, because the reaction time is reduced, the probability of unwanted by-products is reduced, the purity of the obtained product, the reproducibility of the system and the ease of optimization are increased [102]. Regarding the use of MW-assisted synthesis approaches in the synthesis of graphene/MONS-based materials, Foroughi et al. synthesized the graphene/CuO nanocomposite structure on GCE to detect glucose. In the experimental part, the solution prepared by mixing the certain chemicals was sonicated to create a homogeneous solution, and then placed in a household microwave oven and treated for 10 min. Then, the obtained product was annealed at 500 °C for 5 h to remove impurities. In the study, it was noted that graphene/CuO nanocomposite improves the glucose sensing performance of the GCE [104].

3.2.4 Electrochemical Approach

The electrochemical approach is another method used in the synthesis of graphene/MONS-based nanocomposites. This method is a simple, fast, low-cost, stable, reproducible and environmentally friendly approach using ionic liquid electrolytes or aqueous surfactants and an electric current to achieve oxidation or reduction, intercalation, and exfoliation in a piece of graphite. The size, shape, and yield of materials produced by the electrochemical approach can be easily controlled by modifying the electrochemical cell design and conditions, such as the molarity of the electrolytes, the pH of the electrolytic solution, the power supply voltage, and the electrolysis time [105–108]. In a study on graphene/MONS-based nanocomposite was synthesized by Nanakkal et al. The graphene/ZnO nanocomposite structure was produced in three steps by using the electrochemical synthesis method. The first step was the synthesis of graphene by electrolysis, the second step was the modification of the synthesized graphene sheets, and the final step was the functionalization of the modified graphene sheets with ZnO NPs. In the experiment using sodium chloride (NaCl) as the electrolytic, a suitable constant DC voltage was applied to two fresh graphite rod electrodes immersed in the electrolytic solution. After that, it was modified with H_2SO_4 to roughen the graphene-coated electrode surfaces. In the final step, the graphene-deposited graphite rod served as the cathode and zinc sulphate ($ZnSO_4$) solution was used as the electrolyte, and a graphene/ZnO nanocomposite was obtained by depositing ZnO NPs on graphene sheets by electrolysis. Next, the synthesized graphene/ZnO sample was annealed at 500 °C. It was stated that the graphene/ZnO nanocomposite structure obtained in the study showed more advanced photocatalytic activity than ZnO [107].

3.3 *Applications of Carbonaceous/Metal Oxides-Based Nanocomposites*

In recent years, there has been a great interest in producing nanocomposites of largely carbonaceous-based materials with MONS and using them for different applications as depicted in Fig. 7. This section contains studies of the uses of carbonaceous/MONS-based nanocomposites in energy storage, solar cell and catalytic applications, particularly in electrochemical sensors.

Carbonaceous-based materials and nanocomposite materials obtained from MONS are used in many areas. Among these studies, regarding photocatalytic and energy storage applications, Liu et al. synthesized the nanocomposite material from nitrogen-doped MWCNT and V_2O_5 (V_2O_5@N/CNTs) to be used as the positive electrode of a lithium-sulfur battery. The capacity retention rate of the material with improved performance has been determined as 60%. The initial charge capacity of the material reached 1,453 mAh g^{-1} at rate of 0.1 C, and the high discharge specific capacity was determined as 538 mAh g^{-1} even after 200 charge/discharge cycles

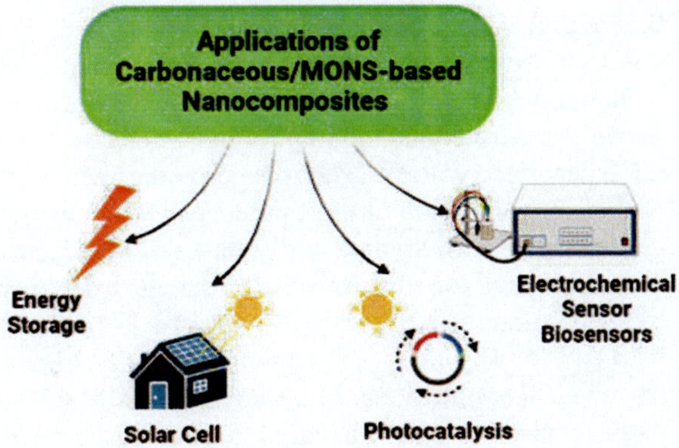

Fig. 7 Schematic illustration of applications of carbonaceous/MONS-based materials

at rate of 0.5 C. It also exhibited a specific discharge capacity of 492 mAh g^{-1} even after 400 charge and discharge cycles at rate of 1.0C [109]. In a study based on catalytic activity, TiO_2 and ZnO were deposited as single and multilayer on MWCNT, yielding CNT–TiO_2, CNT–ZnO, CNT–TiO_2–ZnO and CNT–ZnO–TiO_2 core/shell nanocomposites produced to evaluate the catalytic activity of methyl orange. Structures in which ZnO is the outer layer (CNT–ZnO and CNT–TiO_2–ZnO) showed the most effective catalytic activity in the decomposition of methyl orange. Samples containing multi-layered oxide (CNT–ZnO–TiO_2 and CNT–TiO_2–ZnO) exhibited lower catalytic activity than CNTs containing single-layer oxide due to higher average densities and lower surface areas [110].

Carbonaceous/MONS-based nanocomposites have been widely used in sensor applications recently. Some of the studies conducted in this area are reviewed. Yue et al. synthesized graphene/ZnO nanorods onto GCE to create an electrochemical sensor for the simultaneous detection of sulfamethoxazole (SMX) and trimethoprim (TMP) antibiotics. A synergistic effect occurred between ZnO preventing agglomeration of graphene and graphene increasing the conductivity of semiconductor ZnO. The result demonstrated excellent electrocatalytic performance against SMX and TMP. In addition, the repeatability and stability of the electrode were measured in the study, the electrode was stored at 4 °C for 21 days and then their response to SMX and TMP was compared. The relative standard deviations for both antibiotics on the electrodes were observed as 4.9 and 4.7%, respectively, and it was noted that the electrode had good stability and repeatability [111]. Among the various semiconductor oxides, ZrO_2 is used in sensor applications due to its good catalytic property but is supplemented by carbonaceous-based materials due to its low conductivity. In a study, GQDs@ZrO_2 hybrid nanostructures were used as sensing electrodes for the electrochemical measurement of Ochratoxin A (OTA). The GQDs@ZrO_2 hybrid prevented GQD restacking, and improved sensor performance by providing high active sites, enhanced hydrophilicity, and high electron transport rate. In the study, it was stated that the BSA/anti-OTA/GQDs@ZrO_2/ITO immunosensor has a detection range of 1–20 ng/mL, a sensitivity of 5.62 μA mL/ng cm^2, and a detection limit of

0.38 ng/mL [112]. Tungsten trioxide (WO_3), on the other hand, is an n-type semiconductor with a tunable wide energy band gap (2.4–3.6 eV). It is preferred in sensor technology because of its features such as low toxicity and low cost, high stability in acidic environment and improving the electrochemical performance of materials. GCE decorated with WO_3/graphene synthesized by Ansari et al. was used for detection of the antianginal drug Ranolazine (RZ). Owing to WO_3 NP and graphene, the electrical conductivity and surface area of the electrode increased, which improved the electrocatalytic performance of the electrode, resulting in two linear ranges for RZ determination between 0.2–1.4 and 1.4–14 μM with the low detection limit of 0.13 μM [113]. MnO_2 is also preferred with carbonaceous-based materials in electrochemical applications. In a study, AuNPs/MnO_2/graphene/CNT hybrid structure using graphene and CNT as carbonaceous-based materials were synthesized and used for H_2O_2 determination. In this structure, MnO_2 NPs had excellent electrocatalytic activities against H_2O_2, while Au NPs were preferred because they both increased electrical conductivity and improved supportive sensing performance against H_2O_2. In the study, the hybrid nanostructures exhibited excellent electrocatalytic activity against H_2O_2 with ultrahigh detection sensitivity of 452 μA mM^{-1} cm^{-2} and low detection limit of 0.1 μM (S/N = 3) [114].

4 Electrochemical Sensors

Living organisms use their sense organs to interact with the environment in their daily lives. They process the information coming from their environment through intermediary organs or tissues such as eyes, nose, tongue, skin and respond with their subsequent movements. One of the basic questions to be asked, is there a technological equivalent to these perception and reaction capabilities of living systems? Yes indeed! Sensors.

Sensors are devices that detect environmental changes and process them electrically, optically, biologically or chemically to produce measurable results. Sensors are generally classified according to their operating mechanism (active/passive sensors or absolute/relative sensors), the type of interaction used during the measurement (physical, chemical or biological sensors), and the dimensions of the measured substance (microsensors, nanosensors, etc.).

Chemical sensors are sensors that use chemical reactions to detect and measure a target substance. The best examples of natural chemical sensors are the tongue and nose, which collaborate in taste detection, owing to their highly flexible and sensitive sensing capabilities. The tongue can detect different tastes such as salty, sour, bitter, and sweet by chemical analysis of the substances whereas the nose, which is another natural chemical sensor, can detect and distinguish thousands of different odors. All biological odor detection systems are based on a relatively small number of distinguishable sensors. The odor recognition system is based on pattern matching of the response of different chemical sensors in the nose to various odors [115, 116]. Chemical sensors consist of a recognition element (receiver), a transducer on the receiver part that converts chemical information into a measurable signal, and an electronic system that can reflect this signal [9]. Chemical sensors can be classified as electrochemical sensors, optical chemical sensors, and calorimetric sensors according to the working principle of the transducer. Electrochemical sensors are devices that measure the interactions between the target analyte and the electrode by electrochemical methods with the help of electrodes in solutions called electrolytes [10]. The efficiency of electrochemical sensors can vary depending on the electrode type and electrolyte used, the chemical and physical properties of the electrode surfaces, and the applied potential [11]. The reliability of the sensors is directly proportional to the precision, measurability and reproducibility of the results obtained. It is important to consider the following sensor characteristics in a successful sensor design.

Sensitivity: One of the most important characteristics of a sensor is sensitivity which is usually defined as the ratio of the change in output to the change in input. The sensitivity of an ideal sensor is constant throughout its dynamic range [117].

Linearity: Linearity is a measure of how close the characteristic curve of the sensor is to a straight line. Ideally, the characteristic curve is expected to follow a straight line [10].

Repeatability: Repeatability is the ability of the sensor to produce the same output when exposed to a given measured variable multiple times at different times and under the same conditions [117].

Resolution: The resolution of a sensor is the minimum increase in stimulus it can respond to. For example, on a sensor with a resolution of 0.01 °C, this means that a temperature increase of 0.01 °C produces an easily measurable output [116].

Measuring Range (Dynamic Range): The measuring range of the sensor is the range of measured values for a sensor response [116, 117].

Accuracy: Detection and operating errors in the sensor are parameters that define the accuracy of the device [116].

Selectivity: Selectivity means that the sensor responds only to the relevant measurement [117].

Response Time: Response time is the time it takes for the sensor to reach 90% of the steady state value after the measurand has been entered. Recovery time is the time it takes for the sensor to remain within 10% of its value before being exposed to the measurand [117].

Hysteresis (Delay): Hysteresis is the name given to the deviation of the sensor output at any point when approached from the ascending or descending direction [116].

Drift: Drift is considered as deviations in the output signal. Drift changes the slope of the static sensitivity curve and causes the sensor to lose sensitivity [118].

It is possible to modify the sensor electrodes to improve the sensor properties described above, especially the catalytic performance, sensitivity, analyte selectivity and diffusion kinetics of the sensor [12]. Different methods and materials are used for the modification of electrodes. Nanocomposites made of carbonaceous-based materials with MONS are frequently preferred in the production of modified electrodes due to their properties such as large surface area, high conductivity and fast heterogeneous electron transfer.

4.1 *Classifications of Sensors*

Although sensor classifications are represented in different ways in many different sources, it is possible to divide sensors into several classes. Sensors are generally classified based on their working mechanism (active/passive sensors or absolute/relative sensors), type of interaction used during measurement (physical, chemical, or biological sensors), and dimensions of the measured substance (microsensors, nanosensors, etc.). One of the basic classification of sensors is given with examples for each sensor derivatives in Fig. 8.

4.1.1 Electrochemical Sensors

Analytical chemistry, one of the oldest fields of chemistry, is the science that studies the chemical structure and components of a particular substance. In analytical chemistry, various instrumental techniques are used to study the structure and components of chemicals. Thanks to advanced technology, practical sensors that are sensitive to sample detection began to be developed, and thus chemical sensors emerged. Chemical sensors can be briefly defined as sensors that transform chemical changes between the target substance (analyte) and the recognition element into meaningful

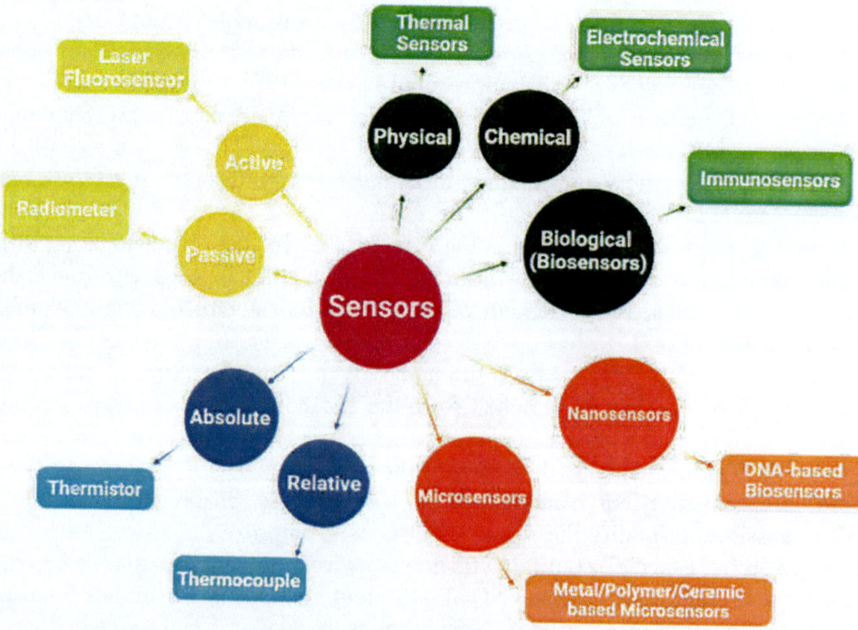

Fig. 8 A fundamental classifications of sensors

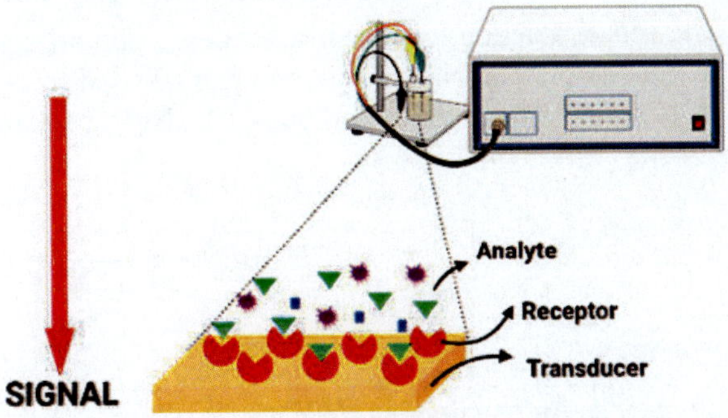

Fig. 9 Schematic illustration of basic chemical sensors

and measurable results. In their simplest design, chemical sensors consist of a recognition element (receptor), a transducer that converts chemical information into a measurable signal in the receptor part, and an electronic system that can reflect this signal, as shown in Fig. 9. The working mechanism of a chemical sensor is based on the fact that the target substance given to the sensor interacts with the recognition elements of the sensor through chemical reactions and gives an output as a result of this reaction. An ideal chemical sensor should respond quickly and operate for a long time, be inexpensive and specific, that is, it should respond to only one analyte, or at least be selective towards a group of analytes [9, 119]. Chemical sensors can be classified as electrochemical sensors, optical chemical sensors and calorimetric sensors according to the working principle of the transducer.

Electrochemical sensors are devices that use electrodes in the electrolyte solutions to measure the interactions between the target analyte and the electrode, the electrical properties of the target analyte, or the effect of the analyte on the electrical properties of another material. These sensors and supporting electronic sensors are generally simple in design and the resulting products are frequently used in demanding applications [10]. The efficiency of electrochemical sensors is affected by the type of electrode and electrolyte used, the chemical and physical properties of the electrode surfaces, the applied potential, and the modifications applied to the electrode surface nanomaterials [11]. As a result of the electrode modifications, more interaction is achieved by increasing the surface area resulting in increasing the catalytic performance, and properties such as selectivity, sensitivity, and diffusion kinetics against the analyte can be improved owing to immobilized functional groups and additives [12]. Electrochemical sensors are generally divided into two groups as potentiometric sensors and voltammetric/amperometric sensors [10].

Potentiometric Chemical Sensors: A potentiometric sensor is a type of chemical sensor that might be used the determination of the concentration of some components of the analyte. The determination of the analyte concentration occurs by measuring the variation potential difference between working and reference electrodes under the conditions of no current flow. The signal of general potentiometric sensors is based on the Nernst equation. The Nernst equation is an equation often used by electrochemists. The equation is a thermodynamic relationship that allows the reduction potential of a reaction to be calculated from the number of electrons involved in the redox reaction, chemical activities, standard electrode potential, and constant temperature. The Nernst equation can be used to determine the cell potential of a cell under non-standard conditions. The Nernst equation is given below in Eq. 1 [10, 120].

$$E = E^{\theta} - \left(\frac{RT}{nF}\right) \ln Q \tag{1}$$

where;

Q stands for reaction quotient
R Universal gas content
n Amount in mol of electrons transferred
F Faraday constant
T Temperature in K.

The working principle of the potentiometric sensor is based on the change of electric potential at the surface of a solid material immersed in a solution where redox reactions take place. Due to the charge separation at the surface, a potential difference occurs between the surface of the solid and the solution. However, this contact potential cannot be measured directly. Therefore, the electrochemical cell is created using electrodes and the potential between the two electrodes can be directly measured. In this type of sensor, it is critical that the potential is ion-specific and that the electrodes are selective towards solutions. These are called ion-specific electrodes (ISEs) or membranes. Also, the potential can vary depending on the number and concentration of ions in the solution [116]. In order to increase selectivity in such sensors, electrodes can be coated with conductive or large surface area nanomaterials or ionic liquids can be applied during the process [121]. ISEs are widely used in potentiometric sensors. It has polymeric membranes containing selective carriers (ionophores) and is widely used for the direct determination of various inorganic and organic ions in medical, environmental and industrial analysis. Compared to other analytical techniques, ISEs have attractive features such as being portable, user-friendly, low energy consumption and low cost [122, 123]. In addition to the ISEs, the ion-selective membrane is also an important potentiometric ion sensor component. This membrane determines the reference to which the sensor responds to the ion of interest in the presence of various other ionic components in the sample [10].

Voltammetric/Amperometric Sensors: The working mechanism of voltammetric/amperometric sensors is based on the application of an increasing and/or decreasing potential to the cell until oxidation and/or reduction of the substance to be analyzed occurs and a sharp rise and/or decrease in the current peaks. The height of the peak current is directly proportional to the concentration of the electroactive material. Since the main operating feature of these sensors is based on the transfer of electrons from the analyte, changes in the concentration of the analyte and the redox reaction affect the current. Selectivity and sensitivity are important factors in these sensors, and the electrodes are modified with various materials to improve these properties

[121]. In these sensors, the electrochemical cell consists of two electrodes immersed in a suitable electrolyte. Sometimes, however, there are more complex systems in which a three-electrode cell is used, with one of the electrodes acting as a reference electrode. The working electrode is the electrode where the reaction of interest occurs, while the reference electrode (e.g. silver/silver chloride (Ag/AgCl), mercury/mercury chloride (Hg/Hg_2Cl_2)) provides a stable potential compared to the working electrode. An inert conductive material (e.g. Pt, graphite) is often used as an auxiliary electrode [119]. Amperometric sensors are required to have a high electrical conductivity. Therefore, conductive materials such as carbon-based materials are used as scaffolds both to immobilize active molecules and to hold the active molecules in monolayers close to the electrode surface.

4.2 *Functionalization of Working Electrodes for Electrochemical Sensor*

Electrochemical sensors are built of three main components: an analyte (sample), a receptor, and a transducer. Electrochemical sensors employ special electrodes to detect chemical reactions or charge transfer caused by the reaction. The electrodes act as the transducer that will either oxidase or reduce the analyte of interest. As a result, the current is produced and converted into meaningful data [124].

Electrochemical sensors use special electrodes where a chemical reaction occurs or charge transfer by the reaction occurs. The most fundamental rule for an electrochemical sensor is that it always requires a closed circuit, meaning that an electric current (DC or AC) must be present to make a measurement. Since the flow of electric current requires a closed loop, electrochemical sensors may contain two-electrode-based systems, three-electrodes-based system, and screen-printed electrode (SPE) system. In three electrode system, there are a working electrode (WE), reference electrode (RE), and counter electrode (CE) whereas two-electrode systems consist of a WE and CE, as shown in Fig. 10 [125]. The WE can be produced from many different materials such as Ag, Au, Pt, glassy carbon (GC), boron-doped, diamond, etc. Depending on the redox reaction, reduction or oxidation may occur so the WE is called the cathode or anode, respectively. The CE is often called as auxiliary electrode which is produced from inert materials such as Ag, Au, Pt, etc. The RE should maintain a constant potential with respect to the analyte, regardless of its concentration or type. Ideally, the RE should not cause any chemical change in the analyte and the analyte should not alter the reference electrode. Commonly used REs are fabricated from Ag/AgCl, normal hydrogen electrode (NHE), and standard hydrogen electrode (SHE) [10].

In order to increase the performance of the WE, this electrode is functionalized in several ways. Modification of WE are usually carried out by using carbonaceous-based materials that increase electrical conductivity, facilitate the adhesion of the

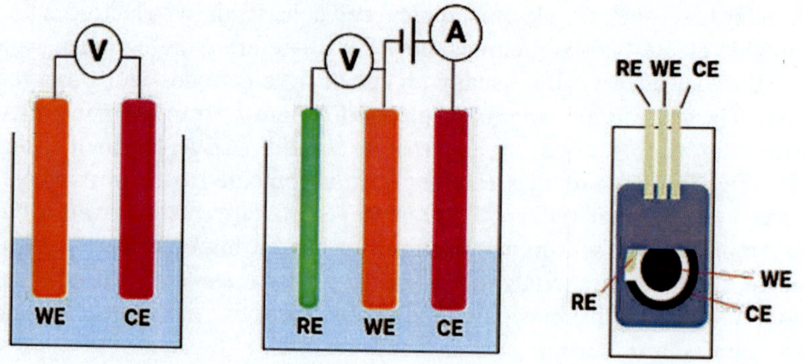

Fig. 10 Illustration of two, three electrodes and SPE electrode systems with WE, CE and RE

substance to be detected, allow specific recognition and enable strong signal transmission. Güneş et al. developed a heterostructured nanocomposite (HNC) electrode modified with graphene foam/hematite (GF/α-Fe_2O_3) nanowire arrays for use in glucose detection. Graphene foams grown on Ni foam by CVD method improved the electrochemical performance of the electrodes by increasing the electron transfer rate, providing a large surface area and high charge transfer capacity. This sensor, in which the electrode whose surface area increased after modification, showed high sensitivity and selectivity against glucose, and the LOD and sensitivity of the sensor were calculated as 71.6 μM and 20.03 μAmM^{-1} cm^{-2}, respectively, according to the active surface area of the electrode [5]. Aykaç et al. modified the ultra-micro electrodes (UME) using ZnO-nanowire (ZnO NW) coated carbon fiber (CF). In the study, ZnO NWs were grown by hydrothermal method on CFs. The UME, which was then covered with the ZnONW/CF structure, was integrated with a metal oxide–semiconductor field-effect transistor (MOSFET) for the construction of an EGFET pH-microsensor. In this study, ZnONW/CF was used to fabricate a pH sensitive UME by inserting it in two a glass micropipette using a micro manipulator. pH of the real samples from 5 to 12 were recorded. The sensitivity has been found 47.4 mW/pH [126].

4.3 *Carbonaceous/Metal Oxides-Based Nanocomposites for Electrochemical Sensors*

Electrochemical sensors developed with the incorporation of carbonaceous/MONS-based nanocomposites gained great interest of scientists recently. These sensors are used for detection of different analytes in many application areas. One of these application areas is pharmaceutical technology. Supraja et al. developed heterogeneous electrodes with MWCNT/ZnO nanocomposite structure for atrazine (ATZ) detection. Thanks to the high conductivity of the nanocomposite structure, 33% increased surface area compared to pure ZnO, and low band gap, the sensitivity, and LOD of the sensor as 21.61 (KΩ μg^{-1} mL^{-1}) cm^{-2} and 5.368 zM for a wide detection range of 10 zM^{-1} μM were recorded. As a result, it was stated that MWCNT/ZnO nanocomposite structure showed higher conductivity than pure ZnO nanofibers [87]. Yang et al. conducted a study considering that dopamine (DA) and uric acid (UA) are difficult to detect due to their similar peak potentials. In this study, ZnO nanofibers were sequentially synthesized on the CF surface using magnetron sputtering and in situ hydrothermal methods. As a result, distinguishing oxidation potential peaks of DA and UA was achieved. The electrode with improved performance showed high sensitivity and selectivity. The results showed a linear range of 6–20 μM and a LOD ~0.402 μM for DA detection [127]. In another study, SPCEs surfaces were modified with MWCNT-sponge (MWCNT-Spng), MWCNT-Spng-AuNPs, and graphene, respectively, and their performance for dopamine (DA) detection was evaluated. According to the study, the modified electrodes showed an improvement in both LOD and sensitivity compared to bare SPCE. Among the modified electrodes, MWCNT-Spng-AuNPs/SPCE showed the highest performance, and the LOD and sensitivity were calculated as 4.96 μM and 526 μA/mM cm^2, respectively [128]. Selvi et al. synthesized graphene/WO_3 nanocomposite structure was coated on SPCE and used for the electrochemical detection of nitrofurantoin (NTF). Gr/WO_3/SPCE electrode with increased electrocatalytic activity has exhibited against NFT with a lower detection limit (0.002 μM), good linear response range (0.01–234 μM), and acceptable sensitivity (2.18 μAμM^{-1} cm^{-2}) [129].

Electrochemical sensors for the detection of food additives and contaminants are used to maintain higher safety and quality standards in processed food and beverages. These sensors perform very effectively for the detection of food contaminants such as hydrazine, bisphenol A, and some commonly used preservatives such as caffeine, ascorbic acid, sulfites and nitrites [130]. In a related study, m-Fe_3O_4/GO structure was obtained using mesoporous magnetite NP (m-Fe_3O_4) and GO by solvothermal synthesis, and then integrated on GCE and used for electrochemical detection of hydrazine (HDZ). The electrocatalytic activity of the electrochemical sensor improved with the increased surface area due to the high electron transfer rate, and the porous structure of both GO and Fe_3O_4 NPs. m-Fe_3O_4/GO/GCE had high sensitivity, low LOD and wide linear detection range compared to GO/GCE. The LOD, sensitivity, and linear detection range values for the new GCE were 59 nM, 27 μA μM^{-1} cm^{-2}, and 1–4400 μM were recorded, respectively [131].

Electrochemical sensors, which are also frequently preferred for the detection of heavy metal ions, show further improved performance by using carbonaceous/MONS-based nanocomposites. A selective Hg^{2+} sensor was designed by preparing CuO/poly(methyl methacrylate) (PMMA) and various carbonaceous-based nanofillers (CuO/PMMA/SWCNT, CuO/PMMA/MWCNT and CuO/PMMA/G) as triple nanocomposites. Owing to nanofillers, the thermal properties of PMMA have improved and CuO/PMMA/SWCNT had highest response to Hg^{2+} ions in a buffer medium (phosphate type) compared to other nanofillers. In addition, the sensor using CuO/PMMA/SWCNT showed good repeatability, fast response time, good linearity, and good stability [132].

5 Conclusion

Environmental pollution and climate change, the growing demand for energy and food as the world's population grows, and the pollution of a limited number of clean water resources are among the most important problems facing humanity. Fortunately, scientists are making great efforts to identify, prevent, and solve these problems. One of the most concrete recent examples of scientific developments in society, as we all know, is the unique efforts of scientists during the COVID-19 pandemic that started at the end of 2019. In particular, the development of rapid diagnostic kits made it possible to diagnose patients in a short time, which helped to isolate them and ultimately prevent the disease from becoming much more devastating. In addition, vaccines developed in a very short time gave hope for an end to the pandemic and helped prevent the transmission of the disease. Vaccines have made a very positive contribution to protecting the elderly population and immunocompromised against the SARS-Cov19 virus. Taking all this into account, the need for simple, fast, practical, cheap and reliable sensors is growing every day, especially for the rapid diagnosis in medicine, for detecting harmful additives in food, for the improvement of polluted clean water, for use in the pharmaceutical industry and many other fields. Although research in this field is progressing in many different areas, the most prominent of these studies have focused on the producing nanocomposites from carbonaceous-based materials, particularly graphene and its derivatives, and producing production of electrodes from these nanocomposites. Graphene has been modified with many different materials and methods in order to achieve superior physical and chemical properties of 2D graphene and its derivatives with different properties and/or to have a suitable design. Among these modifications, the preparation of nanocomposites of graphene and its derivatives with MONS and their widespread use, especially in sensor technologies, has been one of the most common subjects of study in recent research. In this chapter, not only the use of graphene/MONS-based nanocomposites as electrochemical sensors but also carbonaceous/MONS-based nanocomposites are presented, accompanied by some of the most comprehensive, detailed and most recent scientific papers in the field. We have discussed some fascinating applications of these nanocomposites in the field of electrochemical sensors. In this context, the chapter started with a fundamental introduction, followed by a section on graphene and its derivatives, a section on carbonaceous/MONS-based nanocomposites, and finally completed with a section on electrochemical sensors. It should be noted that although the most fundamental and recent studies are covered in this chapter, the studies on the subject are not limited to these, but are broader and more comprehensive.

References

1. J.B. Reece, L.A. Urry, M.L. Cain, P. V Minorsky, R.B. Jackson, *Campbell Biology*, 10th edn (Benjamin Cummings, 2013)
2. C.N.R. Rao, K. Biswas, K.S. Subrahmanyam, A. Govindaraj, Graphene, the new carbon. J. Mater. Chem. **19**, 2457–2469 (2009). https://doi.org/10.1039/B815239J
3. K. Movlaee, M.R. Ganjali, P. Norouzi, G. Neri, Iron-based nanomaterials/graphene composites for advanced electrochemical sensors. Nanomaterials **7**, 406 (2017). https://doi.org/10.3390/nano7120406
4. A. Aykaç, H. Gergeroglu, B. Beşli, E.Ö. Akkaş, A. Yavaş, S. Güler, F. Güneş, M. Erol, An overview on recent progress of metal oxide/graphene/CNTs-based nanobiosensors. Nanosc. Res. Lett. **16**, 1–19 (2021). https://doi.org/10.1186/s11671-021-03519-w
5. F. Güneş, A. Aykaç, M. Erol, Ç. Erdem, H. Hano, B. Uzunbayir, M. Şen, A. Erdem, Synthesis of hierarchical hetero-composite of graphene foam/α-Fe2O3 nanowires and its application on glucose biosensors. J. Alloys Compd. **895**, 162688 (2022). https://doi.org/10.1016/j.jallcom.2021.162688
6. M.S. Chavali, M.P. Nikolova, Metal oxide nanoparticles and their applications in nanotechnology. SN Appl. Sci. **1**, 1–30 (2019). https://doi.org/10.1007/S42452-019-0592-3
7. M. Zhi, C. Xiang, J. Li, M. Li, N. Wu, Nanostructured carbon–metal oxide composite electrodes for supercapacitors: a review. Nanoscale **5**, 72–88 (2013). https://doi.org/10.1039/C2NR32040A
8. H. Jiang, J. Ma, C. Li, Mesoporous carbon incorporated metal oxide nanomaterials as supercapacitor electrodes. Adv. Mater. **24**, 4196 (2012). https://doi.org/10.1002/adma.201290182
9. P. Gründler, Chemical sensors: an introduction for scientists and engineers (2007). https://doi.org/10.1007/978-3-540-45743-5
10. J. Fraden, Handbook of Modern Sensors (2016). https://doi.org/10.1007/978-3-319-19303-8
11. A.C. Power, B. Gorey, S. Chandra, J. Chapman, Carbon nanomaterials and their application to electrochemical sensors: a review. Nanotechnol. Rev. **7**, 19–41 (2018). https://doi.org/10.1515/ntrev-2017-0160
12. S. Tajik, H. Beitollahi, S.Z. Mohammadi, M. Azimzadeh, K. Zhang, Q. Van Le, Y. Yamauchi, H.W. Jang, M. Shokouhimehr, Recent developments in electrochemical sensors for detecting hydrazine with different modified electrodes. RSC Adv. **10**, 30481–30498 (2020). https://doi.org/10.1039/D0RA03288C
13. K.S. Novoselov, A.K. Geim, S.V Morozov, D. Jiang, Y. Zhang, S.V Dubonos, I.V Grigorieva, A.A. Firsov, Electric field effect in atomically thin carbon films. Science (80) **306**, 666–669 (2004). https://doi.org/10.1126/science.1102896
14. C.K. Chua, M. Pumera, Chemical reduction of graphene oxide: a synthetic chemistry viewpoint. Chem. Soc. Rev. **43**, 291–312 (2014). https://doi.org/10.1039/C3CS60303B
15. G. Liu, W. Jin, N. Xu, Graphene-based membranes. Chem. Soc. Rev. **44**, 5016–5030 (2015). https://doi.org/10.1039/C4CS00423J
16. W. Gao, The chemistry of graphene oxide, in *Graphene Oxide* (Springer, 2015), pp. 61–95. https://doi.org/10.1007/978-3-319-15500-5_3
17. C.-N. Yeh, K. Raidongia, J. Shao, Q.-H. Yang, J. Huang, On the origin of the stability of graphene oxide membranes in water. Nat. Chem. **7**, 166–170 (2015). https://doi.org/10.1038/nchem.2145
18. R. Tarcan, O. Todor-Boer, I. Petrovai, C. Leordean, S. Astilean, I. Botiz, Reduced graphene oxide today. J. Mater. Chem. C. **8**, 1198–1224 (2020). https://doi.org/10.1039/C9TC04916A
19. P. Cataldi, A. Athanassiou, I.S. Bayer, Graphene nanoplatelets-based advanced materials and recent progress in sustainable applications. Appl. Sci. **8**, 1438 (2018). https://doi.org/10.3390/app8091438
20. S.-Y. Yang, W.-N. Lin, Y.-L. Huang, H.-W. Tien, J.-Y. Wang, C.-C.M. Ma, S.-M. Li, Y.-S. Wang, Synergetic effects of graphene platelets and carbon nanotubes on the mechanical and

thermal properties of epoxy composites. Carbon N. Y. **49**, 793–803 (2011). https://doi.org/10.1016/j.carbon.2010.10.014
21. S.K. Tiwari, S. Sahoo, N. Wang, A. Huczko, Graphene research and their outputs: status and prospect. J. Sci. Adv. Mater. Dev. **5**, 10–29 (2020). https://doi.org/10.1016/j.jsamd.2020.01.006
22. X. Duan, C. Zhao, W. Liu, X. Zhao, L. Chang, Fabrication of a novel PbO2 electrode with a graphene nanosheet interlayer for electrochemical oxidation of 2-chlorophenol. Electrochim. Acta **240**, 424–436 (2017). https://doi.org/10.1016/j.electacta.2017.04.114
23. M. Li, T. Chen, J.J. Gooding, J. Liu, Review of carbon and graphene quantum dots for sensing. ACS Sens. **4**, 1732–1748 (2019). https://doi.org/10.1021/acssensors.9b00514
24. M. Bacon, S.J. Bradley, T. Nann, Graphene quantum dots. Part. Part. Syst. Charact. **31**, 415–428 (2014). https://doi.org/10.1002/ppsc.201300252
25. X.T. Zheng, A. Ananthanarayanan, K.Q. Luo, P. Chen, Glowing graphene quantum dots and carbon dots: properties, syntheses, and biological applications. Small **11**, 1620–1636 (2015). https://doi.org/10.1002/smll.201402648
26. J. Liu, B.-W. Li, Y.-Z. Tan, A. Giannakopoulos, C. Sanchez-Sanchez, D. Beljonne, P. Ruffieux, R. Fasel, X. Feng, K. Müllen, Toward cove-edged low band gap graphene nanoribbons. J. Am. Chem. Soc. **137**, 6097–6103 (2015). https://doi.org/10.1021/jacs.5b03017
27. M. Pizzochero, G.B. Barin, K. Cernevics, S. Wang, P. Ruffieux, R. Fasel, O.V. Yazyev, Edge disorder in bottom-up zigzag graphene nanoribbons: implications for magnetism and quantum electronic transport. J. Phys. Chem. Lett. **12**, 4692–4696 (2021). https://doi.org/10.1021/acs.jpclett.1c00921
28. A.A. Balandin, S. Ghosh, W. Bao, I. Calizo, D. Teweldebrhan, F. Miao, C.N. Lau, Superior thermal conductivity of single-layer graphene. Nano Lett. **8**, 902–907 (2008). https://doi.org/10.1021/nl0731872
29. K.S. Novoselov, A.K. Geim, S.V. Morozov, D. Jiang, M.I. Katsnelson, I.V. Grigorieva, S.V. Dubonos, A.A. Firsov, Two-dimensional gas of massless Dirac fermions in graphene. Nature **438**, 197–200 (2005). https://doi.org/10.1038/nature04233
30. D.G. Papageorgiou, I.A. Kinloch, R.J. Young, Mechanical properties of graphene and graphene-based nanocomposites. Prog. Mater. Sci. **90**, 75–127 (2017). https://doi.org/10.1016/j.pmatsci.2017.07.004
31. C. Lee, X. Wei, J.W. Kysar, J. Hone, Measurement of the elastic properties and intrinsic strength of monolayer graphene. Science (80) **321**, 385–388 (2008). https://doi.org/10.1126/science.1157996
32. M.S.A. Bhuyan, M.N. Uddin, M.M. Islam, F.A. Bipasha, S.S. Hossain, Synthesis of graphene. Int. Nano Lett. **6**, 65–83 (2016). https://doi.org/10.1007/s40089-015-0176-1
33. B. Lang, A LEED study of the deposition of carbon on platinum crystal surfaces. Surf. Sci. **53**, 317–329 (1975). https://doi.org/10.1016/0039-6028(75)90132-6
34. X. Lu, M. Yu, H. Huang, R.S. Ruoff, Tailoring graphite with the goal of achieving single sheets. Nanotechnology **10**, 269 (1999). https://doi.org/10.1088/0957-4484/10/3/308
35. V.M. Freire Soler, Fabrication and characterization of macroscopic graphene layers on metallic substrates (2014)
36. F. Farjadian, S. Abbaspour, M.A.A. Sadatlu, S. Mirkiani, A. Ghasemi, M. Hoseini-Ghahfarokhi, N. Mozaffari, M. Karimi, M.R. Hamblin, Recent developments in graphene and graphene oxide: properties, synthesis, and modifications: a review. ChemistrySelect **5**, 10200–10219 (2020). https://doi.org/10.1002/slct.202002501
37. B.C. Brodie, Sur le poids atomique du graphite. Ann. Chim. Phys. **59**, e472 (1860)
38. L. Staudenmaier, Verfahren zur darstellung der graphitsäure. Berichte Der Dtsch. Chem. Gesellschaft. **31**, 1481–1487 (1898). https://doi.org/10.1002/cber.18980310237
39. W.S. Hummers Jr., R.E. Offeman, Preparation of graphitic oxide. J. Am. Chem. Soc. **80**, 1339 (1958). https://doi.org/10.1021/ja01539a017
40. A. Adetayo, D. Runsewe, Synthesis and fabrication of graphene and graphene oxide: a review. Open J. Compos. Mater. **9**, 207 (2019). https://doi.org/10.4236/ojcm.2019.92012

41. D.C. Marcano, D.V. Kosynkin, J.M. Berlin, A. Sinitskii, Z. Sun, A. Slesarev, L.B. Alemany, W. Lu, J.M. Tour, Improved synthesis of graphene oxide. ACS Nano **4**, 4806–4814 (2010). https://doi.org/10.1021/nn1006368
42. Z. Sun, D.K. James, J.M. Tour, Graphene chemistry: synthesis and manipulation. J. Phys. Chem. Lett. **2**, 2425–2432 (2011). https://doi.org/10.1021/jz201000a
43. W. Choi, J. Lee, Graphene: synthesis and applications (CRC press, 2019). https://doi.org/10.1201/b11259
44. K.A.I. Yan, L.E.I. Fu, H. Peng, Z. Liu, Designed CVD growth of graphene via process engineering. Acc. Chem. Res. **46**, 2263–2274 (2013). https://doi.org/10.1021/ar400057n
45. X. Li, W. Cai, J. An, S. Kim, J. Nah, D. Yang, R. Piner, A. Velamakanni, I. Jung, E. Tutuc, Large-area synthesis of high-quality and uniform graphene films on copper foils. Science (80) **324**, 1312–1314 (2009). https://doi.org/10.1126/science.1171245
46. N. Tajima, T. Kaneko, J. Nara, T. Ohno, A first principles study on the CVD graphene growth on copper surfaces: a carbon atom incorporation to graphene edges. Surf. Sci. **653**, 123–129 (2016). https://doi.org/10.1016/j.susc.2016.06.012
47. H. Kim, C. Mattevi, M.R. Calvo, J.C. Oberg, L. Artiglia, S. Agnoli, C.F. Hirjibehedin, M. Chhowalla, E. Saiz, Activation energy paths for graphene nucleation and growth on Cu. ACS Nano **6**, 3614–3623 (2012). https://doi.org/10.1021/nn3008965
48. E. Gülsüm, F. Güneş, A. Aykaç, M. Can, Improving the sheet resistance of CVD-graphene films via doping. Eskişehir Tek. Üniversitesi Bilim ve Teknol. Derg. B-Teorik Bilim. **6**, 243–249 (n.d.). https://doi.org/10.20290/aubtdb.464140
49. Q. Yu, J. Lian, S. Siriponglert, H. Li, Y.P. Chen, S.-S. Pei, Graphene segregated on Ni surfaces and transferred to insulators. Appl. Phys. Lett. **93**, 113103 (2008). https://doi.org/10.1063/1.2982585
50. S.J. Chae, F. Güneş, K.K. Kim, E.S. Kim, G.H. Han, S.M. Kim, H. Shin, S. Yoon, J. Choi, M.H. Park, Synthesis of large-area graphene layers on poly-nickel substrate by chemical vapor deposition: wrinkle formation. Adv. Mater. **21**, 2328–2333 (2009). https://doi.org/10.1117/12.828039
51. J. Wang, M. Zhu, R.A. Outlaw, X. Zhao, D.M. Manos, B.C. Holloway, Synthesis of carbon nanosheets by inductively coupled radio-frequency plasma enhanced chemical vapor deposition. Carbon N. Y. **42**, 2867–2872 (2004). https://doi.org/10.1016/J.CARBON.2004.06.035
52. N.G. Shang, P. Papakonstantinou, M. McMullan, M. Chu, A. Stamboulis, A. Potenza, S.S. Dhesi, H. Marchetto, Catalyst-free efficient growth, orientation and biosensing properties of multilayer graphene nanoflake films with sharp edge planes. Adv. Funct. Mater. **18**, 3506–3514 (2008). https://doi.org/10.1002/adfm.200800951
53. T. Zhang, L. Fu, Controllable chemical vapor deposition growth of two-dimensional heterostructures. Chem **4**, 671–689 (2018). https://doi.org/10.1016/j.chempr.2017.12.006
54. Z. Morçimen, Ş Taşdemir, Ç. Erdem, F. Gunes, A. Sendemir, Investıgatıon of the adherence and proliferation characteristics of SH-SY5Y neuron model cells on graphene foam surfaces. Mater. Today Proc. **19**, 40–46 (2019). https://doi.org/10.1016/j.matpr.2019.07.654
55. M. Tabandeh-Khorshid, A. Kumar, E. Omrani, C. Kim, P. Rohatgi, Synthesis, characterization, and properties of graphene reinforced metal-matrix nanocomposites. Compos. Part B Eng. **183**, 107664 (2020). https://doi.org/10.1016/j.compositesb.2019.107664
56. A.K. Jehad, K. Kocabas, M. Yurddaskal, A comparative study for producing few-layer graphene sheets via electrochemical and microwave-assisted exfoliation from graphite powder. J. Mater. Sci. Mater. Electron. **31**, 7022–7034 (2020). https://doi.org/10.1007/s10854-020-03268-z
57. I.H. Lone, J. Aslam, A. Akhter, 3—Characterization of advanced green materials, in *Woodhead Publishing Materials* ed. by S.B.T.-A.G.M. Ahmed (Woodhead Publishing, 2021), pp. 31–41. https://doi.org/10.1016/B978-0-12-819988-6.00003-3
58. K. Kakaei, M.D. Esrafili, A. Ehsani, Chapter 4—Characterization, in Graphene Surfaces eds. by K. Kakaei, M.D. Esrafili, A. Ehsani (Elsevier, 2019), pp. 109–151. https://doi.org/10.1016/B978-0-12-814523-4.00004-6

59. D. Han, S. Hwang, S.-M. Bak, K.-W. Nam, Controlling MoO2 and MoO3 phases in MoOx/CNTs nanocomposites and their application to anode materials for lithium-ion batteries and capacitors. Electrochim. Acta. **388**, 138635 (2021). https://doi.org/10.1016/j.electacta.2021.138635
60. Y.S. Yun, V.-D. Le, H. Kim, S.-J. Chang, S.J. Baek, S. Park, B.H. Kim, Y.-H. Kim, K. Kang, H.-J. Jin, Effects of sulfur doping on graphene-based nanosheets for use as anode materials in lithium-ion batteries. J. Power Sour. **262**, 79–85 (2014). https://doi.org/10.1016/j.jpowsour.2014.03.084
61. M.T.T. Moghadam, M. Seifi, Fabrication and investigation of ZnO-CNT@ Fe3O4/NF as supercapacitor electrode by using a novel preparation method of CNT. Diam. Relat. Mater. **125**, 108962 (2022). https://doi.org/10.1016/j.diamond.2022.108962
62. H. Cao, X. Peng, M. Zhao, P. Liu, B. Xu, J. Guo, Oxygen functional groups improve the energy storage performances of graphene electrochemical supercapacitors. RSC Adv. **8**, 2858–2865 (2018). https://doi.org/10.1039/C7RA12425B
63. B. O'regan, M. Grätzel, A low-cost, high-efficiency solar cell based on dye-sensitized colloidal TiO2 films. Nature **353**, 737–740 (1991). https://doi.org/10.1038/353737A0
64. R. Ishikawa, S. Yamazaki, S. Watanabe, N. Tsuboi, Layer dependency of graphene layers in perovskite/graphene solar cells. Carbon N. Y. **172**, 597–601 (2021). https://doi.org/10.1016/j.carbon.2020.10.065
65. M. Xing, W. Fang, X. Yang, B. Tian, J. Zhang, Highly-dispersed boron-doped graphene nanoribbons with enhanced conductibility and photocatalysis. Chem. Commun. **50**, 6637–6640 (2014). https://doi.org/10.1039/C4CC01341G
66. G. Chen, X. Wang, L. Wang, Application of carbon based material for the electrochemical detection of heavy metal ions in water environment. Int. J. Electrochem. Sci. **15**, 4252–4263 (2020). https://doi.org/10.20964/2020.05.64
67. C. Raril, J.G. Manjunatha, Fabrication of novel polymer-modified graphene-based electrochemical sensor for the determination of mercury and lead ions in water and biological samples. J. Anal. Sci. Technol. **11**, 1–10 (2020). https://doi.org/10.1186/s40543-019-0194-0
68. R. Ziółkowski, A. Uścińska, M. Mazurkiewicz-Pawlicka, A. Małolepszy, E. Malinowska, Directly-thiolated graphene based electrochemical sensor for Hg (II) ion. Electrochim. Acta **305**, 329–337 (2019). https://doi.org/10.1016/j.electacta.2019.03.070
69. X. Zhang, K.-P. Wang, L.-N. Zhang, Y.-C. Zhang, L. Shen, Phosphorus-doped graphene-based electrochemical sensor for sensitive detection of acetaminophen. Anal. Chim. Acta. **1036**, 26–32 (2018). https://doi.org/10.1016/j.aca.2018.06.079
70. L. Qian, A.R. Thiruppathi, R. Elmahdy, J. van der Zalm, A. Chen, Graphene-oxide-based electrochemical sensors for the sensitive detection of pharmaceutical drug naproxen. Sensors **20**, 1252 (2020). https://doi.org/10.3390/s20051252
71. A.H. Oghli, A. Soleymanpour, Polyoxometalate/reduced graphene oxide modified pencil graphite sensor for the electrochemical trace determination of paroxetine in biological and pharmaceutical media. Mater. Sci. Eng. C **108**, 110407 (2020). https://doi.org/10.1016/j.msec.2019.110407
72. L. Fritea, F. Bănică, T.O. Costea, L. Moldovan, C. Iovan, S. Cavalu, A gold nanoparticles-Graphene based electrochemical sensor for sensitive determination of nitrazepam. J. Electroanal. Chem. **830**, 63–71 (2018). https://doi.org/10.1016/j.jelechem.2018.10.015
73. S. Sharma, S.K. Ganeshan, P.K. Pattnaik, S. Kanungo, K.N. Chappanda, Laser induced flexible graphene electrodes for electrochemical sensing of hydrazine. Mater. Lett. **262**, 127150 (2020). https://doi.org/10.1016/j.matlet.2019.127150
74. M. Sun, H. Liu, Y. Liu, J. Qu, J. Li, Graphene-based transition metal oxide nanocomposites for the oxygen reduction reaction. Nanoscale **7**, 1250–1269 (2015). https://doi.org/10.1039/C4NR05838K
75. E. Frackowiak, F. Beguin, Carbon materials for the electrochemical storage of energy in capacitors. Carbon N. Y. **39**, 937–950 (2001). https://doi.org/10.1016/S0008-6223(00)00183-4

76. M. Khan, M.N. Tahir, S.F. Adil, H.U. Khan, M.R.H. Siddiqui, A.A. Al-warthan, W. Tremel, Graphene based metal and metal oxide nanocomposites: synthesis, properties and their applications. J. Mater. Chem. A **3**, 18753–18808 (2015). https://doi.org/10.1039/C5TA02240A
77. S. Mallakpour, E. Khadem, Carbon nanotube–metal oxide nanocomposites: fabrication, properties and applications. Chem. Eng. J. **302**, 344–367 (2016). https://doi.org/10.1016/j.cej.2016.05.038
78. Y. Chen, B. Song, M. Li, L. Lu, J. Xue, Fe3O4 nanoparticles embedded in uniform mesoporous carbon spheres for superior high-rate battery applications. Adv. Funct. Mater. **24**, 319–326 (2014). https://doi.org/10.1002/adfm.201300872
79. D. Seok, Y. Jeong, K. Han, D.Y. Yoon, H. Sohn, Recent progress of electrochemical energy devices: metal oxide–carbon nanocomposites as materials for next-generation chemical storage for renewable energy. Sustainability **11**, 3694 (2019). https://doi.org/10.3390/su11133694
80. A. Didehban, M. Zabihi, M. Faghihi, F. Akbarbandari, H. Akhtarivand, Design and fabrication of core-shell magnetic and non-magnetic supported carbonaceous metal organic framework nanocomposites for adsorption of dye. J. Phys. Chem. Solids. **152**, 109930 (2021). https://doi.org/10.1016/j.jpcs.2020.109930
81. Y. Zhao, L.P. Wang, M.T. Sougrati, Z. Feng, Y. Leconte, A. Fisher, M. Srinivasan, Z. Xu, A review on design strategies for carbon based metal oxides and sulfides nanocomposites for high performance Li and Na ion battery anodes. Adv. Energy Mater. **7**, 1601424 (2017). https://doi.org/10.1002/aenm.201601424
82. A. Rehman, M. Park, S.-J. Park, Current progress on the surface chemical modification of carbonaceous materials. Coatings **9**, 103 (2019). https://doi.org/10.3390/coatings9020103
83. C. Liu, F. Li, L. Ma, H. Cheng, Advanced materials for energy storage. Adv. Mater. **22**, E28–E62 (2010). https://doi.org/10.1002/adma.200903328
84. G. Gryglewicz, J. Machnikowski, E. Lorenc-Grabowska, G. Lota, E. Frackowiak, Effect of pore size distribution of coal-based activated carbons on double layer capacitance. Electrochim. Acta. **50**, 1197–1206 (2005). https://doi.org/10.1016/j.electacta.2004.07.045
85. H. Gergeroglu, S. Yildirim, M.F. Ebeoglugil, Nano-carbons in biosensor applications: an overview of carbon nanotubes (CNTs) and fullerenes (C60). SN Appl. Sci. **2**, 603 (2020). https://doi.org/10.1007/s42452-020-2404-1
86. C. Soldano, Hybrid metal-based carbon nanotubes: Novel platform for multifunctional applications. Prog. Mater. Sci. **69**, 183–212 (2015). https://doi.org/10.1016/j.pmatsci.2014.11.001
87. P. Supraja, V. Singh, S.R.K. Vanjari, S. Govind Singh, Electrospun CNT embedded ZnO nanofiber based biosensor for electrochemical detection of Atrazine: a step closure to single molecule detection. Microsyst. Nanoeng. **6**, 1–10 (2020). https://doi.org/10.1038/s41378-019-0115-9
88. H. Chu, L. Wei, R. Cui, J. Wang, Y. Li, Carbon nanotubes combined with inorganic nanomaterials: preparations and applications. Coord. Chem. Rev. **254**, 1117–1134 (2010). https://doi.org/10.1016/j.ccr.2010.02.009
89. D. Yadav, F. Amini, A. Ehrmann, Recent advances in carbon nanofibers and their applications—A review. Eur. Polym. J. **138**, 109963 (2020). https://doi.org/10.1016/j.eurpolymj.2020.109963
90. R. Rauti, M. Musto, S. Bosi, M. Prato, L. Ballerini, Properties and behavior of carbon nanomaterials when interfacing neuronal cells: how far have we come? Carbon N. Y. **143**, 430–446 (2019). https://doi.org/10.1016/j.carbon.2018.11.026
91. L. Sabantina, M. Klöcker, M. Wortmann, J.R. Mirasol, T. Cordero, E. Moritzer, K. Finsterbusch, A. Ehrmann, Stabilization of polyacrylonitrile nanofiber mats obtained by needleless electrospinning using dimethyl sulfoxide as solvent. J. Ind. Text. **50**, 224–239 (2020). https://doi.org/10.1177/1528083718825315
92. Q. Cao, Y. Zhang, J. Chen, M. Zhu, C. Yang, H. Guo, Y. Song, Y. Li, J. Zhou, Electrospun biomass based carbon nanofibers as high-performance supercapacitors. Ind. Crops Prod. **148**, 112181 (2020). https://doi.org/10.1016/j.indcrop.2020.112181

93. S. Pei, H.-M. Cheng, The reduction of graphene oxide. Carbon N. Y. **50**, 3210–3228 (2012). https://doi.org/10.1016/j.carbon.2011.11.010
94. N.M. El-Shafai, M.E. El-Khouly, M. El-Kemary, M.S. Ramadan, M.S. Masoud, Graphene oxide–metal oxide nanocomposites: fabrication, characterization and removal of cationic rhodamine B dye. RSC Adv. **8**, 13323–13332 (2018). https://doi.org/10.1039/C8RA00977E
95. I. Khalil, S. Rahmati, N.M. Julkapli, W.A. Yehye, Graphene metal nanocomposites—Recent progress in electrochemical biosensing applications. J. Ind. Eng. Chem. **59**, 425–439 (2018). https://doi.org/10.1016/J.JIEC.2017.11.001
96. S. Banerjee, P. Benjwal, M. Singh, K.K. Kar, Graphene oxide (rGO)-metal oxide (TiO2/Fe3O4) based nanocomposites for the removal of methylene blue. Appl. Surf. Sci. **439**, 560–568 (2018). https://doi.org/10.1016/J.APSUSC.2018.01.085
97. S. Bashir, A. Jamil, A. Habib, S.M.I. Shamsah, M. Shahid, Role of carbonaceous matrix and nobel metal doping to boost the electrochemical aptitude of nanostructured MoO3. Ceram. Int. **47**, 35224–35233 (2021). https://doi.org/10.1016/j.ceramint.2021.09.065
98. B. De, P. Sinha, S. Banerjee, T. Pal, K.D. Verma, A. Tyagi, P.K. Manna, K.K. Kar, Transition metal oxide/graphene/reduced graphene oxide composites as electrode materials for supercapacitors, in *Handb. Nanocomposite Supercapacitor Mater. II* (Springer, 2020), pp. 297–331. https://doi.org/10.1007/978-3-030-52359-6_12
99. A. Halder, M. Zhang, Q. Chi, Electrocatalytic applications of graphene—Metal oxide nanohybrid materials, in *Advanced Catalytic Materials: Photocatalysis and Other Current Trends, BoD-Books on Demand* (2016), pp. 379–413. https://doi.org/10.5772/61808
100. J. Song, L. Xu, C. Zhou, R. Xing, Q. Dai, D. Liu, H. Song, Synthesis of graphene oxide based CuO nanoparticles composite electrode for highly enhanced nonenzymatic glucose detection. ACS Appl. Mater. Interfaces. **5**, 12928–12934 (2013). https://doi.org/10.1021/am403508f
101. H.R. Naderi, M.R. Ganjali, A.S. Dezfuli, High-performance supercapacitor based on reduced graphene oxide decorated with europium oxide nanoparticles. J. Mater. Sci. Mater. Electron. **29**, 3035–3044 (2018). https://doi.org/10.1007/s10854-017-8234-2
102. A. Kumar, Y. Kuang, Z. Liang, X. Sun, Microwave chemistry, recent advancements, and eco-friendly microwave-assisted synthesis of nanoarchitectures and their applications: a review. Mater. Today Nano **11**, 100076 (2020). https://doi.org/10.1016/j.mtnano.2020.100076
103. X. Xie, Y. Zhou, K. Huang, Advances in microwave-assisted production of reduced graphene oxide. Front. Chem. **7**, 355 (2019). https://doi.org/10.3389/fchem.2019.00355
104. F. Foroughi, M. Rahsepar, M.J. Hadianfard, H. Kim, Microwave-assisted synthesis of graphene modified CuO nanoparticles for voltammetric enzyme-free sensing of glucose at biological pH values. Microchim. Acta **185**, 1–9 (2018). https://doi.org/10.1007/s00604-017-2558-8
105. M. Alanyalıoğlu, J.J. Segura, J. Oro-Sole, N. Casan-Pastor, The synthesis of graphene sheets with controlled thickness and order using surfactant-assisted electrochemical processes. Carbon N. Y. **50**, 142–152 (2012). https://doi.org/10.1016/j.carbon.2011.07.064
106. P.T. Yin, S. Shah, M. Chhowalla, K.-B. Lee, Design, synthesis, and characterization of graphene–nanoparticle hybrid materials for bioapplications. Chem. Rev. **115**, 2483–2531 (2015). https://doi.org/10.1021/cr500537t
107. A.R. Nanakkal, L.K. Alexander, Photocatalytic activity of graphene/ZnO nanocomposite fabricated by two-step electrochemical route. J. Chem. Sci. **129**, 95–102 (2017). https://doi.org/10.1007/s12039-016-1206-x
108. G. Wang, B. Wang, J. Park, Y. Wang, B. Sun, J. Yao, Highly efficient and large-scale synthesis of graphene by electrolytic exfoliation. Carbon N. Y. **47**, 3242–3246 (2009). https://doi.org/10.1016/J.CARBON.2009.07.040
109. C. Liu, M. Xiang, H. Zhang, S. Feng, J. Xiao, S. Ma, X. Li, Preparation and electrochemical performance of V2O5@ N-CNT/S composite cathode materials. Front. Energy Res. **8**, 615558 (2021). https://doi.org/10.3389/fenrg.2020.615558
110. L.P. Bakos, N. Justh, U.C. Moura da Silva Bezerra da Costa, K. László, J.L. Lábár, T. Igricz, K. Varga-Josepovits, P. Pasierb, E. Färm, M. Ritala, Photocatalytic and gas sensitive multiwalled carbon Nanotube/TiO2–ZnO and ZnO–TiO2 composites prepared by atomic layer deposition. Nanomaterials **10**, 252 (2020). https://doi.org/10.3390/nano10020252

111. X. Yue, Z. Li, S. Zhao, A new electrochemical sensor for simultaneous detection of sulfamethoxazole and trimethoprim antibiotics based on graphene and ZnO nanorods modified glassy carbon electrode. Microchem. J. **159**, 105440 (2020). https://doi.org/10.1016/j.microc.2020.105440
112. P.K. Gupta, D. Chauhan, Z.H. Khan, P.R. Solanki, ZrO2 nanoflowers decorated with graphene quantum dots for electrochemical immunosensing. ACS Appl. Nano Mater. **3**, 2506–2516 (2020). https://doi.org/10.1021/acsanm.9b02598
113. S. Ansari, M.S. Ansari, S.P. Satsangee, R. Jain, WO3 decorated graphene nanocomposite based electrochemical sensor: a prospect for the detection of anti-anginal drug. Anal. Chim. Acta. **1046**, 99–109 (2019). https://doi.org/10.1016/j.aca.2018.09.028
114. S.-J. Li, J.-C. Zhang, J. Li, H.-Y. Yang, J.-J. Meng, B. Zhang, A 3D sandwich structured hybrid of gold nanoparticles decorated MnO2/graphene-carbon nanotubes as high performance H2O2 sensors. Sens. Actuators B Chem. **260**, 1–11 (2018). https://doi.org/10.1016/j.snb.2017.12.184
115. J.S. Wilson, *Sensor Technology Handbook* (Elsevier, 2004). https://doi.org/10.1016/B978-0-7506-7729-5.X5040-X
116. N. Ida, *Sensors, Actuators and their Interfaces* (2012)
117. J. Vetelino, A. Reghu, *Introduction to Sensors* (CRC press, 2017). https://doi.org/10.1201/9781315218274
118. V. Button, Introduction to biomedical variables transducing. Princ. Meas. Transduct. Biomed. Var. 1–24 (2015). https://doi.org/10.1016/B978-0-12-800774-7.00001-5
119. N.R. Stradiotto, H. Yamanaka, M.V.B. Zanoni, Electrochemical sensors: a powerful tool in analytical chemistry. J. Braz. Chem. Soc. **14**, 159–173 (2003). https://doi.org/10.1590/S0103-50532003000200003
120. E. Bakker, K. Chumbimuni-Torres, Modern directions for potentiometric sensors. J. Braz. Chem. Soc. **19**, 621–629 (2008). https://doi.org/10.1590/S0103-50532008000400003
121. H. Karimi-Maleh, F. Karimi, M. Alizadeh, A.L. Sanati, Electrochemical sensors, a bright future in the fabrication of portable kits in analytical systems. Chem. Rec. **20**, 682–692 (2020). https://doi.org/10.1002/tcr.201900092
122. T. Yin, W. Qin, Applications of nanomaterials in potentiometric sensors, TrAC. Trends Anal. Chem. **51**, 79–86 (2013). https://doi.org/10.1016/J.TRAC.2013.06.009
123. J. Bobacka, A. Ivaska, A. Lewenstam, Potentiometric ion sensors. Chem. Rev. **108**, 329–351 (2008). https://doi.org/10.1021/cr068100w
124. A. Harper, M.R. Anderson, Electrochemical glucose sensors—Developments using electrostatic assembly and carbon nanotubes for biosensor construction. Sensors **10**, 8248–8274 (2010). https://doi.org/10.3390/s100908248
125. J. Janata, *Principles of Chemical Sensors* (Springer Science & Business Media, 2010). https://doi.org/10.1007/b136378
126. A. Aykaç, I.D. Tunç, F. Güneş, M. Erol, M. Şen, Sensitive PH measurement using EGFET PH-microsensor based on ZnO nanowire functionalized carbon-fibers. Nanotechnology **32**, 365501 (2021). https://doi.org/10.1088/1361-6528/ac0666
127. C. Yang, C. Zhang, T. Huang, X. Dong, L. Hua, Ultra-long ZnO/carbon nanofiber as free-standing electrochemical sensor for dopamine in the presence of uric acid. J. Mater. Sci. **54**, 14897–14904 (2019). https://doi.org/10.1007/s10853-019-04000-x
128. M. Şen, E. Azizi, İ Avcı, A. Aykaç, K. Ensarioğlu, İ Ok, G.F. Yavuz, F. Güneş, Screen printed carbon electrodes modified with 3D nanostructured materials for bioanalysis. Electroanalysis (2022). https://doi.org/10.1002/elan.202100624
129. S.V. Selvi, R. Rajakumaran, S.-M. Chen, A.M. Rady, S. Veerasankar, T.-W. Chen, S.-P. Rwei, B.-S. Lou, Graphene/Tungsten trioxide (Gr/WO3) composite modified screen-printed carbon electrode for the sensitive electrochemical detection of nitrofurantoin in biological samples. Int. J. Electrochem. Sci. **14**, 6454–6467 (2019). https://doi.org/10.20964/2019.07.67
130. V.S. Manikandan, B. Adhikari, A. Chen, Nanomaterial based electrochemical sensors for the safety and quality control of food and beverages. Analyst **143**, 4537–4554 (2018). https://doi.org/10.1039/C8AN00497H

131. G. Vinodha, P.D. Shima, L. Cindrella, Mesoporous magnetite nanoparticle-decorated graphene oxide nanosheets for efficient electrochemical detection of hydrazine. J. Mater. Sci. **54**, 4073–4088 (2019). https://doi.org/10.1007/s10853-018-3145-z
132. D.F. Katowah, S. Alqarni, G.I. Mohammed, S.Z. Al Sheheri, M.M. Alam, S.H. Ismail, A.M. Asiri, M.A. Hussein, M.M. Rahman, Selective Hg2+ sensor performance based various carbon-nanofillers into CuO-PMMA nanocomposites. Polym. Adv. Technol. **31**, 1946–1962 (2020). https://doi.org/10.1002/pat.4919

Metal Oxide Nanostructures-Based Electronics

Manoj Dey, Nikhilesh Maity, and Abhishek K. Singh

Abstract Metal oxide nanostructures have garnered significant attention in recent years due to their unique physical, chemical, and electrical properties. One of the inherent advantages of oxide nanostructures is their high surface area-to-volume ratios, which enable improved charge carrier transport and device performance. In this chapter, we provide an overview of metal oxide nanostructures and their potential applications in solar cells, displays, water splitting and cleaning, memory devices, bioelectronics, etc. We cover the synthesis and characterization of these materials, in addition to their advantages over traditional nanostructures. The chapter also discusses the challenges of metal oxide nanostructures in micro- and nano-electronics, along with future research goals. Overall, this chapter provides a comprehensive overview of the current state of the field and the potential of metal oxide nanostructures for advancing electronics technology.

Keywords Metal oxides · Nanostructures · Modern electronics · Electronics devices · Quantum effects · Field-effect transistors · CMOS technology

1 Introduction

According to Moore's law, the density of transistors in an integrated circuit becomes two-fold in two years, leading to a corresponding increase in processing power and a decrease in cost per transistor [1]. However, as transistors continue to shrink, it is becoming increasingly difficult and expensive to manufacture them, and there are

M. Dey · N. Maity · A. K. Singh (✉)
Materials Research Centre, Indian Institute of Science, Bangalore, Karnataka 560012, India
e-mail: abhishek@iisc.ac.in

M. Dey
e-mail: manojdey@iisc.sc.in

N. Maity
e-mail: nikhileshm@iisc.ac.in

V. Kumar et al. (eds.), *Optical Properties of Metal Oxide Nanostructures*,
Progress in Optical Science and Photonics 26,
https://doi.org/10.1007/978-981-99-5640-1_12

concerns that Moore's Law may be reaching its limit. Achieving multi-functionality and diversity in electronic devices is the long-term future goal. Therefore, it is timely to investigate the possibility of emerging new materials and even new methods in building electronic devices. In this aspect, metal oxide (MO) nanostructures possess both novel quantum effects and excellent semiconductor properties that make them promising materials for electronics [2]. Moreover, MO nanostructures are deemed a major element in the development and building of many nanoelectronics devices due to their extraordinary electrical, physical, and chemical properties [3]. MO nanostructures have also gained significant attention due to their potential to solve various global issues, particularly in the biomedical and environmental fields [4]. Recent developments in these areas include the use of MO nanostructures in drug delivery systems, gas sensors, and improved medical diagnostic accuracy [5]. Additionally, MO nanostructures have also been used in personal care products [6], energy storage [7], and paint & coating industries [8]. These nanostructures exhibit excellent optical and electrical properties, are capable of being transparent, can be applied to large areas, and possess a high level of mechanical flexibility. Additionally, the properties of MO nanostructures can be tailored to meet the demands of different applications by adjusting their morphology. With environmental concerns at the forefront of research, it is essential to prioritize low-cost, sustainable materials and methods that can still provide high integration and performance levels. In this regard, non-critical metal oxides like ZnO and TiO_2 have demonstrated their ability to meet the requirements of multi-functionality while being environmentally friendly [9]. MO nanostructures possess a wide range of exotic and innovative quantum properties, including ferroelectricity, piezoelectricity, thermoelectricity, and magnetostriction [2]. These properties make MO nanostructures suitable for various applications, such as data storage, actuators, biosensors, wearable electronics, displays, and smart windows [2]. Despite having many advantages, MO nanostructures also have several limitations that restrict their use in modern nano- and micro-electronics. In recent years, researchers have focused on developing novel methods to synthesize MO nanostructures, and using computational methods to study their microstructures and predict macro-level properties [4, 10]. A deeper understanding of the relationship between microstructure and exotic properties can lead to improved performance and functionalities in modern electronic devices. This can also open up the possibility of designing and fabricating MO nanostructures to meet specific electronic applications.

In this chapter, we will discuss recent advancements in the fabrication, analysis, and potential applications of MO nanostructures from both a structural and application perspective in modern electronic devices. We arrange this chapter as follows. In Sect. 2, we will explore several state-of-the-art synthesis and characterization methods for fabricating oxide nanostructures of different dimensions. The exotic properties of oxide nanostructures which make them potential for modern electronics, will be discussed in Sect. 3. In the subsequent Sect. 4, we will review various classes of oxide nanostructures and their applications in electronic devices. Finally, we will discuss several challenges and future directions of electronics in the concluding section. These discussions will serve as a foundation for developing new MO nanostructures with improved performance for electronic device applications.

Additionally, this chapter will provide a platform for translating scientific discoveries into practical multi-functionalities.

2 Synthesis and Fabrication

Device fabrication from metal oxide nanostructures is greatly dependent on their dimensionality, composition, morphology, and uniformity of nanostructures. Therefore, different top-down and bottom-up approaches (see Fig. 1) have been developed to control the growth mechanism over the years. The top-down process starts with macroscopic structures and incorporates external control of nanostructure production. This method utilizes state-of-the-art milling and nanolithography techniques to insert or eliminate thin layers from bulk materials. In contrast, the bottom-up approach involves reducing materials components to the atomic level and then utilizing physical forces to make self-assembly into larger, stable structures. By organizing atomic or molecular components hierarchically, complex nanostructures are formed. For instance, semiconducting oxide nanostructures synthesis in nanowires morphology using chemical vapor deposition (CVD). These experimental techniques offer control over atomic and electronic structures, particle size, microstructures, morphology, and dimensionality. The state-of-the-art synthesis methods and characterization techniques are discussed below for fabricating different dimensional metal-oxide nanostructures.

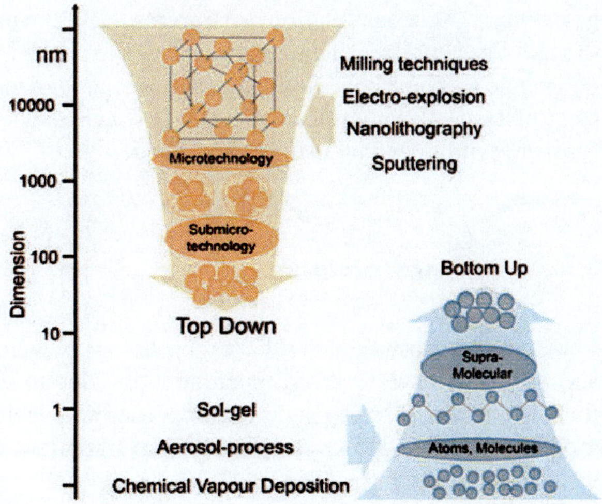

Fig. 1 Top-down and bottom-up approaches for the synthesis of metal oxide nanostructures

2.1 0D Nanostructures

Zero-dimensional (0D) metal-oxide nanostructures include nanoparticles, nanoclusters, and quantum dots with particle sizes within a few nanometers. Device fabrication from 0D nanostructures is advantageous owing to the large surface-to-volume ratio and a low surface area between nanoparticle and surface. Depending upon the preparation technique, nanoparticles can be characterized with different morphologies, such as cube, sphere, octahedron, decahedron, etc. Solvothermal, liquid-phase reduction, and electrochemical deposition techniques are commonly used for 0D nanostructure characterization. The solvothermal method uses solvent under moderate to high temperature and pressure to facilitate the heterogeneous reaction. Water is a widely used solvent, especially for metal-oxide nanostructure characterization, then this method is termed "hydrothermal synthesis". By controlling the growth temperature, concentration, and reaction time, different morphological 0D nanostructures can be characterized by hydrothermal methods. After the arrival of the atoms, nanostructure growth happens in three steps,

(i) surface diffusion, (ii) nucleation, and (iii) layer-by-layer deposition. The hydrothermal method is a simple and cost-effective method for nanostructure characterization. This method is successfully implemented to characterize ZnO quantum-dots and nanocrystals [11], Fe_3O_4 nanocrystals [12]. Metal-oxide nanoparticles can also be directly obtained in liquid-phase reduction by using different soluble precursors in a solvent. Noble metal loaded metal-oxide nanostructures (such as Pt/TiO_2), which are very good catalysts, can be synthesized by liquid-phase reduction techniques [13]. Electrochemical deposition uses an electrode possessing a solid-liquid-gas triphase. Aqueous solution can trap gas pockets while exposing the liquid surface to a solid/gas interface. This electrode helps to create locally high interface pH value independent of bulk conditions, thus, beneficial for depositing metal-oxides in robust electrolytes [14]. Pure phase of Cu_2O [14] and ZnO [15] in the 0D nanostructured form are synthesized by using this method.

2.2 1D Nanostructures

Metal-oxide nanostructure with the morphology of nanowires, nanotubes, nanofibers, and nanobelts have received much attention due to its capability of surface functionalization for tailoring superior properties and device applications. Liquid-phase growth and vapor-phase growth are very important methods for 1D metal-oxide nanostructure synthesis. The liquid-phase deposition technique relies on the chemical equilibrium of the metal-oxide and metal-fluoro complexes. Therefore, uniform 1D nanostructures with large surface areas and complex morphologies are fabricated on different substrates. Moreover, the multi-component 1D nanostructure can also be synthesized as it uses a homogeneous liquid-phase. This process is also known as a template-based synthesis process, as 1D nanostructures can be grown as templates in confined spaces. $CoFe_2O_4$ nanowires [16], Ni nanopillar in TiO_2 surface [17], and VO_2 nanotube arrays [18] are successfully synthesized using the liquid phase technique in the sol-gel template. In general, vapor-phase growth can be classified into physical and chemical deposition processes. Physical vapor deposition (PVD) includes vacuum evaporation and sputtering, in which bulk metal-oxide sublimate into vapor and further condensate into nanostructures. In chemical vapor deposition (CVD), on the other hand, vapor-solid chemical reaction takes place in the vicinity of pre-heated substrate surface. Therefore, the preferable growth of nanostructures can be achieved by controlling the flow rate and precursor partial pressure. CVD is the most widely used technique for the growth of semiconductor nanowires such as Ga_2O_3 [19], Fe_2O_3 [20], SnO_2 [21], In_2O_3 [22], MgO [23], SiO_2 [24].

2.3 2D Nanostructures

Two-dimensional (2D) nanostructures have gained significant attention owing to their intrinsic layered atomic structures and electron confinement. 2D nanostructures play a crucial role in improving the efficiency of surface-enhanced processes, including catalysis and sensing. Pulsed laser deposition (PLD) and magnetic sputtering are the widely used methods for the preparation of 2D metal oxide nanostructures. A high-energy laser pulse is shot in a rotating target material consisting of the desired substrate in preheated or room temperature conditions in the PLD method. The plasma created by the laser ablation process on the substrate then adiabatically expands on the vacuum chamber and further condensates as a thin film. PLD is used to prepare high-quality thin film for device application, especially for gas sensing [25]. Magnetron sputtering processes use high-energy particles to bombard target material, which then eject atoms or molecules. The ejected atoms or molecules are then deposited on the substrate to create superlattices or nanofilms. Magnetron sputtering produces high-purity uniform thin films over a large area due to the high deposition rate and ability to sputter metal, alloy, or compound. These processes are generally termed top-down approaches (as shown in Fig. 1). WO_3 [26], TiO_2 [27], and ZnO [28] nanofilms are examples of metal oxide nanostructures produced by PLD method. Another widely used thin film deposition technique known as atomic layer deposition (ALD) involves sequentially exposing a substrate to a number of gas-phase chemical precursors. These precursors interact with the substrate's surface to create a thin coating of the desired substance. ALD is a thin film deposition technique that is highly reliable and controlled, and it is frequently used to create 2D nanostructures. Atomic layer deposition (ALD) is commonly used to fabricate high-k gate dielectrics (like Al_2O_3, HfO_2, and ZrO_2 [29]). Molecular beam epitaxy (MBE) whereas uses a beam of atoms or molecules to deposit thin films of a material onto a substrate in a vacuum chamber. The beam is directed onto the substrate using magnetic or electric fields, allowing precise control over the film thickness and composition. As a result, MBE is an excellent choice for creating heterostructures, heterointerfaces, and complex oxide films of high quality [30]. Recently, Lithography and self-assembly techniques are also becoming famous, especially for the fabrication of chemical and biomedical oxide nanostructured materials [31].

3 Unique Properties of Metal Oxide Nanostructures

MO nanostructures have a number of unique properties that make them interesting for a wide range of applications, including electronics. Some of the key properties (shown in Fig. 2) of MO nanostructures include:

High surface area: One of the most important properties of oxide nanostructures is their large surface area, which is responsible for many of their distinct characteristics and prospective uses. The high surface area makes oxide nanostructures ideal for catalytic and other surface-based applications, as they provide a large number of active sites for chemical reactions to occur [32]. Furthermore, the high surface area of oxide nanostructures can contribute to their high stability and chemical resistance by allowing for strong chemical bonding on the surface.

Quantum effects: The electronic characteristics of nanostructures vary significantly from their bulk phase when the crystal size is equivalent to the de Broglie wavelength of electrons because of the quantum confinement effect [33]. When the dimension is lowered, the inter-band transition and carrier mobility are impacted, as is the bandgap of the nanostructure. In general, the band gap widens and has an inverse relationship with the dimension of nanostructures in diverse morphologies such as nanowires, nanofibres, quantum dots, etc.

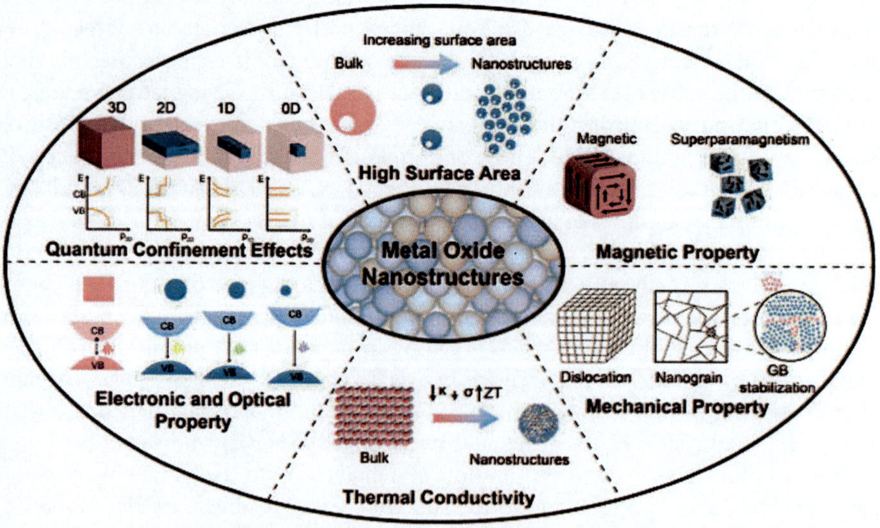

Fig. 2 Schematic representation of several exotic properties of metal-oxide nanostructures

Electrical and Magnetic properties: The electrical and magnetic characteristics of oxide nanostructures highly depend on their size and structural morphologies, and they differ significantly from their bulk counterparts. Oxide nanostructures have distinct electrical characteristics that can be used in a wide range of applications. Some oxide nanostructures, for example, have a high dielectric constant, making them suitable as insulating materials in electronics. Others have low resistance and are suitable for use as conducting materials. The specific electrical properties of an oxide nanostructure depend on its specific chemical composition, size, shape, and surface properties. Similar to this, certain oxides—like iron oxide (magnetite), for example can be used to make magnetic nanostructures since they are inherently magnetic. Other oxides—like alumina (aluminum oxide), can be made magnetic by adding magnetic impurities or by subjecting them to a magnetic field, even if they are not magnetic by nature [34]. This provides a lot of flexibility when designing devices for specific electronic applications.

Thermal conductivity: Due to their distinctive physical and chemical characteristics, oxide nanostructures exhibit exceptional thermal conductivity. At the nanoscale, oxide materials have a greater surface area-to-volume ratio, which can help with phonon scattering and the transmission of heat [35]. Strong covalent bonds are another factor in the heat conduction of many oxide materials. Many different applications, including the creation of electrical devices and heat management systems, frequently employ these materials.

Mechanical properties: Oxide nanostructures have high strength and stiffness, fracture toughness, and low ductility due to their small size, high surface area, and strong covalent bonding [36]. The mechanical properties of oxide nanostructures make them suitable for use in electronic devices that require structural stability, wear resistance, and hardness. However, their low ductility and brittle-ductile transition temperature may limit their use in certain applications.

These characteristics have made oxide nanostructured materials beneficial for a variety of applications, significantly improving the performances of numerous technologies. The following discussion will provide detail about various oxide nanostructures, their characteristics, and their uses in electronic technologies.

4 Applications of Metal Oxide Nanostructures in Electronics

4.1 Graphene Oxide Based Electronics

Graphene, a single-layer or few-layer carbon allotrope, has emerged as a subject of intense research and technological interest due to its extraordinary physical and chemical properties [37]. As a conducting semimetal, graphene has revolutionized electronics, sensing, and energy storage devices. The single or few layer thick graphite oxide and its reduced form i.e., graphene oxide (GO) and reduced graphene

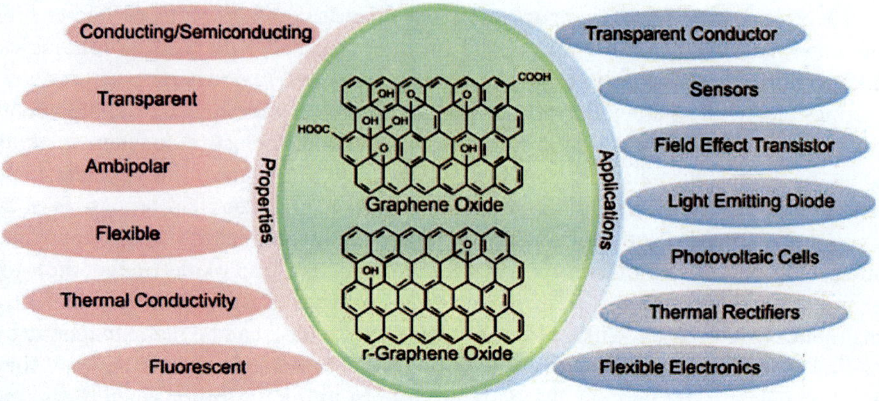

Fig. 3 Properties of graphene oxide and reduced graphene oxide and their applications in modern electronics

oxide (rGO) has also been explored as emerging materials for electronic applications because of their unparalleled advantages over graphene [38, 39]. The GO in pristine form shows low electrical conductivity leading to insulating or semi-conductive behavior depending on the degree of oxidation. However, it can be reduced to form conductive reduced graphene oxide (rGO) using various methods. The exotic properties and applications of graphene oxide and reduced graphene oxide nanostructures are depicted in Fig. 3. Chemical reduction, hydrothermal, plasma, and microwave methods are widely used synthesis techniques for GO and rGO synthesis.

By controlling the degree of oxidation or introducing defects or dopants into the material, GO can also be modified to become a conductor, which makes GO, along with rGO, promising materials for use in transistors [40], sensors [41], and energy storage devices [42]. Additionally, the excellent mechanical properties, including high tensile strength and flexibility [43] of GO and rGO make them ideal materials for use in flexible electronics, such as wearable devices and flexible displays [44]. The chemical stability and biocompatibility of GO are suitable for use in a variety of applications, including drug delivery systems [45] and bioelectronics [46]. The GO can also serve as transparent conductor in electronic devices such as touch screens [47] and solar cells [48]. However, the large-scale production of high-quality graphene derivatives remains a technological challenge. The complexity and expense of synthesizing and purifying graphite, along with controlling the degree of oxidation, composition, and size, are obstacles to the widespread industrialization of graphene oxide-based nanostructure electronics.

4.2 *Ferroelectric Oxide Nanostructures*

Ferroelectric oxide nanostructures, possessing the ability to switch between two thermodynamically stable electrical states of opposing ionic polarization, have been utilized in a wide range of electronic devices such as sensors, actuators, flash drives, and solid-state drives [49]. High dielectric constant of ferroelectric oxide nanostructures enables them to store a large amount of electrical charge. Ferroelectric oxide nanostructures generally include perovskites (such as $BaTiO_3$, $SrTiO_3$, $BiFeO_3$, $PbZr_xTi_{1-x}O_3$) and layered perovskites ($Bi_4La_4Ti_3O_{12}$, $SrBi_2Ta_2O_9$) materials [49]. These materials can be synthesized in the form of thin films, nanoparticles, or other nanostructures through various techniques, including pulsed laser deposition, sol-gel processing, and chemical vapor deposition [51]. Due to their ease of incorporation into nanostructures, ferroelectric materials allow the integration in ultrahigh nanoscale nonvolatile memory devices (NVMDs) such as nonvolatile ferroelectric field effect transistor (FeFET) [52] (see Fig. 4).

In addition to their practical applications, ferroelectric oxide nanostructures are also intriguing to researchers studying fundamental phenomena such as ferroelectricity and ferromagnetism at the nanoscale. The ability to fabricate and study these materials at the nanoscale has opened up new possibilities for understanding and exploiting their unique properties. However, a number of intrinsic difficulties, including fatigue, endurance, and charge retention time, restrict the deployment of ferroelectric nanostructures [53, 54].

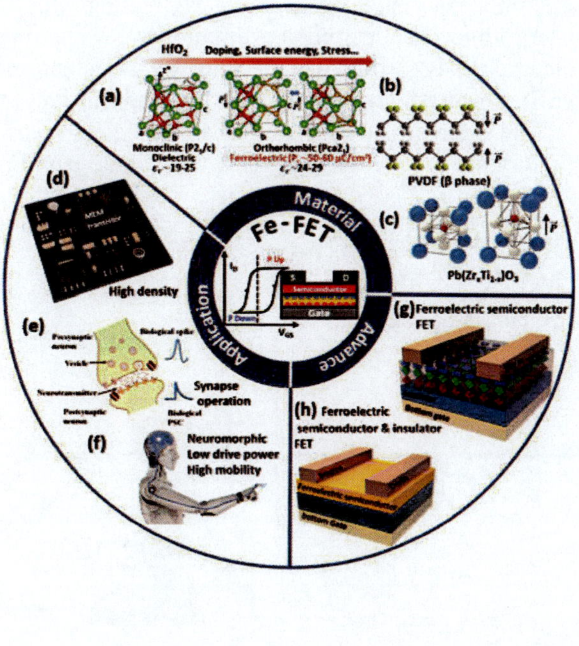

Fig. 4 Applications of Ferroelectric metal oxide nanostructures in FeFET. Reprinted from Ref. [50], with the permission of AIP Publishing

4.3 *Piezotronics*

Piezoelectric semiconductor in low-dimensional nanostructure morphology possesses exceptional mechanical properties and can be incorporated into flexible devices that are capable of withstanding significant strain. The interplay between semiconducting property and piezoelectricity leads to the emergence of unique device characteristics. One of the distinctive features of the piezoelectric oxide nanostructure is its reversibility; materials that show the direct piezoelectric effect when stress is applied also display the reverse piezoelectric effect (generate stress, when an electric field is applied) [55]. Zinc oxide (ZnO), lead zirconate titanate ($Pb(Zr,Ti)O_3$ or PZT), or barium titanate ($BaTiO_3$) have attracted significant attention due to their unique piezoelectric properties at the nanoscale [56, 57].

The extreme sensitivity to mechanical deformation of piezoelectric oxide nanostructures can create a substantial electric charge in response to modest quantities of applied stress, which is one of their key benefits. As a result, they are used in a variety of sensors and actuators, such as pressure sensors [59], acceleration sensors [60], and strain gauges [61] (see Fig. 5). In addition to their high sensitivity, piezoelectric oxide nanostructures also have high piezoelectric coefficient. Therefore, a larger electric charge can be generated in comparison to other piezoelectric materials, making them suitable for use in high-voltage applications [62].

Piezoelectric oxide nanostructures can be used in energy harvesting devices, such as piezoelectric generators [63], which can convert mechanical energy from sources such as vibrations or pressure changes into electrical energy. Piezoelectric oxide in microfiber morphologies offers a fascinating approach for converting mechanical energy into chemical energy through water splitting, a process known as piezoelectric catalysis [64]. This technology holds enormous potential for cutting-edge environmental remediation efforts, making it an area of great interest and excitement.

The strain-sensitive properties of piezoelectric materials have an impact on their performance in electrical devices. Few oxide-based piezoelectric devices have low

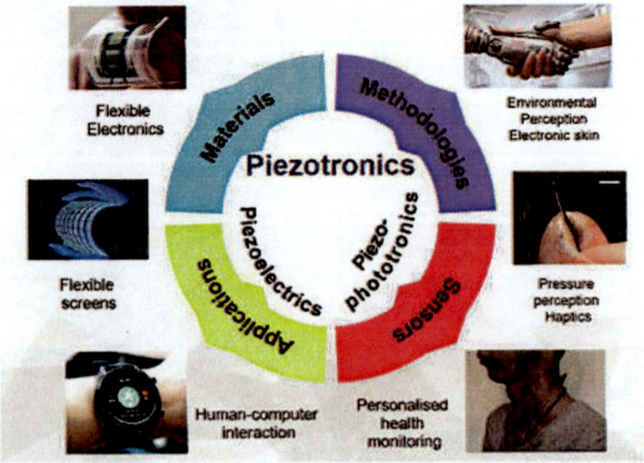

Fig. 5 Metal oxide nanostructures in piezotronics. Reprinted from Ref. [58], Copyright 2021, with permission from Elsevier

output voltage, which affects overall power generation. Some piezoelectric materials are also unstable at high temperatures or humidity, and the synthesis and processing of them can be costly, limiting their usage in specific applications. Therefore, further study is required to fully understand their potential and to create scalable and cost-effective ways for synthesizing and integrating them into practical applications.

4.4 Magnetic Oxide Nanostructures

In recent years, magnetic oxide nanostructures have gained considerable attention owing to their distinctive nanoscale properties (such as superparamagnetism, shape anisotropy, high magnetic moment, etc.) and potential applications in diverse fields like data storage [65] and sensing [66]. The emergence of state-of-the-art synthesis techniques like sputtering, electrodeposition, superconducting quantum interference device (SQUID), and lithography enables the production of oxide nanostructures in diverse shapes and sizes [67].

Traditional data storage technologies, such as hard disk drives, rely on the magnetic properties of materials to store and retrieve data. However, the size of these devices is limited by the minimum size of the magnetic domains that can be used to store data [69]. By using magnetic oxide nanostructures, it is possible to create smaller, and more efficient data storage devices.

Another potential application of magnetic oxide nanostructures is in drug delivery (see Fig. 6). These nanostructures may be tailored to certain cells or tissues in the body by functionalizing them with particular chemicals or medications. The effectiveness of drugs may be increased, and adverse effects may be decreased with this tailored distribution. In drug delivery systems, superparamagnetic iron oxide nanoparticles (SPIONs) are a key component and may be effective MFH agents (magnetic fluid hyperthermia) [70, 71].

Magnetic oxide nanostructures have a wide range of potential applications, however, there are challenges that must be overcome before their full potential can be realized. For instance, creating homogeneous, high-quality nanostructures is difficult. The characteristics of these materials at the nanoscale and how they might be modified for particular applications also need to be better understood.

4.5 Transparent Electronics

Transparent conductive oxide (TCO) nanostructures are ideal conductive materials that exhibit low electromagnetic wave absorption while maintaining high transparency across a broad range of the visible spectrum. These nanostructures are typically composed of materials such as indium tin oxide (ITO) [72], aluminum-doped zinc oxide (AZO) [73], or tin dioxide (SnO_2) [74]. Among them, ITO is widely used in transparent conductive nanostructures owing to its ability to be deposited in a thin

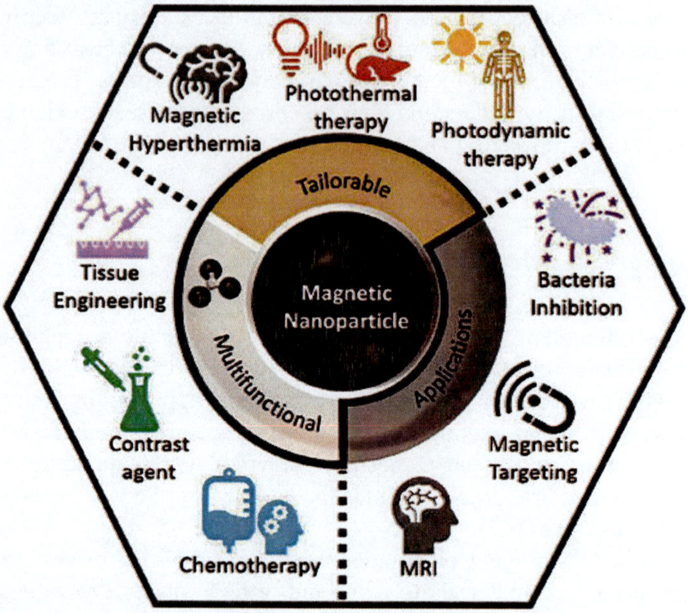

Fig. 6 Magnetic nanoparticles for various biomedical applications. Reprinted with permission from Ref. [68]. Copyright 2021 American Chemical Society

film, besides its electrical conductivity and transparency [72]. While ITO or AZO can be doped n-type, achieving p-type conductivity is very hard in these materials due to heavy hole masses. Thanks to the delafossites (ternary oxide $Cu_xA_yO_z$, which constitute Cu and any other one metal A) and mayenite cage structure type transparent conducting oxides, which typically show p-type semiconducting properties [76]. The ability to transmit a high percentage of visible light makes TCO nanostructures ideal for use in transparent electronic devices [77]. In addition, TCO nanostructures are cost-effective alternative to traditional transparent electrodes owing to their relatively easy fabrication [78]. The high transparency with extensive electrical conductivity makes them useful for various electronic devices, including touch screens [79], LED displays [80], solar cells [81], and smart windows [82]. TCO nanostructures are also being explored in flexible electronics and wearable devices [79] (see Fig. 7). However, TCO nanostructures are not as conductive as metals such as copper or silver, which can limit their use in specific applications. In addition, TCO nanostructures can be brittle and prone to cracking, affecting their long-term durability.

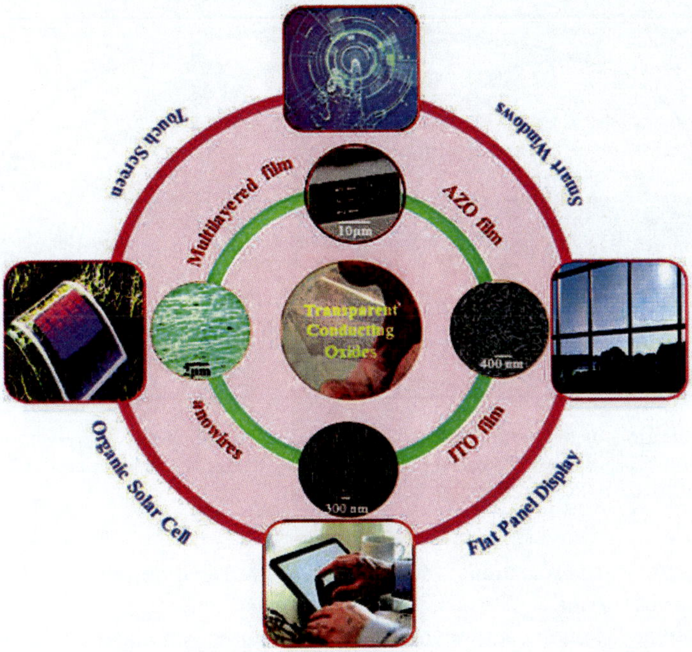

Fig. 7 Schematic representations of transparent conducting oxide nanostructures in modern electronic applications. Reproduced from Ref. [75] with permission from the Royal Society of Chemistry

4.6 *Oxide Nanostructures as Photocatalyst*

Heterogeneous photocatalysis offers the ability to directly transform solar energy into efficient chemical energy, thus providing environment-friendly solutions [83] (see Fig. 8). Due to the ability of charge carrier creation and separation in stimulated phase with adequate energy, oxide nanostructures are potential in electronics. Photocatalytic oxide nanostructures have an advantage in using visible light efficiently for photocatalysis, unlike most metal oxides that can only absorb ultraviolet (UV) light. By creating metal oxides in nanostructure form, their light absorption can be extended to the visible spectrum, enabling them to use more solar energy. In addition to their ability to absorb light, photocatalytic oxide nanostructures also have a large surface area, which allows them to catalyze chemical reactions efficiently. For example, in water purification applications, the large surface area of the nanostructures allows them to effectively remove contaminants from water by catalyzing their breakdown into harmless byproducts. The efficient photocatalyst should have an adequate bandgap with suitable morphology besides a large surface area. Transition metal-based oxide nanostructures like titanium dioxide (TiO_2) [85], zinc oxide (ZnO) [86], or tungsten oxide (WO_3) [87] have been widely studied for their potential

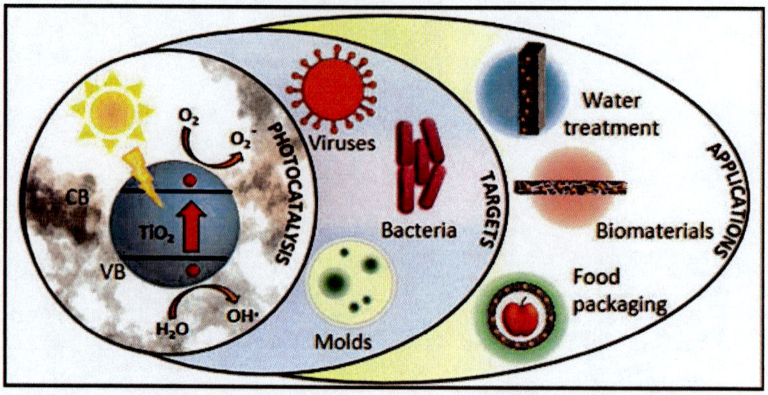

Fig. 8 Applications of photocatalytic TiO_2-based nanostructured materials. Reprinted with permission from Ref. [84], which is licensed under a Creative Commons Attribution 4.0 International License. Licensee MDPI, Basel, Switzerland © 2020

use in a variety of applications, including water purification, air pollution control, and energy conversion.

The stability of photocatalytic oxide nanostructures is a major challenge as they can degrade over time, and their synthesis is difficult as it requires precise control over their size and shape. Moreover, the presence of dopants and defects significantly affects the conductivity and charge carrier mobility in oxide nanostructured photocatalyst and hence the solar-to-hydrogen conversion efficiency. Changing the layer number [88], hetero-structuring [89], varying nanowire diameter [33], and introducing strain [90] in nanostructures are proposed as efficient strategies for engineering defect transition level and hence the device performance. However, oxide nanostructures are susceptible to persistent localized polarons and defects [91], which can have poor charge transport properties [92].

4.7 *Thermoelectric Oxide Nanostructures*

Thermoelectric (TE) materials have the ability to convert heat into electricity, or vice versa, to help overcome global warming and climate change issues with efficient energy utilization and suppression of CO_2 emissions from fossil fuels. Until now, bismuth chalcogenides (Bi_2Te_3, Bi_2Se_3) [94, 95] and lead telluride (PbTe, $PbTe_xSe_{1-x}$) [96] based nanostructured materials are widely used for waste heat recovery and temperature sensing due to their high electrical conductivity (σ) and Seebeck coefficient (S) with low thermal conductivity (κ) resulting in a higher figure of merit ($ZT = S^2\sigma T/\kappa$) for energy conversion. However, these materials are composed of toxic, naturally rare, heavy elements easily oxidizable in air, which restricts the deployment of thermoelectric materials for extensive applications. In contrast, thermoelectric oxide nanostructures are composed of nontoxic, naturally abundant, comparatively light, and cheap elements, thus having the potential for thermoelectric applications [97]. Spinel oxides [98], and oxychalcogenides [99] based oxide nanostructured are proposed as potential thermoelectric materials due to ultralow thermal conductivity and anharmonic phonon scattering.

In addition to their high ZT, thermoelectric oxide nanostructures have good stability and durability, making them suitable for long-term applications. Because of that, these materials are realized in thermoelectric generators [100] and solar thermal energy generation [51] (see Fig. 9). They are also compatible with a wide range of fabrication techniques, which allows them to be easily integrated into various devices.

Despite their potential, the low electrical conductivity of thermoelectric oxide nanostructures can limit their efficiency [101]. While p-type thermoelectric materials have been fabricated mostly Co-based oxides (Na_xCoO_2 [102] and $Ca_3Co_2O_5$ [103]), achieving n-type TE materials is still challenging [104]. Additionally, synthesizing thermoelectric oxide nanostructures is difficult, as it requires precise control over the size and shape of the nanostructures. Several design principles, such as rattling [105], chemical bond hierarchy [99], decoupling atomic contributions [98], and isovalent substitutions have been developed to achieve low thermal conductivity with high ZT. However, a manual search for potential TE with state-of-the-art theoretical and experimental techniques is time-consuming and resource-extensive. In this aspect, a high-throughput search of TE materials combined with machine learning [106, 107]

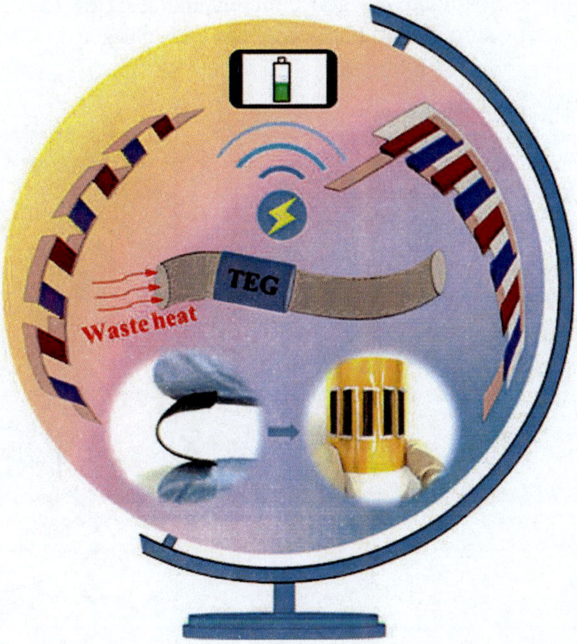

Fig. 9 Schematic depiction of thermoelectric generators (TEGs), which converts waste heat into electrical power. Reprinted with permission from Ref. [93], which is licensed under a Creative Commons Attribution 4.0 International License

has been demonstrated as an efficient approach for accelerated search of viable TE materials, including oxide-based nanostructured materials.

4.8 Oxygen-Deficient Metal Oxide Nanostructures

Metal oxide nanostructures are widely employed in a variety of applications, ranging from energy devices to bio-electronics, due to the benefits over ordinary nanostructures outlined in earlier sections. Although each metal oxide has its own restrictions for particular applications, generally, the low electrical conductivity of metal oxides has been a permanent barrier to their use as electrode materials [108]. As a consequence, oxygen-deficient nanostructures are a class of materials that has attracted substantial interest in recent years due to their ability to modify and fundamentally enhance their electrical characteristics by the controlled insertion of oxygen vacancies into metal oxides.

These nanostructures are highly reactive and have a high surface area, which is often several times larger than that of bulk materials making them suitable for various electronic applications [110] (see Fig. 10). In addition to their high surface area, oxygen-deficient metal oxide nanostructures also exhibit a range of other interesting properties. For example, they can show enhanced electrical conductivity, magnetic properties, and chemical stability compared to their bulk counterparts [111]. These properties make oxygen deficient metal oxide nanostructures attractive for use in energy storage and harvesting devices, gas sensors, catalysts, and fuel cells [2].

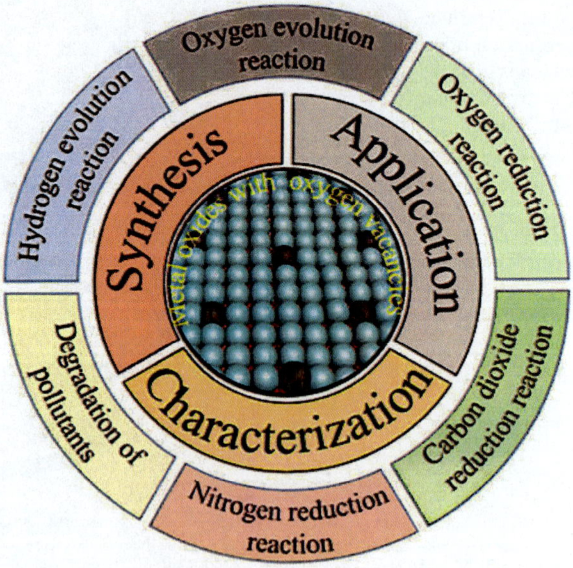

Fig. 10 Applications of oxygen-deficient metal oxide nanostructures semiconductors for energy and environmental catalysis. Reprinted from Ref. [109], Copyright 2021, with permission from Elsevier

One of the most widely studied oxygen-deficient metal oxide nanostructures is cerium dioxide (CeO_2), also known as cerium oxide or ceria. Ceria has a cubic crystal structure and is a good electrical conductor, making it useful in a variety of applications, such as catalysts [112] and gas sensors [113]. It is also a good oxygen storage material and can be used to remove pollutants from exhaust gases [114]. Another important oxygen-deficient metal oxide nanostructure is titanium dioxide (TiO_2), which has a variety of applications in areas such as photocatalysis, solar cells, and coatings [115]. Because of their superior pseudocapacitances, oxygen-deficient metal oxide nanostructures like as manganese oxide (MnO_2), cobalt oxide (Co_3O_4), and nickel oxide (NiO) have also been explored for supercapacitors [116].

Although oxygen-deficient metal oxide nanostructures have promising potential applications, their synthesis and investigation are ongoing areas of research that present many challenges. One of the primary obstacles is the difficulty in controlling the amount of oxygen deficiency in these materials, which can greatly affect their properties. Additionally, there is a need for additional research to comprehensively understand the mechanisms underlying their unique properties and to optimize their manufacturing and processing techniques.

5 Challenges

Metal oxides in nanoarchitecture and hierarchical structure have widespread applications in micro to nanoelectronics, as discussed in the preceding section. Here, we summarize several general challenges and limitations of using metal oxide nanostructures in electronics:

Synthesis and processing: Metal oxide nanostructures with desirable characteristics and in a reproducible manner can be difficult to synthesize and process. The most significant challenge is the cost-effective production of oxide nanostructures. High-quality defect-free nanomaterials are often created using specialized equipment and, under extreme circumstances, restricting their large-scale manufacture. Controlled production of nanomaterials remains a challenging task. More concentrated efforts are needed to create novel synthesis methods that overcome the difficulties associated with traditional methods.

Stability: The chemical composition and structure of the oxide, external variables (such as temperature, humidity, and pressure), and the existence of defects or impurities all influence the stability of oxide nanostructures. Oxide nanostructures with high crystalline perfection are often more stable than those with defects or impurities. This is due to the fact that defects or impurities can serve as nucleation sites for structural alterations or degradation, resulting in the creation of unstable phases or the disintegration of the oxide. External circumstances can potentially affect the stability of oxide nanostructures. High temperatures can cause new phases to develop or the oxide to decompose. In the same way, the presence of moisture or other contaminants can cause instability. Overall, the stability of oxide nanostructures is a complicated and multidimensional topic, and the stability of each single oxide nanostructure will rely on its specific circumstances and features.

Conductivity: Majority of oxide nanostructures have poor conductivity. Due to the high surface-to-volume ratio of nanostructures, surface atoms can behave as scattering centers, obstructing electron pathways and reducing the conductivity of the material. The conductivity of oxide nanostructures is further hindered by impurities and the quantum size effect. Therefore, a future research objective should be to discover strategies to improve the conductivity by lowering the scattering centers.

Compatibility with existing technology: The successful integration of oxide nanostructures in electrical devices is one of the most complex challenges in oxide nanostructure research. It has been shown that nanostructure doping, contact resistance, and surface passivation are plausible approaches to achieving accurate and repeatable positioning of those nanostructures. Despite significant progress in these fields, a comprehensive strategy that addresses all of these challenges at once and builds a solid technical foundation for oxide nanostructure integration is still lacking, at least not in a fashion that is consistent with low-cost and large-area processing.

Toxicity: Some metal oxide nanostructures, like those containing certain transition metals (like Co-based), can be toxic to humans and the environment, which can be a concern when using these materials in electronic devices. Potential risks might arise from oxide nanostructure bio-accumulation. The toxic effects of metal oxide nanostructures can be facilitated by a number of factors, including exposure, size, dissolution, etc. As a result, future research should concentrate on reducing oxide nanostructures' toxicity before incorporating them into electronic systems.

Limited fundamental understanding: The importance of oxide nanostructures in electronic applications is growing at a rapid pace because of various advantages. However, there is still a limited fundamental understanding of the exotic properties and behavior of metal oxide nanostructures, which can make it challenging to design and optimize their use in electronics.

Therefore, by combining advanced experimental techniques, theoretical analysis, and data-driven approaches that span from understanding at the atomic level to device fabrication, we can achieve the long-term goal of creating diverse and multi-functional electronics that are also sustainable.

6 Future of Electronics: CMOS Technology and Beyond

The shrinking of devices owing to advancements in silicon lithography makes the future of electronics exciting yet challenging. While transistors approach closer to the atomic scale and production prices rise, the need for alternate methods has become more essential. Investment in computer architecture and fundamental sciences, such as materials science, is necessary in this circumstance to investigate potential replacement materials and alternative device physics to enable continuous technological advancement.

Metal oxide nanostructures have demonstrated great potential in various electronic applications, as discussed in Sect. 4. In transparent electronics, several metal oxide in nanostructure morphology can serve as a replacement for expensive and fragile indium tin oxide as transparent conductive materials. MO nanostructures have also proven to be useful in energy storage applications, such as lithium-ion batteries and supercapacitors, owing to high specific capacitance and good cycling stability. Due to their great sensitivity, selectivity, and rapid response time, MO nanowires have been employed in sensing applications such as gas sensors, biosensors, and environmental sensors. The interplay of electronic and thermal properties makes them promising for thermoelectric applications, which can be tuned by controlling their size, shape, and doping. MO nanostructures have a bright future in various fields, and their potential for commercial applications is vast. Further research and development in this area will undoubtedly lead to the development of novel materials and devices with improved performance and functionality.

The current complementary metal oxide semiconductor (CMOS) technology has been the backbone of the electronics industry for several decades. By combining p-type and n-type MOS transistors on a single integrated circuit, CMOS technology has effectively resolved the drawbacks associated with the use of separate p-MOS and n-MOS circuits. This innovative approach allows for more versatile circuit designs and significantly reduces circuit noise and complexity, making it an attractive option for a wide range of electronic applications. With CMOS, high-performance circuits can be developed, that are not only efficient but also highly reliable and cost-effective. CMOS has undergone significant advancements over the years to increase the density of transistors on a single chip, which has led to the development of more powerful and efficient processors and other digital devices. However, to overcome the challenges of data movement costs and other limitations, new computing architectures and advanced packaging technologies, such as monolithic three-dimensional integration and photonic co-packaging, are being explored. The development of FinFET technology is also seen as a viable option for continuing to scale CMOS beyond its current limits.

Looking beyond CMOS, there are three axial-pathways that can shape the future of electronics (depicted in Fig. 11). The first axial-pathway involves the exploration of new materials and devices such as carbon nanotubes, graphene, and other two-dimensional materials that have unique electronic properties and have the potential to outperform traditional materials like silicon. The second axial-pathway is focused on developing a new model of computation, such as quantum computing, which has the potential to solve problems beyond the capability of classical computing. Moreover, neuromorphic computing emulates the structure and function of the human brain, leading to more energy-efficient devices that can perform complex tasks with greater accuracy. Further, artificial intelligence-driven machine learning and deep learning along with the Internet of Things (IoT) are developing with great promises for the advancement of future electronics. The third axial pathway focuses on creating more energy-efficient and re-configurable architectures, such as the development of system-on-chip with 3D stacking. In the search for a CMOS replacement, the scalable development of new transistors is a crucial area of research. Promising candidates

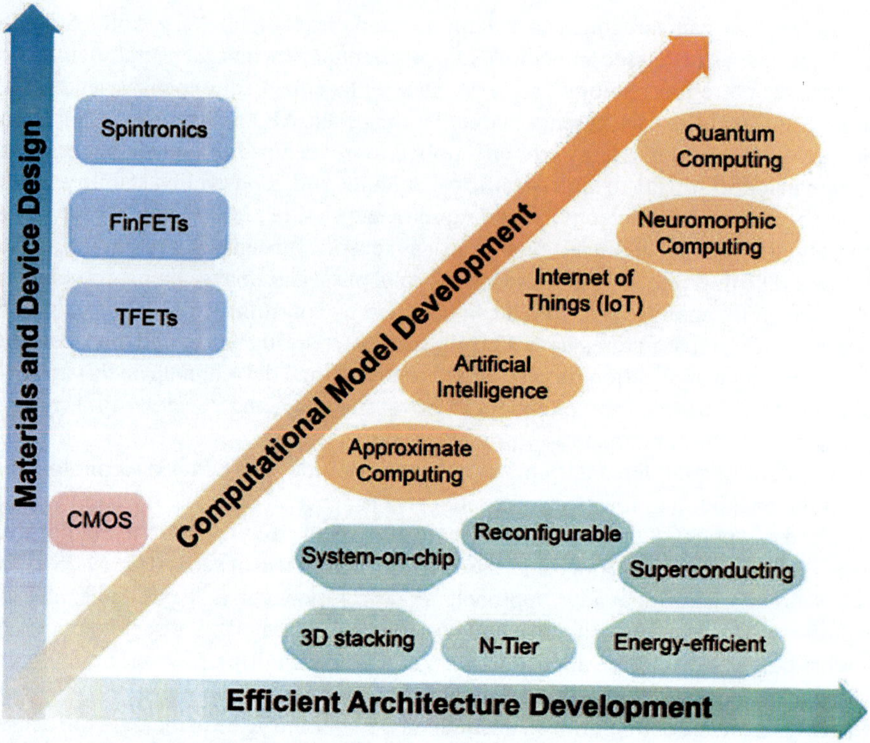

Fig. 11 Three promising avenues for development and progression of future electronic technologies

such as tunnel field-effect transistors, spintronic devices, and nanoelectromechanical systems (NEMS) are being explored to push the limits of electronics beyond what is currently possible.

The future of electronics is bright, with many exciting developments on the horizon. By investing in architecture and the exploration of the three axial pathways, we can push the boundaries of what is possible in electronics and continue to develop smaller, more powerful, and energy-efficient devices that will transform our lives in numerous ways.

7 Conclusions

In summary, this chapter provides an overview of metal-oxide nanostructures, including their synthesis, characterization, and potential applications in modern electronic devices. It highlights the benefits of using MO nanostructures over traditional nanostructures while acknowledging the challenges that still need to be overcome. The chapter also discusses three pathways based on future research goals in the field, such as improving the properties and performance of oxide nanostructures, developing new synthesis and characterization techniques along with efficient architecture development. Overall, the chapter emphasized the importance of metal-oxide nanostructures in progressing the future of electronic applications.

Acknowledgements Authors acknowledge the support from Science and Engineering Research Board (SERB), Dept of Science and Technology, Government of India (File Number: CRG/2021/000633), and DST Nanomission, Department of Science and Technology, Ministry of Science and Technology, Government of India (No. DST/NM/TUE/QM-1/2019). The authors also thank the support from the Institute of Eminence (IoE) scheme of The Ministry of Human Resource Development (MHRD), Government of India.

References

1. Z. Lin Wang, Z.L. Wang, Progress in piezotronics and piezo-phototronics. Adv. Mater. **24**(34), 4632–4646 (2012). https://doi.org/10.1002/adma.201104365
2. M.S. Chavali, M.P. Nikolova, Metal oxide nanoparticles and their applications in nanotechnology. SN Appl. Sci. **1**(6), 1–30 (2019). https://doi.org/10.1007/S42452-019-0592-3
3. A.A. Annu, S. Ahmed, Green synthesis of metal, metal oxide nanoparticles, and their various applications. Handb. Ecomater. **19**, 1–45 (2018). https://doi.org/10.1007/978-3-319-48281-1115-1
4. F. Diao, Y. Wang, Transition metal oxide nanostructures: premeditated fabrication and applications in electronic and photonic devices. J. Mater. Sci. **53**(6), 4334–4359 (2017). https://doi.org/10.1007/s10853-017-1862-3
5. D. Stanicki, T. Vangijzegem, I. Ternad, S. Laurent, An update on the applications and characteristics of magnetic iron oxide nanoparticles for drug delivery. Expert Opin. Drug Deliv. **19**(3), 321–335 (2022). https://doi.org/10.1080/17425247.2022.2047020
6. F. Séby, Metal and metal oxide nanoparticles in cosmetics and skin care products. Compr. Anal. Chem. **93**, 381–427 (2021). https://doi.org/10.1016/bs.coac.2021.02.009
7. J. Jiang, Y. Li, J. Liu, X. Huang, C. Yuan, X. Wen Lou, J. Jiang, J.P. Liu, X.T. Huang, Y.Y. Li, C.Z. Yuan, X.W. Lou, Recent advances in metal oxide-based electrode architecture design for electrochemical energy storage. Adv. Mater. **24**(38), 5166–5180 (2012). https://doi.org/10.1002/adma.201202146
8. A.B. Sengul, E. Asmatulu, Toxicity of metal and metal oxide nanoparticles: a review. Environ. Chem. Lett. **18**(5), 1659–1683 (2020). https://doi.org/10.1007/s10311-020-01033-6
9. A. Lamberti, ZnO- and TiO_2-based nanostructures. Nanomaterials **8**(5), 325 (2018). https://doi.org/10.3390/nano8050325
10. D. Zappa, A. Bertuna, E. Comini, N. Kaur, N. Poli, V. Sberveglieri, G. Sberveglieri, Metal oxide nanostructures: preparation, characterization and functional applications as chemical sensors. Beilstein J. Nanotechnol. **8**, 1205–1217 (2017). https://doi.org/10.3762/bjnano.8.122
11. E.M. Wong, P.G. Hoertz, C.J. Liang, B.M. Shi, G.J. Meyer, P.C. Searson, Influence of organic capping ligands on the growth kinetics of ZnO nanoparticles. Langmuir **17**(26), 8362–8367 (2001). https://doi.org/10.1021/la010944h
12. H. Zeng, J. Li, J.P. Liu, Z.L. Wang, S. Sun, Exchangecoupled nanocomposite magnets by nanoparticle self-assembly. Nature **420**(6914), 395–398 (2002). https://doi.org/10.1038/nature01208
13. H. Eom, I.H. Hwang, D.Y. Lee, S.M. Lee, S.S. Kim, Preparation of liquid-phase reduction method-based Pt/TiO_2 catalyst and reaction characteristics during HCHO room-temperature oxidation. Ind. Eng. Chem. Res. **59**(35), 15489–15496 (2020). https://doi.org/10.1021/acs.iecr.0c02059

14. J. Zhang, X. Sheng, X. Cheng, L. Chen, J. Jin, X. Feng, Robust electrochemical metal oxide deposition using an electrode with a superhydrophobic surface. Nanoscale **9**(1), 87–90 (2016). https://doi.org/10.1039/c6nr07421a
15. A. Zhu, J. Zhang, F. Guan, H. Tang, X. Feng, Green approach for metal oxide deposition at an air-liquid-solid triphase interface with enhanced photocatalytic activity. ACS Omega **4**(2), 3534–3538 (2019). https://doi.org/10.1021/acsomega.8b03234
16. J. Chen, Y. Wang, Y. Deng, Highly ordered $CoFe_2O_4$ nanowires array prepared via a modified sol–gel templated approach and its optical and magnetic properties. J. Alloy. Compd. **552**, 65–69 (2013). https://doi.org/10.1016/j.jallcom.2012.10.073
17. Y. Xu, M. Zhou, L. Wen, C. Wang, H. Zhao, Y. Mi, L. Liang, Q. Fu, M. Wu, Y. Lei, Highly ordered three-dimensional Ni–TiO_2 nanoarrays as sodium ion battery anodes. Chem. Mater. **27**(12), 4274–4280 (2015). https://doi.org/10.1021/acs.chemmater.5b00633
18. C. Huang, L. Chen, G. Xu, L. Miao, Sol-gel template synthesis and characterization of VO_2 nanotube arrays. J. Sol-Gel. Sci. Technol. **63**(1), 103–107 (2012). https://doi.org/10.1007/s10971-012-2769-8
19. T. Terasako, Y. Kawasaki, M. Yagi, Growth and morphology control of β-Ga_2O_3 nanostructures by atmospheric-pressure CVD. Thin Solid Films **620**, 23–29 (2016). https://doi.org/10.1016/j.tsf.2016.07.074
20. D. Peeters, A. Sadlo, K. Lowjaga, O. Mendoza Reyes, L. Wang, L. Mai, M. Gebhard, D. Rogalla, H.W. Becker, I. Giner, G. Grundmeier, D. Mitoraj, M. Grafen, A. Ostendorf, R. Beranek, A. Devi, Nanostructured Fe_2O_3 processing via water-assisted ALD and low-temperature CVD from a Versatile Iron Ketoiminate Precursor. Adv. Mater. Interfaces **4**(18), 1700155 (2017). https://doi.org/10.1002/admi.201700155
21. I. Sayago, M.J. Fernandez, J.L. Fontecha, M.C. Horrillo, J.P. Santos, Synthesis and characterization of SnO_2 nanowires grown by CVD for application as gas sensors, in *2015 10th Spanish Conference on Electron Devices (CDE)* (2015). https://doi.org/10.1109/cde.2015.7087486
22. D.H. Kuo, C.W. Hsu, C.H. Liang, CVD growth of In_2O_3 nanowires using a mixed source of indium and indium chloride. J. Electrochem. Soc. **155**(9), 156 (2008). https://doi.org/10.1149/1.2952822
23. H. Li, M. Li, G. Qiu, C. Li, C. Qu, B. Yang, Synthesis and characterization of MgO nanocrystals for biosensing applications. J. Alloy. Compd. **632**, 639–644 (2015). https://doi.org/10.1016/j.jallcom.2015.01.294
24. H. Lee, J.-H. Park, N. Maity, D. Kim, D. Jang, C. Kim, Y.G. Yoon, A.K. Singh, Y. Han, S.-G. Yoon, Diffusion-enhanced preferential growth of m-oriented GaN micro-domains on directly grown graphene with a large domain size on Ti/SiO_2/Si (001). Mater. Today Commun. **30**, 103113 (2022). https://doi.org/10.1016/j.mtcomm.2021.103113
25. J. Huotari, V. Kekkonen, T. Haapalainen, M. Leidinger, T. Sauerwald, J. Puustinen, J. Liimatainen, J. Lappalainen, Pulsed laser deposition of metal oxide nanostructures for highly sensitive gas sensor applications. Sens. Actuators B Chem. **236**, 978–987 (2016). https://doi.org/10.1016/j.snb.2016.04.060
26. F. Mitsugi, E. Hiraiwa, T. Ikegami, K. Ebihara, R.K. Thareja, WO_3 thin films prepared by pulsed laser deposition. Jpn. J. Appl. Phys. Part 1: Regul. Pap. Short Notes Rev. Pap. **41**(8), 5372–5375 (2002). https://doi.org/10.1143/jjap.41.5372
27. N. Inoue, H. Yuasa, M. Okoshi, TiO_2 thin films prepared by PLD for photocatalytic applications. Appl. Surf. Sci. **197–198**, 393–397 (2002). https://doi.org/10.1016/s0169-4332(02)00347-1
28. F.K. Shan, B.C. Shin, S.W. Jang, Y.S. Yu, Substrate effects of ZnO thin films prepared by PLD technique. J. Eur. Ceram. Soc. **24**(6), 1015–1018 (2004). https://doi.org/10.1016/s0955-2219(03)00397-2
29. T. Gougousi, Atomic layer deposition of high-k dielectrics on III–V semiconductor surfaces. Prog. Cryst. Growth Charact. Mater. **62**(4), 1–21 (2016). https://doi.org/10.1016/j.pcrysgrow.2016.11.001
30. F. Baiutti, F. Wrobel, G. Christiani, G. Logvenov, Oxide molecular beam epitaxy of complex oxide heterointerfaces, in *Metal Oxide-Based Thin Film Structures* (2018), pp. 53–78. https://doi.org/10.1016/b978-0-12-811166-6.00003-0

31. D. Baek, S.H. Lee, B.H. Jun, S.H. Lee, Lithography technology for micro-and nanofabrication. Nanotechnol. Bioappl. **1309**, 217–233. https://doi.org/10.1007/978-981-33-6158-4
32. Y. Du, H. Sheng, D. Astruc, M. Zhu, Atomically precise noble metal nanoclusters as efficient catalysts: a bridge between structure and properties. Chem. Rev. **120**(2), 526–622 (2020). https://doi.org/10.1021/acs.chemrev.8b00726
33. M. Dey, S. Chowdhury, S. Kumar, A. Kumar Singh, Quantum confinement effect on defect level of hydrogen doped rutile VO_2 nanowires. J. Appl. Phys. **131**(23), 235702 (2022). 1063/5.0095834
34. E.C. Nykwest, D. Trujillo, S.P. Alpay, Surface localized magnetism in transition metal doped alumina. Sci. Rep. **11**(1), 1–9 (2021). https://doi.org/10.1038/s41598-021-85791-5
35. M. Ohtaki, Nanostructured oxide thermoelectric materials with enhanced phonon scattering, in *Oxide Thin Films, Multilayers, and Nanocomposites* (2015), pp. 107–122. https://doi.org/10.1007/978-3-319-14478-87
36. Q. Wu, W.S. Miao, Y.D. Zhang, H.J. Gao, D. Hui, Mechanical properties of nanomaterials: a review. Nanotechnol. Rev. **9**(1), 259–273 (2020). https://doi.org/10.1515/ntrev-2020-0021
37. A.K. Geim, K.S. Novoselov, The rise of graphene. Nat. Mater. **6**(3), 183–191 (2007). https://doi.org/10.1038/nmat1849
38. J. Wu, H. Lin, D.J. Moss, K.P. Loh, B. Jia, Graphene oxide for photonics, electronics and optoelectronics. Nat. Rev. Chem. 1–22 (2023). https://doi.org/10.1038/s41570-022-00458-7
39. K. Raidongia, A.T.L. Tan, J. Huang, Graphene oxide: some new insights into an old material. Carbon Nanotub. Graph.: Ed. **2**, 341–374 (2014). https://doi.org/10.1016/B978-0-08-098232-8.00014-0
40. B. Standley, A. Mendez, E. Schmidgall, M. Bockrath, Graphenegraphite oxide field-effect transistors. Nano Lett. **12**(3), 1165–1169 (2012). https://doi.org/10.1021/nl2028415
41. J.T. Robinson, F.K. Perkins, E.S. Snow, Z. Wei, P.E. Sheehan, Reduced graphene oxide molecular sensors. Nano Lett. **8**(10), 3137–3140 (2008). https://doi.org/10.1021/nl8013007
42. D. Dhamodharan, P.P. Ghoderao, V. Dhinakaran, S. Mubarak, N. Divakaran, H.S. Byun, A review on graphene oxide effect in energy storage devices. J. Ind. Eng. Chem. **106**, 20–36 (2022). https://doi.org/10.1016/j.jiec.2021.10.033
43. P. Poulin, R. Jalili, W. Neri, F. Nallet, T. Divoux, A. Colin, S.H. Aboutalebi, G. Wallace, C. Zakri, Superflexibility of graphene oxide. Proc. Natl. Acad. Sci. U. S. Am. **113**(40), 11088–11093 (2016). https://doi.org/10.1073/pnas.1605121113
44. M. Uz, K. Jackson, M.S. Donta, J. Jung, M.T. Lentner, J.A. Hondred, J.C. Claussen, S.K. Mallapragada, Fabrication of high-resolution graphene-based flexible electronics via polymer casting. Sci. Rep. **9**(1), 1–11 (2019). https://doi.org/10.1038/s41598-019-46978-z
45. C.L. Weaver, J.M. Larosa, X. Luo, X.T. Cui, Electrically controlled drug delivery from graphene oxide nanocomposite films. ACS Nano **8**(2), 1834–1843 (2014). https://doi.org/10.1021/nn406223e
46. D. San Roman, R. Garg, T. Cohen-Karni, Bioelectronics with graphene nanostructures. APL Mater. **8**(10), 100906 (2020). https://doi.org/10.1063/5.0020455
47. Y. Zhu, Z. Sun, Z. Yan, Z. Jin, J.M. Tour, Rational design of hybrid graphene films for high-performance transparent electrodes. ACS Nano **5**(8), 6472–6479 (2011). https://doi.org/10.1021/nn201696g
48. K. Parvez, R. Li, K. Müllen, Graphene as transparent electrodes for solar cells, in *Nanocarbons Advanced Energy Conversion* (2015), pp. 249–280. https://doi.org/10.1002/9783527680016.ch10
49. I. Vrejoiu, M. Alexe, D. Hesse, U. Gösele, Ferroelectric nanostructures. J. Vac. Sci. Technol. B: Microelectron. Nanometer Struct. Process. Meas. Phenom. **27**(1), 498 (2009). https://doi.org/10.1116/1.3025907
50. J.Y. Kim, M.J. Choi, H.W. Jang, Ferroelectric field effect transistors: progress and perspective. APL Mater. **9**(2), 021102 (2021). https://doi.org/10.1063/5.0035515
51. L. Liang, X. Kang, Y. Sang, H. Liu, L. Liang, X. Kang, Y. Sang, H. Liu, One-dimensional ferroelectric nanostructures: synthesis, properties, and applications. Adv. Sci. **3**(7), 1500358 (2016). https://doi.org/10.1002advs.201500358

52. X. Liu, Y. Liu, W. Chen, J. Li, L. Liao, Ferroelectric memory based on nanostructures. Nanosc. Res. Lett. **7**(1), 1–16 (2012). https://doi.org/10.1186/1556-276x-7-285
53. R. Gysel, I. Stolichnov, A.K. Tagantsev, S.W.E. Riester, N. Setter, G.A. Salvatore, D. Bouvet, A.M. Ionescu, Retention in nonvolatile silicon transistors with an organic ferroelectric gate. Appl. Phys. Lett. **94**(26), 263507 (2009). https://doi.org/10.1063/1.3158959
54. Y. Kato, Y. Kaneko, H. Tanaka, K. Kaibara, S. Koyama, K. Isogai, T. Yamada, Y. Shimada, Overview and future challenge of ferroelectric random access memory technologies. Jpn. J. Appl. Phys. **46**(4 B), 2157–2163 (2007). https://doi.org/10.1143/jjap.46. 2157
55. S. Trolier-Mckinstry, P. Muralt, Thin film piezoelectrics for MEMS. J. Electroceram. **12**(1–2), 7–17 (2004). https://doi.org/10.1023/B:JECR.0000033998.72845.51
56. J. Briscoe, S. Dunn, Piezoelectric nanogenerators—A review of nanostructured piezoelectric energy harvesters. Nano Energy **14**, 15–29 (2015). https://doi.org/10.1016/j.nanoen.2014.11.059
57. R.K. Pandey, J. Dutta, S. Brahma, B. Rao, C.P. Liu, Review on ZnO-based piezotronics and piezoelectric nanogenerators: aspects of piezopotential and screening effect. J. Phys.: Mater. **4**(4), 044011 (2021). https://doi.org/10.1088/2515-7639/ac130a
58. Y. Wu, Y. Ma, H. Zheng, S. Ramakrishna, Piezoelectric materials for flexible and wearable electronics: a review. Mater. Des. **211**, 110164 (2021). https://doi.org/10.1016/j.matdes.2021.110164
59. J.B. Park, M.S. Song, R. Ghosh, R.K. Saroj, Y. Hwang, Y. Tchoe, H. Oh, H. Baek, Y. Lim, B. Kim, S.W. Kim, G.C. Yi, Highly sensitive and flexible pressure sensors using position- and dimension-controlled ZnO nanotube arrays grown on graphene films. NPG Asia Mater. **13**(1), 1–9 (2021). https://doi.org/10.1038/s41427-021-00324-w
60. C. Ai, X. Zhao, D. Wen, Characteristics research of a high sensitivity piezoelectric MOSFET acceleration sensor. Sensors **20**(17), 4988 (2020). https://doi.org/10.3390/s20174988
61. M. Panth, B. Cook, Y. Zhang, D. Ewing, A. Tramble, A. Wilson, J. Wu, High-performance strain sensors based on vertically aligned piezoelectric zinc oxide nanowire array/graphene nanohybrids. ACS Appl. Nano Mater. **3**(7), 6711–6718 (2020). https://doi.org/10.1021/acsanm.0c01150
62. M.Y. Soomro, I. Hussain, N. Bano, O. Nur, M. Willander, Piezoelectric power generation from zinc oxide nanowires grown on paper substrate. Phys. Status Solidi (RRL)—Rapid Res. Lett. **6**(2), 80–82 (2012). https://doi.org/10.1002/pssr.201105519
63. Z.L. Wang, J. Song, Piezoelectric nanogenerators based on zinc oxide nanowire arrays. Science **312**(5771), 242–246 (2006). https://doi.org/10.1126/science.1124005
64. G. Nie, Y. Yao, X. Duan, L. Xiao, S. Wang, Advances of piezoelectric nanomaterials for applications in advanced oxidation technologies. Curr. Opin. Chem. Eng. **33**, 100693 (2021). https://doi.org/10.1016/j.coche.2021.100693
65. S.M. Yakout, Spintronics: future technology for new data storage and communication devices. J. Supercond. Novel Magn. **33**(9), 2557–2580 (2020). https://doi.org/10.1007/s10948-020-05545-8
66. Y. Slimani, E. Hannachi, Magnetic nanosensors and their potential applications, in *Nanosensors for Smart Cities* (2020), pp. 143–155. https://doi.org/10.1016/b978-0-12-819870-4.00009-8
67. S. Shukla, P.K. Deheri, R.V. Ramanujan, Magnetic nanostructures: synthesis, properties, and applications, in *Springer Handbook of Nanomaterials* (2013), pp. 473–514. https://doi.org/10.1007/978-3-642-20595-812
68. P.M. Martins, A.C. Lima, S. Ribeiro, S. Lanceros-Mendez, P. Martins, Magnetic nanoparticles for biomedical applications: from the soul of the earth to the deep history of ourselves. ACS Appl. Bio Mater. **4**(8), 5839–5870 (2021). https://doi.org/10.1021/acsabm.1c00440
69. A. Gaul, D. Emmrich, T. Ueltzhöffer, H. Huckfeldt, H. Doganay, J. Hackl, M.I. Khan, D.M. Gottlob, G. Hartmann, A. Beyer, D. Holzinger, S. Nemšák, C.M. Schneider, A. Gölzhäuser, G. Reiss, A. Ehresmann, Size limits of magnetic-domain engineering in continuous in-plane exchange-bias prototype films. Beilstein J. Nanotechnol. **9**(1), 2968 (2018). https://doi.org/10.3762/bjnano.9.276

70. T. Vangijzegem, D. Stanicki, S. Laurent, Magnetic iron oxide nanoparticles for drug delivery: applications and characteristics. Expert Opin. Drug Deliv. **16**(1), 69–78 (2018). https://doi.org/10.1080/17425247.2019.1554647
71. C. Comanescu, Magnetic nanoparticles: current advances in nanomedicine. Drug Deliv. MRI Chem. **4**(3), 872–930 (2022). https://doi.org/10.3390/chemistry4030063
72. G. Lavareda, C. Nunes de Carvalho, E. Fortunato, A.R. Ramos, E. Alves, O. Conde, A. Amaral, Transparent thin film transistors based on indium oxide semiconductor. J. Non-Crystal. Solids **352**(2325), 2311–2314 (2006). https://doi.org/10.1016/j.jnoncrysol.2006.03.031
73. M. Wu, D. Sun, C. Tan, X. Tian, Y. Huang, Al-doped ZnO monolayer as a promising transparent electrode material: a firstprinciples study. Materials **10**(4), 359 (2017). https://doi.org/10.3390/ma10040359
74. A.K. Singh, A. Janotti, M. Scheffler, C.G. Van De Walle, Sources of electrical conductivity in SnO_2. Phys. Rev. Lett. **101**(5), 055502 (2008). https://doi.org/10.1103/PhysRevLett.101.055502
75. G.K. Dalapati, H. Sharma, A. Guchhait, N. Chakrabarty, P. Bamola, Q. Liu, G. Saianand, A.M. Sai Krishna, S. Mukhopadhyay, A. Dey, T.K.S. Wong, S. Zhuk, S. Ghosh, S. Chakrabortty, C. Mahata, S. Biring, A. Kumar, C.S., Ribeiro, S. Ramakrishna, A.K. Chakraborty, S. Krishnamurthy, P. Sonar, M. Sharma, Tin oxide for optoelectronic, photovoltaic and energy storage devices: a review. J. Mater. Chem. A **9**(31), 16621–16684 (2021). https://doi.org/10.1039/d1ta01291f
76. A. Stadler, Transparent conducting oxides—An up-to-date overview. Materials **5**(4), 661–683 (2012). https://doi.org/10.3390/ma5040661
77. K.L. Chopra, S. Major, D.K. Pandya, Transparent conductors—A status review. Thin Solid Films **102**(1), 1–46 (1983). https://doi.org/10.1016/0040-6090(83)90256-0
78. S.Q. Li, R.P.H. Chang, L.E. Ocola, TCO Nanostructures as building blocks for nanophotonic devices in the infrared. Proc. SPIE 8809, in *Plasmonics: metallic Nanostructures and Their Optical Properties XI*, vol. 8809 (2013), pp. 186–193 https://doi.org/10.1117/12.2025289
79. C.C. Wu, Highly flexible touch screen panel fabricated with silverinserted transparent ITO triple-layer structures. RSC Adv. **8**(22), 11862–11870 (2018). https://doi.org/10.1039/c7ra13550e
80. H. Liu, V. Avrutin, N. Izyumskaya, U. Ozgr, H. Morkoç, Transparent conducting oxides for electrode applications in light emitting and absorbing devices. Superlattices Microstruct. **48**(5), 458–484 (2010). https://doi.org/10.1016/j.spmi.2010.08.011
81. S.J. Ikhmayies, Transparent conducting oxides for solar cell applications, in *Mediterranean Green Buildings and Renewable Energy* (2017), pp. 899–907. https://doi.org/10.1007/978-3-319-30746-670
82. C.G. Granqvist, Electrochromics for smart windows: oxide-based thin films and devices. Thin Solid Films **564**, 1–38 (2014). https://doi.org/10.1016/j.tsf.2014.02.002
83. A.O. Ibhadon, P. Fitzpatrick, Heterogeneous photocatalysis: recent advances and applications. Catalysts **3**(1), 189–218 (2013). https://doi.org/10.3390/catal3010189
84. I.D. Pasquale, C.L. Porto, M. Dell'Edera, F. Petronella, A. Agostiano, M.L. Curri, R. Comparelli, Photocatalytic TiO_2-based nanostructured materials for microbial inactivation. Catalysts **10**(12), 1382 (2020). https://doi.org/10.3390/catal10121382
85. S.Y. Lee, S.J. Park, TiO_2 photocatalyst for water treatment applications. J. Ind. Eng. Chem. **19**(6), 1761–1769 (2013). https://doi.org/10.1016/j.jiec.2013.07.012
86. T. Shiyani, I. Banerjee, S.K. Mahapatra, A.K. Ray, Flexible zinc oxide photoelectrode for photo electrochemical energy conversion. J. Mater. Sci.: Mater. Electron. **32**(11), 15386–15392 (2021). https://doi.org/10.1007/s10854-021-06088-x
87. Y. Miseki, H. Kusama, H. Sugihara, K. Sayama, Cs-modified WO_3 photocatalyst showing efficient solar energy conversion for O_2 production and Fe (III) ion reduction under visible light. J. Phys. Chem. Lett. **1**(8), 1196–1200 (2010). https://doi.org/10.1021/jz100233w
88. A. Singh, M. Dey, A.K. Singh, Origin of layer-dependent electrical conductivity of transition metal dichalcogenides. Phys. Rev. B **105**(16), 165430 (2022). https://doi.org/10.1103/PhysRevB.105.165430

89. A. Singh, A. Manjanath, A.K. Singh, Engineering defect TransitionLevels through the van der Waals Heterostructure. J. Phys. Chem. C **122**(42), 24475–24480 (2018). https://doi.org/10.1021/acs.jpcc.8b08082
90. X. Yan, P. Li, S.-H. Wei, B. Huang, Universal theory and basic rules of strain-dependent doping behaviors in semiconductors. Chin. Phys. Lett. **38**(8), 087103 (2021). https://doi.org/10.1088/0256-307x/38/8/087103
91. M. Dey, A. K. Singh, Broad photoluminescence from large Frank-Condon relaxation dynamics of hole polarons in $LiGaO_2$. Phys. Rev. B **108**(4), L041201 (2023). https://doi.org/10.1103/PhysRevB.108.L041201
92. M. Dey, A. Singh, A.K. Singh, Formation of a small electron polaron in tantalum oxynitride: origin of low mobility. J. Phys. Chem. C **125**(21), 11548–11554 (2021). https://doi.org/10.1021/acs.jpcc.1c00702
93. Y. Du, J. Xu, B. Paul, P. Eklund, Flexible thermoelectric materials and devices. Appl. Mater. Today **12**, 366–388 (2018). https://doi.org/10.1016/j.apmt.2018.07.004
94. R. Juneja, T. Pandey, A.K. Singh, High thermoelectric performance in n-doped silicon-based chalcogenide Si_2Te_3. Chem. Mater. **29**(8), 3723–3730 (2017). https://doi.org/10.1021/acs.chemmater.7b00728
95. R. Pathak, P. Dutta, A. Srivastava, D. Rawat, R.K. Gopal, A.K. Singh, A. Soni, K. Biswas, Strong anharmonicity-induced low thermal conductivity and high n-type mobility in the topological insulator $Bi_{1.1}Sb_{0.9}Te_2S$. Angew. Chem. **134**(41), 202210783 (2022). https://doi.org/10.1002/ange.202210783
96. J.P. Heremans, V. Jovovic, E.S. Toberer, A. Saramat, K. Kurosaki, A. Charoenphakdee, S. Yamanaka, G.J. Snyder, Enhancement of thermoelectric efficiency in PbTe by distortion of the electronic density of states. Science **321**(5888), 554–557 (2008). https://doi.org/10.1126/science.1159725
97. K. Koumoto, Y. Wang, R. Zhang, A. Kosuga, R. Funahashi, Oxide thermoelectric materials: a nanostructuring approach. Annu. Rev. Mater. Res. **40**, 363–394 (2010). https://doi.org/10.1146/annurev-matsci-070909-104521
98. A. Srivastava, M. Mukherjee, A.K. Singh, Decoupled atomic contribution boosted high thermoelectric performance in mixed cation spinel oxides ACo_2O_4. Appl. Phys. Lett. **120**(24), 243901 (2022). https://doi.org/10.1063/5.0099452
99. M. Mukherjee, A.K. Singh, Strong chemical bond hierarchy leading to exceptionally high thermoelectric figure of merit in oxychalcogenide AgBiTeO. ACS Appl. Mater. Interfaces. **12**(7), 8280–8287 (2020). https://doi.org/10.1021/acsami.9b21358
100. S. Dalola, G. Faglia, E. Comini, M. Ferroni, C. Soldano, D. Zappa, V. Ferrari, G. Sberveglieri, Planar thermoelectric generator based on metal-oxide nanowires for powering autonomous microsystems. Procedia Eng. **47**, 346–349 (2012). https://doi.org/10.1016/j.proeng.2012.09.154
101. M. Mukherjee, A. Srivastava, A.K. Singh, Recent advances in designing thermoelectric materials. J. Mater. Chem. C **10**(35), 12524–12555 (2022). https://doi.org/10.1039/d2tc02448a
102. I. Terasaki, Y. Sasago, K. Uchinokura, Large thermoelectric power in $NaCo_2O_4$ single crystals. Phys. Rev. B **56**(20), 12685 (1997). https://doi.org/10.1103/PhysRevB.56.R12685
103. R. Funahashi, I. Matsubara, H. Ikuta, T. Takeuchi, U. Mizutani, S. Sodeoka, Oxide single crystal with high thermoelectric performance in air. Jpn. J. Appl. Phys. **39**(11 B), 1127 (2000). https://doi.org/10.1143/jjap.39.L1127
104. K. Koumoto, I. Terasaki, R. Funahashi, Complex oxide materials for potential thermoelectric applications. MRS Bull. **31**(3), 206–210 (2006). https://doi.org/10.1557/mrs2006.46
105. R. Juneja, A.K. Singh, Rattling-induced ultralow thermal conductivity leading to exceptional thermoelectric performance in $AgIn_5S_8$. ACS Appl. Mater. Interfaces **11**(37), 33894–33900 (2019). https://doi.org/10.1021/acsami.9b10006
106. R. Juneja, G. Yumnam, S. Satsangi, A.K. Singh, Coupling the high-throughput property map to machine learning for predicting lattice thermal conductivity. Chem. Mater. **31**(14), 5145–5151 (2019). https://doi.org/10.1021/acs.chemmater.9b01046

107. M. Mukherjee, S. Satsangi, A.K. Singh, A statistical approach for the rapid prediction of electron relaxation time using elemental representatives. Chem. Mater. **32**(15), 6507–6514 (2020). https://doi.org/10.1021/acs.chemmater.0c01778
108. C.N.R. Rao, G.V. Subba Rao, Electrical conduction in metal oxides. Phys. Status Solidi (a) **1**(4), 597–652 (1970). https://doi.org/10.1002/pssa.19700010402
109. H. Liu, H. Fu, Y. Liu, X. Chen, K. Yu, L. Wang, Synthesis, characterization and utilization of oxygen vacancy contained metal oxide semiconductors for energy and environmental catalysis. Chemosphere **272**, 129534 (2021). https://doi.org/10.1016/j.chemosphere.2021.129534
110. J.K.H. Orbeck, R.J. Hamers, Surface properties and interactions of transition metal oxide nanoparticles: a perspective on sustainability. J. Vac. Sci. Technol. A: Vac. Surf. Films **38**(3), 031001 (2020). https://doi.org/10.1116/1.5141853
111. O.V. Kononenko, A.N. Red'kin, G.N. Panin, G.N. Baranov, A.A. Firsov, V.I. Levashov, V.N. Matveev, E.E. Vdovin, E.E.: Study of optical, electrical and magnetic properties of composite nanomaterials on the basis of broadband oxide semiconductors. Nanotechnol. Russia **4**(11–12), 822–827 (2009). https://doi.org/10.1134/s1995078009110093
112. T. Montini, M. Melchionna, M. Monai, P. Fornasiero, Fundamentals and catalytic applications of CeO_2-based materials. Chem. Rev. **116**(10), 5987–6041 (2016). https://doi.org/10.1021/acs.chemrev.5b00603
113. J. Hu, Y. Sun, Y. Xue, M. Zhang, P. Li, K. Lian, S. Zhuiykov, W. Zhang, Y. Chen, Highly sensitive and ultra-fast gas sensor based on CeO_2-loaded In_2O_3 hollow spheres for ppb-level hydrogen detection. Sens. Actuators B: Chem. **257**, 124–135 (2018). https://doi.org/10.1016/j.snb.2017.10.139
114. P. Li, X. Chen, Y. Li, J.W. Schwank, A review on oxygen storage capacity of CeO_2-based materials: Influence factors, measurement techniques, and applications in reactions related to catalytic automotive emissions control. Catal. Today **327**, 90–115 (2019). https://doi.org/10.1016/j.cattod.2018.05.059
115. I. Ali, M. Suhail, Z.A. Alothman, A. Alwarthan, Recent advances in syntheses, properties and applications of TiO_2 nanostructures. RSC Adv. **8**(53), 30125–30147 (2018). https://doi.org/10.1039/c8ra06517a
116. M. Kandasamy, S. Sahoo, S.K. Nayak, B. Chakraborty, C.S. Rout, Recent advances in engineered metal oxide nanostructures for supercapacitor applications: experimental and theoretical aspects. J. Mater. Chem. A **9**(33), 17643–17700 (2021). https://doi.org/10.1039/d1ta03857e

Metal Oxide Based Thermoelectric Materials

Olena Okhay and Alexander Tkach

Abstract The energy demands of the mankind are exponentially rising and the worldwide research is focused on the clean and sustainable energy sources during the last years. As one of the possible ways to enable the transformation from a fossil fuel based to a low-carbon socio-economical epoch is the harvesting of unused heat in automotive exhaustion, industrial processes and home heating, etc. Thermoelectric (TE) generators can convert the heat to electrical energy thanks to Seebeck effect when electricity appears between cold and hot ends of usually semiconductor materials. To be interesting for potential commercial use in TE generators the materials need to be of high figure of merit (*ZT*), possessing high electrical conductivity and Seebeck coefficient together with low thermal conductivity. In addition to high-*ZT* chalcogenide- and skutterudite-type heavy-metal-based alloys, including rare-earth and expensive elements, metal oxide TE materials have advantages of low cost, environment friendly manufacturing and chemical stability at high temperatures. However, since they exhibit rather low electrical and high thermal conductivity compared to traditional TE materials, numerous studies have been done to optimise the TE oxide response by doping, nanostructuring, etc. Here, reports on most studied *p*- and *n*-type semiconductor metal oxides as well as the newest research on the enhancement of their TE performance by different methods are reviewed.

Keywords Thermoelectric materials · Metal oxides · Figure of merit · Seebeck coefficient · Thermal conductivity · Electrical conductivity · Nanostructuring · Doping · Inclusion · Graphene

O. Okhay (✉)
TEMA-Center for Mechanical Technology and Automation, Department of Mechanical Engineering, University of Aveiro, 3810-193 Aveiro, Portugal
e-mail: olena@ua.pt

LASI - Intelligent Systems Associate Laboratory, Guimaraes, Portugal

A. Tkach
Department of Materials and Ceramic Engineering, CICECO-Aveiro Institute of Materials, University of Aveiro, 3810-193 Aveiro, Portugal
e-mail: atkach@ua.pt

V. Kumar et al. (eds.), *Optical Properties of Metal Oxide Nanostructures*,
Progress in Optical Science and Photonics 26,
https://doi.org/10.1007/978-981-99-5640-1_13

1 Introduction

In view of the current climate change problem, thermoelectric (TE) energy conversion is considered as particularly attractive for harvesting electricity from solar heat or for recovering waste heat [1–4]. The use of waste heat as source of electrical energy in vehicles is also an attractive option for automotive manufacturers, who actively pursue this technology to implement thermoelectric generators (TEGs) in car exhaust systems [5]. Moreover, it can significantly reduce the CO_2 content in industrial production processes. Furthermore, silent, reliable, and versatile TEGs without moving parts [6] can be used to power portable electronic devices as additional sources of energy. Thus, TEGs hold a tantalizing promise of greater energy efficiency by providing a robust and clean option for waste heat recovery and its conversion into useful electrical energy through the Seebeck effect [7–10]. In addition, their size can be easily varied, since TEGs consist of TE modules (called also as thermocouples or pairs), which include *n*- and *p*-type semiconductors connected electrically in series and thermally in parallel [11–13] (see Fig. 1).

In *n*-type or donor-doped semiconductors, the majority of charge carriers are the negatively charged electrons. In *p*-type or acceptor-doped semiconductors, the majority of charge carriers are the positively charged holes, which can be considered as a lack of electrons at their position in atomic lattice. When *p*-type and *n*-type materials are connected in a manner shown in Fig. 1a, b temperature gradient (the difference in temperature levels) is created on the two sides of the generator, the main charge carriers present at the hot end would be at a high energy level as compared to those present at the cool end side. This means that the hot charge carriers will tend to move towards the cool end under the temperature gradient and create a potential difference between the two ends. The creation of potential difference due to temperature gradient defined as Seebeck effect is thus used in TEGs.

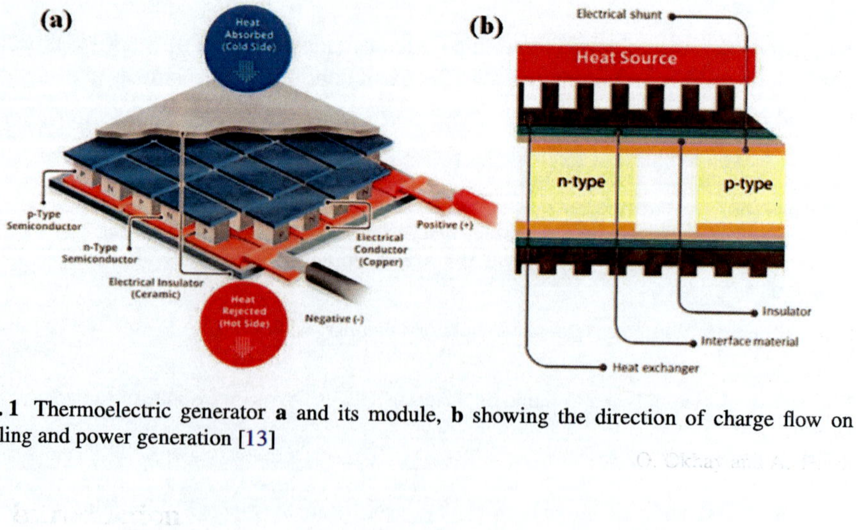

Fig. 1 Thermoelectric generator **a** and its module, **b** showing the direction of charge flow on cooling and power generation [13]

In general, the efficiency of a TE device for generation of electricity is given by η [14]:

$$\eta = \frac{\text{energy provided to the load}}{\text{heat energy absorbed at hot junction}} \tag{1}$$

The maximum efficiency of TEG is:

$$\eta_{\max} = \frac{T_h - T_c}{T_h} \frac{\sqrt{1 + ZT} - 1}{\sqrt{1 + ZT} + \frac{T_c}{T_h}} \tag{2}$$

where T_h and T_c are source temperature and sink temperature, respectively. ZT is called a dimensionless figure of merit of TE material at the absolute temperature T:

$$ZT = \frac{\sigma \times S^2 \times T}{k} = \frac{PF \times T}{k} \tag{3}$$

where S is Seebeck coefficient (also called thermopower), σ is electrical conductivity, and k is thermal conductivity. The ability of a material to produce useful electrical power under a given temperature difference is quantified by its power factor, *PF* ($PF = \sigma \times S^2$).

Accordingly, efficient TE materials should possess a great Seebeck coefficient, S, and a high electrical conductivity, σ, but a low thermal conductivity, κ. Then, a high voltage related to the Seebeck coefficient can be generated by TE material placed in a temperature gradient. The electrical conductivity should be high to help the charge transport to the material ends, reducing the internal resistivity, $\rho = 1/\sigma$, of the material. Thermal conductivity of material should be low to create a large temperature difference between the ends of the TE material. For semiconductor materials, the total κ consists of lattice component κ_L and electronic component κ_e, namely $\kappa = \kappa_L + \kappa_e$ [15]. In majority of the cases, the lattice related mechanism is the dominant one. However, σ is directly proportional to the carrier concentration, n, but S, being proportional to $n^{-2/3}$, according to the Pisarenko relation for doped semiconductors [16], depends on the carrier concentration in the opposite way [17]. Moreover, there is a direct relationship between σ and the electronic contribution to thermal conductivity $\kappa_e = L\sigma T$, according to the Wiedemann-Franz law, where L is the Lorenz number [15]. Thus, an enhancement of the conversion efficiency is a major challenge for the implementation of TE energy conversion. Moreover, TE materials should also be stable under operating conditions, having as well long life cycles, great cost efficiency, and compatibility with existing technologies. Currently, chalcogenides [18], skutterudites [19] and metal oxides [20] are the most widely studied TE materials [7].

At the same time, different TE materials have high *ZT* at different operating temperatures, since each *ZT* component has its own relationship with temperature. Thus, currently TE materials are divided into three groups, which are: low temperature (<500 K), e.g. Bi_2Te_3; middle temperature (500–800 K), e.g. PbTe; and high temperature (>800 K), e.g. SiGe [21]. However, it can be noticed that majority of these materials are typically scarce and expensive. More importantly, some of them contain toxic heavy metals and are readily to cause environmental pollution. Thus, a search for TE materials, which are environmentally friendly and have high efficiency, becomes a hot topic in recent investigations.

Nowadays, both *p*- and *n*-type thermoelectrics can show a high value of *ZT* that strongly depends on the material properties, such as stoichiometry/doping level, charge carrier concentration, and prevailing scattering mechanisms, as well as on processing methods/conditions [12, 22]. The top *ZT* values up to ~2.8 at 773 K [23] and ~2.6 at 923 K [24] were reported for *n*-type Br-doped and *p*-type nanostructured chalcogenide SnSe, respectively. *ZT* up to ~2.2 at 915 K for *n*-type PbTe [25] and ~1.9 at 320 K for *n*-type $Bi_{0.5}Sb_{1.5}Te$ were also reported [26]. All these systems belong to the family of chalcogenide TE materials. However, typically *ZT* values reported for other chalcogenides such as Bi_2Te_3 and Sb_2Te_3 [19], or for skutterudites such as CoSb-family materials [20], or for metal oxides is lower (<1.8) [27]. Among them, metal oxide TE materials possess a considerably smaller figure of merit so far, but they are more high-*T* stable, robust and cost-effective than heavy metal alloys [28–30]. However, since metal oxides are usually wide band gap semiconductors, their electrical resistivity is rather high that results in low *PF* and *ZT*.

Trying to achieve a general and ambitious goal of TE research to produce a material with *ZT* > 3, different methods such as increasing electrical conductivity and Seebeck coefficient or decreasing thermal conductivity by addition of different dopants [31] as well as 2D materials such as graphene [32] and reduced graphene oxide (rGO) [33] are the main directions of the reported research works.

2 *P*-Type Metal Oxide Thermoelectrics

2.1 *Layered Cobalt Oxides*

After the first report by Terasaki et al. about the high Seebeck coefficient of ~100 mV K^{-1} at 300 K obtained for $NaCo_2O_4$ ($Na_{0.5}CoO_2$) [34], different types of cobalt oxide materials with layered crystal structures were studied for the possible TE application (see Fig. 2 [35]). In particular, an *in-plane* thermal conductivity for $Bi_2Sr_3Co_2O_9$, $Ca_3Co_4O_9$ and $NaCo_2O_4$ were measured below room temperature by Satake et al. [35]. It was reported that the charge reservoir block (the block other than the CoO_2 block) dominates the thermal conduction, while a nano-block integration concept is effective for material design. Moreover, at temperature below 300 K the thermal conductivity of $Bi_2Sr_3Co_2O_9$ and $Ca_3Co_4O_9$ was found to be lower than that of layered cobalt oxides with Na (see Fig. 2).

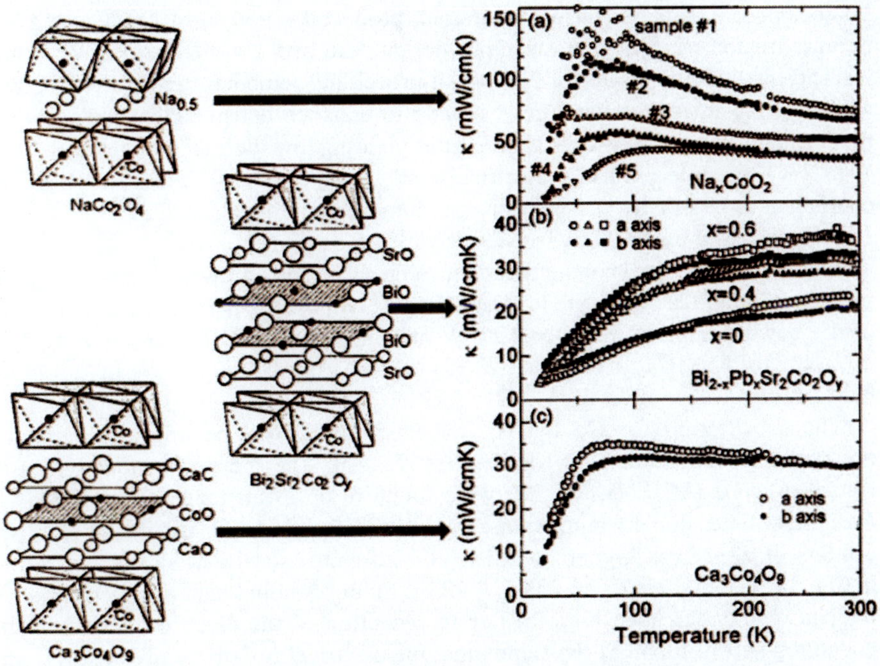

Fig. 2 Crystal structure of the layered cobalt oxides with the corresponding *in-plane* thermal conductivities of layered cobalt oxides: **a** Na_xCoO_2, **b** $Bi_{2-x}Pb_xSr_2Co_2O_y$ ($x = 0$, 0.4 and 0.6), **c** $Ca_3Co_4O_9$. Open and solid symbols represent *a*- and *b*-axis thermal conductivity, respectively. Reprinted from [35] with the permission of AIP Publishing

2.1.1 $NaCo_2O_4$

Na_xCoO_2 consists of the layers of Na-ions plane and CoO_2 plane along the c-axis, with a hexagonal layered crystal structure. In these materials the electrons and holes are transported by passing through the CoO_2 layer and phonons use the path through Na^+ layer as shown in Fig. 2. Such a layered structure results in "phonon glass electronic crystals" properties, which are very promising for TE applications due to high electrical conductivity and low thermal conductivity [36]. High *PF* value of 14×10^{-4} W m^{-1} K^{-2} at 300 K reported for $Na_{0.85}CoO_2$ [37] led to in-depth study of this material for TE power generation. Moreover, in the range of Na content in Na_xCoO_2 crystals of $0.6 < x < 1$ the critical doping $x \sim 0.85$ was reported to result in *Z* value at 80 K ~40 times higher than that for crystals with $x \leq 0.71$ [37].

Concerning ceramic $NaCo_2O_4$, Erdal et al. reported rather low *ZT* values of 6×10^{-7} and 4×10^{-5} at 300 K for ceramics obtained using powders prepared by sol-gel and electrospinning processes, respectively [38]. That was related to the polycrystalline and fiber ceramic microstructure. Moreover, a formation of Na-poor second phases leading to a degradation of $NaCo_2O_4$ above 900–950 °C due to the volatile nature of Na and the slow densification kinetics is rather common. Therefore,

Jakubczyk et al. studied $NaCo_2O_4$ ceramics produced at 950 and 1000 °C using Na-rich pre-treated sol-gel synthesized powder mixture with 2 wt.% NaOH [39]. While insufficient to compensate for Na loss at processing temperatures of 1000 °C and above, at the lower temperature it is able to enhance densification and facilitate the formation of complex crystal structures yielding low thermal conductivity (0.66 $W\,m^{-1}\,K^{-1}$) coupled with high electrical conductivity (3.8×10^3 $S{\cdot}m^{-1}$) and Seebeck coefficient (34.9 $mV\,K^{-1}$). The resultant room-temperature *PF* and *ZT* were 6.19×10^{-6} $W\,m^{-1}\,K^{-2}$ and 0.0026, respectively [39].

Increase of electrical conductivity and Seebeck coefficient was observed after the incorporation of Ca on Na site [40] as well as Cu on Co site [41]. It leads to increase of the power factor from 1.3×10^{-3} $W\,m^{-1}\,K^{-2}$ for $NaCo_2O_4$ to 3.6×10^{-3} $W\,m^{-1}\,K^{-2}$ for $Na_{0.7}Ca_{0.3}Co_2O_4$ at 573 K [40] and from 0.4×10^{-3} $W\,m^{-1}\,K^{-2}$ for $NaCo_2O_4$ to 3.08×10^{-3} $W\,m^{-1}\,K^{-2}$ for $Na(Co_{0.9}Cu_{0.1})_2O_4$ at 1073 K [41].

The substitution for Co by Ti, Rh, Pd in $NaCo_2O_4$ ceramics prepared by hot pressing was studied by Kurosaki et al. [42]. The thermal conductivities of $NaCo_{1.9}M_{0.1}O_4$ (*M* = Ti, Rh, Pd) were found to be lower than that of $NaCo_2O_4$ over the whole studied temperature range. Due to that, *ZT* of $NaCo_{1.9}Rh_{0.1}O_4$ (~0.033 at 723 K) is higher than that of $NaCo_2O_4$ (~0.018 at 723 K) as shown in Fig. 3a. The highest *ZT* = 0.045 at 723 K among studied materials was reported for $NaCo_{1.9}Pd_{0.1}O_4$ (see Fig. 3a) due to reduction of the electrical resistivity by palladium substitution. At the same time, the electrical resistivity of $NaCo_2O_4$ and $NaCo_{1.9}M_{0.1}O_4$ (*M* = Rh and Pd) increased slightly with temperature presenting a typical metallic behavior. For $NaCo_{1.9}Ti_{0.1}O_4$, the electrical resistivity decreased with increasing temperature showing a semiconducting behavior and resulting in very low *ZT* [42].

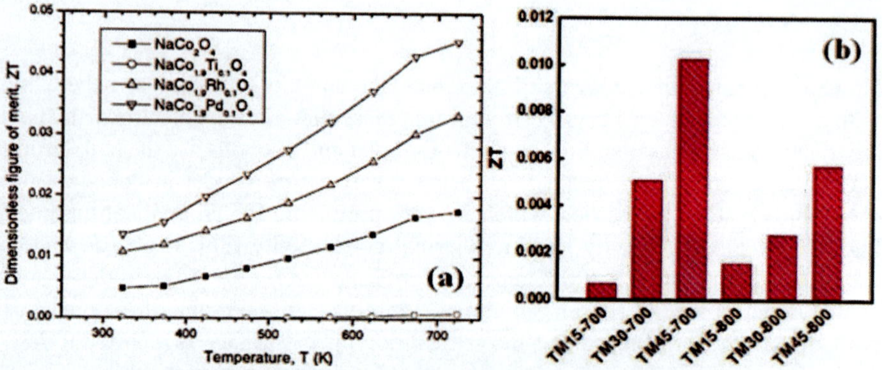

Fig. 3 Temperature dependence of the dimensionless figure of merit for $NaCo_2O_4$ and $NaCo_{1.9}M_{0.1}O_4$ (*M* = Ti, Rh, Pd) (**a**). Reprinted from [42]. Copyright 2001, with permission from Elsevier. Room-temperature figure of merit (*ZT*) of CNF/$NaCo_2O_4$-based composite nanofiber webs fabricated named as TM*x-y*, where x and y denote the content of $NaCo_2O_4$ nanofibers by wt.% and the calcination/carbonization temperatures of 700–800 °C, respectively (**b**) [44]

In addition to the mentioned above replacement of Co by Ti, Rh and Pd, incorporation of Ni and Ni–B on Co site in $NaCo_2O_4$ ceramics prepared using powders synthesized by electrospinning technique was also reported [43]. However, presented values of *ZT* at 300 K as 4.25×10^{-5} (undoped), 5.3×10^{-6} ($NaCo_{1.9}Ni_{0.1}O_4$), 8.6×10^{-5} ($NaCo_{1.7}Ni_{0.3}O_4$) and 9×10^{-6} ($NaCo_{1.6}Ni_{0.3}B_{0.1}O_4$) were rather low and inconsistent [43].

Furthermore, carbon-based materials can also enhance the TE performance of $NaCo_2O_4$ as was reported by Ryu et al. based on the study of composite nanofiber webs, fabricated by dual-electrospinning of polyacrylonitrile (PAN) and sodium cobalt oxide ($NaCo_2O_4$), heat-treated for simultaneous carbonation and calcination [44]. PAN derived carbon nanofiber (CNF) and $NaCo_2O_4$-based ceramic nanofiber webs have shown Seebeck coefficient increasing with $NaCo_2O_4$ content from 15 to 45 wt.%. The increase was from ~26.77 to ~73.28 $\mu V\ K^{-1}$ and from ~14.83 to ~40.56 $\mu V\ K^{-1}$ for webs fabricated at 700 and 800 °C, respectively. On the other hand, the electrical conductivity of the composite nanofiber webs increased with the decrement of the relative content of $NaCo_2O_4$ nanofibers as well as the increment of the heat-treatment temperature, and the highest electrical conductivity of 1.23 S cm^{-1} was obtained for the samples with 15 wt.% $NaCo_2O_4$ in the contrast to 0.2 S cm^{-1} reported for composites with 45 wt.% $NaCo_2O_4$ (both samples were prepared at 800 °C) [44].

Owing to different contributions of $NaCo_2O_4$ nanofibers and CNFs to the Seebeck coefficient, electrical conductivity and thermal conductivity, a maximum power factor of ~5.79 $\mu W\ m^{-1}\ K^{-2}$ and a figure of merit of ~0.01 were attained at room temperature for CNF/$NaCo_2O_4$-based composite nanofiber webs fabricated at 45 wt.% input composition of $NaCo_2O_4$ and at heat treatment temperature of 700 °C (see Fig. 3b) [44].

Significant suppression of the grain growth as well as formation of many new grain boundaries and interfaces to reduce thermal conductivity were reported by Zhang et al. after addition of SiC nanoparticles to $NaCo_2O_4$ [45]. Moreover, the electrical conductivity and Seebeck coefficient were increased in the composites with 0.2 vol.% nano-SiC resulting in the maximum *ZT* of 0.122 at 650 °C, which is approximately twice higher than that of pristine $NaCo_2O_4$ [45].

2.1.2 $Bi_2Sr_3Co_2O_9$

High Seebeck coefficient of ~300 mV K^{-1} at 973 K, electrical resistivity of ~3 mΩ cm and low thermal conductivity of ~2 W $m^{-1}\ K^{-1}$ were reported for single-crystalline whiskers of $Bi_2Sr_2Co_2O_y$ [46]. The resulting dimensionless TE figure of merit *ZT* was more than 1.1 as shown in Fig. 4 [46]. The power factor of ~0.13 mW $m^{-1}\ K^{-2}$ at ~923 K was reported for ceramic $Bi_2Sr_2Co_2O_y$ [47]. Moreover, when Na was added to the ceramics on Sr site the grain density increased and electrical resistivity decreased without significant modification of Seebeck coefficient. Thus, the power factor of $Bi_2Sr_{2-x}Na_xCo_2O_y$ reached the maximum value of 0.21 mW $m^{-1}\ K^{-2}$ at 923 K for $x = 0.075$ [47].

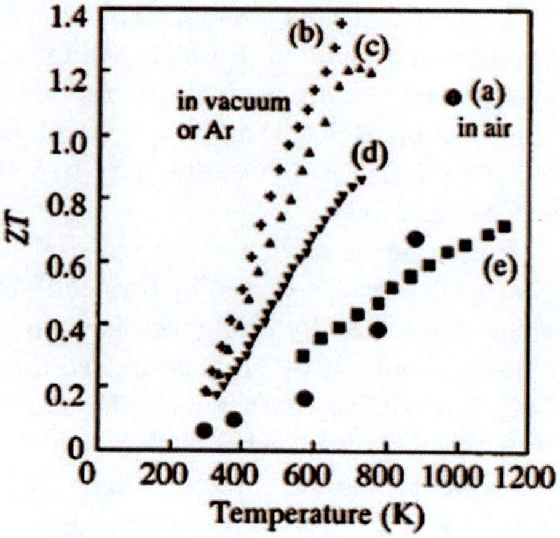

Fig. 4 *ZT* vs *T* plot for $Bi_2Sr_2Co_2O_y$ single-crystalline whisker (**a**). Reprinted from [46] with the permission of AIP Publishing. For comparison, *ZT* is plotted for *p*-b-Zn_4Sb_3 (**b**) [48]; *p*-TeAgGeSb (*p*-TAGS) (**c**) [49]; *p*-$CeFe_4Sb_{12}$ (**d**) [50]; *p*-Si-Ge alloy (**e**) [51] under vacuum or Ar atmosphere

2.1.3 $Ca_3Co_4O_9$

Single crystals of $(Ca_2CoO_3)_{0.7}CoO_2$ with a $Ca_3Co_4O_9$ (Co349) structure grown by modified strontium chloride flux technique presented plate-like single crystals with relatively large $5 \times 10 \times 0.05$ mm^3 size and with *ZT* of ~0.45 at elevated temperature of ~800 K (see Fig. 5a) [52]. Moreover, obtaining *S* of ~240 μV K^{-1} and ρ of ~2.3 $\times$ 10^{-5} Ω m, as well as extrapolating *k* to ~3 W m^{-1} K^{-1} at 973 K, Shikano et al. reported rather high *ZT* of ~0.87 at 973 K [52].

In contrast, polycrystalline $[Ca_2CoO_{3.34}]_{0.614}[CoO_2]$, also known as $Ca_3Co_4O_9$, that was measured by Miyazaki et al. up to 300 K has shown Seebeck coefficient at

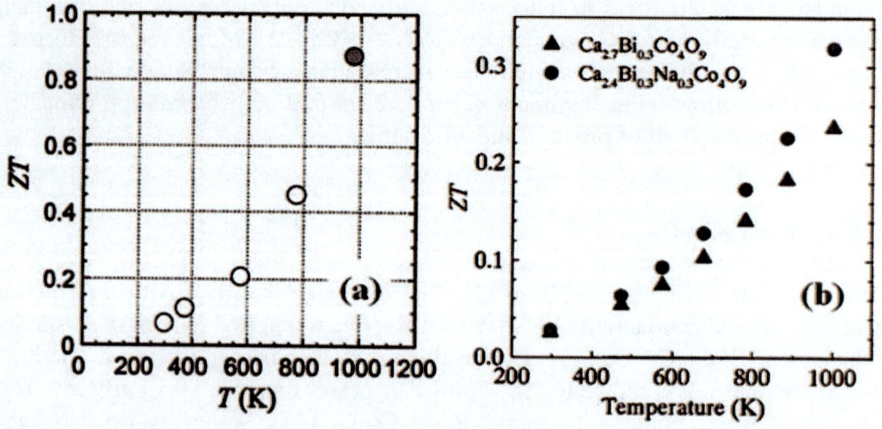

Fig. 5 Temperature dependence of figure of merit *ZT* for $(Ca_2CoO_3)_{0.7}CoO_2$ single crystal (shaded symbol shows evaluation from extrapolated value of thermal conductivity *k*) (**a**). Reprinted from [52] with the permission of AIP Publishing and for polycrystalline $Ca_{2.7}Bi_{0.3}Co_4O_9$ and $Ca_{2.4}Bi_{0.3}Na_{0.3}Co_4O_9$ (**b**). Reprinted from [56] with the permission of AIP Publishing

300 K of ~133 μV K^{-1}, electrical resistivity of 15 mΩ·cm, thermal conductivity of 9.8 mW·cm^{-1} K^{-1} that resulted in rather low $ZT_{300} = 3.5 \times 10^{-2}$ [53].

Madre et al. analyzed the influence of the preparation process on the thermoelectric properties of $Ca_3Co_4O_9$ ceramics produced by a directional melt-grown method using the laser floating zone technique [54]. Among 24, 48, 72 and 96 h durations of the annealing process at 900 °C, the optimum time of 72 h was determined to get the maximum power factor of ~0.42 mW m^{-1} K^{-2} [54].

At the same time Santos et al. compared 74–78% dense $Ca_3Co_4O_9$ ceramics obtained by the cold sintering process at ~150 °C with no, 6 or 24 h further heat treatment at 900 °C and 58% dense ceramics prepared by the conventional sintering at 900 °C for 6 h [55]. The highest *ZT* at 873 K was 0.12 for ceramics prepared by cold sintering process with 24 h heat treatment, while the lowest one was 0.03 for conventionally sintered ceramics.

To increase *ZT*, polycrystalline Bi-, Na- or Bi–Na-substituted $Ca_3Co_4O_9$ were also prepared using a hot-pressing technique and their TE properties were carefully studied in air from room temperature to 1000 K [56]. The substitutions of Bi^{3+} or Na^+ for Ca^{2+}, as well as Bi^{3+} and Na^+ double substitution, cause both the electrical conductivity and thermoelectric power to increase simultaneously. In addition, Xu et al. reported that the double substitution of Bi and Na on Ca site is an effective method to decrease the thermal conductivity leading to high value of *ZT* ~0.32 at 1000 K (see Fig. 5b) [56].

2.2 La_2CuO_4

Materials based on La_2CuO_4 were reported as possible thermoelectrics, although Cu-site doping was not found to enhance the power factor of La_2CuO_4, when the electrical conductivity and thermoelectric power were measured between 300 and 1173 K for solid solutions of $La_2Cu_{0.9}M_{0.1}O_4$ (M = Mn, Fe, Co and Ni) prepared by polymeric precursor synthesis [57]. Room temperature *ZT* of polycrystalline La_2CuO_4 was reported to reach 0.007 [58]. Further, $La_2Ag_xCuO_4$ ceramics synthesized by solid state reaction method were studied by Zhu et al. in the temperature range 10–380 K [58]. The resistivity and Seebeck coefficient of these samples tend to decrease, while the thermal conductivity first decreases and then increases with Ag content varying from 0 to 15 mol.%. Consequently, La_2CuO_4 ceramics with the addition of 5 mol.% Ag presented the maximum *ZT* value of 0.018 at 308 K that is about twice higher than that for undoped La_2CuO_4 as shown in Fig. 6a [58].

At the same time, substituting La by Sr as well as Ba and Pb can lead to increase of *ZT* as it was reported by Liu et al. and can be seen in Fig. 6b for $La_{2-x}Sr_xCuO_4$ ceramics [59]. Moreover, in that work Liu et al. reported that Ba has even better doping effect than Sr and Pb, and presents an optimal carrier concentration of $\sim 10^{20}$ cm^{-3}. Correspondingly, $La_{1.95}Ba_{0.05}CuO_4$ showed a maximum average *ZT* ~0.03 over the

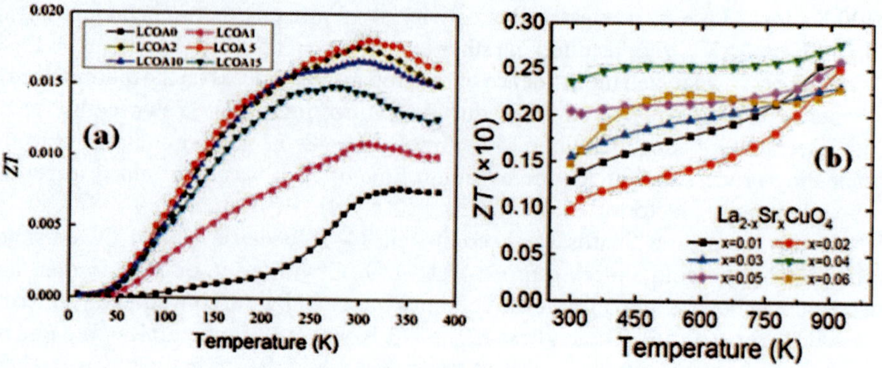

Fig. 6 Temperature dependence of *ZT* for $La_2Ag_xCuO_4$ (**a**). Reprinted from [58]. Copyright 2015, with permission from Elsevier and $La_{2-x}Sr_xCuO_4$ (**b**) [59] ceramics

entire temperature range of 300–923 K that was slightly higher than that reported for $La_{1.95}Pb_{0.05}CuO_4$ or $La_{1.95}Sr_{0.05}CuO_4$ and ~5 times higher than *ZT* of La_2CuO_4 [59].

3 *N*-Type Metal Oxide Thermoelectrics

3.1 *Simple Oxides*

3.1.1 MnO_2

A giant value of Seebeck coefficient higher than 20 mV K^{-1} was measured by Song et al. on MnO_2 powder, suggesting MnO_2 as one of the best TE materials [60]. However, as it was shown by Kumar et al., many factors such as the carrier concentration, electrical conductivity type (*n*- or *p*-) as well as phase (α- or β-) of MnO_2, direction of the measurements, etc. can play significant role on the value of *ZT* [61]. At the same time, Kumar et al. reported the maximum *ZT* of 0.001 at 300 K for the α-MnO_2 sample with *n*-type conduction in the system with a Seebeck coefficient of ~160 μV K^{-1}, an electrical conductivity of ~42 S m^{-1} and a thermal conductivity of 0.65 W m^{-1} K^{-1} synthesized by the hydrothermal route [61]. *ZT* value was also reported to increase with the temperature above 100 K [61]. An order of magnitude higher *ZT* ~0.01 at 300 K was theoretically predicted and reported for α-MnO_2 with electron carrier concentration of 10^{18} cm^{-3} by Chepkoech et al. [62].

An enhanced value of the power factor up to 161.4 μW m^{-1} K^{-2} was reported for MnO_2 with increased electrical conductivity and Seebeck coefficient by addition of conductive polymer polypyrrole (PPy) [63].

In addition, Hedden et al. suggested that TE properties of β-MnO_2 depend heavily on particle size distribution and particle morphology that can be varied by ground time [64]. It was also shown that thermal conductivity can increase with density of the powder compact. The highest *ZT* ~3.28×10^{-4} at 293 K and *PF* ~5.76×10^{-7} W m^{-1} K^{-2} were reported by Hedden et al. for a high density compact pressed with particles of 140 μm average size [64]. Some of the TE properties such as Seebeck coefficient or electrical conductivity or thermal conductivity of MnO_2 in α [61], β [60, 65, 66], or γ [67–69] phases were reported in the literature, although without many details or essential comparative analysis.

3.1.2 TiO_2

Typical *ZT* of pristine TiO_2 ranges from ~0.013 at 873 K [70] to ~0.016 at 800 K [71] depending on the preparation method and conditions. For nonstoichiometric TiO_{2-x} prepared by spark plasma sintering (SPS), addition of Cu powder resulted in a decrease of the electrical resistivity and Seebeck coefficient of the composites together with strong increase of the thermal conductivity. As a result of these variations, the greatest *ZT* value ~0.03 at 873 K was obtained for Cu/TiO_{2-x} composite with Cu volume fraction of 12% (see Fig. 7) [70].

An increase of *ZT* from 3.0×10^{-4} to 6.4×10^{-2} at 1050 K was also reported by Nam et al. after addition of 4 wt.% reduced graphene oxide (rGO) to TiO_2 due to enhancement in electrical conductivity and the reduction in thermal conductivity [72]. Significant increase of *ZT* from 0.016 to 0.24 was observed for SPS nanocomposite of TiO_2 with 0.75 wt.% polyparaphenylene (PPP) [71]. However, the highest *ZT* values were reported for the nonstoichiometric TiO_{2-x} [73] and $Ti_x(O, N)_{2x-1}$ [74].

The lattice defect engineering was sufficiently achieved by Liu et al. for the fabrication of multi-scale hierarchical nonstoichiometric TiO_{2-x} by high pressure and high temperature (HPHT) reactive sintering with appropriate ratio of coarse Ti to nanosized TiO_2 [73]. Lower electrical resistivity, Seebeck coefficient and thermal

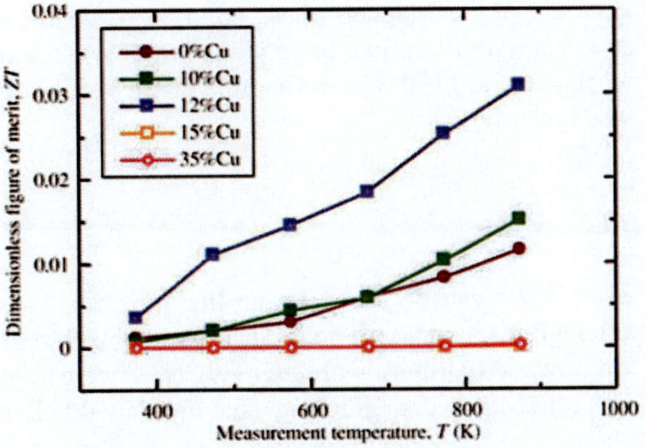

Fig. 7 *ZT* of Cu/TiO_{2-x} composites with different volume fractions of Cu powder as a function of measurement temperature. Reprinted from [70]. Copyright 2013, with permission from Elsevier

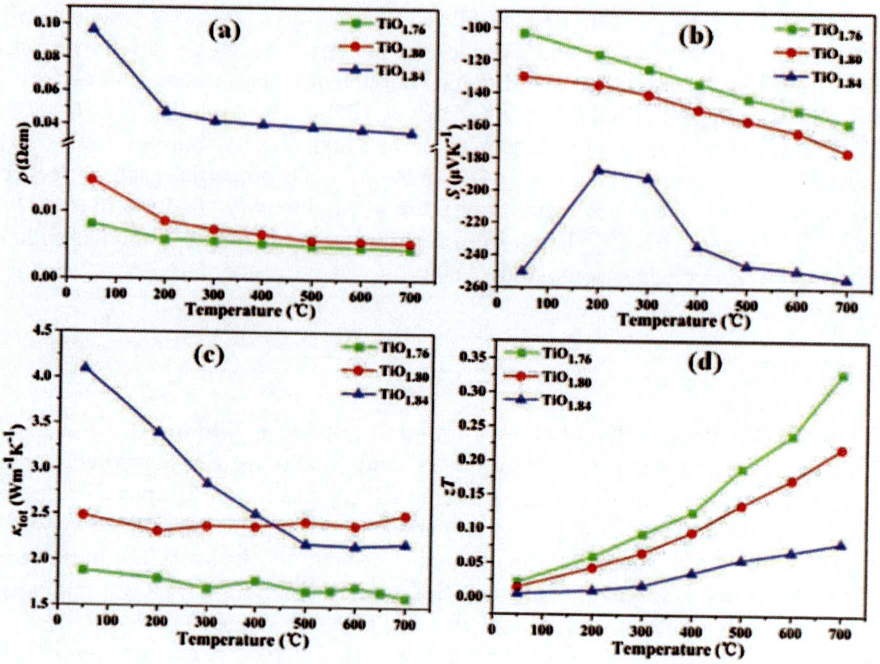

Fig. 8 The temperature dependence of electrical resistivity (**a**), Seebeck coefficient (**b**), power factor (**c**) and thermoelectric figure of merit (**d**) for TiO_{2-x} ($x = 0.24, 0.20, 0.16$) [73]

conductivity but the highest ZT up to 0.33 at 700 °C (see Fig. 8) were reported for the most reduced $TiO_{1.76}$ composition in comparison with that of $TiO_{1.80}$ and $TiO_{1.84}$ [73].

TiO_2 with oxygen defects and Magneli phase $Ti_x(O,N)_{2x-1}$ controlled by the solid-state reaction of TiO_2 and TiN were also investigated by Mikami et al. [74]. The introduction of oxygen deficiency as electron doping resulted in a significant reduction of electrical resistivity (see Fig. 9a). The nanometer-scale periodic planar defect structure of the Magneli phase contributed to the strong enhancement of phonon scattering, resulting in a large reduction of κ (see Fig. 9c). The estimated ZT value reached 0.4 at 1150 K in the sample mainly consisting of the Ti_7O_{13} Magneli phase (see Fig. 9d) [74].

3.1.3 ZnO

$ZT < 0.005$ can be found in the literature for the pure ZnO (see Fig. 10a) [75], although it was possible to be increased by different additives or doping elements. Single walled carbon nanotubes (SWCNTs) were found to have less influence on ZnO TE performance than graphene (see Fig. 10) although the addition of either of these

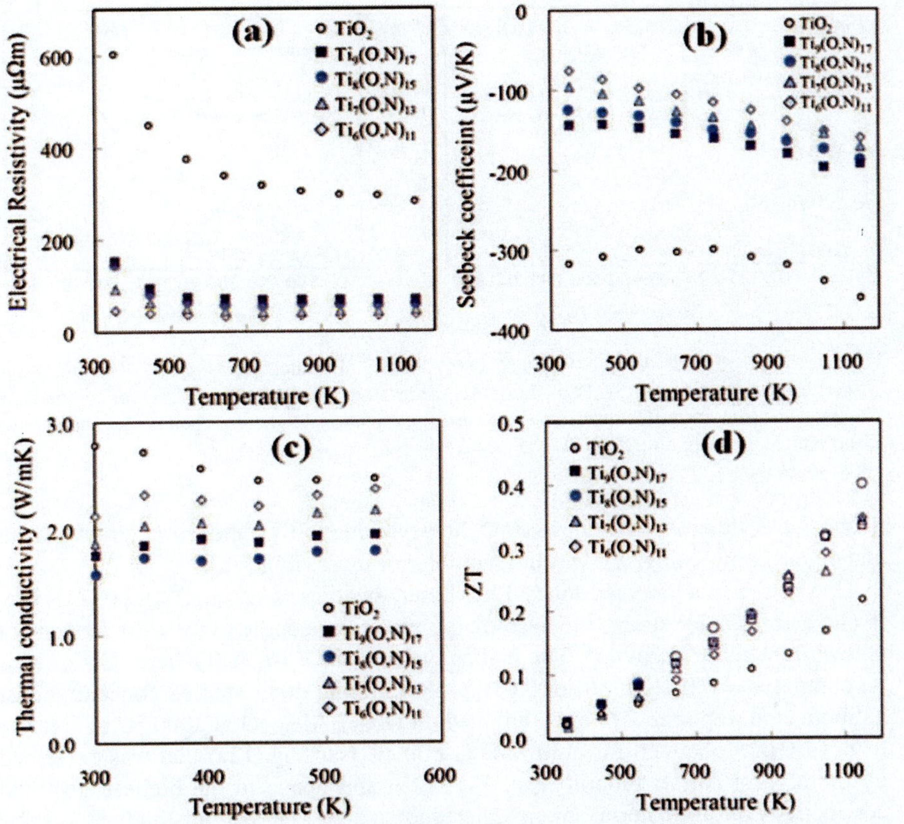

Fig. 9 Temperature dependence of electrical resistivity (**a**), Seebeck coefficient (**b**), thermal conductivity (**c**) and thermoelectric figure of merit *ZT* (**d**) of $Ti_x(O,N)_{2x-1}$ [74]. In **d** above 600 K, the *k* value at 550 K was used for the estimation (marked with open symbols)

low-dimensional carbon materials lead to intrinsically strong transport anisotropy observed in the in-plane and cross-sectional measurement directions [75].

Among doped ZnO, the influence of Ni [76], Ga [77], or In [78] doping on ZnO TE performance was studied. Colder et al. reported that $Zn_{1-x}Ni_xO$ materials with x increasing until the solubility limit of Ni in the wurtzite structure x = 0.03 have significant rise in both electrical conductivity and Seebeck coefficient [76]. The highest power factor of 0.6 mW m^{-1} K^{-2} and *ZT* ~0.09 were reported for $Zn_{0.97}Ni_{0.03}O$ at 1000 K [76]. Ga-doped ZnO ceramics presented the highest *ZT* ~0.025 at 800 °C for a composition with 2 mol.% Ga and the following increase of Ga content lead to decrease of *ZT*. An introduction of dopants with low concentration increased both electron and point defect concentrations, resulting in significantly increased electrical conductivity with reduced thermal conductivity [77]. Jood et al. reported very low thermal conductivity ~3 W m^{-1} K^{-1} and high *ZT* ~0.45 at 1000 K measured for In-doped ZnO (see Fig. 11) [78]. These authors also reported high

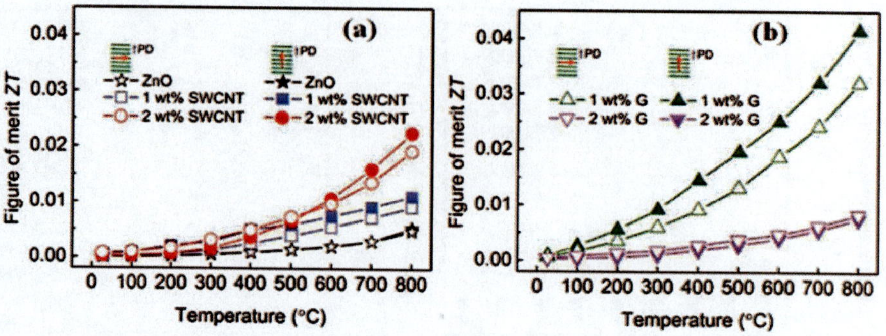

Fig. 10 Figure of merit *ZT* for SWCNT/ZnO (**a**) and G/ZnO nanocomposites (**b**). Reprinted from [75] with permission from RSC Publishing. Also shown are the data for pure ZnO for comparison. The open and filled symbols represent the measuring directions that are perpendicular and parallel to the sample pressing direction (PD), respectively

Seebeck coefficient for Bi-doped ZnO, however due to Bi segregation, grain growth and defect complexing, *ZT* was unfavorable for increasing [78].

One of the most studied today ZnO-based systems is $(Zn_{1-x}Al_x)O$ (AZO) that is characterized by strong increase of the electrical conductivity with Al content (showing metallic behavior). The highest values of *ZT* for AZO were obtained for Al content $x = 0.02$ [79, 80] or 0.03 [81–83] among other studied concentrations. Tsubota et al. reported *ZT* ~0.30 attained on $(Zn_{0.98}Al_{0.02})O$ at 1000 °C in contrast to *ZT* < 0.025 obtained for pure ZnO at 800 °C (see Fig. 12a) that was explained by the highest carrier mobility (see Fig. 12b) and hence to the highest electrical conductivity of this material among other studied ones [79]. Slightly higher *ZT* value up to 0.36 at 1073 K was reported by Zhang for $(Zn_{0.98}Al_{0.02})O$ composite prepared via hydrothermal synthesis and spark plasma sintering [80].

Al-doped ZnO ceramics synthesized via double emulsion method followed by compaction by SPS processing have shown significant decrease of the grain size with Al concentration (see Fig. 12c) [81]. The significantly enhanced electrical conductivity and strikingly reduced lattice thermal conductivity were reported for

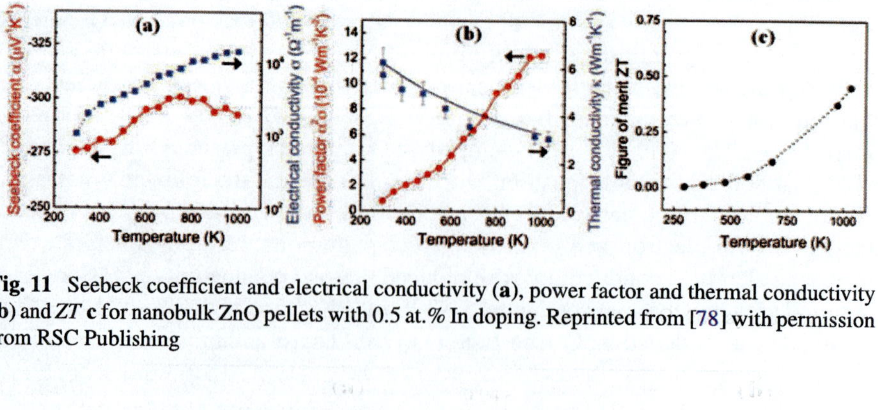

Fig. 11 Seebeck coefficient and electrical conductivity (**a**), power factor and thermal conductivity (**b**) and *ZT* **c** for nanobulk ZnO pellets with 0.5 at.% In doping. Reprinted from [78] with permission from RSC Publishing

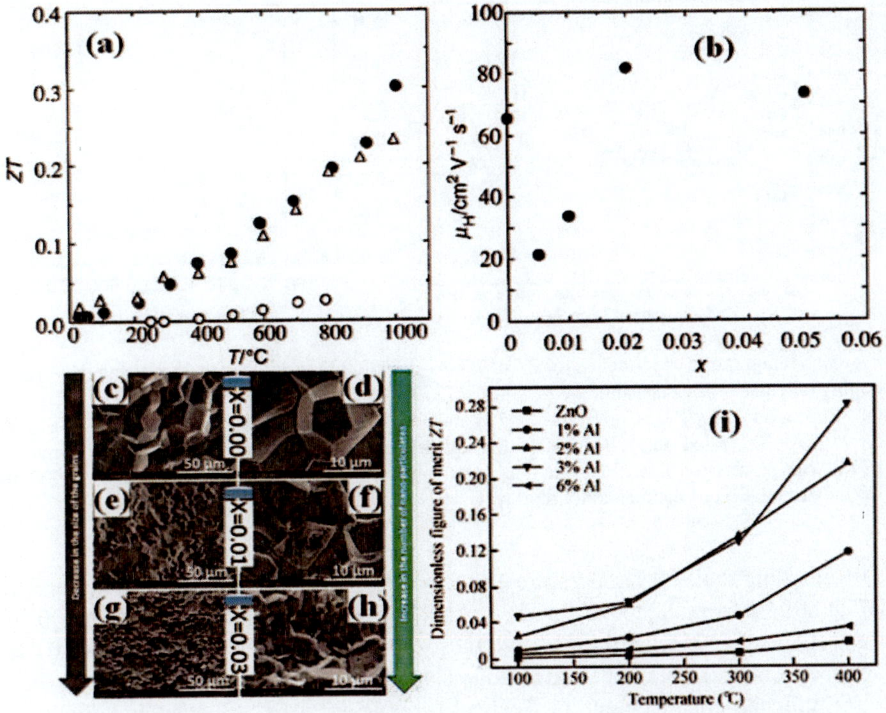

Fig. 12 The dimensionless figures of merit of $Zn_{1-x}Al_xO$ as a function of temperature for $x = 0$ (o), 0.02 (●), 0.05 (Δ) **(a)** and its Hall mobility at room temperature as a function of x **(b)**. Reprinted from [79] with permission from RSC Publishing. Cross-sectional SEM micrographs of sintered $Zn_{1-x}Al_xO$ samples with the nominal composition for $x = 0.00$ (**c, d**), 0.01 (**e, f**), 0.03 (**g, h**), showing a systematic reduction in grain size with Al-content and the presence of $ZnAl_2O_4$ precipitates [81]. Temperature dependence of figure of merit *ZT* of ZnO and Al-doped ZnO (**i**). Reprinted from [83] with permission from IOP Science

these ceramics. $ZT_{max} \sim 0.13$ at 750 K for 3 at. % Al-doped ZnO and $ZT_{max} \sim 0.075$ at 750 K for the undoped ZnO were reported by Virtudazo et al. [81]. *ZT* ~0.28 was reported for $Zn_{1-x}Al_xO$ with $x = 0.03$ made using powder synthesized by hydrothermal method (see Fig. 12d) [83]. The particle size of these samples decreased with increasing Al concentration, although Seebeck coefficient did not vary significantly in opposite to the electrical resistivity that decreased significantly for higher Al content [83].

In addition, Han et al. reported that γ-Al_2O_3 precursor promoted the Al donor doping in ZnO compared with α-Al_2O_3, more effectively shrinking the unit cell volume and increasing the electrical conductivity for 2% Al-doped ZnO. As a result, $Zn_{0.98}Al_{0.02}O$ ceramics prepared by SPS at 1173 K for 5 min using powders calcined at 1173 K showed a *ZT* value of 0.17 at 1173 K for γ-Al_2O_3 doping precursor, which is 27% higher than that for the α-Al_2O_3 doping precursor (see Fig. 13a) [84].

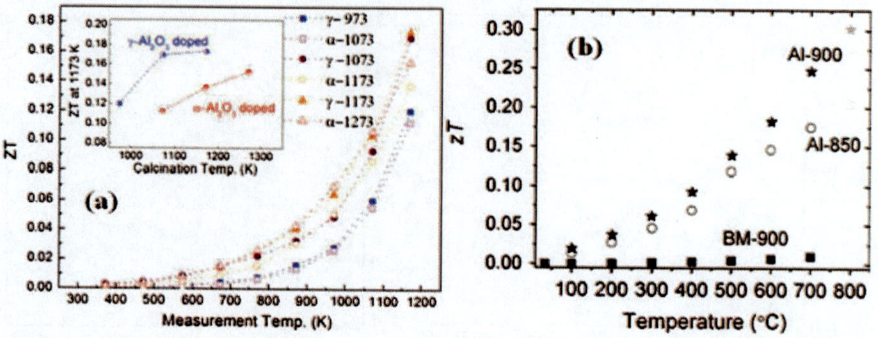

Fig. 13 Temperature dependence for *ZT* of 2% Al-doped ZnO ceramics prepared using a/γ-Al_2O_3 doping precursors (the inset shows *ZT* values at 1173 K as a function of the calcination temperature) (**a**). Reprinted from [84]. Copyright 2013, with permission from Elsevier and for Al-850 (open circles), Al-900 (filled stars), BM-900 (filled squares) samples (**b**). Reprinted from [85]. Copyright 2013, with permission from Elsevier. The gray values in (**b**) for samples Al-850 and Al-900 are from extrapolation of the resistivity to 800 °C

Regarding the SPS temperature effect Sondergaard et al. reported an increase in *ZT* at 800 °C from 0.2 to 0.3 for $Zn_{0.98}Al_{0.02}O$ ceramics sintered at 850 and 900 °C, respectively (see Fig. 13b) [85]. Both these values were much higher than ~0.01 for undoped ZnO ceramics sintered at 900 °C [85].

Significant enhancement in Seebeck coefficient was achieved by dually Al and Ga doping ZnO that resulted in *ZT* ~0.47 at 1000 K and 0.65 at 1247 K for $Zn_{0.96}Al_{0.02}Ga_{0.02}O$ [86]. In addition, the thermal conductivity was drastically decreased by the dual doping, while the decrease in the electrical conductivity was relatively small [86].

The effect of co-doping with transition metals (Fe, Ni, and Sm) on the thermoelectric properties of Al-doped ZnO ceramics was studied by Yamaguchi et al. [87]. Only Ni-co-doped AZO showed significant enhancement in electron density and hence in the electrical conductivity, while the thermal conductivity was decreased after Ni or Sm co-doping of AZO. As a result, the highest *ZT* of 0.126 at 1073 K was obtained for Ni-co-doped AZO, being superior to that of conventional AZO by a factor of 1.5 [87].

Similar to pure ZnO, Al-doped ZnO was also modified by CNT [88] and reduced graphene oxide (RGO) [89, 90]. Addition of rather low amount of 0.1 wt.% CNT to AZO lead to increase of *ZT* from 0.02 to 0.08 at 873 K [88]. At the same time, addition of 1.5 wt.% rGO to AZO powder resulted in *ZT* of 0.28 at 900 °C in comparison to 0.035 for pristine AZO (see Fig. 14a) [89]. Al-doped ZnO quantum dots (QDs) significantly changed *ZT* after RGO addition: from <0.1 to 0.52 at 1100 K as can be seen in Fig. 14b [90]. Al doping was mainly reported to give a rise in electrical conductivity, reducing the thermal conductivity and Seebeck coefficients. However, RGO inclusion not only increased the overall electrical conductivity but also enhanced the Seebeck coefficient through energy filtering in comparison to bare

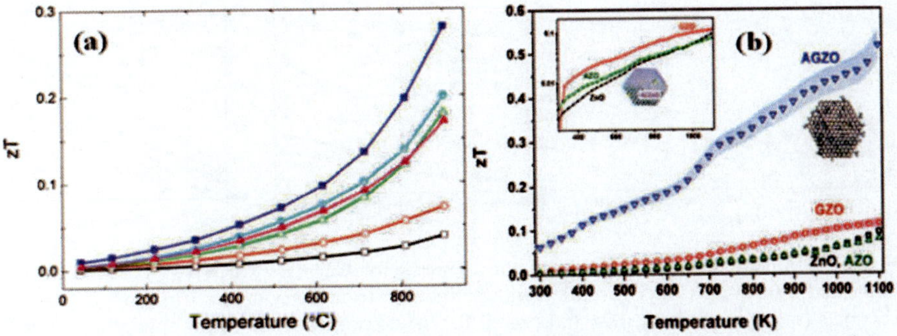

Fig. 14 Temperature-dependent dimensionless ZT for AZO and AZO/rGO composites in the range of 25–900 °C (**a**). Reprinted with permission from [89]. Copyright 2015, American Chemical Society and for undoped ZnO, Al-doped ZnO (AZO), RGO-encapsulated ZnO (GZO), and RGO-encapsulated Al-doped ZnO (AGZO) samples (**b**). Reprinted with permission from [90]. Copyright 2021, American Chemical Society. The inset in (**b**) shows the magnified view of the low ZT values of the samples

Al-doped ZnO. A simultaneous drop in thermal conductivity of this composite is attributed to the selective phonon scattering by point defects and interfaces [90].

3.2 *Double Oxides*

3.2.1 $CaMnO_3$

$CaMnO_3$ with perovskite structure demonstrates temperature dependent electrical resistivity (from ~6000 mΩ·cm at ~320 K to ~1000 mΩ·cm at ~420 K) and Seebeck coefficient (from ~−405 μV K^{-1} at ~320 K to ~−380 μV K^{-1} at ~420 K) [91]. Many different doping elements such as Bi [92, 93], Gd [94], Sm [93, 95], Yb [93], Dy [93], co-doping by Dy with La/Nd/Sm [96], or Dy with Yb [97] or Pr with Yb [98] were incorporated on Ca site to modify the TE properties of $CaMnO_3$. Both the electrical resistivity and absolute value of the Seebeck coefficient decrease with increasing Bi doping level in $Ca_{1-x}Bi_xMnO_3$ indicated an increase in carrier concentration with Bi content [92]. The maximum power factor of 4.67×10^{-4} W m^{-1} K^{-2} at 423 K that is four times larger than that of the undoped $CaMnO_3$ was achieved for $Ca_{0.97}Bi_{0.03}MnO_3$. The lowest thermal conductivity of 1.4 W m^{-1} K^{-1} was obtained at 973 K for the same composition ($x = 0.03$). The maximum obtained ZT value was ~0.25 at 973 K for $x = 0.03$ as can be seen in Fig. 15a [92].

The electrical conductivity of Gd-doped $CaMnO_3$ ceramics prepared using 50–70 nm nanopowders obtained by co-precipitation route was found be lower than that of ceramics prepared from 200 to 400 nm fine-grained powders made by solid-state reaction. However, ceramics prepared by the co-precipitation method have also

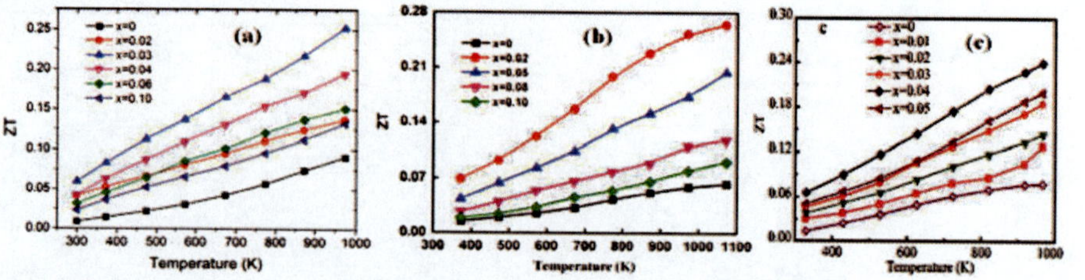

Fig. 15 Temperature dependence of figure of merit for $Ca_{1-x}Bi_xMnO_3$ ($0 \leq x \leq 0.10$) (**a**). Reprinted from [92]. Copyright 2015, with permission from Elsevier; for $Ca_{1-2x}Dy_xYb_xMnO_3$ ceramics ($x = 0$, 0.02, 0.03, 0.04, 0.06 and 0.10) (**b**). Reprinted from [97]. Copyright 2014, with permission from Elsevier; and for $Ca_{1-2x}Pr_xYb_xMnO_3$ with different Pr and Yb contents ($x = 0$, 0.01, 0.02, 0.03, 0.04 and 0.05) [98]

much lower thermal conductivity values of 1.3–1.5 W m^{-1} K^{-1} than that of 2.0–2.4 W m^{-1} K^{-1} for ceramics prepared by the solid-state reaction [94], although κ values of the fine-grained ceramics are still relatively low due to the enhancement of grain boundary scattering. Thus, the highest dimensionless figure of merit $ZT = 0.24$ was obtained at 973 K in the air for nanopowder $Ca_{0.96}Gd_{0.04}MnO_3$, while fine-grained ceramics showed ZT of 0.16 [94].

Among $M =$ Yb, Dy, Sm and Bi substituting for Ca in $Ca_{1-x}M_xMnO_3$ system the highest power factor of 165 μW m^{-1} K^{-2} at 400 °C was reported for $Ca_{0.9}Yb_{0.1}MnO_3$ [93]. At the same time, Zhu et al. reported ZT of 0.2 at 973 K for $CaMnO_3$ with 2% Dy and 2% La co-doping on Ca site ($Ca_{0.96}Dy_{0.02}La_{0.02}MnO_3$) [96]. Both reported ceramics of co-doped $Ca_{1-2x}Dy_xYb_xMnO_3$ ($0 \leq x \leq 0.10$) synthesized by solid state reaction [97] and $Ca_{1-2x}Pr_xYb_xMnO_3$ ($0 \leq x \leq 0.05$) synthesized by co-precipitation method [98] presented significantly enhanced ZT ~0.27 for $x = 0.02$ at 1073 K (see Fig. 15b) and 0.24 for $x = 0.04$ at 973 K (see Fig. 15c), respectively.

Besides Ca-site substitution, the incorporation of Nb [99] or W [100] on Mn site in $CaMnO_3$ was also studied, and increased Seebeck coefficient was reported for Nb doping [99]. However, although more carriers were introduced when Nb-doping content increases, both the electrical conductivity and absolute Seebeck coefficient decrease in heavily Nb-doped $CaMnO_3$ [99]. Moreover, W acted as an electron donor together with the oxygen vacancies formed in the high temperature region that resulted in ZT of 0.25 at 1225 K for $CaMn_{1-x}W_xO_{3-\delta}$ ceramics with $x = 0.04$ in comparison to ~0.15 for $x = 0$ [100].

Simultaneous co-doping by Bi on Ca site and Nb on Mn site was reported by Park et al. [101], but $Ca_{1-x}Bi_xMn_{1-y}Nb_yO_3$ ($0 \leq x = y \leq 0.1$) presented ZT just of ~0.1 at 873 K due to the electrical conductivity decrease as the temperature increased, indicating metallic behavior, although Seebeck coefficient increased with increasing temperature [101].

3.2.2 $SrTiO_3$

Thermoelectric properties of bulk $SrTiO_3$ (STO) were studied since 1964, when Seebeck coefficient up to 800 $\mu V\ K^{-1}$ was reported [102]. However, due to high thermal conductivity ZT value of pure STO is found to be below 0.008 at 760 K, although that can be increased up to 0.1 by oxygen deficiency and RGO addition [103] besides doping.

The substitution for Sr by La was studied by Okuda et al. on $Sr_{1-1.5x}La_xTiO_3$ single crystals with $x \leq 0.1$ that presented ZT value of ~0.1 at 300 K [104]. The optimal La doping content $x = 0.08$ as well as the maximum ZT of 0.22 at 800 K were reported by Zhang et al. for $Sr_{1-1.5x}La_xTiO_3$ ceramics prepared by SPS during 30 min at 1573 K [105]. At the same time Kikuchi et al. prepared $Sr_{0.92}La_{0.08}TiO_{3\pm\delta}$ ceramics by combustion synthesis with post-spark plasma sintering at 1300 °C, finding that the 5-min-sintered ceramics result in ZT of 0.37 at 1045 K as presented in Fig. 16a [106]. Park et al. reported the same ZT of ~0.37 at lower 973 K for similar $Sr_{0.91}La_{0.09}TiO_{3\pm\delta}$ ceramics prepared by SPS at 1400 °C during 5 min. using hydrothermally obtained powder [107]. Moreover, such nanostructured bulk La-doped $SrTiO_3$ ceramics exhibited almost the same ZT before and after the heat treatment at 973 K for 24 h [107]. The highest ZT value for single La-doped STO was reported by Lu et al. to reach 0.41 at 973 K in the case of $Sr_{1-1.5x}La_xTiO_{3\pm\delta}$ ceramics with $x = 0.15$ (see Fig. 16b), conventionally sintered in a flowing 5%H_2–95%N_2 gas at 1500 °C for 6 h [108].

Furthermore, the room-temperature ZT of 0.42 was reported for $Sr_{0.900}La_{0.067}TiO_{3\pm\delta}$ ceramics modified by addition of 0.6 wt.% of reduced graphene oxide (rGO) sheets, whereas ZT at 1023 K was increased from 0.12 for monophasic ceramics to 0.36 for those with graphene nano-inclusions (see Fig. 16c) [110], although rGO itself is not the best TE material [33, 112].

In addition to Sr substitution by La, Kovalevsky et al. reported that among STO ceramics doped with 10 at.% of trivalent ions on the Sr site in $Sr_{0.9}M_{0.1}TiO_{3-\delta}$ (M = La, Ce, Pr, Nd, Sm, Gd, Dy, Y), conventionally sintered in air at 1700 °C for 10 h and further reduced in a 10%H_2–90%N_2 atmosphere at 1500 °C for 10 h, the $Sr_{0.90}Dy_{0.10}TiO_{3\pm\delta}$ composition reveals the highest ZT value of 0.41 at 1073 K and 0.42 around 1200 K (see Fig. 16d) [111].

A slightly lower ZT of 0.35 but at much lower temperature of 773 K was measured by Dehkordi et al. in $Sr_{0.85}Pr_{0.15}TiO_{3\pm\delta}$ using spark plasma sintering at 1400–1500 °C for 5 min [113]. The results were explained by incomplete solid solubility of 15% Pr in STO, and thereby Pr-rich grain boundaries [114].

Thus, the highest ZT values which could be obtained by single Sr site doping in STO, are only in the order of 0.35–0.41 at temperatures of 970–1100 K [106–108, 111] mainly due to the relatively high thermal conductivity.

Moreover, there were several reports on the fact that the electrical conductivity of donor-doped $SrTiO_3$ is sensitive to Sr content, and that the Sr deficient compositions sintered in reduced atmospheres have a much higher electrical conductivity, thereby enhancing TE performance compared to their stoichiometric counterparts [108, 115–118].

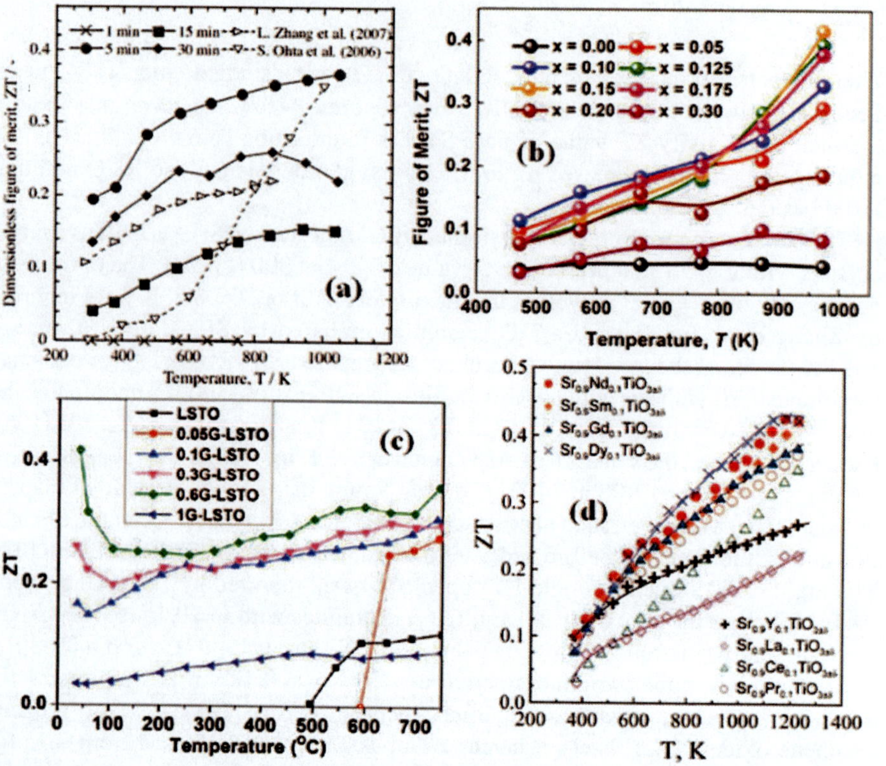

Fig. 16 Temperature dependence of *ZT* for $Sr_{0.92}La_{0.08}TiO_3$ prepared at various holding times (1–30 min.) during sintering (**a**). Reprinted from [106]. Copyright 2010, with permission from Elsevier, in comparison with *ZT* data for Nb-doped $SrTiO_3$ reported by Ohta et al. [109] and for La-doped $SrTiO_3$ reported by Zhang et al. [105]. Temperature dependence of *ZT* for $Sr_{1-1.5x}La_xTiO_{3-\delta}$ ceramics ($x \leq 0.30$) at 1773 K for 6 h (**b**) [108]. Temperature dependence of *ZT* for La-doped $SrTiO_3$ (LSTO) and the nanocomposites with different amount of graphene (0.05–1 g) as a function of temperature (**c**). Reprinted with permission from [110]. Copyright 2015, American Chemical Society. Temperature dependence of *ZT* for $Sr_{0.9}M_{0.1}TiO_{3-\delta}$ ($M =$ La, Ce, Pr, Nd, Sm, Gd, Dy, Y) (**d**). Reprinted from [111] with permission from RSC Publishing

In parallel to Sr-site substitution, incorporation of doping elements onto Nb site was also widely studied. In 2006 Ohta et al. reported a *ZT* value of 0.35 at 1000 K for ~20 μm-grain-size $SrTi_{0.80}Nb_{0.20}O_{3\pm\delta}$ ceramics hot pressed at 36 MPa and 1400 °C for 2 h in an Ar atmosphere, although on cooling toward room temperature *ZT* diminished below 0.01 [109].

ZT values up to 0.36 at 970 K were obtained by Tkach et al. for $SrTi_{0.80}Nb_{0.20}O_{3\pm\delta}$ ceramics, conventionally sintered in air at 1450 °C and further in 10%H_2–90%N_2 at 1400 °C [119]. Such values are observed only when the ceramics were prepared from centrifuged powders having an average particle size of 270 nm, while for the ceramics prepared from powders with 800 nm average particle size *ZT* just reached 0.14 at 970 K (see Fig. 17a). Corresponding enhancement of the TE performance was explained as a result of a bimodal grain size distribution due to abnormal grain growth [119].

ZT ~0.29 at 1160 K was reported by Okhay et al. for nonstoiciometric STO with 2% of Sr vacancy (rGO:$Sr_{0.98}Ti_{0.90}Nb_{0.10}O_{3\pm\delta}$, designed in Fig. 17b as nSTN10 + rGO) ceramics obtained from the sol–gel prepared powder and modified by addition

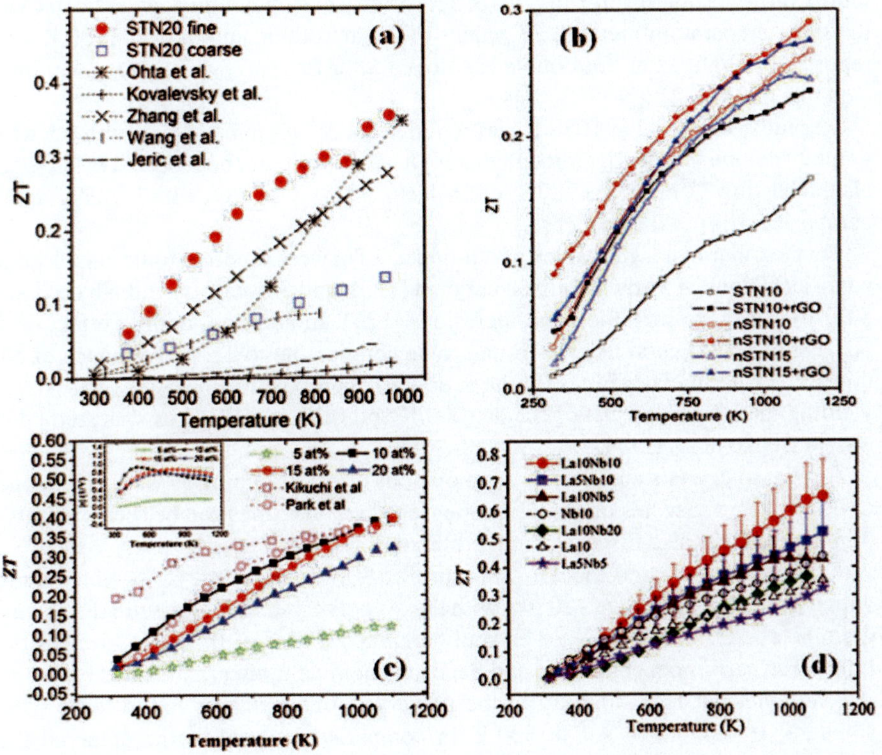

Fig. 17 Figure of merit *ZT* for $SrTi_{0.8}Nb_{0.2}O_{3\pm\delta}$ (STN20) ceramics produced from fine-(solid circles) and coarse-(open squares) particle-sized powders versus temperature (**a**). Reprinted with permission from [119] Copyright 2018, American Chemical Society, in comparison to *ZT* for ceramics with identical composition from Ohta et al. [109], Kovalevsky et al. [115], Zhang et al. [29], Wang et al. [120], and Jeric et al. [121]. Temperature dependence of figure of merit *ZT* of stoichiometric $SrTi_{1-x}Nb_xO_{3\pm\delta}$ (STN_x) and 2% Sr deficient $Sr_{0.98}Ti_{1-x}Nb_xO_{3\pm\delta}$ ($nSTN_x$) ceramics with Nb content $x = 10$ and 15% and their composites with rGO (**b**). Reprinted from [118]. Copyright 2018, with permission from Elsevier. Temperature dependence of *ZT* and thermoelectric compatibility factors (inset) for Nb-doped STO samples with different Nb concentrations (**c**). Reprinted from [29] with permission of RSC Publishing, in comparison to *ZT* for La-doped $SrTiO_3$ reported by Kikuchi et al. [106] and Park et al. [107]. Temperature dependence of *ZT* values for $SrTiO_3$ doped with 10 mol.% Nb (Nb10) or with 10 mol.% La (La10), $SrTiO_3$ co-doped with 5 mol.% La and 5 mol.% Nb (La5Nb5), with 5 mol.% La and 10 mol.% Nb (La5Nb10), with 10 mol.% La and 10 mol.% Nb (La10Nb10), with 10 mol.% La and 5 mol.% Nb (La10Nb5), with 10 mol.% La and 20 mol.% Nb (La10Nb20) (**d**). Reprinted from [122]. Copyright 2017, with permission from Elsevier

of 0.6 wt.% of RGO sheets as nano-inclusions [118], indicating that RGO helps to obtain the TE performance of materials in its best-quality state [33].

Furthermore, Zhang et al. reported that TE performance of Nb-doped STO can be enhanced up to a *ZT* of 0.4 at 1100 K for $SrTi_{0.90}Nb_{0.10}O_{3\pm\delta}$ ceramics (see Fig. 17c and its comparison there with La-doped $SrTiO_3$ reported by Kikuchi et al. [106] and Park et al. [107]), using the hydrothermal method for nano-powder fabrication, and muffle furnace sintering at 1300 °C for 5 h within a bed of carbon powder [29]. Using the same preparation method, *ZT* values of 0.6 at 1000 K and 0.65 at 1100 K were reported by Wang et al. for double site doped $Sr_{0.90}La_{0.10}Ti_{0.90}Nb_{0.10}O_{3\pm\delta}$ ceramics (see Fig. 17d) [122].

Recently, *ZT* ~1.42 at 1050 K was reported for $SrTi_{0.85}Nb_{0.15}O_{3\pm\delta}$ with 0.5 wt.% graphite composite [123], although that high value could not be repeated by anybody else, achieving *ZT* of 0.68 at 980 K for $La_{0.07}Sr_{0.93}Ti_{0.93}Nb_{0.07}O_3$ with 1 wt.% graphite composite [124] in the best case.

On the other hand, *ZT* values of Nb-doped STO were reported to be enhanced by nano-inclusions of yttria stabilised zirconia [125], and those of La and Nb co-doped STO by Cu or Fe metallic nano-inclusions [126]. In addition, doping effect of Ta on Ti site in STO ceramics was found to be comparable to TE performance of Nb-doped STO, particularly in case of the Sr deficient composition $Sr_{0.95}Ti_{0.90}Ta_{0.10}O_{3\pm\delta}$, yielding the *ZT* values just of 0.3 at 1000 K and 0.37 at 1230 K as was studied by Yaremchenko et al. [117].

Similar to crystals and ceramics, an obvious effect of doping as well as influence of the oxygen vacancies on the TE properties of metal oxides can be seen on $SrTiO_3$ thin films as was also discussed in the literature [109, 127, 128].

Ohta et al. compared the TE properties of $SrTiO_3$ doped by 20% Nb ceramics with the average grain size ~20 μm with that of polycrystalline film with the average grain size ~200 nm and single-crystalline epitaxial film of the same composition [109]. The carrier concentration and Seebeck coefficient of ceramics and both films were similar, but Hall mobility of the polycrystalline film was found to be rather small <<10^{-1} cm^2 V^{-1} s^{-1} at 300 K in comparison to that of the epitaxial film and ceramics (~3 cm^2 V^{-1} s^{-1}). However, after heating above 700 °C it drastically increased and exhibited a similar value to that of the epitaxial film and ceramics. As a result, no effect of grain size / boundaries on the figure of merit of 20%-Nb-doped STO ceramics as well as both films was found by Ohta et al. at 1000 K where *ZT* of ~0.35 was reported irrespective of the grain size and microstructure [109].

Regarding other STO-based films for TE applications, Kozuka et al. demonstrated that carrier density increased with Nb content while the electron mobility decreased in $SrTiO_3$ films prepared at 1200–1300 °C by pulsed laser deposition (PLD) on $SrTiO_3$ substrates using an ultrahigh vacuum chamber with an infrared laser heating system (see Fig. 18) [127]. Oxygen vacancies were reported to enhance the low temperature charge mobility of STO thin films. As can be seen in Fig. 18, the electron mobility of $SrTiO_3$ thin film is strongly dependent on the temperature and significantly decrease with the temperature increase in the contrast to carrier density that in almost independent on the temperature [127].

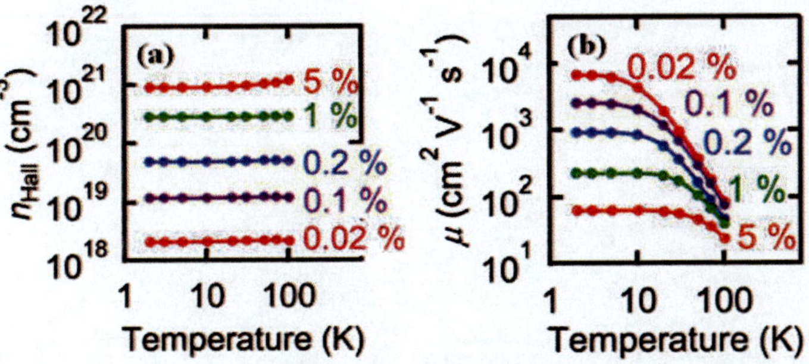

Fig. 18 Carrier density (**a**) and electron mobility (**b**) of $SrTiO_3$ thin films versus temperature for five dopant concentrations, estimated from the Hall effect. Reprinted from [127] with the permission of AIP Publishing

The incorporation of Nb on Ti site in $SrTiO_3$ films deposited by PLD was reported to increase both the electrical conductivity and Seebeck coefficient (see Fig. 19a, b, respectively) similar to ceramics [128]. By using 3ω method for the obtained films with the thickness 300 $\pm$ 10 nm the total cross-plane thermal conductivity of $SrTi_{0.88}Nb_{0.12}O_3$ was found to be 3.1 W m^{-1} K^{-1} at 300 K and the figure of merit was estimated as ZT ~0.29 at 1000 K (see Fig. 19c). Moreover, Seebeck coefficient was reported to decrease in absolute value with carrier concentration due to increase in carrier effective mass m^*/m_0 (m_0 is free electron mass) of the films (see Fig. 19d) [128].

4 Conclusions

Thus, according to the literature review, different approaches can be considered to enhance the TE performance of oxide materials:

- (i) creating lattice defects such as vacancies can involve phonon scattering within the unit cell. Complex crystal structures can be used to separate the electron-crystal from the phonon glass. Here the goal is to achieve a phonon glass without disrupting the crystallinity of the electron-transport region.
- (ii) grain boundary engineering involving nano-inclusions. This approach lies in the phonon scattering at interfaces, via the use of multiphase composites mixed at the nanometre scale and to suppress the grain boundary resistivity as in the case of metallic and graphene-derived nano-inclusions.
- (iii) decrease of the grain size that approaching quantum confinement results in the disruption of the phonon paths, contributing to a decrease of the thermal diffusivity of the material. Indeed, reducing the grain size and nano-structuring has been theoretically reported to lower the value of the thermal conductivity. An increase in the number of such interfaces, as grain boundaries, will increase

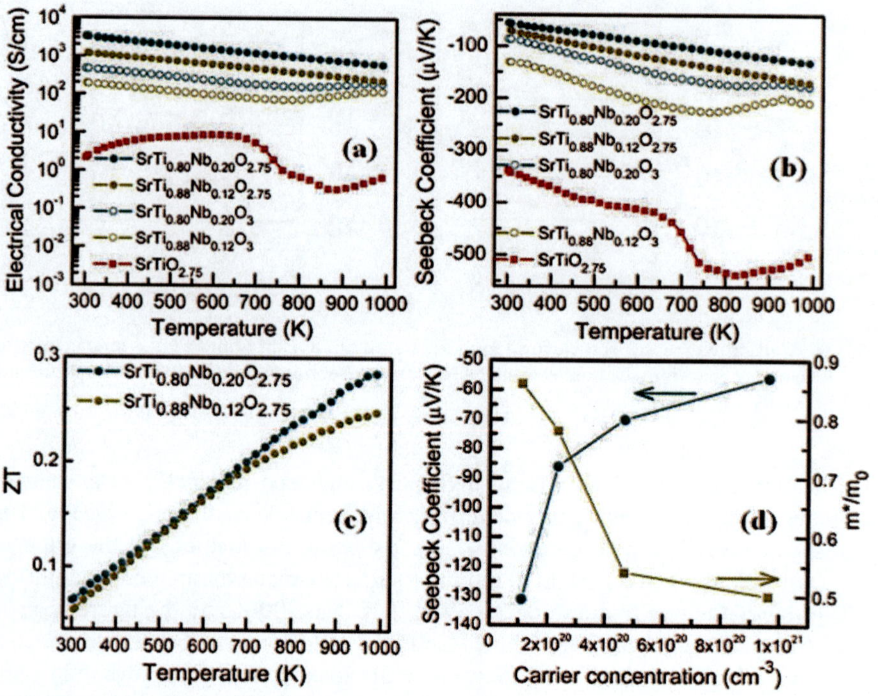

Fig. 19 High temperature electrical conductivity of STO and Nb-doped STO films, measured in Ar/H_2 ambient (the solid symbols correspond to films grown in Ar, while the open symbols correspond to films grown in oxygen (**a**). Seebeck coefficient of STO and Nb-doped STO films as a function of temperature. The solid symbols correspond to films grown in Ar, while the open symbols correspond to films grown in oxygen (**b**). Estimated high temperature *ZT* of Nb-doped STO films grown in Ar (**c**). Dependence of Seebeck coefficient and carrier effective mass on carrier density of the films (**d**). Reprinted with permission from [128]. Copyright 2013 American Chemical Society

the phonon scattering in the polycrystalline structure, leading to a decrease of κ. Additionally, the reduction of the grain size will increase the scattering of the phonons with shorter mean free paths at the grain boundaries, creating a wider range of scattered phonons. However, it can be also problematic, since reducing the grain size leads to an increase in the electrical resistivity, which is not favourable for enhancing the *ZT* values. To overcome such dual effect of grain size, high-performance TE ceramics may need to be a combination of coarse and fine grains in an individual material.

Acknowledgements This work was supported by national funds, through FCT (Fundacao para a Ciencia e a Tecnologia) in the scope of the framework contract foreseen in numbers 4,5 and 6 of article 23 of the Decree Law 57/2016, of 29 August, UIDB/00481/2020 and UIDP/00481/2020; and CENTRO-01-0145-FEDER-022083—Centro Portugal Regional Operational Programme (Centro2020), under the PORTUGAL 2020 Partnership Agreement, through the European Regional Development Fund and developed within the scope of the project CICECO-Aveiro Institute of Materials, UIDB/50011/2020, UIDP/50011/2020 & LA/P/0006/2020, financed by national funds through the FCT/MEC (PIDDAC) as well as within FCT independent researcher grant 2021.02284.CEECIND.

References

1. X.F. Zheng, C.X. Liu, Y.Y. Yan, Q. Wang, A review of thermoelectrics research—Recent developments and potentials for sustainable and renewable energy applications. Renew. Sust. Energ. Rev. **32**, 486–503 (2014). https://doi.org/10.1016/j.rser.2013.12.053
2. M.A. Alghoul, S.A. Shahahmadi, B. Yeganeh, N. Asim, A.M. Elbreki, K. Sopian, S.K. Tiong, N. Amin, A review of thermoelectric power generation systems: roles of existing test rigs/prototypes and their associated cooling units on output performance. Energy Convers. Manag. **174**, 138–156 (2018). https://doi.org/10.1016/j.enconman.2018.08.019
3. L.C. Ding, A. Tan, A review of power generation with thermoelectric system and its alternative with solar ponds. Renew. Sust. Energ. Rev. **81**, 799–812 (2018). https://doi.org/10.1016/j.rser.2017.08.010
4. M.Q. Khan, S. Malarmannan, G. Manikandaraja, Power generation from waste heat of vehicle exhaust using thermos electric generator: a review. Mater. Sci. Eng. **402**, 012174 (2018). https://doi.org/10.1088/1757-899X/402/1/012174
5. G.J. Snyder, Small thermoelectric generators. Electrochem. Soc. Interface **17**, 54–56 (2008). https://doi.org/10.1149/2.F06083IF
6. S. Bensaid, M. Brignone, A. Ziggiotti, S. Specchia, High efficiency thermo-electric power generator. Int. J. Hydrog. Energy **37**, 1385–1398 (2012). https://doi.org/10.1016/j.ijhydene.2011.09.125
7. M.A. Zoui, S. Bentouba, J.G. Stocholm, M. Bourouis, A review on thermoelectric generators: progress and applications. Energies **13**, 3606 (2020). https://doi.org/10.3390/en13143606
8. T.J. Seebeck. *Magnetische Polarisation der Metalle und Erze durch Temperatur-Differenz (Magnetic polarization of metals and minerals by temperature differences)*. Abhandlungen der Königlichen Akademie der Wissenschaften zu Berlin (Treatises of the Royal Academy of Sciences in Berlin) (1825), pp. 265–373
9. T.J. Seebeck, Ueber die Magnetische Polarisation der Metalle und Erze durch Temperatur-Differenz (On the magnetic polarization of metals and minerals by temperature differences). Annalen der Physik und Chemie **6**, 1–20 (1826)
10. G.J. Snyder. Thermoelectric power generation: efficiency and compatibility, in *Thermoelectrics Handbook: Macro to Nano*, ed. by D.M. Rowe (CRC Press, Taylor & Francis Group, Boca Raton, FL, USA, 2006)
11. S. Lv, Z. Qian, D. Hu, X. Li, W. He, A comprehensive review of strategies and approaches for enhancing the performance of thermoelectric module. Energies **13**, 3142 (2020). https://doi.org/10.3390/en13123142
12. G. Snyder, E. Toberer, Complex thermoelectric materials. Nat. Mater. **7**, 105–114 (2008). https://doi.org/10.1038/nmat2090
13. P.M. Kumar, V.J. Babu, A. Subramanian, A. Bandla, N. Thakor, S. Ramakrishna, H. Wei. The design of a thermoelectric generator and its medical applications. Designs **3**, 22 (2019) https://doi.org/10.3390/designs3020022
14. K. Gaurav, S.K. Pandey, Efficiency calculation of a thermoelectric generator for investigating the applicability of various thermoelectric materials. J. Renew. Sustain. Energy **9**, 014701 (2017). https://doi.org/10.1063/1.4976125
15. H.S. Kim, Z.M. Gibbs, Y. Tang, H. Wang, J. Snyder, Characterization of Lorenz number with Seebeck coefficient measurement. APL Mater. **3**, 41506 (2015). https://doi.org/10.1063/1.4908244

16. A.F. Ioffe, *Physics of Semiconductors* (Academic Press, New York, NY, USA, 1960)
17. T. Zhu, Y. Liu, C. Fu, J.P. Heremans, J.G. Snyder, X. Zhao, Compromise and synergy in high-efficiency thermoelectric materials. Adv. Mater. **29**, 1605884 (2017). https://doi.org/10.1002/adma.201605884
18. Y. Yu, M. Cagnoni, O. Cojocaru-Mirédin, M. Wuttig, Chalcogenide thermoelectrics empowered by an unconventional bonding mechanism. Adv. Funct. Mater. **30**, 1904862 (2020). https://doi.org/10.1002/adfm.201904862
19. G. Rogl, P. Rogl, Skutterudites, a most promising group of thermoelectric materials. Curr. Opin. Green Sustain. Chem. **4**, 50–57 (2017). https://doi.org/10.1016/j.cogsc.2017.02.006
20. J. Lian. Metal oxide-based thermoelectric materials, in *Metal Oxides in Energy Technologies* (2018), pp. 49–72. https://doi.org/10.1016/B978-0-12-811167-3.00003-1
21. K. Koumoto, R. Funahashi, E. Guilmeau, Y. Miyazaki, A. Weidenkaff, Y. Wang, C. Wan, X.-D. Zhou, Thermoelectric ceramics for energy harvesting. J. Am. Ceram. Soc. **96**, 1–23 (2013). https://doi.org/10.1111/jace.12076
22. R. Freer, A.V. Powel, Realising the potential of thermoelectric technology: a roadmap. J. Mater. Chem. C **8**, 441–463 (2020). https://doi.org/10.1039/c9tc05710b
23. C. Chang, M. Wu, D. He, Y. Pei, C.-F. Wu, X. Wu, H. Yu, F. Zhu, K. Wang, Y. Chen, L. Huang, J.-F. Li, J. He, L.-D. Zhao, 3D charge and 2D phonon transports leading to high out-of-plane ZT in n-type SnSe crystals. Science **360**, 778–783 (2018). https://doi.org/10.1126/science.aaq1479
24. L.-D. Zhao, S.-H. Lo, Y. Zhang, H. Sun, G. Tan, C. Uher, C. Wolverton, V.P. Pravid, M.G. Kanatzidis, Ultralow thermal conductivity and high thermoelectric figure of merit in SnSe crystals. Nature **508**, 373–377 (2014). https://doi.org/10.1038/nature13184
25. K. Biswas, J. He, I.D. Blum, C.-I. Wu, T.P. Hogan, D.N. Seidman, V.P. Dravid, M.G. Kanatzidis, High-performance bulk thermoelectrics with all-scale hierarchical architectures. Nature **489**, 414–418 (2012). https://doi.org/10.1038/nature11439
26. S.I. Kim, K.H. Lee, H.A. Mun, H.S. Kim, S.W. Hwang, J.W. Roh, D.J. Yang, W.H. Shin, X.S. Li, Y.H. Lee, G.J. Snyder, S.W. Kim, Dense dislocation arrays embedded in grain boundaries for high-performance bulk thermoelectrics. Science **348**, 109–114 (2012). https://doi.org/10.1126/science.aaa4166
27. D. Beretta, N. Neophytou, J.M. Hodges, M.G. Kanatzidis, D. Narducci, M. Martin-Gonzalez, M. Beekman, B. Balke, G. Cerretti, W. Tremel, A. Zevalkink, A.I. Hofmann, C. Müller, B. Dörling, M. Campoy-Quiles, M. Caironi, Thermoelectrics: from history, a window to the future. Mater. Sci. Eng. R. Rep. **138**, 100501 (2019). https://doi.org/10.1016/j.mser.2018.09.001
28. H.H. Wang, F. Chen, S.Y. Dai, T. Zhao, H.B. Lu, D.F. Cui, Y.I. Zhou, Z.H. Chen, G.Z. Yang, Sb-doped $SrTiO_3$ transparent semiconductor thin films. Appl. Phys. Lett. **78**, 1676 (2001). https://doi.org/10.1063/1.1355992
29. B. Zhang, J. Wang, T. Zou, S. Zhang, X. Yaer, N. Ding, C. Liu, L. Miao, Y. Lia, Y. Wu, High thermoelectric performance of Nb-doped $SrTiO_3$ bulk materials with different doping levels. J. Mater. Chem. C **3**, 11406–11411 (2015). https://doi.org/10.1039/C5TC02016F
30. H. Muta, K. Kurosaki, S. Yamanaka, Thermoelectric properties of rare earth doped $SrTiO_3$. J. Alloys Compd. **350**, 292–295 (2003). https://doi.org/10.1016/S0925-8388(02)00972-6
31. Z. Ma, J. Wei, P. Song, M. Zhang, L. Yang, J. Ma, W. Liu, F. Yang, X. Wang, Review of experimental approaches for improving zT of thermoelectric materials. Mater. Sci. Semicond. Process. **121**, 105303 (2021). https://doi.org/10.1016/j.mssp.2020.105303
32. A.K. Geim, K.S. Novoselov, The rise of graphene. Nat. Mater. **6**, 183–191 (2007). https://doi.org/10.1038/nmat1849
33. O. Okhay, A. Tkach. Impact of graphene or reduced graphene oxide on performance of thermoelectric composites **C 7**, 37 (2021). https://doi.org/10.3390/c7020037
34. I. Terasaki, Y. Sasago, K. Uchinokura, Large thermoelectric power in $NaCo_2O_4$ single crystals. Phys. Rev. B **56**, R12685 (1997). https://doi.org/10.1103/PhysRevB.56.R12685
35. A. Satake, H. Tanaka, T. Ohkawa, T. Fujii, I. Terasaki, Thermal conductivity of the thermoelectric layered cobalt oxides measured by the Harman method. J. Appl. Phys. **96**, 931 (2004). https://doi.org/10.1063/1.1753070

36. A. Nag, V. Shubha, Oxide thermoelectric materials: a structure-property relationship. J. Electron. Mater. **43**, 962–977 (2014). https://doi.org/10.1007/s11664-014-3024-6
37. M. Lee, L. Viciu, L. Li, Y. Wang, M.L. Foo, S. Watauchi, R.A. Pascal Jr., R.J. Cava, N.P. Ong, Large enhancement of the thermopower in Na_xCoO_2 at high Na doping. Nat. Mater. **5**, 537–540 (2006). https://doi.org/10.1038/nmat1669
38. M.O. Erdal, M. Koyuncu, I. Uslu, The effect of synthesis technique on thermoelectric properties of nanocrystalline $NaCo_2O_4$ ceramics. J. Nanopart. Res. **16**, 2715 (2014). https://doi.org/10.1007/s11051-014-2715-5
39. E.M. Jakubczyk, A. Mapp, C.C. Chung, C.L. Sansom, J.L. Jones, R.A. Dorey, Enhancing thermoelectric properties of $NaCo_2O_4$ ceramics through Na pre-treatment induced nano-decoration. J. Alloys Compd. **788**, 91–101 (2019). https://doi.org/10.1016/j.jallcom.2019.02.199
40. E. Ermawan, S. Poertadji, Thermoelectric properties enhancement of $NaCo_2O_4$ through Ca atom partial substitution on Na atom. Adv. Mater. Res. **1123**, 140–144 (2015). https://doi.org/10.4028/www.scientific.net/AMR.1123.140
41. K. Park, K.U. Jang, H.-C. Kwon, J.-G. Kim, W.-S. Cho, Influence of partial substitution of Cu for Co on the thermoelectric properties of $NaCo_2O_4$. J. Alloys Compd. **419**, 213–219 (2006). https://doi.org/10.1016/j.jallcom.2005.08.081
42. K. Kurosaki, H. Muta, M. Uno, S. Yamanaka, Thermoelectric properties of $NaCo_2O_4$. J. Alloys Compd. **315**, 234–236 (2001). https://doi.org/10.1016/S0925-8388(00)01277-9
43. M.O. Erdal, M. Koyncu, M.L. Aksu, I. Uslu, S. Koçyiğit, Thermoelectric properties of Nickel and Boron Co-substituted $NaCo_2O_4$ prepared by electrospinning technique. Nano Hybrids Compos. **19**, 34–45 (2018). https://doi.org/10.4028/www.scientific.net/NHC.19.34
44. K.M. Ryu, Y.H. Kang, S.Y. Cho, S.H. Lee, Y.C. Choi, M.S. Kim, Y.G. Jeong, Microstructure and thermoelectric characterization of composite nanofiber webs derived from polyacrylonitrile and sodium cobalt oxide precursors. Sci. Rep. **10**, 9633 (2020). https://doi.org/10.1038/s41598-020-66667-6
45. W. Zhang, K. Zhu, J. Liu, J. Wang, K. Yan, P. Liu, Y. Wang, Enhanced thermoelectric properties of nano-SiC dispersed $NaCo_2O_4$ composites. Funct. Mater. Lett. **12**, 1950009 (2019). https://doi.org/10.1142/S1793604719500097
46. R. Funahashi, M. Shikano, $Bi_2Sr_2Co_2O_y$ whiskers with high thermoelectric figure of merit. Appl. Phys. Lett. **81**, 1459 (2002). https://doi.org/10.1063/1.1502190
47. G.Ç. Karakaya, B. Özçelik, O. Nane, A. Sotelo, S. Rasekh, M.A. Torres, M.A. Madre, Improvement of $Bi_2Sr_2Co_2O_y$ thermoelectric performances by Na doping. J. Electroceram. **40**, 11 (2018). https://doi.org/10.1007/s10832-017-0078-x
48. T. Caillat, J.-P. Fleurial, A. Borshchevsky, Preparation and thermoelectric properties of semi-conducting Zn_4Sb_3. J. Phys. Chem. Solids **58**, 1119 (1997). https://doi.org/10.1016/S0022-3697(96)00228-4
49. E. Skrabek, D.S. Trimmer. Properties of the general tags system, in *Handbook of Thermoelectrics*, ed. by D.M. Rowe (CRC, Boca Raton, FL, 1995)
50. B.C. Sales, D. Mandrus, B.C. Chakoumakos, V. Keppens, J.R. Thompson, Filled skutterudite antimonides: electron crystals and phonon glasses. Phys. Rev. B **56**, 15081–15089 (1997). https://doi.org/10.1103/PhysRevB.56.15081
51. J.P. Dismukes, L. Ekstron, E.F. Steigmeier, I. Kudman, D.S. Beers, Thermal and electrical properties of heavily doped Ge-Si alloys up to 1300 K. J. Appl. Phys. **35**, 2899 (1964). https://doi.org/10.1063/1.1713126
52. M. Shikano, R. Funahashi. Electrical and thermal properties of single-crystalline $(Ca_2CoO_3)_{0.7}CoO_2$ with a $Ca_3Co_4O_9$ structure. Appl. Phys. Lett. **82**, 1851(2003). https://doi.org/10.1063/1.1562337
53. Y. Miyazaki, K. Kudo, M. Akoshima, Y. Ono, Y. Koike, T. Kajitani. Low-temperature thermoelectric properties of the composite crystal $[Ca_2CoO_{3.34}]_{0.614}[CoO_2]$. Jpn. J. Appl. Phys. **39**, L531–L533 (2000). https://doi.org/10.1143/JJAP.39.L531
54. M.A. Madre, F.M. Costa, N.M. Ferreira, A. Sotelo, M.A. Torres, G. Constantinescu, S. Rasekh, J.C. Diez, Preparation of high-performance $Ca_3Co_4O_9$ thermoelectric ceramics produced by

a new two-step method. J. Eur. Ceram. Soc. **33**, 1747–1754 (2013). https://doi.org/10.1016/j.jeurceramsoc.2013.01.029

55. A.M. dos Santos, D. Thomazini, M.V. Gelfuso, Cold sintering and thermoelectric properties of $Ca_3Co_4O_9$ ceramics. Ceram. Int. **46**, 14064 (2020). https://doi.org/10.1016/j.ceramint.2020.02.206
56. G. Xu, R. Funahashi, M. Shikano, I. Matsubara, Y. Zhou, Thermoelectric properties of the Bi- and Na-substituted $Ca_3Co_4O_9$ system. Appl. Phys. Lett. **80**, 3760 (2002). https://doi.org/10.1063/1.1480115
57. S.I. Seok, D.S. Park, S.J. Park, The effect of Cu-sites doping on electrical properties of La_2CuO_4 as thermoelectric material. J. Mater. Sci.: Mater. Electron. **6**, 140–143 (1995). https://doi.org/10.1007/BF00190544
58. C. Zhu, J. Gong, Z. Li, H. Zhou, G. Tang, Investigation of the thermoelectric properties of La_2CuO_4 ceramics with the addition of Ag. Phys. B **456**, 26 (2015). https://doi.org/10.1016/j.physb.2014.08.016
59. S. Liu, B. Qin, D. Wang, L.-D. Zhao. Investigations on the thermoelectric transport properties in the hole-doped La_2CuO_4. Z. Anorg. Allg. Chem. e202200036 (2022). https://doi.org/10.1002/zaac.202200036
60. F.F. Song, L. Wu, S. Liang, Giant Seebeck coefficient thermoelectric device of MnO_2 powder. Nanotech **23**, 085401 (2012). https://doi.org/10.1088/0957-4484/23/8/085401
61. A. Kumar, K. Kumari, C.V. Tomy, A.D. Thakur, Magnetic and thermoelectric properties of α-MnO_2 nanorods. AIP Conf. Proc. **2220**, 080008 (2020). https://doi.org/10.1063/5.0001333
62. M. Chepkoech, D.P. Joubert, G.O. Amolo, First principles calculations of the thermoelectric properties of α-MnO_2 and β-MnO_2. Eur. Phys. J. B. **91**, 301 (2018). https://doi.org/10.1140/epjb/e2018-90321-4
63. A. Bibi, A. Shakoor, Electrical, structural, and thermo-electric power studies of polypyrrole-MnO_2 composites. Polym. Sci. Ser. B **63**, 606 (2021). https://doi.org/10.1134/S1560090421050018
64. M. Hedden, N. Francis, J.T. Haraldsen, T. Ahmed, C. Constantin, Thermoelectric properties of nano-meso-micro β-MnO_2 powders as a function of electrical resistance. Nanosc. Res. Lett. **10**, 292 (2015). https://doi.org/10.1186/s11671-015-1000-6
65. S. Walia, S. Balendhran, P. Yi, D. Yao, S. Zhuiykov, M. Pannirselvam, R. Weber, M.S. Strano, M. Bhaskaran, S. Sriram, K. Kalantar-zadeh, MnO_2-based thermopower wave sources with exceptionally large output voltages. J. Phys. Chem. C **117**, 9137 (2013). https://doi.org/10.1021/jp401731b
66. A.K.M.F.U. Islam, R. Islam, K.A. Khan. Studies on the thermoelectric effect in semiconducting MnO_2 thin films. J. Mater. Sci.: Mater. Electron. **16**, 203–207 (2005). https://doi.org/10.1007/s10854-005-0766-1
67. X. Xia, H. Li, Z.-H. Chen, The study of semiconduction properties of γ-MnO_2 with different degrees of reduction. J. Electrochem. Soc. **136**, 266 (1989). https://doi.org/10.1149/1.2096600
68. E. Preisler, Semiconductor properties of manganese dioxide. J. Appl. Electrochem. **6**, 311–320 (1976). https://doi.org/10.1007/BF00608916
69. S.K. Bhargande, P.S. Patil, Studies on structural, optical and electrical properties of spray deposited manganese oxide (MnO_2) thin films. Res. Rev.: J. Pure Appl. Phys. **1**, 11–16 (2013)
70. Y. Lu, K. Sagara, Y. Matsuda, L. Hao, Y.R. Jin, H. Yoshida, Effect of Cu powder addition on thermoelectric properties of Cu/TiO_{2-x} composites. Ceram. Int. **39**, 6689–6694 (2013). https://doi.org/10.1016/j.ceramint.2013.01.107
71. Z.-H. Wu, H.-Q. Xie, Y.-B. Zhai, L.-H. Gan, J. Liu, Enhanced thermoelectric performance of TiO_2-based hybrid materials by incorporating conducting polymer. Chin. Phys. B **24**, 034402 (2015). https://doi.org/10.1088/1674-1056/24/3/034402
72. W.H. Nam, Y.S. Lim, W. Kim, H.K. Seo, K.S. Dae, S. Lee, W.-S. Seo, J.Y. Lee, A gigantically increased ratio of electrical to thermal conductivity and synergistically enhanced thermoelectric properties in interface-controlled TiO_2-RGO nanocomposites. Nanoscale **9**, 7830 (2017). https://doi.org/10.1039/c7nr00911a

73. H. Liu, H. Ma, T. Su, Y. Zhang, B. Sun, B. Liu, L. Kong, B. Liu, X. Jia, High-thermoelectric performance of TiO_{2-x} fabricated under high pressure at high temperatures. J. Mater. **3**, 286–292 (2017). https://doi.org/10.1016/j.jmat.2017.06.002
74. M. Mikami, K. Ozaki, Thermoelectric properties of nitrogen-doped TiO_{2-x} compounds. J. Phys.: Conf. Ser. **379**, 012006 (2012). https://doi.org/10.1088/1742-6596/379/1/012006
75. X. Liang, Y. Yang, F. Dai, C. Wang, Orientation dependent physical transport behavior and the micro-mechanical response of ZnO nanocomposites induced by SWCNTs and graphene: importance of intrinsic anisotropy and interface. J. Mater. Chem. C **7**, 1208–1221 (2019). https://doi.org/10.1039/c8tc05148h
76. H. Colder, E. Guilmeau, C. Harnois, S. Marinel, R. Retoux, E. Savary, Preparation of Ni-doped ZnO ceramics for thermoelectric applications. J. Eur. Ceram. Soc. **31**, 2957–2963 (2011). https://doi.org/10.1016/j.jeurceramsoc.2011.07.006
77. X. Liang, Thermoelectric transport properties of naturally nanostructured Ga-ZnO ceramics: effect of point defect and interfaces. J. Eur. Ceram. Soc. **36**, 1643–1650 (2016). https://doi.org/10.1016/j.jeurceramsoc.2016.02.017
78. P. Jood, R.J. Mehta, Y. Zhang, T. Borca-Tasciuc, S.X. Dou, D.J. Singh, G. Ramanath, Heavy element doping for enhancing thermoelectric properties of nanostructured zinc oxide. RSC Adv. **4**, 6363 (2014). https://doi.org/10.1039/c3ra46813e
79. T. Tsubota, M. Ohtaki, K. Eguchi, H. Arai, Thermoelectric properties of Al-doped ZnO as a promising oxide material for high temperature thermoelectric conversion. J. Mater. Chem. **7**, 85–90 (1997). https://doi.org/10.1039/A602506D
80. D.-B. Zhang, H.-Z. Li, B.-P. Zhang, D.-D. Liang, M. Xia, Hybrid-structured ZnO thermoelectric materials with high carrier mobility and reduced thermal conductivity. RSC Adv. **7**, 10855 (2017). https://doi.org/10.1039/c6ra28854e
81. R.V.R. Virtudazo, B. Srinivasan, Q. Guo, R. Wu, T. Takei, Y. Shimasaki, H. Wada, K. Kuroda, S. Bernik, T. Mori, Improvement in the thermoelectric properties of porous networked Al-doped ZnO nanostructured materials synthesized via an alternative interfacial reaction and low-pressure SPS processing. Inorg. Chem. Front. **7**, 4118 (2020). https://doi.org/10.1039/d0qi00888e
82. K.F. Cai, E. Müller, C. Drašar, A. Mrotzek, Preparation and thermoelectric properties of Al-doped ZnO ceramics. Mater. Sci. Eng. B **104**, 45 (2003). https://doi.org/10.1016/S0921-5107(03)00280-0
83. S. Jantrasee, P. Moontragoon, S. Pinitsoontorn, Thermoelectric properties of Al-doped ZnO: experiment and simulation. J. Semicond. **37**, 092002 (2016). https://doi.org/10.1088/1674-4926/37/9/092002
84. L. Han, N.V. Nong, L.T. Hung, T. Holgate, N. Pryds, M. Ohtaki, S. Linderoth, The influence of a- and c-Al_2O_3 phases on the thermoelectric properties of Al-doped ZnO. J. Alloys. Compd. **555**, 291–296 (2013). https://doi.org/10.1016/j.jallcom.2012.12.091
85. M. Søndergaard, E.D. Bøjesen, K.A. Borup, S. Christensen, M. Christensen, B.B. Iversen, Sintering and annealing effects on ZnO microstructure and thermoelectric properties. Acta Mater. **61**, 3314–3323 (2013). https://doi.org/10.1016/j.actamat.2013.02.021
86. M. Ohtaki, K. Araki, K. Yamamoto, High thermoelectric performance of dually doped ZnO ceramics. J. Electron. Mater. **38**, 1234–1238 (2009). https://doi.org/10.1007/s11664-009-0816-1
87. H. Yamaguchi, Y. Chonan, M. Oda, T. Komiyama, T. Aoyama, S. Sugiyama, Thermoelectric properties of ZnO Ceramics Co-doped with Al and transition metals. J. Electron. Mater. **40**, 723–727 (2011). https://doi.org/10.1007/s11664-011-1529-9
88. C. Dreβler, R. Löhnert, J. Gonzalez-Julian, O. Guillon, J. Töpfer, S. Teichert, Effect of carbon nanotubes on thermoelectric properties in $Zn_{0.98}Al_{0.02}$. J. Electron. Mat. **45**, 1459–1463 (2016). https://doi.org/10.1007/s11664-015-4070-4
89. D. Chen, Y. Zhao, Y. Chen, B. Wang, H. Chen, J. Zhou, Z. Liang, One-step chemical synthesis of ZnO/graphene oxide molecular hybrids for high-temperature thermoelectric applications. ACS Appl. Mater. Interfaces **7**, 3224 (2015). https://doi.org/10.1021/am507882f

90. S. Biswas, S. Singh, S. Singh, S. Chattopadhyay, K. Kanishka H. De Silva, M. Yoshimura, J. Mitra, V.B. Kamble. Selective enhancement in phonon scattering leads to a high thermoelectric figure-of-merit in graphene oxide-encapsulated ZnO nanocomposites. ACS Appl. Mater. Interfaces **13**, 23771–23786 (2021). https://doi.org/10.1021/acsami.1c04125
91. S. Paengson, T. Seetawan, P. Muthitamongkol, C. Thanachayanont, Characterization p-$Ca_3Co_4O_9$ and n-$CaMnO_3$. Adv. Mater. Res. **802**, 209–212 (2013). https://doi.org/10.4028/www.scientific.net/AMR.802.209
92. R. Kabir, R. Tian, T. Zhang, R. Donelson, T.T. Tan, S. Li, Role of Bi doping in thermoelectric properties of $CaMnO_3$. J. Alloys Compd. **628**, 347–351 (2015). https://doi.org/10.1016/j.jallcom.2014.12.141
93. M. Mouyane, B. Itaalit, J. Bernard, D. Houivet, J.G. Noudem, Flash combustion synthesis of electron doped-$CaMnO_3$ thermoelectric oxides. Powder Technol. **264**, 71–77 (2014). https://doi.org/10.1016/j.powtec.2014.05.022
94. J. Lan, Y.-H. Lin, H. Fang, A. Mei, C.-W. Nan, Y. Liu, S. Xu, M. Peters, High-temperature thermoelectric behaviors of fine-grained Gd-doped $CaMnO_3$ ceramics. J. Am. Ceram. Soc. **93**, 2121–2124 (2010). https://doi.org/10.1111/j.1551-2916.2010.03673.x
95. A. Chik, R. Mohd Zaki, A.A. Adewale, F. Che Pa, C.K. Yeoh, Thermoelectric properties of Sm doped $CaMnO_3$ using density functional theory method. Mater. Sci. Forum. **1010**, 334–338 (2020). https://doi.org/10.4028/www.scientific.net/MSF.1010.334
96. Y. Zhu, C. Wang, W. Su, J. Liu, J. Li, Y. Du, X. Zhang, Y. Qin, L. Mei. Enhanced thermoelectric response of $Ca_{0.96}Dy_{0.02}Re_{0.02}MnO_3$ ceramics (Re = La, Nd, Sm) at high temperature. J. Electron. Mater. **44**, 414–419 (2015). https://doi.org/10.1007/s11664-014-3450-5
97. Y.-H. Zhu, W.-B. Su, J. Liu, Y.-C. Zhou, J. Li, X. Zhang, Y. Du, C.-L. Wang, Effects of Dy and Yb co-doping on thermoelectric properties of $CaMnO_3$ ceramics. Ceram. Int. **41**, 1535–1539 (2015). https://doi.org/10.1016/j.ceramint.2014.09.089
98. C. Li, Q. Chen, Y. Yan, Effects of Pr and Yb dual doping on the thermoelectric properties of $CaMnO_3$. Materials **11**, 1807 (2018). https://doi.org/10.3390/ma11101807
99. Y. Li, J. Liu, J.-C. Li, Y.-F. Chen, X.-M. Zhang, X.-J. Wang, F.-N. Wang, W.-B. Su, L.-L. Zhao, C.-L. Wang, Electron localization in niobium doped $CaMnO_3$ due to the energy difference of electronic states of Mn and Nb. Phys. Chem. Chem. Phys. **20**, 20571 (2018). https://doi.org/10.1039/c8cp02783h
100. P. Thiel, J. Eilertsen, S. Populoh, G. Saucke, M. Döbeli, A. Shkabko, L. Sagarna, L. Karvonen, A. Weidenkaff, Influence of tungsten substitution and oxygen deficiency on the thermoelectric properties of $CaMnO_{3-\delta}$. J. Appl. Phys. **114**, 243707 (2013). https://doi.org/10.1063/1.4854475
101. J.W. Park, D.H. Kwak, S.H. Yoon, S.C. Choi, Thermoelectric properties of Bi, Nb co-substituted $CaMnO_3$ at high temperature. J. Alloys Compd. **487**, 550–555 (2009). https://doi.org/10.1016/j.jallcom.2009.08.012
102. H.P.R. Frederikse, W.R. Thurber, W.R. Hosler, Electronic transport in strontium titanate. Phys. Rev. **134**, A442–A445 (1964). https://doi.org/10.1103/PhysRev.134.A442
103. X. Feng, Y. Fan, N. Nomura, K. Kikuchi, L. Wang, W. Jiang, A. Kawasaki, Graphene promoted oxygen vacancies in perovskite for enhanced thermoelectric properties. Carbon **112**, 169–176 (2017). https://doi.org/10.1016/j.carbon.2016.11.012
104. T. Okuda, K. Nakanishi, S. Miyasaka, Y. Tokura. Large Thermoelectric Response of Metallic Perovskites: $Sr_{1-x}La_xTiO_3$ ($0 < x < 0.1$). Phys. Rev. B **63**, 113104 (2001). https://doi.org/10.1103/PhysRevB.63.113104
105. L. Zhang, T. Tosyo, N. Okinaka, T. Akiyama, Thermoelectric properties of combustion-synthesized lanthanum-doped strontium titanate. Mater. Trans. **48**, 1079–1083 (2007). https://doi.org/10.2320/matertrans.48.1079
106. A. Kikuchi, N. Okinaka, T. Akiyama, A large thermoelectric figure of merit of La-doped $SrTiO_3$ prepared by combustion synthesis with post-spark plasma sintering. Scr. Mater. **63**, 407–410 (2010). https://doi.org/10.1016/j.scriptamat.2010.04.041
107. K. Park, J.S. Son, S.I. Woo, K. Shin, M.-W. Oh, S.-D. Park, T. Hyeon, Colloidal synthesis and thermoelectric properties of La-doped $SrTiO_3$ nanoparticles. J. Mater. Chem. A **2**, 4217–4224 (2014). https://doi.org/10.1039/C3TA14699E

108. Z. Lu, H. Zhang, W. Lei, D.C. Sinclair, I.M. Reaney, High-figure-of-merit thermoelectric La-doped A-site-deficient $SrTiO_3$ ceramics. Chem. Mater. **28**, 925–935 (2016). https://doi.org/10.1021/acs.chemmater.5b04616
109. S. Ohta, H. Ohta, M. Hirano, K. Koumoto, Grain size dependence of thermoelectric performance of Nb-doped $SrTiO_3$ polycrystals. J. Ceram. Soc. Jpn. **114**, 102–105 (2006). https://doi.org/10.2109/jcersj.114.102
110. Y. Lin, C. Norman, D. Srivastava, F. Azough, L. Wang, M. Robbins, K. Simpson, R. Freer, I.A. Kinloch, Thermoelectric power generation from lanthanum strontium titanium oxide at room temperature through the addition of graphene. ACS Appl. Mater. Interfaces **7**, 15898–15908 (2015). https://doi.org/10.1021/acsami.5b03522
111. A.V. Kovalevsky, A.A. Yaremchenko, S. Populoh, P. Thiel, D.P. Fagg, A. Weidenkaff, J.R. Frade, Towards a high thermoelectric performance in rare-earth substituted $SrTiO_3$: effects provided by strongly-reducing sintering conditions. Phys. Chem. Chem. Phys. **16**, 26946–26954 (2014). https://doi.org/10.1039/C4CP04127E
112. O. Okhay, G. Gonçalves, C. Dias, J. Ventura, E.M.F. Vieira, L.M.V. Gonçalves, A. Tkach, Tuning electrical and thermoelectric properties of freestanding graphene oxide papers by carbon nanotubes and heat treatment. J. Alloys Compd. **781**, 196–200 (2019). https://doi.org/10.1016/j.jallcom.2018.12.033
113. A.M. Dehkordi, S. Bhattacharya, J. He, N. Alshareef, Significant enhancement in thermoelectric properties of polycrystalline Pr-doped $SrTiO_3$ ceramics originating from nonuniform distribution of Pr dopants. Appl. Phys. Lett. **104**, 193902 (2014). https://doi.org/10.1063/1.4875925
114. A.M. Dehkordi, S. Bhattacharya, T. Darroudi, J.W. Graff, U. Schwingenschlogl, H.N. Alshareef, T.M. Tritt, Large thermoelectric power factor in Pr-doped $SrTiO_{3-\delta}$ ceramics via grain-boundary-induced mobility enhancement. Chem. Mater. **26**, 2478–2485 (2014). https://doi.org/10.1021/cm4040853
115. A.V. Kovalevsky, A.A. Yaremchenko, S. Populoh, A. Weidenkaff, J.R. Frade, Effect of a-site cation deficiency on the thermoelectric performance of donor-substituted strontium titanate. J. Phys. Chem. C **118**, 4596–4606 (2014). https://doi.org/10.1021/jp409872e
116. C. Chen, T. Zhang, R. Donelson, T.T. Tan, S. Li. Effects of yttrium substitution and oxygen deficiency on the crystal phase, microstructure, and thermoelectric properties of $Sr_{1-1.5x}Y_xTiO_{3-\delta}$ $(0 \leq x \leq 0.15)$. J. Alloys Compd. **629**, 49–54 (2015). https://doi.org/10.1016/j.jallcom.2014.12.125
117. A.A. Yaremchenko, S. Populoh, S.G. Patricio, J. Macias, P. Thiel, D.P. Fagg, A. Weidenkaff, J.R. Frade, A.V. Kovalevsky, Boosting thermoelectric performance by controlled defect chemistry engineering in Ta-substituted strontium titanate. Chem. Mater. **27**, 4995–5006 (2015). https://doi.org/10.1021/acs.chemmater.5b01389
118. O. Okhay, S. Zlotnik, W. Xie, K. Orlinski, M. J. Hortiguela Gallo, G. Otero-Irurueta, A.J.S. Fernandes, D.A. Pawlak, A. Weidenkaff, A. Tkach. Thermoelectric performance of Nb-doped $SrTiO_3$ enhanced by reduced graphene oxide and Sr deficiency cooperation. Carbon **143**, 215–222 (2019). https://doi.org/10.1016/j.carbon.2018.11.023
119. A. Tkach, J. Resende, K.V. Saravanan, M.E. Costa, P. Diaz Chao, E. Guilmeau, O. Okhay, P.M. Vilarinho. Abnormal grain growth as a method to enhance the thermoelectric performance of Nb-doped $SrTiO_3$ ceramics. ACS Sustain. Chem. Eng. **6**, 15988–15994 (2018). https://doi.org/10.1021/acssuschemeng.8b03875
120. J. Wang, X. Ye, X. Yaer, B. Zhang, W. Ma, L. Miao, High thermoelectric performance of niobium-doped strontium titanate bulk material affected by all-scale grain boundary and inclusions. Scr. Mater. **99**, 25–28 (2015). https://doi.org/10.1016/j.scriptamat.2014.11.018
121. M. Jeric, J. de Boor, J. Zavasnik, M. Ceh. Lowering the thermal conductivity of $Sr(Ti_{0.8}Nb_{0.2})O_3$ by SrO and CaO doping: microstructure and thermoelectric properties. J. Mater. Sci. **51**, 7660–7668 (2016). https://doi.org/10.1007/s10853-016-0048-8
122. J. Wang, B.Y. Zhang, H.J. Kang, Y. Li, X. Yaer, J.F. Li, Q. Tan, S. Zhang, G.H. Fan, C.Y. Liu, L. Miao, D. Nan, T.M. Wang, L.-D. Zhao, Record high thermoelectric performance in bulk $SrTiO_3$ via nano-scale modulation doping. Nano Energy **35**, 387–395 (2017). https://doi.org/10.1016/j.nanoen.2017.04.003

123. M. Acharya, S.S. Jana, M. Ranjan, T. Maiti, High performance (ZT > 1) n-type oxide thermoelectric composites from earth abundant materials. Nano Energy **84**, 105905 (2021). https://doi.org/10.1016/j.nanoen.2021.105905
124. S.S. Jana, T. Maiti, Enhanced thermoelectric performance in oxide composites of La and Nb codoped $SrTiO_3$ by using graphite as the electron mobility booster. ACS Appl. Mater. Interfaces **14**, 14174–14181 (2022). https://doi.org/10.1021/acsami.1c24625
125. N. Wang, H. Chen, H. He, W. Norimatsu, M. Kusunoki, K. Koumoto, Enhanced thermoelectric performance of Nb-doped $SrTiO_3$ by nano-inclusion with low thermal conductivity. Sci. Rep. **3**, 3449 (2013). https://doi.org/10.1038/srep03449
126. D. Srivastava, C. Norman, F. Azough, M.C. Schafer, E. Guilmeau, R. Freer, Improving the thermoelectric properties of $SrTiO_3$-based ceramics with metallic inclusions. J. Alloys Compd. **731**, 723–730 (2018). https://doi.org/10.1016/j.jallcom.2017.10.033
127. Y. Kozuka, M. Kim, H. Ohta, Y. Hikita, C. Bell, H.Y. Hwang, Dramatic mobility enhancements in doped thin films by defect management. Appl. Phys. Lett. **97**, 222115 (2010). https://doi.org/10.1063/1.3457994
128. S.R.S. Kumar, A.Z. Barasheed, H.N. Alshareef, High temperature thermoelectric properties of strontium titanate thin films with oxygen vacancy and niobium doping. ACS Appl. Mater. Interfaces **5**, 7268–7273 (2013). https://doi.org/10.1021/am4015956

Memristors and Resistive Switching in Metal Oxides

Charu Singh and Nirat Ray

Abstract The term memristor, derived from the words memory and resistor, refers to a memristor's desirable ability to remember its history. The memristor was the fourth fundamental passive electronic element, following the resistor, capacitor, and inductor, and was theoretically predicted by Leon Chua in 1971. Despite numerous publications in the field, the memristor remains a "mysterious" electronic element that is not yet in widespread use. One important reason for this is a lack of understanding of the physical phenomena that govern the occurrence of the switching effect. Metal oxides such as HfO_x, TaO_x, and TiO_x have shown promise in memristive applications. In this chapter, we review the working mechanisms and models of resistive switching memristors, as well as an overview of the metal oxides that are used. We investigate the various emerging applications of memristors, such as resistive switching memory (RRAM), analogue and digital logic circuit applications, and neuromorphic applications.

Keywords Resistive switching · Memory · Hysteresis · MIM junction

1 Introduction and Motivation

The term "memristor" is derived from the words "memory" and "resistor" and it accurately describes a device's ability to remember its history. The memristor was theoretically predicted by Leon Chua in 1971 [1], and research into the subject is still ongoing today. A memristor is often a two-terminal device whose resistance is affected by both the amplitude and polarity of the voltage given to it, as well as the duration for which the voltage was applied. Figure 1a illustrates a typical current–voltage characteristic highlighting the hypothetical memristor's behaviour. Unlike a linear plot observed for a resistor, the plots, which resemble bow ties, maps the

C. Singh · N. Ray (✉)
Department of Materials Science and Engineering, Indian Institute of Technology, Delhi, New Delhi, India
e-mail: nirat@iitd.ac.in

V. Kumar et al. (eds.), *Optical Properties of Metal Oxide Nanostructures*, Progress in Optical Science and Photonics 26,
https://doi.org/10.1007/978-981-99-5640-1_14

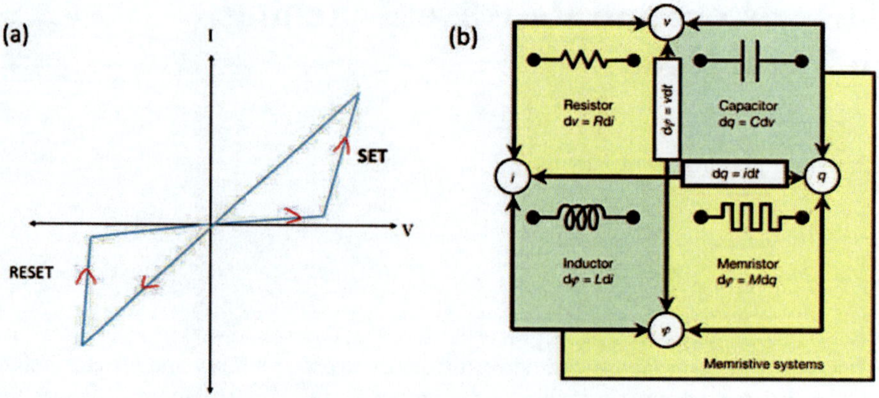

Fig. 1 **a** A typical I-V characteristics of the memristor. **b** The four elements of the circuit model explained by Chua. Reprinted with permission from [3] © 2008 Springer Nature

switching behaviour of the device. The memristor begins in a high resistance state, and the current increases as the applied voltage increases. The resistance decreases as the applied voltage increases, causing the slope of the current–voltage characteristics to change. This process continues until a maximum is reached. If the voltage is reduced beyond this maximum, the current decreases slowly as the resistance decreases, resulting in an on-switching loop [2]. When a negative voltage is applied, an off-switching loop is observed. The bottom line is that the applied voltage changes the resistance of the memristor.

The resistor, inductor, and capacitor are the three most common passive elements in a circuit. The memristor is the family's fourth member. Figure 1b depicts the entire circuit model, including all four elements. Chua [1] considered four fundamental variables in a circuit in his original proposal of memristors: electric current I, voltage V, charge q, and magnetic flux, ϕ. The three basic components of a two-terminal circuit; a resistor with resistance R (dV = R di), a capacitor with capacitance C (dq = CdV), and an inductor with inductance L (d = L di)—are defined by the relationship between these fundamental variables. The missing relationship between the charge q and flux ϕ, can be used to define the memristance, M, such that dϕ = M dq. Thus, the relationship between charge and flux (d = M dq) could be used to find the "missing" element [3]. In cases where M is a constant, the memristance is the same as the resistance. However, if M is a function of q, resulting in a nonlinear circuit element, the situation becomes more interesting. This device's current voltage characteristics are generally a frequency-dependent Lissajous figure, and the non-linear properties of memristors cannot be defined by any relation of resistive, capacitive and inductive components [3].

An interesting analogy for memristive behaviour is with water flow in a pipe [4]. Here the flow of water is analogous to the flow of electric charge. The pipe provides resistance to flow of water based on the diameter of the pipe. The memristor can then be thought of as a pipe that can change the diameter of the pipe with respect to the amount and the direction of water flowing through it. Further, the memristor also remembers the diameter corresponding to the last water flow event, and if the water flow is abruptly stopped, the diameter of the pipe is sustained until the water is turned back on.

Memristors gained much interest of the researchers worldwide as an alternative for non-volatile memory devices with over hundreds of research papers in the late 1960s. While the phenomenon was interesting, a lack of explanation about the origin and the underlying mechanism stalled research in the area. It was only after the year 2008, that it again came to the forefront as Hewlett-Packard provided an explanation of the hysteresis behaviour and implemented a circuit model based on the mathematical equations described by Chua. The structure was a crossbar array; essentially a mesh of perpendicular wires, with every wire crossing connected by a switch. The switch here was made with a 40 nm cube of TiO_2. The cube consisted an insulating top layer of stoichiometric TiO_2 and a metallic conductive layer of non-stoichiometric TiO_{2-x}. The conducting behaviour in the metallic non-stoichometric layer was attributed to oxygen vacancies, and the idea was to open and close the switches by playing with oxygen stoichometry and thereby form a connection between the vertical and horizontal wires, by applying a voltage. This could also be thought of as a data storage system, with an open switch representing 0 and a closed switch representing 1. When the memristor experiences the positive volatge, it repels the (positive) oxygen deficiencies in the TiO_{2-x} layer of the upper metallic part and sends them into the insulating TiO_2 layer below [5]. This shifts down the boundary between the two material, such that the percentage of conducting TiO_{2-x} increases and thereby increasing the conductivity of the entire memristor switch. If the positive voltage is increased, conductive also increases. On the other hand, if a negative voltage is applied voltage on the switch, it attracts the positively charged oxygen bubbles such that they get pulled out of the TiO_2. This results in increasing the amount of resistive (insulating) TiO_2, thereby making the entire switch a resistive switch. Thus, an increased negative voltage will result in lesser conducting switch or a more resistive switch [6]. The switch is memristive because when the voltage is turned off, from a positive or negative value, the oxygen vacancies stay frozen, and so does the boundary between the two TiO_2 layers. The switch essentially remembers how much voltage was last applied. Since then, the memristors took the hot seat in the academia for memory devices and the research continues to replace the current existing silicon based non-volatile devices by resistive random-access memory (RRAM) devices. Moreover, silicon-based memory devices have almost reached to its limit of scalability hence the speed. Nonetheless, with technology engulfing almost every aspect of our lives and increased dependency on artificial intelligence, the demand for data storage needs to be fulfilled in a much faster and economical way without compromising the reliability. Along with data storage, data processing is another aspect which decides the speed and functioning of the memory. This leads to the 'in-memory' concept where data storage and data processing occur simultaneously unlike the present Von-Newman architecture which consumes power and by default have delay issue due to separate data processing of data and its storage. In the mid-1980s, Carver Mead, a Caltech professor proposed a brain mimicking memory concept termed as neuromorphic memory [7] which could function like a human brain (recognition, analysis, association). Such a concept would serve with faster processing, increased density of data storage capacity with low power consumption. In this regard, memristor was seen as a potential candidate to serve as the building blocks of neuromorphic memory or neuromorphic computing. It could serve as the synapse of the brain which controls the flow of charge between neurons and hence, form a complex neural network for computing. It sounds interesting but still a through grasp of underlying principle and charge transport mechanism is needed and becomes essential for memristive devices to explore and utilize its full potential to be deployed into a memory technology.

Nonetheless, we define the memristors as an emerging and promising element which could update our existing technology by far. It needs deeper insight into several aspects of its functioning from its conduction mechanisms to modelling into a memory device, structural and materials aspects, etc. Throughout the years, the study of memristors moved from micro regime to nano regime which not only serves to scale the memristive technology but also a stage for intrusive study. This chapter will try to cover some of the important aspects in this area such that the readers could well acquaint themselves with the memristors.

2 Mechanism of Resistive Switching

The heart of memristive devices lies in the ability of the device to show multiple values of resistances and switching between them; also known as resistive switching (RS). Several mechanisms have been identified for change of resistance, and memristors have been classified based on active switching region and geometry into filament type, barrier type, and bulk type. We discuss each of these in detail.

2.1 Filament Type Resistors

The TiO_2 based memristive switch discussed earlier is an example of filament type switching. In filament type memristors, the repeated formation and rupture of conductive filaments (CFs) causes switching when an external electric field is applied [8, 9]. A memristive system ideally has a high resistance (HRS) and a low resistance state (LRS). While the current flows uniformly in the layer in HRS, it is limited in the LRS and follows a local path within the insulating layer. There have been many reports of filamentary conducting paths in literature, and it was seen that the filament type can vary drastically from one material to another, such as filaments of metallic nano-bridges [10], conductive channels made of oxygen vacancies [11], and dislocations [12]. In the case of oxides, the drift of oxygen vacancies leads to the formation of the conducting filament in the presence of strong electric fields. Following the development of the oxygen vacancy filament, the LRS would then be independent of the device area, whereas the resistance would scale with size in the case of the HRS [13]. Thus, the size of the conducting filament would determine the device's scaling limit. The size of the device was then advised to be roughly double that of the filament as a rule of thumb [14]. Valence-change memory (VCM) or anion-based memory are the terms used to describe the devices that are controlled by oxygen ion migration's effects on the valence change of the metals in metal oxides, which results in the development of conductive filament (CF) [15]. VCM filamentary switching includes devices made of HfO_x, TaO_x, TiO_x, etc. [2, 16].

Another type of filament is formed by a redox reaction, which results in metallic filaments. The RRAM, mediated by the redox reaction formed filament, is defined as electrochemical metallization (ECM) cell forming a conducting bridge [8]. A solid electrolyte thin film, a metal electrode (electrochemically active such as Ag, Cu, or Ni), and another metal electrode that must be electrochemically inert such as Pt, Au, W, or Ir make up the ECM cell. Under the applied voltage, the ECM cell undergoes a redox reaction in which the active metal is oxidised at the interface, as represented by $M \rightarrow M^{n+} + ne^-$, where M^{n+} represents the metal cation. The mobile metal cations then move through the solid electrolyte layer towards the inert electrode and are reduced at the cathode, which is represented as $M^{n+} + ne^- \rightarrow M$. Such cation movement and reduction results in the accumulation of metal atoms at the cathode, forming a metal protrusion that grows and reaches the anode, forming a highly conductive filament and switching the device to a LRS. A negative voltage, on the other hand, will reset the device into HRS by causing dissolution of the film at the weakest point of the conducting filament (thinnest part of the CF near the anode). The CF's break point becomes the starting point for the next cycle. Repetition of the preceding procedure causes the device to switch between high and low resistance states in a reproducible manner.

Another possible way to form the CF in a memristor device, could be initiated by the dielectric breakdown of the insulating layer. Possibly, voltage could lead to partial dielectric breakdown such discharge filament modifies the material due to Joule heating. If a limit is set for the current (compliance current), then, a weak conductive filament is formed having its resistance controlled. The so formed filament possibly comprises of the transported electrode metal into the insulator, carbon from residual organics or decomposed insulator material such as sub-oxides [15].

Dislocations may also be used to drive resistive switching. Here, the conducting filament is formed by the dislocations within the material and dislocations cause current crowding induced localized heating. $SrTiO_3$ has been extensively explored for the study of dislocation initiated resistive switching [17]. It has been seen that the dislocations related with the formation of slip lines are caused by the plastic deformation of the device, induced by the stress from thermal gradients, created by the high-power dissipation [18]. Dislocation formations are generally impacted by the size of device, current compliance levels, and concentration of vacancy [17]. The advantage of dislocation-based resistive switching is that dislocations can be scaled down to the size of a single dislocation and switched as a result of the scaling process [18]. However, these defects could result in uncontrollable ion transport which will then require transistors for neural analog computing without which the device would be impractical for resistive switching [19]. In this regard, SiGe was explored to incorporate defect selective CF formation such that by confining the filaments in dislocations gives exceptionally low temporal variation [19]. This low variation allows to have a pattern which becomes feasible to study and recognize when dealing with neuromorphic networks [19] discussed later in the chapter.

2.1.1 Ex-Situ and In-Situ Observation of Filaments

The conducting filament can be observed either during the real-time operation of memristors under practical conditions ('in-situ' observation) or during external conditioning to form the conducting filament ('ex-situ' observation). Scanning tunnelling electron microscopy was used to observe the electroforming process in Au/VO_2/Ge vertical devices, where the morphology and structural changes were observed and confirmed that the conducting filament was formed with V_5O_9. It was clearly demonstrated that oxygen deficiency is critical in oxides for the formation of conducting filament [20]. Because of its unrivalled spatial resolution, in-situ TEM is a highly preferred technique for capturing the dynamic evolution of morphology and moving paths at a nanometer scale; thus, the system can directly map the structural changes and transition during resistive switching [21]. Most in situ techniques can characterise the ion migration region and its morphological evolution while also characterising the device switching. Correlating their electric behaviour and morphological changes allows us to infer the variation in memristive parameters [22].

Atom probe tomography (APT) is another technique used to analyze the composition and structure of materials at the atomic scale by measuring the positions and identities of individual atoms within a sample, using the principle of selectively ionizing and detecting the atoms by a combination of a high-voltage electric field and a time-of-flight mass spectrometer. Atom probe tomography (APT) analysis of Ag/TiO_2/Pt devices revealed a three-dimensional elemental distribution map of CFs and revealed that the conduction path was formed due to doping/removal of mobile Ag ions rather than the connection/disconnection of the filament [23]. The combination of electron energy loss spectroscopy (EELS) and transmission electron microscopy (TEM) could be used to observe oxygen evolution in the switching layer [24]. The ability to visualise morphological changes in memristors aids in characterising their functions and, as a result, is critical in the development of switching characteristics [15].

A non-invasive optical microscopy technique has been used to examine the movement of oxygen vacancies within the TiN/$SrTiO_3$/AuNP with strontium titanate ($SrTiO_3$) as the switching material. The morphological changes within the switching material can be well tracked by the information provided by the optical spectrum of the plasmonic hotspot in the junction where it shows the creation and destruction of conductive channels between the electrodes [25]. Scanning transmission X-ray microscopy (STXM) in situ with TEM enables non-destructive study of formation of conduction filaments in Pt/HfO_2/Hf/Pt device with ring-like feature [26]. The main focus of in-situ investigations has been to capture the filament rupture and formation. Other stages in the process, more specifically ion migration during pre- and post-rupture formation remain largely unexplored [22]. Ex-situ techniques could be used to shed light on these aspects.

Ex situ transmission electron microscopy (TEM), while being a destructive technique, enables high-resolution visualization of filaments and metal clusters and also provides morphology of the switching material before and after the switching [15]. These studies have helped underline the dynamics of filament growth such as geometry and apparently the origin of the filament. It is worthy to mention that geometry of the filament is pivotal in the operation of resistive memories as it helps to identify the thinnest part of the filament, which is the hotspot for the dissolution and recreation, during SET/RESET process [15]. Figure 2 shows the before and after pictures of the conductive filaments of Ag formed in a SiO_2 based memristor using transmissions electron microscopy technique [27]. Nonetheless, ex-situ characterization may also miss the significant filaments regions which could hinder the understanding of the complete filament growth process which is the crucial part for the materials research.

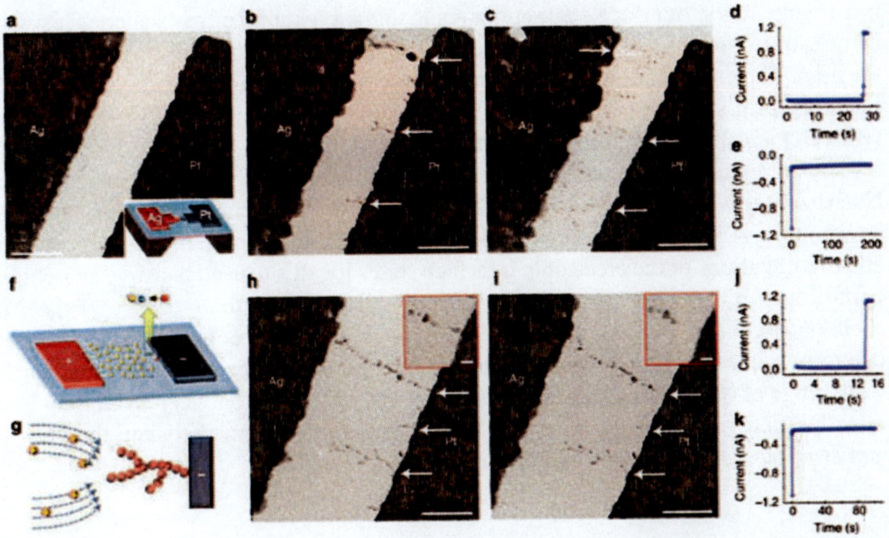

Fig. 2 **a** Transmission electron micrograph of a SiO_2-based planar device fabricated on a silicon nitride membrane. Inset shows a schematic of the device. **b** Transmission electron micrograph of the same device after filament formation. **c** Transmission electron micrograph of the same device after filament rupture. **d** Current as a function of time during the forming process and **e** rupture process at an applied voltage of 8 V. Reprinted with permission from [27] © 2012 Springer Nature

2.2 Barrier Type Memristors

Barrier type memristors differ from filament type memristors in the way they function and the internal structure of the device. Barrier type memristors experience interfacial switching due to the modulation of the charge transport barrier at the electrode/switching layer [28]. The height of the barrier at the interface, as an example, could govern the electrical characteristics [9]. The memristance of barrier type memristors is area dependent, unlike filamentary memristors, because the entire area participates in the switching phenomenon. Another distinguishing feature is that, unlike conductive filament based switching, the resistance changes uniformly and gradually due to interface effects of barrier memristors [29]. As a result, this RRAM type is more suitable for analog applications.

While several models have been proposed to explain the switching mechanism, the most commonly used example is the change in the Schottky contact at the metal-oxide interface [29]. A Schottky barrier gives rise to a high contact resistance due to the formation of a space-charge layer in which the majority carriers are strongly depleted [30]. Another possibility is the activation of ions within the active layer such that there is a change in the interface characteristics leading to change of resistance. This was demonstrated in Nb-doped $SrTiO_3$ where the resistance was inversely proportional to the area. Thus, the scaling limit in barrier type memristors are area dependent [31]. Such uniform switching is reported in perovskite oxides, such as Bismuth Ferrite ($BiFeO_3$) and Praseodymium Calcium manganite ($Pr_{1-x}Ca_xMgO_3$) [32, 33]. High resistance retention time was observed in a double barrier memristor devised with $Nb/Al/Al_2O_3/Nb_xO_y/Au$ where Nb_xO_y is the ultra-thin memristive layer lying in between the Al_2O_3 tunnel barrier and a Schottky-like contact. Diffusion of oxygen and modifications of the electronic interface states locally, caused resistive switching within the Nb_xO_y layer, which influenced the properties at the interface of the Au (Schottky) contact and also the Al_2O_3 tunnel barrier, respectively [28]. Another emerging candidate for interfacial memristors is Strontium ferrite, $SrFeO_x$, having a wide range of oxygen non-stoichiometry and easy transition between insulating and metallic (perovskite) phase because of lower energy difference between the phases and also lower energy for oxygen vacancy formation [34].

2.3 Mathematical Modelling of Resistive Switching

There are two primary equations governing the description of a memristive systems, and these equations provide a comprehensive understanding of how a memristive system works. The first equation is the memristor version of the Ohm's law, $v = M(x, i)i$, where v and i are the voltage and current in the circuit, w can be a set of state variables, and $M(x, i)$ is the memristance. This equation is significant because it highlights an important characteristic of memristive systems, and that is that no current flows through the memristive system when the voltage drop across it is zero.

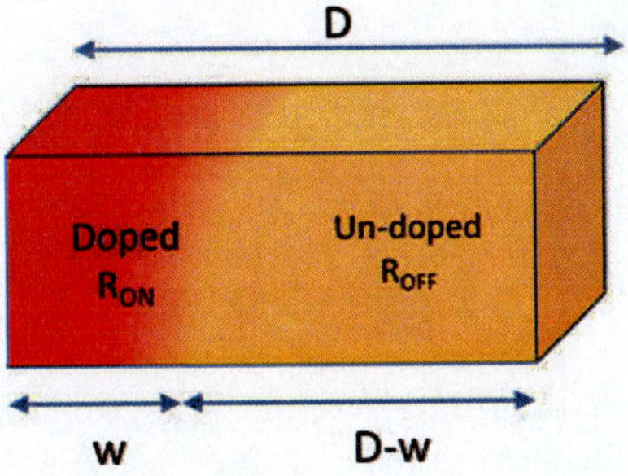

Fig. 3 Schematic illustrating the doped (red) and undoped (orange) regions in a device of width D, as described within the linear model

The second equation is the state equation, $\frac{dw}{dt} = f(w, i)$. In this equation the function f can in general be an explicit function of time. This equation describes how the state of the system changes over time, based on the current flowing through it and the internal state variables of the system. The mathematical models for memristive behavior may further be classified into linear, non-linear, and exponential.

2.3.1 Linear Ion Drift Model

The linear drift model [35] begins with the assumption that a device of width D, consists of two regions; one side of width w being doped with (typically oxygen) ions and the other being undoped, as shown in Fig. 3. Here, the doped region will have low resistance R_{ON}, and correlate with the ON state, while the undoped will have high resistance, R_{OFF} for the OFF state [35]. The size of the doped region changes with applied voltage, and this in turn modifies the resistance of the memristor, and the memristance can be expressed as $M(x) = R_{\text{ON}}x + R_{\text{OFF}}(1 - x)$.

The memristor version of Ohm's law and the state equation can now be expressed as:

$$v = \left(R_{\text{ON}}\frac{w(t)}{D} + R_{\text{OFF}}\left(1 - \frac{w(t)}{D}\right)\right) \cdot i(t)$$

$$\frac{dw}{dt} = \mu_V \frac{R_{\text{ON}}}{D} i(t)$$

where μ_V is the average velocity of ions.

2.3.2 Nonlinear Model

The first level of non-linearity can be added to the models by using window functions to the linear ion drift model. The state variable w, must be zero at the bounds, namely 0 and D. There have been several suggested window functions that vary depending on the device structure. The window function $f(w) = 1 - (2w - 1)^{2p}$, proposed in the Joglekar model [36], has a control parameter p that is a positive integer. The non-linearity of the model is likewise controlled by this control parameter. Another window function, developed by Biolek [37], uses a current dependant step function, $f(w) = 1 - (w - stp(-i))^{2p}$, where stp(i) is equal to 1 for i $\geq$ 0 and 0 for i $<$ 0.

2.3.3 Exponential Model

Experimental current–voltage characteristics of memristive devices often show non-linearity, and the addition of a simple window function to the linear drift model is not sufficient to explain the experimental observations. Non-linear effects may become particularly dominant at large electric fields, and over thin junctions. Wang et al. [5] fitted experimentally measured IV characteristics using:

$$i(t) = w(t)^n \beta \sinh(\alpha v(t)) + \chi\left[\exp(\gamma v(t)) - 1\right]$$

where α, β, γ and χ represents fitting constants, and n is a free parameter. In the OFF-state, the current voltage characteristic resembles that of a diode, and this is the origin of the exponential term in the equation. In the ON-state the curve follows a tunneling process and that is descripted by the hyperbolic function. The state equation is modeled by the nonlinear equation:

$$\dot{w} = a\sinh(bv) f(w, i)$$

where $f(w, i)$ is an appropriate window function.

2.3.4 Other Models

A more accurate physical model for asymmetric switching characteristics was proposed by Simmons tunnel barrier model where a resistor isn't alone but in series with an electron tunnel barrier [38]. The state variable x can be differentiated considering it as the oxygen drift velocity and may be presented by:

$$\frac{dx(t)}{dt} = \begin{Bmatrix} C_{\text{off}} \sinh\left(\frac{i}{i_{\text{off}}}\right)\exp\left[-\exp\left(\frac{x-a_{\text{off}}}{w_e} - \frac{|i|}{b}\right) - \frac{x}{w_e}\right], i > 0 \\ C_{\text{on}} \sinh\left(\frac{i}{i_{\text{on}}}\right)\exp\left[-\exp\left(\frac{x-a_{\text{on}}}{w_e} - \frac{|i|}{b}\right) - \frac{x}{w_e}\right], i < 0 \end{Bmatrix}$$

A modification in the Simmons model is captured by Threshold adaptive memristor model (TEAM) [39]

$$\frac{dx(t)}{dt} = \begin{Bmatrix} k_{\text{off}} \cdot \left(\frac{i(t)}{i_{\text{off}}} - 1\right)^{\alpha_{\text{off}}} f_{\text{off}}(x), & 0 < i_{\text{off}} < i \\ k_{\text{on}} \cdot \left(\frac{i(t)}{i_{\text{on}}} - 1\right)^{\alpha_{\text{on}}} f_{\text{on}}(x), & 0 < i_{\text{on}} < i \\ 0, & \text{otherwise} \end{Bmatrix}$$

Any change in the tunnel barrier width leads to memristive changes, which change exponentially, resulting in a current–voltage relationship given by:

$$v(t) = R_{\text{on}} e^{\left(\frac{\lambda}{x_{\text{off}}} - x_{\text{on}}\right)(x - x_{\text{on}})} i(t).$$

The memristor practically acts like a resistor within high frequencies for inversely related flux and frequency [35].

2.4 First Principles Calculations

Mathematical models can be used to understand mechanical motion, time evolution, and error estimation in complex dynamical systems. However, these models may not always be solvable, necessitating the use of numerical solutions. Further, these may be inadequate for understanding the mechanism due to the random formation of conducting filaments. Therefore, it is crucial to conduct theoretical calculations at the atomic scale to study the physical mechanism of the memristors. To obtain the ground state electronic properties and the transition energies of the stable states, first-principles (or ab initio) calculations, based on quantum mechanical principles, are generally used [36].

There are a few aspects that may easily be understood using ab-initio calculations. The formation energy of various neutral or charged defects in the material which acts as the active layer of the memristor. The study of low concentrations of defects becomes challenging as the calculations will require a large number of atoms. Once created and optimized, the contribution of the defect states to the density of states and electrical conductivity can be explored. Another aspect that can be understood are the barrier heights for migration through the material [35].

However, most ab-initio calculations would require large supercells to capture the basic physics, and while these enable direct simulation of systems without the introduction of model-related parameters, they end up being computationally very intensive. For simple dynamics investigations, such as the migration of a single oxygen atom, charging and discharging [36], and resistance volatility [38], the calculations are extremely effective. While these calculations are simple for single crystals, they become more difficult when dealing with polycrystalline, amorphous, or disordered structures. Figure 4a–e depicts a graphical representation of different first principles strategies [31].

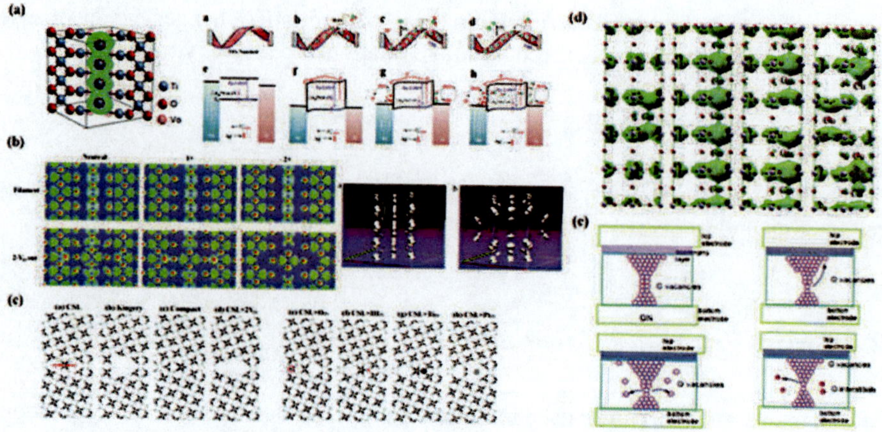

Fig. 4 Summary of strategies employed to understand memristive behaviour using first principles calculations. **a** Oxygen vacancy properties. **b** Electron localization functions showing the formation of conducting filament in TiO_2 as Ti atoms overlap and form a delocalized electron distribution. The rupture in the filament is caused by the disruption of charge distribution by oxygen vacancies. Three distinct charge states (neutral, 1+, and 2+) are considered here. **c** Grain boundary models in HfO_2. **d** Iso-surface chart of the partial charge density related with the defect states for Ta_2O_5, with **a–d** separately representing a single interstitial Cu, direct path Cu, defective path Cu, and deformed path Cu at 0.04 e $Å^{-3}$, respectively. **e** Schematic flow for switching processes: a-LRS, b-HRS (vacancies movement toward electrodes), c-HRS (vacancies scattered into the insulating matrix) and d-HRS (vacancies recombination with O interstitials in the matrix). Reprinted with permission from [35] © 2013 Royal Society of Chemistry

3 Materials and Devices

3.1 Overview of Metal Oxides Used for Memristive Applications

Metal oxides are the heart of the memristors which serve the active layer. Oxide based insulators have simple atomic structure, enhanced thermal stability, CMOS compatibility with optimal switching characteristics [40]. Initially, studies were done on Ti based oxides and then extensive research followed to study switching mechanisms in wide variety of simple binary transition metal oxides like ZnO, NiO, MnO, CuO to perovskites e.g. $SrTiO_3$, $Ba_{1-x}Sr_xTiO_3$, $SrZrO_y$, and $BiFeO_3$ and transparent conducting oxides, such as indium tin oxide (ITO) and SnO_2 [40]. The fundamental mechanism in these oxides is the oxygen anion species migration under an external electric field [40]. Rapid diffusion channels are created by grain boundaries existing inside the germanium-oxide GeO_x for ion movement [41].

High Entropy Oxides (with offering high dielectric constant [42]) have emerged as the super oxides for memristive applications, offering tunability of migration of oxygen vacancy through enhanced lattice distortion and poor diffusion effects [43]. Homogeneous distribution of oxygen vacancies [44] occurs to maintain the charge neutrality when elements with different charge valences are mixed [45]. Zr, Hf, Nb, Ta, Mo, and W based oxides are high entropy oxides and their memristors devices show forming-free operation, gradual conductance modulation, low device and cycle variability, long retention (~10^3 s) and high endurance (~70,000 cycles) [46].

Oxides could be differentiated based on polarity of switching as of unipolar, bipolar and both. Oxides such as TiO_x, ZrO_x, MoO_x, AlO_x, and HfO_2 can show atypical co-existent bipolar and unipolar resistive switching. The device shows bipolar switching mode most probably at low current and unipolar switching mode is achieved usually at high current accompanied by Joule heating [40]. Certain low resistance oxides such as ITO could be used to improve switching characteristics (retention time, resistance range, and switching speed [47] if used as a thin buffer layer between electrode and the main oxide layer. Capping layer promotes the diffusivity of oxygen species or vacancy carriers because the affinity of the capping material towards oxygen anions could dictate the extent of vacancy migration and hence dictates the speed at which conductive filaments are formed [48]. Ti and Al have been reported to enhance the switching process owing to its oxygen affinity. Capping layer of Ti in HfO_x forms TiO_x/HfO_x bi-layer and thereby increased the resistance change [49] and also improved the overall switching process [40].

3.2 Solution Processed Memristors

Integration of organic solutions and oxides for electrical switching and memristive behaviour was emphasized five decades ago [50]. Ever since, the researchers have been working to obtain the synergistic effect of oxides with organic solutions or polymers. Solution based memristors offer a low-cost synthesis route, and may easily integrated with flexible substrates, thereby paving the way for flexible ReRAM devices.

Spin coating has been widely used thin-film fabrication techniques for fabricating network of memristors to form artificial synaptic devices (discussed in the next section). Simple fabrication of large-scale film is possible with controllable thickness in ambient condition. Film thickness is dependent on the solution viscosity, spin coating speed, and solvent volatilization speed [51]. TiO_2 layer of 60 nm was deposited by spinning a titanium isopropoxide solution on the flexible plastic substrate to get a 2 mm × 2 mm $Al/TiO_2/Al$ [52]. Gel-form TiO_2 structure contains defects due to amorphous state and thus these memristors can switch without undergoing forming process [53]. Changing the aluminium electrode with the gold electrode as the anode causes oxygen evolution and electrode deformation [53]. Also, an interplay with the electrode could lead to both unipolar and bipolar switching [53]. With spin coated aluminium nitrate nonahydrate ($Al(NO_3)_3{\cdot}9H_2O$), AlO_x layer is obtained to get $Ni/AlO_x/Pt$ and $TiN/AlO_x/Pt$ RRAM devices which show bipolar resistive switching with operation voltages as low as 1 V and larger than 10^3 ON/OFF ratio. On annealing $Ni/AlO_x/Pt$ RRAM devices offered better stability at lower voltages. Apparently, such variation in Ni and TiN as top electrode might be related to the work function difference between the top and bottom electrode [51]. Stacked layers of Ag/GaO_x/AlO_x/ITO memristive device could perform with voltage as low as ±0.6 V, showed ON/OFF ratio of ~2 × 10^4, >2 × 10^4 s retention time and >1000 endurance cycles [54].

Other spin coated active layers are tantalum oxide (TaO_x) [55], SnO_2 [56], and NiO [57] with electrodes such as gold, silver, platinum, ITO (where the ITO layer acts as current limiting resistor) [58]. Introducing quantum dot (QD) based solution processed memristors, allows another layer of complexity. Citric acid quantum dot (CA QD)-PVP composite film was used as the active layer with Ag-Nanowires as electrodes for flexible, water soluble memristors which showed unipolar resistive switching [59]. A memristor with CdSe/ZnS QDs in the polymethyl methacrylate (PMMA) matrix showed ON/OFF ratio of ratio of 10^4 and a retention time over 10^4 s, with endurance of 400 cycles. The device showed extra trap states after the HRS/LRS switch cycles which degraded the performance and invoking the need for surface stabilization. QDs due to their relatively small bandgap, are effective charge trappers, when embedded in the polymer [60]. Another advantage of QD was seen in MgO-graphene oxide quantum dots (MgO-GO-QDs) based memristor, developed with Al as top electrode and ITO as bottom electrode on a polyethylene terephthalate (PET) flexible substrate. Here, the introduction of GO-QDs enhanced the switching characteristics and lowered the power consumption. Interconnected NWs offers MIM junction by forming core-shell structure of metal being the core and an insulating layer over it. The insulating layer could either be a metal oxide (same metal or different) or polymer like for example PVP such that the interconnection of NWs form MIM structure. ReRAM using single NW includes Au–NiO–Au (NiO NW 250 nm diameter) [28] and Ni–Ag_2Se–Ni NWs (200 nm diameter) showed bipolar switching. ZnO NW memristor with Ag electrodes form a conducting filament of Ag atoms [21]. On the other hand, biocompatible and biodegradable materials based memristive devices can be easily prepared by simple drop cast method [51]. Ag/a–TiO_2/Ag memristor was fabricated with fully inkjet printing process forming a dense network to be used for brain mimicking computing (neuromorphic computing) [61]. A glucose-based RRAM device was fabricated on thin-polyimide (PI) film as well as on glass and rice paper substrates with a Mg electrode. It exhibits non-volatile bipolar resistive switching behaviour and has perceived its way towards biocompatibility for the implantable bioelectronic devices [62]. Flexible conductors devised using metal nanowires, conducting polymers, graphene and carbon nanotubes serve as potential substitutes for metal and metal oxides using solution processes [59].

4 Applications of Memristors

Memristors have potential to be the first in line for the future alternatives to the present CMOS technology. Better scalability, high density, low power consumption are the benefits offered by the memristor based technology. There are several applications that can utilize memristor technology are memory, analog/digital, or neuromorphic circuits [63].

4.1 Resistive Switching Random Access Memory (RRAM)

Random access memory (RAM) can be volatile or non-volatile. When the power supply is removed, volatile memory loses data, whereas non-volatile memory retains data. As we move toward a digital age, the need to store data has increased dramatically, and a new class of memory is being investigated with the ambition of combining the switching speeds observed in volatile memory with non-volatility. An ideal memory device has a lower operating voltage (>1 V), a longer cycling endurance (($>10^{17}$ cycles), a longer data retention time (>10 years), a lower energy consumption (fJ/bit), and superior scalability (10 nm) [64]. Resistive random access memory (RRAM), which uses resistive switching mechanisms, appears to be a promising technology in these aspects.

RRAM device performance can be measured using key performance metrics such as endurance and retention. Endurance is calculated by counting the number of times an RRAM device can be switched between high and low resistance states, as each switching event may potentially degrade the memory. RRAM endurance is typically measured by performing a series of current–voltage sweeps and noting the resistance of the ON and OFF states as a function of sweep number. There are RRAM devices with excellent performance characteristics, such as a 30 nm cell size HfO_x-based RRAM with a retention time of approximately 10^5 s [65], and the device array stability can be improved further by layering the bottom electrode with Al_2O_3. The stability of the ON and OFF states as a function of time, which also relates to the non-volatility of the memory, is used to calculate retention in an RRAM device. This can be measured experimentally by applying a small constant voltage and measuring the current as a function of time. Because switching occurs in the active layer, obtaining a long retention time in the OFF state is typically difficult. Retention is high in the ON state as this is the device's natural state [66]. HfO_x-based RRAM have shown a cycle endurance of 10^7 cycles [65]. It was discovered that the pulse width and amplitude had a significant impact on the endurance of TaO_x-based RRAM; for example, increasing the pulse width and amplitude degraded endurance performance, as seen in the Ta/Ta_2O_5/TiN RRAM structure [67]. However, another cause of endurance deterioration is the reaction of oxygen ions with the TiN electrode. This can be solved by reducing the oxygen reaction by shrinking the Ta_2O_5 layer to 3 nm, which increases the endurance to 10^9 switching cycles with a pulse width of 5 ns [68].

Ta_2O_5 was studied for large-scale array performance and showed endurance of 10^5 cycles [69] and low power consumption as the cell current for LRS falls below 50 μA. Higher endurance of $>10^9$ cycles have been reported in different types of RRAM cells like NiO which is a unipolar device and bipolar devices like tantalum oxide (TaO_x)-based switching mediums [70]. Also, tantalum oxide-based RRAM devices seem to be exhibiting the retention time of ten years as well. Several RRAM devices based on metal oxides are mentioned in Table 1 for the comparison. The biggest challenge with RRAM technologies, however, is the poor uniformity of various device characteristics, and this is also a limiting factor for scaling up of manufacturing. The devices are prone not only to cycle to cycle variations but also device to device fluctuations stemming from the random nature of filament rupture and formation.

In previous sections, we have discussed the memristors and their underlying mechanisms, which are the principles behind RRAM devices. The sandwiched metal–metal oxide-metal structure of memristors makes them simple and robust. With the increasing dependence of our modern lives on technology, there is a growing demand for high connectivity, fast processing, and memory storage. However, current technology struggles to keep up with these demands due to limitations such as low endurance and high voltage requirements, as well as the approaching scalability limit of silicon-based memory technology. In contrast, the scalability limit of memristors is dependent on the filament length or atomic redox reaction, making RRAM an attractive option as a new generation of low power consuming CMOS compatible ultra-fast, and highly scalable non-volatile memory devices (Fig. 5).

In terms of architecture, RRAM can be configured in a crossbar layout or integrated with CMOS technology using one memristor and one transistor, also known as the 1R–1 T architecture [82]. A MOS transistor (1 T) is connected in series with a memristor (1R) in the 1 T–1R cell architecture, so that all the drains of the transistor cells in the same column are connected to the bit lines, with the bottom electrode of the memristor connected to the drain. Similarly, all of the transistor's gates are connected to the same line, which is then connected to the metal word line. Transistor sources are thus connected to the power supply line, allowing memristor devices to be accessed via the top electrode and the source terminal, as well as by using the gate

Table 1 Examples of metal-oxide based RRAM devices with performance characteristics

S. No	Device	Endurance	Retention	References
1.	Ideal RRAM	$>10^{17}$ cycles	>10 years	[64]
2.	NiO	10^{12} cycles	2×10^7 s	[71]
3.	TaO_x	10^9 cycles	=10 years	[72]
4.	MgO/CoO_x	10^8 cycles	$>10^2$ s	[73]
5.	HfO_x	5×10^7 cycles	10^5 s	[65]
6.	Cu doped SiO_2	10^7 cycles	5×10^4 s	[74]
7.	TiO_x/HfO_2	$>10^6$ cycles	10 years	[75]
8.	$WSiO_x$	10^5 cycles	10^5 s	[76]
9.	GdO_x	10^5 cycles	3×10^4 s	[77]
10.	TiO_2	10^4 cycles	10^4 s	[78]
11.	$HfLaO_x$	10^4 cycles	10^4 s	[79]
12.	AlO_x	10^4 cycles	10^5 s	[80]
13.	ZrO_2	10^3 cycles	10^4 s	[81]

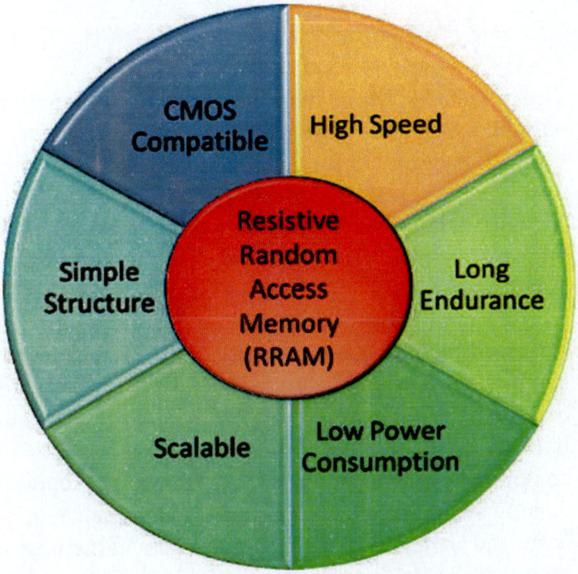

Fig. 5 Useful device and performance characteristics of the Resistive Random Access Memory RRAM

to control the transistor's ON and OFF states. In this manner, if a voltage is selectively applied on the word line and current is read on the bit line, then any cell can be read [82]. The feature of compliance current is associated with the transistor, where the gate could be used to control the current that passes from the memristor, thus acting as an access point for controlling the memristors. During the Set operation in 1R–1T memory configuration in TiN/Al:HfO_2/Ti/TiN memristor device, stacked with NMOS transistor, the compliance current was changed from a single high resistance state to three different low resistance states and vice versa [83]. The use of a transistor also helps with the issue of sneak path current in memristors based memory arrays. Apparently, the low current finds an alternate path to the other memristors in the array, resulting in faulty memory function because the detected current no longer reflects the correct resistance state of the memristors.

Although the 1R–1 T structure is simple, it is space-consuming because the connection with the transistor increases the effective structure area of the memory devices. Crossbar structures are prominent in terms of scalability because no transistor is attached. A typical crossbar memory array system has parallelly running metal lines that are perpendicular to each other and are referred to as word lines and bit lines, respectively. Thus, at the intersections of word and bit lines, two-terminal memristor devices are formed [84].

4.2 Analog and Digital Circuits

The resistance change in the memristors can be either 'gradual' or 'abrupt'. Based on such resistive switching behaviour, it can be categorized as 'ana-log' for gradual behavior and'digital' for abrupt circuits. The differences in the behaviour of the resistive switching could manipulate its usefulness in the applications. Analog memristors lays the foundation of neural networks, while digital memristors involve the on-chip logic operations. Also, the digital memristors are used during the initialization procedure required to operate memristor arrays [85]. This complexity can be reduced by the single memristor device which could operate both analog and digital modes to switch to digital mode, it involves the random rupture/formation of the filaments leading to high variability and low reliability in the devices [86]. Digital memristors can run between multiple limited resistance states because of which it has potential to achieve high-density, low-power storage [85]. The transition from digital to analog in a rapid fashion could help the device to own better control and functioning. Analog switching shares similarities with the adaption in biological synapses of the brain and are responsible for the adaption of memory and learning [87]. Hence, these memristors acts as artificial synapses which could perform synaptic functions like short term plasticity (STP), long term plasticity (LTP) and spike time dependent plasticity (STDP) [88]. The study of these functions could help improve the information processing performance of the digital world. It was found that by inserting reduced graphene oxide layer between $SrTiO_3$ and FTO, it modulated the switching performance of rupture/formation of Ag filaments responsible for the digital switching in the device with Ag as the top electrode [86]. Basically, the resistance change at the interface under the applied voltage causes the resistive switching from digital to analog. A digital–analog integrated memristor was realized using nanoparticles and nanowires forming ZnO NPs/CuO NWs heterostructure [85]. Digital circuits based on memristors were made as a hybrid CMOS-memristor circuit. This hybrid CMOS-memristor circuits have higher density, enhanced performance and reduced power dissipation. The memristors are augmented on the top of the CMOS base such that there are reduced interconnect wires and has clear advantage in terms of reduced area and possibly the higher speed [89].

4.3 Neuromorphic Applications

Neuromorphic memory is a new generation memory that aims to develop circuits that compute in the same way as the human brain, integrating memory and processing at the same location [14]. However, creating an artificial computer that can function similarly to the human brain is a difficult task. When an electrical signal is sent to a pre-synaptic neuron, it causes an impulse in the post-synaptic neurons, which causes the release of chemicals known as neurotransmitters, resulting in an electrical spike known as an action potential [90]. Neuromorphic circuits use the analogue characteristics of memristors to represent synapse whose synaptic weight (represented by memristor conductance) can be manipulated by an electrical signal. It is possible to imagine the device's synaptic weight being altered by either electrical signal transmission or a back propagating signal following a training event. These behaviours are roughly related to brain functioning, and with extensive research in memristive functions, could potentially perform memory, recognition, cognition, analysis [14], and other functions to supplement traditional von Neumann architectures.

Artificial synapses and resistive switching are two key components in neuromorphic computing using metal oxides. Artificial synapses can be used to model the behavior of biological synapses in a neural network. They can be implemented using metal oxides that exhibit memristive behavior, where the resistance of the material changes based on the history of applied voltages. This allows the artificial synapses to store and recall information based on their resistance state.

Artificial synapse requires analog resistance changes, synapse weight regulation (i.e. strength between both neurons), and workable synaptic plasticities. The resistance change is a change in synaptic weight. There are numerous difficulties associated with creating a cognitive computing system, such as scale constraints, system complexity, and so on. For example, the human brain contains ~10^{11} neurons and 10^{15} synapses, with each neuron having more than 1000 synapse connections to other neurons as a result, such a network can be created artificially through the use of electronic neurons. A memristor allows you to build an electronic synapse in a simple structure that uses very little energy. It could provide a 'in-memory' concept in which memory and processing occur within the same structure, as opposed to current computing systems, and thus reduce power consumption. At the device level, neuromorphic capabilities can be attributed broadly to electronic transport mechanisms mediated by nanoscale geometric confinement, typically across a metal-insulator-metal (MIM) junction [8]. Memristors can be used as artificial synapses because of their small size, straightforward structure, high compatibility, and ability to simulate changes in synaptic weight by varying their resistance values.

It is worth noting that sneak current has less of an impact on neuromorphic computing and machine learning than it does on crossbar arrays of sequential read-and-write isolated memristors. Because of their self-rectifying behaviour, Pt/TaO_y/nanoporousTaO_x/Ta memristor synapses can suppress unwanted leakage pathways [91]; bilayer cerium oxide memristors are forming-free, have a low voltage (|0.8 V|), and are energy-efficient (full on/off switching at 8 pJ with 20 ns pulses, intermediate states switching at fJ) [92].

5 Conclusions

In conclusion, this chapter explored the development of memristors from being a missing element to an emerging and promising element which could update our existing technology. The heart of memristive devices lies in the ability of the device to show multiple values of resistances and switching between them; also known as resistive switching. Different resistive switching mechanisms including filament formation and rupture, and barrier type memristors were explored. Key mathematical linear and non-linear models were discussed. The role of first-principles investigations, particularly in single crystalline materials, to understand memristive behaviour was also highlighted. The integration of memristive devices and flexible electronics has been made possible with the development of solution processed memristors offering a low-cost synthesis route, in addition to the ease of integration with flexible substrates. Memristor technology can be utilized directly for memory devices, analog and digital circuits as well as neuromorphic computing. We surveyed the key device parameters such as endurance and retention and found promising numbers from NiO and TaO based memristors. Furthermore, as current silicon-based memory technology approaches the scalability limit, the filament size limit in memristors may offer a new route to alternative memory technologies. We believe that memristors, with metal oxides as the core component, have potential to be the first in line for the future alternatives to the present CMOS technology, offering unique benefits such as better scalability, high density and low power consumption.

References

1. L. Chua, Memristor-the missing circuit element. IEEE Trans. Circuit Theory **18**(5), 507–519 (1971). https://doi.org/10.1109/TCT.1971.1083337
2. I. Daniele, R. Waser, *Resistive Switching: from Fundamentals of Nanoionic Redox Processes to Memristive Device Applications* (Wiley-VCH, 2016)
3. D.B. Strukov, G.S. Snider, D.R. Stewart, R.S. Williams, The missing memristor found. Nature **453**(80), 7191 (2008)
4. G. Johnsen, An introduction to the memristor: a valuable circuit element in bioelectricity and bioimpedance. J. Electr. Bioimpedance **3** (2012). https://doi.org/10.5617/jeb.305
5. J.J. Yang, M.D. Pickett, X. Li, D.A.A. Ohlberg, D.R. Stewart, R.S. Williams, Memristive switching mechanism for metal/oxide/metal nanodevices. Nat. Nanotechnol. **3**(7), 429–433 (2008). https://doi.org/10.1038/nnano.2008.160
6. M. Shahsavari, P. Boulet, Memristor nanodevice for unconventional computing: review and applications (2017). http://arxiv.org/abs/11703.00331
7. C. Mead, Neuromorphic electronic systems. Proc. IEEE **78**(10), 1629–1636 (1990). https://doi.org/10.1109/5.58356
8. R. Waser, R. Dittmann, G. Staikov, K. Szot, Redox-based resistive switching memories—Nanoionic mechanisms, prospects, and challenges. Adv. Mater. **21**(25–26), 2632–2663 (2009). https://onlinelibrary.wiley.com/doi/pdf/https://doi.org/10.1002/adma.200900375. https://doi.org/10.1002/adma.200900375
9. F. Pan, C. Chen, Z.-S. Wang, Y.-C. Yang, J. Yang, F. Zeng, Nonvolatile resistive switching memories-characteristics, mechanisms and challenges. Progr. Nat. Sci.: Mater. Int. **20**, 1–15 (2010). https://doi.org/10.1016/S1002-0071(12)60001-X
10. Y.C. Yang, F. Pan, Q. Liu, M. Liu, F. Zeng, Fully room-temperature-fabricated nonvolatile resistive memory for ultrafast and high-density memory application. Nano Lett. **9**(4) (2009)
11. N. Xu, L. Liu, X. Sun, X. Liu, D. Han, Y. Wang, R. Han, J. Kang, B. Yu, Characteristics and mechanism of conduction/set process in tinznopt resistance switching random-access memories. Appl. Phys. Lett. **92**(23), 232112 (2008). https://doi.org/10.1063/1.2945278

12. K. Szot, W. Speier, G. Bihlmayer, R. Waser, Switching the electrical resistance of individual dislocations in single-crystalline SrTiO3. Nat. Mater. **5**(4), 312–320 (2006)
13. S. Zhang, S. Long, W. Guan, Q. Liu, Q. Wang, M. Liu, Resistive switching characteristics of MnOsubix/i/sub-based ReRAM. J. Phys. D: Appl. Phys. **42**(5), 055112 (2009). https://doi.org/10.1088/0022-3727/42/5/055112
14. R. Waser, R. Dittmann, M. Salinga, M. Wuttig, Function by defects at the atomic scale—New concepts for non-volatile memories. Solid-State Electron. **54**, 830–840 (2010). https://doi.org/10.1016/j.sse.2010.04.043
15. R. Waser, M. Aono, Nanoionics-based resistive switching memories. Nat. Mater. **6**(11), 833–840 (2007). https://doi.org/10.1038/nmat2023
16. R. Dittmann, J.P. Strachan, Redox-based memristive devices for new computing paradigm. APL Mater. **7**(11), 110903 (2019). https://doi.org/10.1063/1.5129101
17. R.J. Kamaladasa, M. Noman, W. Chen, P.A. Salvador, J.A. Bain, M. Skowronski, Y.N. Picard, Dislocation impact on resistive switching in single-crystal srtio3. J. Appl. Phys. **113**(23), 234510 (2013). https://doi.org/10.1063/1.4811525
18. W. Jiang, R.J. Kamaladasa, Y.M. Lu, A. Vicari, R. Berechman, P.A. Salvador, J.A. Bain, Y.N. Picard, M. Skowronski, Local heating-induced plastic deformation in resistive switching devices. J. Appl. Phys. **110**(5), 054514 (2011). https://doi.org/10.1063/1.3633271
19. S. Choi, S.H. Tan, Z. Li, Y. Kim, C. Choi, P.-Y. Chen, H. Yeon, S. Yu, J. Kim, Sige epitaxial memory for neuromorphic computing with reproducible high performance based on engineered dislocations. Nat. Mater. **17**(4), 335–340 (2018). https://doi.org/10.1038/s41563-017-0001-5
20. S. Cheng, M.-H. Lee, X. Li, L. Fratino, F. Tesler, M.-G. Han, J. del Valle, R.C. Dynes, M.J. Rozenberg, I.K. Schuller, Y. Zhu, Operando characterization of conductive filaments during resistive switching in Mott Vo_2. Proc. Natl. Acad. Sci. **118**(9), 2013676118 (2021). https://www.pnas.org/doi/pdf/https://doi.org/10.1073/pnas.2013676118, https://doi.org/10.1073/pnas.2013676118
21. Y. Yang, R. Huang, Probing memristive switching in nanoionic devices. Nat. Electron. **1**(5), 274–287 (2018). https://doi.org/10.1038/s41928-018-0069-1
22. W. Sun, B. Gao, M. Chi, Q. Xia, J.J. Yang, H. Qian, H. Wu, Understanding memristive switching via in situ characterization and device modeling. Nat. Commun. **10**(1), 3453 (2019). https://doi.org/10.1038/s41467-019-11411-6
23. B.-G. Chae, J.-B. Seol, J.-H. Song, K. Baek, S.-H. Oh, H. Hwang, C.-G. Park, Nanometer-scale phase transformation determines threshold and memory switching mechanism. Adv. Mater. **29**(30), 1701752 (2017). https://onlinelibrary.wiley.com/doi/pdf/https://doi.org/10.1002/adma.201701752, https://doi.org/10.1002/adma.201701752
24. D. Cooper, C. Baeumer, N. Bernier, A. Marchewka, C. La Torre, R.E. Dunin-Borkowski, S. Menzel, R. Waser, R., Dittmann, Anomalous resistance hysteresis in oxide ReRAM: oxygen evolution and reincorporation revealed by in situ TEM. Adv. Mater. **29**(23) (2017)
25. G. Di Martino, A. Demetriadou, W. Li, D. Kos, B. Zhu, X. Wang, B. de Nijs, H. Wang, J. MacManus-Driscoll, J.J. Baumberg, Real-time in situ optical tracking of oxygen vacancy migration in memristors. Nat. Electron. **3**(11). https://doi.org/10.1038/s41928-020-00478-5
26. S. Kumar, Z. Wang, X. Huang, N. Kumari, N. Davila, J.P. Strachan, D. Vine, A.L.D. Kilcoyne, Y. Nishi, R.S. Williams, Conduction channel formation and dissolution due to oxygen thermophoresis/diffusion in hafnium oxide memristors. ACS Nano (2016)
27. Y. Yang, P. Gao, S. Gaba, T. Chang, X. Pan, W. Lu, Observation of conducting filament growth in nanoscale resistive memories. Nat. Commun. **3**(1), 732 (2012). https://doi.org/10.1038/ncomms1737
28. M. Hansen, M. Ziegler, L. Kolberg, R. Soni, S. Dirkmann, T. Mussenbrock, H. Kohlstedt, A double barrier memristive device. Sci. Rep. **5**(1), 13753 (2015). https://doi.org/10.1038/srep13753
29. E. Mikheev, B.D. Hoskins, D.B. Strukov, S. Stemmer, Resistive switching and its suppression in Pt/Nb:SrTiO3 junctions. Nat. Commun. **5**, 3990 (2014)
30. M. Das, S. Kumar, Effect of surface variations on resistive switching, in *Memristor*, ed. by Y.-F. Chang (IntechOpen, Rijeka, 2021, Chap. 2). https://doi.org/10.5772/intechopen.97562

31. H. Sim, H. Choi, D. Lee, M. Chang, D. Choi, Y. Son, E.-H. Lee, W. Kim, Y. Park, I.-K. Yoo, H. Hwang, Excellent resistance switching characteristics of pt/srtio/sub 3/schottky junction for multi-bit nonvolatile memory application, in *IEEE International Electron Devices Meeting, 2005* (IEDM Technical Digest., 2005), pp. 758–761. https://doi.org/10.1109/IEDM.2005.1609464
32. S. Slesazeck, T. Mikolajick, Nanoscale resistive switching memory devices: a review. Nanotechnology **30**(35), 352003 (2019)
33. A. Sawa, Resistive switching in transition metal oxides. Mater. Today **11**(6), 28–36 (2008). https://doi.org/10.1016/S1369-7021(08)70119-6
34. J. Rao, Z. Fan, L. Hong, S. Cheng, Q. Huang, J. Zhao, X. Xiang, E.-J. Guo, H. Guo, Z. Hou, Y. Chen, X. Lu, G. Zhou, X. Gao, J.-M. Liu, An electroforming-free, analog interface-type memristor based on a srfeox epitaxial heterojunction for neuromorphic computing. Mater. Today Phys. **18**, 100392 (2021). https://doi.org/10.1016/j.mtphys.2021.100392
35. L. Gao, Q. Ren, J. Sun, S.-T. Han, Y. Zhou, Memristor modeling: challenges in theories, simulations, and device variability. J. Mater. Chem. C **9**, 16859–16884 (2021). https://doi.org/10.1039/D1TC04201G
36. Y.N. Joglekar, S.J. Wolf, The elusive memristor: properties of basic electrical circuits. Europ. J. Phys. **30**(4), 661 (2009). https://doi.org/10.1088/0143-0807/30/4/001
37. D. Biolek, V. Biolková, Z. Biolek, Spice model of memristor with nonlinear dopant drift. Radioengineering **18**, 210–214 (2009)
38. M.D. Pickett, D.B. Strukov, J.L. Borghetti, J.J. Yang, G.S. Snider, D.R. Stewart, R.S. Williams, Switching dynamics in titanium dioxide memristive devices. J. Appl. Phys. **106**(7), 074508 (2009). https://doi.org/10.1063/1.3236506. https://doi.org/10.1063/1.3236506
39. S. Kvatinsky, E.G. Friedman, A. Kolodny, U.C. Weiser, Team: threshold adaptive memristor model. IEEE Trans. Circuits Syst. I: Regul. Pap. **60**(1), 211–221 (2013). https://doi.org/10.1109/TCSI.2012.2215714
40. B. Mohammad, M.A. Jaoude, V. Kumar, D.M.A. Homouz, H.A. Nahla, M. Al-Qutayri, N. Christoforou, State of the art of metal oxide memristor devices. Nanotechnol. Rev. **5**(3), 311–329 (2016). https://doi.org/10.1515/ntrev-2015-0029
41. S. Maikap, D.S.Z. Rahaman, Bipolar resistive switching memory characteristics using al/cu/geox/w memristor. ECS Trans. **45**, 257 (2012). https://doi.org/10.1149/1.3700961
42. D. Bérardan, S. Franger, D. Dragoe, A.K. Meena, N. Dragoe, Colossal dielectric constant in high entropy oxides. Phys. Status Solidi (RRL)—Rapid Res. Lett. **10**(4), 328–333 (2016). https://onlinelibrary.wiley.com/doi/pdf/https://doi.org/10.1002/pssr.201600043, https://doi.org/10.1002/pssr.201600043
43. J.-W. Yeh, Recent progress in high-entropy alloys. Euro. J. Control **31**, 633–648 (2006)
44. A. Sarkar, C. Loho, L. Velasco, T. Thomas, S.S. Bhattacharya, H. Hahn, R. Djenadic, Multicomponent equiatomic rare earth oxides with a narrow band gap and associated praseodymium multivalency. Dalton Trans. **46**, 12167–12176 (2017). https://doi.org/10.1039/C7DT02077E
45. Z. Rak, C.M. Rost, M. Lim, P. Sarker, C. Toher, S. Curtarolo, J.-P. Maria, D.W. Brenner, Charge compensation and electrostatic transferability in three entropy-stabilized oxides: results from density functional theory calculations. J. Appl. Phys. **120**(9), 095105 (2016) https://doi.org/10.1063/1.4962135
46. M. Ahn, Y. Park, S.H. Lee, S. Chae, J. Lee, J.T. Heron, E. Kioupakis, W.D. Lu, J.D. Phillips, memristors based on (zr, hf, nb, ta, mo, w) high-entropy oxides. Adv. Electron. Mater. **7**(5), 2001258 (2021). https://onlinelibrary.wiley.com/doi/pdf/https://doi.org/10.1002/aelm.202001258, https://doi.org/10.1002/aelm.202001258
47. E.O. Filatova, A.P. Baraban, A.S. Konashuk, M.A. Konyushenko, A.A., Selivanov, A.A., Sokolov, F. Schaefers, V.E. Drozd, Transparent-conductive-oxide (tco) buffer layer effect on the resistive switching process in metal/tio2/tco/metal assemblies. New J. Phys. **16**(11) (2014). https://doi.org/10.1088/1367-2630/16/11/113014
48. Y. Hou, B. Chen, B. Gao, Z.Y. Lun, Z. Xin, R. Liu, L.F. Liu, D.D. Han, Y. Wang, X.Y. Liu, J.F. Kang, Self-compliance multilevel resistive switching characteristics in tin/hfox/al/pt rram devices, in *2013 IEEE International Conference of Electron Devices and Solid-state Circuits* (2013), pp. 1–2. https://doi.org/10.1109/EDSSC.2013.6628050

49. H.Y. Lee, Y.S. Chen, P.S. Chen, T.Y. Wu, F. Chen, C.C. Wang, P.J. Tzeng, M.-J. Tsai, C. Lien, Low-power and nanosecond switching in robust hafnium oxide resistive memory with a thin Ti cap. IEEE Electron. Device Lett. **31**(1), 44–46 (2010). https://doi.org/10.1109/LED.2009.2034670
50. F. Argall, Switching phenomena in titanium oxide thin films. Solid-State Electron. **11**(5), 535–541 (1968). https://doi.org/10.1016/0038-1101(68)90092-0
51. Z. Shen, C. Zhao, Y. Liu, Y. Qi, I.Z. Mitrovic, L. Yang, C. Zhao, Performance variation of solution-processed memristor induced by different top electrode. Solid-State Electron. **186**, 108132 (2021). https://doi.org/10.1016/j.sse.2021.108132
52. N. Gergel-Hackett, B. Hamadani, B. Dunlap, J. Suehle, C. Richter, C. Hacker, D. Gundlach, A flexible solution-processed memristor. IEEE Electron. Device Lett. **30**(7), 706–708 (2009). https://doi.org/10.1109/LED.2009.2021418
53. E. Gale, D. Pearson, S. Kitson, A. Adamatzky, B. de Lacy Costello, The effect of changing electrode metal on solution-processed flexible titanium dioxide memristors. Mater. Chem. Phys. **162**, 20–30 (2015). https://doi.org/10.1016/j.matchemphys.2015.03.037
54. Z. Shen, C. Zhao, T. Zhao, W. Xu, Y. Liu, Y. Qi, I.Z. Mitrovic, L. Yang, C.Z. Zhao, Artificial synaptic performance with learning behavior for memristor fabricated with stacked solution-processed switching layers. ACS Appl. Electron. Mater. **3**(3), 1288–1300 (2021) https://doi.org/10.1021/acsaelm.0c01094
55. M. Sophocleous, N. Mohammadian, L.A. Majewski, J. Georgiou, Solution-processed, low voltage tantalum-based memristive switches. Mater. Lett. **269**, 127676 (2020). https://doi.org/10.1016/j.matlet.2020.127676
56. Y. Pan, T. Wan, H. Du, B. Qu, D. Wang, T.-J. Ha, D. Chu, Mim- icking synaptic plasticity and learning behaviours in solution processed sno2 memristor. J. Alloy. Compd. **757**, 496–503 (2018). https://doi.org/10.1016/j.jallcom.2018.05.092
57. S. Zhang, S. Long, W. Guan, Q. Liu, Q. Wang, M. Liu, Resistive switching characteristics of mnox-based reram. J. Phys. D Appl. Phys. **42**, 055112 (2009). https://doi.org/10.1088/0022-3727/42/5/055112
58. X. Wu, Z. Xu, Z. Yu, T. Zhang, F. Zhao, T. Sun, Z. Ma, Z. Li, S. Wang, Resistive switching behavior of photochemical activation solution-processed thin films at low temperatures for flexible memristor applications. J. Phys. D Appl. Phys. **48**(11), 115101 (2015). https://doi.org/10.1088/0022-3727/48/11/115101
59. Z. Zhou, H. Mao, X. Wang, T. Sun, Q. Chang, Y. Chen, F. Xiu, Z. Liu, J. Liu, W. Huang, Transient and flexible polymer memristors utilizing full-solution processed polymer nanocomposites. Nanoscale **10**, 14824–14829 (2018). https://doi.org/10.1039/C8NR04041A
60. M. Kim, S. Oh, S. Song, J. Kim, Y.-H. Kim, Solution-processed memristor devices using a colloidal quantum dot-polymer composite. Appl. Sci. **11**(11) (2021). https://doi.org/10.3390/app11115020
61. B. Salonikidou, T. Yasunori, B. Le Borgne, J. England, T. Shizuo, R.A. Sporea, Toward fully printed memristive elements: a-tio2 electronic synapse from functionalized nanoparticle ink. ACS Appl. Electron. Mater. **1**(12), 2692–2700 (2019). https://doi.org/10.1021/acsaelm.9b00701
62. S.P. Park, Y.J. Tak, H.J. Kim, J.H. Lee, H. Yoo, H.J. Kim, Analysis of the bipolar resistive switching behavior of a biocompatible glucose film for resistive random access memory. Adv. Mater. **30**(26), 1800722 (2018). https://onlinelibrary.wiley.com/doi/pdf/https://doi.org/10.1002/adma.201800722, https://doi.org/10.1002/adma.201800722
63. C. Pearson, J.H. Ahn, M.F. Mabrook, D.A. Zeze, M.C. Petty, K.T. Kamtekar, C. Wang, M.R. Bryce, P. Dimitrakis, D. Tsoukalas, Electronic memory device based on a single-layer Fluorene-containing organic thin film. Appl. Phys. Lett. **91**(12), 123506 (2007). https://doi.org/10.1063/1.2783481
64. S. Yu, Resistive random access memory (RRAM), in *Resistive Random Access Memory* (2016)
65. B. Govoreanu, G.S. Kar, Y.-Y. Chen, V. Paraschiv, S. Kubicek, A. Fantini, I.P. Radu, L. Goux, S. Clima, R. Degraeve, N. Jossart, O. Richard, T. Vandeweyer, K. Seo, P. Hendrickx, G. Pourtois, H. Bender, L. Altimime, D.J. Wouters, J.A. Kittl, M. Jurczak, $10 \times 10\ nm^2$ Hf/HfO_x crossbar

resistive ram with excellent performance, reliability and low-energy operation, in *2011 International Electron Devices Meeting* (2011), pp. 31–613164. https://doi.org/10.1109/IEDM.2011.6131652
66. F. Zahoor, T.Z. Azni Zulkifli, F.A. Khanday, Resistive random access memory (RRAM): an overview of materials, switching mechanism, performance, multilevel cell (MLC) storage, modeling, and applications. Nanosc. Res. Lett. **15**(1), 90 (2020). https://doi.org/10.1186/s11671-020-03299-9
67. C.Y. Chen, L. Goux, A. Fantini, A. Redolfi, S. Clima, R. Degraeve, Y.Y. Chen, G. Groeseneken, M. Jurczak, Understanding the impact of programming pulses and electrode materials on the endurance properties of scaled Ta_2O_5. RRAM Cells. https://doi.org/10.1109/IEDM.2014.7047049
68. L. Goux, A. Fantini, A. Redolfi, C.Y. Chen, F.F. Shi, R. Degraeve, Y.Y. Chen, T. Witters, G. Groeseneken, M. Jurczak, Role of the Ta scavenger electrode in the excellent switching control and reliability of a scalable low-current operated $TiN_2\backslash O_5\backslash Ta$ RRAM device, in *2014 Symposium on VLSI Technology (VLSI-Technology): digest of Technical Papers* (2014), pp. 1–2. https://doi.org/10.1109/VLSIT.2014.6894401
69. Y. Hayakawa, A. Himeno, R. Yasuhara, W. Boullart, E. Vecchio, T. Vandeweyer, T. Witters, D. Crotti, M. Jurczak, S. Fujii, S. Ito, Y. Kawashima, Y. Ikeda, A. Kawahara, K. Kawai, Z. Wei, S. Muraoka, K. Shimakawa, T., Mikawa, S. Yoneda, Highly reliable TaO_x ReRAM with centralized filament for 28-nm embedded application, in *2015 Symposium on VLSI Technology (VLSI Technology)* (2015), pp. 14–15
70. C. Hsu, I. Wang, C. Lo, M. Chiang, W. Jang, C. Lin, T.-H. Hou, Self-rectifying bipolar TaO_x/TiO_2 RRAM with superior endurance over 10^{12} cycles for 3d high-density storage-class memory, in *2013 Symposium on VLSI Technology, VLSIT 2013—Digest of Technical Papers. Digest of Technical Papers—Symposium on VLSI Technology* (2013), pp. 166–167; Conference date: 11-06-2013 Through 13-06-2013
71. I.-G. Baek, M.S. Lee, S. Seo, M. Lee, D.H. Seo, D. Suh, J.C. Park, S.O. Park, H.-S. Kim, I.-K. Yoo, U.-I. Chung, J.-T. Moon, Highly scalable nonvolatile resistive memory using simple binary oxide driven by asymmetric unipolar voltage pulses, in *IEDM Technical Digest. IEEE International Electron Devices Meeting, 2004* (2004), pp. 587–590
72. Z. Wei, Y. Kanzawa, K. Arita, Y. Katoh, K. Kawai, S. Muraoka, S. Mitani, S. Fujii, K. Katayama, M. Iijima, T. Mikawa, T. Ninomiya, R. Miyanaga, Y. Kawashima, K. Tsuji, A. Himeno, T. Okada, R. Azuma, K. Shimakawa, H. Sugaya, T. Takagi, R. Yasuhara, K. Horiba, H. Kumigashira, M. Oshima, Highly reliable TaO_x ReRAM and direct evidence of redox reaction mechanism, in *2008 IEEE International Electron Devices Meeting* (2008), pp. 1–4. https://doi.org/10.1109/IEDM.2008.4796676
73. T. Yanagida, Memristive switching phenomena in a single oxide nanowire, in *2011 IEEE Nanotechnology Materials and Devices Conference* (2011), pp. 28–31. https://doi.org/10.1109/NMDC.2011.6155307
74. C. Schindler, S.C.P. Thermadam, R. Waser, M.N. Kozicki, Bipolar and unipolar resistive switching in cu-doped SiO2. IEEE Trans. Electron. Devices **54**(10), 2762–2768 (2007). https://doi.org/10.1109/TED.2007.904402
75. H. Kim, P.C. McIntyre, C. On Chui, K.C. Saraswat, S. Stemmer, Engineering chemically abrupt high-k metal oxidesilicon interfaces using an oxygen-gettering metal overlayer. J. Appl. Phys. **96**(6), 3467–3472 (2004) https://doi.org/10.1063/1.1776636. https://doi.org/ https://doi.org/10.1063/1.1776636
76. Y.-E. Syu, T.-C., Chang, T.-M. Tsai, G.-W. Chang, K.-C. Chang, Y.-H. Tai, M.-J. Tsai, Y.-L. Wang, S.M. Sze, Silicon introduced effect on resistive switching characteristics of WOX thin films. Appl. Phys. Lett. **100**(2), 022904 (2012). https://doi.org/10.1063/1.3676194
77. D. Jana, M. Dutta, S. Samanta, S. Maikap, Rram characteristics using a new $Cr/GdO_x/TiN$ structure. Nanosc. Res. Lett. **9**(1), 680 (2014). https://doi.org/10.1186/1556-276X-9-680
78. Y.H. Do, J.S. Kwak, Y.C. Bae, J.H. Lee, Y. Kim, H. Im, J.-P. Hong, Tin electrode-induced bipolar resistive switching of TiO_2 thin films. Curr. Appl. Phys. **10** (2010)

79. L. Chen, Y. Xu, Q.-Q. Sun, P. Zhou, P.-F. Wang, S.-J. Ding, D. Zhang, Atomic-layer-deposited HfLAO-based resistive switching memories with superior performance. Electron. Device Lett. IEEE **31**, 1296–1298 (2010). https://doi.org/10.1109/LED.2010.2069081
80. Y. Wu, Y. Chai, H. Chen, S. Yu, H. Wong, Resistive switching ALO_x-based memory with CNT electrode for ultra-low switching current and high density memory application, in *2011 Symposium on VLSI Technology, VLSIT 2011—Digest of Technical Papers. Digest of Technical Papers—Symposium on VLSI Technology* (2011), pp. 26–27; Conference date: 14-06-2011 Through 16-06-2011
81. B. Sun, Y.X. Liu, L.F. Liu, N. Xu, Y. Wang, X.Y. Liu, R.Q. Han, J.F. Kang, Highly uniform resistive switching characteristics of TiN/ZrO2/Pt memory devices. J. Appl. Phys. **105**(6), 061630 (2009) https://doi.org/10.1063/1.3055414
82. G. Molas, M. Harrand, C. Nail, P. Blaise, Advances in oxide-based conductive bridge memory (cbram) technology for computing systems, in *Advances in Non-Volatile Memory and Storage Technology, Woodhead Publishing Series in Electronic and Optical Materials*, 2nd edn, eds. by B. Magyari-Köpe, Y. Nishi (Woodhead Publishing, 2019), pp. 321–364. https://doi.org/10.1016/B978-0-08-102584-0.00010-3. https://www.sciencedirect.com/science/article/pii/B9780081025840000103
83. E. Pérez-Bosch Quesada, R. Romero-Zaliz, E. Pérez, M. Kalishettyhalli Mahadevaiah, J. Reuben, M.A. Schubert, F. Jiménez-Molinos, J.B. Roldán, C. Wenger, Toward reliable compact modeling of multilevel 1T-1R RRAM devices for neuromorphic systems. Electronics **10**(6) (2021). https://doi.org/10.3390/electronics10060645
84. H. Li, S. Wang, X. Zhang, W. Wang, R. Yang, Z. Sun, W. Feng, P. Lin, Z. Wang, L. Sun, Y. Yao, Memristive crossbar arrays for storage and computing applications. Adv. Intell. Syst. **3**(9), 2100017 (2021). https://onlinelibrary.wiley.com/doi/pdf/https://doi.org/10.1002/aisy.202100017. https://doi.org/10.1002/aisy.202100017
85. Y. Wang, W. Wang, C. Zhang, H. Kan, W. Yue, J. Pang, S. Gao, Y. Li, A digital–analog integrated memristor based on a ZnO NPs/CuO NWs heterostructure for neuromorphic computing. ACS Appl. Electron. Mater. **4**(7), 3525–3534 (2022). https://doi.org/10.1021/acsaelm.2c00495
86. T. Wan, B. Qu, H. Du, X. Lin, Q. Lin, D.-W. Wang, C. Cazorla, S. Li, S. Liu, D. Chu, Digital to analog resistive switching transition induced by graphene buffer layer in strontium titanate based devices. J. Colloid Interface Sci. **512**, 767–774 (2018). https://doi.org/10.1016/j.jcis.2017.10.113
87. J.-D. Kim, Y.-J. Baek, Y. Choi, C.J. Kang, H.H. Lee, H.-M. Kim, K.-B. Kim, T.-S. Yoon, Investigation of analog memristive switching of iron oxide nanoparticle assembly between Pt electrodes. J. Appl. Phys. **114**, 224505–224505 (2013). https://doi.org/10.1063/1.4846759
88. Z.Q. Wang,H.Y. Xu, X.H. Li, H. Yu, Y.C. Liu, X.J. Zhu, Synaptic learning and memory functions achieved using oxygen ion migration/diffusion in an amorphous ingazno memristor. Adv. Funct. Mater. **22**(13), 2759–2765 (2012) https://onlinelibrary.wiley.com/doi/pdf/https://doi.org/10.1002/adfm.201103148. https://doi.org/10.1002/adfm.201103148
89. D. Sacchetto, G.D. Micheli, Y. Leblebici, Multiterminal memristive nanowire devices for logic and memory applications: a review. Proc. IEEE **100**(6), 2008–2020 (2012). https://doi.org/10.1109/JPROC.2011.2172569
90. P.R. Mickel, A.J. Lohn, M.J. Marinella, Memristive switching: physical mechanisms and applications. Modern Phys. Lett. B **28**(10), 1430003 (2014). https://doi.org/10.1142/S0217984914300038
91. S. Choi, S. Jang, J.-H. Moon, J.C. Kim, H.Y. Jeong, P. Jang, K.-J. Lee, G. Wang, A self-rectifying TaO_y/nanoporous TaO_x memristor synaptic array for learning and energy-efficient neuromorphic systems. NPG Asia Mater. **10**(12), 1097–1106 (2018). https://doi.org/10.1038/s41427-018-0101-y
92. C.-C. Hsieh, A. Roy, Y.-F. Chang, D. Shahrjerdi, S.K. Banerjee, A sub-1-volt analog metal oxide memristive-based synaptic device with large conductance change for energy-efficient spike-based computing systems. Appl. Phys. Lett. **109**(22), 223501 (2016) arXiv:1603.03979. https://doi.org/10.1063/1.4971188

Metal Oxides-Based Photodetectors and Sensors

Savita Chowdhury and Anirban Mitra

Abstract Recently, metal oxide-based photodetectors and sensors have attracted the attention of researchers due to the wide range of applications in modern society. Metal oxides have received a significant interest in recent decades due to their versatile and remarkable structural, optical, and electronic properties. This chapter discusses the parameters of photodetectors, the recent progress on the different types of photodetectors such as photoconductor, self-powered, plasmonic etc. Moreover, different types of the metal-oxide based sensors have also been reviewed. Consequently, we describe the present approaches and future research directions for metal oxide based semiconductors for next-generation photodetectors and sensors.

Keywords Metal oxide · Photodetector · Self-powered · Sensor · Biosensor

1 Introduction

Metal oxides and their alloys have garnered considerable attention in recent decades due to their diverse and remarkable structural, optical, and electrical characteristics [1, 2]. Metal oxides have a variety of functional properties those are determined by their crystal structure and, most importantly, the bonding between the metal cation and the oxygen atoms. The electronic configuration and oxidation state of the metal cation primarily provide exceptional magnetic and attractive transport properties [3]. Metal oxide thin films are used for charge creation and recombination as they have a high work function and can display good optical transparency in the visible region [7]. Transition metal oxides have distinct geometrical, electrical, and magnetic properties due to the role of d-orbital electrons in bond formation [4–6]. Because of these properties, metal oxide is a particularly appealing candidate for use in a

S. Chowdhury · A. Mitra (✉)
High Power Laser Laboratory, Department of Physics, Indian Institute of Technology, Roorkee, Uttarakhand 247667, India
e-mail: anirban.mitra@ph.iitr.ac.in

V. Kumar et al. (eds.), *Optical Properties of Metal Oxide Nanostructures*, Progress in Optical Science and Photonics 26,
https://doi.org/10.1007/978-981-99-5640-1_15

wide range of applications, including gas sensing, photocatalysis, solar cells, and photodetectors [7].

Researchers are presently concentrating their efforts on the development of smart, environmentally friendly Internet of Things technology (IoT). According to statistics, the number of sensors used in IoT technology has topped 50 billion, with that figure predicted to exceed 200 billion by 2025. As a result, the IoT market's demand for highly efficient sensors is expanding at an unanticipated rate [8, 9]. Photodetectors (PDs), one of these sensors, show great promise for use in future nanodevices. Photodetectors (PDs) are optoelectronic devices those detect optical signals using electronic processes. They are widely used in a variety of applications, including fibre optic communication networks, digital cameras, medical imaging, and scientific instruments. The three stages of a typical photodetector operation are:

1. Generation of charge carriers by incident light
2. Carrier movement and/or multiplication
3. Collection of charge carriers to generate the output signal.

Photodetectors are designed to optimize the performance of each step of operation in order to fulfil the demanding requirements of high photoresponse, quick response times, and low noise. They should also be small, self-biased, and stable while in operation, which can perform through the use of optimal materials, improved fabrication techniques, and accurate control of device characteristics.

The human body is equipped with a variety of sensors. We can sense taste, heat, light, and pressure. As the name implies, sensors sense or monitor a chemical or physical quantity. The development of sensors based on metal oxide has been relatively moderate. However, given the rising need for security, environmental management, health improvement, and other factors, sensors are predicted to play a more significant role and develop more quickly soon.

In this chapter, we have reviewed the recently published papers on metal oxide-based photodetector and sensors. In this study sections have been organized as follows. Section 2 discusses the key parameters of photodetector. In Sect. 3 different kinds of photodetector based on metal oxide film have been enlisted. Section 4 describes the different types of sensors based on metal oxide. The summary of this chapter is presented in Sect. 5.

2 Photodetector Characteristics

A range of factors can describe the performance of a photodetector. These performance traits show how a detector performs. When detecting a wavelength, the detector's reaction should be excellent and the additional noise should be minimal. High response speeds are necessary to see fluctuations in the input optical signal. The parameters of the photodetectors are outlined in the subsections that follow.

2.1 Responsivity (R)

Responsivity of a photodetector is a crucial characteristic for determining its performance. It is defined as the ratio of the output photocurrent to the incident optical power. The responsivity of a photodetector is affected by a number of parameters, including the detector's material qualities, the wavelength of the incident light, the detector's temperature, and the applied bias voltage.

The metal oxide material of the detector has properties of reflection and absorption that vary with wavelength, and as a result, responsivity varies with wavelength. The responsivity may be expressed as:

$$R = \frac{I_{ph}}{P_{inc}} \tag{1}$$

where P_{inc} is the optical power of incident light (W), and I_{ph} is the photocurrent (A) of the photodetector. The responsivity may also be expressed as:

$$R = \frac{\eta \lambda q}{hc} = \frac{\eta \lambda (\mu m)}{1.24} (A/W) \tag{2}$$

where λ is the wavelength of incident light, and η is the external quantum efficiency. The designer often specifies responsivity, which is a crucial characteristic. The manufacturer can determine via responsivity how much output the detector needs for a particular application.

2.2 Quantum Efficiency

The ratio of an electron production rate to an incidence photon rate is known as a photodetector's quantum efficiency (η). The following equation relates quantum efficiency to photodetector responsivity:

$$\eta = \frac{I_{ph}/q}{P_{inc}/hv} = \frac{I_{ph}}{q} \cdot \frac{hv}{P_{inc}} \tag{3}$$

2.3 Detectivity

The detectivity (D) of a device is a figure of merit that describes the ability of the device to detect weak signals. It is defined as:

$$D^* = R(A/2qI_d)^{1/2} \tag{4}$$

It is important to define the wavelength of the incident radiation, any bias current that has been applied to the device, and the detector's area when the detectivity is used to describe the detector. The detectivity of a photodetector refers to its ability to detect weak light signals.

2.4 *Response Time*

The viability of a device for a given application will frequently be determined by the response time T, which is critical for all detectors. This parameter is explained by the time taken by it to respond with respect to a rapid change in the input signal. The response time of a photodetector is defined as the time taken by the output photocurrent to rise from 10 to 90% or fall from 90 to 10% of its maximum value in response to a change in the incident light. It measures how quickly the detector responds to the variations in light intensity. The rise and fall times of the detector can be so short that they appear as a square wave. This is because the speed of carrier transport and recombination processes within the device restrict the photodetector's ability to respond.

3 Photoconductor

The photoconductive detector (photoconductor) fundamentally functions as a radiation-sensitive resistor and has resistor-like fundamental characteristics. Xu et al. fabricated a photoconductive ultraviolet (UV) detector using zinc oxide (ZnO) thin film deposited on a quartz substrate by radio frequency (RF) sputtering. The I–V curve of the ZnO-based UV photodetector is exhibited in Fig. 1, which offers linear nature. The responsivity reported 18 A/W at 5 V applied voltage under exposure of 365 nm wavelength light. The device displays a fast response speed with rise and decay times of 100 ns and 1.5 μs [10]. Xie et al. studied the photoconductivity of ZnO NRs at room temperature, which was synthesized through clean room technology. They explore how the photoresponse of these rods changes with temperature and the environment by analyzing their spectrum and frequency dependency. It can be observed that beyond bandgap light responds quickly but below-bandgap light responds slowly, exhibiting clear indications of persistent photoconductivity [11]. The photoconductor device based on ZnO single nanowires was studied in air and vacuum conditions by Bao and his research group [12]. They observed that the photoconductivity of the device was higher in a vacuum compared to air, which was explained by the decomposition of ZnO.

Tawfik et al. fabricated cobalt doped CuO and pure films using AC and DC reactive magnetron sputtering methods. The doping ratio of the films was controlled by the RF power of the AC sputtering technique. Ohmic contacts and increments in the conductivity values with doping are observed in I–V measurements. Photoresponse

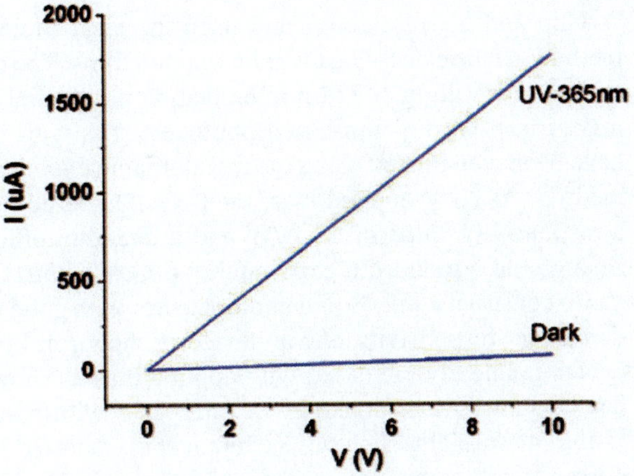

Fig. 1 I–V measurement of ZnO photodetector under dark and illumination of 365 nm wavelength, reprinted with permission from [10]

studies have revealed that cobalt-doped CuO films show higher photoresponse than pure CuO films [13].

Manga et al. fabricated a hybrid TiO_2 graphene-based photoconducting matrix using a solution-processable technique. The device based on TiO_2-graphene shows higher photoconductive gain, higher detectivity, and faster response speed than the TiO_2-based device [14]. The photoconductivity of monocrystalline V_2O_5 nanowires prepared by the physical vapor deposition (PVD) technique has been examined by Chen and his research group. They also reported that V_2O_5 film-based device fabricated by the PVD method shows two times of photoconversion efficiency than the device fabricated by the hydrothermal method. The responsivity and gain values of the device based on V_2O_5 (NWs) reported around 7900 AW^{-1} and 30,000, respectively [15].

3.1 *Metal-Semiconductor-Metal Photodetector*

A photodiode with two Schottky metal contacts placed back-to-back on a coplanar surface is referred to as an MSM (Metal-Semiconductor-Metal) photodetector. Metal contacts are commonly interdigitated stripes, and the semiconductor material absorbs light in the space between the metal contacts. This type of photodetector has the advantage of being easy to construct and fabricate, with a simple design. At low applied voltages, photocurrent in MSM photodetectors typically increases with voltage because of the rise in the depletion area, which improves the device's internal quantum efficiency. However, the photocurrent eventually saturates at higher voltages.

Guo and his research group used the laser molecular beam epitaxy (L-MBE) method to fabricate β-Ga_2O_3 epitaxial thin films. The device displays a clear photoresponse to UV light at 254 nm, indicating a potential use in solar-blind photodetectors [16]. β-Ga_2O_3 film-based photodetectors with interdigitated metal electrodes have been constructed using the metalorganic chemical vapor deposition (MOCVD) method. At 20 V applied bias, the β-Ga_2O_3 based photodetector device displayed a responsivity value of 17 A/W and a quantum efficiency of almost 8228% [17]. Zhang et al. produced a good-quality β-Ga_2O_3 thin film using N_2O as the reaction gas to construct a solar-blind photodetector with good performance. The wavelength-dependent responsivity of the device is shown in Fig. 2a. The responsivity of the device is linearly increased with applied bias as shown in the inset of Fig. 2a. The device showed a responsivity of 26.1 A/W with rise/fall time of 0.48/0.18 s [18]. Using the sol-gel technique, Kokubun et al. reported MSM photodetectors based on β-Ga_2O_3 with a maximum responsivity of 8×10^{-5} A/W [19]. Zn-doped β-Ga_2O_3 (ZnGaO) and undoped β-Ga_2O_3 epitaxial films based on solar-blind photodetectors were fabricated and characterized by Alema et al. The peak response of the ZnGaO detector, which is significantly greater than that of the β-Ga_2O_3 device, was 210 A/W at 232 nm [20]. Ga_2O_3-based photodetector was constructed by Jubu et al. using the reducing-ambient CVD process. Figure 2b exhibited the responsivity and detectivity variation with an applied bias for Ga_2O_3-based MSM device. The measured values of responsivity and detectivity for the MSM photodetector are 38.161 A/W and 8.3916×10^9 Jones, respectively [21].

Sol-gel fabrication was used to fabricate anatase TiO_2 MSM photodetectors on a Si substrate. At a bias of 5 V, the spectral response of TiO_2 MSM photodetectors with Au Schottky contact is presented in Fig. 2c. The device's photoresponse was uniform across the cut-off wavelength of 330 nm. The device displayed 1.9 nA and 199 A/W values of dark current and responsivity at 5 V applied bias under the illumination of 260 nm wavelength, respectively. However, the device was slow to react, with the response and recovery times of 6 s and 15 s, respectively [22].

Kong et al. reported a TiO_2-based MSM photodetector prepared by the sol-gel method, which shows the responsivity value of 889.6 A/W at 5 V bias upon exposure

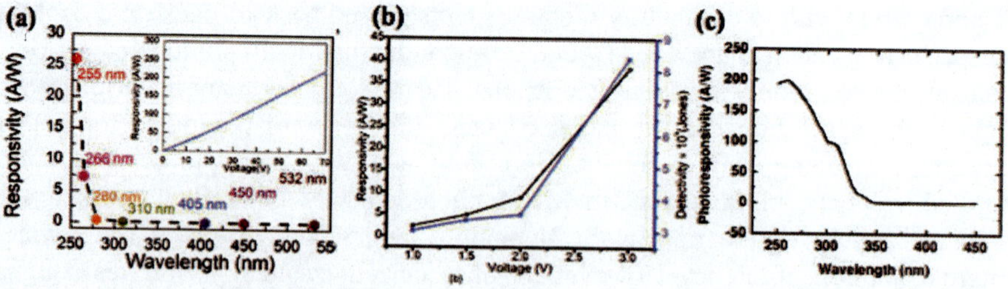

Fig. 2 **a** Spectral response of the photodetector at 10 V applied bias. The responsivity vs applied voltage is depicted in the inset, reprinted with permission from [18], **b** Applied bias dependent detectivity and responsivity, reprinted with permission from [21], **c** Wavelength dependent photoresponsivity curve of MSM device based on TiO_2, reprinted with permission from [22]

of 260 nm wavelength. The spectral-dependent responsivity of the $Ni/TiO_2/Ni$-based device at 5 V applied bias is shown in Fig. 3a [23]. Radio frequency (RF) magnetron sputtering was used by Shih et al. to fabricate TiO_2 MSM photodetectors with oxygen plasma modifications. Various O_2 plasma treatment times were used to examine the responsivities of the TiO_2-based MSM photodetectors. The peak responsivities of the TiO_2-based photodetectors were determined to be 36 A/W, 144 A/W, 153 A/W, and 53 A/W, with O_2 plasma modifications of 0 min, 1 min, 2 min, and 3 min, respectively. [24].

Deniz Caliskan examined the performance of a TiO_2-based MSM photodetector produced by RF magnetron sputtering, which has a responsivity value of 1.73 A/W at 50 V applied voltage [25]. The wavelength-dependent photoresponse of the device is shown in Fig. 3b. The responsivity of the device is decreased below 275 nm wavelength, which is explained by the absorption plot shown in the inset of Fig. 3b.

Zhao et al. prepared a Mg-doped NiO thin film on a quartz substrate via electron beam evaporation, with a peak response of 0.147 mA/W at 5 V applied voltage

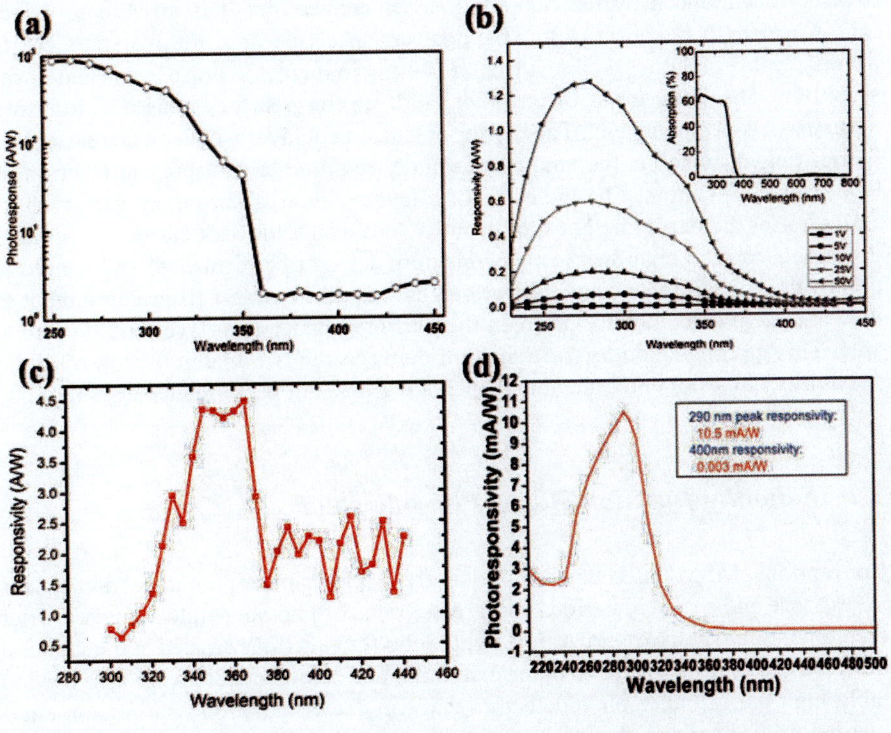

Fig. 3 **a** The wavelength dependent response of the $Ni/TiO_2/Ni$ structure-based photodetector, reprinted with permission from [23], **b** Responsivity plot with wavelength for MSM type photodiode. The wavelength dependent absorption of the TiO_2 layer is shown in the inset, reprinted with permission from [25], **c** Spectral response of the Au/NiO/Au structure-based UV photodetector at 5 V applied voltage, reprinted with permission from [26], **d** wavelength dependent responsivity of the device based on $Mg_{0.47}Zn_{0.53}$ O at 5 V applied voltage, reprinted with permission from [27]

and 320 nm illumination [28]. Ahmed and his group fabricated an MSM photodetector based on NiO/Si heterostructure using RF magnetron sputtering technique. The variation in responsivity with spectral region from 300 to 440 nm of Au/NiO/Au structure-based device at 5 V applied bias is shown in Fig. 3c. The maximum responsivity and photosensitivity values of the MSM UV device were 4.5 A/W and $6.4 \times 10^3\%$ under exposure to 365 nm wavelength [26].

Zhao et al. demonstrated $Mg_{0.47}Zn_{0.53}O$ thin film based MSM photodetector, which was prepared by RF magnetron co-sputtering. With a cutoff at 312 nm under a 5 V bias, a maximum responsivity value of 10.5 mA/W under the exposures of 290 nm wavelength was measured. The spectral-dependent photoresponse of the device based on $Mg_{0.47}Zn_{0.53}O$ is shown in Fig. 3d. The detector's response time exhibited a fast component with a response and recovery time of 10 ns and 30 ns, respectively [27]. There has been a lot of reported research on doped ZnO that is used for UV photodetection, such as Cu-doped ZnO NWs [29], Ga-doped ZnO [30], La-doped ZnO NWs [31], and Sb-doped ZnO NWs [32].

The fundamental shortcomings of the MSM-based photodetector are the large dark current induced by the Schottky-barrier connection. It is difficult to detect long-wavelength radiation in low-bandgap materials because the materials create a considerable quantity of dark current, resulting in increased noise levels and lower sensitivity. The dark current, on the other hand, may be greatly decreased by inserting a barrier-enhancement layer. The barrier-enhancement layer works by increasing the barrier height, which is the amount of energy required to transport an electron or hole over the junction. The inclusion of a layer with a larger energy gap, such as GaAs, raises the barrier height significantly. As a result, the dark current has greatly decreased, causing significant enhancement in detector performance. The composition of the barrier-enhancement layer can be graded to avoid carrier entrapment at the band-edge discontinuity between the barrier-enhancement layer and the active area. This grading smooths the transition between the two materials, lowering the possibility of carrier entrapment and enhancing detector performance even more.

3.2 *Schottky Junction-Based Photodetector*

Conventional MSM-based devices frequently display slower photodetection speeds in response and recovery periods and greater rates of photogenerated charge carrier recombination when exposed to UV light. Schottky junctions created at the junction between conductors (metal or other materials) and semiconductors are also used to build high-performance photodetectors that can detect radiation with photon energies below the bandgap of semiconductors through internal photoemission. Schottky junction-based UV PDs feature a relatively low dark current, a rapid response, and flexible photoresponse performance when driven externally [33, 34]. Schottky PDs have many benefits over photoconductive-type PDs, including low dark current value, fast response times, high UV-visibility rejection ratio, and perhaps even the ability to be self-powered.

Similarly, Nie et al. fabricated a Schottky junction UV photodetector based on a graphene film/ZnO NR array. The photoconductive gain and responsivity of the device are obtained as 385 and 113 A/W at −1 V bias, respectively [35]. Zhang and the research group demonstrated the construction of high-performance graphene-ZnO Schottky junction-based UV photodetectors. The responsivity and gain of the device have been decreased with reducing bias voltage, which is shown in Fig. 4a. The responsivity, detectivity, and gain of the device were calculated to be 3×10^4 A/W, 4.33×10^{14} cm $Hz^{1/2}W^{-1}$, and 10^5, respectively [36]. Through localized Schottky effects, Au nanoparticle (NP) decorating of single ZnO nanowire (NW) photodetectors (PDs) can effectively increase both photocurrent and photoconductive gain, which has been analyzed by Chen and his research group [37]. The excitation power-dependent photoconductive gain for the ZnO NW and the Au-decorated ZnO NW photodetectors are shown in Fig. 4b.

An Au/β-Ga_2O_3 nanowire array film-based Schottky photodiode was developed using a straightforward thermal partial oxidation technique. The Schottky photodiode

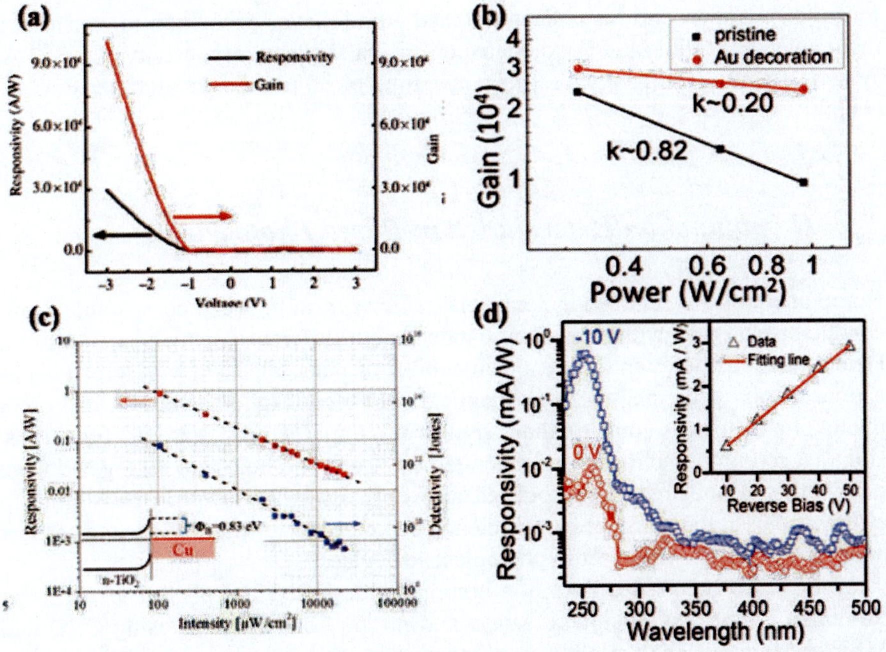

Fig. 4 **a** Responsivity and gain of the UV photodetector at various applied voltages, reprinted with permission from [36], **b** photoconductive gain dependent on excitation power for the pristine and the Au NP-decorated ZnO NW PDs, reprinted with permission from [37], **c** responsivity and detectivity of the Cu/TiO_2 Schottky photodetector vary with light intensity, reprinted with permission from [39], **d** responsivity of the device at 0 and −10 V bias. The inset exhibits the responsivity of the photodetector under the illumination of 254 nm wavelength as a result of applied bias, reprinted with permission from [38]

formed of Au/β-Ga_2O_3 nanowires array film has good rectifying characteristics and a low dark current of just 10 pA at an applied bias of 30 V. The wavelength-dependent responsivity of the device at 0 V and 10 V reverse bias are shown in Fig. 4d. The variation in the responsivity with a reverse bias of the device under exposure of 254 nm wavelength is exhibited in the inset of Fig. 4d. Photoresponse study revealed that device is a suitable solar-blind photodetector with a cutoff edge at about 270 nm and can function without bias. The response and recovery time of the device were ~1 and 64 μs, respectively [38].

According to Patel et al., TiO_2 based Schottky photodetectors were fabricated using the DC sputtering technique. The values of peak responsivity and detectivity for devices were determined to be 0.897 A/W and 4.5×10^{12} Jones at −1 V applied bias, respectively. The variation in the responsivity with illumination intensity for the photodiode is displayed in Fig. 4c. The photodetector also exhibited a very fast response, with response/recovery times of 1.12/1.46 ms, [39]. Chang et al. developed a TiO_2 MIS photodetector by sandwiching a polyvinyl phenol (PVP) layer between the electrodes and the TiO_2 nanowires. This photodetector exhibits response values of 2.46×10^3 A/W under 390 nm at 5 V and detectivity values of 8.86×10^{11} Jones [40]. Akbari and his colleagues used atomic layer deposition to construct a monolayer TiO_2 film-based photodetector, which shows a responsivity of 0.352 A/W at 1 V applied voltage with a response time of 30 μs and decay time of 63 μs [41].

3.3 *Homojunction/Heterojunction-Based Photodetector*

Homojunctions are semiconductor interfaces that form between comparable semiconductor materials with equivalent bandgaps but different doping concentrations. Heterojunctions between different semiconductor materials have widespread application in solar cells, photo-electrochemical (PEC) cells, photodetectors, and other photovoltaic devices, due to their significant impacts on electronic properties. Contrary to homojunction, photogenerated electrons and holes are segregated from one another by both internally produced and externally applied electric fields. Built-in electric fields at the interface can efficiently separate photogenerated charge carriers and regulate carrier transport in photoelectric devices.

Wang et al. reported a ZnO p–n homojunction-based UV photodetector using Sb-doped p-type ZnO nanowire arrays and n-type ZnO thin film using CVD and MBE techniques. The variation in photocurrent measurement with wavelength at different applied voltages has been displayed in Fig. 5b. Photocurrent measurements revealed a very positive response to UV light irradiation [42]. Leung and co-authors reported on the fabrication of a ZnO p–n homojunction using the sol-gel method. The device showed a high responsivity of 4 A/W when exposed to a wavelength of 384 nm and an applied voltage of −3 V. The wavelength-dependent photoresponse of the device with different applied biases is shown in Fig. 5a [43]. Various UV photodetectors based on ZnO have also been described, for example, p–n ZnO NW homojunction [44], Cu-doped ZnO/ZnO NRs homojunction [45], p-$Mg_{0.2}Zn_{0.8}O$/n-ZnO homojunction [46], ZnO microwire p–n homojunction [47], ZnO sheet-like nanorod p–n homojunction [48], and ZnO NW p–n homojunction [49].

Tsai and co-authors studied NiO/ZnO heterostructure-based UV photodetector. The current density versus voltage plot of the NiO/ZnO structure-based device in dark and under light illumination conditions is shown in Fig. 5c [50]. ZnO/SnO_2 heterojunction-based UV PD was constructed by Tian and co-authors [51]. The photodetector showed the highest sensitivity of 5×10^3 when exposed to light at 300 nm. The photodetector's observed response and recovery time were 32.2 s and 7.8 s, respectively. A high-performance UV PD using a ZnS/ZnO biaxial NB heterostructure was developed by Fang's group. Due to the built-in electric field, this heterostructure demonstrated more effective photodetection characteristics than pure ZnO or ZnS nanostructures [52]. This device at 5 V bias exhibited a high responsivity of 5.0×10^5 A/W with rise and decay time of 0.3 and 1.5 s, respectively.

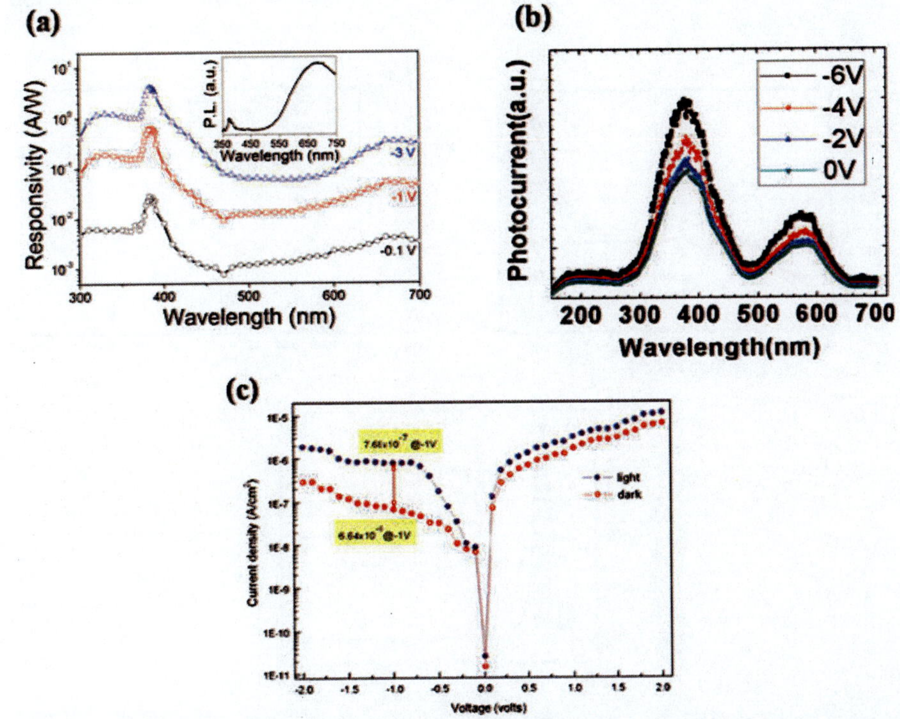

Fig. 5 **a** Responsivity characteristics at the reverse voltage of 0.1, 1, and 3 V. The inset exhibits the PL spectra of the ZnO film under 325 nm excitation, reprinted with permission from [43], **b** photocurrent spectra under different reverse biases, reprinted with permission from [42], **c** The J–V curve of the NiO/ZnO heterojunction measured both in the dark and under 365 nm wavelength illumination conditions with the power density of 0.3 mW/cm^2, reprinted with permission from [50]

Therefore, it can be inferred that due to synergistic effects and inbuilt internal electric fields, junction-based UV photodetectors exhibit greater photodetection behavior than their pristine equivalents. Mahesh et al. have studied the photoresponse characteristics of a Cu-doped ZnO/Si heterojunction device. I-V measurements exhibit rectifying behavior similar to a diode in dark and light illumination conditions, which is displayed in Fig. 6a [53]. The responsivity of the device was 6.2 A/W with response/recovery time of 0.14 s/0.11 s, respectively. Heterostructures of ZnO with other metal oxide semiconductor materials have been presented, e.g. ZnO/Si heterostructure [54], ZnO NWs/p-GaN heterojunction [55], NiO/ZnO p–n junction [56], n-Ga:ZnO NRs/p-GaN heterostructure [57], ZnO/GaN heterojunction [58], p-GaN/n-ZnO NRs heterostructure [59], ZnO NR/graphene heterostructure [60], and ZnO NP-graphene core-shell heterostructure [61].

A van der Waals interaction demonstrated a mixed-dimensional ultrawide bandgap heterojunction consisting of an n-type exfoliated-Ga_2O_3 nanolayer and a p-type diamond substrate [64]. The device showed outstanding photoresponse properties, including responsivity (12 A/W), photo-to-dark-current ratio (3900), rejection

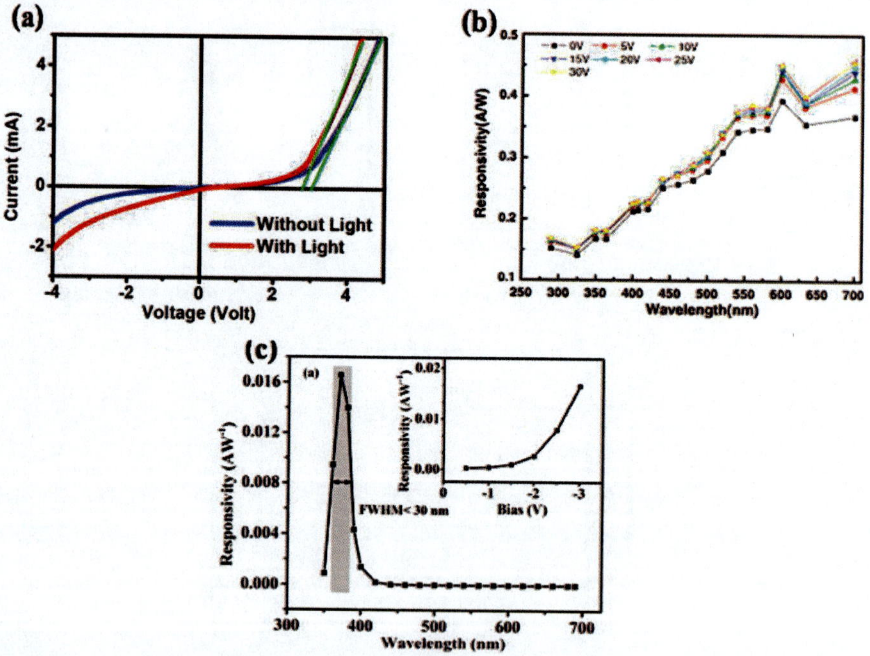

Fig. 6 **a** Current–voltage measurement of n-CZO/p-Si heterostructure under dark and white light illumination within the voltage range of −4 to +5 V, reprinted with permission from [53], **b** spectral responsivity curves obtained in the range of 290–700 nm at different reverse voltage such as 0, 5, 10, 15, 20, 25, and 30 V, reprinted with permission from [62], **c** responsivity with the wavelength of the ITO/H-NiO/IGZO/ITO photodetector at −3 V bias. The inset exhibits the responsivity versus the reverse bias plot, reprinted with permission from [63]

ratio (8.5×10^3), and quick response/decay characteristics while demonstrating no persistent photoconductivity. GaN/Ga_2O_3 dual-band photodetector fabrication was proposed by Weng et al. [65]. Qu et al. employed laser molecular beam epitaxy to fabricate β-Ga_2O_3 thin films on n-type 4H-SiC substrates [66]. When Au/Ti was replaced as the top electrode with a single graphene layer, the responsivity value improved from 7.14×10^{-4} AW^{-1} to 0.18 AW^{-1}, and the response time reduced from 2.14 to 0.65 s. The deep UV solar-blind photodetector was constructed by Guo et al. using laser molecular beam epitaxy to prepare β-Ga_2O_3 films on Si [67]. The responsivity was 370 A/W at −3 V bias under the illumination of 254 nm, and the associated external quantum efficiency was more than 1.8×10^5%, which may be ascribed to charge carrier multiplication. Furthermore, the response and recovery times were 1.79 and 0.27 s, respectively.

An easy solution procedure was used by Sun et al. to construct a Fe-doped TiO_2/n-Si heterostructure-based self-powered UV-visible photodetector [68]. At 0 V, the device had a responsivity of 60 mA/W under the irradiation of 600 nm wavelength with the light intensity of 0.5 mW/cm^2, with rise and fall time of less than 10 ms and 15 ms, respectively. Zhang et al. fabricated TiO_2 thin films on $SrTiO_3$ (001) substrate using the sol-gel technique [69]. The value of responsivity was 46.1 A/W at 260 nm with response and decay time of 3.5 ms and 1.4 s, respectively. Then, their team created an n-TiO_2/p-NiO heterostructure-based UV photodetector, which showed an overall performance development (responsivity value was 181.9 AW^{-1} with a recovery time was 598 ms at 280 nm) compared to their earlier devices [70]. Due to its internal electric field (E_{bi}) in the dark, the p–n heterojunction can deplete the bulk of carriers, potentially reducing device noise and I_{dark}. Liu et al. used radio frequency magnetron sputtering to deposit an ultrathin TiO_2 layer on the surface of a single crystal diamond to create heterojunction photodetectors [71]. The photodetector exhibited two peaks of responsivity at about 225 nm (3 mA/W) and 290 nm (23.6 mA/W), which were caused by the responses of diamond and TiO_2, respectively. Avijit et al. investigated a comparative study of the photodetection behavior of TiO_2 film grown on chemically textured Si and pristine Si. They obtained maximum responsivity (0.22 A/W at −2 V) under UV light for textured-Si/TiO_2 compared to the pristine-Si/TiO_2 (0.11 A/W at −2 V) [72].

Avijit and his research group studied TiO_2/Si heterojunction-based photodetector, in which TiO_2 films at various thicknesses were grown on a Si. Photoresponse of TiO_2/Si device is improved by decreasing the thickness of TiO_2 films [73].

NiO film has mainly been applied in the fabrication of p–n heterostructure-based UV photodetectors with some other n-type semiconductors, including Silicon, indium gallium zinc oxide, ZnO, and TiO_2, due to its p-type nature. Choi et al. constructed a p–n heterojunction by evaporating the NiO layer on the Si substrate, and the constructed photodetectors had a responsivity of 0.15 AW^{-1} at 0 V under the exposures of a 290 nm wavelength light source. The wavelength-dependent responsivity was displayed at different reverse biases in Fig. 6b [62]. A magnetron sputtering technique was used by Li et al. to construct an n-IGZO/p-NiO photodetector which exhibited maximum responsivity of 0.016 A/W at 3 V bias under exposure of 370 nm wavelength. The responsivity versus wavelength plot for the device based on high conductive NiO film at a reverse bias of 3 V is displayed in Fig. 6c [63].

The NiO/Si heterostructure-based photodiode prepared by Zhang and his research group demonstrated a maximum responsivity of 0.815 A/W at a reverse bias of 2 V. The p-NiO thin film's conductivity significantly affected the photodetector's behavior, which could be controlled by adjusting the O_2 working pressure during NiO film deposition [74].

3.4 *Self-powered Photodetector*

In a typical photodetector, an external driving force is required to separate and move photogenerated charge carriers for photoconduction. Internal heat generated by the presence of an external bias voltage in typical photodetectors might result in higher power consumption and lower device performance. Nevertheless, junction-based photodetectors do not require an external bias voltage and instead rely on the charge carrier separation and movement provided by the internal built-in electric field. In a p–n junction-based photodetector, for example, the p-type and n-type areas create a built-in electric field that separates photo-generated electrons and holes, generating a measurable photocurrent. Similarly, the barrier height between the metal and semiconductor in a Schottky barrier photodetector provides an internal electric field that separates the charge carriers and generates a photocurrent.

As a result, considerable study has been done on the preparation of self-powered photodetectors in the present scientific society. Self-powered photodetectors provide various advantages over conventional photodetectors, such as extremely low power usage, a decrease in unwanted energy-related issues, and decreased photodetector size. Furthermore, self-powered photodetectors exhibit quicker photoconduction behavior under light illumination due to photogenerated charge carriers' rapid separation and movement by a built-in electric field.

3.4.1 Schottky Junction-Based Self-Powered Photodetectors

The graphene/ZnO:Al NR Schottky junction-based self-biased photodetectors developed by Duan and coworkers. The use of ZnO:Al NRs as the active layer provides a large surface area for light absorption, while the graphene layer acts as a transparent electrode, allowing light to pass through and reach the ZnO:Al NRs. These photodetectors have potential applications in low-power and self-powered electronics, as well as in UV sensing and imaging [75]. Zhang and his research group created an Sb-doped ZnO nanobelt ability of self via bridging a Schottky contact with Au and an Ohmic contact with Ag [76]. This device demonstrated self-powered photodetection performance, with current increasing from 1 to 23 nA without an applied voltage when exposed to UV light. High donor impurity density caused by Sb doping in ZnO improves the electric field and enhances self-powering behaviour. This photodetector demonstrated a significantly quicker photodetection speed of less than 100 ms.

Asymmetric Schottky junctions have been shown to be effective in enhancing the performance of self-driven UV photodetectors. Chen et al. used Au asymmetric electrodes to regulate charge carrier movement and demonstrated devices with various electrode width ratios. The device with a width ratio of 20:1 (wide finger: narrow finger) showed a responsivity of 20 mA/W, a response time of 710 ns, and a recovery time of 4 ms. Additionally, the piezo-phototronic phenomenon of ZnO was coupled to enhance self-powered photodetector performance by developing piezo potentials across ZnO that allowed for the separation and movement of photogenerated charge carriers. Zhang's group studied a self-powered photodetector based on a ZnO/Au structure, and its photodetection performance was enhanced by introducing an external strain [77].

3.4.2 Heterojunction-Based Self-powered Photodetectors

p-GaN/n-Sn:Ga_2O_3 heterostructure-based self-driven UV photodetector was prepared by growing an Sn-doped Ga_2O_3 film on the GaN film. The responsivity and detectivity at 254 nm are obtained as 3.05 AW^{-1} and 1.69×10^{13} cm $Hz^{1/2}$ W^{-1}, respectively, without a power supply [78].

Zheng et al. presented an economically feasible large-scale heterojunction structure based on n-TiO_2 nano wells and p-NiO mesoporous nanosheets. The hybrid PD based on p–n junctions functioned excellently in a self-bias mode, with excellent responsivity, significant wavelength selectivity, and good stability [79].

The thermal evaporation method was applied to construct V_2O_5/n-Si heterostructure-based photodetector successfully. The responsivity versus wavelength plot for V_2O_5/n-Si heterostructures is shown in Fig. 7a, in which the Si surface is passivated by H and CH_3 groups. The methyl group passivated device exhibited the highest responsivity compared to the H-passivated device. The effect of V_2O_5 thickness and interface passivation on heterojunction-based photodetector performance was thoroughly investigated. The photodetector displayed a strong on/off ratio of 1.4×10^4, a quick response speed of 9.5 s, and a high specific detectivity of 1.34×10^{12} Jones when illuminated with 650 nm light [80].

Optoelectronic properties of a NiO/Si heterostructure-based photodiode for broadband photodetection that was made by employing the PLD technique to deposit a NiO thin film over a commercial Si substrate. The wavelength-dependent responsivity of the NiO/Si-based device is shown in Fig. 7b. Under 0 V conditions, the photodiode shows good responsivity values of 24.6 mA/W with irradiation of 485 nm wavelength, respectively. The rise and fall time at 1.695 kHz were calculated to be 0.17 ms and 0.05 ms, respectively, under 485 nm light [81].

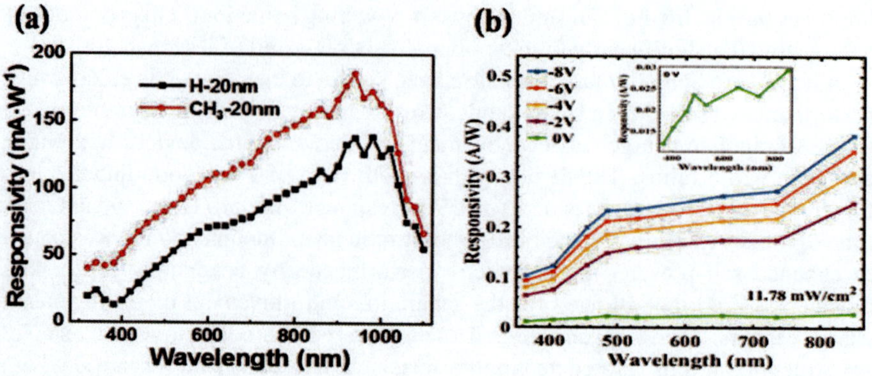

Fig. 7 **a** Responsivity with wavelength dependent of the two heterojunctions, reprinted with permission from [80], **b** the wavelength-dependent responsivity at different applied voltages, reprinted with permission from [81]

3.5 *Plasmonic Based Photodetector*

Venugopal et al. fabricated Ag (nanoisland)/n-Al:ZnO/p-Si heterostructure using pulsed laser deposition (PLD) and thermal evaporation. They observed that introducing Ag in the device exhibited an increment in the current compared to the bare Al: ZnO/Si device [82]. Gurpreet et al. fabricated Al:ZnO/Ag(nanoisland)/Cu_2O heterostructure using a PLD technique. Enhancement in current is observed after introducing the Ag nanoisland at the interface of Al: ZnO and Cu_2O, which is explained by localized surface plasmon resonance [83]. Mitra's group [84] published a comparison study of the self-biased photodetection characteristics of NiO/Si and NiO/Ag-NPs/Si heterostructures-based devices. Current–voltage curves measured, in the dark and irradiation (UV and AM1.5G), for NiO/Ag-NP/Si heterostructure device is shown in Fig. 8a. In comparison to the NiO/Si structure with zero bias, we were able to obtain large responsivity (6.58 mAW^{-1}), significant detectivity (2.1×10^{10} $cmHz^{1/2}W^{-1}$), and quick response for the Ag-NPs incorporated NiO/Si heterostructure. Joshna et al. [85] exhibited a self-powered photodetector based on p-NiO/n-rGO heterostructure. Current–voltage measurement of the Al:ZnO/Ag/Cu_2O and Al:ZnO/Cu_2O structures-based solar cells in illumination (1.5 AMU) is shown in Fig. 8b, and the wavelength-dependent IPCE curve was displayed in the inset of Fig. 8b. Ag NPs incorporated p-NiO/n-rGO heterostructure device demonstrated a 10^4-fold increase in photocurrent and a significant improvement in UV photoconductivity compared to the pristine NiO/rGO. Ag NPs included devices that exhibited improved electrical characteristics, including efficiency, responsivity, and a high on/off ratio.

rGO/Si structure-based UV photodetector with incorporated Au nanoparticles was constructed [86]. The photoresponse characteristics confirmed that the Au-rGO/Si heterostructure had a considerably larger UV photoresponse activity than rGO/Si. By dissolving the base Ag foil and integrating it to produce Schottky junctions,

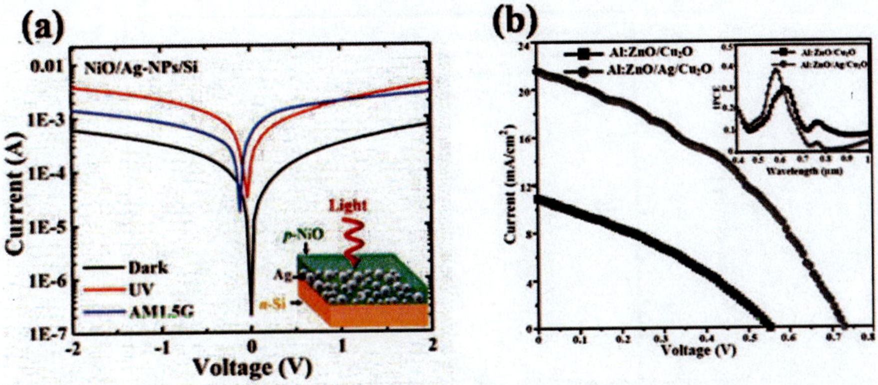

Fig. 8 **a** Current–voltage curves measured, in the dark and irradiation (UV and AM1.5G), for NiO/Ag-NP/Si heterostructure device, reprinted with permission from [84], **b** current–voltage measurement of the Al:ZnO/Cu_2O and Al:ZnO/Ag/Cu_2O heterostructures based solar cells in light exposure (1.5 AMU). The wavelength-dependent IPCE curve was displayed in the inset, reprinted with permission from [83]

Ayhan showed the fabrication of Ag-NPs on graphene [87]. The photoresponse of the device was 122 mA/W at 550 nm, respectively. The enhanced photoresponse was induced by light interaction with the plasmonic Ag-NPs and effective Schottky junction formation.

4 Sensor

4.1 Thermal Sensor

Thermal sensors detect temperature changes and transform them into an electrical signal that may be measured or displayed. Depending on the type of sensor, these sensors operate on different principles such as resistance change, voltage change, or infrared radiation detection.

They are used in a variety of industries, including industrial, medical, and automotive, for temperature monitoring, control, and safety. As the temperature is an essential characteristic for measuring the different properties of materials, it behaves as a suitable temperature sensor due to the various functional properties of ZnO. Figure 9a depicts the relationship between sensitivity and operating temperature of as-deposited AZO films under a 1000 ppm CO environment [88]. Copper oxides have particular importance in the application of temperature sensors [89]. The CuO nanowires-based temperature sensing materials exhibit high sensitivity and high current response, and they are potentially strong candidate for gas sensors and other electronic devices.

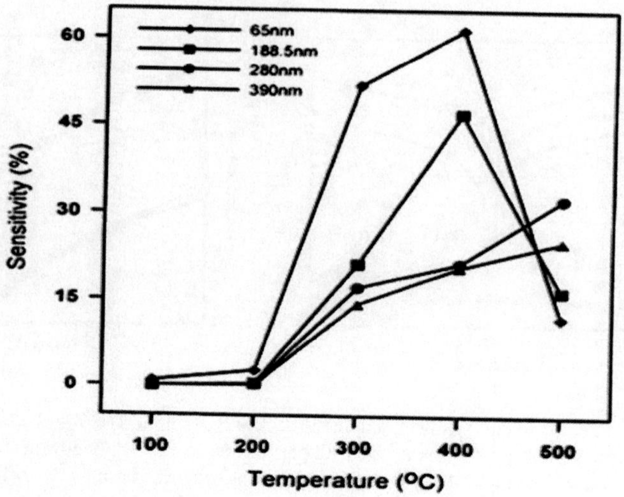

Fig. 9 The relationship between sensitivity and operating temperature of as-deposited AZO films in a 1000 ppm CO environment, reprinted with permission from [88]

4.2 Chemical Sensor

Chemical or biological information of a quantitative or qualitative kind is converted into an analytically useful signal by a chemical reaction taking place between the analyte gas and liquid in a small device known as a chemical sensor. The signal from a sensor is frequently electronic and appears as a change in impedance, conductance, or voltage as a result of electron exchange.

4.3 Gas Sensor

Chemical gas sensors can change their properties based on how the surface interacts with the surrounding atmosphere. Metal oxide-based sensors are particularly useful in this regard due to their abundance, ease of fabrication, and low cost. These sensors operate based on the surface phenomena between the material, the adsorbed air oxygen ions, and the target gaseous molecule, which controls the electrical resistance of the material. Metal oxide-based sensors have been effectively employed in various applications, including safety, climate, security, process monitoring, and medical equipment like breath analyzers.

Metal oxide-based gas sensors offer several advantages, including high sensitivity, selectivity, and stability over time. They can also operate at room temperature, making them energy-efficient and suitable for portable applications. As a result, metal oxide-based sensors are becoming increasingly popular for various gas-sensing applications, and ongoing research is focused on further improving their performance and extending their applications.

Gas sensors based on ZnO [90], WO_3 nanoparticles [91], TiO_2 nanotubes [92], SnO_2 [93], and In_2O_3 nanofibers [94] have been reported to react to the gases of formaldehyde, NO_2, H_2S, and CO. Ahn et al. examined ZnO-nanowire based gas sensors demonstrating using a NO_2 gas. Response versus NO_2 concentration curve is shown in Fig. 10a, which exhibits linear dependence in the range of 0.5–3 ppm and then the indication of slight saturation nature at higher concentrations [95].

Researchers are increasingly concentrating on investigating metal oxide heterojunction-based high-performance sensors. The hydrothermal approach was used to synthesize the ordered mesoporous WO_3/ZnO nanocomposites, which showed an improved sensor sensitivity of 168.7 to 1 ppm NO_2 at 150 °C working temperature, nearly 10 times greater than that of the pure WO_3 [96]. The sensor response of NiO-modified CeO_2 nanostructures was studied to be 1570, much more than that of pure CeO_2. The reaction and recovery time of the NiO-modified CeO_2 sensor were 15 and 19 s, respectively; these times were likewise faster than those of

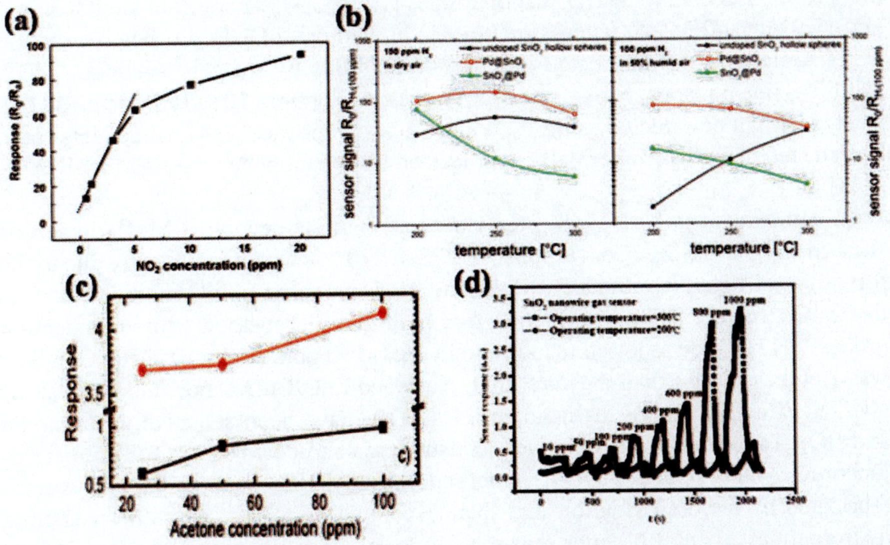

Fig. 10 **a** Response versus NO_2 concentration curve, which exhibits linear dependence in the range of 0.5–3 ppm and then the indication of slight saturation nature at higher concentration, reprinted with permission from [95], **b** sensor signal (R_0/R_{gas}) for all three samples when exposed to 100 ppm H_2 as a function of the operating temperature in both dry air (left side) and 50% relative humidity (right side), reprinted with permission from [104], **c** at 40% relative humidity, 20 °C ambient temperature, and 400 °C operating temperature, the behavior of thick wires (black) and nanowires (red) is displayed in relation to acetone concentration, reprinted with permission from [105], **d** Sensor response of the SnO_2 nanowire to hydrogen at 200 and 300 °C, in continuous operation mode, at varied concentrations of hydrogen diluted in the air from 10 to 1000 ppm, reprinted with permission from [106]

pure CeO_2 (96 and 118 s, respectively) [97]. The sensors based on MoS_2/ZnO [98] and SnO_2–CuO [99] heterojunction showed highly improved performance compared to the ZnO and SnO_2, respectively. The gas sensing capabilities of ZnO-decorated SnO_2 hollow spheres for ethanol, synthesized using a two-step hydrothermal process, were reported by Lu et al. [100]. At their optimized operating temperature of 225 °C, the ZnO-decorated SnO_2 hollow spheres were found to have a sensor response of 34.8 toward 30 ppm ethanol, significantly 5.7 times greater than the bare SnO_2. The SnO_2/SiO_2 heterojunction was fabricated using a magnetron sputtering process and showed good H_2 sensing capability at room temperature [101]. Compared to solid nanospheres, hybrid Co_3O_4/SnO_2 core-shell nanospheres fabricated using a hydrothermal process showed a measured response of 13.6–100 ppm NH_3 at 200 °C [102]. By combining a hydrothermal procedure with the wet impregnation method, CeO_2-decorated ZnO nanosheets were fabricated, and they demonstrated an improved sensor response for 90–100 ppm ethanol at 310 °C [103].

ZnO–SnO_2 nanofibers were created by Kim et al. using an electrospinning approach to detect CO efficiently [104]. Many researchers reported the improving performance of the gas sensor based on following metal oxides-based heterojunctions: ZnO/SnO_2 heterojunction [107], ZnO/$La_{0.8}Sr_{0.2}Co_{0.5}Ni_{0.5}O_3$ heterojunction [108], ZnO nanorods/TiO_2 nanoparticles [109], Ga_2O_3-core/ZnO-shell nanorod [110], α-MoO_3/0D ZnO heterostructure [111], and α-MoO_3/Fe_2O_3 heterostructure [112]. Galatsis et al. fabricated α-MoO_3–WO_3 films by sol-gel technique, which exhibits a linear response to O_2 concentrations ranging from 10 to 10,000 ppm [113]. Acetone might be detected using WO_3 and SnO_2 composites [114], while NH_3 could be detected using hydrothermally synthesized ZnO nanoplates modified with WO_3 [115].

In comparison to bare MoO_3, the CuO nanoparticle-decorated MoO_3 nanorods showed a higher H_2S sensor response of 272 at 270 °C, which was mostly due to the formation of heterojunctions [116]. Zhang et al. prepared a PdO-modified flower-like ZnO structure using a surfactant-free hydrothermal method with an annealing process [117]. They reported response of a 35.4–100 ppm ethanol at 320 °C, which was 10 times greater than the bare ZnO. Al-doped ZnO/CuO nanocomposites [118] and ZnO/Co_3O_4 composites nanoparticles [119] have been sensitive to ammonia and NO_2. Yang et al. [120] effectively constructed nano-coaxial Co_3O_4/TiO_2 heterojunctions using a standard two-step method. Initially, TiO_2 nanotubular arrays were fabricated by anodic oxidation and then decorated with Co_3O_4 synthesized by a hydrothermal process. A better sensor response of 40 to 100 ppm ethanol at 260 °C with rise and decay time of 1.4 and 7.2 s was shown by the nano-coaxial Co_3O_4/TiO_2 heterojunction-based sensor. The electrospinning process was used with annealing at a working voltage of 15 kV to prepare the SnO_2–Co_3O_4 composite nanofibers [121]. Additionally, sensing materials comprised of CuO/ZnO heterostructural nanorods [122], CuO-decorated SnO_2 nanowires[123], flower-like p-CuO/n-ZnO nanorods [124], CuO nanoparticles decorated ZnO flowers [125], NiO@ZnO heterostructured nanotubes [126], Co_3O_4 decorated flower-like SnO_2 nanorods [127] and ZnO/NiO heterojunction nanofibers [128]. TiO_2 composited with ZnO [129], MoS_2 [130], MoO_3 [131], V_2O_5 [132], and WO_3 [133] have also been prepared, demonstrating improvement in gas detecting performances for ethanol, alcohol, NO_2, and ammonia. To construct the high-performance gas sensors, In_2O_3 composited with WO_3 [134], Fe_2O_3 [135], TiO_2 [136], and SnO_2 [137] were also synthesized. It has been reported that an octahedral-like ZnO/CuO composite [138] and SnO_2@TiO_2 heterostructure [139] have improved gas sensing characteristics for acetone, formaldehyde, and H_2S, respectively. Co_3O_4 composites with SiO_2 [140], In_2O_3 [141], and TiO_2 [142] were also synthesized using thermal conversion, hydrothermal, and simple nanoscale coordination polymer techniques, respectively and they displayed better gas sensing characteristics to those of pure Co_3O_4. Additionally, TiO_2-decorated Co_3O_4 [143], In_2O_3-decorated CuO [144] and rod-like α-Fe_2O_3/NiO heterojunction nanocomposites [145] were developed to research their enhanced gas-sensing characteristics. Thermal evaporation and sputter deposition techniques were used in combination to construct TeO_2/CuO core-shell nanorods by Lee et al. [146]. The sensor response for TeO_2/CuO core-shell nanostructure and TeO_2 was obtained to be 4.25 and 2, respectively, for 10 ppm NO_2 at 150 °C, which concludes that the core-shell nanostructure shows higher response compared to pure TeO_2.

SnO_2 is a significant conductance-type semiconductor used for CO and O_2 gas sensing [106]. The SnO_2 nanowires-based hydrogen gas sensor is prepared by the self-assembly method. The nanodevices can detect hydrogen as low as 10 ppm with great sensitivity. With increasing hydrogen content and operating temperature, the sensor response of the SnO_2 nanowire improves, which is shown in Fig. 10d [106].

Hubner et al. developed SnO_2@Pd and Pd@SnO_2 core@shell nanocomposites to examine the effect of the location of Pd in core@shell nanocomposites on CO and H_2 sensing performance in dry and humid conditions at different temperatures. The sensor signal for samples, when exposed to 100 ppm H_2 as a function of the operating temperature in both dry air and humidity condition, are shown in Fig. 10b [104]. From pure In_2O_3 powders, In_2O_3 wires were formed by thermal evaporation and condensation, and the contribution of metallic nanoparticles grown on substrates before evaporation was examined. In order to customize the density distribution and the lateral dimensions of the wires, which are the primary factors governing the gas sensing capabilities, it is crucial to choose the substrate seeding appropriately. The response of thick wires and nanowires as a function of acetone concentration is exhibited in Fig. 10c [105].

Wang et al. fabricated a flower-like structure of SnO_2, which having a BET surface area of 36 m^2/g; this material shows good sensor performance for carbon monoxide, methane, methanol, and ethanol [147]. The faceted SnO_2 polyhedrons self-assembled synthesized using a sodium dodecyl sulfate (SDS)-assisted hydrothermal method. The as-obtained SnO_2 nanopolyhedrons were sensitive to all chemicals at 1 ppm, had a rapid response and recovery period, and were highly selective for acetone [148]. Lupan et al. investigated the preparation of a single ZnO nanorod-based hydrogen sensor using a focused ion beam in-situ lift-out technique [149].

Rai et al. reported the shape-controlled synthesis of ZnO and corresponding gas sensing properties [150]. Under various substrate temperatures, normal and oblique angle sputtering deposition methods were used to construct nano-structured ZnO for gas sensing applications [151]. Undoped and Co-doped ZnO nanorod sensors have been made using a low-temperature, mass-production approach that is both cost-effective and efficient. Fast reaction and excellent sensitivity are displayed by ZnO-based nanorod sensors [152]. This research investigated the ethanol detecting characteristics of sensors based on undoped and Au-doped ZnO nanowires prepared by the oxidation reaction process [153].

4.4 Biosensor

Biosensors play an essential role in environmental processes, food safety, medical care, energy-efficient systems, chemical and agricultural sectors. Biosensors have been suggested as an effective instrument for quick measurement and analysis since continuous onsite monitoring is required and must have versatile and dependable properties. Adapting the materials of biosensors for different applications (screening techniques, quality control, safety gear, and environmental evaluation) is a significant research area with several obstacles to overcome.

Wang et al. prepared a biosensor based on MoS_2/g–C_3N_4/b-TiO_2 heterojunction to detect miRNA, and the biosensor improved with Au nanoparticles [154]. The TiO_2-based photoelectrochemical biosensor was developed for detecting heme and glucose by Cakiroglu et al. [155] and Rajendran et al. [156], respectively. The photocurrent vs. concentration graph for TiO_2-based biosensor is shown in Fig. 11a, and the linear portion of graph is exhibited in the inset of the figure. A TiO_2 film-based glucose detector with a 0.7 μM limit of detection (LOD) was functionalized using glucose oxidase (GO_x). Zhu et al. prepared a polyaniline-TiO_2 nanotube composite to detect glucose, showing a better LOD value (0.5 μM) than TiO_2 film-based glucose sensor [157]. TiO_2 nanorods are fabricated via a facile method to detect glucose sensor, which shows a 0.002 mM LOD value [158]. Liu et al. [159] fabricated bag-like TiO_2 microspheres using hydrothermally to detect H_2O_2 which functionalized with hemoglobin. They reported a LOD value of 10 nM obtained by amperometric measurements. Kafi et al. reported a LOD value of 0.08 μM to detect H_2O_2 obtained by amperometry. The GC/TiO_2/Hb/CH modified electrodes were prepared by the TiO_2 nanotube co-immobilized with hemoglobin (Hb) and chitosan (CH). Another electrode GC/TiO_2/Hb/MB/CH is prepared with TiO_2 nanotubes, Hb, CH and methylene blue (MB). The current versus H_2O_2 concentration plot for all modified electrodes is shown in Fig. 11b [160]. There is also work reported based on TiO_2 for cholesterol [161, 162] and breast cancer cell detection [163, 164].

SnO_2 nanowires-based H_2O_2 sensor was constructed using the thermal evaporation method [165]. The electrospinning technique was also used to produce SnO_2 nanowires for amperometric glucose detection [166]. SnO_2 nanoparticles were synthesized by precipitation [167] and microwave irradiation [168], which are used for L-cysteine and H_2O_2 detection, respectively.

WO_3 nanowire-based biosensor was fabricated by a simple casting technique, which was functionalized with hemoglobin [169]. The hydrothermal process was designed to make WO_3 sensors with various geometries: flower-like for aflatoxin B1

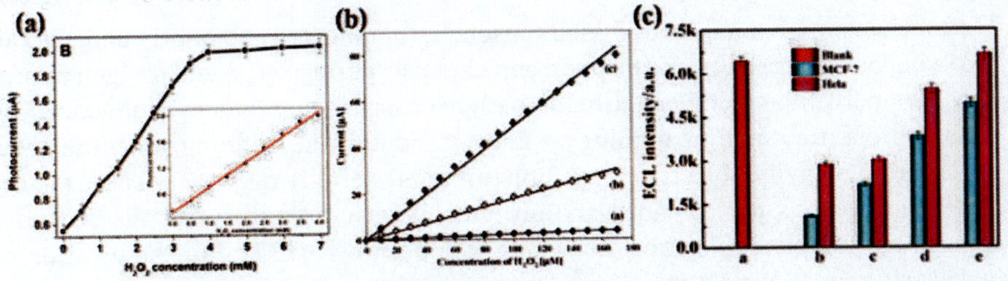

Fig. 11 **a** The plot of photocurrent with concentration (inset: the linear part of the graph, Error bars indicate the standard deviation of repeated 3 measurements), reprinted with permission from [155]. **b** Corresponding calibration plots of the **a** GC/Hb/CH. **b** GC/TiO_2/Hb/CH. **c** GC/TiO_2/Hb/MB/CH-modified electrode, reprinted with permission from [160], and **c** Analysis of the miRNA-21 in different tumor cells detection, **a** blank; **b** 10^6 cells; **c** 10^5 cells; **d** 10^4 cells; **e** 10^3 cells of MCF-7 and Hela. Reprinted with permission from [174]

[170], nanorods for bisphenol A [171], and nanocomposite for cardiac biomarker Troponin I [172]. Santos et al. [173] fabricated WO_3 nanoparticle-based biosensors by using the hydrothermal method, which functionalized with cytochrome c nitrite reductase.

ZnO nanostars are fabricated using the chemical bath deposition (CBD) technique to detect microRNA-21 in cancer cells. The variation in ECL response signal with blank and different tumor cells of MCF-7 and Hela, are displayed in Fig. 11c [174]. A similar process was used to create ZnO nanoparticles for detecting the Zika virus in undiluted urine [175]. The Zika virus is spread by mosquito bites and causes headaches, arthralgia, myalgia, and conjunctivitis. The LOD was 1.00 pg/mL. Early identification of the Zika virus is possible with this biosensor. Hydrothermally produced ZnO nanorods were employed in biosensors for glucose [176], phosphate [175], and G Immunoglobulin [176] detection. The biosensor based on ZnO nanorods exhibited a detection limit of 1.0 μM for glucose concentration. ZnO nanocone arrays were fabricated for dopamine detection using the hydrothermal method [177].

5 Conclusions

Metal oxide-based devices offer distinct properties that make them useful for a wide range of applications, including environmental monitoring, defense technology, medical analysis, and fire prevention. Metal oxide semiconductors are ideal for use in various applications due to their large bandgap, solution processability, affordable, and environmental friendliness. Nevertheless, much more study is needed to increase the performance of photodetectors and sensors based on metal oxide. One area of study is p-type doping, which is difficult in metal oxide but necessary for the creation of high-performance devices. Another area of investigation is the improvement of material quality, which is crucial for boosting device performance. Device structure design is also critical for enhancing the performance of metal oxide-based devices.

The mechanism of metal oxide-based sensors is complicated and poorly understood, involving charge transfers, adsorption and chemisorption stages, and oxygen migration. The performance of metal oxide-based sensors is significantly influenced by surface properties such as topology and crystallinity. The defining properties of a sensor are sensitivity, selectivity, reaction time, and stability over time. These factors affect the sensor's performance and are crucial for confirming that the sensor is appropriate for the application for which it is designed [177]. For the creation of high-performance metal oxide-based sensors that can satisfy the requirements of diverse applications, it is essential to improve these properties.

In summary, this chapter provides a systematic study on the recent advances in metal-oxide based photodetector and sensor application. Metal oxides provide a broad range of adaptable and outstanding structural, optical, and electrical characteristics that may be used in a range of optoelectronic devices. In this chapter, metal-oxide based different kinds of photodetector are discussed in detailed. The working mechanism behind the photodetection is also studied. Further, the sensor based on metal oxide is covered and studied in detail. These devices based on metal oxide have received significant attention in different areas such as defense, medical treatments, astronomical observation, etc. due to their availability, large bandgap, low-cost fabrication, and stability. There is still a lot of work to be required in the development of photodetectors and sensor based on metal oxide, such as to reduce the weight, decrease the size, improve the response, rapid reaction time, and so on.

References

1. C.N.R. Rao, Annu. Rev. Phys. Chem. **40**, 291 (1989)
2. T. Guo, M.-S. Yao, Y.-H. Lin, C.-W. Nan, CrystEngComm **17**, 3551 (2015)
3. B. Raveau, J. Eur. Ceram. Soc. **25**, 1965 (2005)
4. L.G. Gerling, S. Mahato, A. Morales-Vilches, G. Masmitja, P. Ortega, C. Voz, R. Alcubilla, J. Puigdollers, Sol. Energy Mater. Sol. Cells **145**, 109 (2016)
5. M. Bonomo, J. Nanoparticle Res. **20**, 222 (2018)
6. A.A. Ahmed, M.R. Hashim, T.F. Qahtan, M. Rashid, Ceram. Int. (2022)
7. A.M. Soleimanpour, in (2013)
8. O.A. Saraereh, A. Alsaraira, I. Khan, B.J. Choi, Sensors **20**, 407 (2020)
9. M. Haras, T. Skotnicki, Nano Energy **54**, 461 (2018)
10. Q.A. Xu, J.W. Zhang, K.R. Ju, X.D. Yang, X. Hou, J. Cryst. Growth **289**, 44 (2006)
11. Y. Xie, M. Madel, M. Feneberg, B. Neuschl, W. Jie, Y. Hao, X. Ma, K. Thonke, Mater. Res. Express **3**, 045011 (2016)
12. J. Bao, I. Shalish, Z. Su, R. Gurwitz, F. Capasso, X. Wang, Z. Ren, Nanosc. Res. Lett. **6**, 404 (2011)
13. W.Z. Tawfik, Z.S. Khalifa, M.S. Abdel-wahab, A.H. Hammad, J. Mater. Sci. Mater. Electron. **30**, 1275 (2019)
14. K.K. Manga, S. Wang, M. Jaiswal, Q. Bao, K.P. Loh, Adv. Mater. **22**, 5265 (2010)
15. R.-S. Chen, W.-C. Wang, C.-H. Chan, H.-P. Hsu, L.-C. Tien, Y.-J. Chen, Nanosc. Res. Lett. **8**, 443 (2013)
16. D. Guo, Z. Wu, P. Li, Y. An, H. Liu, X. Guo, H. Yan, G. Wang, C. Sun, L. Li, W. Tang, Opt. Mater. Express **4**, 1067 (2014)

17. G.C. Hu, C.X. Shan, N. Zhang, M.M. Jiang, S.P. Wang, D.Z. Shen, Opt. Express **23**, 13554 (2015)
18. D. Zhang, W. Zheng, R.C. Lin, T.T. Li, Z.J. Zhang, F. Huang, J. Alloys Compd. **735**, 150 (2018)
19. Y. Kokubun, K. Miura, F. Endo, S. Nakagomi, Appl. Phys. Lett. **90**, 031912 (2007)
20. F. Alema, B. Hertog, O. Ledyaev, D. Volovik, G. Thoma, R. Miller, A. Osinsky, P. Mukhopadhyay, S. Bakhshi, H. Ali, W.V. Schoenfeld, Phys. Status Solidi **214**, 1600688 (2017)
21. P.R. Jubu, F.K. Yam, Sens. Actuators A Phys. **312**, 112141 (2020)
22. H. Xue, X. Kong, Z. Liu, C. Liu, J. Zhou, W. Chen, S. Ruan, Q. Xu, Appl. Phys. Lett. **90**, 201118 (2007)
23. X. Kong, C. Liu, W. Dong, X. Zhang, C. Tao, L. Shen, J. Zhou, Y. Fei, S. Ruan, Appl. Phys. Lett. **94**, 123502 (2009)
24. W.S. Shih, S.J. Young, L.W. Ji, W. Water, T.H. Meen, H.W. Shiu, IEEE Sens. J. **11**, 3031 (2011)
25. D. Çalışkan, B. Bütün, Ş. Özcan, E. Özbay, J. Vac. Sci. Technol. B Nanotechnol. Microelectron. Mater. Process. Meas. Phenom. **31**, 020606 (2013)
26. A.A. Ahmed, M. Devarajan, N. Afzal, Sens. Actuators A Phys. **262**, 78 (2017)
27. Y. Zhao, J. Zhang, D. Jiang, C. Shan, Z. Zhang, B. Yao, D. Zhao, D. Shen, ACS Appl. Mater. Interfaces **1**, 2428 (2009)
28. Y. Zhao, J. Zhang, D. Jiang, C. Shan, Z. Zhang, B. Yao, D. Zhao, D. Shen, J. Phys. D. Appl. Phys. **42**, 092007 (2009)
29. N. Kouklin, Adv. Mater. **20**, 2190 (2008)
30. H.-M. Chiu, J.-M. Wu, J. Mater. Chem. A **1**, 5524 (2013)
31. C.-L. Hsu, H.-H. Li, T.-J. Hsueh, ACS Appl. Mater. Interfaces **5**, 11142 (2013)
32. T.-H. Kao, J.-Y. Chen, C.-H. Chiu, C.-W. Huang, W.-W. Wu, Appl. Phys. Lett. **104**, 111909 (2014)
33. D.Y. Guo, Z.P. Wu, Y.H. An, X.C. Guo, X.L. Chu, C.L. Sun, L.H. Li, P.G. Li, W.H. Tang, Appl. Phys. Lett. **105**, 023507 (2014)
34. S. Liang, H. Sheng, Y. Liu, Z. Huo, Y. Lu, H. Shen, J. Cryst. Growth **225**, 110 (2001)
35. B. Nie, J.-G. Hu, L.-B. Luo, C. Xie, L.-H. Zeng, P. Lv, F.-Z. Li, J.-S. Jie, M. Feng, C.-Y. Wu, Y.-Q. Yu, S.-H. Yu, Small **9**, 2872 (2013)
36. T.-F. Zhang, G.-A. Wu, J.-Z. Wang, Y.-Q. Yu, D.-Y. Zhang, D.-D. Wang, J.-B. Jiang, J.-M. Wang, L.-B. Luo, Nanophotonics **6**, 1073 (2016)
37. M.-W. Chen, C.-Y. Chen, D.-H. Lien, Y. Ding, J.-H. He, Opt. Express **18**, 14836 (2010)
38. X. Chen, K. Liu, Z. Zhang, C. Wang, B. Li, H. Zhao, D. Zhao, D. Shen, ACS Appl. Mater. Interfaces **8**, 4185 (2016)
39. D.B. Patel, K.R. Chauhan, W.-H. Park, H.-S. Kim, J. Kim, J.-H. Yun, Mater. Sci. Semicond. Process. **61**, 45 (2017)
40. S.-J. Chang, T.-Y. Tsai, Z.-Y. Jiao, C.-J. Chiu, W.-Y. Weng, S.-H. Wang, C.-L. Hsu, T.-J. Hsueh, S.-L. Wu, IEEE Electron. Device Lett. **33**, 1577 (2012)
41. M. Karbalaei Akbari, Z. Hai, S. Depuydt, E. Kats, J. Hu, S. Zhuiykov, IEEE Trans. Nanotechnol. **16**, 880 (2017)
42. G. Wang, S. Chu, N. Zhan, Y. Lin, L. Chernyak, J. Liu, Appl. Phys. Lett. **98**, 041107 (2011)
43. Y.H. Leung, Z.B. He, L.B. Luo, C.H.A. Tsang, N.B. Wong, W.J. Zhang, S.T. Lee, Appl. Phys. Lett. **96**, 053102 (2010)
44. P.-J. Li, Z.-M. Liao, X.-Z. Zhang, X.-J. Zhang, H.-C. Zhu, J.-Y. Gao, K. Laurent, Y. Leprince-Wang, N. Wang, D.-P. Yu, Nano Lett. **9**, 2513 (2009)
45. C.-L. Hsu, Y.-D. Gao, Y.-S. Chen, T.-J. Hsueh, ACS Appl. Mater. Interfaces **6**, 4277 (2014)
46. Y.F. Li, B. Yao, R. Deng, B.H. Li, J.Y. Zhang, Y.M. Zhao, D.Y. Jiang, Z.Z. Zhang, C.X. Shan, D.Z. Shen, X.W. Fan, Y.M. Lu, J. Phys. D. Appl. Phys. **42**, 105102 (2009)
47. L. Shi, F. Wang, B. Li, X. Chen, B. Yao, D. Zhao, D. Shen, J. Mater. Chem. C **2**, 5005 (2014)
48. W. Ko, S. Lee, N. Myoung, J. Hong, J. Mater. Chem. C **4**, 142 (2016)
49. H.D. Cho, A.S. Zakirov, S.U. Yuldashev, C.W. Ahn, Y.K. Yeo, T.W. Kang, Nanotechnology **23**, 115401 (2012)

50. S.-Y. Tsai, M.-H. Hon, Y.-M. Lu, Solid. State. Electron. **63**, 37 (2011)
51. W. Tian, T. Zhai, C. Zhang, S.-L. Li, X. Wang, F. Liu, D. Liu, X. Cai, K. Tsukagoshi, D. Golberg, Y. Bando, Adv. Mater. **25**, 4625 (2013)
52. L. Hu, J. Yan, M. Liao, H. Xiang, X. Gong, L. Zhang, X. Fang, Adv. Mater. **24**, 2305 (2012)
53. M. Saini, R. Singh, A. Mitra, T. Som, Sol. Energy **207**, 228 (2020)
54. S.K. Singh, P. Hazra, S. Tripathi, P. Chakrabarti, Superlattices Microstruct. **91**, 62 (2016)
55. C.-H. Chen, S.-J. Chang, S.-P. Chang, M.-J. Li, I.-C. Chen, T.-J. Hsueh, C.-L. Hsu, Chem. Phys. Lett. **476**, 69 (2009)
56. N. Park, K. Sun, Z. Sun, Y. Jing, D. Wang, J. Mater. Chem. C **1**, 7333 (2013)
57. L. Yang, H. Zhou, M. Xue, Z. Song, H. Wang, Sens. Actuators A Phys. **267**, 76 (2017)
58. H. Zhu, C.X. Shan, B. Yao, B.H. Li, J.Y. Zhang, D.X. Zhao, D.Z. Shen, X.W. Fan, J. Phys. Chem. C **112**, 20546 (2008)
59. L.S. Vikas, K.A. Vanaja, P.P. Subha, M.K. Jayaraj, Sens. Actuators A Phys. **242**, 116 (2016)
60. H. Chang, Z. Sun, K.Y.-F. Ho, X. Tao, F. Yan, W.-M. Kwok, Z. Zheng, Nanoscale **3**, 258 (2011)
61. D. Shao, M. Yu, H. Sun, T. Hu, J. Lian, S. Sawyer, Nanoscale **5**, 3664 (2013)
62. J.-M. Choi, S. Im, Appl. Surf. Sci. **244**, 435 (2005)
63. H.K. Li, T.P. Chen, S.G. Hu, X.D. Li, Y. Liu, P.S. Lee, X.P. Wang, H.Y. Li, G.Q. Lo, Opt. Express **23**, 27683 (2015)
64. H. Kim, S. Tarelkin, A. Polyakov, S. Troschiev, S. Nosukhin, M. Kuznetsov, J. Kim, ECS J. Solid State Sci. Technol. **9**, 045004 (2020)
65. W.Y. Weng, T.J. Hsueh, S.J. Chang, G.J. Huang, H.T. Hsueh, IEEE Sens. J. **11**, 1491 (2011)
66. Y. Qu, Z. Wu, M. Ai, D. Guo, Y. An, H. Yang, L. Li, W. Tang, J. Alloys Compd. **680**, 247 (2016)
67. X.C. Guo, N.H. Hao, D.Y. Guo, Z.P. Wu, Y.H. An, X.L. Chu, L.H. Li, P.G. Li, M. Lei, W.H. Tang, J. Alloys Compd. **660**, 136 (2016)
68. L. Sun, C. Wang, T. Ji, J. Wang, G.-C. Yi, X. Chen, RSC Adv. **7**, 51744 (2017)
69. M. Zhang, H. Zhang, K. Lv, W. Chen, J. Zhou, L. Shen, S. Ruan, Opt. Express **20**, 5936 (2012)
70. D. Zhang, C. Liu, R. Xu, B. Yin, Y. Chen, X. Zhang, F. Gao, S. Ruan, Nanotechnology **28**, 365505 (2017)
71. Z. Liu, J.-P. Ao, F. Li, W. Wang, J. Wang, J. Zhang, H.-X. Wang, Mater. Lett. **188**, 52 (2017)
72. A. Dewasi, A. Mitra, J. Mater. Sci. Mater. Electron. **29**, 9209 (2018)
73. A. Dewasi, A. Mitra, J. Mater. Sci. Mater. Electron. **28**, 18075 (2017)
74. D. Zhang, S. Nozaki, K. Uchida, J. Vac. Sci. Technol. B Microelectron. Nanom. Struct. **32** (2014)
75. L. Duan, F. He, Y. Tian, B. Sun, J. Fan, X. Yu, L. Ni, Y. Zhang, Y. Chen, W. Zhang, ACS Appl. Mater. Interfaces **9**, 8161 (2017)
76. Y. Yang, W. Guo, J. Qi, J. Zhao, Y. Zhang, Appl. Phys. Lett. **97**, 223113 (2010)
77. S. Lu, J. Qi, S. Liu, Z. Zhang, Z. Wang, P. Lin, Q. Liao, Q. Liang, Y. Zhang, ACS Appl. Mater. Interfaces **6**, 14116 (2014)
78. D. Guo, Y. Su, H. Shi, P. Li, N. Zhao, J. Ye, S. Wang, A. Liu, Z. Chen, C. Li, W. Tang, ACS Nano **12**, 12827 (2018)
79. L. Zheng, F. Teng, Z. Zhang, B. Zhao, X. Fang, J. Mater. Chem. C **4**, 10032 (2016)
80. Y. Fu, Y. Liu, K. Ma, Z. Ji, W. Mai, C. Zhao, J. Alloys Compd. **819**, 153063 (2020)
81. S. Chaoudhary, A. Dewasi, R.P.S, V. Rastogi, R.N. Pereira, A. Sinopoli, B. Aïssa, A. Mitra, Nanotechnology **33**, 255202 (2022)
82. N. Venugopal, G. Kaur, A. Mitra, Appl. Surf. Sci. **320**, 30 (2014)
83. G. Kaur, K.L. Yadav, A. Mitra, Appl. Phys. Lett. **107** (2015)
84. S. Chaoudhary, A. Dewasi, V. Rastogi, R.N. Pereira, A. Sinopoli, B. Aïssa, A. Mitra, J. Mater. Sci. Mater. Electron. **33**, 10761 (2022)
85. P. Joshna, S.R. Gollu, P.M.P. Raj, B.V.V.S.N.P. Rao, P. Sahatiya, S. Kundu, Nanotechnology **30**, 365201 (2019)
86. T. Chandrakalavathi, K.R. Peta, R. Jeyalakshmi, Mater. Res. Express **5**, 025011 (2018)
87. M.E. Ayhan, G. Kalita, M. Kondo, M. Tanemura, RSC Adv. **4**, 26866 (2014)

88. J.F. Chang, H.H. Kuo, I.C. Leu, M.H. Hon, Sens. Actuators B Chem. **84**, 258 (2002)
89. B.C. Yadav, A.K. Yadav, Int. J. Green Nanotechnol. Mater. Sci. Eng. **1**, M16 (2009)
90. M.-W. Ahn, K.-S. Park, J.-H. Heo, J.-G. Park, D.-W. Kim, K.J. Choi, J.-H. Lee, S.-H. Hong, Appl. Phys. Lett. **93**, 263103 (2008)
91. R. Ionescu, Sens. Actuators B Chem. **104**, 132 (2005)
92. S. Lin, D. Li, J. Wu, X. Li, S.A. Akbar, Sens. Actuators B Chem. **156**, 505 (2011)
93. Y.-J. Choi, I.-S. Hwang, J.-G. Park, K.J. Choi, J.-H. Park, J.-H. Lee, Nanotechnology **19**, 095508 (2008)
94. S.K. Lim, S.-H. Hwang, D. Chang, S. Kim, Sens. Actuators B Chem. **149**, 28 (2010)
95. M.-W. Ahn, K.-S. Park, J.-H. Heo, D.-W. Kim, K.J. Choi, J.-G. Park, Sens. Actuators B Chem. **138**, 168 (2009)
96. J. Sun, L. Sun, N. Han, J. Pan, W. Liu, S. Bai, Y. Feng, R. Luo, D. Li, A. Chen, Sens. Actuators B Chem. **285**, 68 (2019)
97. N. Jayababu, M. Poloju, J. Shruthi, M.V.R. Reddy, RSC Adv. **9**, 13765 (2019)
98. S. Zhao, G. Wang, J. Liao, S. Lv, Z. Zhu, Z. Li, Appl. Surf. Sci. **456**, 808 (2018)
99. S. Bai, W. Guo, J. Sun, J. Li, Y. Tian, A. Chen, R. Luo, D. Li, Sens. Actuators B Chem. **226**, 96 (2016)
100. J. Liu, T. Wang, B. Wang, P. Sun, Q. Yang, X. Liang, H. Song, G. Lu, Sens. Actuators B Chem. **245**, 551 (2017)
101. C. Ling, Q. Xue, Z. Han, H. Lu, F. Xia, Z. Yan, L. Deng, Sens. Actuators B Chem. **227**, 438 (2016)
102. L. Wang, Z. Lou, R. Zhang, T. Zhou, J. Deng, T. Zhang, ACS Appl. Mater. Interfaces **8**, 6539 (2016)
103. G. Hui, M. Zhu, X. Yang, J. Liu, G. Pan, Z. Wang, Mater. Lett. **278**, 128453 (2020)
104. M. Hübner, A. Sackmann, F. Gyger, C. Feldmann, P. Bockstaller, D. Gerthsen, U. Weimar, N. Barsan, Procedia Eng. **47**, 208 (2012)
105. A. Vomiero, S. Bianchi, E. Comini, G. Faglia, M. Ferroni, G. Sberveglieri, Cryst. Growth Des. **7**, 2500 (2007)
106. B. Wang, L.F. Zhu, Y.H. Yang, N.S. Xu, G.W. Yang, J. Phys. Chem. C **112**, 6643 (2008)
107. H. Huang, H. Gong, C.L. Chow, J. Guo, T.J. White, M.S. Tse, O.K. Tan, Adv. Funct. Mater. **21**, 2680 (2011)
108. K.-C. Hsu, T.-H. Fang, S.-H. Chen, E.-Y. Kuo, Ceram. Int. **45**, 8744 (2019)
109. C.W. Zou, J. Wang, W. Xie, J. Colloid Interface Sci. **478**, 22 (2016)
110. C. Jin, S. Park, H. Kim, C. Lee, Sens. Actuators B Chem. **161**, 223 (2012)
111. J. Li, H. Liu, H. Fu, L. Xu, H. Jin, X. Zhang, L. Wang, K. Yu, J. Alloys Compd. **788**, 248 (2019)
112. D. Jiang, W. Wei, F. Li, Y. Li, C. Liu, D. Sun, C. Feng, S. Ruan, RSC Adv. **5**, 39442 (2015)
113. K. Galatsis, Y. Li, W. Wlodarski, K. Kalantar-zadeh, Sens. Actuators B Chem. **77**, 478 (2001)
114. L. Zhu, W. Zeng, Y. Li, Mater. Lett. **231**, 5 (2018)
115. D.D. Nguyen, D.T. Do, X.H. Vu, D.V. Dang, D.C. Nguyen, Adv. Nat. Sci. Nanosci. Nanotechnol. **7**, 015004 (2016)
116. T.-S. Wang, Q.-S. Wang, C.-L. Zhu, Q.-Y. Ouyang, L.-H. Qi, C.-Y. Li, G. Xiao, P. Gao, Y.-J. Chen, Sens. Actuators B Chem. **171–172**, 256 (2012)
117. Z. Lou, J. Deng, L. Wang, L. Wang, T. Fei, T. Zhang, Sens. Actuators B Chem. **176**, 323 (2013)
118. M. Poloju, N. Jayababu, M.V. Ramana Reddy, Mater. Sci. Eng. B **227**, 61 (2018)
119. S. Park, S. Kim, H. Kheel, C. Lee, Sens. Actuators B Chem. **222**, 1193 (2016)
120. Y.Q. Liang, Z.D. Cui, S.L. Zhu, Z.Y. Li, X.J. Yang, Y.J. Chen, J.M. Ma, Nanoscale **5**, 10916 (2013)
121. J.-H. Kim, J.-H. Lee, A. Mirzaei, H.W. Kim, S.S. Kim, Sens. Actuators B Chem. **248**, 500 (2017)
122. Q. Xu, D. Ju, Z. Zhang, S. Yuan, J. Zhang, H. Xu, B. Cao, Sens. Actuators B Chem. **225**, 16 (2016)
123. G.-J. Sun, S.-W. Choi, A. Katoch, P. Wu, S.S. Kim, J. Mater. Chem. C **1**, 5454 (2013)

124. Y.-B. Zhang, J. Yin, L. Li, L.-X. Zhang, L.-J. Bie, Sens. Actuators B Chem. **202**, 500 (2014)
125. J. Huang, Y. Dai, C. Gu, Y. Sun, J. Liu, J. Alloys Compd. **575**, 115 (2013)
126. L. Xu, R. Zheng, S. Liu, J. Song, J. Chen, B. Dong, H. Song, Inorg. Chem. **51**, 7733 (2012)
127. H. Wang, M. Chen, Q. Rong, Y. Zhang, J. Hu, D. Zhang, S. Zhou, X. Zhao, J. Zhang, Z. Zhu, Q. Liu, Nanotechnology **31**, 255501 (2020)
128. K. Lokesh, G. Kavitha, E. Manikandan, G.K. Mani, K. Kaviyarasu, J.B.B. Rayappan, R. Ladchumananandasivam, J. Sundeep Aanand, M. Jayachandran, M. Maaza, IEEE Sens. J. **16**, 2477 (2016)
129. S. Ng, P. Kuberský, M. Krbal, J. Prikryl, V. Gärtnerová, D. Moravcová, H. Sopha, R. Zazpe, F.K. Yam, A. Jäger, L. Hromádko, L. Beneš, A. Hamáček, J.M. Macak, Adv. Eng. Mater. **20**, 1700589 (2018)
130. P.X. Zhao, Y. Tang, J. Mao, Y.X. Chen, H. Song, J.W. Wang, Y. Song, Y.Q. Liang, X.M. Zhang, J. Alloys Compd. **674**, 252 (2016)
131. Y.-J. Chen, G. Xiao, T.-S. Wang, F. Zhang, Y. Ma, P. Gao, C.-L. Zhu, E. Zhang, Z. Xu, Q. Li, Sens. Actuators B Chem. **155**, 270 (2011)
132. Y. Wang, Y. Zhou, C. Meng, Z. Gao, X. Cao, X. Li, L. Xu, W. Zhu, X. Peng, B. Zhang, Y. Lin, L. Liu, Nanotechnology **27**, 425503 (2016)
133. Y. Yao, J. Yuan, X. Chen, L. Tan, Q. Gu, W. Zhao, J. Chen, J. Mater. Res. Technol. **8**, 3580 (2019)
134. C. Feng, X. Li, J. Ma, Y. Sun, C. Wang, P. Sun, J. Zheng, G. Lu, Sens. Actuators B Chem. **209**, 622 (2015)
135. X. Chi, C. Liu, L. Liu, S. Li, H. Li, X. Zhang, X. Bo, H. Shan, Mater. Sci. Semicond. Process. **18**, 160 (2014)
136. S. Park, J. Alloys Compd. **696**, 655 (2017)
137. H. Du, J. Wang, Y. Sun, P. Yao, X. Li, N. Yu, Sens. Actuators B Chem. **206**, 753 (2015)
138. Z. Yang, L.-M. Li, Q. Wan, Q.-H. Liu, T.-H. Wang, Sens. Actuators B Chem. **135**, 57 (2008)
139. H.-K. Chang, D.-S. Ko, D.-H. Cho, S. Kim, H.-N. Lee, H.S. Lee, H.-J. Kim, T.J. Park, Y.M. Park, Ceram. Int. **47**, 5985 (2021)
140. S. Mandal, M. Rakibuddin, R. Ananthakrishnan, ACS Omega **3**, 648 (2018)
141. S. Shi, F. Zhang, H. Lin, Q. Wang, E. Shi, F. Qu, Sens. Actuators B Chem. **262**, 739 (2018)
142. B. Zhang, M. Li, Z. Song, H. Kan, H. Yu, Q. Liu, G. Zhang, H. Liu, Sens. Actuators B Chem. **249**, 558 (2017)
143. L. Zhang, Z. Gao, C. Liu, Y. Zhang, Z. Tu, X. Yang, F. Yang, Z. Wen, L. Zhu, R. Liu, Y. Li, L. Cui, J. Mater. Chem. A **3**, 2794 (2015)
144. X. Li, X. Li, N. Chen, X. Li, J. Zhang, J. Yu, J. Wang, Z. Tang, J. Nanomater. **2014**, 1 (2014)
145. Z. Wang, K. Zhang, T. Fei, F. Gu, D. Han, Sens. Actuators B Chem. **318**, 128191 (2020)
146. S. Park, S. Kim, G.-J. Sun, W. Lee, K.K. Kim, C. Lee, Nanosc. Res. Lett. **9**, 638 (2014)
147. Q. Wang, L.-S. Zhang, J.-F. Wu, W.D. Wang, W.-G. Song, W. Wang, J. Phys. Chem. C **114**, 22671 (2010)
148. H. Huang, Y.C. Lee, O.K. Tan, W. Zhou, N. Peng, Q. Zhang, Nanotechnology **20**, 115501 (2009)
149. O. Lupan, G. Chai, L. Chow, Microelectron. Eng. **85**, 2220 (2008)
150. P. Rai, Y.-T. Yu, Sens. Actuators B Chem. **161**, 748 (2012)
151. J. Chu, X. Peng, M. Sajjad, B. Yang, P.X. Feng, Thin Solid Films **520**, 3493 (2012)
152. Y.-J. Li, K.-M. Li, C.-Y. Wang, C.-I. Kuo, L.-J. Chen, Sens. Actuators B Chem. **161**, 734 (2012)
153. N. Hongsith, C. Viriyaworasakul, P. Mangkorntong, N. Mangkorntong, S. Choopun, Ceram. Int. **34**, 823 (2008)
154. M. Wang, H. Yin, Y. Zhou, C. Sui, Y. Wang, X. Meng, G.I.N. Waterhouse, S. Ai, Biosens. Bioelectron. **128**, 137 (2019)
155. B. Çakıroğlu, M. Özacar, Biosens. Bioelectron. **141**, 111385 (2019)
156. S. Rajendran, D. Manoj, K. Raju, D.D. Dionysiou, M. Naushad, F. Gracia, L. Cornejo, M.A. Gracia-Pinilla, T. Ahamad, Sens. Actuators B Chem. **264**, 27 (2018)
157. J. Zhu, X. Liu, X. Wang, X. Huo, R. Yan, Sens. Actuators B Chem. **221**, 450 (2015)

158. Z. Yang, Y. Tang, J. Li, Y. Zhang, X. Hu, Biosens. Bioelectron. **54**, 528 (2014)
159. P. Liu, X. Huo, Y. Tang, J. Xu, X. Liu, D.K.Y. Wong, Anal. Chim. Acta **984**, 86 (2017)
160. A.K.M. Kafi, G. Wu, P. Benvenuto, A. Chen, J. Electroanal. Chem. **662**, 64 (2011)
161. S. Cao, L. Zhang, Y. Chai, R. Yuan, Biosens. Bioelectron. **42**, 532 (2013)
162. B. Batra, V. Narwal, Sumit, J. Ahlawat, M. Sharma, Sens. Int. **2**, 100111 (2021)
163. F. Zanghelini, I.A.M. Frías, M.J.B.M. Rêgo, M.G.R. Pitta, M. Sacilloti, M.D.L. Oliveira, C.A.S. Andrade, Biosens. Bioelectron. **92**, 313 (2017)
164. M. Safavipour, M. Kharaziha, E. Amjadi, F. Karimzadeh, A. Allafchian, Talanta **208**, 120369 (2020)
165. L. Li, J. Huang, T. Wang, H. Zhang, Y. Liu, J. Li, Biosens. Bioelectron. **25**, 2436 (2010)
166. S. Alim, A.K.M. Kafi, J. Rajan, M.M. Yusoff, Int. J. Biol. Macromol. **123**, 1028 (2019)
167. Y. Dong, J. Zheng, J. Mol. Liq. **196**, 280 (2014)
168. N. Lavanya, S. Radhakrishnan, C. Sekar, Biosens. Bioelectron. **36**, 41 (2012)
169. H. Liu, C. Duan, C. Yang, X. Chen, W. Shen, Z. Zhu, Mater. Sci. Eng. C **53**, 43 (2015)
170. J. Feng, Y. Li, Z. Gao, H. Lv, X. Zhang, Y. Dong, P. Wang, D. Fan, Q. Wei, Sens. Actuators B Chem. **270**, 104 (2018)
171. Y. Zhou, L. Yang, S. Li, Y. Dang, Sens. Actuators B Chem. **245**, 238 (2017)
172. D. Sandil, S. Srivastava, B.D. Malhotra, S.C. Sharma, N.K. Puri, J. Alloys Compd. **763**, 102 (2018)
173. L. Santos, C.M. Silveira, E. Elangovan, J.P. Neto, D. Nunes, L. Pereira, R. Martins, J. Viegas, J.J.G. Moura, S. Todorovic, M.G. Almeida, E. Fortunato, Sens. Actuators B Chem. **223**, 186 (2016)
174. X. Zhang, W. Li, Y. Zhou, Y. Chai, R. Yuan, Biosens. Bioelectron. **135**, 8 (2019)
175. A.M. Faria, T. Mazon, Talanta **203**, 153 (2019)
176. X. Zong, R. Zhu, Sens. Actuators B Chem. **255**, 2448 (2018)
177. A. Dey, Mater. Sci. Eng. B **229**, 206 (2018)

Thermo-Optical Properties of Metal Oxide Nanoparticles and Their Applications

Victor K. Pustovalov

Abstract This part presents a comprehensive analysis of previous and recent research and advances related to the thermo-optical properties of metal oxide nanoparticles. Modern nanotechnologies produce various nanoparticles from different oxides with unique properties. Metal oxide nanoparticles have significant advantages over other nanoparticles due to their special optical properties and high oxide melting and evaporation temperatures, which allows them to be stable in a high-temperature surrounding. The thermo-optical properties of oxide nanoparticles have recently received significant attention due to their various applications in solar energy conversion, laser processing, photocatalytic applications, photothermal therapies. The development of modern high-temperature photonics, photocatalysis, and laser technologies in recent years requires the study and use of the optical parameters of metal oxide nanoparticles. The analysis of thermo-optical properties of oxide Al_2O_3, TiO_2, NiO, ZnO and other conventional oxide nanoparticles and their dependences on temperature has been carried out. These results can be used for the development and application of various optical and laser devices and technologies based on the use of thermo-optical properties of metal oxide nanoparticles, as well as for high-temperature nanophotonics and nanotechnology.

Keywords Nanoparticles · Metal oxide · Optical · Properties · Applications

1 Introduction

Metal oxide nanoparticles (NPs) are widely used for various purposes, such as heterogeneous photocatalysis, laser biotechnology and medicine, photonics, solar energy conversion, etc. [1–4]. The synthesis of oxides with desired properties is carried out by various synthetic methods, including hydrothermal, calcination, emulsion, spray pyrolysis, plasma and inert-atmosphere growth methods to obtain and modify

V. K. Pustovalov (✉)
Belarusian National Technical University, Pr. Independency, 65, 220013 Minsk, Belarus
e-mail: pustovalovv@mail.ru

V. Kumar et al. (eds.), *Optical Properties of Metal Oxide Nanostructures*,
Progress in Optical Science and Photonics 26,
https://doi.org/10.1007/978-981-99-5640-1_16

numerous compositions, phases, structures [5–8]. In particular, a mechanism has been developed for the synthesis of nanomaterials in a spray flame and gas-phase synthesis for the production and industrial applications of tailored materials for gas sensors, catalytic combustion of CO and CH_4, energy storage, catalytic activity for selective CO oxidation and H_2 production [8]. All these procedures make it possible to achieve high productivity and use all oxides in powder form.

In recent years, the synthesis of nanocolloids based on laser ablation in a liquid [9–12] at room temperature and atmospheric pressure has been used as a versatile approach to obtaining oxide NPs. This method uses a laser beam directed at a solid target immersed in a liquid to generate a dispersion of NPs in the liquid by means of laser ablation of the solid surface in the liquid [9–12]. One way is to obtain oxide NPs from a bulk oxide target using short wavelengths in the near UV. On the other hand, laser ablation synthesis of oxide NPs in a liquid solution is reported, starting from a pure metal bulk target. In this case, the matter extracted from the solid target collides with liquid solution molecules, water vapor, or gas microbubbles consisting of H_2, O_2, etc. [12]. Therefore, water molecules (H_2O) or additives (atmospheric O_2) will react with the metal, which will promote its oxidation, eventually forming oxide structures. Laser irradiation of preliminary prepared metal powder in a liquid is also used to obtain metal oxide NPs by laser oxidation [10, 11].

2 Processes of Light Interaction with Nanoparticles

For the use and processing of NPs two main optical sources are used: laser and solar radiation. In fact, the production and subsequent control of the size (reduction or increase), morphology (shapes) and structure of colloidal NPs are based on a combination of processes such as NP absorption, heating, oxidation, melting, evaporation, etc. under the action of optical and laser radiation. It is worth noted that for the first time complex theoretical study of NP heating, melting, evaporation, bubble formation, etc. was carried out [13]. The influence of the optical properties of NPs on the processes of solar and laser interaction with NPs will be briefly considered.

The absorption of NPs of laser or solar radiation [14] and the release of heat are the first and main processes that will determine the dynamics and results of all subsequent thermal and other processes with NPs and their surroundings. The optical properties of metallic NPs are widely reported in the literature [14–20]. The use of thermal laser applications in nanotechnology is based on laser heating of NPs, heat transfer to the environment and the initiation and implementation of subsequent processes. It is very important to take into account the dependences of the thermo-optical parameters and properties of NPs on their temperature [21]. The heating of NPs by laser or solar radiation has been experimentally and theoretically studied [20, 22–28]. The conversion of light into heat and the heating of single NPs, their ensembles, and the environment under the action of laser pulses and solar radiation were studied [20, 24].

Heating of NP by laser radiation can be described by the equation [22, 24]

$$\rho_0 c_0 V_0 \frac{dT_0}{dt} = \frac{1}{4} I K_{abs}(r_0, \lambda, T_0) S_0 - J_C S_0 . T_0(t = 0) = T_\infty \quad (1)$$

where $S_0 = 4\pi r_0{}^2$ and $V_0 = 4/3\pi r_0{}^3$ are the NP surface area and volume, c_0 is the heat capacity and ρ_0 is the density of NP material, I is the intensity of radiation with wavelength λ, r_0 and T_0 are the radius and temperature of NP, J_C is the heat flux density from NP due to thermal conductivity in a solid or liquid medium, K_{abs} (r_0, λ, T_0) is the factor of absorption efficiency of radiation with wavelength λ by a spherical particle of radius r_0 and with temperature T_0 [14], T_∞ is the initial temperature of NP and medium, and t is time.

The analytical solution of Eq. (1) with constant value of $K_{abs} = K_{abs}^{\infty}$ with quasi-stationary heat conduction around NP under laser action has been obtained [22, 24]

$$T_0 = T_\infty + \frac{I K_{abs}^{\infty} r_0}{4k_m}\left[1 - \exp\left(-\frac{3k_m t}{c_0 \rho_0 r_o^2}\right)\right] \quad (2)$$

K_m constant coefficient of thermal conduction.

The heating of NP by solar radiation is determined by the equation [20]

$$\rho_0 c_0 V_0 \frac{dT_0}{dt} = \pi r_0^2 \int_{\lambda_1}^{\lambda_2} I_S(\lambda) K_{abs}(r_0, \lambda, T_0) d\lambda - J_C S_0 \quad (3)$$

$I_S(\lambda)$ is the intensity of solar radiation in the studied wavelength interval $\lambda_2 - \lambda_1$.

The main factor determining NP heating (1–3) and subsequent thermal processes is the radiation absorption efficiency factor of K_{abs} (r_0, λ, T_0), which will be considered here. Laser or solar heating of NPs and the surrounding liquid (water) leads to vaporization and steam generation around NPs. Bubble formation is created of water vapor (gas oxidants) inside hot surrounding liquids and contributes to the formation of oxide NPs. The thresholds of the bubble formation under the laser pulses have been experimentally and theoretically determined [29–33].

The NP oxidation is the main process of the formation of oxide NPs under the action of radiation. The rate of oxidation and the thermochemistry of hydroxide formation in water increase in the presence of water vapor [34]. Metals actively react with oxygen in the air to produce metal oxides. The processes of high temperature oxidation of metal NPs, as a rule, are governed by the following chemical reaction involving oxygen O_2 molecules and metal M atoms [35]:

$$\mathrm{nM} + 1/2\mathrm{kO_2} = M_n O_k$$

n, k are the number of metal atoms and oxygen molecules involved in the chemical oxidation reaction. For example, magnesium oxide is formed by the reaction of magnesium with oxygen when heated in air:

$$2\text{Mg} + \text{O}_2 \rightarrow 2\text{MgO}$$

In this case, as for any thermally activated kinetic process, the chemical reaction rate K_0 strongly depends on temperature, and this dependence of K_0 on the oxidation temperature can be expressed by the Arrhenius equation [35]:

$$K_0 = A \exp\left(-\frac{E_a}{RT}\right)$$

where A is the pre-exponential factor, E_a is the apparent activation energy, R is the universal gas constant, and T is the absolute temperature. The value of T is close to the NP temperature T_0 near the NP surface. Characteristic oxidation temperatures are experimentally determined as 1300–1700 K in tests for isothermal oxidation [36, 37].

Some of the obtained oxidized NPs were formed on the bulk surface under the action of a laser pulse. The second part of the oxidation of the formed NPs is carried out under the action of laser radiation after leaving from the bulk surface. The oxidation rate makes it possible to control the composition and structure of oxide NPs, ranging from full oxide NP to core–shell structures where only an outer shell of oxide is formed around a pure metal core. In some cases, NPs with a metal core and an oxide shell are formed when the concentration of reactive oxygen molecules is low enough to avoid complete oxidation of the metallic core [38]. Theoretically, the combustion of a particle in air on exposure of laser radiation was considered taking into account heating, evaporation of the metal and oxide shell, oxidation according to the Arrhenius equation and diffusion of oxidant molecules to the metal core through the condensed oxide shell [39]. Chemical oxidizing agents in the solution promote the reaction of target atoms with oxygen, as shown by laser ablation of bulk Cu in pure water and aqueous solutions, resulting in the formation of Cu_2O or CuO nanocrystals, respectively [40]. The optical properties of core-shell NPs are approximately equal to pure oxide NPs made from shell materials when the shell thickness is greater than the core radius r_0 [41].

Processing of oxide NPs with laser pulses is accompanied by their melting, fusion and spheroidization. Melting, fusion and spheroidization of NPs under the action of laser pulses have been theoretically and experimentally studied [42–48]. Nanometer and submicron spherical oxide particles can be obtained by laser melting using materials with sufficient optical absorption and whose melting temperature is lower than evaporation temperature for metallic [42–44] and oxide [45–47] NPs. After NP melting due to the absorption of laser radiation, surface stress forces pull the particle to a spherical shape. Appropriate choice of laser wavelength and fluence is important, and significant NP optical absorption is required to fabricate spherical oxide particles. Most experiments on laser melting are carried out by irradiating a colloidal suspension of raw particles with a pulsed laser in the frequency mode. The pulsed lasers required for melting have nanoseconds or picoseconds pulse width, and their pulsed fluences are in the range from 10 to 200 mJ/cm^2, depending on the material [10, 48]. Various insulating oxides have a band gap larger than 4–5 eV, and shorter laser wavelengths are required to provide optical absorption at the level required for melting. To melt of oxide NPs from the materials with large band gap, it is necessary to use the fluence of about ~1 J/cm^2 and a laser wavelength of 355 or 532 nm [10]. Laser melting in a liquid can be used to fuse NP aggregates into larger nanospheres or to melt micrometric powders into submicron spheres.

Laser evaporation or fragmentation of NPS in a liquid is carried out under the action of intense pulses. The laser beam is focused inside a liquid dispersion of nanometric powders for their photoreduction and fragmentation into smaller NPs. A significant decrease in the size of colloidal NPs under the action of high-intensity laser pulses was studied experimentally and theoretically [48–56]. Significant ("complete") evaporation, and in some cases fragmentation of NP, is a process of laser impact on NP, as a result of which large NP disappeared and many small NPs with sizes ~10 times smaller are formed. To explain the disappearance of large NPs under laser irradiation, two different terms are used—a significant size reduction due to evaporation [50–53] or fragmentation [54, 55], which means one final result of laser irradiation and the impossibility in some cases to distinguish between these two processes. The formation of small NPs can occur due to the thermal ejection of molecules or small clusters from the surface of NPs or the condensation of vapors of the NP material (metal) in the environment (water). A significant reduction in the size can be considered as thermal evaporation of atoms from the surface of NPs. A change in the shape and a significant reduction of NPs occur due to melting and, accordingly, evaporation of gold NPs. These results can be interpreted in the frame of thermal model of size reduction. As a rule, the evaporation (fragmentation) NP by a laser turned out to be an effective method for obtaining small (~10 nm) and ultra-small (<3 nm) NPs, starting from the large NPs or microparticles dispersed in a liquid.

3 Optical Properties of Selected Homogeneous Oxide Nanoparticles

The optical absorbing properties of oxide NPs are the main parameters, that determine the processes and efficiency of the interaction of NPs with laser and optical radiation, and have been actively studied in recent years. These studies have focused on the UV, VIS and near IR spectral intervals for laser and optical (solar) sources and their applications. The optical data available today from the literature has been provided with a focuse on oxide nanostructures and claimed applications. It is possible to achieve excellent optical properties of NPs by manipulating of NP plasmon resonance, that should ensure maximum heating of NPs. However, essential information is currently limited, and only separate results for various oxide NPs, including pure oxide phases, will be presented here. These materials have shown several useful properties that are discussed with particular emphasis on photocatalytic, biomedical and optical application. We will restrict ourselves to consideration of conventional nano-oxides and will not consider the unconventional oxide structures with doped materials, non-equilibrium compounds, metal-oxide core–shells, ligands, additives, etc.

Modern nanotechnologies make it possible to obtain various homogeneous and other types of NPs from various metals and materials with unique thermo-optical properties. The tunable optical properties of oxide NPs, including absorption, scattering, and field enhancement, differ from most other nanostructures due to adaptable coupling to the NP structure. The design of NPs with selected and tuned thermo-optical and plasmonic properties is possible by choosing the materials, structure, shape and geometric dimensions of NP. This fact can play a decisive role in increasing the efficiency of laser and optical radiation for various applications. Optical properties are determined using the optical refractive indices of NP oxides for calculation within Mie or other theories. Optical refractive indices at the initial temperature are presented in various publications [57–62].

The main optical parameters that describe the electromagnetic interaction with single NP and their optical properties are the efficiency factors of absorption K_{abs}, scattering K_{sca}, extinction K_{ext} of radiation [14]. It is used the parameter P_1, which describes the correlation between the absorption and scattering of radiation by NP and characterizes their contribution [18]:

$$P_1 = \frac{K_{abs}}{K_{sca}} \tag{4}$$

The parameter P_1 makes it possible to select the spectral intervals in which NP absorption dominates scattering, $P_1 > 1$, and this situation can be used for efficient absorption of radiation by NPs. The opposite case, when radiation scattering by NP prevails over absorption, $P_1 < 1$, is not suitable for thermal applications and can be used for efficient radiation scattering.

It is interesting to compare the optical properties of homogeneous oxide NPs upon absorption of solar radiation. NPs of TiO_2, NiO oxides were chosen for subsequent analysis based on a comparison of their optical properties. Figure 1 shows the dependences of solar irradiance I_S, absorption efficiency factor K_{abs} and parameter P_1 for homogeneous TiO_2, NiO NPs with various NP radii r_0 on λ [20]. The refractive indices of oxides [61, 62] were used for calculations. TiO_2 and NiO NPs have significant absorption only in the spectral interval 300–500 nm. Their K_{abs} values sharply decrease in the IR spectral interval $\lambda > 700$ nm up to ~3–4 orders of magnitude, which means almost insignificant absorption of solar radiation by TiO_2 and NiO NPs in this spectral interval. The optical properties of nanofluid (NF), based on TiO_2 NPs and water, were experimentally and theoretically studied in the spectral range 250–1100 nm [63].

Figure 2 shows the scattering indicatrixes of radiation with the wavelengths $\lambda =$ 300 (UV) nm, 560 (VIS) nm, 1000 (IR) nm for homogeneous TiO_2 NPS with various radii r_0 [20]. The scattering indicatrix shows the relative radiation intensity scattered by a NP in the range of scattering angles from 0 to 360°. The indicatrixes for $\lambda =$ 300, 560 nm are elongated forward in the direction of radiation propagation. The

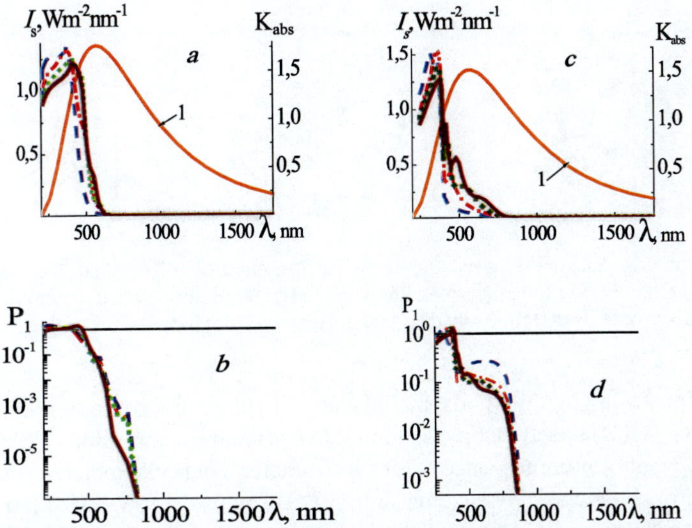

Fig. 1 Dependences of the absorption factor K_{abs} and parameter P_1 of homogeneous TiO_2 **a**, **b** and NiO, **c**, **d** NPs with $r_0 = 50$ (dashed, blue), 75 (dashed-dotted, red), 100 (dotted, green), 125 (solid, brown) nm and solar irradiance I_S (1, solid, orange, a–d) on wavelength λ. Horizontal solid line (black) indicates the value $P_1 = 1$. Adopted from [20]

amount of scattered radiation decreases significantly with wavelength increasing, especially for small NP sizes. The scattering indicatrixes of radiation for NiO NPs with the same wavelengths λ exhibit similar features presented here. The radiation absorbed by a NPs is extracted from subsequent optical processes, while scattered radiation by NPs propagates further and participates in subsequent acts of absorption and scattering by NPs. Preferential scattering of radiation into the rear hemisphere is realized for rather small wavelengths (for example, 300, 560 nm) and for all given values of the NP radii. During the interaction of IR light with NP, the scattering of radiation (1000 nm) will be approximately equally distributed in all directions. These data make it possible to understand the distribution of scattered radiation after the first collision of a photon with an NP for its subsequent application.

In general, TiO_2 and NiO NPs are therefore not suitable for absorption in solar thermal applications due to the extremely low absorption in the VIS and IR regions of the spectrum and their $P_1 << 1$ parameters. An increase in the radiation wavelength leads to a more uniform distribution of scattered radiation in space, with approximately equal scattering in the rear and frontal directions. This situation will be more suitable for the absorption of solar radiation by the NF. TiO_2 NPs are used in material science, solar cells, photocatalysis, and engineering, as well as, in various industrial applications [1–4].

Let us briefly present the experimental results on the optical properties of oxide NPs. In experimental studies, the structure, morphology and optical properties of prepared NPs were studied by various methods used in the publications presented

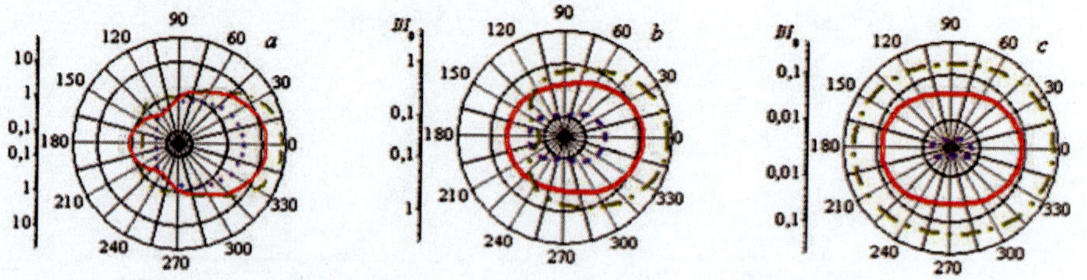

Fig. 2 Scattering indicatrixes of radiation with wavelengths λ = 300 (**a**), 560 (**b**), 1000 (**c**) nm for TiO_2 NPs with r_0 = 50 (dotted, purple), 75 (solid, red), 100 (dashed-dotted, green) nm. Radiation is propagated from left to right (from180° to 0°). Adopted from [20]

below. The following methods include atomic force microscopy (AFM), X-ray diffraction (XRD) spectroscopy, scanning (SEM) and transmission (TEM) electron microscopy, photoluminescence (PL) spectrometer, energy dispersive spectroscopy (EDS), proton-induced X-ray emission (PIXE) analysis, UV-VIS ultraviolet and visible spectroscopy, field emission scanning electron microscope (FESEM), Fourier transform infra-red (FTIR) and Raman spectroscopy.

The optical properties of the synthesized nanocrystalline copper oxide (CuO) nanopowder were studied by UV-VIS absorption and photoluminescence spectroscopy [64, 65]. X-ray diffraction study shows that the size of the CuO NPs is about15 nm, as determined from Debye-Scherrer formula, and the particles are almost irregular in shape, with some nanorod-like structures. The thermo-optical properties of copper oxide NFs with average NP diameters of less than 40 nm were studied for direct absorption of solar radiation in the spectral interval 200–2500 nm [66] (see Figs. 3a and 4). The volume fractions of CuO NF samples are equal accordingly $C_5 = 12.5$, $C_4 = 25$, $C_3 = 50$, $C_2 = 75$, $C_1 = 100$ ppm. CuO NPs are used as a catalyst in several oxidation processes, in photoconductive and photothermal applications.

Recently, zinc oxide (ZnO) NPs have attracted the attention of researchers due to their optical properties and applications in optoelectronic devices [67–72]. Zinc oxide NPs are the semiconductor materials with a band gap of 3.37 eV at room temperature. In addition, it is non-toxic, environment friendly and transparent in the visible spectrum. Electrical and optical properties are highly dependent on both the shape and size of the NPs. ZnO NPs were synthesized by the sol gel method and the UV-visible spectrum showed the transparency of NPs in the entire visible range [67]. The FTIR analysis confirms the formation of zinc oxide NPs with an average size of 14.36 nm. The sample absorbs the radiations in the UV range up to 361.75 nm, and almost all of the visible spectrum radiations are transmitted by the ZnO nanoparticles. The band gap energy is approximately equal 3.3 eV.

The morphology of zinc oxide NPs with crystal sizes between 19 and 28 nm synthesized by the solvothermal method and their optical properties (photoluminescence emission spectra) were studied [68]. A study of the optical and morphological properties of zinc oxide NPs was made [69]. The absorption spectrum of 4.0 nm ZnO particles showed maximum between 300 and 350 nm (4.14 and 3.55 eV) followed by a sharp decrease. Structural and optical properties of ZnO NPs were investigated

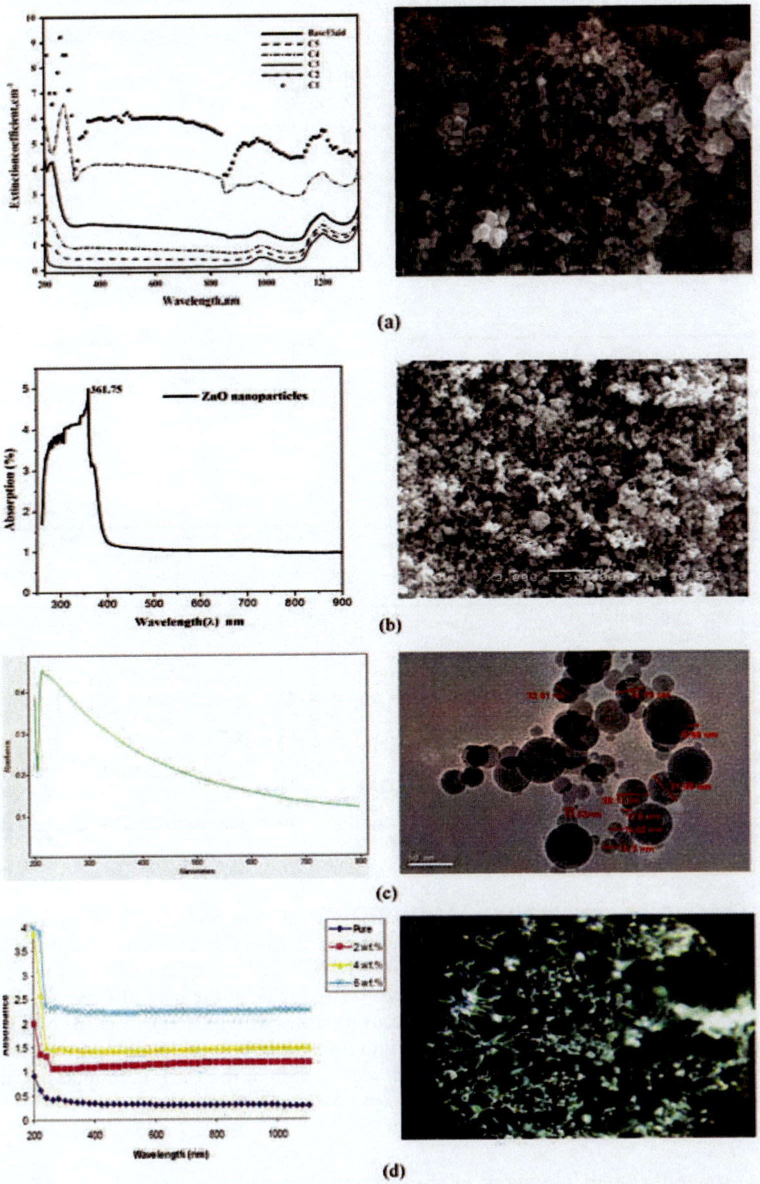

Fig. 3 Spectral extinction coefficient for CuO NF samples at different volume fractions (left) and SEM image of CuO NPs with an average diameter less than of 40 nm (right). Adopted from [66] with the permission from Elsevier (**a**); UV and VIS spectra (left) and SEM image of ZnO NPs with average size of 14.36 nm (right). Adopted from [67] based on Open access condition of journal (**b**). UV-VIS absorption spectrum (left) and TEM images of Al_2O_3 NPs with an average particle size of 19.43 nm (right) (**c**). Variation of absorbance with wavelength (left) and photomicrographs (x40) for (PVA-PEO-MnO_2) nanocomposites for 2 wt.% MnO_2 NPs (right) (**d**). Adopted from [73] (**c**) and [74] (**d**) on the base of a Creative Commons Attribution 4.0 International License

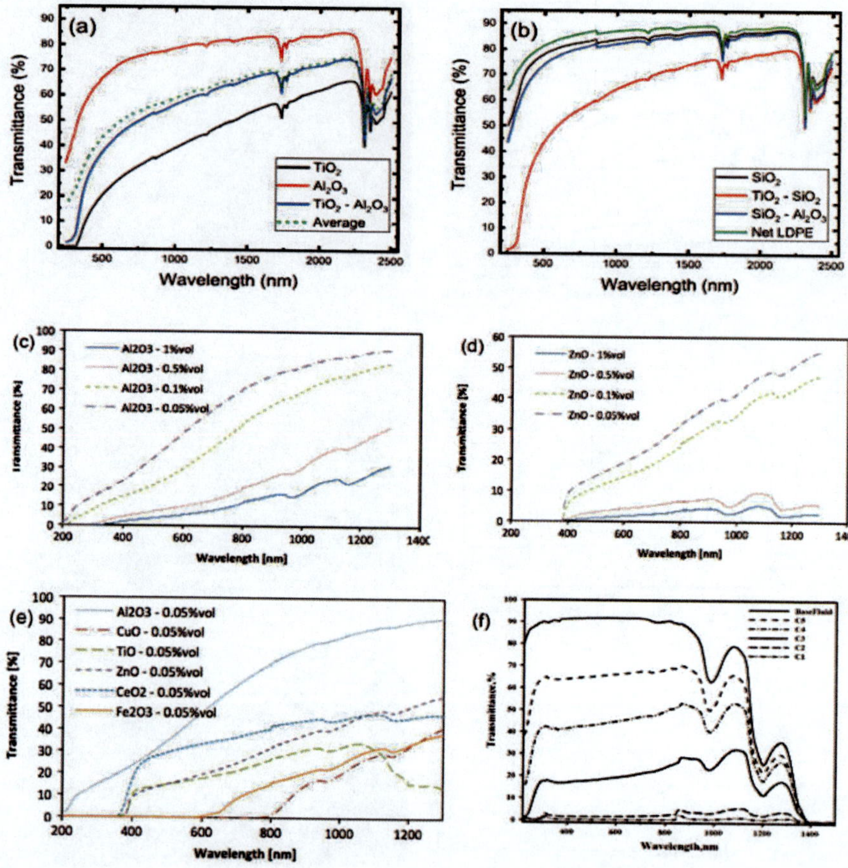

Fig. 4 Optical transmittance of TiO_2, Al_2O_3, TiO_2–Al_2O_3 and SiO_2, TiO_2–SiO_2, SiO_2–Al_2O_3 NPs embedded in low density polyethylene LDPE. The dashed line marks the calculated average of the TiO_2 and Al_2O_3 curves (**a**). Transmittance of NPs embedded in LDPE and the curve of net LDPE is given for comparison (**a, b**). Adopted from [82] on the base of Creative Commons Attribution 4.0 International license. Optical transmittances of Al_2O_3 (**c**), ZnO (**d**) NPs in water (NFs) and several NFs as a function of wavelength for various NPs (**e**). Adopted from [85] on the base of the permission from Elsevier. Transmittance spectra of the CuO NF samples at different volume fractions (**f**). Adopted from [66] on the base of the permission from Elsevier

[70]. The average particle sizes of the prepared samples are 9, 12, 25, 45 nm and the morphology of the particles can be observed using TEM analysis. The optical properties were studied by using UV-VIS absorption and photoluminescence spectroscopy. The optical band gap value of the prepared ZnO NP was ~3.36 eV. It can be noted that the absorption peak shifts with decrease in crystallite size. The absorption edge of the nanocrystalline ZnO is shifted in blue region compared to the edge of the corresponding bulk.

A review on the influence of processing parameters on the control of the morphology and optical properties of sol-gel synthesized ZnO nanoparticles was presented [71]. The UV-vis absorption spectra of the samples were recorded in the wavelength range from 200 to 800 nm at room temperature, and the data were transformed using the Kubelka-Munk method. The thermal radiating properties of the ZnO NFs for the selective absorption of solar radiation were characterized [72]. The optical properties of ZnO NPs are the determining factors for successful and efficiency applications in the above-mentioned studies. The absorbing properties and TEM, SEM images of various NPs and NFs [66, 67, 73, 77] are shown in Fig. 3. ZnO NPs can be applied in gas sensors, optoelectronic devices, in sunscreen creams, and as catalysts for numerous types of organic reactions.

Alumina (Al_2O_3) NPs synthesized from leaf extract are stable, good semiconductors and showed good optical properties with a particle size ranging from 18 to 25 nm [73]. Using UV-VIS spectroscopy, the energy band gap is 2.55 eV. The UV-VIS spectrum was used to study optical properties, and the absorbance was measured by varying wavelength of light (Fig. 3c). The sample showed a strong absorption peak at 230 nm in the UV region. This is associated with photo excitation of electrons from the valence band to the conduction band. The effect of adding the aluminum oxide NPs with different weight percentages on the absorption spectra of polymer composites in the wavelength range 220–800 nm was studied [74]. The results showed that the absorption and extinction coefficients, real and imaginary part of dielectric constant increased with increasing concentrations of Al_2O_3 NPs. The change in the optical absorption of the PVA-PEG-Al_2O_3 nanocomposite depending on the wavelength and the investigation of the optical properties of alumina NFs with different aggregation properties were studied [76]. The optical properties of alumina NFs were studied experimentally and theoretically in the spectral range 250–1100 nm [63].

Polymer-manganese oxide nanocomposites were prepared with different concentrations of the blend and MnO_2 NPs which were added to the blend by weight percentages 2, 4 and 6 wt.% [77]. Optical measurements have shown, that the absorption and extinction coefficients, real and imaginary dielectric constants of nanocomposites increase with increasing concentrations of MnO_2 NPs.

The photothermal efficiency of iron oxide (Fe_3O_4, Fe_2O_3) NPs with a variable composition, achieved by partial replacement of iron by zinc atoms has been systematically studied [75]. A series of zinc ferrites NPs with size of ~12 nm and variable composition, $Zn_xFe_{(3-x)}O_4$ ($x = 0.2, 0.5, 0.8$) have been synthesized. Laser radiation was used at a wavelength of 800 nm with variable power density (up to 1.4 Wcm^{-2}). The efficiency of light-to-heat conversion gradually increases with the iron content, which indicates the dominant role of iron related transitions in heating processes. The experimental extinction coefficient in the spectral interval 250–1100 nm for magnetite NF Fe_3O_4 with NP sizes of 15 nm was higher than the calculated values [19].

Nanostructured particles of bismuth oxide (Bi_2O_3) were synthesized by pulsed laser ablation on irradiation of a solid bismuth target in 3% H_2O_2 solution with laser radiation at a wavelength of 355 nm generated by a third harmonic Nd:YAG laser [78]. The XRD pattern of the prepared sample was indexed as bismuth oxide

α-Bi_2O_3 and the particle size was estimated to be ~4 nm. The UV–visible spectrum showed, that the absorption edge was shifted to the blue side, and the band gap of the prepared sample was estimated at 3.8 eV. The Raman spectrum exhibited peaks at the characteristic bands of α-Bi_2O_3.

Nanomaterials based on tungsten oxide (WO_3) were synthesized using the modified method of plasma-arc gas condensation with the aspect ratio from 1.1 for equiaxed particles to 12.7 for rods [79]. Tungsten oxide nanomaterials were characterized using TEM, UV-VIS and PL spectroscopes. Compared to NPs, nanorods exhibited redshift in the UV-visible spectrum, blue emission in the PL spectrum, and good performance in field emission.

Tellurite glasses based on amorphous TeO_2 have a high refractive index and optical nonlinearity, which make them attractive for a number of applications [80]. Particles have a small exhibit scattering of visible light and potential applications as polymeric nanocomposites with linear and nonlinear refractive index and as host matrices for rare-earth ions that can act as IR-to-VIS upconverters.

Figure 4 shows the transmittances of various samples of metal oxides (NFs) [66, 82, 85].

The cerium oxide (CeO_2) NPs were synthesized by the hydroxide mediated method in a cubic structure with size ~6.4 nm [81]. The optical absorption spectra of cerium oxide NPs were recorded on a UV-VIS spectrophotometer in the range of 320–600 nm and photoluminescence spectra in the range of 400–540 nm. The absorption peak observed at 349 nm depends on many factors such as band gap, oxygen deficiency, grain size, impurity centers, lattice strain and surface roughness. The CeO_2 NPs were also obtained by pulsed electron evaporation of ceramic oxide targets with vapor condensation in vacuum [84]. It has been established, that cerium oxide exhibit catalase-mimic activity, as well as differential cytotoxicity against tumor and healthy cell cultures, that is associated with the presence both Ce^{3+} and Ce^{4+} ions on the nanocrystal surface. CeO_2 NPs can be used in the fields of catalysis, gas sensing, electrochemistry, biomedical, and materials chemistry.

Optical absorption measurements were carried out on several water-based NFs (Al_2O_3, CuO, TiO_2, ZnO, CeO_2, and Fe_2O_3) in the wavelengths range of 200–1300 nm [66] depending on the concentration of NPs. For all NFs, volumetric concentrations were studied: 0.05, 0.1, 0.5 and 1%. The results showed different optical behaviors of the NFs depending on the material and concentration of NPs. The transmittance increased from the visible to the infrared region, and the extinction distance increased to values exceeding the typical diameter of a solar receiver, when the NPs concentration is too low. The results can be used for the solar absorption devices.

Of great interest is the use of NPs oxide embedded in polymer films to create various devises. The nanocomposite films presented here can be widely used in various sectors of various constructions, screen/TV technology, photography, etc. A bulk scale process for the production of nanostructured film composites with dispersed metal oxide NPs was implemented. The Al_2O_3, SiO_2 and TiO_2 NPs were processed in such a way that they could be suspended at the optimal size and composition [82] (Fig.4e, f). Silica nanopowder with an average particle size of 12 nm, titanium (IV) oxide anatase-nanopowder with a particle size of <25 nm, and aluminum oxide with a particle size of 40–50 nm was used. Transmission and reflection measurements in the UV, VIS and IR ranges were performed to determine the optical properties of the nanostructured films. The optical properties of the films can be controlled depending on the concentration, size, shape of the NPs and the types of polymers, and the degree of NP dispersion in the polymer matrix.

Thin films of aluminum oxide Al_2O_3 NPs were obtained by UV-irradiation and by measuring the structural and optical properties [85]. The AFM results showed that all films have a grain size of about 25.97–65 nm. The energy of the optical band gap was reduced from 5.19 to 4.68 eV by increasing the thickness. Absorbance in the wavelength range 200–800 nm increases with thickness. Sensitivity to gas is proportional to film thickness, so aluminum oxide NPs are considered good gas sensors for NH_3 gas.

A composite of PVA optical polymer and Co_3O_4 NPs with diameters in the range of 30–40 nm was prepared by adding NPs to a 6% PVA solution [86]. Optical properties of films with thicknesses of hundreds of nanometers were studied. The refractive index and absorption coefficient were measured, and two direct band gaps (1.38 and 2.0 eV) were determined from the absorption spectrum. The nonlinear transmissions of radiation with a wavelength of 523.5 nm with a pulse width of about 10 ns through three samples was studied depending on the illuminating fluence. Saturable absorption was observed when the films were illuminated with different fluences. The transmission of PVA:Co_3O_4 films decreases with increasing laser fluence, reaches a minimum at 600 mJ/cm^2, and then increases with increasing fluence due to saturable absorption of Co_3O_4 caused by the band filling effect. Experimental results shown that the Co_3O_4 NP/PVA composite is a promising material for nonlinear optical devices in the visible range and has the excellent mechanical and process properties.

4 Temperature Dependences of the Thermo-Optical Properties of Oxide Nanoparticles

Recent development of modern high-temperature technologies such as laser ablation and processing of NPs (their melting, evaporation, fragmentation, etc.), nanomedicine, and NPs applications in solar photo-processes such as photocatalysis, absorption and conversion of solar energy using solid nanostructured absorbers, etc. are implemented at high NP temperatures of the order of 1000 K and more. For the effective implementation of these processes, it is necessary to study and use the temperature dependences of the thermo-optical parameters of various oxides, dielectrics and materials.

The effect of temperature on the optical characteristics of NPs determines the dynamics and results of high-temperature processes involving NPs under the action of intense laser radiation. Establishing the temperature dependence of the optical properties of metal NPs is of crucial importance for the development of high temperature applications in various emerging fields. Optical properties that depend on temperature can be realized as thermo-optical properties of NPs. To date, there is no broad understanding of the temperature dependences of the thermo-optical parameters of various materials of NPs and the environment, and significant information on the effect of temperature on optical properties of oxide NPs is currently limited. Below some results are presented on the dependences of the thermo-optical parameters of the selected materials of NPs and surrounding media for the spectral interval ~200 to 2500 nm. The analysis will be limited to consideration of the currently available optical parameters of bulk oxides and the properties of NPs, which are significant for thermal processes of intensive laser or optical interactions with NPs. This part presents the results on the influence of the temperature dependences of the refractive indices of oxides on the optical properties of NPs.

Publications [87–92] present the results of the temperature dependences of the thermo-optical parameters of various oxides, such as Al_2O_3, SiO_2, TiO_2, etc. Optical refractive n and absorption k indices of TiO_2 thin films were studied by spectroscopic ellipsometry in the spectral 300–800 nm and in the temperature 293–533 K ranges [87]. Thermal expansion and electron–phonon interaction can be introduced to explain decrease of the refractive indices with increasing temperature. The results of this study can be applied to design and manufacture photonic devices and solar cell based on TiO_2, used at various temperatures.

The dependences of the indices n, k of fused silica (quartz) on the wavelength λ in the spectral range ~300 to 3000 nm for temperatures $T = 299$, 744, 1101 K were studied [88]. Changes of n, k in the presented ranges can affect the optical properties of fused silica. These results agree with the data of [89]. The dependences of the refractive n and absorption k indices of alumina Al_2O_3 on wavelength λ in the range T = 300–2300 K were studied [90, 91]. It should be noted that the optical indices of both oxides increase with temperature. It should also be noted the work [92] with the development of first-principles studies of the temperature dependence of the optical properties of rutile TiO_2 (Fig. 5).

Measurements of the optical absorption of NPs of oxides ZnO, CeO_2, Fe_2O_3 in the temperature range 25–500 C (298–773 K) were experimentally carried out [93, 94]. The transmission spectra of ZnO, Fe_2O_3 CeO_2 NPs were measured in air in the range 20–500 °C. Iron and cerium oxide NPs have a spherical shape with an average diameter of ~4.3 nm and ~3.0 nm, respectively.

The absorption spectra of CeO_2 film are shown depending on the temperature of the sample in the range 25–500 °C, and an increase in temperature does not change the absorption edges in this sample at Fig. 6. The change in the absorption spectra of iron oxide (Fe_2O_3) is seen in the inset of the Fig. 6, where room temperature measurements made before (dashed line) and after (continuous line) the annealing process with an integrated sphere. The observed red-shift is related to the structural phase transition of NPs induced by high temperature. The phase of nanocrystals irreversibly transforms into α-Fe_2O_3 (hematite) at high temperature, which make them more stable. The absorption coefficient of CeO_2 oxide does not change significantly in this temperature range. Measurements of the absorption spectra of Fe_2O_3 and ZnO [94] showed a more significant temperature dependence. Unfortunately, these results can only be used as

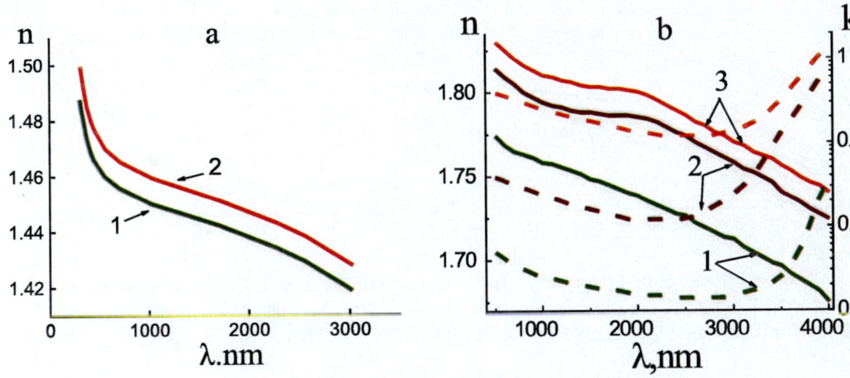

Fig. 5 Dependences of the refractive n index of fused silica SiO_2 for temperatures $T = 299$ (1, green), $T = 1101$ (2, red) K (**a**) and the refractive n and absorption k indices of alumina Al_2O_3 for $T = 300$ (1, green), 1700 (2, brown), 2300 (3, red) K (**b**) on wavelength λ. Adopted from [91]

qualitative ones for application in gas-based or nanostructured solid-based absorbers in direct absorption solar power systems.

Dependence of K_{abs} on temperature can be presented as [21]

$$K_{abs} = K_\infty (T_0/T_\infty)^b \quad \text{(a)} \quad K_{abs} = K_\infty \exp[\varepsilon_0 (T_0 - T_\infty)] \quad \text{(b)} \qquad (5)$$

$K_\infty = K_{abs}(T_0 = T_\infty)$, parameters $b = \text{const}$, $\varepsilon_0 = \text{const}$. Dependence (5b) can be used for the oxides (dielectrics) as materials of NPs, and the parameter $1/\varepsilon_0$ can be possibly considered as material band gaps. There is no dependence of K_{abs}

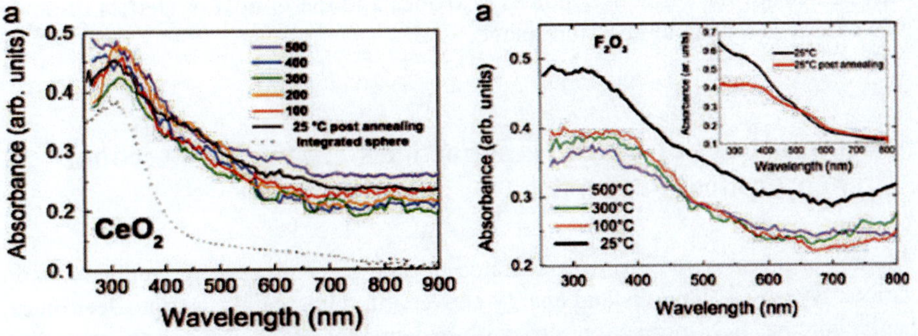

Fig. 6 Absorption spectra of CeO_2 (left) and Fe_2O_3 (right) NPs at room temperature and in the range 25–500 °C. The absorption spectrum at room temperature was also measured after sample cooling (grey line). The absorption spectrum obtained with the integrated sphere at room temperature is also shown (dashed line). Inset in Fe_2O_3: the absorption spectra obtained at room temperature, before (dashed line) and after (continuous line) the measurements at high temperature. Adopted from [94] with the permission from Elsevier

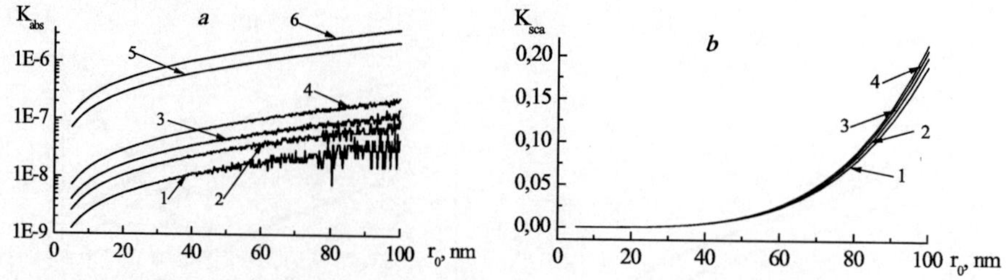

Fig. 7 The dependences of factors K_{abs} (**a**) and K_{sca} (**b**) for a wavelength 800 nm on r_0 in the range $r_0 = 5$–100 nm of spherical Al_2O_3 NPs for $T_0 = 300$ (1), 700 (2), 1100 (3), 1500 (4), 1900 (5), 2300 (6) K (**a**) and $T_0 = 300$ (1), 1100 (2), 1900 (3), 2300 (4) K (**b**). Adopted from [21]

on temperature T_0 at $b = 0$, $\varepsilon_0 = 0$. The temperature dependences of the optical efficiency factors of aluminum oxide NPs were studied [21].

On Fig. 7 shows the calculated dependences of the efficiency factors K_{abs} and K_{sca} of radiation with a wavelength 800 nm by spherical Al_2O_3 NPs in air on r_0 in the range $r_0 = 5$–100 nm on temperature T_0 in the range 300–2300 K. Dependences of the optical refraction and absorption indices on temperature for Al_2O_3 were taken from [90]. The significant influence of temperature on the absorbing properties of Al_2O_3 NPs can lead to a change in the dynamics of the interaction processes between a laser and NP. The effect of temperature on the values of K_{sca} is small. Expression (5b) with $K_\infty \approx 5 \times 10^{-9}$, $\varepsilon_0 \approx 2.3 \times 10^{-3}$ K^{-1} approximates the dependence K_{abs} on T_0 and r_0 (Fig. 7a) with a relative deviation of 10–30% and can be used for estimations. Noteworthy is the study of the temperature dependence of the scattering and absorption for aluminum oxide [95]. The studies of the temperature dependences of the optical properties of metal NPs [96–100] should be noted. Successful applications in high-temperature nanotechnologies require further studies of the temperature dependences of the refractive indices of oxides and the optical properties of oxide NPs oxides in a wide temperature range.

5 Applications of Oxide Nanoparticles Taking into Account Their Optical Properties

The use of metal oxide NPs is associated to their laser processing, applications in various types of solar cells and energy conversion, photocatalysis, optoelectronics, nanomedicine, light protection, etc. with exceptional properties and functionality. These applications are also based on the fact that metal oxide and semiconductors are highly stable materials, environmentally friendly, and cheap. The optical properties of oxide NPs are the main parameters of NPs that determines the results and efficiency of their applications. The applications of oxide NPs in optical and laser technologies are briefly considered.

Laser processing and synthesis of oxide NPs is currently a promising technology. Absorption of laser radiation by NP is the main process that will determine all subsequent thermal and other processes with NPs. Therefore, especially the optical properties of oxide NPs are of importance in their laser processing [9, 10]. Laser synthesis of oxides NPs in liquids can be considered as a qualitative method for obtaining pure oxide NPs with desired properties and their stabilizing for various applications.

Oxide NPs are now used in real solar absorbers for energy conversion. Solar cells can generate charge carriers in a semiconductor that absorbs light and transfer them into conductive materials that transmit electricity. The application of metal oxide NPs in solar cells is interesting topic, covering a wide variety of research and emerging developments. Metal oxide NPs have been used as photoelectrodes in photovoltaics solar cells based on dyes [100]. Several metal oxides with a large bandgap range are easily tuned and applicable as photo-harvesters [101]. Photovoltaic solar cells are the novel energy conversion technology due to their fabrication capability on various substrates, high photovoltaic output, excellent transparency, and low cost. High performance cells can be fabricated with suitable functionalization of oxides NPs, using novel inorganic materials, organic conjugated polymers, etc. [102]. The role of NPs of possible metal oxides as an electron transport layer in dye-sensitized solar cells was studied [103]. A promising layer based on TiO_2 NPs and its ability to be modified are studied.

Stable binary oxide phases with different copper oxidation states are included Cu_2O (cuprous oxide), CuO (copper oxide), and Cu_4O_3 (paramelaconite) which have been already successfully synthesized. Binary copper oxides are a promising class of semiconductors for solar energy applications due to their optical and electrical properties, non-toxicity and elemental content. Copper oxides can be applied in photovoltaic cells as absorbers based on the physical properties of mentioned above phases, including their optical and electrical parameters, band structures, thermodynamics [104]. Optical and structural studies of nanostructured copper oxides showed their effectiveness as a perspective solar cell material. UV-vis spectroscopy and atomic force microscopy have been used to characterize nano-structure powder-thin film with increased efficiency [105]. Raman spectroscopic and photoluminescence studies showed that multi-phonon scattering can play an important role for this improvement. The use of CuO NPs as hybrid heterojunction active layer indicated a noticeable improvement in solar cells characteristics [106]. When CuO NPs are added, the morphology of the photoactive layer undergoes significant changes. The incorporation of CuO NPs increased the power-conversion efficiency and photocurrent density of organic solar cells containing inorganic NPs. TiO_2 NPs were also used to easily produce hybrid solar cells [107]. The effect of surface modification of ZnO NPs on the characteristics of hybrid solar cells with a bulk heterojunction was studied [108].

Metal oxides (MgO, ZnO) NPs were employed as photocatalysts for oxidation catalysis [109, 110] and TiO_2 NPs for photocatalytic decomposition of pollutants [111]. TiO_2, WO_3, NiO were used in the heterogeneous photocatalytic air and water purification. Light-induced generation and migration of the carrier to the nanostructure surface, and reactive processes at the interfaces are the consequent processes for a photocatalytic process. Light-absorption efficiency determine the amount of photogenerated charge carriers and consequently the photocatalytic efficiency. Among binary oxides, TiO_2 NPs are the most investigated photocatalysts due to their low cost, inertness, and photostability [112]. The photocatalytic activity of TiO_2 is determined by its density of surface hydroxyl groups, surface area, crystal structure and adsorption/desorption characteristics. The activity of several titania colloids, modified by nanosecond pulsed laser irradiation, for photocatalytic water splitting was measured [113]. The properties of the NPs were studied and shown that laser irradiation contributes to the observed enhanced photocatalytic activity. The results showed an increase in hydrogen production up to three times under UV irradiation and two times with visible irradiation compared to the unirradiated titania. Thus, its efficiency under visible irradiation is low and it can only absorb UV radiation [114].

The optical properties of oxide nanostructures have stimulated their investigation and application in photonics and optics and in addition to biophotonics and photocatalysis. A strong second-harmonic signal was obtained from ZnO NPs, suggesting realization of the nonlinear processes [115]. Size-dependent optical applications of oxide particles with their appreciable sub-micrometre size range and monodispersity were examined for ZnO random lasing [116] and using TiO_2 for solar cells back reflectors [117]. Recently, Al_2O_3 NPs were studied as a antireflection coating and passivating for silicon photodiodes [118].

Indium tin oxide (ITO) has two main properties: electrical conductivity and optical transparency and can be used as transparent conducting oxides. ITO films exhibit good spectral response and anti-reflection properties in the blue visible region. A film reflecting near IR range was fabricated by spin coating method using ITO sol solution with NP sizes in the range of 15–30 nm. Reflectance on the surface of ITO film was determined by near-IR spectrometer [119]. Optimizing of solution-processed ITO NPs film was used to study efficient transparent heater with oxygen vacancy control [120]. High transparency over 96% was achieved by solution-treated ITO films in the visible spectral interval (500–730 nm). This material is applied as a thin film in polymer-based electronics, photovoltaics, and for glass windows to conserve energy both in research and industry. The ITO material can be customized for the wide applications by controlling the composition, shape, size, and crystallinity of nanostructures, such properties as absorptivity, band gap, conductivity, surface to volume ratio can be influenced on the result. The applications of the metallic oxide (ZnO, TiO_2) NPs as parts of photodetectors, solar-cells, solid-state lightning, and transparent conducting layers in optoelectronics were conducted [121].

Metal oxide NPs were applied in nanobiomedicine as biomedical materials in regenerative medicine, immunotherapy, diagnosis, tissue therapy and wound healing and biosensing platforms. One of the most popular imaging and therapeutic agents in nanomedicine are superparamagnetic iron oxide NPs [122, 123]. Combined infrared optical and magnetic heating of a single nanostructure, using the ability of iron magnetic NPs to efficiently generate heat, makes it possible to develop advanced therapeutic procedures. They can combine the synergetic effects between these two heat sources which, can be joined with high penetration magnetic imaging. Magnetic iron oxide NPs have attracted considerable attention due to their applications, including hyperthermia and the use them as contrast agents for magnetic resonance imaging due to their nontoxic and biodegradable nature [124]. Current challenges in the use of NPs as a contrast agent are presented [125].

Titanium dioxide TiO_2 NPs are common agents for biomedical applications and in tissue engineering, due to its ability to induce cytotoxic effect on melanoma cells and cell migration. Such properties are primarily determined by their optical properties. In vivo experiments with unmodified TiO_2 irradiated with 365 nm light showed suppression of tumor growth in mice with glioma [126]. Nitrogen-doped TiO_2 NPs revealed cytotoxic effect on melanoma cells with UV-induced death of melanoma cells due to higher absorption of light [127]. TiO_2 NPs triggered the death of bladder cancer cells under the action of IR light with a wavelength of 808 nm [128]. It should be noted, that TiO_2 and ZnO NPs have high optical absorption in the UV spectrum, which is important for UV protection in cosmetics and their antibacterial response [129].

Solar radiation absorption and energy conversion are widely used and applied in recent times. The direct absorption of solar energy by NF, determined by its optical properties, is a novel method to increase fluid temperature and to lead to higher collector efficiency. The direct absorption solar collector with alumina NF was tested [130]. The efficiency of a direct absorption collector with Al_2O_3 NF was experimentally studied and NF has increased the efficiency [131]. The optical properties of NiO, Mo_2O_3 NPs and NFs were studied for direct absorption of solar radiation [132]. The collection of solar energy by cobalt oxide Co_3O_4 NF was studied [133]. The effect of water-based TiO_2 and ZnO NFs on the characteristics of a flat plate collector was studied [132]. Optical absorption measurements were carried out on several water-based NFs (Al_2O_3, CuO, TiO_2, ZnO, CeO_2, and Fe_2O_3) [85]. These measurements are important for evaluating the possibility of using the aforementioned metal-oxide NPs in liquid-based NFs for direct absorption flat solar collector. The idea of using water-based NFs with metal oxide foams in the direct absorption solar collectors was proposed [135] and collector is expected to have improved thermal performance.

One of the main problems for the peoples of hot countries is the extraction of fresh water. The amount of water obtained by solar desalination has been significantly increased by adding copper and aluminum oxide NPs [136]. The effects of synthesis parameters on the optical properties of CeO_2 NPs [27] and particle size on the electronic and optical properties of magnesium oxide [97] were studied. On the whole, it can be concluded that knowledge of the optical properties of oxide NPs is important for the development and implementation of the above technologies.

6 Conclusions

A brief review of recent experimental results related to the absorption optical properties of oxide NPs, their changes under heating, and the use of NPs for collection of radiation is presented. This review focuses on the latest results of a study of the optical properties of oxide NPs, of course, taking into account the basic works of previous years. The dynamics and results of laser action on NPs are determined, primarily, by their optical properties. Metal oxide NPs have become significant components for use in solar absorbers and cells, photocatalysis, energy conversion, optoelectronic, electronics, medicine, and cosmetics. Metal oxide NPs with their nanoarchitecture and unique properties have proven to be valuable substances for creating various devices with excellent properties. The problem with oxide NPs is that the light collection efficiency is limited by the oxide large bandgap, and energy can be wasted either due to the low energy of the passing photons or as heat from high-energy photons. Therefore, multiple excitations, up-conversion of nanostructures and tuning of the band gap are of primary importance, which have a direct effect on photon absorption and photocurrent generation.

The significance of the most relevant applications of oxide NPs in high-temperature laser and optical nanotechnologies is considered, and special attention is paid to accounting the temperature dependences of the optical properties of NPs. The need to continue research in this promising direction is emphasized in connection with the unique possibilities of oxide nanostructured materials and nanocolloids in devices and technologies of the future.

References

1. M.S. Chavali, M.P. Nikolova, Metal oxide nanoparticles and their applications in nanotechnology. SN Appl. Sci. **1**, 6 (2019)
2. O. Diwald, T. Berger (eds.), *Metal Oxide Nanoparticles: formation, Functional Properties, and Interfaces* (Wiley, New Jersey, 2021)
3. M.P. Nikolova, M.S. Chavali, Metal oxide nanoparticles as biomedical materials. Biomimetics **5**, 27 (2020)
4. T. Naseem, T. Durrani, The role of some important metal oxide nanoparticles for wastewater and antibacterial applications. Environ. Chem. Ecotoxicol. **3**, 59 (2021)
5. I. Bretos, R. Jimenez, J. Ricote, M.L. Calzada, Low temperature crystallization of solution-derived metal oxide thin films assisted by chemical processes. Chem. Soc. Rev. **47**, 291 (2018)
6. S. Diodati, P. Dolcet, M. Casarin, S. Gross, Low-temperature crystallization of solution-derived metal oxide thin films assisted by chemical processes. Chem. Rev. **115**, 11449 (2015)
7. H. Li, Y.-J. Zhu, Liquid-phase synthesis of iron oxide nanostructured materials and their applications. Chem. Eur. J. **26**, 9180 (2020)
8. G. Schulz, S. Li (eds), Virtual special issue of recent advances in gas-phase synthesis of functional materials for energy. Energy Fuels **35**(7) (2021)
9. H. Palneedi, J.H. Park, D. Maurya, M. Peddigari, G.T. Hwang, V. Annapureddy, J.W. Kim, J.J. Choi, B.D. Hahn, S. Priya, K.J.L.J. Ryu, Laser irradiation of metal oxide films and nanostructures: applications and advances. Adv. Mater. **30**, 1705148 (2018)

10. V. Amendola, D. Amans, Y. Ishikawa, N. Koshizaki, S. Scirè, G. Compagnini, S. Reichenberger, Room-temperature laser synthesis in liquid of oxide, metal-oxide core-shells, and doped oxide nanoparticles. Chem. A Euro. J. **26**, 9206 (2020)
11. S. Singh, H. Zeng, C. Guo, W. Cai (eds.), *Nanomaterials. Processing and Characterization with Lasers* (Wiley-ECH Verlag GmbH, Berlin, 2014)
12. M.R. Kalus, V. Reimer, S. Barcikowski, B. Gokce, Discrimination of effects leading to gas formation during pulsed laser ablation in liquids. Appl. Surf. Sci. **465**, 1096 (2019)
13. V.K. Pustovalov, A.S. Smetannikov, V.P. Zharov, Photothermal and accompanied phenomena of selective nanophotothermolysis with gold nanoparticles and laser pulses. Las. Phys. Lett. **5**, 775 (2008)
14. C. Bohren, D. Huffman, *Absorption and Scattering of Light by Small Particles* (Wiley-Interscience publication, New York, 1983)
15. V.K. Pustovalov, V.A. Babenko, Optical properties of gold nanoparticles at laser radiation wavelengths for laser applications in nanotechnology and medicine. Las. Phys. Lett. **1**, 516 (2004)
16. M. Quinten, *Optical Properties of Nanoparticle Systems: Mie and Beyond* (Wiley-VCH Verlag GmbH & Co, New York, 2011)
17. A. Trugler, *Optical Properties of Metallic Nanoparticles* (Springer, Heidelberg, 2016)
18. L. Astafyeva, V. Pustovalov, W. Fritzsche, Characterization of plasmonic and thermo-optical parameters of spherical metallic nanoparticles. Nano-Struct. Nano-Obj. **12**, 57 (2017)
19. T.B. Gorji, A.A. Ranjbar, A review on optical properties and application of nanofluids in direct absorption solar collectors (DASCs). Renew. Sustain. Ener. Rev. **72**, 10 (2017)
20. V.K. Pustovalov, Modeling and analysis of optical properties of nanoparticles and nanofluids for effective absorption of solar radiation and their heating. Springer Nat. Appl. Sci. **1**, 356 (2019)
21. V. Pustovalov, Modeling of the processes of laser-nanoparticle interaction taking into account temperature dependences of parameters. Las. Phys. **21**, 906 (2011)
22. V.K. Pustovalov, Theoretical study of heating of spherical nanoparticle in media by short laser pulses. Chem. Phys. **308**, 103 (2005)
23. A. Plech, V. Kotaidis, Laser-induced heating and melting of gold nanoparticles studied by time-resolved x-ray scattering. Phys. Rev. **70**, 195423 (2004)
24. V.K. Pustovalov, Light-to-heat conversion and heating of single NPs, their assemblies, and surrounding medium under laser pulses. RSC Adv. **6**, 81266 (2016)
25. L.I. Trakhtenberg, V.A. Astapenko, S.V. Sakhno, M.A. Kozhushner, V.S. Posvyanskii, O.J. Ilegbusi, Absorption of infrared radiation by an electronic subsystem of semiconductor nanoparticles. J. Phys. Chem. C **120**, 23851 (2016)
26. S. Bruzzone, M. Malvaldi, Local field effects on laser-induced heating of metal nanoparticles. J. Phys. Chem. **113**, 15805 (2009)
27. A.M. Schrand, B.M. Stacy, S Payne, L. Dosser, S.M. Hussain. Fundamental examination of nanoparticle heating kinetics upon near infrared irradiation. ACS Appl. Mater. Inter. **3**, 3971 (2011)
28. A.M. Fales, W.C. Vogt K.A. Wear, T.J. Pfefer, I.K. Ilev, Experimental investigation of parameters influencing plasmonic nanoparticle-mediated bubble generation with nanosecond laser pulses. J. Biomed. Opt. **24**, 065003 (2019)
29. S. Wang, L. Fu, Y. Zhang, J. Wang, Z. Zhang Quantitative evaluation and optimization of photothermal bubble generation around overheated nanoparticles excited by pulsed lasers. J. Phys. Chem. C **122**, 24421 (2018)
30. V.K. Pustovalov, Model for estimations of laser threshold fluencies for photothermal bubble generation around nanoparticles. Appl. Phys. A **126**, 196 (2020)
31. S. Ishii, R.P. Sugavaneshwar, K. Chen, T.D. Dao, T. Nagao, Solar water heating and vaporization with silicon NPs at Mie resonances. Opt. Mater. Exp. **6**, 640 (2016)
32. M. Amjad, G. Raza, Y. Xin, S. Pervaiz, J. Xu, X. Du, D. Wen, Volumetric solar heating and steam generation via gold nanofluids. Appl. Energy **206**, 393 (2017)

33. S.R.J. Saunders, M. Monteiro, F. Rizzo, The oxidation behaviour of metals and alloys at high temperatures in atmospheres containing water vapor: a review. Progr. Mater. Sci. **53**, 775 (2008)
34. K.J. Laidler, *Chemical Kinetics* (Harper and Row, New York, 1987)
35. B. He, G. Xu, M. Zhou, Q. Yuan, Effect of oxidation temperature on the oxidation process of silicon-containing steel. Metals **6**, 137 (2016)
36. E. Chicardi, J.M. Córdoba, F.J. Gotor, Kinetics of high-temperature oxidation of (Ti, Ta) (C, N)-based cermets. Corr. Sci. **102**, 168 (2016)
37. A. Yabuki, S. Tanaka, Oxidation behavior of copper nanoparticles at low temperature. Mater. Res. Bull. **46**, 2323 (2011)
38. V.K. Pustovalov, D.S. Bobuchenko, Heating, evaporation and combustion of a solid aerosol particle in a gas exposed to optical radiation. Int. J. Heat Mass Trans. **32**, 3 (1989)
39. M.A. Gondal, T.F. Qahtan, M.A. Dastageer, T.A. Saleh, Y.W. Maganda, D.H. Anjum, Pulsed laser ablation in liquid synthesis of ZnO/TiO_2 nanocomposite catalyst with enhanced photovoltaic and photocatalytic performance. Appl. Surf. Sci. **286**, 149 (2013)
40. V.K. Pustovalov, L.G. Astafyeva, Influence of shell parameters on optical properties of spherical metallic core-oxide shell nanoparticles. J. Nanomater. **2015**, 812617 (2015)
41. A. Fales, W. Vogt, J. Pfefer, I. Ilev Quantitative evaluation of nanosecond pulsed laser-induced photomodification of plasmonic gold nanoparticles. Sci. Rep. **7**, 15704 (2017)
42. S. Inasawa, M. Sugiyama, S. Noda, Y. Yamaguchi, Spectroscopic study of laser-induced phase transition of gold nanoparticles on nanosecond time scales and longer. J. Phys. Chem. B **110**, 3114 (2006)
43. T. Tsuji, S. Sakaki, H. Fujiwara, H. Kikuchi, M. Tsuji, Y. Ishikawa, N. Koshizaki, Stabilizer-concentration effects on the size of gold submicrometer-sized spherical NPs prepared using laser-induced agglomeration and melting of colloidal nanoparticles. J. Phys. Chem. C **122**, 21659 (2018)
44. M. Sugiyama, H. Okazaki, S. Koda, Size and shape transformation of TiO_2 nanoparticles by irradiation of 308-nm laser beam. Jpn. J. Appl. Phys. **41**(1), 4666 (2002)
45. M.S.F. Lima, F.P. Ladario, R. Riva, Microstructural analyses of the nanoparticles obtained after laser irradiation of Ti and W in ethanol. Appl. Surf. Sci. **252**, 4420 (2006)
46. Y. Ishikawa, N. Koshizaki, A. Pyatenko, N. Saitoh, N. Yoshizawa, Y. Shimizu, Nano- and submicrometer-sized spherical particle fabrication using a submicroscopic droplet formed using selective laser heating. J. Phys. Chem. C **120**, 2439 (2016)
47. V.K. Pustovalov, Investigation of threshold laser intensities for melting and evaporation of spherical and spheroidal nanoparticles in media by short laser pulses. Chem. Phys. Lett. **421**, 142 (2006)
48. V.K. Pustovalov, A.N. Chumakov, Laser melting and evaporation of nanoparticles: a simplified model for estimation of threshold fluence. Opt. Las. Techn. **126**, 106082 (2020)
49. A. Takami, H. Kurita, S. Koda, Laser-induced size reduction of noble metal nanoparticles. J. Phys. Chem. B **103**, 1226 (1999)
50. H. Muto, K. Miyajima, F. Mafune, Mechanism of laser-induced size reduction of gold NPs as studied by single and double laser pulse excitation. J. Phys. Chem. B **112**, 5810 (2008)
51. R. Cavicchi, D. Meier, C. Presser, V. Prabhu, S. Guha, Single laser pulse effects on suspended Au- nanoparticle size distributions and morphology. J. Phys. Chem. C **117**, 10866 (2013)
52. M.J. Rivera-Chaverra, E. Restrepo-Parra, C.D. Acosta-Medina, A. Mello, R. Ospina, Synthesis of oxide iron nanoparticles using laser ablation for possible hyperthermia applications. Nanomaterials **10**, 2099 (2020)
53. F. Giammanco, E. Giorgetti, P. Marsili, A. Giusti, Experimental and theoretical analysis of photofragmentation of Au NPs by picosecond laser radiation. J. Phys. Chem. C **114**, 3354 (2010)
54. A.R. Ziefuss, S. Reich, S. Reichenberger, M. Levantino, A. Plech, In situ structural kinetics of picosecond laser-induced heating and fragmentation of colloidal gold spheres. Phys. Chem. Chem. Phys. **22**, 4993 (2020)

55. V.K. Pustovalov, Influence of temperature dependence of optical properties of gold nanoparticles and their heating, melting and evaporation by laser pulses. High Temper. Mater. Process. **26**, 41 (2022)
56. Refractive index database. http://refractiveindex.info/
57. V.M. Zolotarev, V.N. Morozov, E.V. Smirnova. *Optical Constants of Nature and Technical Media* (Chemistry publ., Leningrad, 1984, in Russian)
58. I.H. Malitson, Inter specimen comparison of the refractive index of fused silica. J. Opt. Soc. Am. **55**, 1205 (1965)
59. R. Kitamura, L. Pilon, M. Jonasz, Optical constants of silica glass from extreme ultraviolet to far infrared at near room temperature. Appl. Opt. **46**, 8118 (2007)
60. J.R. Devore, Refractive indices of rutile and sphalerite. J. Opt. Soc. Am. **41**, 416 (1951)
61. S. Mahmoud, S. Alshomer, M. Tarawnh, Structural and optical characterization of spayed nickel oxide thin films. J. Modern Phys. **2**, 1178 (2011)
62. Z. Said, R. Saidur, N. Rahim, Optical properties of metal oxides based nanofluids. Int. Commun. Heat Mass Transf. **59**, 46–54 (2014)
63. N.C. Horti, M.D. Kamatagi, N.R. Patil, M.S. Sannaikar, S.R. Inamdar, Synthesis and optical properties of copper oxide nanoparticles: effect of solvents. J. Nanophoton. **14**, 046010 (2020)
64. A. Chauhan, R. Verma, K.M. Batoo, S. Kumari, R. Kalia, R. Kumar, M. Hadi, E.H. Raslan, A. Imran, Structural and optical properties of copper oxide nanoparticles: a study of variation in structure and antibiotic activity. J. Mater. Res. **36**, 1496 (2021)
65. M. Karami, M. Akhavan-Behabadi, M.R. Dehkordi, S. Delfani, Thermo-optical properties of copper oxide nanofluids for direct absorption of solar radiation. Sol. En. Mater. Sol. Cells **144**, 136 (2016)
66. S.S. Kulkarni, M.D. Shirsat, Optical and structural properties of zinc oxide nanoparticles. Int. J. Adv. Res. Phys. Sci. **2**, 14 (2015)
67. W. Widiyastuti, S. Machmudah, T. Nurtono, S. Winardi, R. Balgis, T. Ogi, K. Okuyama, Morphology and optical properties of zinc oxide nanoparticles synthesized by solvothermal method. Chem. Engin. Trans. **56**, 955 (2017)
68. R.V. Quevedo-Robles, A.R. Vilchis-Nestor, P.A. Luque, Study of optical and morphological properties of nanoparticles semiconductors of zinc oxide synthesized using *Mimosa tenuiflora* extract for photodegradation of methyl orange. Opt. Mater. **128**, 112450 (2022)
69. J. Jassi, V. Biju, A. Shajan, Structural and optical properties of zinc oxide nanoparticles. AIP Conf. Proc. **2263**, 8244066 (2020)
70. S. Arya, P. Mahajan, S. Mahajan, A. Khosla, R. Datt, V. Gupta, S.-J. Young, S.K. Oruganti, Influence of processing parameters to control morphology and optical properties of sol-gel synthesized ZnO nanoparticles. ECS J. Solid State Sci. Technol. **10**, 023002 (2021)
71. Q. Zhu, Y. Cui, L. Mu, L. Tang, Characterization of thermal radiative properties of nanofluids for selective absorption of solar radiation. Int. J. Thermophys. **34**, 2307 (2013)
72. P. Sharma, N. Sharma, N. Neelam, Optical properties of aluminium oxide nanoparticles synthesized by leaf extract of *Ocimum sanctum*. J. Nanosci. Technol. **5**, 817 (2019)
73. H. Hakim, N. Al-Garah, A. Hashim, The Effect of aluminum oxide nanoparticles on optical properties of (polyvinyl alcohol-polyethylene glycol) blend. J. Industr. Engin. Res. **1**, 94 (2015)
74. M. Rivero, J. Hu, D. Jaque, M. Cañete, J. Sánchez-Marcos, A. Muñoz-Bonilla, Compositional tuning of light-to-heat conversion efficiency and of optical properties of superparamagnetic iron oxide nanoparticles. J. Phys. Chem. C **122**, 16389 (2018)
75. D. Song, Y. Wang, D. Jing, J. Geng, Investigation and prediction of optical properties of alumina nanofluids with different aggregation properties. Int. J. Heat Mass Transf. **96**, 430 (2016)
76. A. Hashim, M. Habeeb, Q. Jebur, Structural, dielectric and optical properties for (polyvinyl alcohol–polyethylene oxide-manganese oxide) nanocomposites. Egypt. J. Chem **62**, 735 (2019)
77. M.A. Gondal, Tawfik A. Saleh, Q. Drmosh, Optical properties of bismuth oxide nanoparticles synthesized by pulsed laser ablation in liquids. Sci. Adv. Mater. **4**, 507 (2012)

78. C.-Y. Su, H.-C. Lin, T.-K. Yang, C.-K. Lin, Structure and optical properties of tungsten oxide nanomaterials prepared by a modified plasma arc gas condensation technique. J. Nanopart. Res. **12**, 1755 (2010)
79. H. Zhang, M.T. Swihart, Synthesis of tellurium dioxide nanoparticles by spray pyrolysis. Chem. Mater. **19**, 1290 (2007)
80. M.M. Ali, H.S. Mahdi, A. Parveen, A. Azam. Optical properties of cerium oxide (CeO_2) nanoparticles synthesized by hydroxide mediated method. AIP Conf. Proc. 1953, id.030044 (2018)
81. N.C. Angastiniotis, S. Christopoulos, K.C. Petallidou, A.M. Efstathiou, A. Othonos, L. Koutsokeras, Controlling the optical properties of nanostructured oxide-based polymer films. Sci. Rep. **11**, 16009 (2021)
82. I.H. Hilal, H.l. Abdullah, A.M. Ibraheem, Effect of aluminum oxide nanoparticles films thickness on optical properties. Int. J. Scien. Res. Manag. **6**, 116 (2018)
83. I.N. Bazhukova, S.Y. Sokovnin, V.G. Ilves, A.V. Myshkina, R.A. Vazirov, N. Pizurova, V.V. Kasyanova, Luminescence and optical properties of cerium oxide nanoparticles. Opt. Mater. **92**, 136 (2019)
84. M. Milanese, G. Colangelo, A. Creti, M. Lomascolo, F. Iacobazzi, A. de Risi, Optical absorption measurements of oxide NPs for application as NF in direct absorption solar power systems. Sol. Energy Mater. Sol. Cells **147**, 315 (2016a)
85. X. Zhu, J. Wang, D. Nguyen, J. Thomas, R.A. Norwood, N. Peyghambarian, Linear and nonlinear optical properties of Co_3O_4 nanoparticle-doped polyvinyl-alcohol thin films. Opt. Mater. Exp. **2**, 103 (2012)
86. F. Zhang, R.-J. Zhang, D.-X. Zhang, Z.-Y. Wang, J.-P. Xu, Y.-X. Zheng, L.-Y. Chen et al., Temperature-dependent optical properties of titanium oxide thin films TiO_2 studied by spectroscopic ellipsometry. Appl. Phys. Exp. **6**, 121101 (2013)
87. J.H. Wray, J.T. Neu, Refractive index of several glasses as a function of wavelength and temperature. J. Opt. Soc. Amer. **59**, 774 (1969)
88. C.Z. Tan, J. Arndt, Temperature dependence of refractive index of glassy SiO_2 in the infrared wavelength range. J. Phys. Chem. Solids **61**, 1315 (2000)
89. Y.K. Lingart, V.A. Petrov, N.A. Tikhonova, Optical properties of alumina under high temperatures. Termophys. High Temperat. **20**, 872–880 (1982, in Russian)
90. V. Pustovalov, Optical properties of nanoparticles dispersed in ambient medium and their dependences on temperature. Curr. Nanomater. **8**, 233–258 (2023)
91. Y.-N. Wu, W.A. Saidi, P. Ohodnicki, B. Chorpening, Y. Duan, First-principles investigations of the temperature dependence of electronic structure and optical properties of rutile TiO_2. J. Phys. Chem. C **122**, 22642 (2018)
92. A.M. Cretì, A. Epifani, M. Taurino, F. Catalano, M. Casino, M. Lomascolo, A. Milanese, A. de Risi, Optical absorption measurements at high temperature (500 °C) of oxide nanoparticles for application as gas-based nanofluid in solar thermal collector systems. Adv. Mater. Res. **773**, 80 (2013)
93. M. Milanese, G. Colangelo, A. Cretì, M. Lomascolo, F. Iacobazzi, A. De Risi, Optical absorption measurements of oxide nanoparticles for application as nanofluid in direct absorption solar power systems—Part II: ZnO, CeO_2, Fe_2O_3 nanoparticles behavior. Sol. Energy Mater. Sol. Cells **147**, 321 (2016b)
94. G. Plass, Temperature dependence of the Mie scattering and absorption cross section for aluminum oxide. Appl. Opt. **4**, 1616 (1965)
95. O.A. Yeshchenko, I.S. Bondarchuk, V.S. Gurin, I.M. Dmitruk, A.V. Kotko, Temperature dependence of the surface plasmon resonance in gold nanoparticles. Surf. Sci. **608**, 281 (2013)
96. V.K. Pustovalov, L.G. Astaf'eva, Temperature dependences of the optical properties of metal nanoparticles in various media. Opt. Spectr. **129**, 345 (2021)
97. O. Yeshchenko, I. Dmitruk, A. Alekseenko, A. Kotko, J. Verdal, A. Pinchuk. Size and temperature effects on the surface resonance in silver nanoparticles. Plasmonics **7**, 685 (2012)
98. R. Jose, V. Thavasi, S. Ramakrishna, Metal oxides for dye sensitized solar cells. J. Am. Ceram. Soc. **92**, 289 (2009)

99. O.K. Varghese, M. Paulose, C.A. Grimes, Long vertically aligned titania nanotubes on transparent conducting oxide for highly efficient solar cells. Nat. Nanotechnol. **4**, 592 (2009)
100. D.K. Kumar, J. Kriz, N. Bennett, B. Chen, H. Upadhayaya, K.R. Reddy, V. Sadhu, Functionalized metal oxide nanoparticles for efficient dye-sensitized solar cells (DSSCs): a review. Mater. Sci. Eng. Technol. **3**, 472 (2020)
101. S. Akin, S. Sonmezoglu, Metal oxide nanoparticles as electron transport layer for highly efficient dye-sensitized solar cells. Chapter 2. 39–79, in *Emerging Materials for Energy Conversion and Storage*. eds. by K.Y. Cheong, M.A. Fraga, G. Impellizzeri (Elsevier, Amsterdam, 2018)
102. Y. Wang, J.F. Pierson, Binary copper oxides as photovoltaic absorbers: recent progress in materials and applications. J. Phys. D: Appl. Phys. **54**, 263002 (2021)
103. A. Bhaumik, A. Haque, P. Karnati, K. Ghosh, Copper oxide based nanostructures for improved solar cell efficiency. Thin Solid Films **572**, 126 (2014)
104. H. Siddiqui, M.R. Parra, P. Pandey, M.S. Qureshi, F.Z. Haque, Utility of copper oxide nanoparticles (CuO-NPs) as efficient electron donor material in bulk-heterojunction solar cells with enhanced power conversion efficiency. J. Sci.: Adv. Mater. Dev. **5**, 104 (2020)
105. V. Korstgens, S. Proller, T. Buchmann, D. Mosegui Gonzalez, L. Song, Y. Yao, W. Wang, J. Werhahn, G. Santoro, S.V. Roth, H. Iglev, R. Kienberger, P. Meller-Buschbaum, Laser-ablated titania nanoparticles for aqueous processed hybrid solar cells. Nanoscale **7**, 2900 (2015)
106. S. Shao, K. Zheng, K. Zidek, P. Chabera, T. Pullerits, F. Zhang, Optimizing ZnO nanoparticle surface for bulk heterojunction hybrid solar cells. Sol. Eng. Mat. Sol. Cells **118**, 43 (2013)
107. R. Schlogl, Heterogeneous Catalysis. Angew. Chem. **127**, 3531 (2015)
108. R. Schlögl, Material Science for catalysis: Quo vadis? Z. Anorg. Allg. Chem. **647**, 1998 (2021)
109. S.W. Verbruggen, TiO_2 photocatalysis for the degradation of pollutants in gas phase: from morphological design to plasmonic enhancement. J. Photochem. Photobiol. C **24**, 64 (2015)
110. M. Zimbone, M.A. Buccheri, G. Cacciato, R. Sanz, G. Rappazzo, S. Boninelli, R. Reitano, L. Romano, V. Privitera, M.G. Grimaldi, Photocatalytic and antibacterial activity of TiO_2 nanoparticles obtained by laser ablation in water. Appl. Catal. B **165**, 487 (2015)
111. S. Filice, G. Compagnini, R. Fiorenza, S. ScirH, L. D'Urso, M.E. Fragal, P. Russo, E. Fazio, S. Scalese, Laser processing of TiO_2 colloids for an enhanced photocatalytic water splitting activity. J. Coll. Interface Sci. **489**, 131 (2017)
112. A. Bagabas, A. Al Shammari, M. Aboud, H. Kosslick, Room temperature synthesis of zinc oxide nanoparticles in different media and their application in cyanide photo-degradation. Nanosc. Res. Let. **8**, 516 (2013)
113. I. Rocha-Mendoza, S. Camacho-Lopez, Y.Y. Luna-Palacios, Y. Esqueda-Barrjn, M.A. Camacho-Lopez, M. Camacho-Lopez, G. Aguilar, Second-harmonic generation of ZnO nanoparticles synthesized by laser ablation of solids in liquids. Opt. Las. Technol. **99**, 118 (2018)
114. R. Niyuki, H. Fujiwara, T. Nakamura, Y. Ishikawa, N. Koshizaki, T. Tsuji, K. Sasaki, Double threshold behavior in a resonance-controlled ZnO random laser. APL Photonics **2**, 036101 (2017)
115. H. Wang, M. Miyauchi, Y. Ishikawa, A. Pyatenko, N. Koshizaki, Y. Li, L. Li, X. Li, Y. Bando, D. Golberg, Single-crystalline rutile TiO_2 hollow spheres: room-temperature synthesis, tailored visible-light-extinction, and effective scattering layer for quantum dot-sensitized solar cells. J. Am. Chem. Soc. **133**, 19102 (2011)
116. R.A. Ismail, S.A. Zaidan, R.M. Kadhim, Preparation and characterization of aluminum oxide nanoparticles by laser ablation in liquid as passivating and anti-reflection coating for silicon photodiodes. Appl. Nanosci. **7**, 477 (2017)
117. J.E. Song, Y. Hwan, K. Young, S. Kang, Preparation of indium tin oxide nanoparticles and their application to near IR-reflective film. Curr. Appl. Phys. **6**, 791 (2006)
118. C. Kim, J.-W. Park, J. Kim, S. J. Hong, M. Jung Lee, A highly efficient indium tin oxide nanoparticles (ITO-NPs) transparent heater based on solution-process optimized with oxygen vacancy control. J. Alloys Compd. **726**, 712 (2017)

119. A. Sosna-Głębska, N. Szczecińska, K. Znajdek, M. Sibiński, Review on metallic oxide nanoparticles and their application in optoelectronic devices. Acta Innov. **30**, 5 (2019)
120. F. Benyettou, J.A. Ocadiz Flores, F. Ravaux, R. Rezgui, M. Jouiad, S.I. Nehme, R.K. Parsapur, J.C. Olsen, P. Selvam, A. Trabolsi, Mesoporous γ-iron oxide nanoparticles for magnetically triggered release of doxorubicin and hyperthermia treatment. Chem. A Eur. J. **22**, 17020 (2016)
121. E.C. Abenojar, S. Wickramasinghe, J. Bas-Concepcion, A.C.S. Samia, Structural effects on the magnetic hyperthermia properties of iron oxide nanoparticles. Prog. Nat. Sci. Mater. Int. **26**, 440 (2016)
122. Z. Shen, A. Wu, X. Chen, Iron oxide nanoparticle based contrast agents for magnetic resonance imaging. Mol. Pharm. **14**, 1352 (2017)
123. M. Jeon, A. Mackenzie V. Halbert, Z.R. Stephen, M. Zhang, Iron oxide nanoparticles as T_1 contrast agents for magnetic resonance imaging: fundamentals, challenges, applications, and perspectives. Adv. Mat. **33**, 1906539 (2021)
124. C. Wang, S. Cao, X. Tie, B. Qiu, A. Wu, Z. Zheng, Induction of cytotoxicity by photoexcitation of TiO_2 can prolong survival in glioma-bearing mice. Mol. Biol. Rep. **38**, 523 (2011)
125. P.F. Zeni, D.P.D. Santos, R.R. Canevarolo, J.A. Yunes, F.F. Padilha, J.R.L.C. de Albuquerque, S.M. Egues, M.L. Hernández-Macedo, Photocatalytic and cytotoxic effects of nitrogen-doped TiO2 nanoparticles on melanoma cells. J. Nanosci. Nanotechnol. **18**, 3722 (2018)
126. W. Ni, M. Li, J. Cui, Z. Xing, Z. Li, X. Wu, E. Song, M. Gong, W. Zhou, 808 nm light triggered black TiO2 nanoparticles for killing of bladder cancer cells. Mater. Sci. Eng. C **81**, 252 (2017)
127. M. Kulkarni, A. Mazare, E. Gongadze, Š Perutkova, V. Kralj-Iglič, I. Milošev, P. Schmuki, A. Iglič, M. Mozetič, Titanium nanostructures for biomedical applications. Nanotechnology **26**, 062002 (2015)
128. M. Sajid, Z. Said, R. Saidur, M. Sabri, Applicability of alumina nanofluid in direct absorption solar collectors. Appl. Mech. Mater. **699**, 366 (2015)
129. H. Kumar, B. Gupta, G.D. Agrawal, J. Mathur, Experimental study of water-based Al_2O_3 nanofluid flow in direct absorption solar collector. Soft Mater. **357**, 30 (2015)
130. V.K. Pustovalov, L.G. Astafyeva, W. Fritzsche, Optical properties of nanoparticles and nanofluids for direct absorption of solar radiation. Nanotechn. Environ. Eng. **3**, 15 (2018)
131. V. Bhalla, H. Tyagi, Solar energy harvesting by cobalt oxide nanoparticles in nanofluid absorption based system. Sustain. Eng. Technol. Asses. **24**, 45 (2017)
132. M. Hamdan, M. Sarsour, Effect of nanoparticles on the performance of solar flat plate collectors. J. Ecol. Eng. **19**, 1 (2018)
133. M. Esmaeili, M. Karami, S. Delfani, Performance enhancement of a direct absorption solar collector using copper oxide porous foam and nanofluid. Int. J. Energy Res. **44**, 5527 (2020)
134. M.A. Hamdan, A.M. Al Momani, O. Ayadi, A.H. Sakhrieh, F. Manzano-Agugliaro, Enhancement of solar water desalination using copper and aluminum oxide nanoparticles. Water **13**, 1914 (2021)
135. A.L. Chibac-Scutaru, V. Podasca, I.A. Dascalu, V. Melinte, Exploring the influence of synthesis parameters on the optical properties for various CeO_2 nanoparticles. Nanomaterials **12**, 1402 (2022)
136. M.A. Zwijnenburg, The effect of particle size on the optical and electronic properties of magnesium oxide nanoparticles. Phys. Chem. Chem. Phys. **23**, 21579 (2021)